AutoSketch

FOR WINDOWS

RELEASE 2

Brian L. Duelm

Technical Writer and CAD Consultant

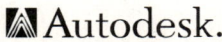

Autodesk.
Registered Author/Publisher

Publishers
The Goodheart-Willcox Company, Inc.
Tinley Park, Illinois

708-687-5000

Copyright 1996

by

The Goodheart-Willcox Company, Inc.

All rights reserved. No part of this book may be reproduced, stored in a retrieval system, or transmitted in any form or by any means, electronic, mechanical, photocopying, recording, or otherwise, without prior written permission of The Goodheart-Willcox Company, Inc. Manufactured in the United States of America.

Library of Congress Catalog Card Number 94-1438
International Standard Book Number 1-56637026-4

1 2 3 4 5 6 7 8 9 10 96   99 98 97 96 95

AutoSketch for Windows / by Brian L. Duelm.

    p.   cm.
ISBN 1-56637-026-4
    1. Computer graphics.   2. AutoSketch for Windows   3. Computer-aided design.   I. Title.
T385.D84   1996   94-1438
604.2'0285'5369—dc20   CIP

Content Notice

This book contains the most complete and accurate information that could be obtained from various authoritative sources at the time of publication. Goodheart-Willcox cannot assume responsibility for any changes, errors, or omissions, nor will it assume responsibility for damage to hardware or software resulting from the use or misuse of the program AutoSketch for Windows.

AutoCAD and AutoSketch are registered in the U.S. Patent and Trademark Office by Autodesk, Inc.

Microsoft, Microsoft Word, Windows, Windows for Workgroups, and Windows 95 are registered in the U.S. Patent and Trademark Office by Microsoft Corporation.

Pagemaker is registered in the U.S. Patent and Trademark Office by Aldus Corporation.

Quark Xpress is registered in the U.S. Patent and Trademark Office by Quark, Inc.

Corel Draw is registered in the U.S. Patent and Trademark Office by Corel Systems Corporation.

INTRODUCTION

AutoSketch for Windows is an innovative text written to show you how to apply AutoSketch to typical drafting and design tasks. This text covers computer hardware, industry practices, and the technical language of computer-aided drafting (CAD). These topics are important for readers who are preparing for a career as a drafter or designer, and for readers who want to use CAD for their own convenience.

For the home user, this book will help take your ideas from thoughts in your head to a hardcopy. For experienced manual drafters, the topics and activities will make your transition from the drafting board to the computer much smoother, and more enjoyable. For experienced CAD users switching to AutoSketch software, this text will help you learn how AutoSketch differs from other CAD programs. Whether you are a new computer user, an experienced manual drafter, or an experienced CAD user, this text will be a valuable resource.

The topics in *AutoSketch for Windows* are covered in an easy-to-learn sequence. Topics also progress in a way that allows you to become comfortable with the commands as your knowledge builds from one chapter to the next. Each chapter includes the following items.

- Objectives at the beginning of each chapter stating what will be learned in the chapter.
- Step-by-step use of AutoSketch commands.
- Tool icons in the margin next to sections discussing that particular tool.
- "Shortcuts." These are tips on how to use AutoSketch more productively.
- In-chapter exercises that help you practice AutoSketch commands as soon as they are presented. Many of the exercises build on previously learned material.
- End-of-chapter lists of Important Terms and New Tools and Commands presented in the chapter. These will help you identify AutoSketch commands and industry language.
- Review questions that help you study important concepts and drafting practices.
- Activities that allow you to apply AutoSketch commands to drafting problems. The Activities section is based on commands presented in the chapter, but may contain activities that also build on previously learned material.

AutoSketch for Windows not only helps you learn how to use AutoSketch commands, but also makes you familiar with other aspects of CAD. Some of the additional material covered includes the following.
- Computer hardware used with AutoSketch.
- Benefits of computer-aided drafting over manual drafting.
- Coordinate systems used by AutoSketch and other drafting programs.
- Drawing and editing techniques that increase your productivity.
- Accepted dimensioning practices.
- Creating working drawings common in mechanical and architectural drafting.
- Creating three-dimensional pictorial drawings.
- Exporting drawings for desktop publishing.
- Creating drawings that can be transferred to computer numerical control (CNC) machines and computer-aided machining systems.

Brian L. Duelm

ACKNOWLEDGEMENTS

The author would like to especially thank the many companies that provided illustrations for this textbook. The following is a list of the companies that provided illustrations.

    ABA
    Autodesk, Inc.
    Barber-Colman
    CAD Technology Corp.
    CADAM
    Compaq Computer Corporation
    Cincinnati Milacron
    E. Henry Fitzgibon, Architect
    GMD, Inc.
    Houston Instruments, A Summagraphics Company
    Hyster Company
    IBM
    Koh-I-Noor Rapidograph
    Kurta
    Light Machines Corporation
    Monarch Sydney

In addition, the author would like to thank the following companies for providing drawings, user manuals, and software to make this book possible.

    Autodesk, Inc. for the latest versions of AutoSketch, AutoCAD, and AutoCAD LT.
    CNC Software Inc. for information regarding computer-aided machining.
    Light Machines Corp. for information regarding computer-aided machining.

The author would also like to thank Compaq Computer Corporation for providing the computer and monitor for the cover photo.

CONTENTS

Using This Text   10
How to Use This Text   10
Installing AutoSketch for Windows Release 2   11

Chapter 1
Introduction to Computer-aided Drafting   15
Introduction to Drafting   15
What is Computer-aided drafting?   16
Drafting, Design, and the Application of Computers   18
Tools of CAD   24
Drafting Positions and Qualifications   39
Important Terms   42
Review Questions   43
Activities   44

Chapter 2
Introduction to AutoSketch   47
Why Use AutoSketch?   47
Your AutoSketch System   49
Starting AutoSketch   49
AutoSketch Window Layout   50
Using the Cursor   55
Selecting Menus, Tools, and Pull-Down Menus   56
Getting Help   60
Basic Drawing with AutoSketch   62
Summary   66
Important Terms   67
New Commands and Tools   67
Review Questions   67
Activities   69

Chapter 3
Starting and Saving Drawings   71
Starting a New Drawing   73
Saving a New Drawing   74
Closing a Saved Drawing   76
Opening a Stored Drawing   76
Saving a Drawing   78
Deleting a Drawing   79
Working with Multiple Files On-screen   80
Ending Your Drawing Session   83
Summary   83
Important Terms   83
New Commands and Tools   83
Review Questions   84
Activities   85

Chapter 4
Drawing and Erasing Objects   87
Setting Object Color, Width, and Linetype   88
Drawing Points   92
Exercise 4-1   93
Drawing Lines   93
Exercise 4-2   93
Drawing Boxes   94
Exercise 4-3   95
Drawing Circles   95
Exercise 4-4   101
Drawing Polygons   101
Exercise 4-5   106
Drawing Arcs   106
Exercise 4-6   112
Drawing Ellipses   112
Exercise 4-7   117
Drawing Curves   117
Exercise 4-8   121
Drawing Polylines   122
Exercise 4-9   125
Erasing Objects   126

Exercise 4-10   128
Undoing and Redoing Your Commands   129
Exercise 4-11   129
Summary   129
Important Terms   130
New Commands and Tools   130
Review Questions   130
Activities   132

Chapter 5
Drawing Precisely with Coordinates and Drawing Aids   133

Cartesian Coordinate System   133
Typing in Coordinate Locations   134
Exercise 5-1   136
Exercise 5-2   137
Exercise 5-3   138
Exercise 5-4   140
Using /lpoint to Enter Location   141
Exercise 5-5   142
Changing the Origin of the Drawing   142
Exercise 5-6   142
Drawing Precisely with the Cursor   143
Coordinate Display   143
Exercise 5-7   145
Grid and Snap   145
Exercise 5-8   148
Ortho Tool   149
Attaching to Objects   149
Exercise 5-9   152
Rulers   152
Summary   153
Important Terms   154
New Commands and Tools   154
Review Questions   154
Activities   155

Chapter 6
Modifying and Editing Objects   163

Choosing Objects to Edit   163
Exercise 6-1   165
Exercise 6-2   169
Moving Objects   170
Exercise 6-3   172
Copying Objects   172
Exercise 6-4   174
Rotating Objects   175
Exercise 6-5   177
Mirroring Objects   177
Exercise 6-6   178

Scaling Objects   179
Exercise 6-7   180
Modifying Object Properties   181
Exercise 6-8   184
Summary   184
Important Terms   185
New Commands and Tools   185
Review Questions   185
Activities   186

Chapter 7
Changing Views of Your Drawing   189

Zoom Tools   190
Exercise 7-1   190
Exercise 7-2   192
Exercise 7-3   193
Exercise 7-4   194
Exercise 7-5   195
Panning and Scrolling   195
Exercise 7-6   197
Aerial View   197
Exercise 7-7   200
Redrawing the View   200
Exercise 7-8   201
Using View Tools within Other Commands   201
Changing What is Seen with View Preferences   201
Summary   205
Important Terms   206
New Commands and Tools   206
Review Questions   206
Activities   207

Chapter 8
Advanced Editing and Measuring Tools   209

Breaking Objects   209
Exercise 8-1   211
Exploding Objects   211
Exercise 8-2   212
Trimming Objects   213
Exercise 8-3   215
Extending Objects   215
Exercise 8-4   217
Drawing Fillets and Chamfers   217
Exercise 8-5   222
Stretching Objects   223
Exercise 8-6   225
Drawing a Box Array   225
Exercise 8-7   232

Drawing a Ring Array   232
Exercise 8-8   238
Editing Curves   238
Exercise 8-9   242
Measuring Distance, Angle, and Area   243
Exercise 8-10   247
Summary   249
Important Terms   250
New Commands and Tools   250
Review Questions   250
Activities   251

Chapter 9
Adding and Editing Text   255

Benefits of CAD Text   255
Selecting Text Features   256
Exercise 9-1   262
Placing Text on the Drawing   262
Exercise 9-2   263
Exercise 9-3   266
Exercise 9-4   267
Editing Text   267
Exercise 9-5   267
Exercise 9-6   269
Summary   270
Important Terms   270
New Commands and Tools   270
Review Questions   270
Activities   272

Chapter 10
Developing New Drawings and Templates   275

The Design Process   275
Exercise 10-1   279
Sketching Designs   279
Exercise 10-2   281
Setting Up a New Drawing   282
Defining the Unit of Measurement   282
Setting up a Drawing Sheet   284
Exercise 10-3   289
Exercise 10-4   290
Sorting Information with Layers   290
Exercise 10-5   294
Setting Object Properties, Drawing Aids, and Text   294
Adding a Border and Title Block   294
Exercise 10-6   295
Using Templates   296
Exercise 10-7   298

Tips for Developing Drawings   298
Summary   299
Important Terms   300
New Commands and Tools   300
Review Questions   300
Activities   301

Chapter 11
Applying AutoSketch   307

Mechanical Drafting   307
Exercise 11-1   314
Exercise 11-2   318
Exercise 11-3   320
Architectural Drafting   320
Desktop Publishing   334
Summary   338
Important Terms   339
Review Questions   339
Activities   341

Chapter 12
Drawing Pattern Fills and Section Views   343

Elements of a Section View   343
Drawing a Basic Section View with AutoSketch   345
Exercise 12-1   350
Editing Pattern Fills   351
Exercise 12-2   354
Types of Sections   354
Other Uses for Pattern Fills   359
Exercise 12-3   361
Summary   361
Important Terms   362
New Tools and Commands   362
Review Questions   362
Activities   363

Chapter 13
Dimensioning Drawings   365

Size and Location Dimensions   365
Elements in Dimensioning   367
Dimensioning with AutoSketch   371
Exercise 13-1   373
Exercise 13-2   374
Dimensioning Linear Distance   374
Linear Dimensions with Leaders   378
Exercise 13-3   379

Dimensioning Angles   379
Exercise 13-4   381
Dimensioning Circles and Arcs   382
Exercise 13-5   386
Leaders   387
Exercise 13-6   388
Associativity   388
Exercise 13-7   390
Editing Dimensions   392
Changing Dimension Text   392
General Rules for Dimensioning   393
Typical Dimensioning Practices   393
Summary   399
Important Terms   399
New Tools and Commands   400
Review Questions   400
Activities   401

Chapter 14
Creating and Using Symbols   405

Symbols   405
Drawing and Saving a Symbol   406
Exercise 14-1   410
Saving an Entire Drawing as a Symbol   410
Exercise 14-2   411
Inserting a Symbol in a Drawing   411
Exercise 14-3   413
Exercise 14-4   414
Exercise 14-5   415
Editing Symbols   415
Symbols Library Management   416
Summary   416
Important Terms   416
New Tools and Commands   416
Review Questions   417
Activities   418

Chapter 15
Drawing Pictorial Views   421

Oblique Drawing   421
Exercise 15-1   422
Perspective Drawing   422
Isometric Drawing   424
Exercise 15-2   425
Using the Isometric Mode of AutoSketch   426
Exercise 15-3   430
Exercise 15-4   431
Exercise 15-5   432
Exercise 15-6   435
Exercise 15-7   436

Summary   436
Important Terms   437
New Tools and Commands   437
Review Questions   437
Activities   438

Chapter 16
Printing and Plotting Drawings   441

Print Setup   442
Exercise 16-1   447
Sheet Setup   447
Exercise 16-2   450
Exercise 16-3   453
Print Preview   453
Printing and Plotting Your Drawing   455
Plotter Media, Pens, and Tips   458
Summary   461
Important Terms   461
New Commands   461
Review Questions   461
Activities   463

Chapter 17
Sharing Drawings with Other Programs   465

Exchanging Drawings with AutoCAD   465
Exchanging Drawings with Other CAD Programs   466
Copying and Pasting Objects with the Windows Clipboard   469
Printing Files to Export to Desktop Publishing   471
Making and Viewing Slides   475
Summary   476
Important Terms   478
New Tools and Commands   478
Review Questions   478
Activities   479

Chapter 18
Drafting for Computer-aided Machining (CAM)   481

What is CAM?   481
Basic Machining Processes   481
The CAD/CAM/CNC Process   485
Creating Drawings for CAM   487
Using Computer-aided Machining Software to Generate NC Code   490

Summary   494
Important Terms   494
Review Questions   495
Activities   496

Chapter 19
Automating AutoSketch with Macros   497

Macro Files   498
Macro Tools   498
Recording and Playing Macros   499
Exercise 19-1   500
Exercise 19-2   501
Exercise 19-3   503
*Creating Custom Toolboxes and
    Macro Buttons   503*
Exercise 19-4   506
Exercise 19-5   509
*Writing and Editing Macros Using the AutoSketch
    Macro Language   510*
Exercise 19-6   515
Advanced Macro Language Topics   515
Exercise 19-7   518
Summary   519
Important Terms   519
New Tools and Commands   520
Review Questions   520
Activities   521

Chapter 20
Managing Files in Windows and DOS   523

*Understanding Directories, Files, and
    Extensions   523*
Using Windows File Manager   525
Using DOS Commands   538
Summary   545
Important Terms   545
New Tools and Commands   546
Review Questions   546
Activities   548

Chapter 21
AutoSketch and Windows 95   549

How Is Windows 95 Different?   549
Installing AutoSketch in Windows 95   552
Using AutoSketch in Windows 95   555
Managing Files Using Windows 95   560
Summary   565
Important Terms   565
New Tools and Commands   565
Review Questions   565
Activities   566

Appendices   567

Function Keys and Key Combinations   567
AutoSketch Fonts   568
AutoSketch Pattern Fills   571
AutoSketch Pull-down Menus   573
Macro Variables and Commands   574

Glossary   580

Index   586

USING THIS TEXT

How to Use This Text

What makes this text important? How is it different from the *User's Guide* and *Getting Started* manuals that come with your AutoSketch software? The answer is that this text teaches you how to apply your AutoSketch software to typical drafting tasks. The manuals that come with AutoSketch just teach you how to use the software. *AutoSketch for Windows* focuses on learning computer-aided drafting, expressing your design ideas, and getting the most out of AutoSketch. There are plenty of drawing exercises to help you develop a wide range of drafting skills.

AutoSketch for Windows introduces AutoSketch commands in the order you will most likely need them. It is not arranged in alphabetical order or in the order of menus. This book assumes that you have very little knowledge of computers or the AutoSketch software. Topics follow a step-by-step approach, and each chapter builds on the previous one.

The text also presents topics on drafting in the same way you would learn traditional drafting instruction. Each chapter shows how to apply AutoSketch to typical drawing tasks. Once you master all of these skills, the text serves as an excellent reference tool.

Using the Chapters of This Text

Each chapter of this text begins with learning objectives. Review the objectives so that you know what important topics to look for in the chapter. Refer to the illustrations as you read the text. Each one points out an important feature of AutoSketch or a drafting practice. As you learn AutoSketch, you will also learn skills required by industry. If you have access to an AutoSketch system, work through the in-chapter exercises as you read through the chapter.

Different type faces are used throughout each chapter to define terms and identify AutoSketch commands. Important terms are always printed in ***bold-italic face, serif*** type and are defined. These terms are then listed at the end of the chapter for your review.

AutoSketch menus, commands, dialog box names, and tool buttons are printed in **bold-face, sans serif** type. Filenames, directory names, paths, and keyboard-entry items appear in the body of the text in Roman, sans serif type. Keyboard keys are shown inside of square brackets and appear in Roman, sans serif type. For example, [Enter] means to press the Enter key.

In addition, commands, menus, and dialog boxes related to Microsoft Windows appear in Roman, sans serif type. Two exceptions are Chapter 20 and Chapter 21. In these chapters, Windows-related terms appear in **bold-face, sans serif** type.

You will also find "icons" in the margin of most chapters. These icons are related to the different buttons that can be used to access AutoSketch commands from toolboxes. An icon will appear next to the paragraph or section where that tool is being discussed. In Chapter 20, these icons are related to Windows for Workgroups 3.11, not AutoSketch. In addition, the icons in Chapter 21 are related to Windows 95. For this reason, the icons in Chapter 20 and Chapter 21 appear in green, not black as in all other chapters.

As you progress through the topics in the text, you will find instances where you will be asked to supply information. On-screen prompts appear in the text like the following example.

Line Enter point:

Here, you must enter a point location, or pick a location on-screen using your mouse and cursor. Look for these prompts as you work through the chapters. Text and numeric values you must enter are shown in the prompt in **bold-face** type. The symbol ↵ is used in prompts to indicate pressing the [Enter] key. In the following example, you would enter R(3,3) using the keyboard and then you would press the [Enter] key.

Line Enter point: **R(3,3)**↵

When finished with the chapter, answer the Review Questions and complete the Activities. The review questions test your comprehension of the chapter topics. The activities include drawing problems for you to apply learned skills using AutoSketch.

Installing AutoSketch for Windows Release 2

There is a minimum amount of computer equipment and software that you will need to use AutoSketch for Windows Release 2. The following sections cover these requirements.

Hardware

The hardware is the physical components of the computer system. The following hardware is required.

Computer

A Compaq, IBM, or compatible computer based on the 386DX, 386SX, 486DX, 486SX, or Pentium microprocessor is required. If you have a 386-based computer, you can increase the performance of AutoSketch by installing an Intel 80387 math coprocessor. A 486DX-based computer (not 486SX) has the math coprocessor integrated into the 486 chip. The Pentium microprocessor also has a math coprocessor integrated into the chip.

Memory

AutoSketch requires that you have 4 megabytes (MB) of memory installed. Additional memory increases the performance of AutoSketch, especially if you work with large drawings. More memory is also key if you use more than one program with Windows at the same time you use AutoSketch. Memory prices have steadily declined and adding memory takes very little effort.

Hard disk storage

Installing AutoSketch will take about 9MB of hard disk storage. The clip art files take up another 2MB. Plus, you will need disk storage for your drawings. As a general rule, have another 2MB of available disk storage.

If you install the AutoSketch Symbol Libraries, up to 20MB more disk space is needed. Finally, if you do not already have Microsoft Windows installed on your computer, plan for another 6MB to 8MB of disk space. Windows 95 may take up over 70MB when installed on your computer.

In total, this takes up about 40MB of disk storage (perhaps near 100MB with Windows 95) when all of the Windows and AutoSketch software pieces are installed.

Monitor

At a minimum, use a video graphics array (VGA) monitor and graphics card. This will give you the resolution needed to view your drawings. Super VGA (SVGA) monitors may give you a larger and/or clearer view of your drawing. Your graphics card that drives the monitor should also be Windows-compatible. Most cards are Windows-compatible.

Mouse

Although it is one of the smallest parts of your computer system, a mouse is very important while using AutoSketch. You will use the mouse to pick commands, draw objects, and edit objects.

Printer or plotter

To print your drawings, you will need to have a printer or plotter installed, and configured in Windows. Windows supports almost every printer and plotter available today.

Software

Software is the written instructions for a computer. The following software is required.

MS-DOS

Most computers come with the Microsoft Disk Operating System (MS-DOS) installed. Version 5.0 or higher is recommended with AutoSketch. However, versions 3.3 and above will work.

Microsoft Windows

Windows is the graphic environment that AutoSketch operates in. Version 3.1 or later is required. It was stated earlier that your hardware must be "Windows-compatible." This simply means that Windows must offer a driver that knows how to communicate with your hardware. Windows offers hundreds of drivers for popular hardware. Refer to the Windows documentation and use the Windows Setup program to configure your hardware.

Microsoft Windows 95

Windows 95 is an operating system that combines the functions of Windows and DOS. AutoSketch will work with Windows 95. For more information, refer to Chapter 21 *AutoSketch and Windows 95*.

Installation Procedure

Before installing AutoSketch, you should have already installed MS-DOS and Microsoft Windows or Windows 95. Follow the instructions that come with Windows to make sure your mouse, monitor, and printer or plotter are installed and working properly. Next, follow these steps.

Note: The following procedure is for installing AutoSketch in Windows 3.1 or Windows for Workgroups 3.11 *only*. If you are installing AutoSketch for Windows in the Windows 95 environment, refer to Chapter 21 *AutoSketch and Windows 95* for the specific installation procedure.

1. Start Windows. This is usually done by typing WIN at the DOS prompt. The DOS prompt is usually shown as C:\ and appears after you start your computer. Many computers "boot" directly into Windows, so you may not need to manually start it. Once Windows is started, you will be in the Windows Program Manager. Here, you can start programs by clicking on their icon.
2. Place Disk 1 of the AutoSketch for Windows installation disks in your floppy drive.
3. From the File menu, choose Run ... to display the Run dialog box.
4. Enter A:SETUP. If the drive where you place the disk is the B: drive on your computer, enter B:SETUP. Then, press [Enter] or click on the OK button.
5. Follow the instructions on-screen.

The AutoSketch installation process is very simple. You will be asked which files you want to install, and which hard disk drive and directory to place them. Accept the defaults whenever you're not sure how to answer them. The installation program will also ask for your name, company, and registration number (printed on your AutoSketch package). Enter this information carefully. Finally, you are asked for the default unit of measurement to start your drawings with.

When the installation is complete, an AutoSketch program group is created, and the AutoSketch icon appears in this group. Start AutoSketch by double-clicking on the icon.

Installing the Symbol Library

Symbols are drawings that you can insert into other drawings to save you time. An example might be a bolt head that you use over and over in a mechanical drawing. Instead of drawing each bolt head, you simply insert a predrawn symbol each time you need it. AutoSketch saves you even more time by providing hundreds of predrawn symbols that you can use or customize to fit your needs.

To install the Symbol Library, you must first have installed AutoSketch. Then, follow these steps.

Note: These steps are for installing the Symbol Library in Windows 3.1 and Windows for Workgroups 3.11 *only*. If you are installing the Symbol Library in Windows 95, refer to Chapter 21 *AutoSketch and Windows 95* for the specific procedure.

1. Place Disk 1 of the Symbol Library in your floppy drive.
2. From the File menu of the Windows Program Manager, choose Run ... to display the Run dialog box.
3. Enter A:SETUP or B:SETUP and press the [Enter] key or click on the OK button.
4. Select the symbol libraries to install. A check mark should appear in the box next to each library you select. When finished choosing, click on OK.
5. Select the hard disk drive where you installed AutoSketch for Windows.
6. Enter the path where AutoSketch is located. If you installed AutoSketch on your C: drive using the default directory name, enter C:\WSKETCH. (Note: Depending on the shipping version you have, this directory may be \WSKETCH2.)
7. Follow the instructions on-screen to insert other disks and complete the installation.

CHAPTER 1

Introduction to Computer-aided Drafting

Objectives
After studying this chapter, you will be able to:
- Explain how drawings help to document an idea.
- Describe how computer-aided drafting is different from manual drafting.
- Identify how computer-aided drafting is applied to the various fields of drafting and graphic design.
- Specify hardware that makes up a CAD system.
- Recognize the different drafting positions.
- List the qualifications needed to be a CAD drafter.

Introduction to Drafting

Drafting is a language used to communicate ideas with drawings. Most products you see began as a sketch on paper or on a computer screen. Completed *working drawings* show the information needed to build, machine, assemble, install, or service a product, Figure 1-1. Drafting applies to the hobbyist and homeowner as much as it does to industry. If you want to design a new chair, a good way to show your ideas is with a drawing.

Why draw rather than give written or spoken directions? One reason is words cannot describe everything. Have you ever tried to give someone directions to a place, only to have the person get lost? Your directions may have been accurate. However, words can mean different things to different people. A simple sketch may have solved the problem. A well-prepared drawing does the same thing. A drawing shows a single product idea that should not be misunderstood.

In this text, you will learn that drafting is a universal language. Lines, measurements, notes, and symbols describe the size, shape, material, finish, and assembly of a product. Drafting uses standard symbols and measurements. This means that all drafters will read the same drawing in the same way. The person who must build or assemble the product will also read the drawing in the same way. *Standards organizations* set the guidelines that companies should follow to prepare their drawings. Two such organizations are the American National Standards Institute (ANSI) and International Standards Organization (ISO).

Figure 1-1.
Working drawings provide the information needed to manufacture parts. (Autodesk, Inc.)

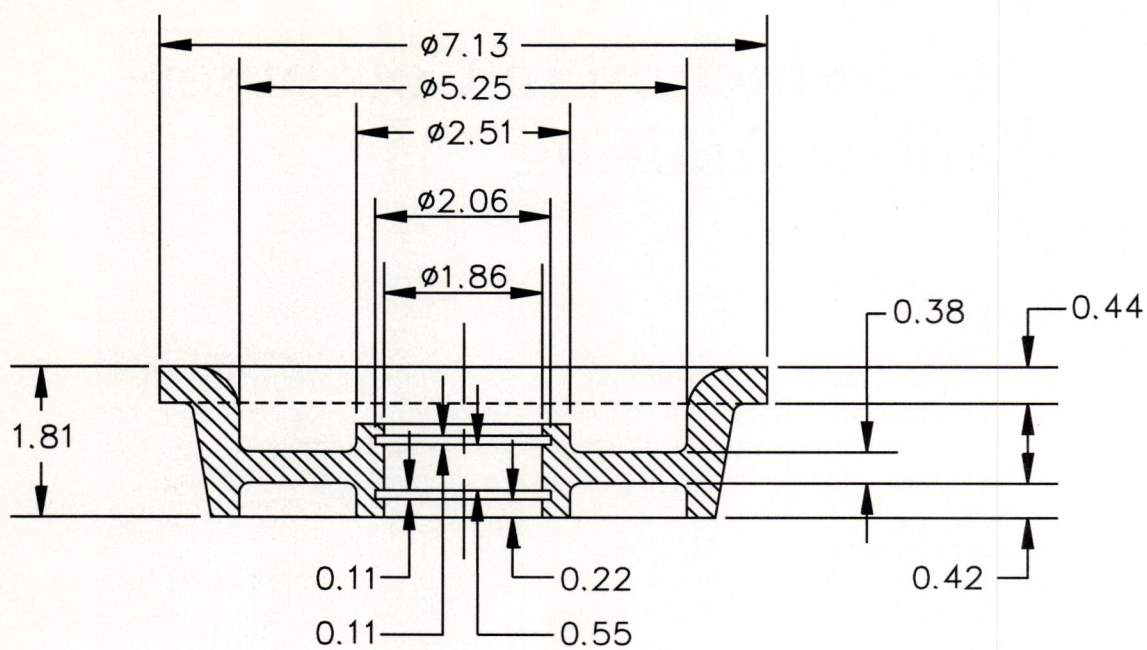

What is Computer-aided Drafting?

To understand computer-aided drafting, you should first know a little bit about *manual drafting*. Manual drafting is the process of creating a drawing on paper using drafting instruments. Manual drafting tools include pencils or lead holders, T-squares or drafting machines, scales, dividers, compasses, erasers, erasing shields, irregular (or French) curves, and triangles. Some of these tools are shown in Figure 1-2. With these tools, you can draw accurate lines, circles, and arcs.

Computer-aided drafting (CAD) is the process of creating a drawing on-screen using a computer and drafting software, Figure 1-3. AutoSketch is a product for CAD. AutoSketch is used with a personal computer to create drawings and designs. Companies are getting rid of their pencils, drawing boards, and scales. Personal computers and CAD software help drafters prepare drawings quicker and more accurately. The key word is *help*. No CAD program, including AutoSketch, can create a drawing by itself. The process of drafting still requires the knowledge and experience of the drafter.

Figure 1-2.
Typical manual drafting tools include a drafting board and machine, scale, triangle, pencil, and eraser.

Figure 1-3.
A typical computer-aided drafting workstation has a computer, monitor, keyboard, pointing device, and CAD software.

Benefits of CAD

You might ask "What makes CAD so much better than manual drafting?" Here are several answers.

- It is simply easier to draw objects using a computer. A manual drafter must create all images by moving a pencil. With CAD, you enter a command and pick one or two points. For example, to draw a circle just select the **Circle** tool, pick the center of the circle, and pick one other point on the circle. Drawing circles, lines, and other objects is discussed in Chapter 4.

- You never have to draw the same shape twice using the computer. Identical shapes are drawn one at a time in manual drafting. With AutoSketch, however, simply select the **Copy** tool, pick the objects to copy, and place as many copies as needed. Copying objects is discussed in Chapter 6.
- CAD drawings are more accurate and readable. The drawings are more accurate because the computer stores drawings in very precise computer code. The drawings are more readable because you cannot smear or tear a drawing stored in computer memory. The paper that a drawing is on can be torn or smeared. Also, the CAD system outputs perfect line widths on paper. Taking drawings from computer memory and putting them on paper, called *printing* or *plotting,* is discussed in Chapter 16.
- Making changes to the drawing is easy. Revisions are often made to a drawing. Making changes on paper is time-consuming. You have to tape the drawing to the drafting board, erase the mistakes, and draw the changes. With CAD, the drawing can be changed easily on-screen. Modifying and editing drawings are discussed in Chapters 6 and 8.
- A CAD system improves communication. The main purpose of drafting is to communicate an idea. When another person needs to use a drawing, you can send it using a computer. The drawing can be stored in a main computer called a s*erver.* Different departments can access the drawing through a network. A *network* is simply a connection of individual computers. A drawing file may also be sent across long distances using a computer modem and the telephone lines.
- A CAD system promotes the consistent use of standard drawing practices. The drafting department supervisor may set up a model drawing called a ***template.*** This template drawing has a border and a title block. It is set up with the proper sheet size, text features, linetypes, and standard symbols. All drafters within the drafting department begin new drawings using this template.

Drafting, Design, and the Application of Computers

Every person who has reason to draw can benefit from CAD. Each year, people find new ways to use computers for drafting and design. Any drawing done manually can be drawn faster and more precisely using a computer and CAD software. Applications described in this section cover only a few of the many areas where you might use AutoSketch or other CAD software.

Technical Drafting

Technical drafting is the process of preparing working drawings. Technical drafting is used in several different industries. Some of these industries include manufacturing, construction, electronics, and land development.

An important area of technical drafting is accurate dimensioning. A feature of most CAD programs is automatic dimensioning. In automatic dimensioning, the drafter simply points to the distance to dimension. The CAD program responds by drawing the dimension lines and adding the measurements. Dimensioning is discussed in Chapter 13.

Mechanical drafting

Mechanical drafting involves making drawings that describe the shape and size of manufactured products, Figure 1-4. The product may be as simple as a hinge or as complex as a car. Mechanical applications were the first use of CAD. These applications remain the biggest use of CAD. Most companies have found that replacing manual drafting tools with computers makes mechanical drafters two to three times more productive.

In many industries, the CAD system is connected to a *computer-integrated manufacturing (CIM)* system. A CIM system uses computers to control manufacturing machinery and the production of products. A CIM system uses the data stored in the CAD drawing as the basis for the manufacturing process. This data is stored in the engineering database and is used by many departments, Figure 1-5. A CIM system speeds the manufacturing process, reduces costs, and improves quality.

Figure 1-4.
Mechanical drafting involves making drawings that describe the shape and size of manufactured products. (Autodesk, Inc.)

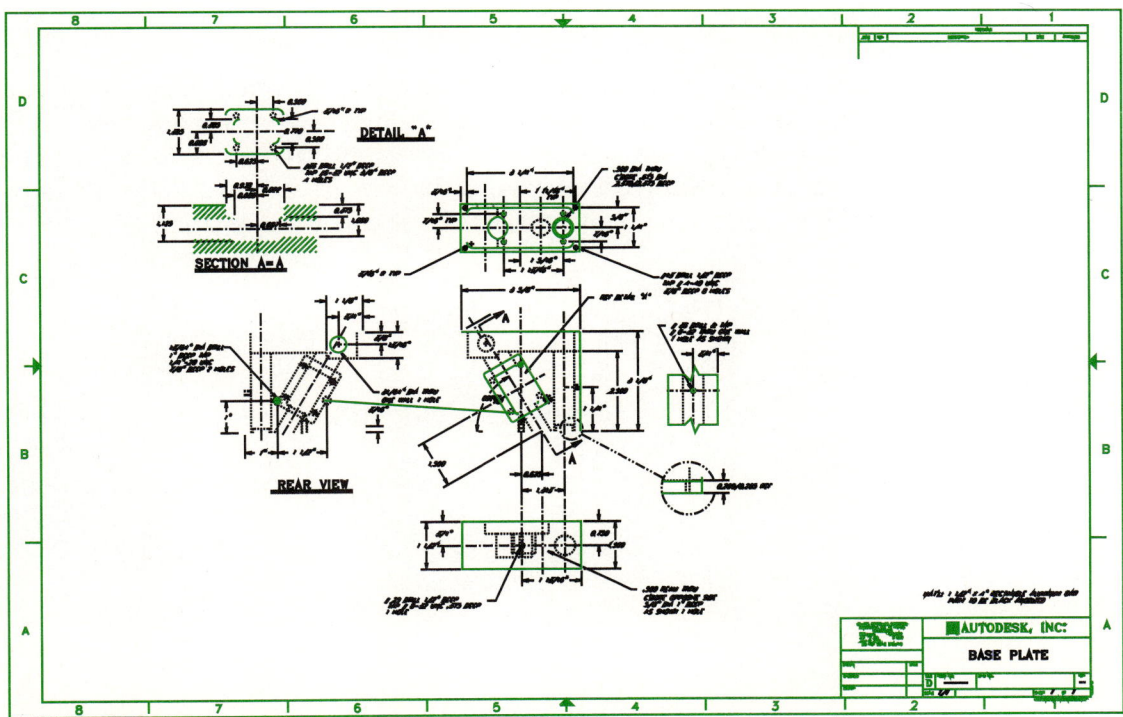

Figure 1-5.
A—This robot is part of a computer-integrated manufacturing system. (Cincinnati Milacron) B—A CIM system improves quality and productivity because the same data can be used by many departments. (GMD, Inc.)

Architectural drafting

Architectural drafting involves preparing drawings that describe the structure and materials for buildings, Figure 1-6. These buildings may be residential or commercial. There are several different types of architectural drawings. These drawings are called *plans.* Several plans make up a set of working architectural drawings. The following list describes some common architectural plans.

- *Plot plans* describe the parcel (piece) of land, its perimeter, elevations, and the building location.
- *Foundation plans* describe the concrete, block, beams, and other materials needed to support the structure.
- *Floor plans* show how the space is divided. Floor plans also show the location of walls, doors, and windows.
- *Electrical plans* show the location of electrical fixtures and switches. Electrical plans also explain the types of fixtures, switches, wiring, and other items necessary to distribute electricity.
- *Heating, ventilation, and air conditioning (HVAC) plans* show how the building environment will be controlled. Controlling the environment means controlling the temperature, humidity, and air flow in the building.

Figure 1-6.
Architects use section views to explain construction details. (Autodesk, Inc.)

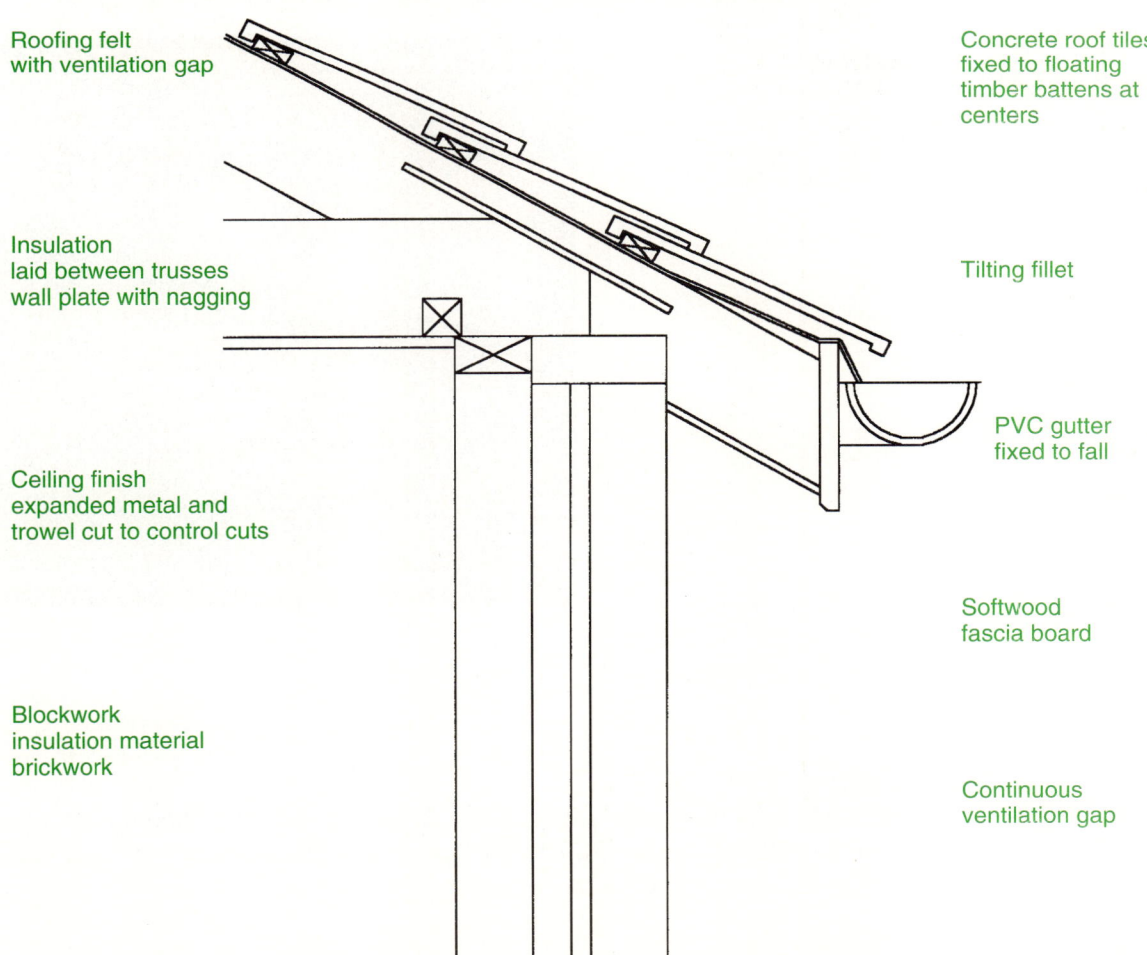

- *Plumbing plans* show where plumbing fixtures will be. These plans trace the flow of water and waste.
- *Elevation plans* show how the building looks. These drawings can be described as "side views" of the building.
- *Landscape plans* show the layout of trees, shrubs, and other ground cover surrounding a building.

Architectural drawings contain many symbols that represent doors, windows, lights, sinks, and appliances. AutoSketch, and most CAD programs, allow you to store important shapes and symbols on disk. Rather than drawing each symbol, you simply insert them on the drawing from disk storage. Drawing, storing, and inserting parts (symbols) are covered in Chapter 14.

Electronics drafting

Electronics drafting involves the design and layout of circuits, parts, and wiring for electrical and electronic products, Figure 1-7. The initial layout is often done by inserting predrawn symbols that represent standard electronic parts. A drafter can manually add the electrical lines that connect parts or have it done automatically by routing software.

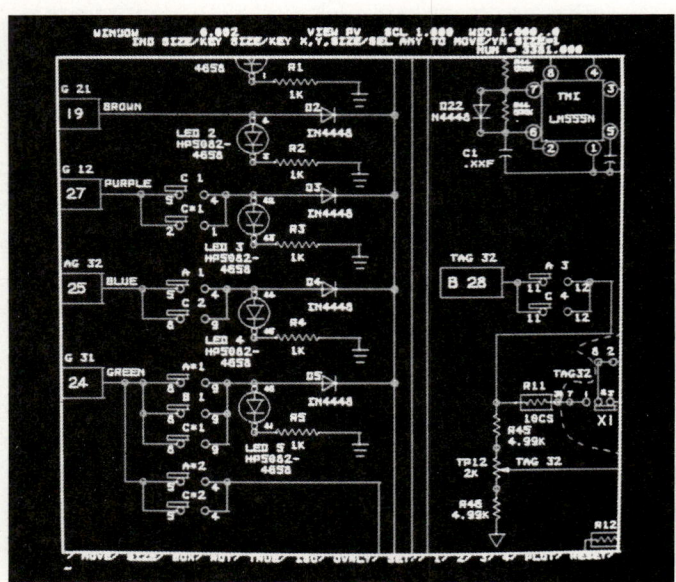

Figure 1-7.
Electronics drafting involves designing circuits for electrical and electronic products. (CADAM)

Civil drafting

Civil drafting is the process of making drawings that describe land terrain, road systems, and utility systems, Figure 1-8. Surveys of property are typical civil drawings. AutoSketch curves can be used to draw the contour lines of different elevations.

Figure 1-8.
Roads, intersections, and parking lot layouts are all part of civil drafting. (Autodesk, Inc.)

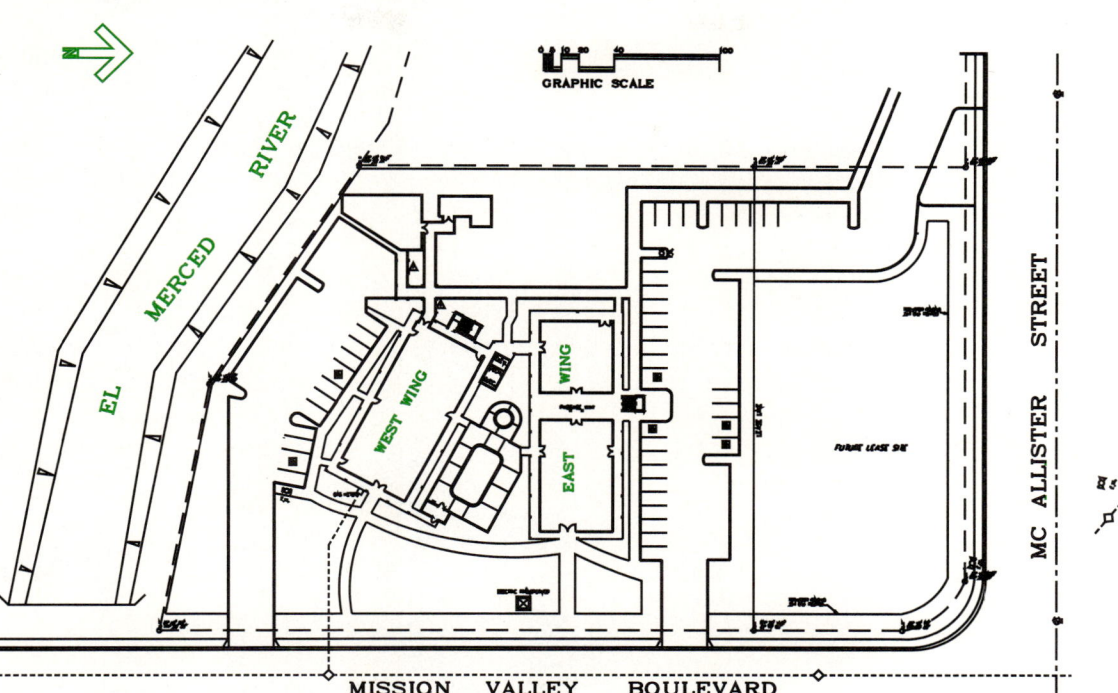

Technical Illustration

Technical illustration involves making two-dimensional or three-dimensional (pictorial) drawings for assembly or presentation, Figure 1-9. A technical illustrator draws the part and may use shading or color to make it look realistic. The drawing may be an exploded view to show how parts fit together. Each part may be identified by name and number. An illustration for presentation usually shows the product assembled and in color. An illustration for a manual is typically a black and white line drawing.

Graphic Design

Many advertising firms, publishing companies, and corporate printing departments have replaced pencils with computers for graphic design. The special effects computers can create have become a trademark of modern graphic design, Figure 1-10.

Publishing firms use CAD systems to create artwork for books and magazines. Much of the art in this book was drawn using a CAD system. In the past, artwork was drawn by hand using technical pens. When corrections were needed, the artist would carefully "white-out" and redraw the art. This process is time-consuming. Art can be quickly created on a computer. Changes can also be made quickly using editing commands. Many times the art is sent to the printer as a computer file on disk. (The "printer" in this case is the place that actually prints the books or documents, not the device connected to your computer.)

Figure 1-9.
A technical illustration shows the assembly of components. (Compaq Computer Corporation)

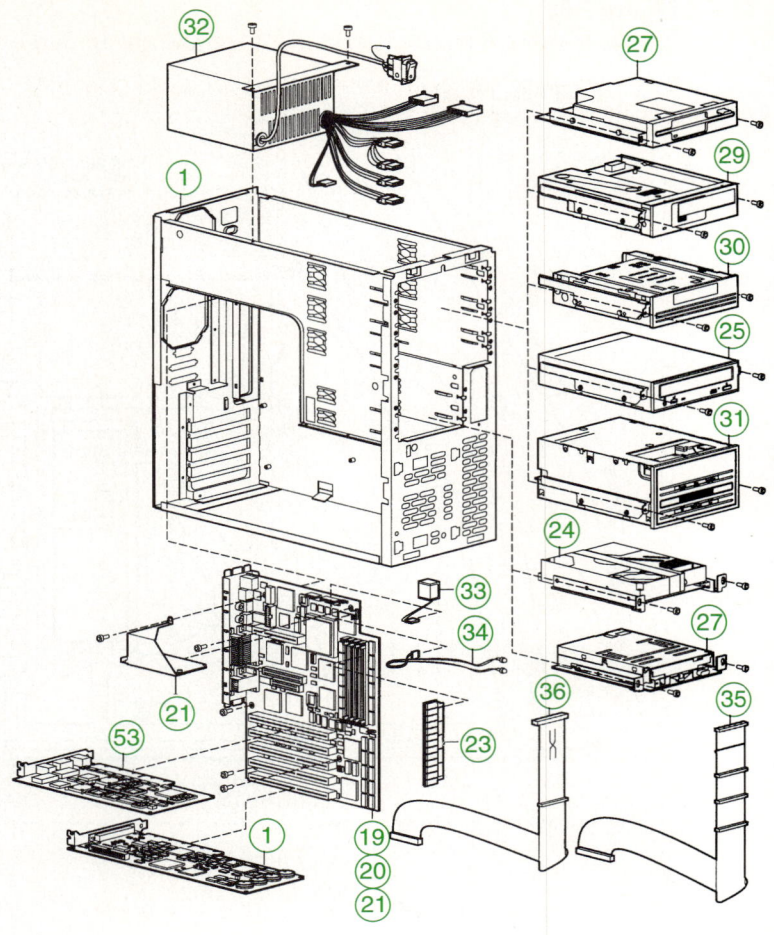

A benefit of graphics created on computer is that they can be brought directly into desktop publishing programs. Desktop publishing programs combine existing text and graphics to create pages for brochures, advertisements, and books. The text can be created in any word processing program. The graphics can come from paint or draw programs, scanned photographs, or CAD programs such as AutoSketch. Exporting drawings for desktop publishing is covered in Chapter 17.

Business Graphics

Companies use CAD to create charts and graphs, Figure 1-11. Many companies also use CAD to create product drawings and illustrations. These drawings may include diagrams of production flows and process sequences. Charts also present sales and marketing data clearly. Colors added to the graph help explain the information.

Tools of CAD

Earlier in this chapter, you were introduced to some manual drafting tools. Most drafters find that a computer is simply a more efficient drawing tool. Your computer-aided drafting system is a combination of software and hardware working together.

Figure 1-10.
Graphic designers have turned to computers to create interesting designs. (Autodesk, Inc.)

Figure 1-11.
Business graphics can be created with a CAD program. Color can be added to help explain details.

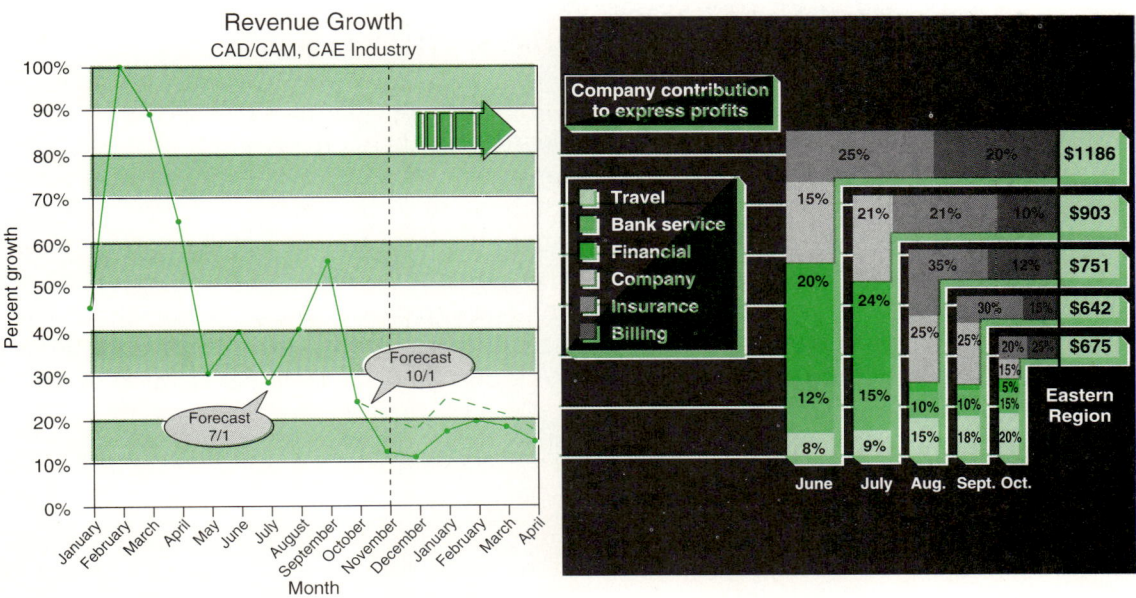

Software is the electronic instructions for a computer. This is also called a program. The software for the purpose of this text is AutoSketch for Windows.

Hardware is the physical parts of a computer system. Hardware includes the following items.

- The *computer* is the "brains" of your CAD system. The computer runs the CAD software. There are many different kinds of computers. AutoSketch runs on a Compaq, IBM, or compatible personal computer.
- A *storage device* stores the CAD program and your drawings. The AutoSketch software is loaded from floppy disks onto a hard disk drive located inside your computer. Drawings can be stored on the hard disk drive or on floppy disks. You can also backup the contents of the hard disk drive onto a tape backup system.
- The *monitor,* or display screen, is where you view the drawing. It shows the drawing held in computer memory. The monitor also shows AutoSketch menus and commands.
- *Input devices* allow you to enter commands and draw on-screen. With AutoSketch, you will use a keyboard to enter text and a mouse to pick commands and draw.
- A *hardcopy device* makes a paper copy of the drawing shown on the monitor. This paper copy is called a *print* or a *plot.* The most popular hardcopy devices are laser printers and pen plotters.
- A *peripheral device* is simply another piece of hardware used with the computer. Peripheral devices include printers, modems, and scanners. A CAD system may include one or all of these devices.

Computers

The *computer* is the center of activity in your CAD system. A computer is an electronic machine that performs operations on information. The information may be a research paper or your CAD drawing. AutoSketch runs on a personal computer (PC). A *personal computer* is a computer that uses a single computer chip to process data, Figure 1-12. Although there are many brands of personal computers, AutoSketch works only on Compaq, IBM, and compatible personal computers.

The most important parts of the computer you should learn about are the microprocessor, math coprocessor, memory, ports, and expansion slots. A computer also contains data storage devices that are important to learn about.

Microprocessor

The *microprocessor* is a single integrated circuit, or "chip." This chip executes the software program. There are many types of microprocessors used in personal computers, Figure 1-13. Most are made by Intel Corporation. Each new microprocessor is faster and more powerful than the one before. The speed of the microprocessor is measured in megahertz (MHz). One megahertz is one million cycles, or operations, per second. Microprocessors range in speed from 4.77 MHz to 100 MHz. The data bus width is how many "bits" of information the microprocessor can work on in one cycle. You should run AutoSketch for Windows on a 386SX or better microprocessor.

Figure 1-12.
A typical personal computer.

Math coprocessor

The *math coprocessor* is an optional chip that speeds up the computer. It is a "number cruncher" that allows the microprocessor to do more important things. AutoSketch will perform faster with a math coprocessor installed. PCs with 386 or 486SX processors will be helped with a math coprocessor. Personal computers that use the Intel 486DX, 486DX2, 486DX4, or Pentium chip have a math coprocessor built into the microprocessor.

Memory

Memory is the place where your drawing is stored while you work on it. Memory is often called random access memory (RAM). RAM is a collection of computer chips that is able to hold information *while the computer is on.* The contents of memory are lost when the computer is turned off. You must save your drawing to either a floppy disk or to the hard drive. Otherwise, the drawing will be lost when the computer is turned off. It is important to remember that *memory* is not the same as *disk space*.

The amount of memory is expressed in bytes. A *byte* is the memory needed to store one character. Memory is usually added in increments of one million bytes. One million bytes is one megabyte (MB). Most personal computers are sold with 2 to 8 megabytes of

Microprocessor	Clock Speed (rated in MHz)	Data Bus Width (rated in bits)	Integrated Math Coprocessor
8088	4.77	8	No
8086	4.77 to 8	16	No
286	8 to 12	16	No
386SX	16 to 33	16	No
386 (or 386DX)	16 to 33	32	No
486SX	25 to 33	32	No
486 (or 486DX)	25 to 50	32	Yes
486DX2	33 to 66	32	Yes
486DX4	100	32	Yes
Pentium	60 to 133	64	Yes

Figure 1-13.
The power of a microprocessor is expressed as clock speed and data bus width.

RAM. Your computer must have 4 megabytes of memory to run AutoSketch for Windows. Memory is typically added with *Single Inline Memory Modules (SIMMs)* that fit into special sockets in the computer.

Ports

Ports are connections on the back of the computer, Figure 1-14. Input and output devices are "hooked-up" to the computer through these ports. All personal computers have a keyboard port, monitor port, parallel port, and serial port. Most also have a mouse port. Most printers are connected to the parallel port. You can connect a plotter, modem, or mouse to the serial port. The mouse should be connected to the mouse port if the computer has one. This will allow you to use the serial port for another device. Each input and output device requires its own port.

Sometimes you need to add more devices to the computer. To do this, you must insert special circuit boards into vacant slots inside the computer, Figure 1-15. These slots are called *expansion slots.* The circuit board you insert is often called a *card.* New devices may be attached to a connector on the card that shows through the back of the computer. Some devices may be attached to the card inside the computer.

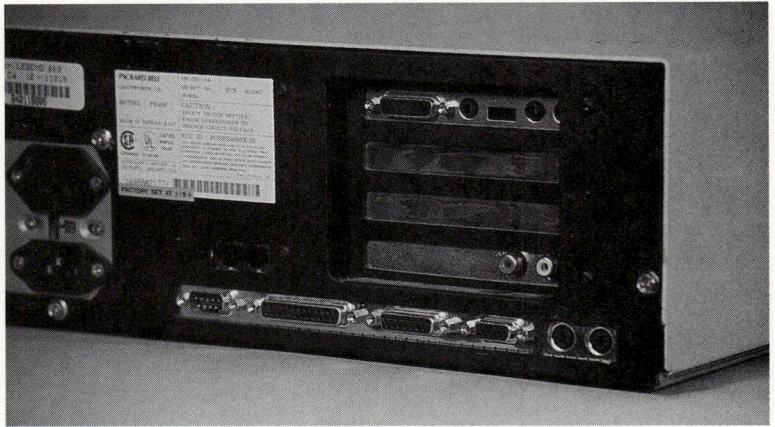

Figure 1-14.
Ports are connections on the back of the computer where you hook up input and output devices. There are several shapes of ports. Each shape has a different number and/or pattern of pins.

Data Storage Devices

Data storage devices save information for later use. These devices include hard drives, disk drives, and tape drives. When you work with AutoSketch, the computer holds the drawing in memory. If the computer is turned off accidentally, all your work is lost. This is why you should save your drawing often. Choosing the **Save** command tells AutoSketch to save the information from memory to the storage device. Also, when you are ready to leave AutoSketch or start a new drawing, save your drawing.

Hard disk drives

Hard disk drives are mass storage devices that have become standard on all personal computers, Figure 1-16. Hard disk drives are also called *hard drives* or *fixed disk drives.* Most hard drives cannot be removed. Hard drives store more information and retrieve it faster than floppy disks. In fact, AutoSketch for Windows must be run from a hard disk because the program needs more than 10MB of storage. Your drawings, however, should fit on a floppy disk.

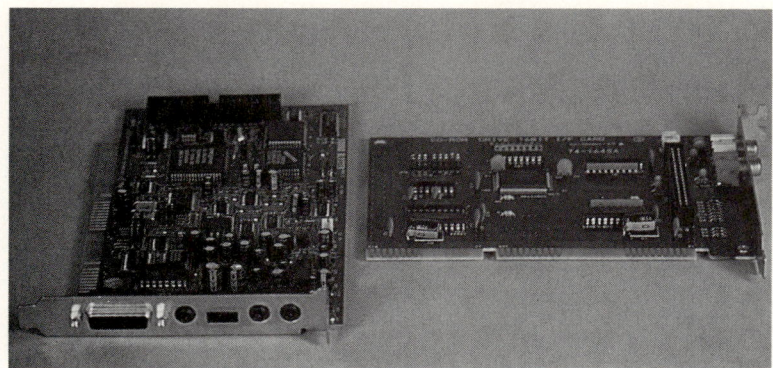

Figure 1-15.
A—Different expansion cards are available to control graphics, sound, CD-ROM drives, and modems. B—Expansion cards fit inside the computer in slots called expansion slots.

A hard drive consists of several stacked aluminum disks called *platters*. These platters are coated with magnetic particles. Several read/write heads store and load information between RAM and the platters. The hard drive is sealed in an airtight enclosure. All you see on the front of the computer is a small light. When the light is on, the hard disk is saving or looking for information.

Hard drives can store from 20 to over 2000 megabytes of information. A floppy disk can only store about 1.5 megabytes of information. To install both Windows and AutoSketch for Windows, the hard drive must have at least 40 megabytes of free space. However, a hard drive with more than 40MB free is recommended.

There are standard naming conventions for drives. The primary hard drive is named the C drive. If a second hard drive is installed in the computer, it is typically named the D drive.

Diskettes

Diskettes are 5.25" or 3.5" plastic disks. These disks are coated with magnetic particles and enclosed in a vinyl jacket or plastic case for protection. The diskette drive is usually called a *disk drive* or a *floppy drive.*

The diskette drive has a read/write head that saves, or "writes," data by charging magnetic particles on the disk's surface. The drive loads data into computer memory by "reading" the charges of the particles. To use the disk, simply insert it into the drive. If the drive has a drive lever or door, turn the lever or close the door to lock the disk in place.

Figure 1-16.
A—From the outside of the computer, all you see is the light that indicates the hard disk drive is storing or reading data. B—Inside a hard disk drive are platters where read/write heads store information. (IBM)

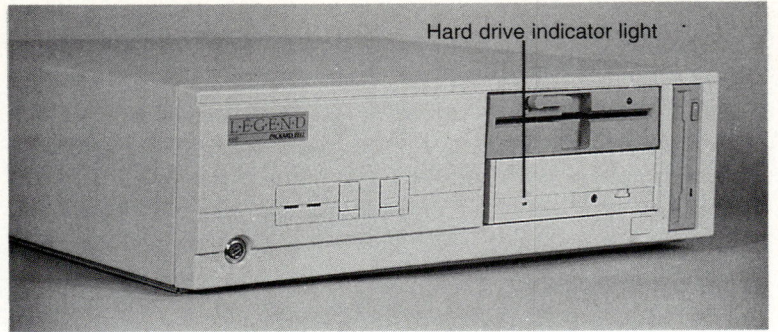

The features of a 5.25" diskette are shown in Figure 1-17. The write-protect notch determines whether data can be written onto the disk. If you cover the notch with a write-protect tab, the disk drive cannot store data on the disk.

A 3.5" diskette has a write-protect switch rather than a notch. Make sure the switch is set to WRITE DATA if you plan to store drawings on the disk. Push the switch to the READ ONLY setting to prevent you or someone else from accidentally erasing files or storing more data.

Not all diskettes are the same, Figure 1-18. There are double-density and high-density disks. The density is the compactness of the magnetic particles on the disk's surface. A high-density 5.25" diskette stores 1.2 megabytes. A high-density 3.5"

Figure 1-17.
A 5.25" high-density diskette stores 1.2 megabytes of data. The different features are shown here.

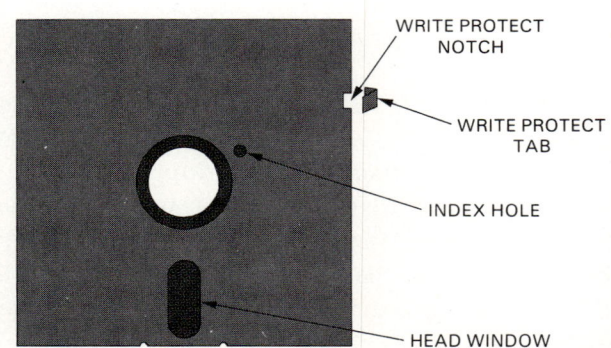

diskette stores 1.44 megabytes. A double-density 5.25" diskette stores 360 kilobytes. A double-density 3.5" diskette stores 720 kilobytes. Most computers now have drives that can read both double-density and high-density diskettes.

There are standard naming conventions for diskette drives. The primary diskette drive is named the A drive. If a second diskette drive is installed in the computer, it is named the B drive.

Formatting diskettes

Before you can store drawings on diskettes, you must carefully select the disk

Figure 1-18.
Although they look similar, 3.5" diskettes can be either high-density or double-density. High-density diskettes (on the right) hold 1.44MB of information. These disks typically have the letters HD printed on the case. Double-density diskettes (on the left) hold 720K of information.

type and prepare it by formatting. Do this before you start AutoSketch. Most computers that can run AutoSketch for Windows will have a high-density drive. These drives can read and write either high-density or double-density disks. If you have doubts, consult your instructor, supervisor, or a computer store salesperson. (Note: Many disks can be purchased preformatted.)

Formatting prepares the surface of new disks into a format readable by the computer. This can be done from the File Manager application in Windows. You can also do this from the MS-DOS system prompt.

To format a disk from the Windows File Manager, insert a new disk in the A: drive (or B: drive). Then, select the **Disk** pull-down menu. Next, select **Format Disk...** from the pull-down menu. Select the correct drive and disk capacity in the dialog box. When the selections are correct, select **OK.**

To format a disk from the DOS prompt, insert a new disk in the A: drive (or B: drive). From the hard drive DOS prompt, type the following command and press the [Enter] key. (Note: This command is for formatting a disk in the A: drive.)

C:**Format A:** ↵

Follow the computer's instructions. It takes only a few minutes to prepare the disk. After the process is done, make sure that the values for "total disk space" and "bytes available on disk" are the same. If these numbers are not the same, you may have a defective disk.

Care of diskettes

Diskettes must be handled carefully. Follow each of these precautions.

- Do not bend. Although 5.25" diskettes are flexible, bending the disk will damage the coating that stores data. Be careful when inserting the diskette in the drive.
- Do not touch the portion of the disk visible through the head window. The surface of the disk is very sensitive. Any dirt or oil from your hand may prevent data from being read or saved.
- Protect from magnetic sources. Since data is stored on diskettes using magnetism, placing the disk on a speaker, TV, or other magnetic appliance may erase the disk.
- Protect from extreme temperature or moisture. Diskettes are sensitive to temperature and moisture. Do not leave them in cold areas, sunlight, or near moisture.
- Write on the diskette label with a felt-tip pen. Do not use a ballpoint pen. Place labels on diskettes to keep track of the drawings.
- Make frequent backups. The File Manager in Windows can be used to copy an entire disk. The MS-DOS program DISKCOPY also allows you to do this. Make frequent backups in case your original disk is damaged. Making backups is discussed in Chapter 20.

Tape drives

Tape drives store information on a length of plastic tape coated with magnetic particles. The tape is wound around two reels enclosed in a plastic cartridge. These tapes are about the same size as an audio cassette tape. The tape passes against a read/write head that sends data back and forth between the computer and the tape.

Tape drives are a popular way to backup hard drives, Figure 1-19. Many tape drives can hold the entire contents of a hard disk drive. The tape cartridge can be stored in a safe place. This is important since you cannot remove most hard drives to store elsewhere. If your hard drive should fail, you can use the tape drive to restore your data to a new drive.

Monitors

With manual drafting, you create and view your drawing on paper. With a CAD system, you view the drawing on a *monitor*. A monitor is also called a display. The screen becomes your "paper" for a CAD drawing. As you create and edit a drawing, the lines appear and change on-screen. You never physically touch the drawing. Instead, the monitor shows the drawing held in computer memory.

Figure 1-19.
Tape cartridge drives are used primarily to backup the data on your hard drive. Tapes are about the same size as an audio cassette.

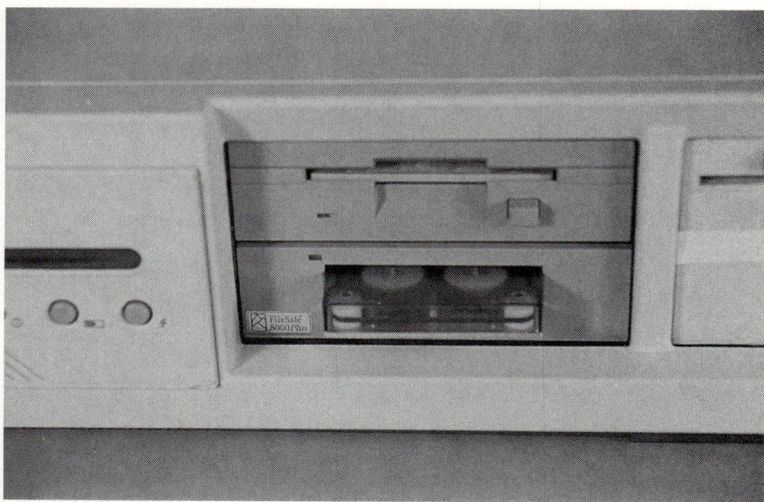

Types of monitors

There are many types of monitors, Figure 1-20. Monitors may be large and bulky, or small and flat. Their size is given in inches measured diagonally between opposite corners. The size of most monitors is from 13" to 21". Some monitors are monochrome (one color) and others show many colors. Color monitors help you separate different parts of the drawing by drawing with several colors.

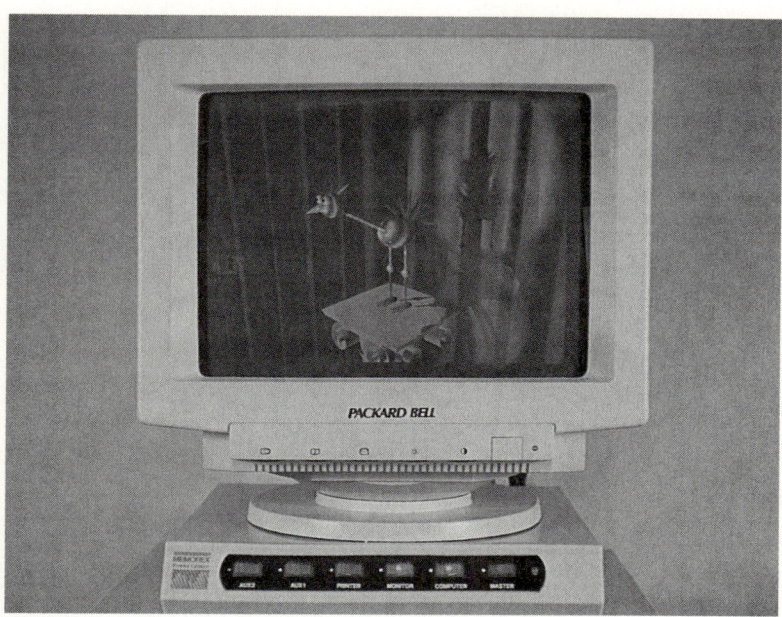

A

B

Figure 1-20.
Displays vary in size. Typical displays for home use are 14" or 15", while the displays on laptop computers can be much smaller.

Resolution

The most important difference between monitors is resolution. *Resolution* refers to how sharp an image is. Resolution is measured by the number of pixels horizontally and vertically. *Pixels* (an abbreviation for picture elements) are small dots or rectangles used to create an image. If you look closely at the screen, you can see the pixels. High resolution monitors make the drawing easier to see. Common resolutions are 640 x 480, 1024 x 768, and 1280 x 1024.

Graphics controller

Monitors require a compatible graphics controller installed in the computer. The *graphics controller* is a circuit board or chip integrated into the computer. This board controls the monitor and determines its resolution and number of colors supported. The most popular controllers are VGA (video graphics array) and SVGA (super video graphics array). Your monitor must support the type of controller installed in the computer. AutoSketch for Windows requires either a VGA or SVGA monitor and graphics controller card.

Input Devices

Input devices allow you to draw on-screen as well as enter text and commands. A cursor appears on-screen to show you where the input device will interact. The cursor might appear as an arrow, hand, crosshairs, or an underline. To select a point or command, you press a *pick button* on the input device.

There are three basic types of input devices: keyboards, mice, and digitizers. Mice and digitizers are often called *pointing devices.* AutoSketch for Windows requires a mouse and a keyboard.

Keyboard

A *keyboard* is found with every computer system. You use the keyboard to enter text and numeric values. A typical personal computer keyboard has alphanumeric keys, function keys, and cursor keys, Figure 1-21.

Figure 1-21.
This is a typical personal computer keyboard. The important groups of keys are labeled in this photo.

Alphanumeric keys

Alphanumeric keys are letters of the alphabet, numbers, and symbols like $, %, and &. These keys let you add text to a drawing. After typing in text, you must press the [Enter] key so that the computer knows to process the data.

Function keys

Function keys provide instant access to important commands. These keys are labeled F1 to F12. Pressing a function key might display a grid or erase an object.

Ctrl, Alt, Shift

Keys labeled Alt (alternate), Ctrl (control), and Shift change the meaning of character keys and function keys when held. For example, pressing function key [F3] will erase an object. However, if you hold [Alt] down while pressing [F3], AutoSketch will perform the **Arc** command.

Cursor keys.

Cursor keys are labeled with arrows and the words Home, End, PgUp, and PgDn. These keys are used to move the cursor around the screen. These keys are typically used when working with text.

Mouse

A *mouse* is a hand-held device that is rolled around a flat surface to move the cursor on-screen, Figure 1-22. The mouse has a ball under it that senses motion when moved. A mouse shows movement, not absolute position. You can pick it up and set it down again in another place without moving the cursor on-screen.

A mouse has one or more buttons. The ***pick button*** is usually the left-hand button (if the mouse has more than one button). Push the pick button to select a command or place a point. This process is called *clicking.* Other buttons on the mouse may be used to perform commands instantly. These buttons work just like function keys.

Trackball

A device that works like a mouse is the ***trackball.*** A trackball is simply an upside-down mouse. You move the ball with your thumb. The buttons are located on the top or side of the trackball enclosure.

Figure 1-22.
Moving a mouse across a flat surface or special pad moves the cursor shown on the display.

Joystick

A device that performs the same functions as a mouse is a joystick. A *joystick* is a small box with a movable shaft. Pushing the shaft in one direction moves the screen pointer in that direction. Pressing the pick button selects commands and points. The pick button is found on the box or shaft. Although once very popular, most joysticks have been replaced by mice. AutoSketch for Windows does not support the use of a joystick. However, most joysticks have a "mouse mode" that might allow you to use the joystick with AutoSketch.

Digitizer

A *digitizer* consists of a rectangular plastic tablet and a stylus or puck, Figure 1-23. Moving the stylus or puck over the tablet moves the screen pointer. A digitizer can be used to trace existing drawings. The most popular sizes of digitizers are 9" x 9", 12" x 12", and 12" x 18".

Inside a digitizer tablet is a grid of closely spaced wires. The puck or stylus senses the intersection of these wires. The screen pointer is shown in the absolute position of the puck or stylus. A *puck* looks like a small box with a plastic extension. Inside the plastic extension are metal crosshairs that sense position on the digitizer. One button on the puck is the pick button. The other buttons may perform other commands. The *stylus* looks like a ballpoint pen. Moving it close to the tablet moves the screen pointer. If you press down on the stylus, you activate the pick function.

Although AutoSketch does not support digitizers, you may need to use one for other CAD programs. The advantage of a digitizer compared to a mouse is that you can trace an existing drawing. To do so, tape the drawing to the tablet surface. Then, select the proper commands to trace the lines, circles, arcs, and curves.

Hardcopy Devices

A *hardcopy device* takes your drawing from memory and puts it on paper. Popular hardcopy devices include laser printers, pen plotters, electrostatic plotters, inkjet printers, thermal printers, and dot matrix printers.

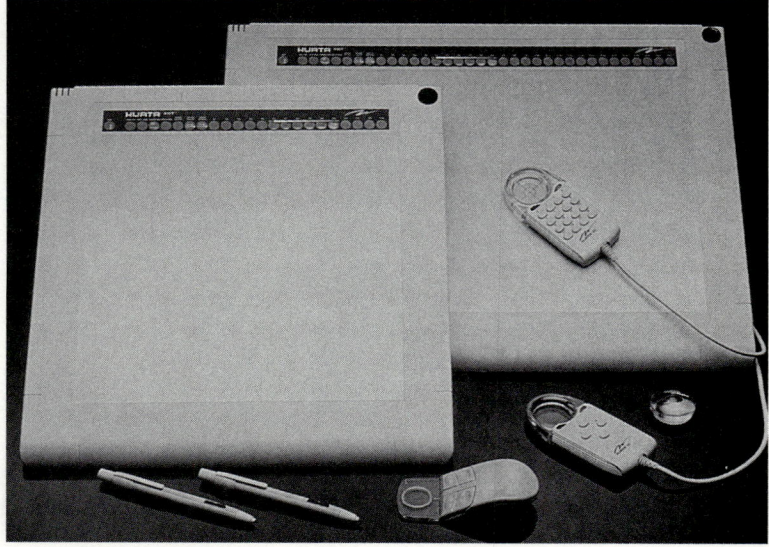

Figure 1-23.
A digitizer is a common input device for CAD systems. A stylus or puck attached to the digitizer is used to enter commands and draw. (Kurta)

Laser printers

Laser printers have become a very popular hardcopy device. Laser printers create a good image on regular copy paper. In fact, the technology of laser printers is very similar to a photocopy machine. Laser printers form an image by exposing a photosensitive drum to light. This light is either a laser or a light-emitting diode (LED). The light beam draws the image onto the drum. The drum becomes electrically charged by the light. As the drum rolls against a toner cartridge, the toner dust is attracted to the drum. The toner is then transferred to paper and heated so that the toner bonds to the paper. The drum is erased after each drawing and re-exposed to make a new drawing. Although color laser printers are available, the most common laser printers print black toner on white paper.

The quality of laser printers is measured in resolution. *Resolution* of a printer is the number of dots placed per inch. The more dots per inch, the sharper the image will be. Most laser printers have resolutions of at least 300 dots per inch. Laser printer resolution can range from 300 to 1200 dots per inch.

One limitation of laser printers is the paper size. Most laser printers only support regular letter-sized paper, (8 1/2" x 11"). This is also called A-size sheets. Some support legal size paper, and others even support B-size sheets (11" x 17"). Still, most technical drawings require larger paper so that a complex drawing is large enough to be readable. AutoSketch will print a large drawing across several sheets of paper. The sheets can then be taped together to make the large print.

Pen plotters

Pen plotters move a pen across a drafting medium such as paper, vellum, or film to create a hard copy of your drawing, Figure 1-24. With a *flatbed pen plotter,* you tape the paper to the flat surface. The paper remains fixed while the pen moves. A *microgrip pen plotter* is somewhat different. Pinch rollers hold the paper on the plotter surface. Both the pen and paper move to draw the image. The pen moves in one direction along a rail. The pinch rollers move the paper to produce the other direction. The advantage of a microgrip plotter is that they support C-size and D-size sheets of paper (17" x 22" and 24" x 36"), where flatbed plotters support A-size and B-size sheets.

Pen plotters are either single-pen or multiple-pen. Both can plot drawings using more than one pen. For a single-pen plotter to do this, the plotter will stop after a certain linetype or color is finished. Then, you insert a different pen. Multipen plotters hold more than one pen in a rack or carriage. The plotter selects and exchanges pens automatically.

There are several types of pens. Wet ink, or liquid ink, pens feed ink through a narrow metal tip. Felt-tip and plastic-tip pens are like markers. Pressurized roller ball and ballpoint pens are very similar to ballpoint writing pens. Pens are discussed in more detail in Chapter 16.

Electrostatic plotter

An *electrostatic plotter* is very similar to a laser printer. It does not draw one line at a time like the pen plotter does. Instead, it forms images with a pattern of small dots. However, that is where the similarity to a laser printer ends. Rather than attracting toner to a drum and then transferring it to paper, an electrostatic plotter applies an electrical charge to the paper. The paper is then fed through toner. Most electrostatic plotters support C-size and D-size paper.

Figure 1-24.
Various pen plotters. A—One flatbed plotter and three microgrip plotters for A-size and B-size drawings. B—A large format multipen microgrip plotter for drawings up to D-size. (Houston Instruments, A Summagraphics Company)

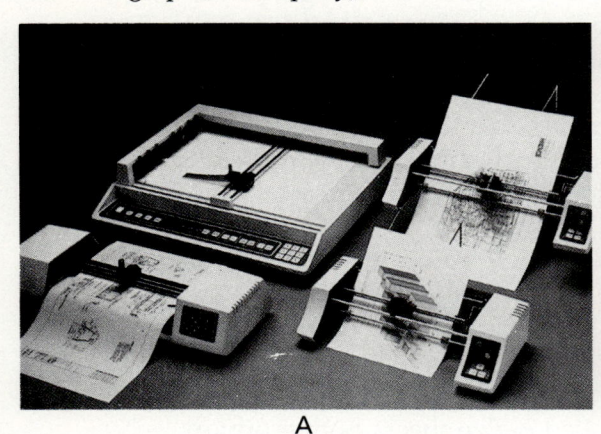

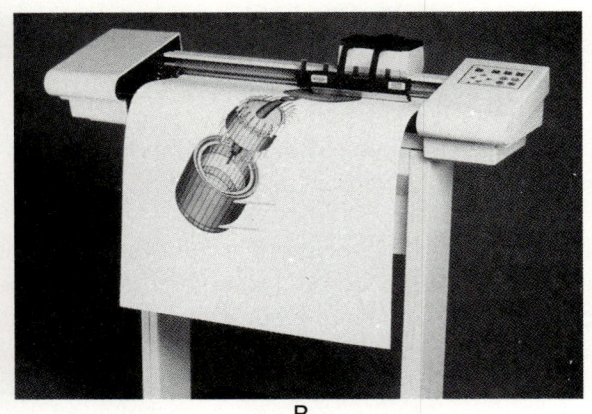

Inside the plotter is a writing head that moves across the paper to apply tiny electrostatic charges. The writing head places the charges on paper in the dot pattern. The paper then passes through toner. The toner sticks to the paper where the charges were placed. Most plots are made with black toner on white paper. However, some plotters make colored drawings by using red, blue, and yellow toner. By combining the colors, the plotter can create multicolor drawings.

The advantage of electrostatic plotters and laser printers over pen plotters is speed. A pen plotter takes a long time to draw each individual line of a complex drawing. An electrostatic plotter can place its dot pattern much quicker, usually 20 times as fast. The disadvantage of electrostatic plotters over laser printers is cost. Most are 10 times as expensive as pen plotters and laser printers.

Ink jet printers

Ink jet printers form images by shooting ink droplets onto paper. The printer guides the drops to hit the paper at precise locations. Like laser printers and electrostatic plotters, ink jet printers use a dot pattern to form images. Most have a resolution of 200 to 400 dots per inch. Unlike laser printers, an ink jet printer prints one line at a time. After completing one line, the printer moves the paper up and prints the next line. This process can be slower than a laser printer. However, many ink jet printers can print in color.

Thermal printers

Thermal printers have become more popular because they can make high-quality multicolor prints. The plotter works by melting ink on a ribbon or transfer sheet and applying the hot ink to paper. It is possible to have ribbons or sheets that carry several colors. The colored ink is deposited in several layers to produce a multicolor image. Like most devices, thermal printers use a dot pattern to form images. Most have a resolution of 300 dots per inch, but support only A-size paper.

Dot matrix printer

The *dot matrix printer* is an inexpensive way to print documents, Figure 1-25. It can also print a low-quality to medium-quality drawing. The image is formed by the impact of 9 or 24 steel wires mounted in a print head. The wires push out and impact the paper through an inked ribbon.

Figure 1-25.
Although dot matrix printers are mostly used to print text, they can be used to plot low-resolution drawings. (IBM)

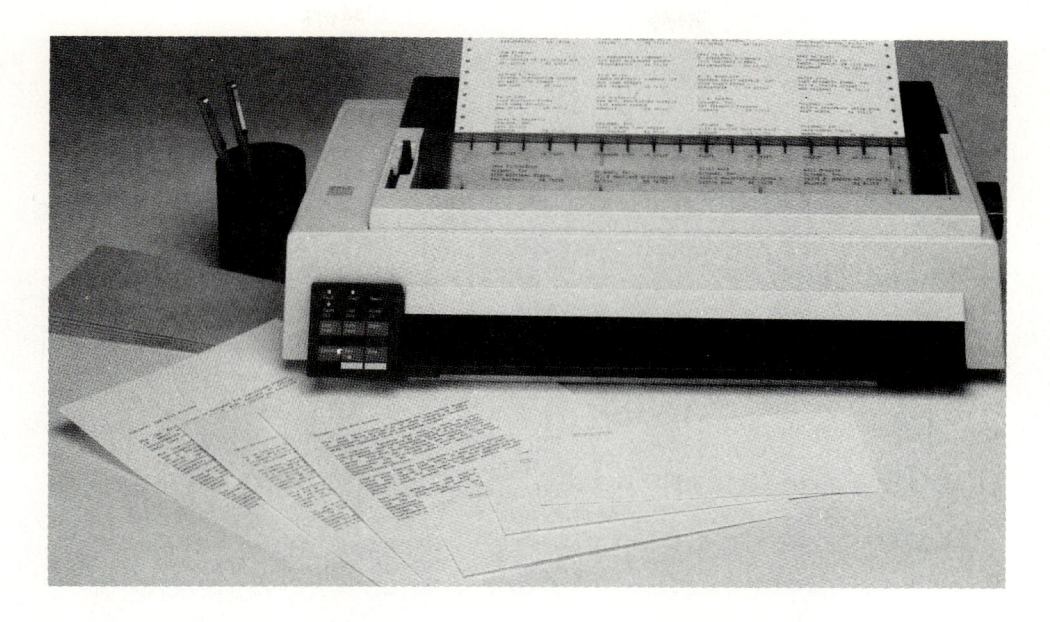

The printer makes one line at a time. The print head moves across the paper. The paper then moves up one line and the print head prints another line. The resolution can be as high as 150 dots per inch.

Drafting Positions and Qualifications

Drafters spend hours converting engineering sketches and specifications into working drawings. These drawings are used to build or manufacture products. Most drafters have specialized training in a single field, such as electronics, mechanical, architectural, or civil drafting. A drafter should know a great deal about their drafting field. The introduction of computer-aided design has affected many drafting-related occupations. Tedious tasks have been greatly reduced. This makes a single drafter much more productive.

Traditional Drafting Occupations

The following are the most common job titles in the field of technical drafting.

- An *apprentice drafter* redraws, revises, and repairs existing drawings. They may also develop simple drawings under the close supervision of a drafter.
- A *drafter*, or *detailer*, develops working drawings from specifications received from sketches, from notes made by an engineer or designer, and from verbal instructions, Figure 1-26.
- A *checker* reviews prepared drawings and looks for errors that were not caught by the original drafter. The checker usually has worked as a drafter for some time.
- A *chief drafter*, or *senior drafter*, supervises the work of the drafting department. This person also may develop complex working drawings. They usually have production experience as well as drafting skills.

Figure 1-26.
Drafters often have to work with both manual drafting equipment and CAD systems. (IBM)

- A *technical illustrator* prepares two-dimensional and three-dimensional (pictorial) drawings for assembly or presentation. This may involve adding colors to the drawings to make them look more realistic. A technical illustrator has both technical and artistic skills.
- A *designer* prepares sketches and writes specifications to develop new products. These materials are turned over to a drafter for detailing. Designers should know processes, materials, standards, and codes of their field. An industrial designer develops product concepts or works to improve the function and appearance of existing products. A tool and die designer develops tools and devices needed to manufacture industrial products. This could include cutting tools, fixtures, jigs, and dies.
- An *engineer* develops ideas into practical designs, including the process used to manufacture a product. This position requires college training, practical experience, and usually a professional license. A product engineer designs products used directly by a consumer. A manufacturing engineer develops parts, machines, and processes used to make products.
- *Architects* design plans for residential, commercial, and industrial structures. A professional architect typically has a four-year degree and a license, and may have a master's degree.

Positions related to CAD

Job titles have been added to most drafting departments that use computers. The duties of apprentice and detail drafters have remained much the same. However, some of the job titles have changed.

- Drafters using a CAD system may be called *CAD drafters* or *CAD operators*. This does *not* mean that they know more about CAD than their field of drafting. Remember, CAD is only a tool, like a pencil and T-square.

- A *systems manager* supervises the entire CAD system. The tasks include loading software, scheduling system use, and reporting hardware problems and software errors.
- A *CAD programmer* develops and maintains the functions of the CAD software. As new releases of software become available, the programmer updates the system. A CAD programmer may also write programs (called *macros*) that automate common tasks to make drafters more productive.

Training

Training to become a CAD drafter involves several skills. These skills include drafting techniques, use of software, and use of hardware.

Drafting skills

The most important skills you can have are those related to drafting techniques. You will need to show competence in developing drawings, dimensioning, and writing specifications. You should also know the materials and processes of your field. Your knowledge of drafting standards is more important than the tools you use to draw. Your education may include training beyond high school at a technical school or college.

Software skills

Many of the skills you learn using AutoSketch apply to other CAD systems. Once you take a drafting position, you should master the functions of the software you will be expected to use. These skills are learned over time. Your drafting speed will increase as you become familiar with the system.

Hardware skills

As a CAD drafter, you will be expected to understand different hardware equipment. Learn the basic functions of computers, input devices, monitors, and output devices. Also know the proper care of this equipment.

The Drafting Environment

Computers require special care to maintain smooth operation. The temperature and humidity level of the room should be controlled. Also, static electricity must be eliminated. The power source should be protected of electrical spikes and surges. These surges can damage delicate computer circuits.

Drafters sit for long periods of time. The seating position in front of a computer is more flexible than a manual drafter's. You are not required to sit above the drafting table on a stool. Choose an adjustable cushioned chair and position your eye level at a convenient angle to the display screen. Also, certain devices are made to help reduce the risk of carpal tunnel syndrome when using a mouse or keyboard.

Another comfort factor is the monitor. Change the brightness and contrast, so you do not have to squint. Make sure the image is clear. This reduces eyestrain.

Job Outlook

The U.S. Department of Labor indicates a need for future CAD drafters. You can find information on jobs in the *Occupational Outlook Handbook* or the *Dictionary of Occupational Titles*. These and other books on jobs can be found at your local library.

While computers have increased productivity, they have not reduced the number of available jobs. Human talents will continue to be vital. The computer will become more of a coworker than a tool. In the future, using a computer will require a higher level of education and skill. This trend will make jobs available in education, as well as in drafting rooms.

Important Terms

Alphanumeric keys
Apprentice drafter
Architects
Architectural drafting
Byte
CAD drafters
CAD programmer
Card
Checker
Chief drafter
Civil drafting
Clicking
Computer
Computer-aided drafting (CAD)
Computer-integrated manufacturing (CIM)
Cursor keys
Data storage devices
Designer
Detailer
Digitizer
Disk drive
Diskettes
Dot matrix printer
Drafter
Drafting
Electrical plans
Electronics drafting
Electrostatic plotter
Elevation plans
Engineer
Expansion slots
Flatbed pen plotter
Floor plans
Floppy drive
Formatting
Foundation plans
Function keys
Graphics controller
Hard disk drives
Hardcopy device
Hardware
Heating, ventilation, and air conditioning (HVAC) plans
Ink jet printers
Input devices
Joystick
Keyboard
Landscape plans
Laser printers
Macros
Manual drafting
Math coprocessor
Mechanical drafting
Memory
Microgrip pen plotter
Microprocessor
Monitor
Mouse
Network
Pen plotters
Peripheral device
Personal computer
Pick button
Pixels
Plans
Platters
Plot
Plot plans
Plotting
Plumbing plans
Pointing devices
Ports
Print
Printing
Puck
Resolution
Senior drafter
Server
Single Inline Memory Modules (SIMMs)
Software
Standards organizations
Storage device
Stylus
Systems manager
Technical drafting
Technical illustration
Technical illustrator
Template
Thermal printers
Trackball
Working drawings

Review Questions

Give the best answer for each of the following quesions.
1. The process of creating a drawing on paper using drafting instruments is called _____.
2. The process of creating a drawing on-screen using a computer and drawing software is _____.
3. List five benefits of CAD over manual drafting.
   A. _____
   B. _____
   C. _____
   D. _____
   E. _____
4. Preparing drawings that describe the shape, size, and features of manufactured products is called _____.
5. List eight plans that typically make up a set of working architectural drawings.
   A. _____
   B. _____
   C. _____
   D. _____
   E. _____
   F. _____
   G. _____
   H. _____
6. The occupation that involves making two-dimensional or three-dimensional (pictorial) drawings for assembly or presentation is _____.
7. List five hardware components that make up a computer-aided drafting system.
   A. _____
   B. _____
   C. _____
   D. _____
   E. _____
8. AutoSketch for Windows runs on what type of computer? _____
9. The single integrated circuit, or "chip," that processes data in a personal computer is called a(n) _____.
10. Personal computer memory is known as _____.
11. The connections on the back of the computer where you connect input and output devices are called _____.
12. Removable disks often used with personal computers are called _____.
13. The _____ is a mass storage device used with personal computers. This device is typically nonremovable.
14. What type of storage device is often used for backing up the information on hard drives? _____
15. With a CAD system, you view your drawing on a(n) _____.
16. The clearness of a displayed image depends on the _____ of the monitor.
17. List the two input devices used with AutoSketch.
    A. _____
    B. _____
18. What is a pointing device? _____

19. Which types of hardcopy devices print a drawing using a pattern of dots?
   _____
20. The person who develops detail and assembly drawings from specifications received from an engineer or designer is a(n) _____.
21. Training to become a CAD drafter typically involves skills in what three areas?
   A. _____
   B. _____
   C. _____

Activities

Complete the following activities. Use a separate piece of paper if you need more room.
1. Identify the manufacturer and model of the computer you, your school, or company uses for computer-aided drafting.
   _____
2. Identify the microprocessor found in the computer you use.
3. Identify how much memory (RAM) your computer has.
4. Identify the speed, in megahertz (MHz), of the microprocessor found in your computer.
5. Is your computer connected to a network?
6. What types of disk storage are used by your system?
7. What is the storage capacity of the floppy used in your computer?
8. What is the storage capacity of the hard drive in your computer?
9. Determine the type of monitor you have. Is the monitor monochrome or color?
   _____
10. Where are the contrast and brightness controls on your monitor?
11. Determine the graphics controller of your computer (such as VGA or SVGA).
12. Determine the resolution of you monitor.
13. Identify the types of input devices used with your CAD workstation.
   _____
14. Determine the type of hardcopy device(s) attached to your computer for printing or plotting drawings.
   _____
15. Identify the resolution (in dots per inch) of your hardcopy device. If you have a pen plotter, is it a single-pen or multipen plotter?
16. Identify two magazines that provide CAD-related topics. Give the titles of two articles found in each magazine.
17. Contact a firm using computer-aided drafting. Interview the supervisor of the drafting department. Identify four positions in the drafting department and list the qualifications of each.
   _____
   _____
   _____
   _____
   _____
   _____

18. Contact a firm using computer-aided drafting. Find out these items:
    A. Identify the computer(s) the company is using.
    _____

    B. Identify the data storage devices the company is using.
    _____

    C. Identify the type monitor and its resolution.
    _____

    D. Identify the input devices used with the CAD workstation.
    _____

    E. Identify the hardcopy devices the company is using.
    _____

    F. Identify what type(s) of CAD software they use.
    _____

The AutoSketch drawing editor.

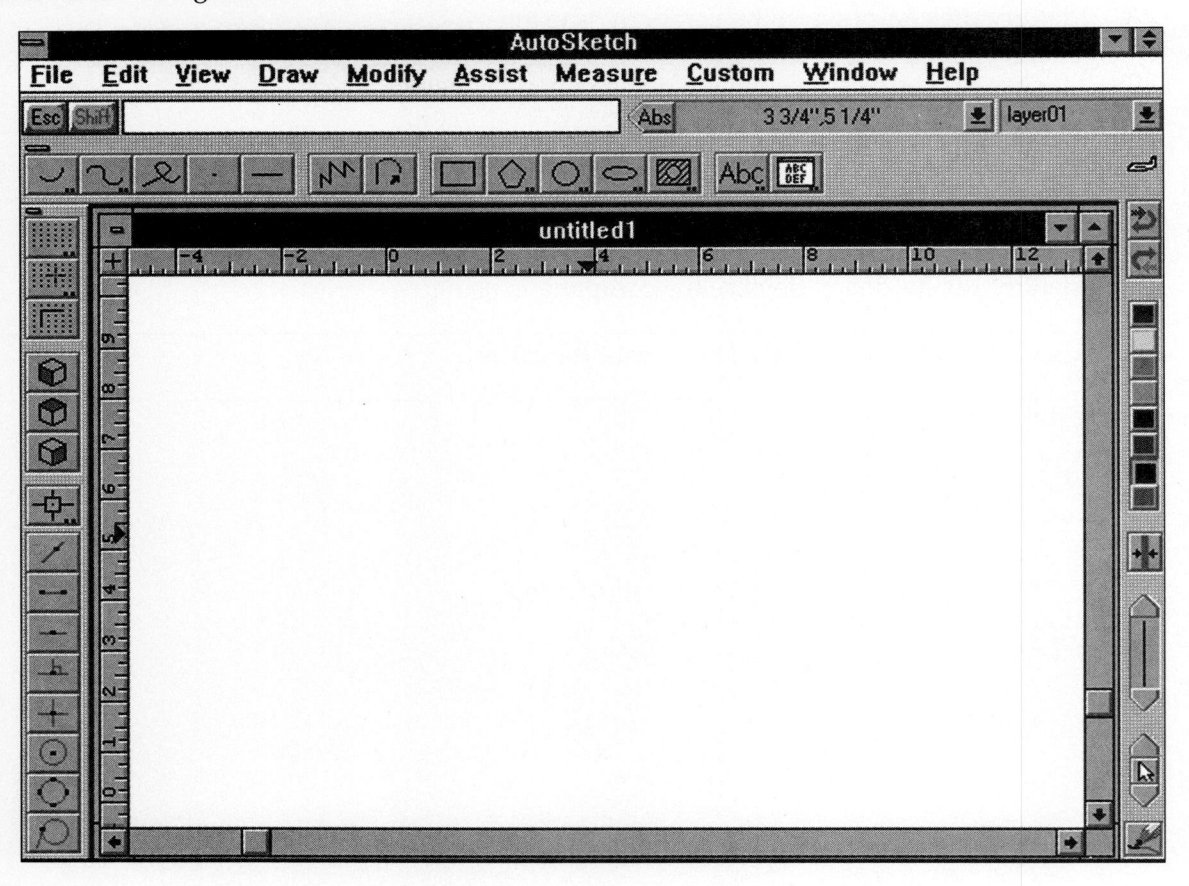

C H A P T E R   2

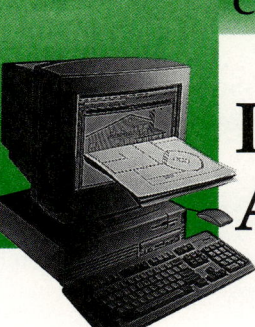

Introduction to AutoSketch

Objectives

After studying this chapter, you will be able to:
- Explain the difference between AutoSketch and other CAD programs.
- Explain the difference between AutoSketch and paint or draw software.
- List the steps taken to start AutoSketch.
- Identify the features of the AutoSketch window.
- Use the cursor to select point locations and commands.
- Select commands using tools and toolboxes.
- Change between toolboxes and pull-down menus.
- Select commands using pull-down menus and function keys.
- Explain how AutoSketch communicates to the user with prompts and dialog boxes.
- Obtain help with AutoSketch functions.

Why Use AutoSketch?

AutoSketch is a unique computer program. It is both an easy-to-use drawing tool and an excellent program for technical drafting and illustration. Instead of using pencils and paper, you edit a drawing held in computer memory. This makes it easy to draw and revise with great precision. AutoSketch is a great tool for students, drafters, engineers, and home users. All the benefits of computer-aided drafting discussed in Chapter 1 are offered by AutoSketch.

AutoSketch Versus Other CAD Programs

AutoSketch is not the most powerful CAD program. It is not meant to be. The makers of AutoSketch purposely designed the program with fewer functions than its big brother, AutoCAD. This has many advantages for new users.

AutoSketch is easy to learn. It has fewer and simpler commands than other programs, so you will learn fast.
- AutoSketch still contains the main functions often found in more powerful computer-aided drafting software. This makes it easy for you to transfer your skills to a more powerful program such as AutoCAD.
- AutoSketch is inexpensive. The price makes it affordable for schools, the home user, and industry.

- AutoSketch is well suited for sketches, schematics, charts, and simple drawings. It helps you "jot down" ideas quickly and accurately.
- AutoSketch does not require that you remember command names to type in. All commands are shown as icons (pictures) or listed in on-screen menus. You simplypick an icon of the command to perform.

AutoSketch Versus Paint and Draw Software

You may have worked with other graphics programs before trying AutoSketch. Many of these programs found today are paint or draw programs.

Paint programs

Paint programs create and store images as dot patterns. The dots align with the pixels of your display screen. These dots may be black, white, or colored. You can recognize paint programs because they allow you to "spray paint" color on-screen and use a "brush."

The main drawback to paint graphics is that the resolution is fixed. This is because your drawing is created with dots. For example, paint programs do not recognize that a group of dots represents a circle or line. If you enlarge the image, the dots become bigger and more noticeable. The image also becomes jagged as it is enlarged, Figure 2-1.

Figure 2-1.
Paint software store images as dots. If you enlarge a "painted" shape, the dots become more visible.

Draw programs

Draw programs are object oriented. This means that they store each *object* mathematically as a single item. These objects include lines, circles, and arcs. For example, a circle is a single object remembered by its center point and radius. It is not remembered by the individual dots that form the circle on-screen.

Draw programs allow you to enlarge, stretch, or reduce the graphic without losing detail. However, most draw programs do not allow you to place dimensions or other items necessary for drafting.

CAD programs

A *CAD program* is a draw program with special functions needed for drafting. AutoSketch allows you to draw lines, circles, and boxes just like a draw program does. However, AutoSketch also allows you to draw ellipses, complex curves, and add dimensions. These features are necessary to create a technical drawing.

All CAD programs offer the same general set of commands to create drawings. AutoSketch refers to items you add to your drawing as *objects.* Other CAD drawings may call them *entities.*

Your AutoSketch System

Chapter 1 discussed the hardware that makes up a typical computer-aided drafting system. AutoSketch does not support every computer, monitor, and hardcopy device currently manufactured. However, because AutoSketch for Windows is based on the Windows operating environment, any computer that can run Windows can also run AutoSketch. The AutoSketch *Getting Started* manual lists hardware that AutoSketch supports. It also tells how to install and configure AutoSketch to work with your equipment. Make sure that you, your instructor, or supervisor has installed both Windows and AutoSketch for Windows properly before proceeding further with this book.

The Windows Environment

AutoSketch for Windows is an application designed for the Windows operating environment. Windows is a graphical user interface that uses icons to represent commands and functions. You will use a mouse to move a cursor across the screen. This cursor is used to navigate and select commands. For example, rather than typing in commands at the DOS prompt, you simply "double-click" on an icon to start the program. (To double-click is to click the pick button twice.)

In addition, Windows allows you to have more than one program running at the same time. You might have AutoSketch and a word processing program loaded at the same time. The ability to run more than one program at the same time is called *multitasking.*

Windows applications are also very consistent. You will always find a **File** menu where you start and save your work. You will also find a **Window** menu to help you organize your windows, and a **Help** menu to get on-line help. You do not need to be an expert with Windows to use AutoSketch. However, learning more about Windows will help you use Windows-based applications better.

Starting AutoSketch

Before you can use AutoSketch, you must boot the computer. *Booting* is the process of turning on the computer and loading the operating system. The *operating system* is a special program that tells the computer how to interact with the hardware and your files. The operating system used with IBM and compatible computers is called MS-DOS. This stands for Microsoft Disk Operating System.

To boot, turn on peripheral devices first, and then the computer. The hard drive will "buzz" and "whir" while the operating system loads automatically. When the computer is ready, a system prompt, shown as C:\↵,appears. The *prompt* tells you that the computer is ready for your instructions.

Once the DOS prompt appears, you must start Windows by typing WIN and pressing [Enter].

C:\> **WIN**↵

Note: Many computers boot directly into Windows. If your system boots into Windows, you will not need to start Windows from the DOS prompt.

Once Windows is loaded, the **Program Manager** window appears. You will see the AutoSketch program group, Figure 2-2. The group will either be open or appear as an icon.

Figure 2-2.
The AutoSketch icon appears in the AutoSketch for Windows program group. If the program group is not open, it will appear as an icon. The icons at the bottom of the screen represent program groups that are not open. Double-click on a program group icon to open it. Double-click on the AutoSketch icon in the AutoSketch for Windows program group to start the AutoSketch program.

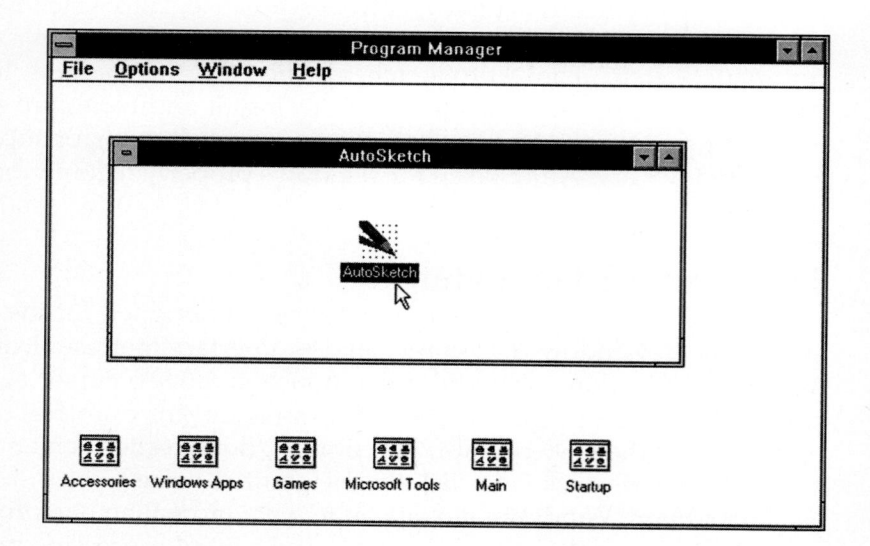

If the group is an icon, double-click on the icon to display the AutoSketch for Windows program group. The AutoSketch icon will appear in the program group. Double-click on the AutoSketch icon to start AutoSketch.

Double-click means to press the pick button on the mouse twice very fast. Remember, the pick button in most cases will be the left mouse button. Double-click is an important term to know. Almost every application in Windows will require you to double-click on icons.

AutoSketch Window Layout

When you start AutoSketch, the **Select Template** dialog box appears, Figure 2-3. This lets you choose a template to start a new drawing. A *template* is a file of preset AutoSketch options. Templates are discussed in more detail in Chapter 10. For the purposes of this chapter, simply pick the Normal template and then **OK**.

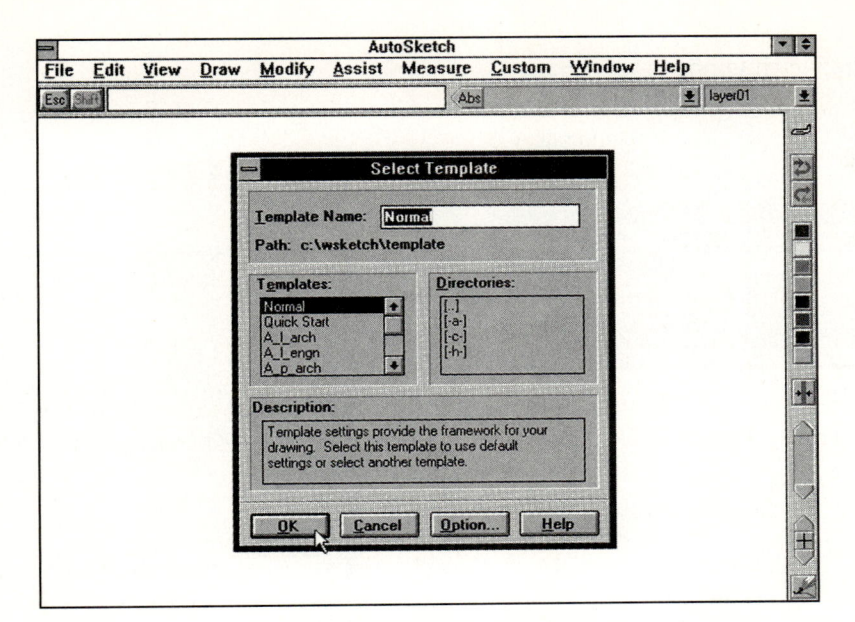

Figure 2-3.
The **Select Template** dialog box appears when you start AutoSketch or begin a new drawing.

If you do not want the **Select Template** dialog box to display each time you start AutoSketch, pick the **Option...** button. Then pick **Show template dialog on startup** to deselect it. Then, click on **OK**. However, it is recommended that you leave this option selected (check mark appears).

Once you pick **OK**, the default AutoSketch window appears. *Default* means the layout that AutoSketch shows the first time you use AutoSketch after installing it. This is important to know, since AutoSketch allows you to customize the layout of the screen. You may later decide to modify the AutoSketch window with different commands that are more appropriate for how you use the program.

Main Window

There are several items in the main AutoSketch window that are useful. Some of these items are shown in Figure 2-4. The following sections define the items shown in Figure 2-4.

Quick Help

As you pass the cursor over any tool in a toolbox, **Quick Help** shows the name of the tool and its purpose.

Menu Bar

The *Menu Bar* contains a list of menus. When you choose a menu name, AutoSketch displays either a pull-down menu or toolbox. Each pull-down menu or toolbox offers a list of AutoSketch drawing functions you might choose.

Horizontal Status Bar

The *Horizontal Status Bar* provides a prompt box, the location of your cursor on the drawing, and the current layer. It also contains the **Esc** and **Shift** buttons.

Figure 2-4.
There are several different features of the AutoSketch window.

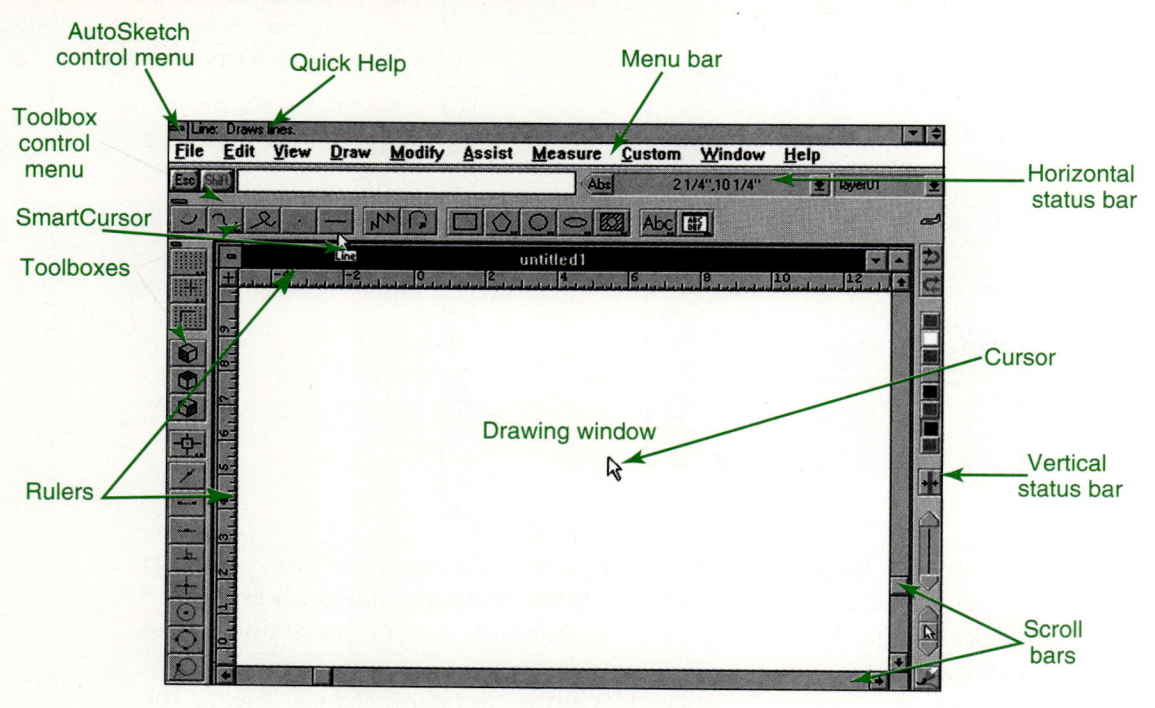

Labels on figure: AutoSketch control menu, Quick Help, Menu bar, Toolbox control menu, Horizontal status bar, SmartCursor, Toolboxes, Cursor, Drawing window, Vertical status bar, Rulers, Scroll bars

Toolbox
A *toolbox* shows AutoSketch commands as icons. An *icon* is a small picture that represents an AutoSketch function, such as **Line, Circle,** and **Copy.** When you first use AutoSketch, two toolboxes will appear. The *horizontal toolbox* reflects the menu name you choose. The *vertical toolbox* contains the **Attach** toolbox.

Drawing window
The *drawing window* is the space where you will create your drawing. The title bar shows the file name of the drawing. It will say "untitled1" for a new drawing until you save the drawing using a specific name.

Cursor
The *cursor* shows your position on-screen, and allows you to choose commands, draw, and select items to edit.

SmartCursor
The *SmartCursor* shows the name of each tool as you move the cursor over it. With the **SmartCursor**, you do not have to memorize the function of each icon.

Rulers
Rulers along the top and left of the drawing area show measurements and your current position on the drawing.

Scroll bars

Scroll bars along the bottom and right of the drawing area allow you to move across the drawing. Click on an arrow to move a short distance. Drag the box in the scroll bar to move any distance horizontally or vertically. Scroll bars are useful when you need to view a different part of the drawing.

Vertical Status Bar

The *Vertical Status Bar* provides settings for the drawing color, width, linetype, and cursor. It also contains the **Undo, Redo,** and **Redraw** buttons.

AutoSketch control menu

The *AutoSketch control menu* provides Windows functions that let you resize, move, or close the entire AutoSketch window. This menu is a standard menu found in all Windows applications. Simply click on the small bar in the upper left-hand corner of the screen to open the menu.

Toolbox control menus

The *Toolbox control menus* allow you to copy the toolbox to another location on-screen. These menus also allow you to change the toolbox to a pull-down menu. Simply click on the small bar at the left or top of the toolbox to open the menu.

Horizontal Status Bar

The following sections define the items in the *Horizontal Status Bar*. Refer to Figure 2-5 as you go through these sections.

Figure 2-5.
The horizontal status bar provides useful information about your drawing.

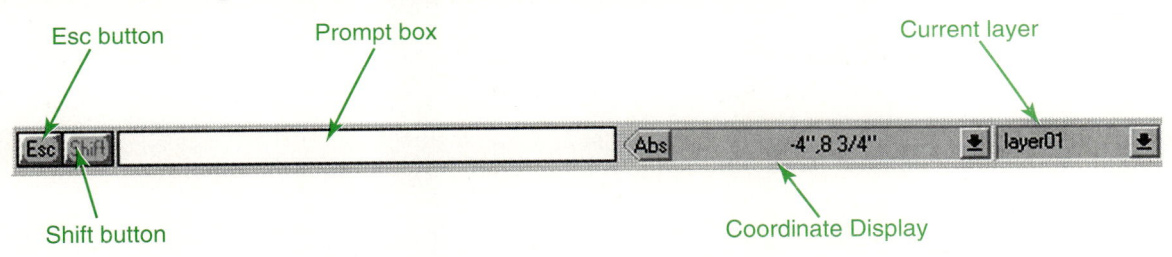

Esc button

The **Esc** button (or icon) cancels a command, ends the current command, or closes a dialog box. You can also press the [Esc] key on the keyboard.

Shift button

The **Shift** button (or icon) allows you to select multiple objects to edit. You can also hold down the [Shift] key to perform the same function.

Prompt box

When AutoSketch needs you to supply information or perform an action, it will ask so in the *prompt box.*

Coordinate Display box

The *Coordinate Display* shows your current location on-screen, or your location relative to the most recent point you placed. It also allows you to select the coordinate type to draw with. Coordinates are discussed in Chapter 5.

Current Layer box

The *Current Layer box* shows the current layer. You can also change, add, or delete layers using this box. Layers are discussed in Chapter 10.

Vertical Status Bar

The following sections define the items in the vertical status bar. Refer to Figure 2-6 as you go through these sections.

Figure 2-6.
The vertical status bar provides quick access to several commonly used items.

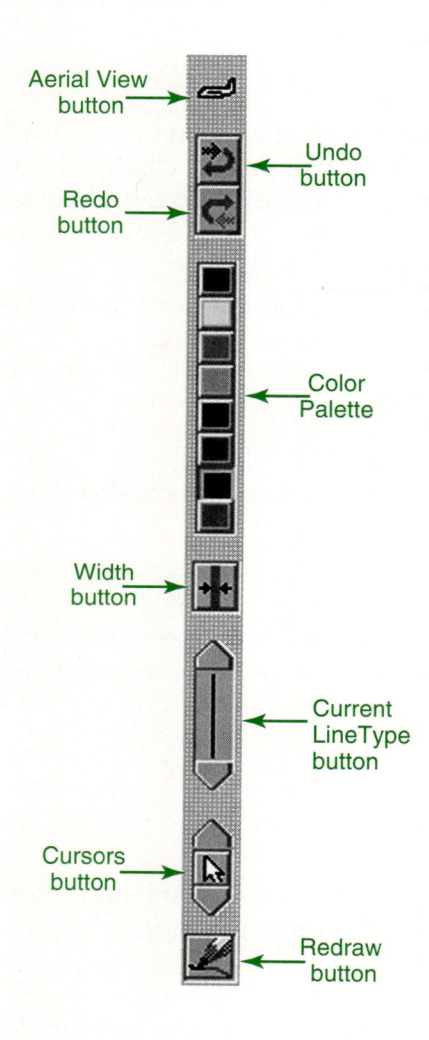

Aerial View button

The *Aerial View button* displays a pop-up window that lets you navigate large drawings. Viewing drawings is discussed in Chapter 7.

Undo and Redo buttons

The *Undo button* reverses the effect of the previous drawing or editing task. Continue picking **Undo** to cancel changes one at a time to the first change you made after you opened or last saved the drawing. The *Redo button* reverses the effect of **Undo**. The **Redo** button will be greyed out until you have used the **Undo** button.

Color Palette

The *Color Palette* lets you choose the color of lines, circles, and other objects you add to the drawing.

Width button

The *Width button* lets you choose the width of lines, circles, and other objects you add to the drawing.

Current LineType button

The *Current LineType button* lets you choose the linetype of lines, circles and other objects you add to the drawing. Some linetype options are solid, dashed, and dotted.

Cursors button

The *Cursors button* allows you change the cursor size and shape for drawing. You can change the cursor shape between an arrow, crosshairs, a box, or crosshairs and a box. The size can be increased or decreased by clicking on the upper and lower triangular buttons.

Redraw button

The *Redraw button* clears the drawing and redraws the same view. This refreshes the view and removes marks or holes left by some drawing and editing commands.

Using the Cursor

The *cursor* allows you to choose menus, tools, to draw, and to select items to edit. In many ways, it is your pencil and eraser. You can move it anywhere on-screen by moving your mouse. You can change the cursor shape to a crosshairs, box, or crosshairs and box by picking on the **Cursors** button, Figure 2-7. You can change the size of the crosshairs by picking the triangular buttons above and below the **Cursors** button.

In this text, you will come across the terms pick, click, and double-click quite often. *Pick* and *click* both mean to locate the cursor on-screen and press the pick button on your input device. This is usually the left mouse button. You might pick a tool to use, a point to place, or an object to edit. *Double-click* means to press the pick button on the mouse twice very fast. You will double-click on tools to change their settings.

Figure 2-7.
Picking the **Cursors** button changes the cursor shape. Picking the triangular button above the **Cursors** button increases the size. Picking the triangular button below the **Cursors** button decreases the size.

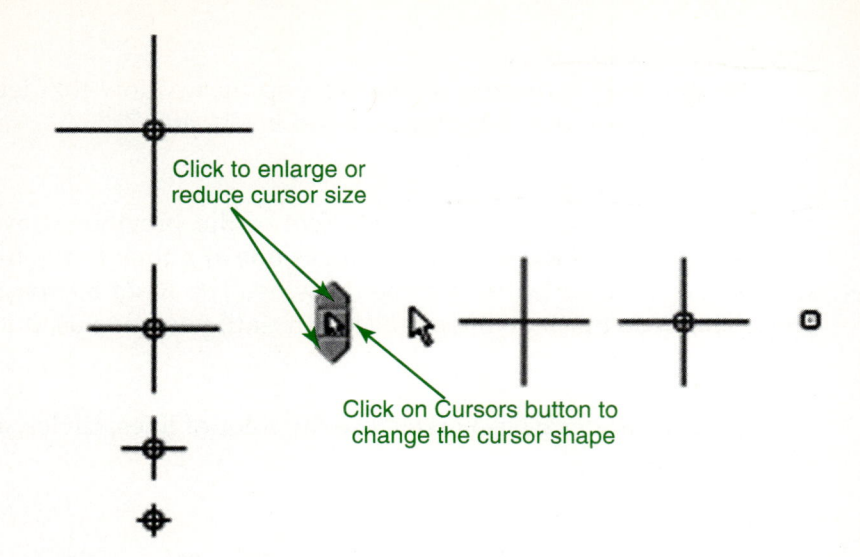

Click to enlarge or reduce cursor size

Click on Cursors button to change the cursor shape

Selecting Menus, Tools, and Pull-down Menus

To perform any AutoSketch function, you must choose a command. Typical AutoSketch commands are **Line, Circle, Erase, Move,** and **Copy.** The commands are either shown as a word in a pull-down menu, or as a tool. To see a group of commands, you must pick the menu name where the commands are contained. A *menu* contains a list of commands, just like a restaurant menu is a list of food. The menus within AutoSketch are defined in the next section.

Menus

There are several menus found in AutoSketch. These are explained in the next sections. Refer back to Figure 2-4 as you go through the next sections.

File

The **File** menu provides commands to let you start, save, close, delete, import, export, and print drawings.

Edit

The **Edit** menu offers commands that let you group objects together. You can also change properties such as color and line width with the **Edit** menu.

View

The **View** menu has commands that determine how much and what part of the drawing is shown on-screen.

Draw

The **Draw** menu contains a list of objects you can add to the drawing. Objectsinclude lines, points, circles, arcs, curves, boxes, polygons, ellipses, pattern fills, and text.

Modify
The **Modify** menu contains a list of commands that let you change objects you have drawn. You might want to erase, copy, move, or rotate an object. AutoSketch provides more than 15 ways to change objects.

Assist
The **Assist** menu contains drawing aids that help you draw precisely. For example, you might want to attach the endpoint of a line exactly at the midpoint of another line. Without **Assist** commands, this would be nearly impossible.

Measure
The **Measure** menu contains commands to let you measure objects and add dimensions to your drawing.

Custom
The **Custom** menu includes commands to help you set up your drawing, customize the AutoSketch window, and record or play macros.

Window
The **Window** menu lets you select which drawing to work on, and how multiple drawings should be displayed on-screen. AutoSketch allows you to have up to five drawings open at the same time.

Help
The **Help** menu gives you access to on-line help. Rather than refer to a printed manual, you can get help about any AutoSketch function using on-line help. You will refer to the help function often as you learn AutoSketch.

Selecting Commands Using Toolboxes and Tools
After you install AutoSketch, the default is to have all menus, except **File**, **Window**, and **Help** menus, shown as toolboxes. To start a command, simply pick the menu and then the tool. A *tool* is an icon picture of the command. It is best to use tools, rather than pull-down menus, for the following reasons.
- Tools remain visible as you draw and edit. This gives you instant access to important commands. When you use pull-down menus, the menu covers your drawing. Tools do not. Therefore, you can select commands without blocking your view of the drawing.
- Tools allow you to select other tools while using a command. For example, while using the **Line** command, you can select any of the commands from the **View** menu to change the drawing view. Doing so does not interrupt the **Line** command.
- Tools allow you to access settings dialog boxes with one double-click. Tools that have settings appear with two dots at the lower-right corner of the tool. By double-clicking on that tool, you access the settings dialog box. For example, if you double-click on the **Circle** tool, the **Circle Settings** dialog box appears, Figure 2-8. In this dialog box, you set how you want to draw the circle. When using pull-down menus, you have to select **Drawing Settings...** from the **Custom** menu. Next, you must choose **Circle...** to access the **Circle Settings** dialog box.

Figure 2-8.
Double-clicking on a tool with two dots brings up the settings dialog box for that tool.

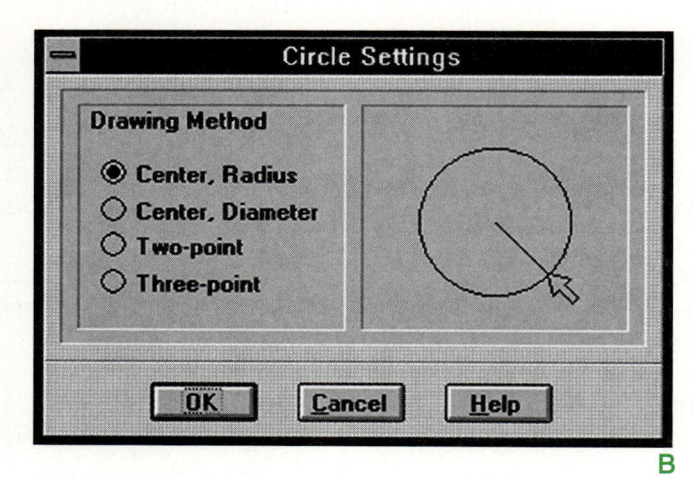

Circle tool
A

B

- You can display more than one toolbox at a time. By default, AutoSketch displays both horizontal and vertical toolboxes. The **Assist** toolbox is used so much that it is displayed all the time by default. You can customize AutoSketch to have more toolboxes at the top, bottom, sides, or even to float around the screen. With pull-down menus, only one menu can be open at any time. Picking a new menu or pressing the [Esc] key closes the previous one.

For the purposes of this text, we will assume that you use the default setup of AutoSketch. The default has all menus, except **File, Window,** and **Help** menus, shown as toolboxes. However, if you choose to use pull-down menus, simply select the command described in the text rather than the tool. Appendix B shows all of the menus and tools.

Selecting Commands from Pull-down Menus

For anyone not comfortable with toolboxes, you can also select commands from *pull-down menus.* To change toolboxes to pull-down menus pick the **Toolbox Control Menu** button, Figure 2-9. Then pick **Pulldown Menu.** Now, all commands within that menu are listed in a pull-down menu, Figure 2-10. To activate a command, simply pick the menu and then the command in the pull-down menu. To change a pull-down menu back to a toolbox, select **Toolbox** at the bottom of the pull-down menu. Appendix B shows all of the pull-down menus and commands.

Selecting Commands Using Function Keys

The most common AutoSketch commands can also be chosen using function keys. *Function keys* are the keys labeled [F1] through [F12] along the top of your keyboard. Some keyboards have function keys grouped along either the right or left side of the keyboard. The [Alt] and [Ctrl] keys give extra options for many function keys. In Figure 2-10, the function key combinations for commands in the **Draw** menu can be seen. Appendix B shows the commands associated with each key combination.

Figure 2-9.
The **Toolbox Control** menu allows you to switch between a toolbox and a pull-down menu. To access the menu, pick the menu button.

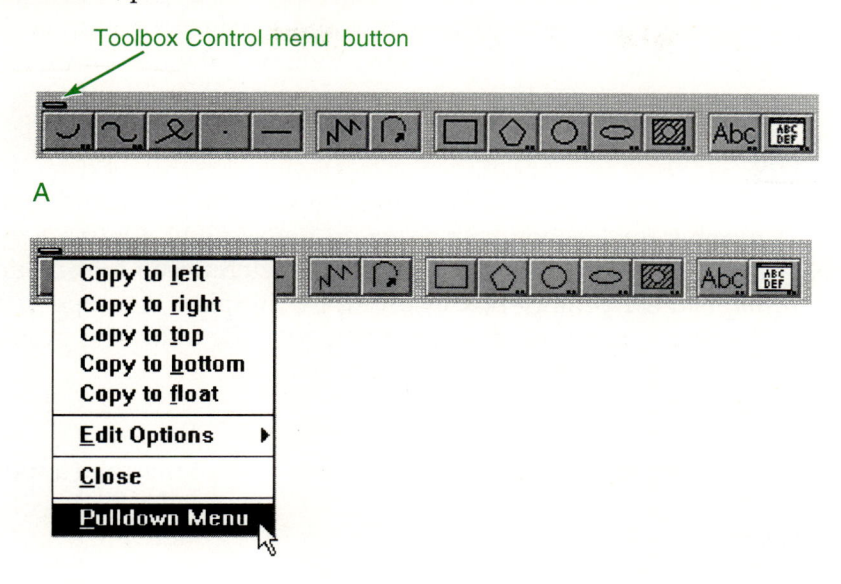

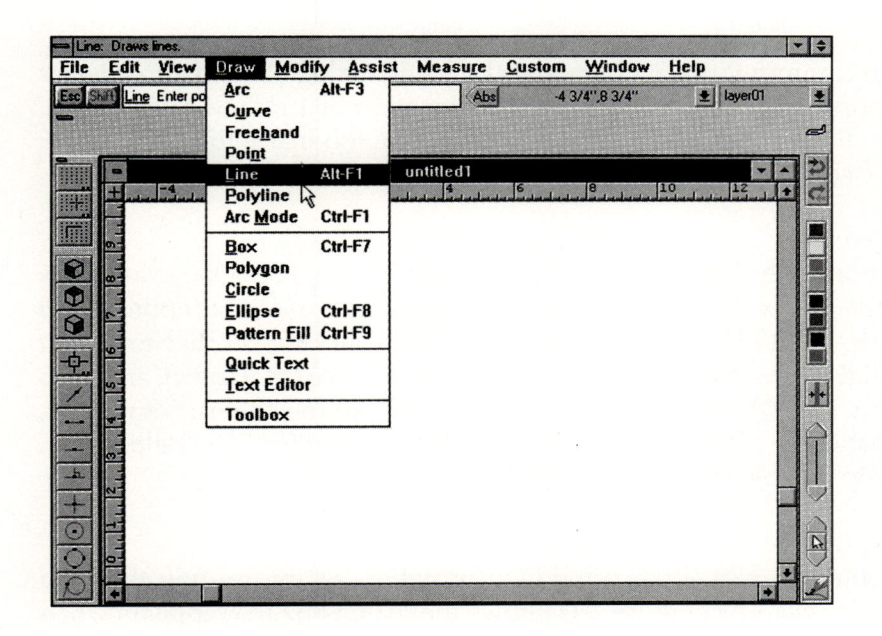

A

B

Figure 2-10.
Some commands in a pull-down menu can be selected with function keys. You can also hold the [Alt] key and press the underlined character (the "hot" key) to activate a menu or command.

Selecting Commands with Hot Keys

There is one more way to select menus and commands. If you are using pull-down menus, each menu and command has one character underlined. These characters represent *hot keys*. To access a menu, press the [Alt] key and then type the underlined letter on the menu you want to access. To execute a command in a menu, press the underlined letter of the command once the menu is open.

For example, to select the **Draw** pull-down menu, press the [Alt] key and then the letter D. To access the **Circle** command once the **Draw** menu is open, simply press the letter C. Refer back to Figure 2-10 to see how commands in the **Draw** menu are underlined.

Repeating Commands

All commands automatically repeat. For example, after drawing a line, AutoSketch will ask you for the first endpoint of another line. This feature reduces the number of times you have to pick menus and tools or commands. To exit the command you are using, simply select another tool, pick the **Esc** button, or press the [Esc] key.

Prompts

Most commands require that you enter some type of information. You may need to type a number, pick a point, or select an item to edit. Only a few commands act immediately without your input. When AutoSketch needs information, it asks for it with a *prompt* in the prompt box. A prompt is simply a message or question to you from AutoSketch. When you select the **Circle** tool, the following prompt appears.

Circle Center point:

After you pick the circle's center, the following prompt message appears.

Circle Radius:

You can either pick a point for the radius or enter the numeric value. Always refer to the prompt box when you're not quite sure what to do.

Dialog Boxes

AutoSketch also communicates with the user by using dialog boxes. *Dialog boxes* are windows that pop up on-screen to request information or provide information. There are dozens of dialog boxes that request information, too many to discuss here. However, there are two general types of dialog boxes you will use.

Settings dialog boxes

The most common type of dialog box is the *settings dialog box.* One example is the **Circle Settings** dialog box shown in Figure 2-8. Tools that have settings appear with two dots at the lower-right corner of the tool. By double-clicking on that tool, you access the settings dialog box. Most "settings" dialog boxes have **OK, Cancel,** and **Help** buttons at the bottom. Pick **OK** or press the [Enter] key to accept the values. Sometimes, you might choose **Cancel** to clear the box without changing a setting. The **Help** button brings up on-line help for that particular dialog box.

Informational dialog boxes

Informational dialog boxes either provide information, or inform you that you have made an error. For example, Figure 2-11 shows the dialog box that appears when you measure distance. To close the dialog box, pick **OK.**

Getting Help

You've already seen the **Quick Help** and **SmartCursor** features of AutoSketch. These provide the name of the tool and its purpose. The prompt box also tells you what action to take. However, what if you need more help? The AutoSketch documents are very complete, but it may be difficult to find the right help you need. An eas-

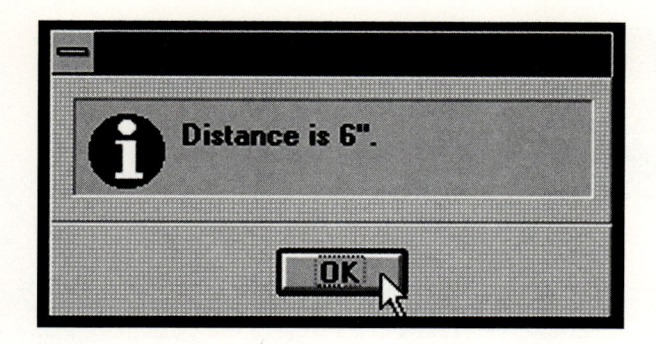

Figure 2-11.
Informational dialog boxes provide the user with information, such as a measured distance.

ier way to get assistance is with on-line help. *On-line help* provides information on every tool, menu, and feature in AutoSketch. You can access on-line help in several ways.

Information About a Tool

For information about a tool, locate the cursor over the tool and press function key [F1]. Or, press [Shift] and [F1] to display the help cursor. Then choose a tool or any portion of the AutoSketch window for help. Refer to Figure 2-12.

Search for Information

You can access the **Help** menu to display a list of help functions. To see the contents of on-line help, pick **Contents,** Figure 2-13. To search for help on a topic, pick **Search** from the **Help** menu, or select the **Search** button after selecting **Contents.**

With the **Search** window on-screen, type the first few characters of the topic you need help on, Figure 2-14. You can also pick the scroll bar and scroll arrows at the right side of the box to move through the list. When the topic appears in the window, double-click on it. Either the information will appear, or a list of subtopics will appear. Double-click on one of the subtopics for information, Figure 2-15.

Moving around in Help

From the main Help window, you can select the **Search** button to search for a specific topic. You can also move the cursor to the topics listed in green. The cursor becomes a pointing finger and you can click to see subtopics.

The **Back** button returns you to the previous help screen. The **History** button shows the list of help topics that you have viewed. This allows you to return to a previous topic quicker than using **Back.**

Finally, you can mark a location for future reference using the **Bookmark** menu. Using this option lets you define places in the help file that you use most often.

When finished with the **Help** window, pick **File** and **Exit**.

Figure 2-12.
A—Press [Shift] and [F1] to display the help cursor. Then choose a tool for help on that tool. B—The AutoSketch for Windows help application provides information on a selected item.

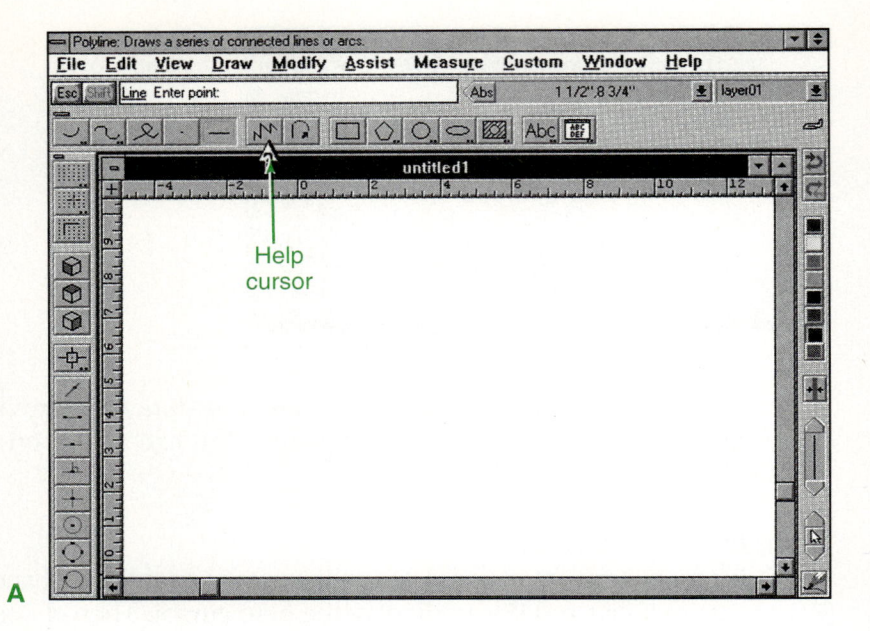

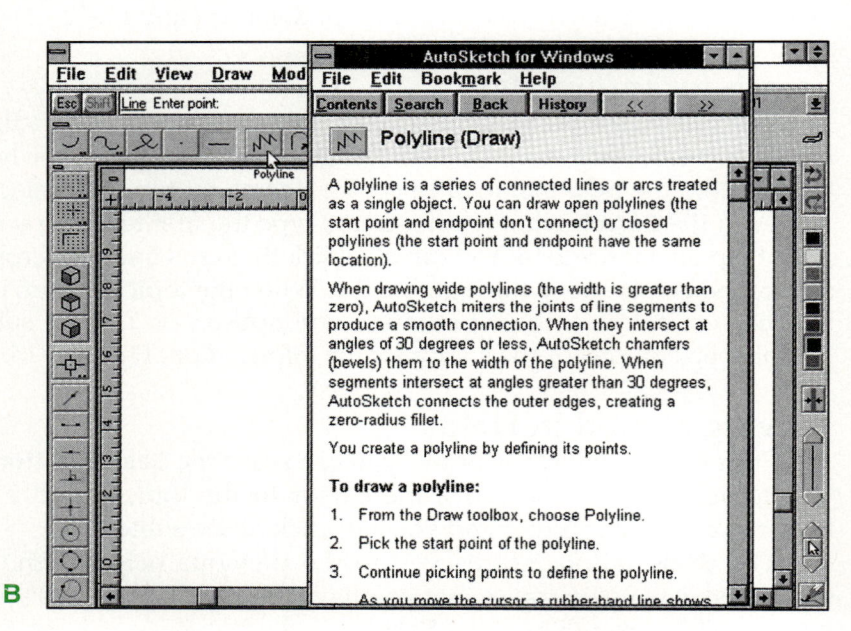

Basic Drawing with AutoSketch

Future chapters will teach you to draw with AutoSketch. To let you experiment, this section will show how to use the **Line** command. Refer to Figure 2-16 as you go through the following instructions.

1. Start AutoSketch by double-clicking on the AutoSketch icon in the AutoSketch for Windows program group.
2. Pick the **Normal** template and then **OK** in the **Select Template** dialog box.

Figure 2-13.
A—Pick the **Contents** command in the **Help** menu to see the contents of the help file. B—The AutoSketch for Windows help application provides a list of topics with help information.

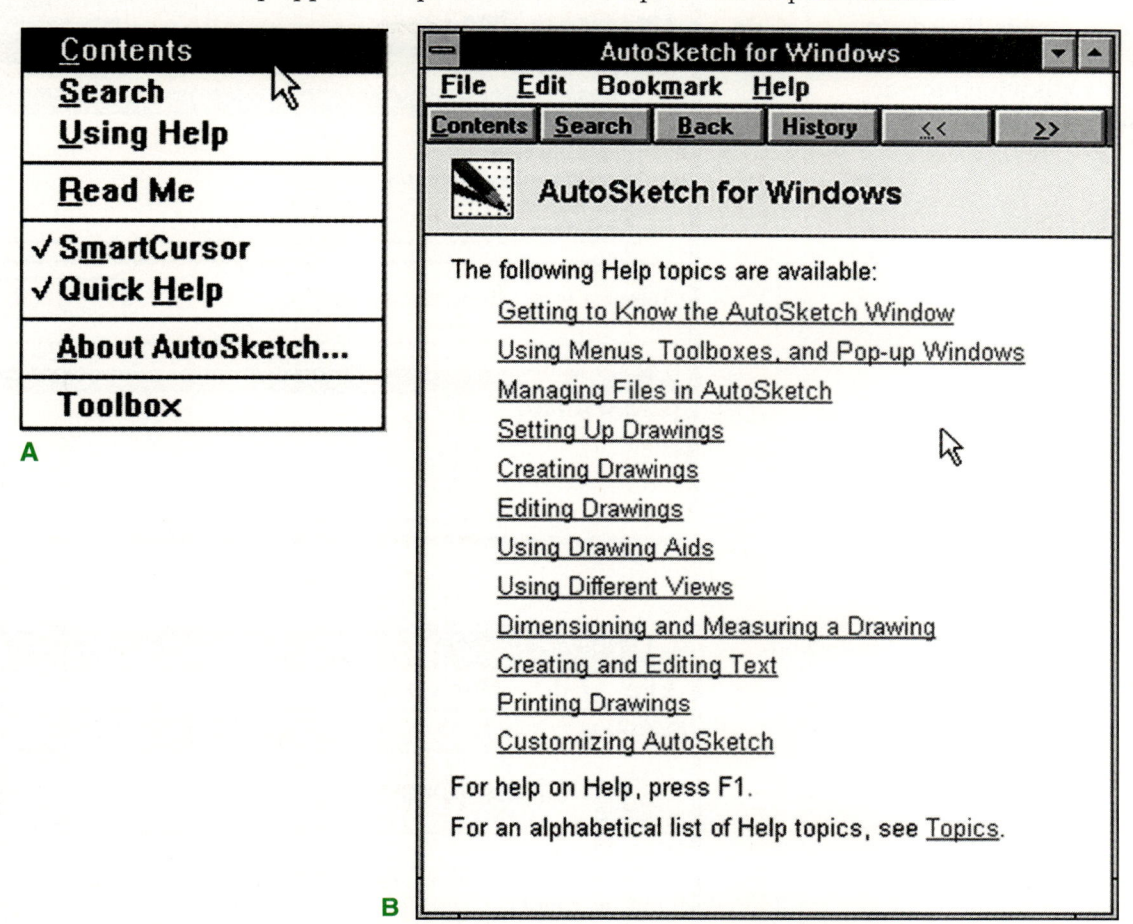

A

B

Figure 2-14.
A—Pick the **Search** command in the **Help** menu to search the help files for a specific topic.
B—The **Search** window allows you to enter a topic or scroll through the list of topics.

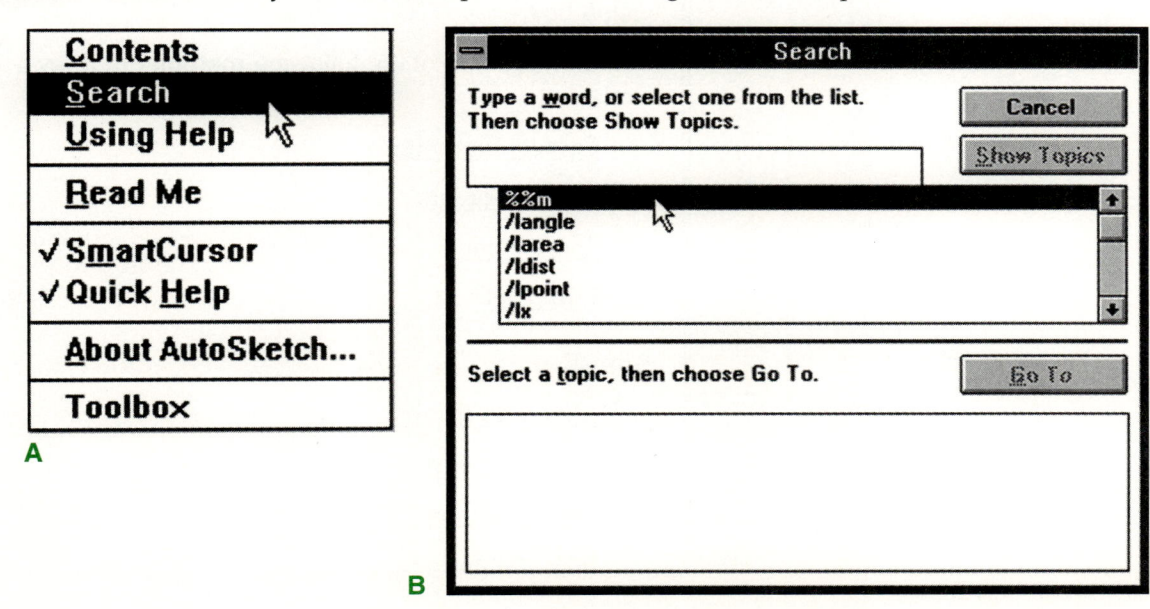

A

B

Figure 2-15.
A—Type a few letters of the topic you need help with and double-click on the topic. Next, double-click on the subtopic. B—The AutoSketch for Widows help application provides information on that topic.

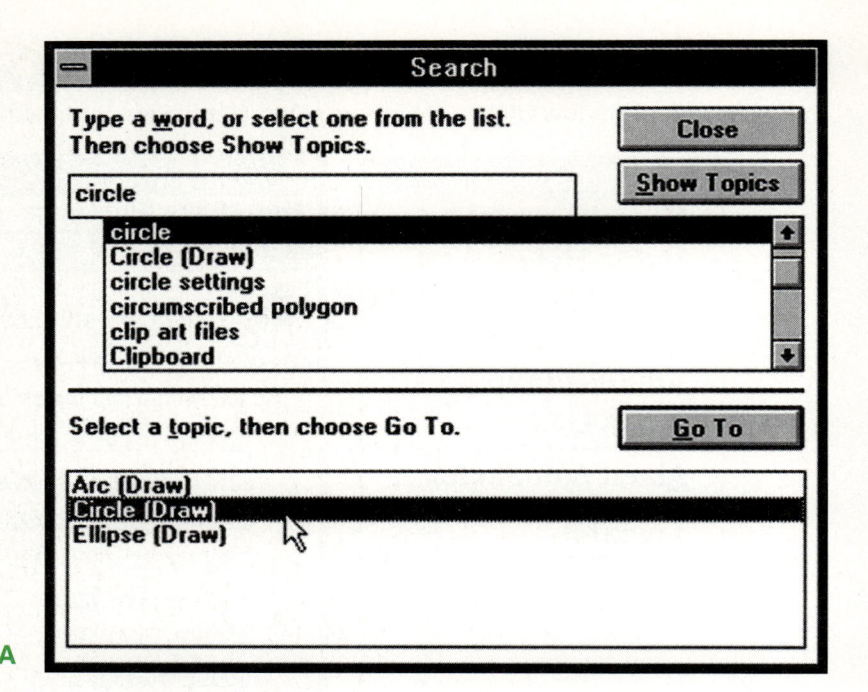

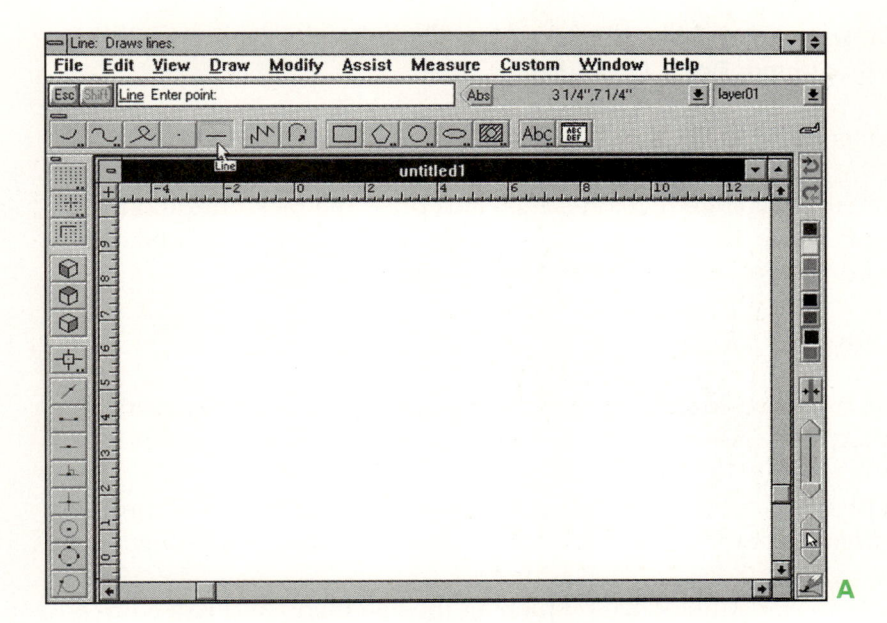

Figure 2-16.
Drawing a line with AutoSketch. A—Pick the **Line** tool from the **Draw** toolbox. B—Pick the first and second endpoints of the line.

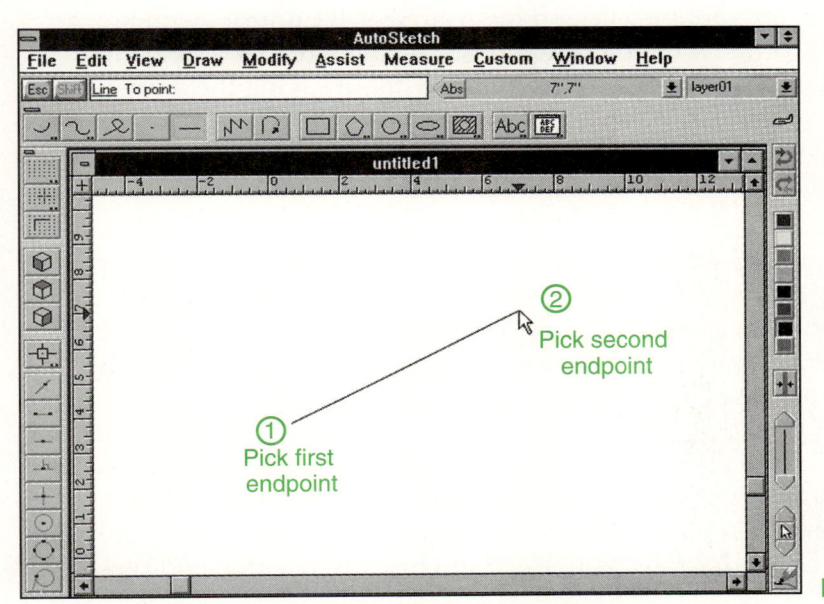

3. Pick the **Draw** menu and then the **Line** tool. The following prompt will appear in the **Prompt Box.**

   Line Enter point:

4. Move your cursor somewhere on-screen and press the pick button on your input device. This action selects the first endpoint of the line. AutoSketch will then prompt you to provide a second point.

   Line To point:

5. Move your cursor on-screen. As you do this, a temporary line appears. It helps you locate the second endpoint to place the line. Select a second point.
6. After you place a line, the **Line** command repeats. AutoSketch will prompt you for a first endpoint again. Continue to draw as many single lines as you want.
7. When finished, exit AutoSketch by picking the **File** menu and **Exit** command. You can also double-click on the AutoSketch Control menu box in the upper-left corner. A dialog box appears asking whether you want to save changes. Pick **No** to leave AutoSketch without saving the drawing.

Summary

AutoSketch is a unique, easy-to-use drawing tool. AutoSketch replaces pencils and paper for making charts, diagrams, illustrations, and working drawings. Unlike paint programs, AutoSketch stores each object mathematically as a single item. This allows objects to be enlarged without distorting the image. Also, unlike draw programs, AutoSketch provides all the functions needed for drafting.

Once AutoSketch is loaded from the Windows Program Manager, the main window appears. This window contains the menu bar, horizontal and vertical status bars, toolboxes, and the drawing window. Using the cursor, you can pick a menu to see a list of tools or commands in a pull-down menu. To perform a command, simply pick the tool or command name. To draw, you use your cursor to position and pick point locations on-screen.

AutoSketch communicates with the user by displaying prompts and dialog boxes. Prompts are messages that appear in the **Prompt Box.** Dialog boxes are windows that pop up on-screen. Dialog boxes may provide you with information or ask you to provide information. Always look to the prompt line. Also be sure to read any dialog box when one is given.

Important Terms

Booting
CAD program
Click
Cursor
Default
Dialog boxes
Double-click
Draw programs
Drawing Window
Entities
Function keys
Horizontal status bar

Horizontal toolbox
Hot keys
Icon
Informational dialog boxes
Menu
Menu bar
Multitasking
Objects
On-line help
Operating system
Paint programs
Pick

Prompt
Pull-down menus
Quick Help
Rulers
Scroll bars
Settings dialog box
Template
Tool
Toolbox
Vertical Status Bar
Vertical toolbox

New Commands and Tools

Aerial View button
Assist menu
AutoSketch Control menu
Back
Bookmark menu
Cancel
Color Palette
Contents
Coordinate Display
Current Layer box
Current LineType button
Cursors button

Custom menu
Draw menu
Drawing Settings...
Edit menu
E**x**it
File
Help
Help menu
Histo**r**y, **Line** tool
Measu**re** menu
Modify menu
OK

Option...
Prompt box
Pulldown Menu
Redo button
Redraw button
Search
SmartCursor
Toolbox
Toolbox Control menu
Undo button
View menu
Width button
Window menu

Review Questions

Give the best answer for each of the following questions.

1. Explain why AutoSketch is an excellent program to learn CAD.

2. Describe the difference between a paint program, a draw program, and a CAD program.

3. List the steps taken to start AutoSketch.
   A. _____
   B. _____
   C. _____

4. List the menus available in AutoSketch.
   A. _____
   B. _____
   C. _____
   D. _____
   E. _____

F. _____
G. _____
H. _____
I. _____
J. _____

5. In what ways will you use the cursor?

6. What tools are found in the horizontal toolbox?

7. What tools are found in the vertical toolbox?

8. What options are found in the **Toolbox Control** menu?

9. Why is it best to use tools rather than pull-down menus?

10. How can you use the keyboard to select menus and commands?

11. List four ways that AutoSketch provides help during your drawing session.
    A. _____
    B. _____
    C. _____
    D. _____

12. Since AutoSketch commands automatically repeat, how do you exit the command you are using?

13. When AutoSketch needs information, it will ask for it with a(n) _____ or a(n) _____.

14. List the two general types of dialog boxes.
    A. _____
    B. _____

15. List the seven steps taken to start a drawing, draw a line, and exit AutoSketch without saving the drawing.
    A. _____
    B. _____
    C. _____
    D. _____
    E. _____
    F. _____
    G. _____

Activities

Here is a mixture of drawing activities to develop your skills using AutoSketch. Complete all of the following activities in order without exiting and restarting AutoSketch.

1. Identify the names and locations of reference manuals that accompany AutoSketch.
2. In Appendix D in the back of this book, find the diagram showing AutoSketch menus and commands. Study it carefully, so that you are familiar with the command structure.

3. Boot your computer and start Windows. Then start AutoSketch and select the **Normal** template.
4. Pick each menu and review the tools and commands in that menu. Record which menus display toolboxes and which display pull-down menus.
5. Pick the **Draw** menu and **Circle** tool. Record the prompt that appears in the prompt box.
6. For each menu that displays a toolbox, use the toolbox control menu to change the toolbox to a pull-down menu.
7. Press the [Alt] key and then press the letter M to activate the **Modify** menu. Select the **Mirror** command by pressing the letter I. Press the [Esc] key to end the command.
8. Change all of the menus so that they display toolboxes rather than pull-down menus.
9. Change the menus so that the **File, Window,** and **Help** menus display pull-down menus. This is the default configuration of AutoSketch.
10. Double-click on the **Arc** tool in the **Draw** menu to open the **Arc Settings** dialog box. Review the options. Pick **OK** when finished.
11. Pick the **Cursors** tool and change the cursor to a crosshairs. Increase the size of the crosshairs until it fills the drawing window. Return the cursor to an arrow when finished.
12. Use the help feature of AutoSketch to display the contents of on-line help. Select one of the underlined topics for more information. Exit the help window when finished.
13. Use the help feature of AutoSketch to search for help on the **Circle** tool. Click on one of the topics to see information on drawing circles. Exit the help window when finished.
14. Draw several lines by picking the **Draw** menu and **Line** tool.
15. Pick **Exit** in the **File** menu. Select **No** to exit without saving your changes.

Saving an AutoSketch drawing.

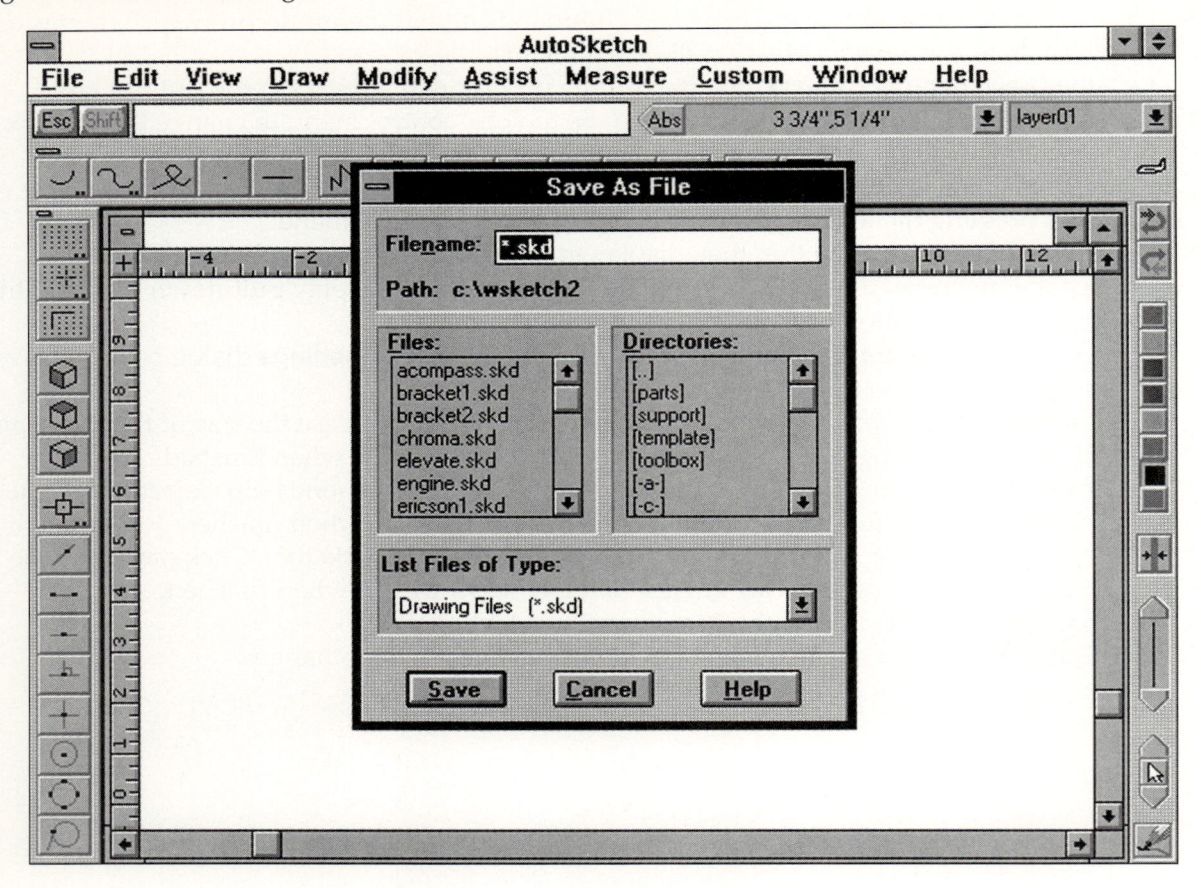

CHAPTER 3

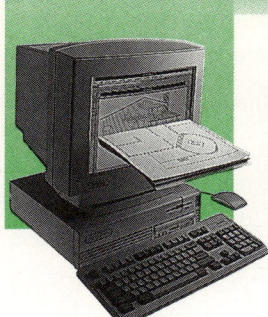

Starting and Saving Drawings

Objectives
After studying this chapter, you will be able to:
- Start a new drawing.
- Save a new drawing.
- Close a drawing.
- Open an existing drawing.
- Save changes made to an existing drawing.
- Save an existing drawing under a new name.
- Open several drawings at the same time.
- Save part of a drawing as a separate drawing file.
- Exit AutoSketch.

Drawings created using AutoSketch are stored as computer files, called *drawing files.* During a drawing session, AutoSketch holds the drawing file in computer memory for you to work on. When finished with the drawing, you must save it from memory to disk storage. Then you can start another drawing or exit AutoSketch. The commands to start new drawings, open existing drawings, and save drawings are found in the **File** menu, Figure 3-1.

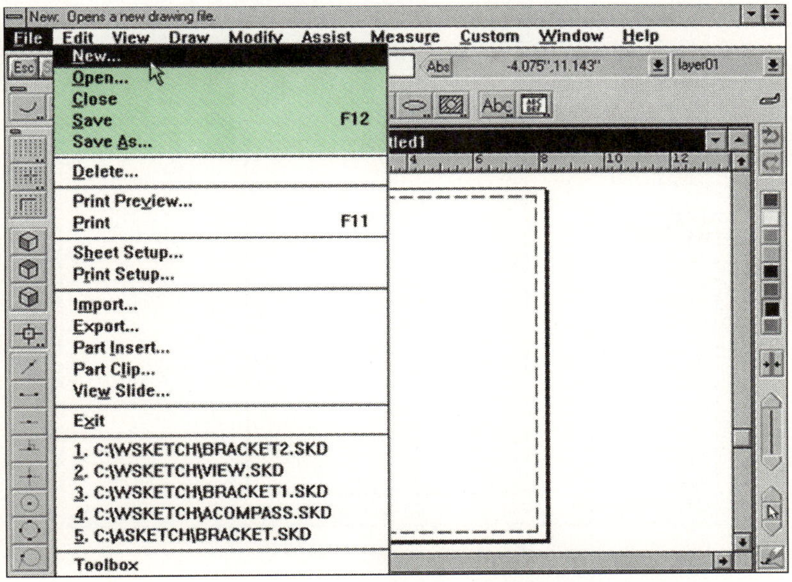

Figure 3-1.
The commands to load and save drawing files are found in the **File** menu.

71

By default, the **File** menu is shown as a pull-down menu. However, by selecting **Toolbox** from the pull-down menu, you can display this menu as a toolbox, Figure 3-2. To return to the pull-down menu, pick the **Toolbox Control Menu** button and select **Pulldown Menu,** Figure 3-3.

Figure 3-2.
Selecting **Toolbox** in the **File** menu displays the **File** toolbox.

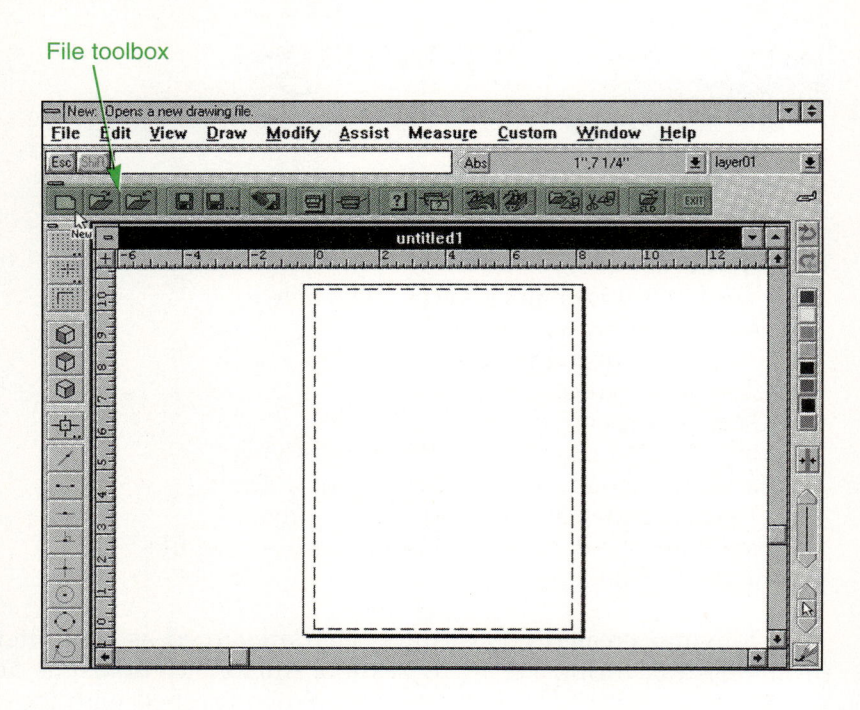

Figure 3-3.
Select **Pulldown Menu** from the **File Toolbox Control** menu to return to the default pull-down menu.

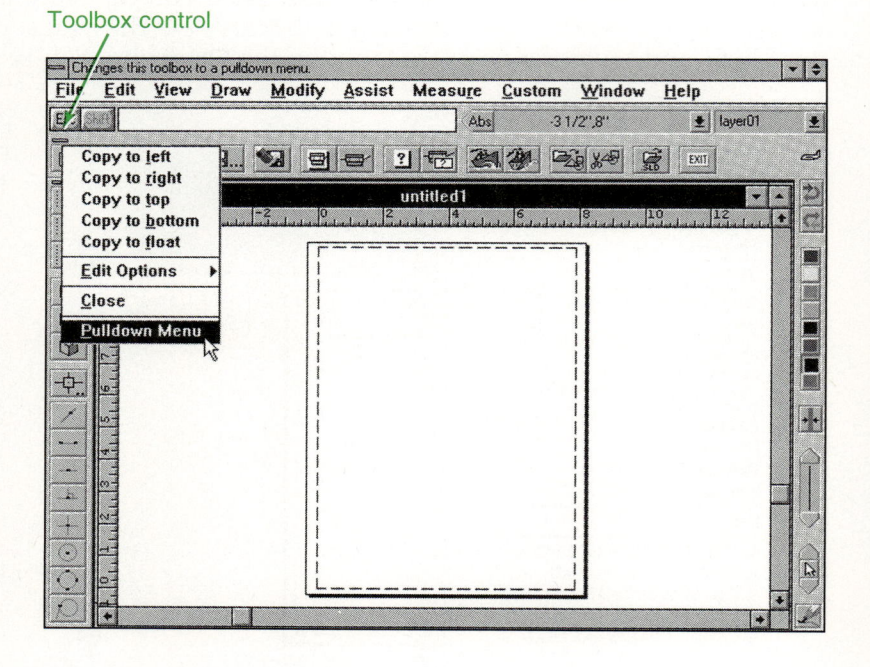

Starting a New Drawing

When you double-click on the icon to start AutoSketch, AutoSketch opens an untitled copy of an AutoSketch template. The **Select Template** dialog box appears, Figure 3-4. A *template* is a drawing file that contains a complete drawing setup, usually including a border and title block. AutoSketch provides templates to save you the time of setting up each and every drawing you start. A number of templates are provided for each drafting situation and every paper size. Setting up a drawing and saving those settings as a template are discussed in Chapter 10. You can double-click on the **Normal** template, or pick the **Cancel** button, in this dialog box to start an AutoSketch drawing using the default values. If you select **Quick Start,** AutoSketch will prompt you for the units of measure and size of the drawing area. These steps are part of setting up a drawing. Setting up a drawing is discussed in Chapter 10.

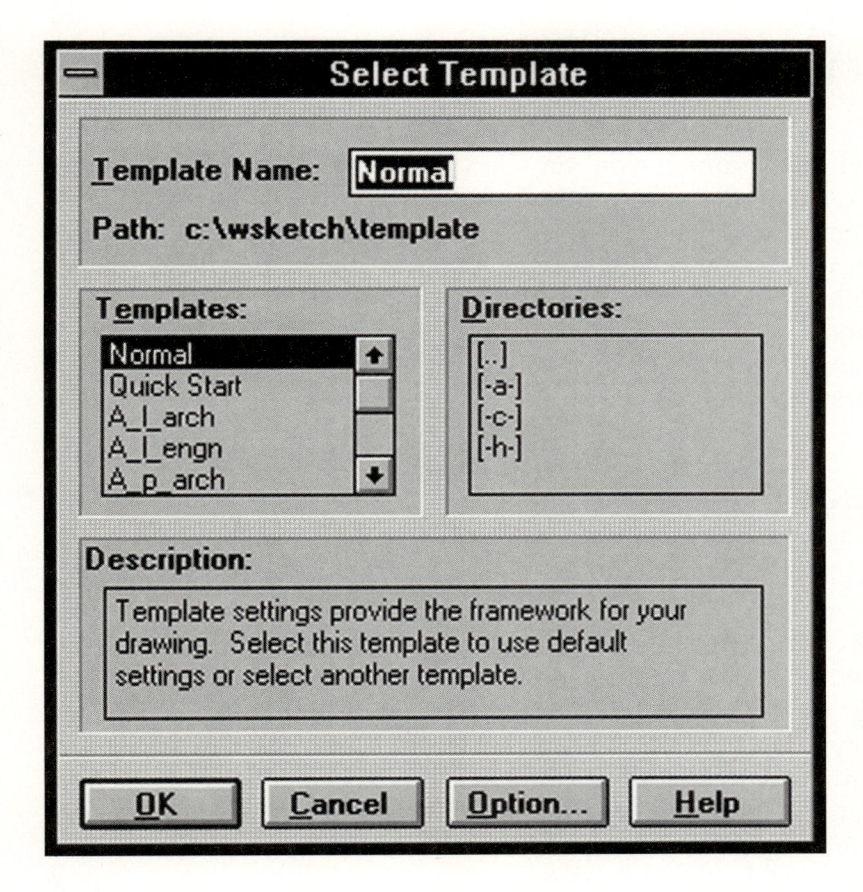

Figure 3-4.
When you start AutoSketch, or start a new drawing, the **Select Template** dialog box allows you to select a pre-saved drawing setup.

Once you double-click on a template, or highlight a template and pick the **OK** button, AutoSketch provides you with a new drawing named "untitled1" to work on, Figure 3-5. Your new drawing will remain unnamed until you select **Save** in the **File** menu to give it a name.

Figure 3-5.
All new drawings are temporarily named "untitled1" until you save the drawing under a filename.

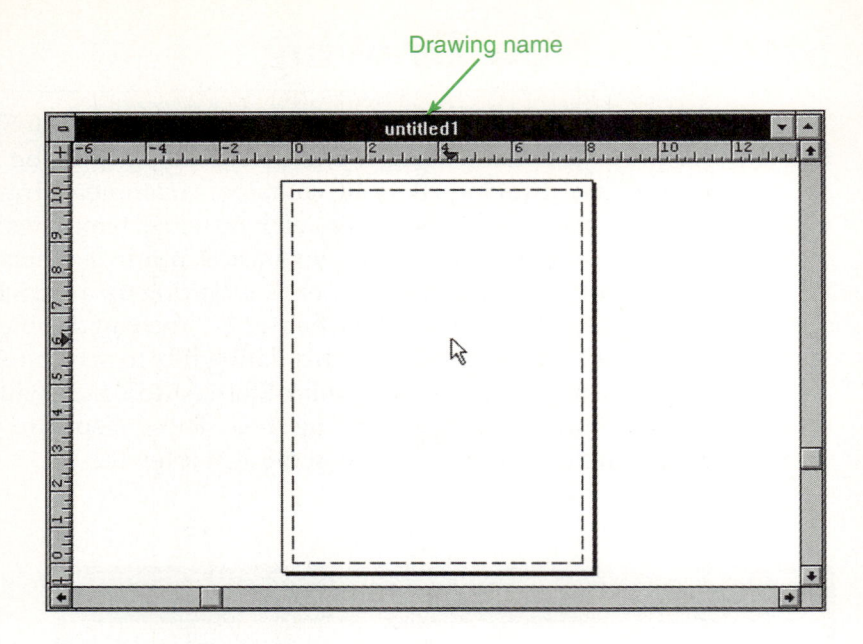

If AutoSketch is already running, you can select **New...** in the **File** menu to start a new drawing. AutoSketch allows you to have up to five drawings active at a time. Thus, you can start a new drawing even if an existing drawing is on-screen. However, each drawing takes up computer memory and processor power. Your performance may slow down significantly. Also, you may soon run out of memory with as few as two drawings on-screen if each drawing contains many objects. It is best to have only one drawing open at any time unless you need to copy objects between two drawings. Opening several drawings at the same time is discussed later in this chapter.

Saving a New Drawing

Once you have added objects to a new drawing, save it to disk under a name. The drawing may not be complete. Most drawings take hours, days, or even weeks to finish. Yet, it is best to save your work before going too far into a drawing session. This will protect your drawing in case a power outage erases the drawing from memory. Saving your drawing not only saves the objects, but any settings in the drawing. Settings for color, linetype, and drawing aids are all saved.

To save a new drawing, select **Save** from the **File** menu. This will bring up the **Save As File** dialog box, Figure 3-6. Your cursor will already be positioned in the **Filename:** box. Type in the drawing name and pick the **Save** button or press the [Enter] key. AutoSketch automatically adds the .SKD extension. This will save the drawing in the current directory of the hard drive. Once you save the drawing, the filename appears in the top border of your drawing window, Figure 3-7.

You can use up to eight characters (letters and numbers) to name a drawing. You can also use the characters &, _, and - in the drawing name. The name should reflect the topic of the drawing. Names such as DRAWING1 and DRAWING2 don't give much information about the drawing. Give your drawings descriptive names. For example, the name INTAKE1 might mean the first design for an intake valve.

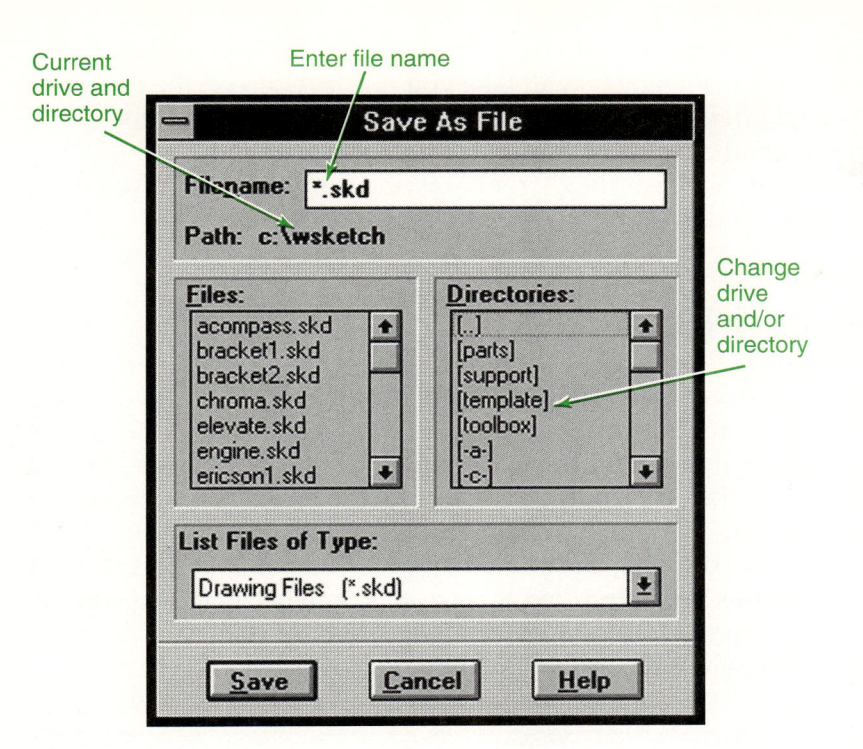

Figure 3-6.
To save a new drawing, or save a drawing under a different name, select the drive and directory and enter a filename.

Current drive and directory
Enter file name
Change drive and/or directory

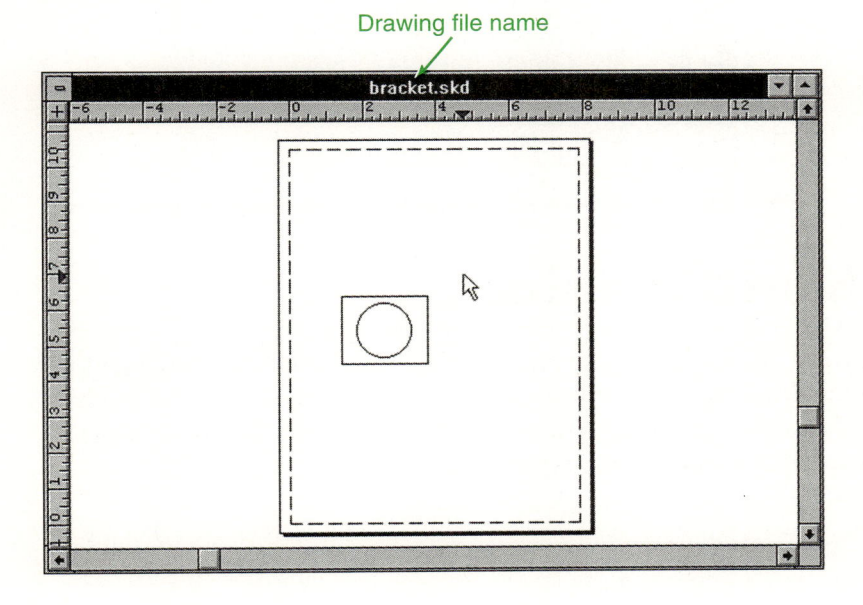

Figure 3-7.
After saving a new drawing, its name appears in the drawing title bar.

Drawing file name

A new drawing will be saved to the default directory on the hard drive. This is the directory where AutoSketch for Windows is installed. This directory will typically be \WSKETCH. To save the drawing to a different directory or to a floppy disk, you can do either of two things.

- In the **Filename:** box, enter the disk drive letter and any directory path before the drawing name. For example, to save the drawing as INTAKE1 on the A: diskette drive, type A:INTAKE1. Make sure you have a formatted disk in the drive before trying to save a drawing to that drive. To save drawings to a different directory on the hard drive, type the drive letter, the directory, and the filename. These must be separated by backslashes. An example might be C:\DRAWINGS\INTAKE1.

- In the **Directories** box, pick the drive letter and a directory by double-clicking on the appropriate selections. Picking [..] will take you one level up on the directory path. When satisfied with the drive and directory, place your cursor in the **Filename:** box and enter the drawing name.

The **Save** command does not clear the current drawing from the screen. You must use the **Close** command to do this.

Closing a Saved Drawing

Suppose after saving a drawing, you no longer plan to work on it. The drawing will remain on-screen until you close it. This consumes computer memory. To close a drawing, select the **Close** command from the **File** menu. You can also double-click on the drawing control bar. The drawing control bar is a short horizontal bar found in the upper left-hand corner of your drawing window.

If you have not saved the drawing since making changes, AutoSketch will warn you with a dialog box, Figure 3-8. Select **Yes** to save the drawing before closing it. Select **No** to close the drawing without saving the changes. Select **Cancel** to return to the drawing without closing it. Closing a drawing will not close AutoSketch.

Figure 3-8.
AutoSketch will warn you if you attempt to close a drawing that has not been saved.

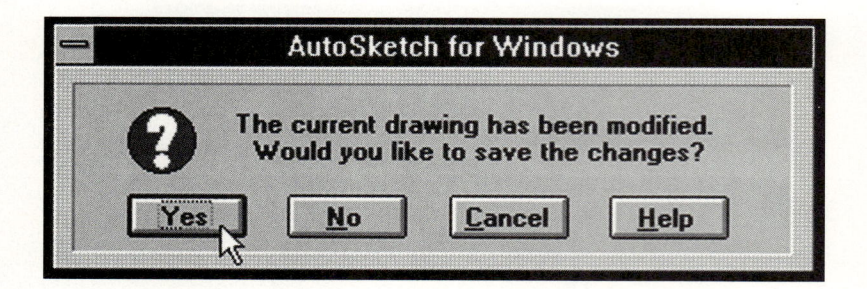

Opening a Stored Drawing

To work on a drawing you have already created and stored on disk, you must load it into memory. Once the drawing is in memory, you can add objects, edit, or otherwise change the drawing. To load a drawing, select the **File** menu and **Open...** command. The **Open Drawing File** dialog box appears, Figure 3-9.

Figure 3-9.
The **Open Drawing File** dialog box shows stored drawings in the current drive and directory as icons.

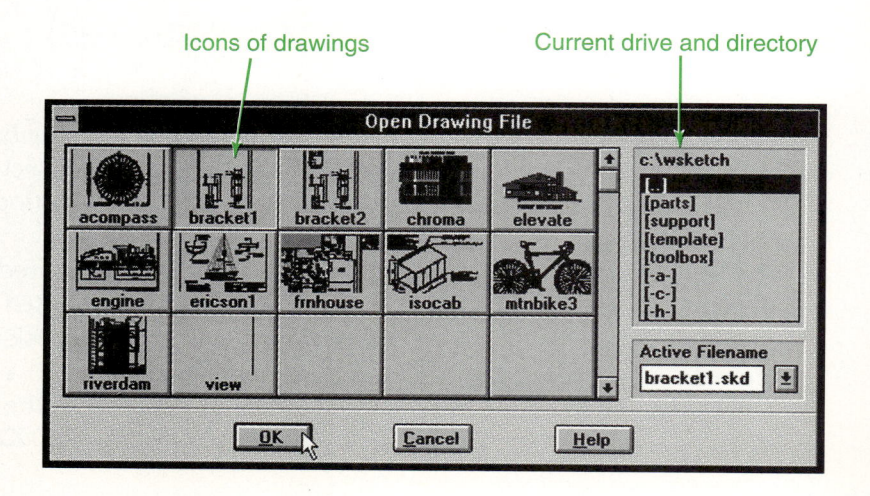

Drawings stored in the current directory appear as *icons.* Icons are miniature pictures. The current drive and directory are shown in the box at the right side of the **Open Drawing File** dialog box. If there are more than fifteen drawings in the current directory, use the scroll bar on the right. You can pick the up or down arrows, or pick the scroll box in the scroll bar.

To open a drawing, either double-click on the icon, or click on the icon and then **OK**. The drawing then appears in a drawing window.

To see drawings stored on another disk drive, double-click on the drive and/or directory until you reach the location where your drawings are stored. Icons of your stored drawings will then appear. For example, to see drawings stored on your diskette in drive A, double-click on [-a-]. The drawings stored on that disk will appear in the file list.

Recent Drawings

AutoSketch allows you to quickly open the 5 most recent drawings. Those drawing names appear toward the bottom of the **File** menu. Simply pick the drawing name to open and the drawing appears on-screen. Refer back to Figure 3-1 for an example.

Drawings from Different Versions of AutoSketch

With AutoSketch for Windows Release 2, you can open drawings created in earlier releases of AutoSketch. These earlier versions include versions 1.0, 2.0, and 3 for DOS, as well as Release 1.0 for Windows. Drawings made with version 1.0 for DOS will not have icons, but the drawing name will appear.

Earlier versions of AutoSketch do not store the specific unit of measurement. When you open a drawing created in an earlier version, a dialog box will appear, Figure 3-10. Click on the scroll arrow and pick the unit type that represents how your drawing was created. To print the drawing, you will also need to set up the sheet size. These settings are not converted from the older versions. Also, once you save this drawing, it can no longer be used with a previous version of AutoSketch.

Figure 3-10.
When loading a drawing from a previous version of AutoSketch, you must specify a unit of measurement.

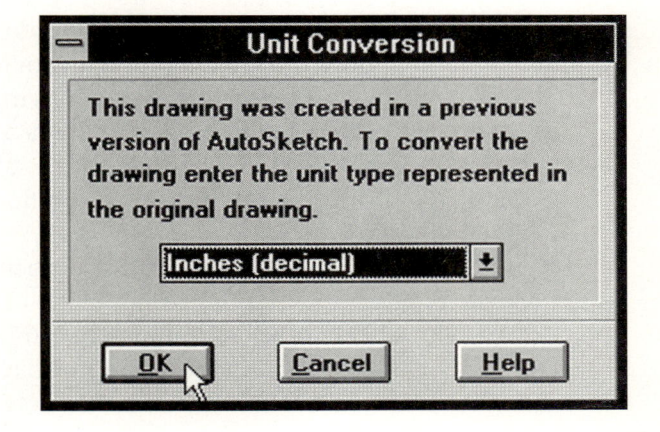

AutoSketch for Windows Release 2.0 stores drawings with greater precision than in previous releases. You may notice that some connected objects will not align precisely after the conversion.

Saving a Drawing

As you work on a drawing, save it often. You never know when a power outage will wipe it from memory. Generally, you should select the **Save** command from the **File** menu about every 5 to 10 minutes. AutoSketch will not prompt you with questions. The drawing file is simply updated on disk. Your drawing remains on-screen in the drawing window.

Saving a Drawing Under a New Name

> **SHORTCUT**
> You can use the [F12] key to quickly save the current drawing.

You learned earlier that saving a new drawing allows you to assign a name to the new drawing. You can also save a copy of the current drawing under another name. By doing so, you keep both the original drawing and the revised drawing under separate names. For example, suppose you create a chart design and name it CHART. Then, you make changes that you want to keep, but also want to keep the original. The **Save As...** command will let you do this.

Select the **Save As** command in the **File** menu. The **Save As File** dialog box appears. Enter a new name, such as NEWCHART. Now you have both versions stored on disk. You can also select a different disk or directory just as when saving new drawing.

Saving Part of a Drawing

AutoSketch allows you to save part of your drawing as a new drawing file. This may be important if only a few objects from your drawing are needed. For example, suppose you just drew a new jig base that needs to be machined. Your drawing includes dimensions and notes about the jig. When you send this drawing to a computer-aided machining system, only the objects that make up the part should be sent. Dimensions and text should not be sent. The **Part Clip...** command can be used to save only the part geometry. A new AutoSketch drawing file will be created.

To clip part of the drawing, select the **File** menu and **Part Clip...** command. The **Part Clip File** dialog box appears asking for a filename, Figure 3-11. You may want to select a new disk or directory to store the new file. Enter the name of the part clip file and pick the **OK** button. The following prompt appears in the **Prompt Box.**

Part Clip Part base location:

The location you pick becomes the origin of the new drawing. Drawing origins are discussed in more detail in Chapter 10.

Once the origin is selected, AutoSketch gives the following prompt.

Part Clip Select object:

You can select single objects, or select multiple objects using a crosses/window box or selection set. These selection techniques are discussed in Chapter 6.

A crosses/window box appears when you select a point that is not an object. Drag the box around the objects you want to select.

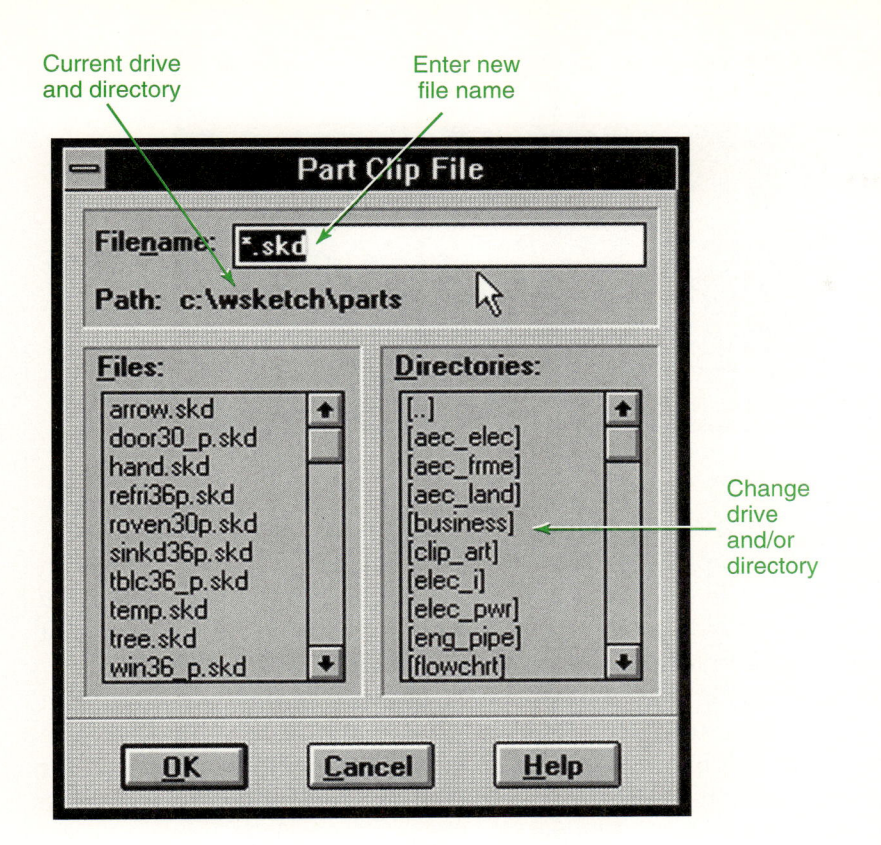

Figure 3-11.
Enter a filename in the **Part Clip File** dialog to save a portion of the drawing under a different filename.

Current drive and directory — Path: c:\wsketch\parts

Enter new file name — Filename: *.skd

Change drive and/or directory

> **SHORTCUT**
>
> It is quicker to hold down the [Shift] key on the keyboard to create the selection set.

To create a selection set, click on the **Shift** button. Then, pick the objects you want to clip. After all objects are selected, click on the **Shift** button again. The objects are saved to the drawing file.

Deleting a Drawing

AutoSketch allows you to delete drawings and backup files from disk. This should only be done after you are certain these files are no longer needed.

Deleting a Drawing File

To delete a drawing file, select the **Delete...** command from the **File** menu. The **Delete Drawing File** dialog box will appear, Figure 3-12. This dialog box looks similar to the **Open Drawing File** dialog box. The drawings in the current directory appear as icons. Click on the icon of the drawing to delete. Then, pick the **Delete** button. A warning message will appear asking if you really want to delete the file. Select either **Yes** or **No**. Note: You cannot double-click on an icon to delete a drawing file. This is a safety precaution to help prevent accidentally deleting a file.

Deleting a Backup Drawing File

The delete function also allows you to delete backup files. With the **Delete Drawing File** dialog box on-screen, select the **.BAK Files** option. Only backup files will appear. Of course, if you delete both your drawing file and backup file, there is no way to retrieve your drawing. The only time you may need to delete backup files is if you are running low on disk space. Backup files take up the same amount of space as drawing files. Deleting them may provide space for more drawings.

> **SHORTCUT**
>
> If you are deleting all backup files, use the Windows File Manager. You can delete all backup files at one time.

Figure 3-12.
To delete a drawing, select the drawing to delete, and then pick the **Delete...** button.

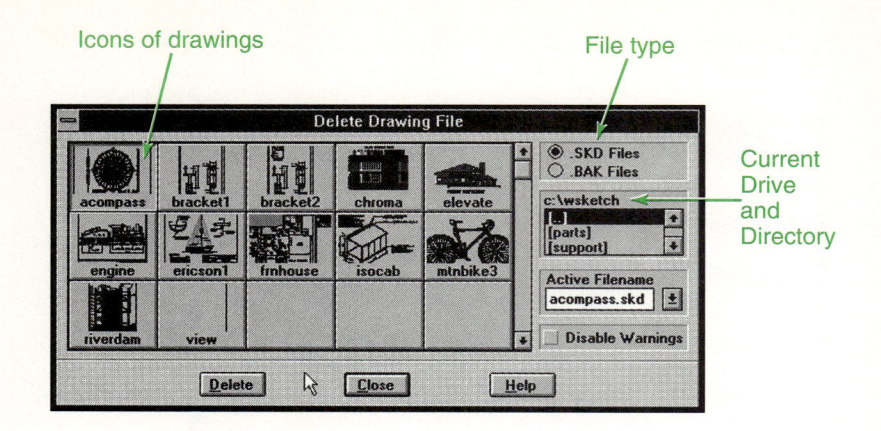

Icons of drawings — *File type* — *Current Drive and Directory* (Delete Drawing File dialog)

Retrieving a Backup Drawing File

If you accidentally delete a drawing file, you may be able to retrieve some of it. AutoSketch creates a backup file each time you save an edited drawing. The *backup file* contains the previously saved version of the drawing. If the backup file has not been deleted, you will be able to restore to a previous version of the drawing.

To restore a backup drawing file, you must first change the file extension from .BAK to .SKD. This can be done using the **Rename...** command in the Windows File Manager application. This command is found in the **File** pull-down menu. After changing the extension, you can return to AutoSketch and open the file using the **Open...** command from the AutoSketch **File** menu.

Working with Multiple Files On-screen

AutoSketch allows you to have up to five drawings open at one time. Remember, though, that each drawing takes up computer memory. You may soon run out of memory with as few as two drawings on-screen. However, if you have plenty of memory, having multiple drawing files open has some advantages. Two advantages are listed below.

- You can view and compare two versions of the same design with two windows open.
- You can copy objects from one open drawing file to another open drawing file.

To open more than one drawing, select the **Open...** command from the **File** menu for each drawing. This is the same procedure as opening a single drawing file. The drawings are now all open, however you will only see one on-screen.

SHORTCUT
You can cycle through open drawings very quickly by holding the [Ctrl] key pressing the function key [F6].

Arranging Drawing Windows

With more than one drawing open, you can arrange the drawings several different ways. You move between active drawings by selecting the drawing you want to view from the **Window** menu.

Tile drawings

When you select **Tile Windows** from the **Window** menu, all drawings fit on-screen next to each other, Figure 3-13. You can move around each drawing using the scroll bars.

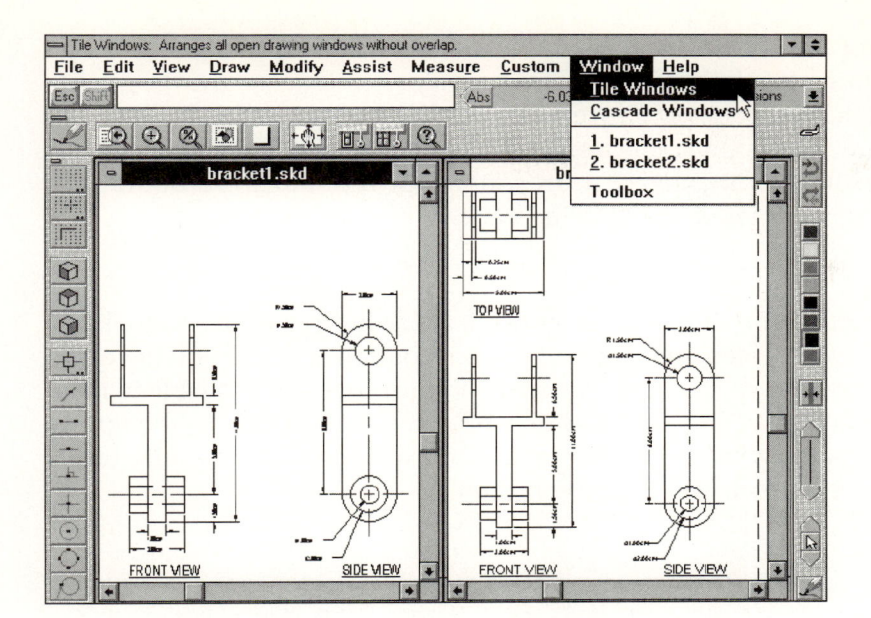

Figure 3-13.
The **Tile Windows** command fits all drawings side-by-side on-screen (as shown here).

Cascade drawings

When you select **Cascade Windows** from the **Window** menu, all drawings overlap each other on-screen, Figure 3-14. You can select a drawing by clicking on it. You can move around the active drawing using the scroll bars.

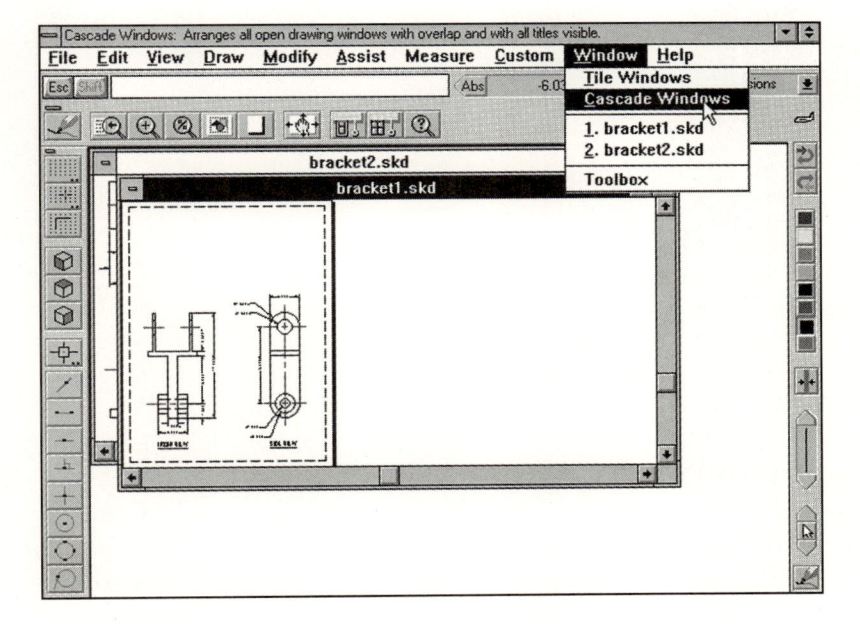

Figure 3-14.
The **Cascade Windows** command fits all drawings by overlapping them on-screen (as shown here).

Maximize a drawing

Maximizing a drawing enlarges the drawing window so that the drawing takes up the entire AutoSketch workspace. However, any other open drawings will not be seen. This is the default view of a single drawing. This is also the default view when you start a new drawing. To maximize a drawing, pick the maximize button (up arrow) in the upper-right corner of the drawing window.

Minimize a drawing

When you *minimize* a drawing, an icon appears at the bottom of the AutoSketch workspace with the drawing filename, Figure 3-15. This option is best if you have two drawings open, but want one drawing maximized and the other still available to maximize. To minimize a drawing, pick the minimize button (down arrow) in the upper-right corner of the drawing window. To minimize a maximized drawing, first tile or cascade the drawing. Then, pick the minimize button.

Figure 3-15. When you minimize a drawing, an icon appears with the drawing name. Double-click on the icon to restore the drawing.

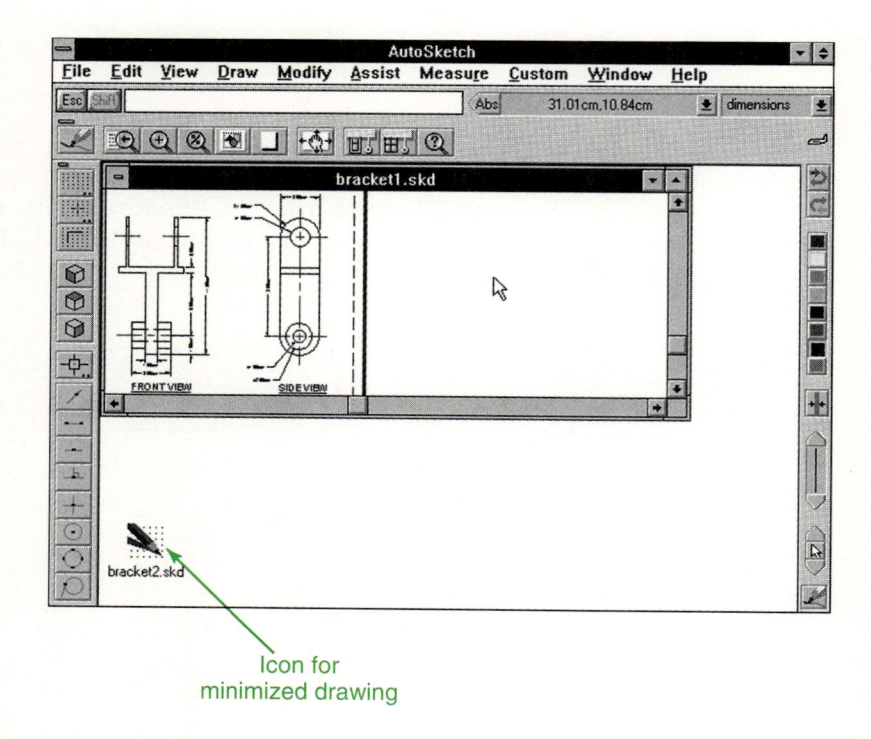

Working with Duplicate Windows

It is possible to have the same drawing open in more than one window. Simply use the **Open...** command in the **File** menu and select the same file. This allows you to work on different views of the same drawing. However, you should do this with caution for two reasons.

- Changes made to one drawing window may not be updated in the view shown in the other windows. The changes are made to the other windows, but they may not be immediately visible.
- Saving changes made in one drawing window are applied to all open windows for that drawing. This means that if you are working in two windows, using the **Save** command keeps overwriting changes made in the other file.

When using duplicate windows, you should do all your work in one window. Use the other window only to view other parts of the drawing.

Ending Your Drawing Session

When you are finished with AutoSketch, select **Exit** from the **File** menu. If all your open drawings are saved, the drawing screen will immediately clear, you will leave AutoSketch, and Windows Program Manger will reappear. If you have made changes since last saving the drawing, a warning will appear for each unsaved drawing.

Never turn the computer off to leave AutoSketch. Always select the **Exit** command and then close Windows to return to a DOS prompt. Also, make backup copies of your drawing files. If your drawings are stored on the hard drive, copy them to a floppy or use a tape backup system. If your drawings are stored on a diskette, make a copy of the disk. This can be done using the **Copy Disk...** command in the Windows File Manager application or by using the DISKCOPY command at a DOS prompt. These commands are discussed in Chapter 20.

> **SHORTCUT**
>
> The quickest way to end AutoSketch is to double-click on the **AutoSketch Control Menu** button.

Summary

The drawings you create using AutoSketch are computer files. The commands to open, save, and start drawings are found in the **File** menu. After starting AutoSketch and selecting a template, you start with a blank, untitled drawing screen. After drawing some objects, select the **Save** command to store the drawing under a specific name. Continue to save your drawing about every 5 to 10 minutes. When you are finished for the day, select **Save, Close,** and then **Exit** to end your drawing session.

To work on the drawing again, you must load it into memory using the **Open...** command. After adding or changing objects, you might want to save the changes as a new drawing. To do so, select the **Save As...** command.

While working on drawings, there are other options that may assist you. Using the **Tile Windows** and **Cascade Windows** commands, you can view multiple drawings on-screen. Using the **Delete...** command you can delete drawing files and backup files from disk storage. Using the **Part Clip...** command, you can save a portion of your drawing as a different drawing file.

Important Terms

Backup file	Icons	Minimizing
Drawing files	Maximizing	Template

New Commands and Tools

Cascade Windows	File**n**ame: box	**S**ave
Close	**N**ew...	Save **A**s...
Delete...	**O**pen	**T**ile Windows
Directories box	Part C**l**ip...	

Review Questions

Give the best answer for each of the following questions.

1. What is the first action you must take upon starting AutoSketch? _____

2. When does the drawing name "untitled1" appear? _____

3. AutoSketch allows you to have up to _____ drawings active at any given time.

4. What are the two ways to save a drawing to a disk or directory other than the default disk and directory?
   A. _____
   B. _____

5. True or false. If the power goes out, any changes you have made to the drawing on-screen are automatically saved to disk. _____

6. You can use up to _____ characters (letters and numbers) to name a drawing.

7. The _____ command clears the active drawing from the screen and memory.

8. How do you save a new drawing to a disk drive other than the default drive? _____

9. What guidelines should you follow when naming drawings? _____

10. After choosing **Open...**, how do you select a drawing to load into memory? _____

11. How can you open one of the five most recently saved drawings without having to use the **Open...** command? _____

12. True or false. You cannot load drawings created in a previous version of AutoSketch. _____

13. Generally, you should save your drawing about every _____ to _____ minutes.

14. Why might you want to save a drawing under a new name? _____

15. What command allows you to save part of your drawing as a new drawing file? _____

16. What must you select in the **Delete Drawing File** dialog box to view backup files? _____

17. What two commands allow you to see all open drawings?
    A. _____
    B. _____

18. Why might you want to open two copies of the same drawing? _____

19. Why is it best not to open two copies of the same drawing? _____

20. If you select **Exit** and have made changes since the last **Save** command, a _____ will appear.

Activities

Here is a mixture of drawing activities to develop your skill with starting and saving drawings. Before starting this activity, format a new, blank diskette where you will store your drawings. Label the disk with your name and the words "AutoSketch Work Disk."

1. Start AutoSketch and select the **Normal** template. Draw several lines by picking the **Draw** menu and **Line** tool. Draw several lines. Save this drawing as C3A1 on your work diskette. Close the drawing.
2. Open the file named C3A1. Draw several new lines. Save the revised drawing on your file disk under a new name, C3A2. Leave this drawing open.
3. Use the **Part Clip...** command to save one line under the filename C3A3. After doing so, clear all of the drawings from the screen using the **Close** command.
4. Open drawings C3A1, C3A2, and C3A3 using the **Open...** command. After doing so, clear all of the drawings from the screen using the **Close** command.
5. Open drawings C3A1, C3A2, and C3A3 by clicking on their names near the bottom of the **File** pull-down menu. Use the **Tile Windows** command to view all drawings on-screen. Add one line to each drawing. After doing so, clear all of the drawings from the screen using the **Close...** command and select **Yes** to save the changes to each drawing.
6. Delete all backup files from your diskette.
7. Exit AutoSketch to end your drawing session.

This drawing contains many different objects. In addition, many objects were created to help construct others, then were later erased. (Autodesk, Inc.)

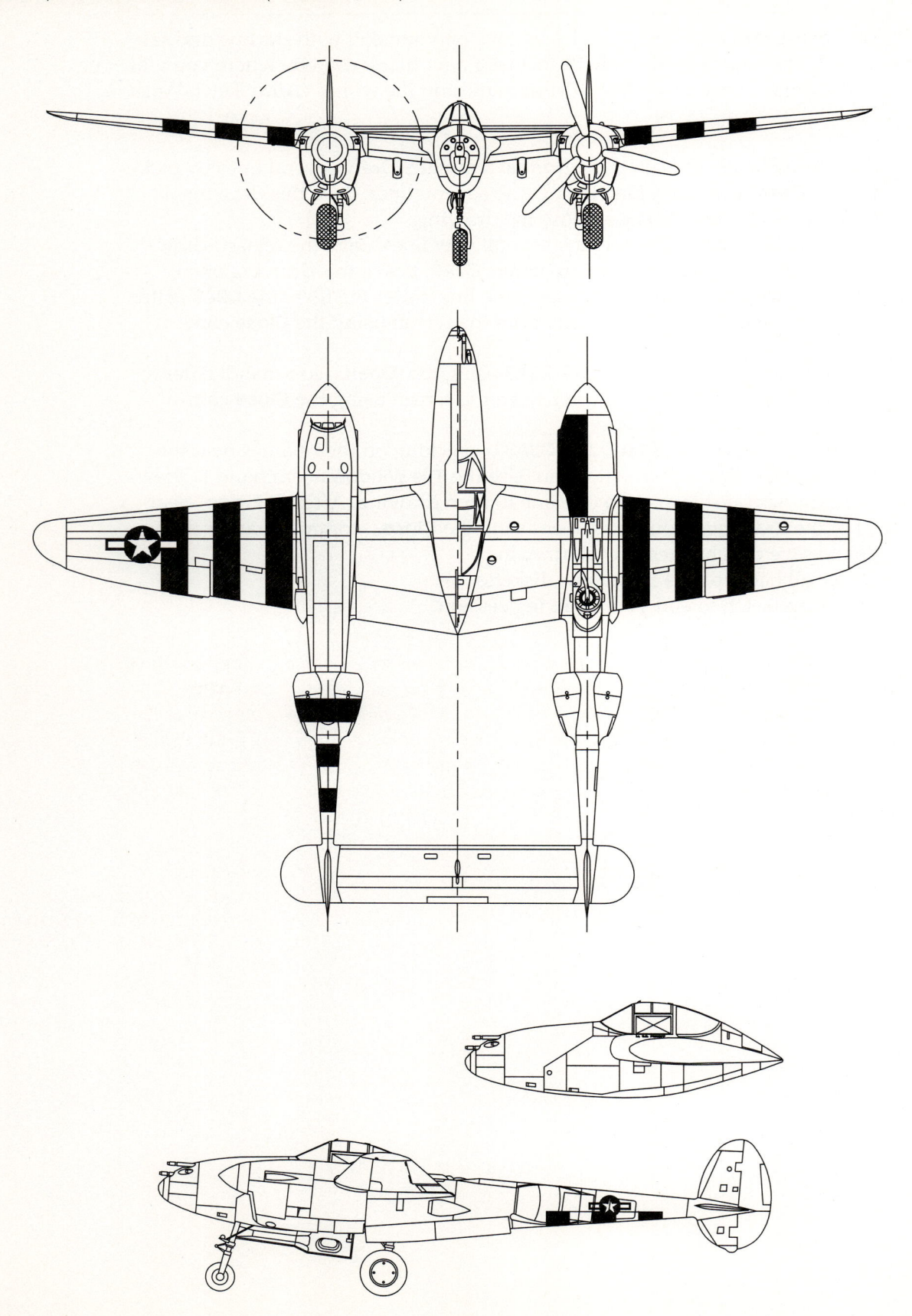

CHAPTER 4

Drawing and Erasing Objects

Objectives

After studying this chapter, you will be able to:
- Set the object color, width, and linetype.
- Draw points.
- Draw lines.
- Draw boxes.
- Draw circles using any of four options.
- Draw polygons using any of three methods.
- Draw arcs using any of four options.
- Draw ellipses using any of three options.
- Draw Bezier, B-spline, and freehand curves.
- Draw polylines with both straight-line and arc segments.
- Erase objects that you have drawn.
- Undo and redo your commands.

The purpose of a drawing is to communicate information using graphics. For example, a technical drawing might show details for a product to be manufactured or constructed. An illustration might show how to assemble a new product. Each of these drawings are composed of one or more objects. *Objects* refer to lines, boxes, polygons, circles, arcs, and other elements that AutoSketch can create. Each of these items are found in the **Draw** menu, Figure 4-1. Although you can also select drawing commands from pull-down menus, this chapter assumes that you are using tools.

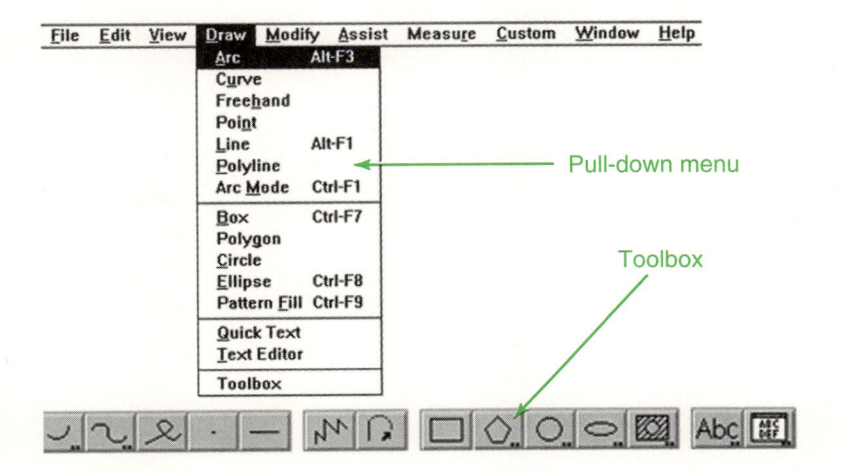

Figure 4-1.
Commands to draw AutoSketch objects are found in the **Draw** menu. You can use a toolbox or pull-down menu.

87

In this chapter, you will learn how to add objects to your drawing. This chapter is not concerned with exact size or position. Therefore, for this chapter, you can create objects without regard to size or placement. Chapter 5 will discuss how to enter coordinates to define exact size or location. Chapter 5 will also discuss how to place objects at precise places using a grid, or attach new objects to existing objects.

There are several exercises in this chapter. Remember, after you place an object, the command repeats. To exit the command, you must pick the **Esc** button, press the [Esc] key, or select a different drawing or editing tool. If you make a mistake, you can always pick the **Undo** button in the vertical status bar.

Setting Object Color, Width, and Linetype

In technical drafting, color, linetype, and width have certain meanings. These settings are called *object properties.* A solid, .20" (.5 mm) black line might outline the shape of an object. A dashed, .15" (.4 mm) yellow line might represent a hidden edge of an object. Chapter 11 discusses standards for using color, linetype, and width.

The settings for color, linetype, and width are found in the vertical status bar. After changing a setting, new objects added to the drawing are given that new setting. However, not all objects are affected by color, width, and linetype, Figure 4-2.

If you draw an object, and later decide to changes its properties, you can do so using the **Change Properties** tool in the **Edit** menu.

Figure 4-2.
The color of all objects can be changed. However, not all objects can have the width or linetype changed.

Settings That Affect Objects			
Object	Color	Wide Line Widths	Multiple Linetypes
Arc	X	X	X
Box	X	X	X
Circle	X	X	X
Curve	X		
Ellipse	X	X	X
Line	X	X	X
Pattern Fill	X		
Point	X		
Polygon	X	X	X
Polyline	X	X	X
Text	X		
Wide Objects	X	X	

Setting Object Color

It is helpful to assign colors to objects, especially if you have a color monitor. Even if you have a monochrome monitor, you may still want to set colors. When you plot the drawing to a pen plotter, the object's color determines which pen is used. The pens may have different ink colors or different tip widths.

Selecting a color

You can use up to 255 colors in your AutoSketch. Eight of the 255 are shown in the **Color Palette**. The **Color Palette** is found in the vertical status bar. To select a color, simply pick its button in the palette. All newly drawn objects will have the new color. All existing objects will remain the same color.

Changing the Color Palette

To select new colors for the **Color Palette**, double-click any button on the **Color Palette**. The **Color Settings** dialog box appears, Figure 4-3. The current palette is displayed on the right of the dialog box. Pick a color from the current palette that you want to change. Then, select a color from the **Available Colors** to replace the color you selected. Repeat this step to change other colors in the palette and pick **OK** when done. The palette in the vertical status bar now displays the new colors.

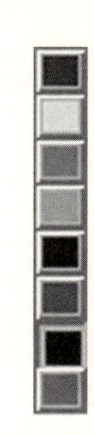

SHORTCUT

Try to use no more than eight colors on a drawing. More than that may "clutter" the drawing and make it hard to read.

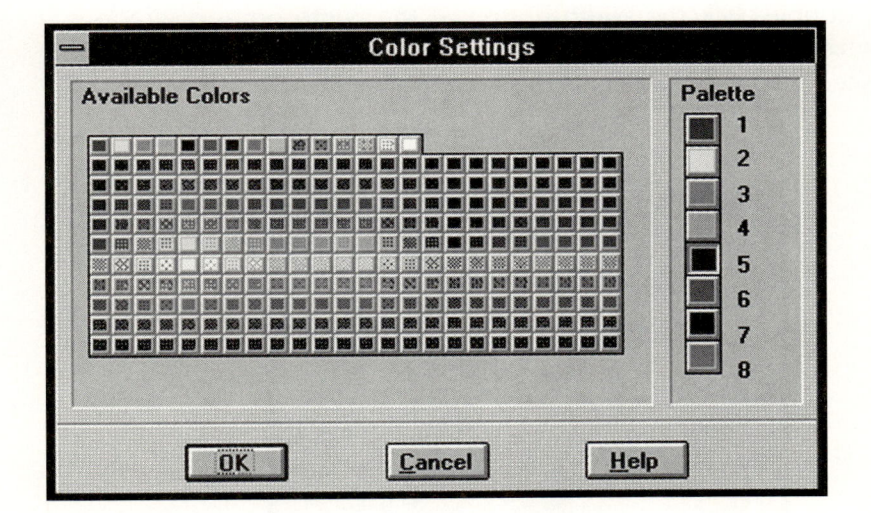

Figure 4-3.
Use the **Color Settings** dialog box to change the colors shown in the palette. Click on a color in the palette to change, and then select one of the available colors.

Setting Object Width

The *width* is the thickness of the lines you draw. These lines define the objects you add to a drawing. Picking the **Width** button opens the **Width** pop-up box. There are 6 different width settings that you can use, Figure 4-4. The *hairline* width is the thinnest line you can draw. To select a new width, simply pick the thickness you want, and then pick the **Width** button again or pick **Close** to close the dialog box. You can also leave the **Width** pop-up box visible while you draw. This will give you instant access to width settings. Once you change line width, all new objects you draw will have the new line width. (An exception is objects that do not support widths larger than hairline. See Figure 4-2.)

Figure 4-4.
Double-click on the **Width** button to open the **Width** pop-up box. Select the object width from the **Width** pop-up box. Six different line widths are displayed in this box. To change a line width, select **Customize.**

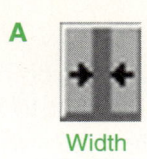

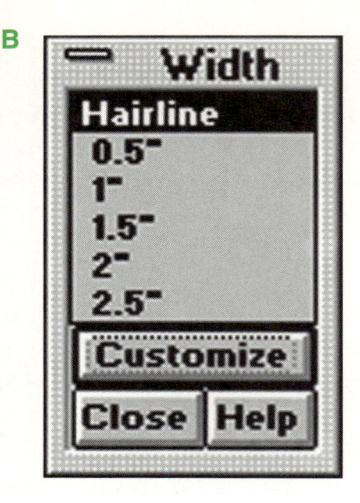

A Width button

B

It is best to use only the hairline width while constructing and editing a drawing. Many editing tools do not work well with wide lines. In addition, wide objects are shown and printed as a solid linetype, regardless of the current linetype setting. When finished with the drawing, you can change line widths as needed.

Customizing Widths

The default width settings for AutoSketch are not appropriate for most technical drawings. Therefore, you should customize the line widths. To access the **Width Settings** dialog box, double-click the **Width** button. If the **Width** pop-up window is open, choose **Customize** or double-click one of the widths, Figure 4-5. You can change any default width except hairline.

> **SHORTCUT**
>
> To save time, draw all objects with a hairline width. Once all constructions and editing are done, change any lines that need to be different line widths.

Figure 4-5.
Use the **Width Settings** dialog box to change the widths shown in the **Width** pop-up box. You can change any of the line widths *except* hairline.

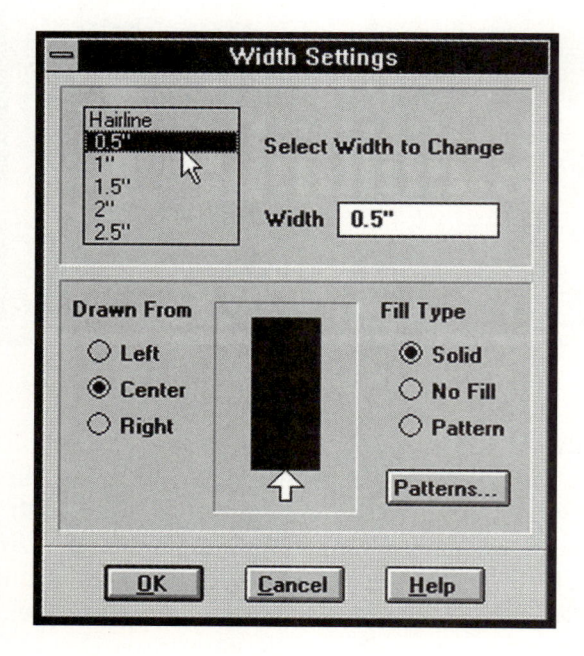

To change a width, first pick the width to change in the **Width Settings** dialog box. Then, enter the new setting for the **Width** option and press the [Enter] key. The **Drawn From** option lets you choose how lines are drawn from the points you pick to define the object. Line widths for technical drawings are typically very thin, less than .025". Therefore, the **Center** option is usually chosen. The **Fill Type** option determines how objects are filled. Most often, you will select **Solid**. Choose **OK** when finished.

Setting Object Linetype

The *linetype* refers to the line pattern of objects. The line may be solid, dashed, or a combination of solid and dashed segments. When you first start AutoSketch, the default linetype is solid. The **Current LineType** button is found in the vertical status bar. The **Previous LineType** arrow is above the **Current LineType** button, and the **Next LineType** arrow is below it. Pick either of these arrows to change the linetype. Continue picking an arrow until the linetype you need is shown.

Changing linetype by name

It may be difficult to see which linetype you have chosen. To choose a linetype by name, double-click on the **Current LineType** button. The **Line Type Settings** dialog box appears, Figure 4-6. Here you can pick a linetype by name. Many of the names reflect the names given in the alphabet of lines for technical drafting. All new objects will have the new linetype. All existing objects will remain the same. (An exception is objects that do not support nonsolid linetypes. See Figure 4-2.)

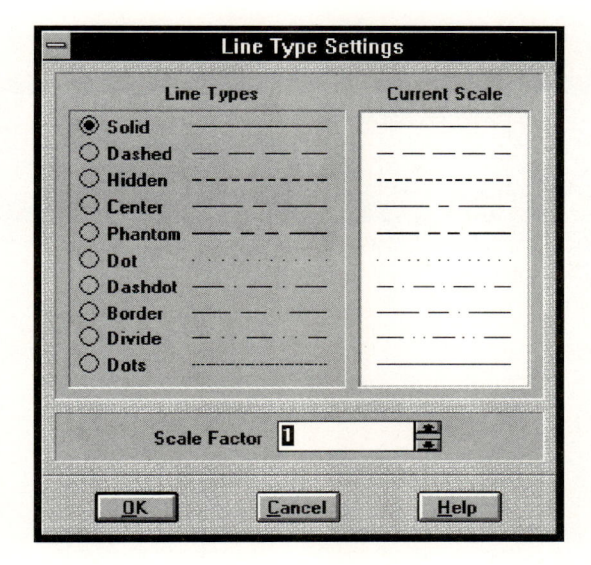

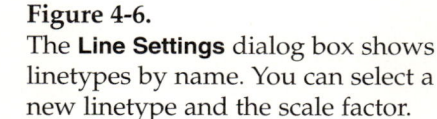

Figure 4-6.
The **Line Settings** dialog box shows linetypes by name. You can select a new linetype and the scale factor.

Changing linetype scale factor

In the **Line Type Settings** dialog box, you can also change the scale factor. The **Scale Factor** option determines how often the line pattern repeats. On a very large drawing, set the scale larger. On a small drawing, set the scale smaller. Otherwise, even a patterned linetype will appear solid.

The default scale factor is 1. To change the factor, pick one of the arrows or enter the new factor in the box. Pick **OK** when you are done.

Experiment with different scale factors to see how the line appears. Resetting the scale factor affects all objects, *including* those already drawn.

Drawing Points

A *point* marks an exact position on the drawing. It has no size, only location. Points are helpful as a reference for placing other objects. To add a point to the drawing, select the **Draw** menu and **Point** tool, Figure 4-7. The following prompt will appear.

Point Enter point:

Move your cursor to the desired place on the drawing and pick that location with the mouse. A small dot will appear on-screen in the current color to mark the point's location.

Figure 4-7.
After choosing the **Point** tool, pick locations on the drawing. Dots appear to represent points.

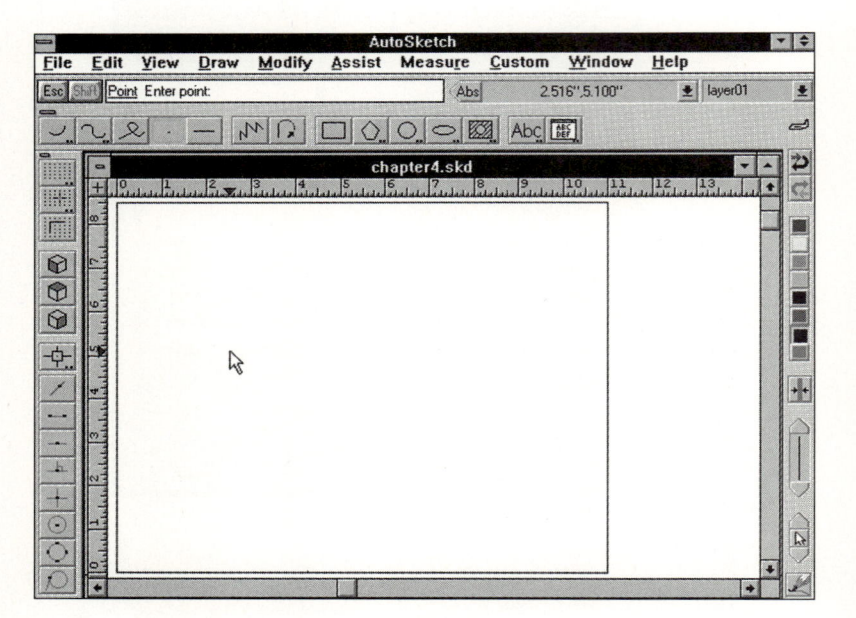

Preparing for Exercises and Activities in This Chapter

Before starting the exercises or activities in this chapter, set your printer for landscape orientation. You can do this in the Windows Control panel. You can also use the following procedure.
1. Opening a new drawing with the **Quick Start** template.
2. Select the **File** menu and **Print Setup...** command.
3. Highlight the printer and select the **Setup...** button.
4. Select **Landscape** for the paper orientation and pick **OK**.
5. Pick **OK** to leave the **Print Setup** dialog box.
6. Close the drawing without saving your changes.

Doing these steps allows your drawing sheet to fit on a single A-size sheet of paper horizontally. You will only need to do this step once, unless you use another Windows application and change the orientation back to portrait.

Exercise 4-1

1. Start a new drawing using the **Normal** template.
2. Use the **Point** tool to place four points on your drawing.
3. Save the drawing as C4E1 on your work diskette and then close the drawing.

Drawing Lines

A *line* is a straight object drawn by picking two endpoints. To draw a line, select the **Line** tool from the **Draw** menu. The following prompt will appear.

Line Enter point:

Locate your cursor and pick the first endpoint of the line. AutoSketch then prompts you for the second endpoint.

Line To point:

Move your cursor to place the second endpoint. As you do this, a "rubber band" will appear, connecting your cursor to the first endpoint, Figure 4-8. This temporary line helps you locate the second endpoint. When satisfied with the line's position, pick the second endpoint.

AutoSketch does not automatically connect multiple lines together. You can connect lines by carefully picking the endpoint at the end of an existing line. However, you can draw connected lines easier using the **Polyline** tool discussed later in this chapter.

> **SHORTCUT**
>
> To access the **Line** tool, you can also hold the [Alt] key and press the function key [F1].

Exercise 4-2

1. Start a new drawing using the **Normal** template.
2. Use the **Line** tool to draw two lines in opposite corners of your drawing.
3. Change the linetype to hidden and the color to red.
4. Use the **Line** tool to draw a box with four lines. Try to draw straight horizontal and vertical lines so that the endpoints touch.
5. Save the drawing as C4E2 on your work diskette and then close the drawing.

Figure 4-8.
To draw a line, pick the first and second endpoints. As you locate the second endpoint, a "rubber band" helps locate the line's position.

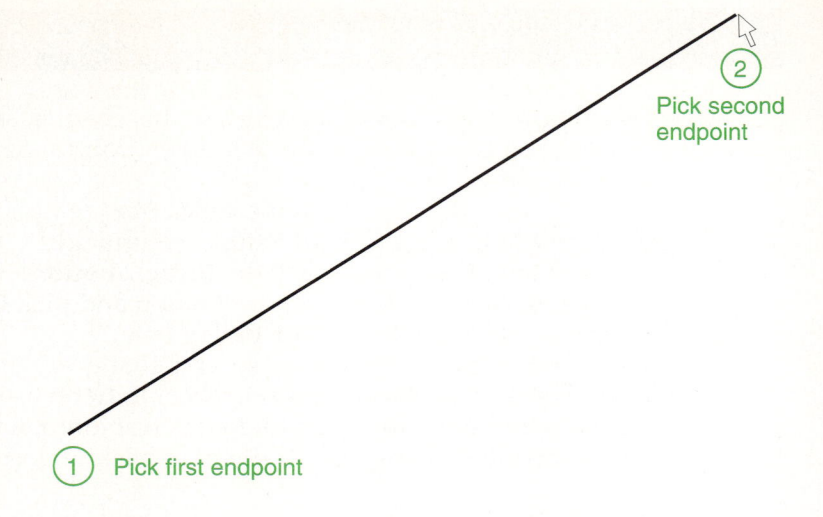

② Pick second endpoint

① Pick first endpoint

Drawing Boxes

The **Box** tool lets you draw rectangles and squares. These shapes are used often on a drawing. They might be the walls of a house or the outline of a computer chip. To draw a box, pick the **Draw** menu and the **Box** tool, Figure 4-9. The following prompt appears.

Box First corner:

> **SHORTCUT**
>
> To access the **Box** tool, you can also hold the [Ctrl] key and press the function key [F7].

Pick one corner of the box. It can be any corner. AutoSketch then asks you for the second corner of the box.

Box Second corner:

Locate the cursor to the opposite corner of the box. As you move the cursor, a temporary box expands and shrinks. This temporary box is similar to the "rubber band" used when drawing a line. When satisfied with the box's size and position, pick the opposite corner.

Figure 4-9.
To draw a box, pick opposite corners. After you pick the first point, a temporary box will appear to help you locate the second point.

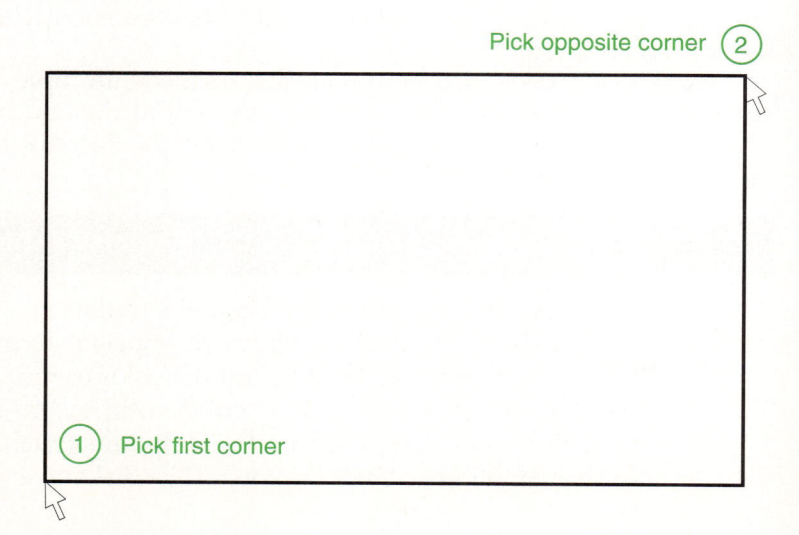

Pick opposite corner ②

① Pick first corner

You can draw boxes using the **Line** tool. However, a box is a single object and, therefore, all four lines are automatically connected. Since a box is a single object, you only need to pick it once to erase, move, or copy it. If a box is made of four single lines, you have to pick each line individually. You can also draw boxes using the **Polygon** or **Polyline** tool. However, the **Box** tool is simply the easiest to use.

Exercise 4-3

1. Start a new drawing using the **Normal** template.
2. Use the **Box** tool to draw a square box on your drawing.
3. Use the **Box** tool again to draw a rectangular box.
4. Save the drawing as C4E3 on your work diskette and then close the drawing.

Drawing Circles

Circles are often used in a drawing to show holes and round objects. The features of a circle are shown in Figure 4-10. The distance from the center to a point on the circle is called the *radius*. The distance across the circle through the center is the *diameter*. The actual distance around a circle is called its *circumference*. AutoSketch lets you draw circles using several options.

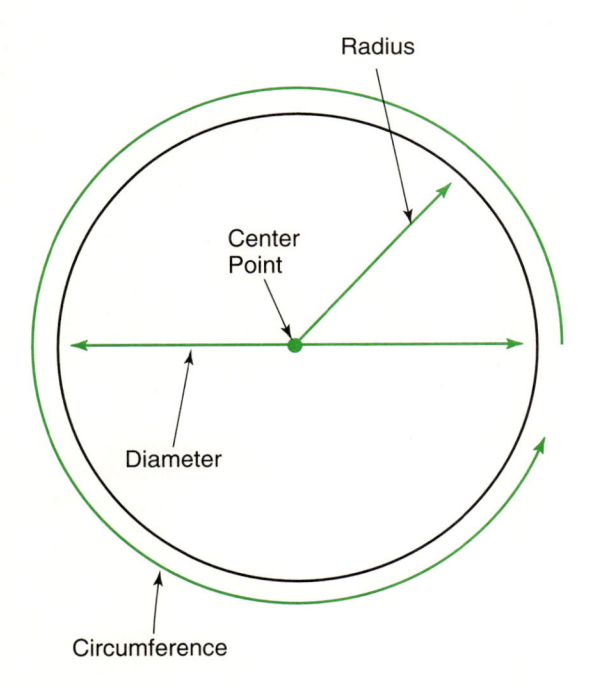

Figure 4-10.
The different parts of a circle are shown here.

Setting Circle Settings

To access the **Circle Settings** dialog box, double-click on the **Circle** tool in the **Draw** menu, Figure 4-11. The sample in the dialog box will adjust to show the setting you choose. After selecting an option, pick **OK** to close the dialog box. The options are discussed in the next sections.

Figure 4-11.
Select the method to draw circles in the **Circle Settings** dialog box. The preview on the right will change to help show you how each option draws a circle.

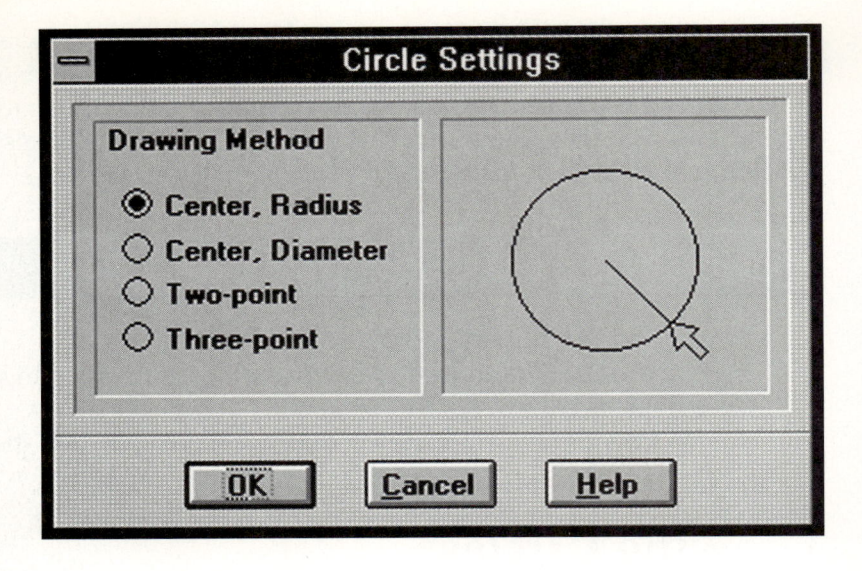

Center point and radius
One option for drawing a circle is the **Center, Radius** option. This option is helpful when you know where the center of the circle should be, and either the exact radius or another location the circle should touch.

Center point and diameter
Another option for drawing a circle is **Center, Diameter.** This option can be used when you know where the center is and the exact diameter of the circle.

Two points
A circle can be drawn with the **Two-point** option if you know two points on the circle. This option is helpful when you know that the circle should touch two other objects in precise places.

Three points
If you know three points on the circle, you can use the **Three-point** option to draw the circle. This option is helpful when you know that the circle should touch three other objects in precise places.

Drawing a Circle Using Center, Radius
To draw a circle by picking the center and radius, choose the **Center, Radius** option in the **Circle Settings** dialog box, Figure 4-12. Then, pick the **Circle** tool in the **Draw** menu. The following prompt appears.

Circle Center point:

Pick where you want the center of the circle to be. AutoSketch then asks you for the radius.

Circle Radius:

Locate the cursor to a radius point. As you move the cursor, a temporary circle will enlarge and shrink. The radius of the circle will also appear in the prompt box. When satisfied with the size of the circle, pick again.

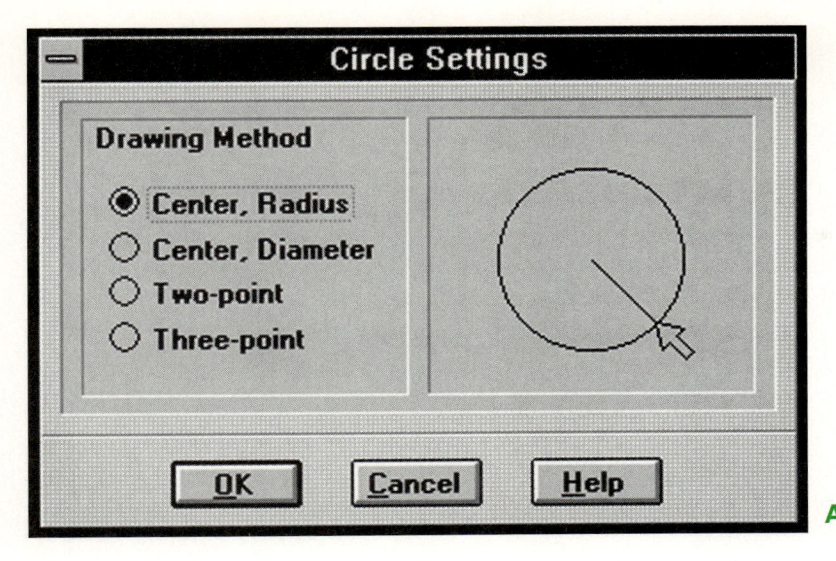

Figure 4-12.
Drawing a circle using the **Center, Radius** option.

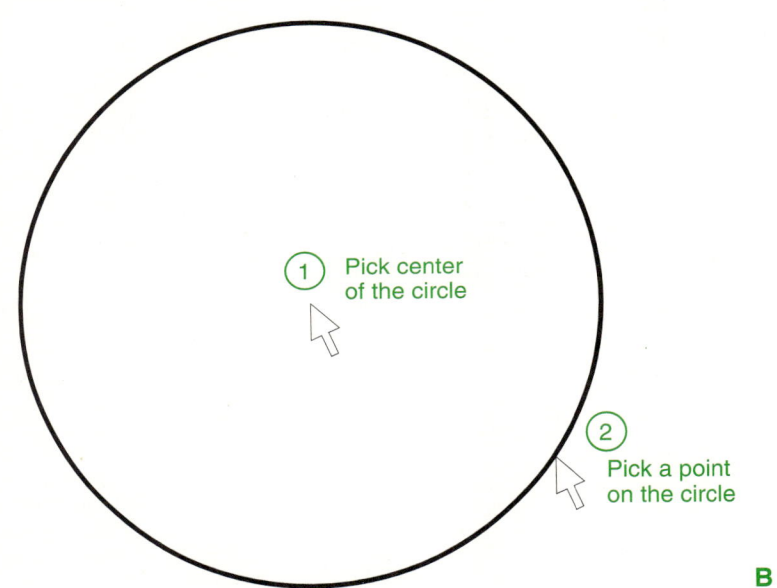

You can also enter an exact radius in the prompt box. After picking the center point, simply type the radius. The numbers will appear in the prompt box. Then press [Enter].

<u>Circle</u> Radius: **2.75**↵

Drawing a Circle Using **Center, Diameter**

To draw a circle by picking the center and a point representing the diameter, choose the **Center, Diameter** option in the **Circle Settings** dialog box, Figure 4-13. Pick **OK** to close the box. The following prompt appears.

<u>Circle</u> Center point:

Figure 4-13.
Drawing a circle using the **Center, Diameter** option.

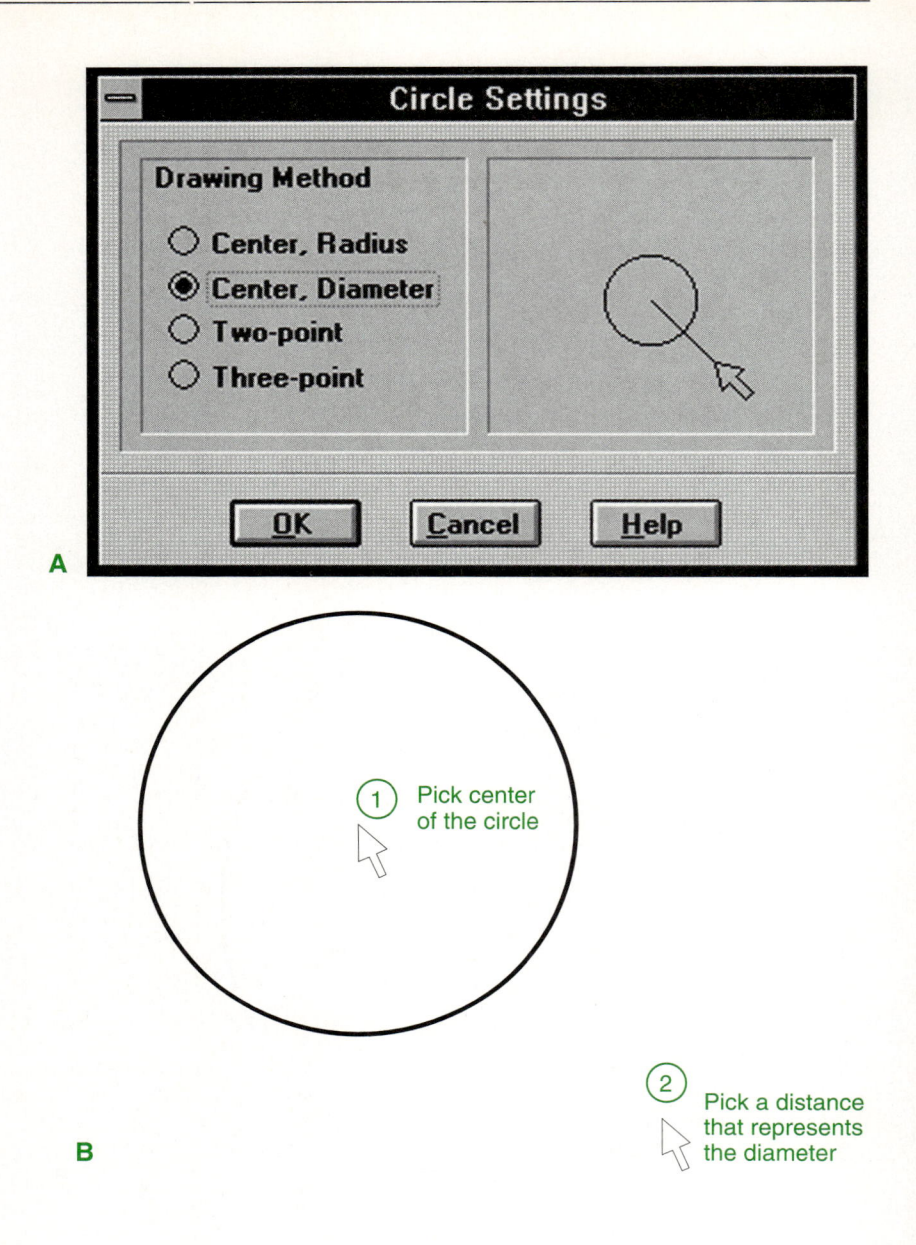

A — Circle Settings dialog with Drawing Method: Center, Radius; Center, Diameter (selected); Two-point; Three-point. OK, Cancel, Help buttons.

B — 1: Pick center of the circle. 2: Pick a distance that represents the diameter.

Pick where you want the center of the circle to be. AutoSketch then asks you for the diameter.

Circle Diameter:

Locate the cursor to a point that is the endpoint of the diameter. As you move the cursor, a temporary circle will enlarge and shrink. The diameter of the circle will also appear in the prompt box. When satisfied with the size of the circle, pick again.

You can also enter an exact diameter in the prompt box. After picking the center point, simply type the diameter. The numbers will automatically be placed in the prompt box.

Circle Diameter: **5.5**↵

Drawing a Two-point Circle

To draw a circle by picking two points, choose the **Two-point** option in the **Circle Settings** dialog box, Figure 4-14. Pick **OK** to close the box. The following prompt appears.

<u>Circle</u> First point:

Pick where you want one endpoint of the circle diameter to be. AutoSketch then asks for the second endpoint.

<u>Circle</u> Second point:

Locate the cursor to a second point that is the endpoint of the diameter. As you move the cursor, a temporary circle will enlarge, shrink, and rotate. The diameter of the

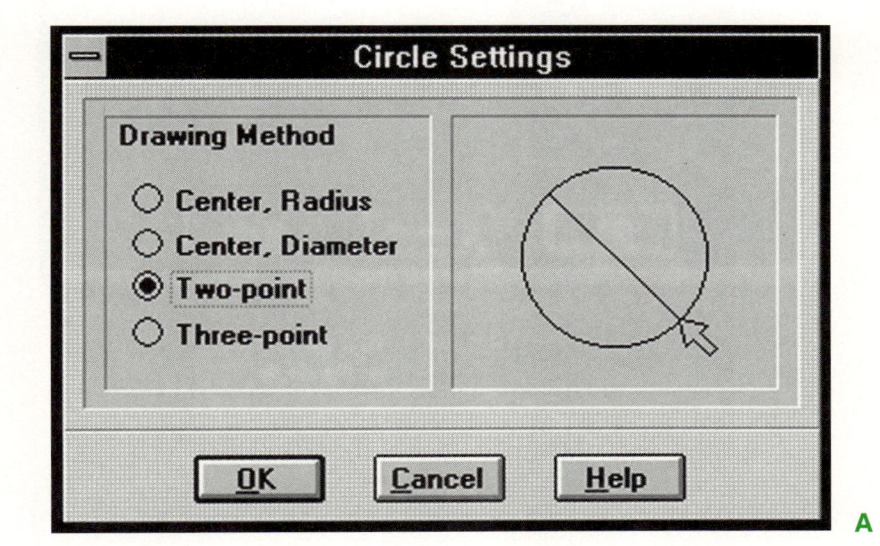

Figure 4-14.
Drawing a circle using the **Two-point** option.

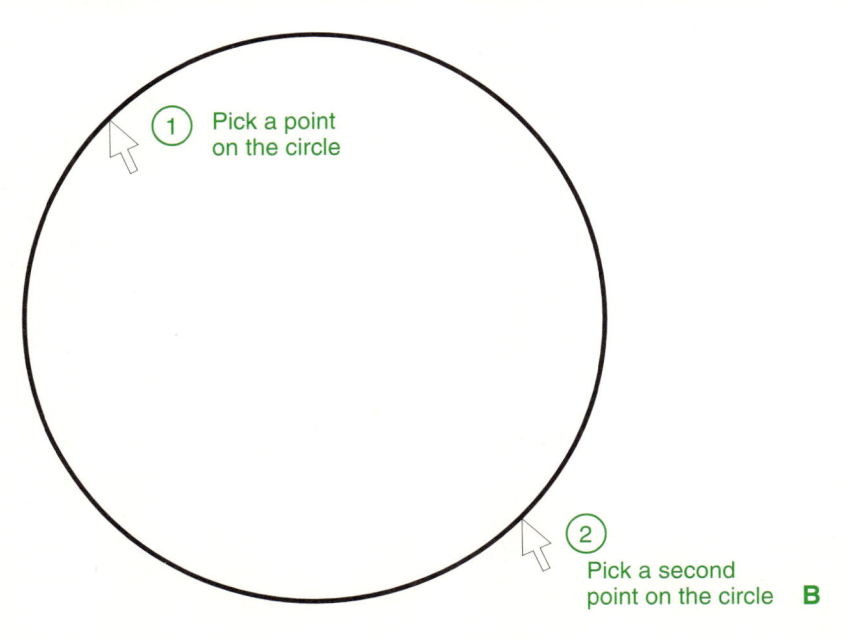

circle will also appear in the prompt box. When satisfied with the size and rotation of the circle, pick again.

Drawing a Three-point Circle

To draw a circle by picking three points, choose the **Three-point** option in the **Circle Settings** dialog box, Figure 4-15. Pick **OK** to close the box. The following prompt appears.

Figure 4-15.
Drawing a circle using the **Three-point** option.

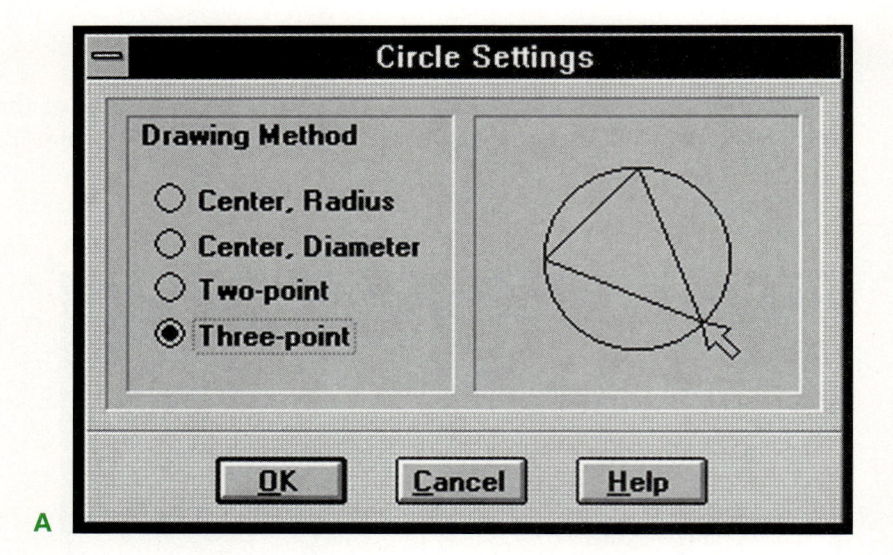

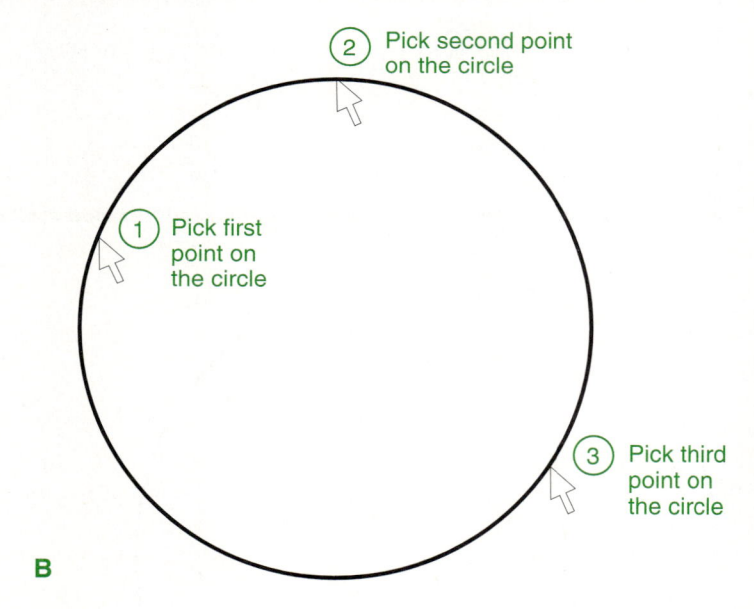

Circle First point:

Pick where you want one point on the circle circumference to be. AutoSketch then asks you for the second point.

Circle Second point:

Pick where you want second point on the circle circumference to be. You are then asked for the third point.

Circle Third point:

Locate the cursor to a third point on the circumference. As you move the cursor, a temporary circle will enlarge and shrink. The radius of the circle will also appear in the prompt box. When satisfied with the size of the circle, pick again.

Exercise 4-4

1. Start a new drawing using the **Normal** template.
2. Use the **Center, Radius** option of the **Circle** tool to draw a circle. Pick both points with your cursor.
3. Use the **Center, Radius** option of the **Circle** tool to draw a circle. Pick the center with your cursor. Enter a radius of 1.25.
4. Use the **Center, Diameter** option of the **Circle** tool to draw a circle. Pick both points with your cursor.
5. Use the **Center, Diameter** option of the **Circle** tool to draw a circle. Pick the center with your cursor. Enter a diameter of 2.5.
6. Draw a line. Then, use the **Two-point** option of the **Circle** tool to draw a circle where the circle touches each endpoint.
7. Draw three lines. Then, use the **Three-point** option of the **Circle** tool to draw a circle where the circle touches one endpoint of each line.
8. Save the drawing as C4E4 on your work diskette and then close the drawing.

Drawing Polygons

A *polygon* is a closed shape made with three or more connected straight lines. The **Polygon** tool draws regular polygons. *Regular polygons* have equal sides and equal internal angles. Squares, hexagons, and octagons are examples of regular polygons. Regular polygons are commonly found in mechanical drawings. For example, a regular polygon may be used to represent the head of a bolt or nut. Before actually drawing the polygon, you need to determine the polygon settings.

Setting Polygon Settings

To access the **Polygon Settings** dialog box, double-click on the **Polygon** tool in the **Draw** menu, Figure 4-16. The sample in the dialog box will adjust to show the setting you choose. After selecting an option, pick **OK** to close the dialog box. The different options are described in the next sections.

Figure 4-16.
Select the method to draw polygons in the **Polygon Settings** dialog box. The preview on the right will change to help show you how each option draws a polygon. Also select the number of sides for the polygon in this dialog box.

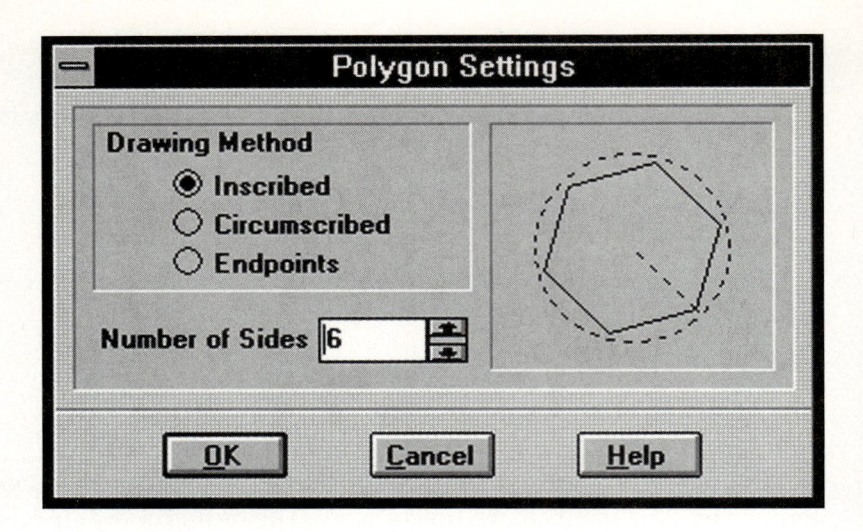

Inscribed

Inscribed means drawn inside of something. An inscribed polygon is drawn by picking a center point and vertex point. A *vertex* is a point on the polygon where two sides meet. This option can be helpful when you want the vertices of the polygon touching a circle.

Circumscribed

Circumscribed means drawn outside of something. A circumscribed polygon is drawn by picking a center point and a midpoint on one of the sides of the polygon. This option can be helpful when you want the sides of the polygon tangent to a circle.

Endpoints

When using the **Endpoints** option, you draw the polygon by picking the two endpoints of one side of the polygon. This option is helpful if you need each side to be an exact length.

Number of sides

The number of sides of a polygon created in AutoSketch can range from 3 to 99. A three-sided polygon is a triangle, a four-sided polygon is a square, and a five-sided polygon is a pentagon. In AutoSketch, you must select how you want to draw the polygon and then choose the number of sides for the polygon.

Drawing an Inscribed Polygon

To draw an inscribed polygon, open the **Polygon Settings** dialog box by double-clicking on the **Polygon** tool in the **Draw** menu. Choose the **Inscribed** option and enter the number of sides for the polygon, Figure 4-17. Then, pick **OK** to close the dialog box. The following prompt appears.

Polygon Enter point:

Pick where you want the center of the polygon to be. AutoSketch then asks you for a vertex.

Polygon Vertex point:

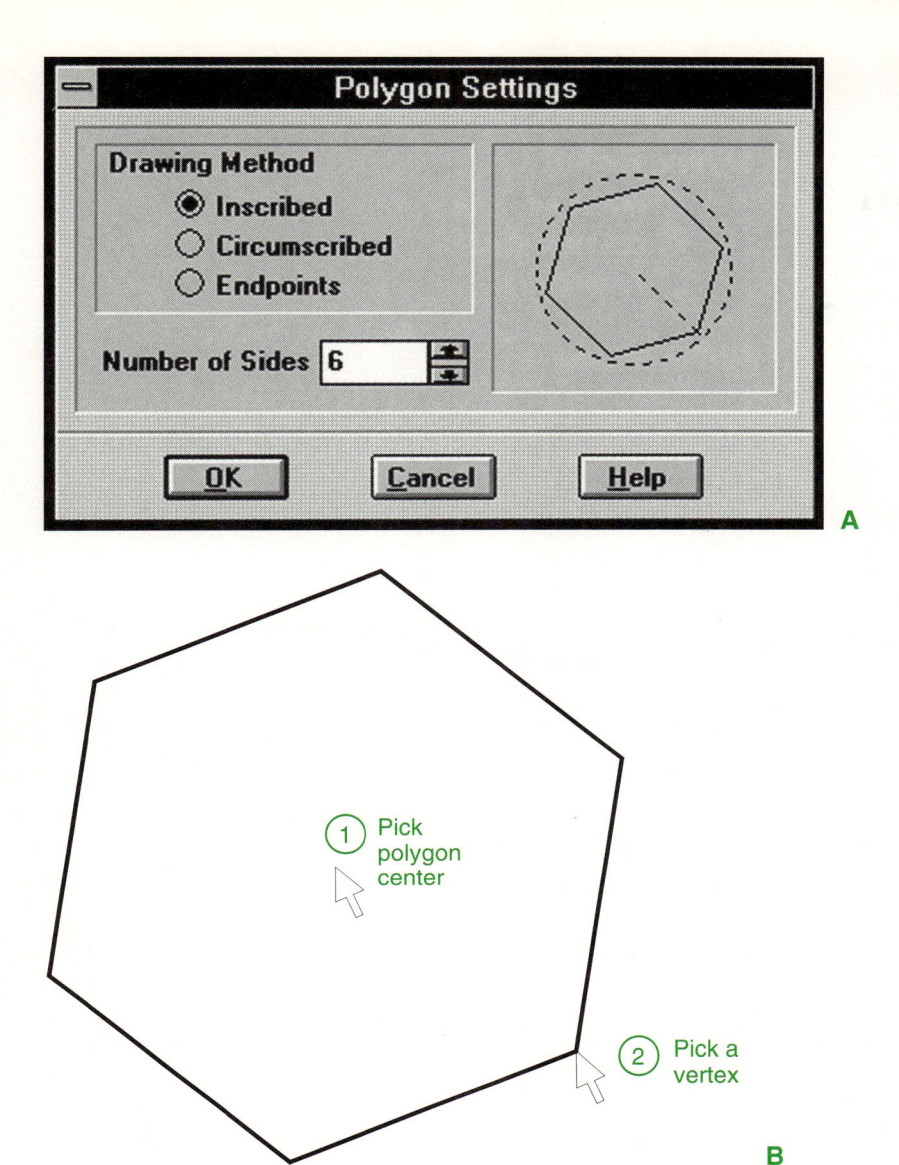

Figure 4-17.
Drawing an inscribed, six-sided polygon.

A

B

Locate the cursor to a vertex point on the polygon. As you move the cursor, a temporary polygon will enlarge, shrink, and rotate. When satisfied with the size and rotation of the polygon, pick again.

Drawing a Circumscribed Polygon

To draw a circumscribed polygon, open the **Polygon Settings** dialog box by double-clicking on the **Polygon** tool in the **Draw** menu. Choose the **Circumscribed** option and enter the number of sides for the polygon, Figure 4-18. Then, pick **OK** to close the dialog box. The following prompt appears.

Polygon Enter point:

Pick where you want the center of the polygon to be. AutoSketch then asks you for the midpoint of a side.

Polygon Midpoint of side:

Figure 4-18.
Drawing a circumscribed, six-sided polygon.

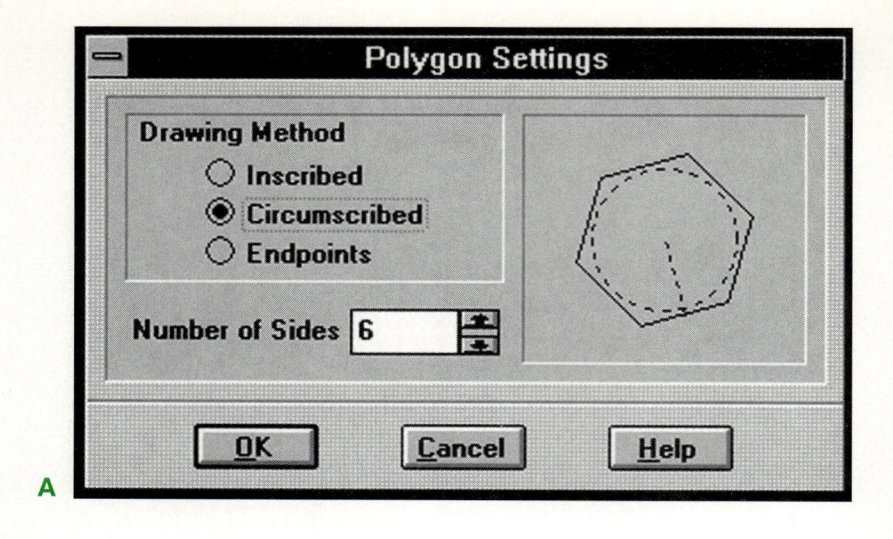

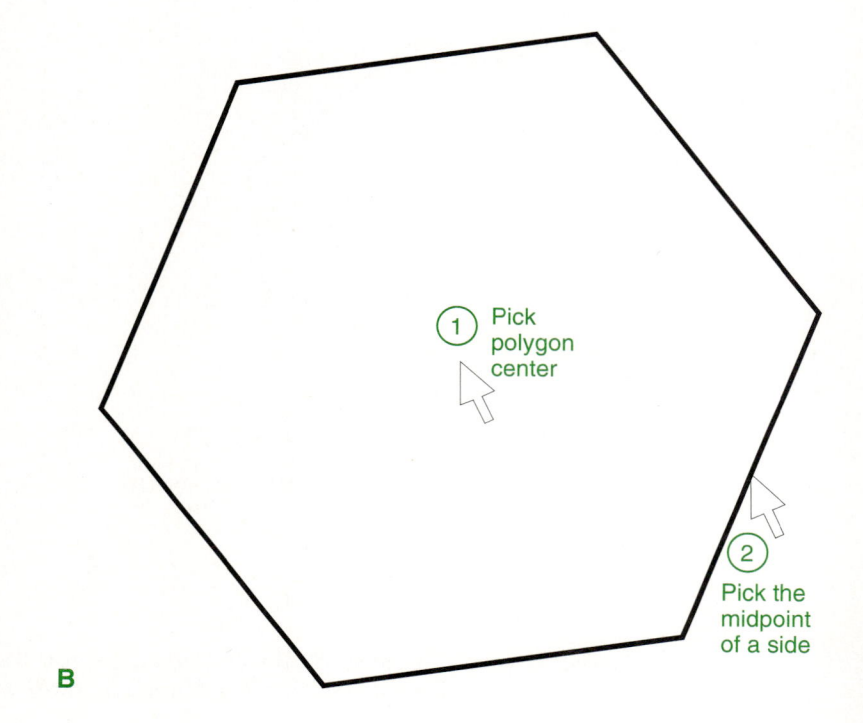

Locate the cursor to the midpoint of one side of the polygon. As you move the cursor, a temporary polygon will enlarge, shrink, and rotate. When satisfied with the size and rotation of the polygon, pick again.

Drawing a Polygon Using Endpoints

To draw a polygon using endpoints, open the **Polygon Settings** dialog box by double-clicking on the **Polygon** tool in the **Draw** menu. Choose the **Endpoints** option and enter the number of sides for the polygon, Figure 4-19. Then, pick **OK** to close the dialog box. The following prompt appears.

Polygon Enter point:

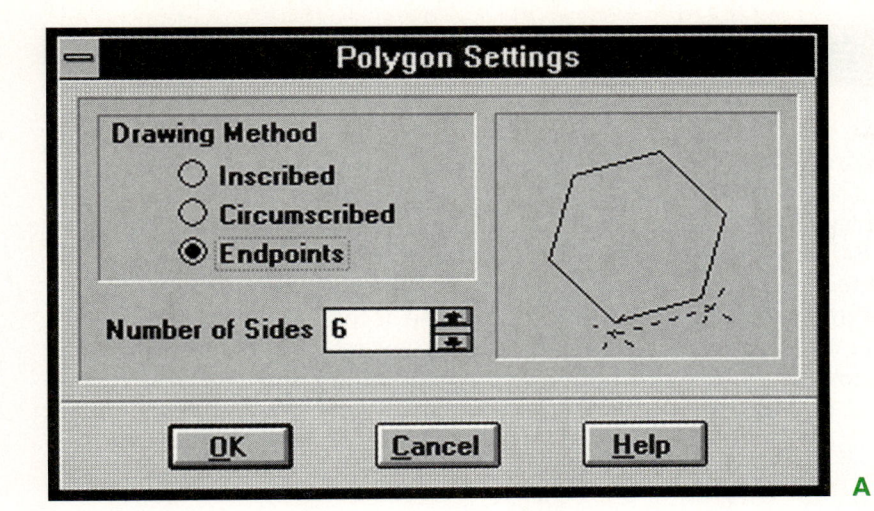

Figure 4-19.
Drawing a six-sided polygon by selecting endpoints.

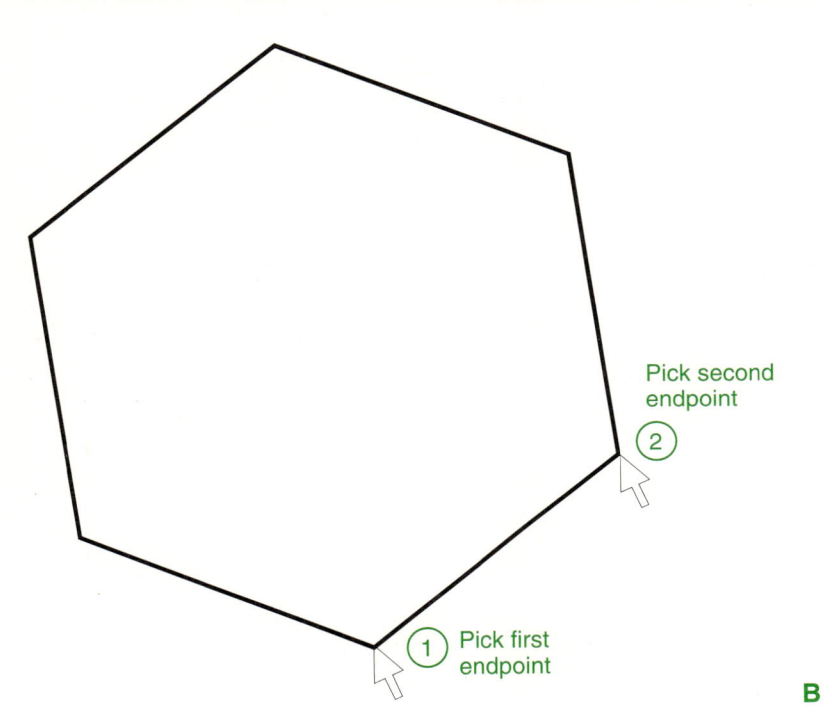

Pick where you want the endpoint of one side of the polygon to be. AutoSketch then asks you for the second endpoint.

<u>Polygon</u> Second endpoint of edge:

Locate the cursor to the second endpoint of one side of the polygon. As you move the cursor, a temporary polygon will enlarge, shrink, and rotate. When satisfied with the size and rotation of the polygon, pick again.

Exercise 4-5

1. Start a new drawing using the **Normal** template.
2. Draw two 3" diameter circles and a square box.
3. Use the **Polygon** tool to draw a five-sided inscribed polygon. Place the center of the polygon at one circle's center. Place the second point on the circle.
4. Use the **Polygon** tool to draw a six-sided circumscribed polygon. Place the center of the polygon at the second circle's center. Place the second point on the circle.
5. Use the **Polygon** tool to draw a three-sided polygon by picking the endpoints of one side of the polygon on the end points of one side of the box.
6. Save the drawing as **C4E5** on your work diskette and then close the drawing.

Drawing Arcs

Arcs are partial circles. Like circles, they have a center point and radius. However, arcs also have a start point, an endpoint, and an included angle, Figure 4-20. Like the **Circle** tool, AutoSketch lets you draw arcs using several different options.

Figure 4-20.
The different parts of an arc are shown here.

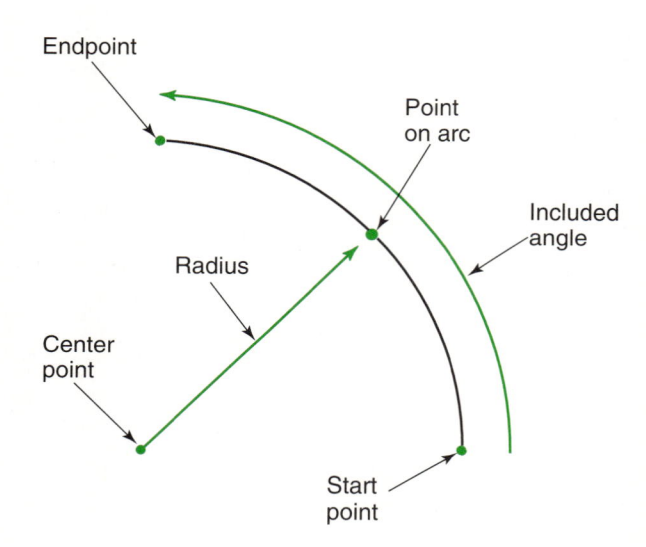

Arc Settings

To access the **Arc Settings** dialog box, double-click on the **Arc** tool in the **Draw** menu, Figure 4-21. The sample in the dialog box will adjust to show you the setting you choose. After selecting an option, pick **OK** to close the dialog box. The different options are covered in the next sections.

SHORTCUT
To access the **Arc** tool, you can hold down the [Alt] key and press the function key [F3].

Start, Point, End

With this option, pick the start point, a point on the arc, and the endpoint in that order. This option is helpful when you know where the arc should start and where it should touch another object.

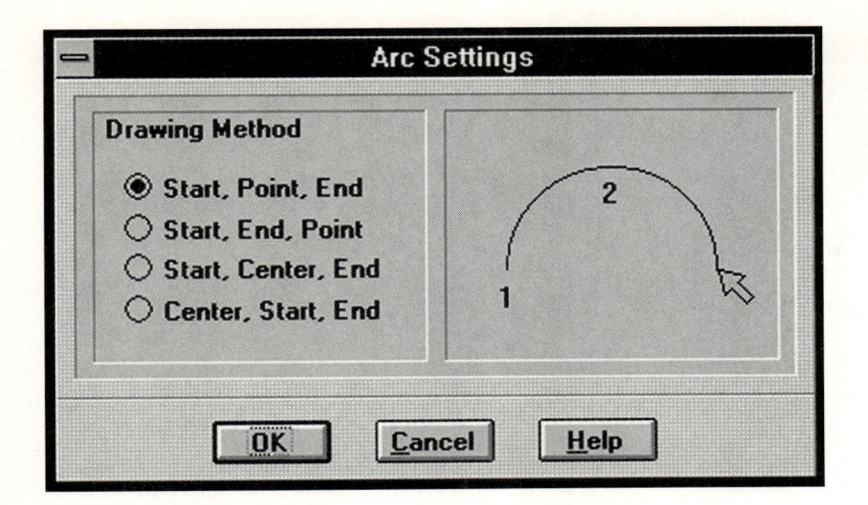

Figure 4-21.
Select the method to draw arcs in the **Arc Settings** dialog box. The preview on the right will change to help show you how each option draws an arc.

Start, End, Point
With this option, pick the start point, the endpoint, and a point on the arc in that order. This option is helpful when you want the start and end points to be an exact distance apart.

Start, Center, End
With this option, pick the start point, the center of the arc, and the endpoint in that order. This option is helpful when you know where the arc should start and where the center should be.

Center, Start, End
With this option, pick the center of the arc, the start point, and the endpoint in that order. This option is helpful when you know that the arc should be centered at a certain location, but also want the freedom to adjust the radius of the arc.

Drawing an Arc Using Start, Point, End
To draw an arc using the **Start, Point, End** option, open the **Arc Settings** dialog box by double-clicking on the **Arc** tool in the **Draw** menu. Choose the **Start, Point, End** option in the **Arc Settings** dialog box, Figure 4-22. Then, pick **OK** to close the dialog box. The following prompt appears.

　　Arc Start point:

Pick where you want the arc to start. AutoSketch then asks you for a point on the arc.

　　Arc Point on arc:

Pick another point or object where you want the arc to pass through. AutoSketch then asks you for the endpoint.

　　Arc Endpoint:

Locate the cursor to where you want the arc to end. As you move the cursor, a temporary arc will enlarge, shrink, flip, and rotate. When satisfied with the size and rotation of the arc, pick again.

Figure 4-22.
Drawing an arc using the **Start, Point, End** option.

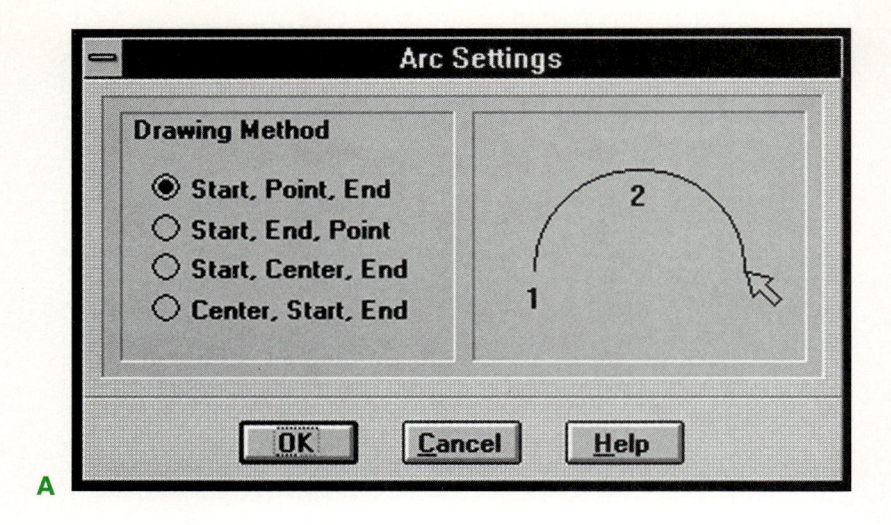

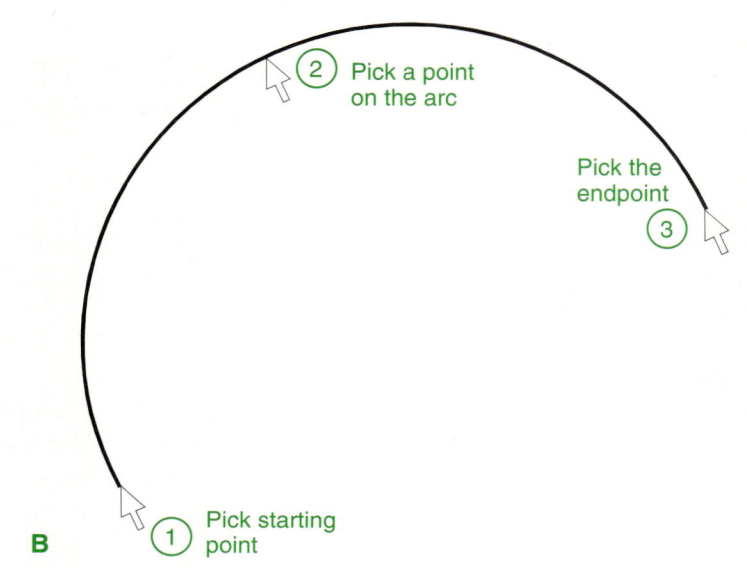

Drawing an Arc Using Start, End, Point

To draw an arc using the **Start, End, Point** option, open the **Arc Settings** dialog box by double-clicking on the **Arc** tool in the **Draw** menu. Choose the **Start, End, Point** option, Figure 4-23. Then, pick **OK** to close the dialog box. The following prompt appears.

Arc Start point:

Pick where you want the arc to start. AutoSketch then asks you for the endpoint of the arc.

Arc Endpoint:

Pick where you want the arc to end. AutoSketch then asks you for a point on the arc.

Arc Point on arc:

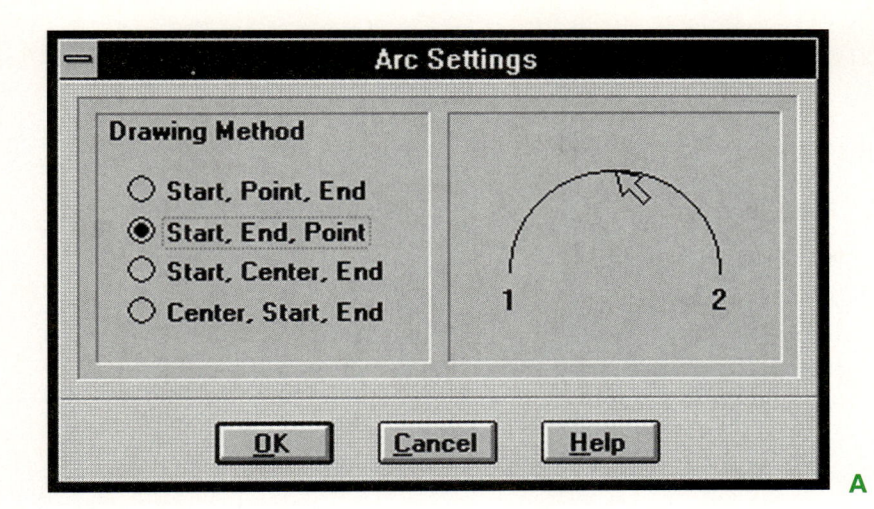

Figure 4-23.
Drawing an arc using the **Start, End, Point** option.

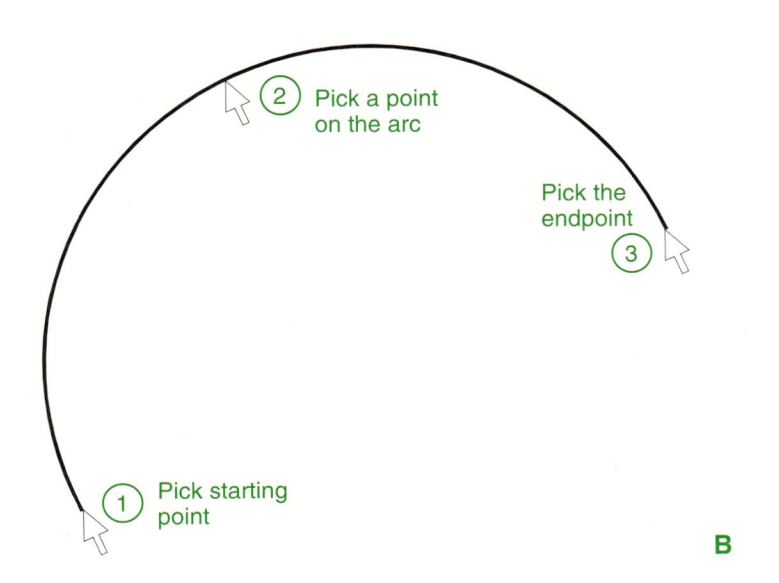

Locate the cursor to a point or object where you want the arc to pass through. As you move the cursor, a temporary arc will stretch and flip. When satisfied with the size and rotation of the arc, pick again.

Drawing an Arc Using Start, Center, End

To draw an arc using the **Start, Center, End** option, open the **Arc Settings** dialog box by double-clicking on the **Arc** tool in the **Draw** menu. Choose the **Start, Center, End** option, Figure 4-24. Then, pick **OK** to close the dialog box. The following prompt appears.

Arc Start point:

Pick where you want the arc to start. AutoSketch then asks you for the center of the arc.

Arc Center of arc:

Figure 4-24.
Drawing an arc using the **Start, Center, End** option.

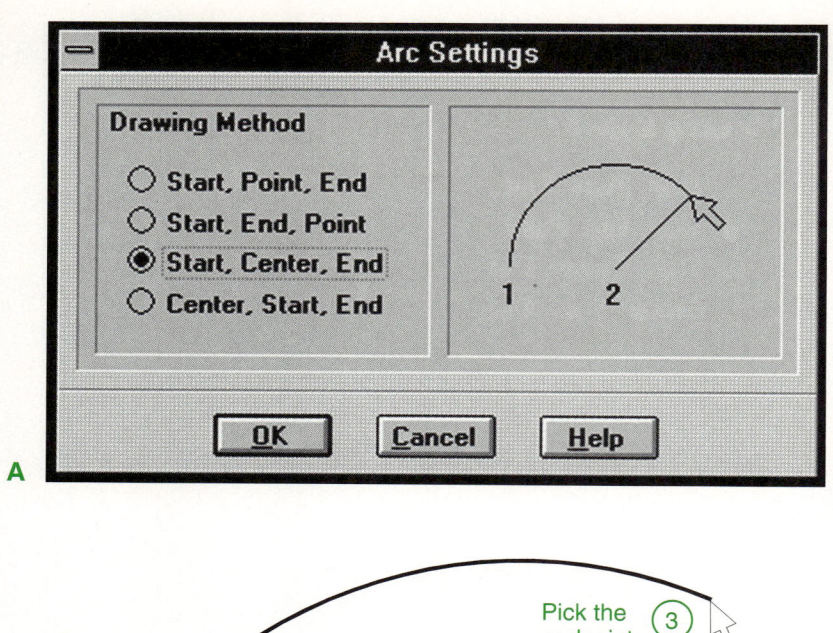

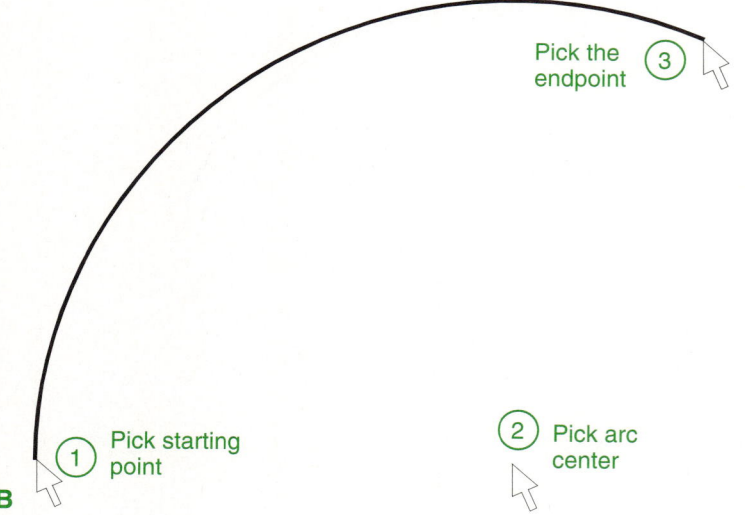

A

B

Pick the endpoint ③

① Pick starting point

② Pick arc center

Pick where you want the center of arc to be. AutoSketch then asks for the endpoint of the arc or the included angle.

Arc Endpoint or included angle:

Locate the cursor to another point to indicate the endpoint. As you move the cursor, a temporary arc extends from the start point. When satisfied with length of the arc, pick again.

You can also enter an included angle. The *included angle* is the angle between two lines that would connect the start and end points of the arc to the center point. For example, entering 45 would create a quarter arc. Entering 270 would create a three-quarter arc. A positive value draws the arc counter-clockwise from the start point. A negative value, such as -45, draws the arc clockwise from the start point. The included angle cannot be 0 degrees (no arc at all) or 360 degrees (a circle). To enter the angle, simply type the number. The number will appear in the prompt box. Press [Enter] to complete the arc.

Drawing an Arc Using Center, Start, End

To draw an arc using the **Center, Start, End** option, open the **Arc Settings** dialog box by double-clicking on the **Arc** tool in the **Draw** menu. Choose the **Center, Start, End** option, Figure 4-25. Then, pick **OK** to close the dialog box. The following prompt appears.

Arc Center of arc:

Pick where you want the center of arc to be. AutoSketch then asks your for the starting point of the arc.

Arc Start point:

Pick where you want the arc to start. AutoSketch then asks you for the endpoint or the included angle.

Arc Endpoint or included angle:

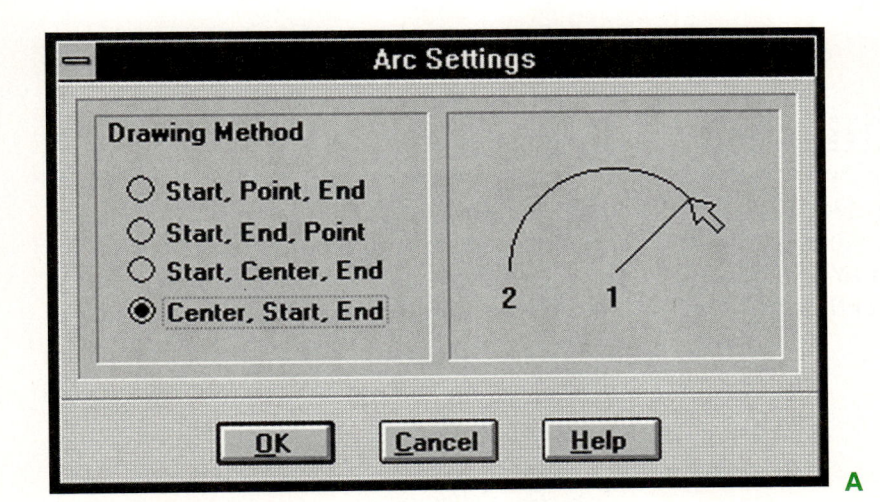

Figure 4-25.
Drawing an arc using the **Center, Start, End** option.

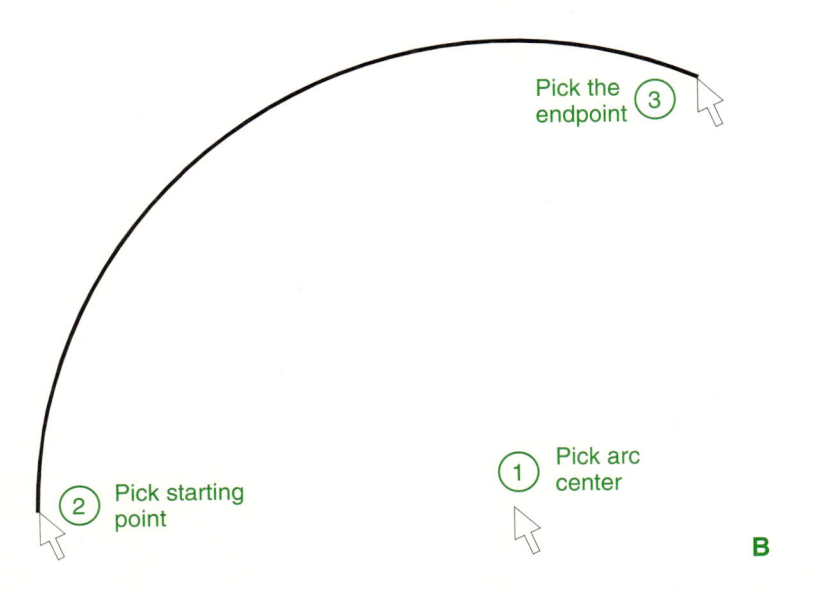

Locate the cursor to another point to indicate the endpoint. As you move the cursor, a temporary arc extends out from the start point. When satisfied with length of the arc, pick again.

Exercise 4-6

1. Start a new drawing using the **Normal** template.
2. Use the **Start, Point, End** option of the **Arc** tool to draw a arc.
3. Draw a line. Then, use the **Start, End, Point** option of the **Arc** tool to draw a arc where the endpoints of the arc touch the endpoints of the line.
4. Use the **Start, Center, End** option of the **Arc** tool to draw a arc. Use your cursor to pick the endpoint.
5. Use the **Start, Center, End** option of the **Arc** tool to draw a arc. Enter a 30 degree included angle.
6. Use the **Center, Start, End** option of the **Arc** tool to draw a arc. Use your cursor to pick the endpoint.
7. Use the **Center, Start, End** option of the **Arc** tool to draw a arc. Enter a 90 degree included angle.
8. Save the drawing as C4E6 on your work diskette and then close the drawing.

Drawing Ellipses

An *ellipse* is formed when a circle is viewed at an angle. It is defined by a center, a major axis, a minor axis, and two focal points, Figure 4-26. (Note: Foci is the plural of focus, or focal.) AutoSketch allows you to draw ellipses using three methods. These methods are **Center and Both Axes, Axis and Planar Rotation,** and **Two Foci and Point.**

Figure 4-26.
Parts of an ellipse.

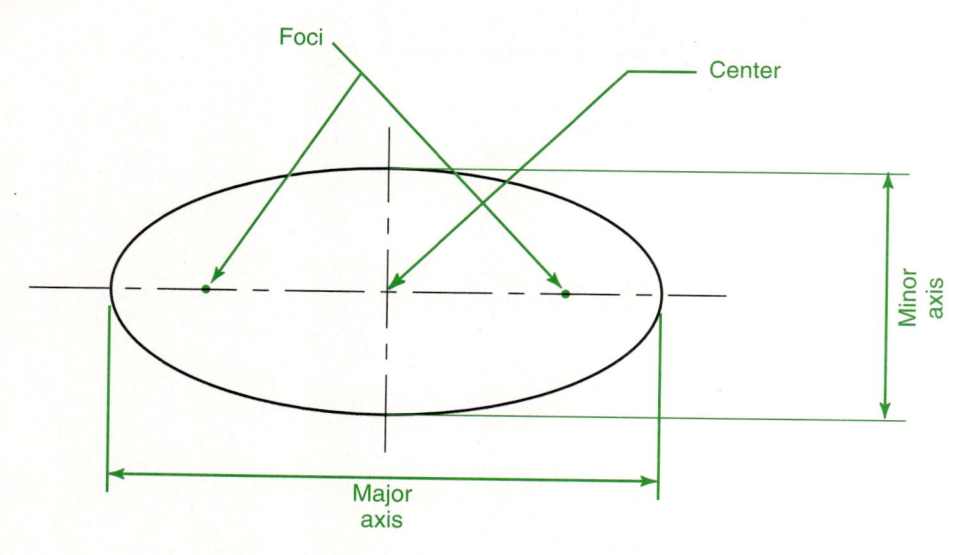

Setting Ellipse Settings

To access the **Ellipse Settings** dialog box, double-click on the **Ellipse** tool in the **Draw** menu, Figure 4-27. The sample in the dialog box will adjust to show you the setting you choose. If you change the ellipse settings while drawing an ellipse, the new settings won't take effect until you draw another ellipse. After selecting an option, pick **OK** to close the dialog box. The different options are covered in the next sections.

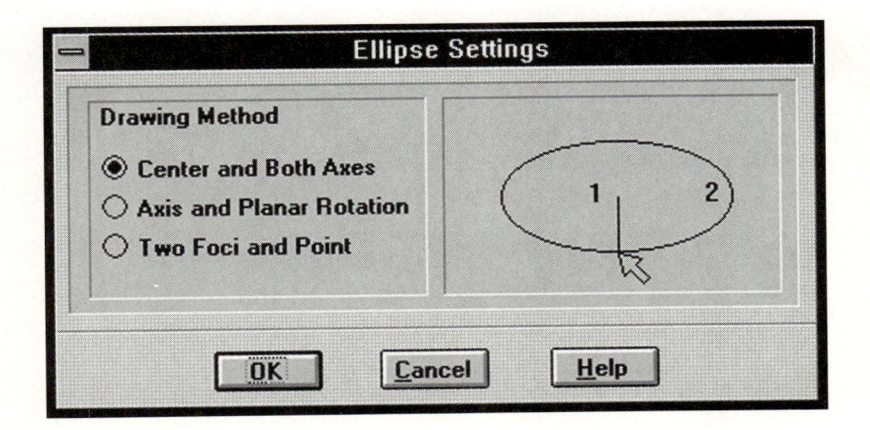

Figure 4-27.
Select the method to draw ellipses in the **Ellipse Settings** dialog box. The preview on the right will change to help show you how each option draws an ellipse.

Center and Both Axes
To draw an ellipse with this option, pick the center point, and one endpoint of each ellipse axis. This option is helpful when you know where the center of the ellipse should be, and two points where the axis should touch.

Axis and Planar Rotation
To draw an ellipse with this option, pick the center point, one axis endpoint, and the rotation of the ellipse. This option is helpful when you know where the center of the ellipse should be, and at what angle the ellipse is viewed.

Two Foci and Point
To draw an ellipse with this option, pick the two focal points and any point on the ellipse. This is useful when you need an ellipse to touch an object and need the foci to be in exact locations.

Drawing an Ellipse Using Center and Both Axes
First, open the **Ellipse Settings** dialog box by double-clicking on the **Ellipse** tool in the **Draw** menu. To draw an ellipse by picking the center and one endpoint of each axis, choose the **Center and Both Axes** option, Figure 4-28. Then, pick **OK** to close the dialog box. The following prompt appears.

Ellipse Center of ellipse:

Pick where you want the center of ellipse to be. AutoSketch then asks you for the endpoint of the first axis.

Ellipse Axis endpoint:

> **SHORTCUT**
> To access the **Ellipse** tool, you can hold down the [Ctrl] key and press the function key [F8].

Figure 4-28.
Drawing an ellipse using the **Center and Both Axes** option.

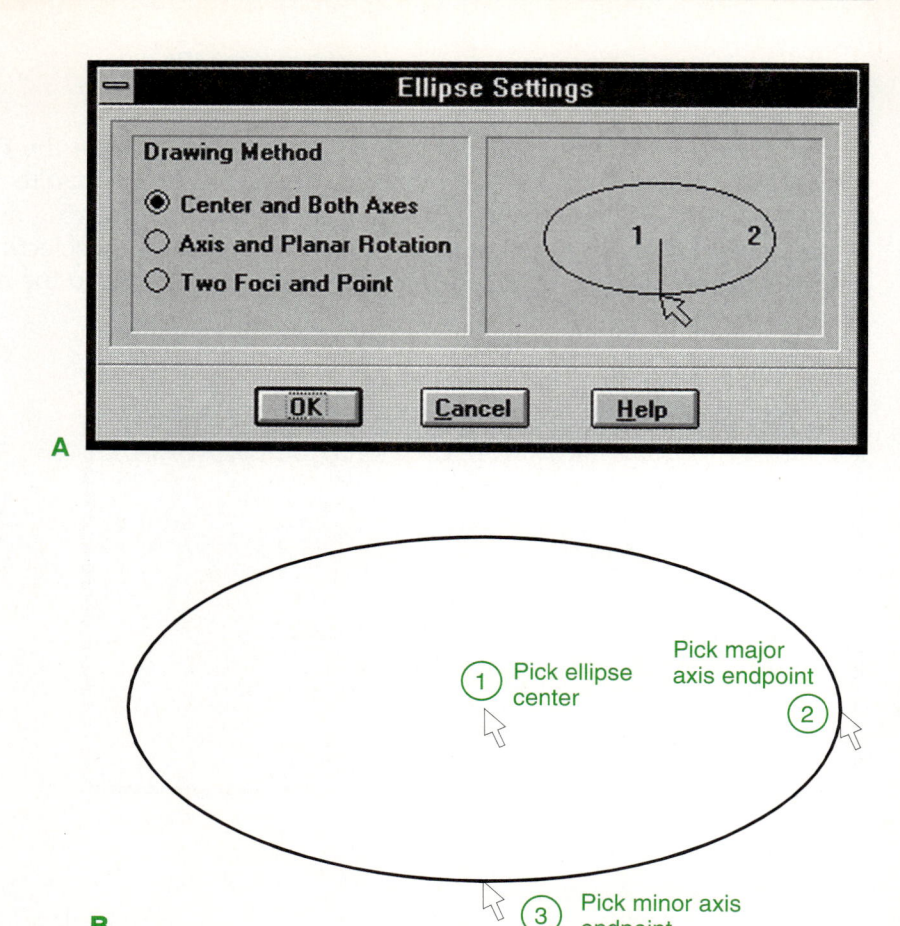

A

B

Pick where you want one axis to end. The angle between the center and this point determines the rotation of the ellipse. AutoSketch then asks you for the endpoint of the second axis.

Ellipse Other axis distance:

Locate the cursor to indicate the other axis endpoint. As you move the cursor, a temporary ellipse extends out from the center point and passes through first axis endpoint. Because the second axis is always at a 90 degree angle from the first axis, the point you choose will not necessarily be on the ellipse. What you are actually doing is defining a distance from the center point. That distance will be one half the length of the second axis. When satisfied with size of the ellipse, pick again to finish the ellipse.

Drawing an Ellipse Using Axis and Planar Rotation

To draw an ellipse by picking the axis and planar rotation, open the **Ellipse Settings** dialog box by double-clicking on the **Ellipse** tool in the **Draw** menu. Choose **Axis and Planar Rotation,** Figure 4-29. Then, pick **OK** to close the dialog box. The following prompt appears.

Ellipse Center of ellipse:

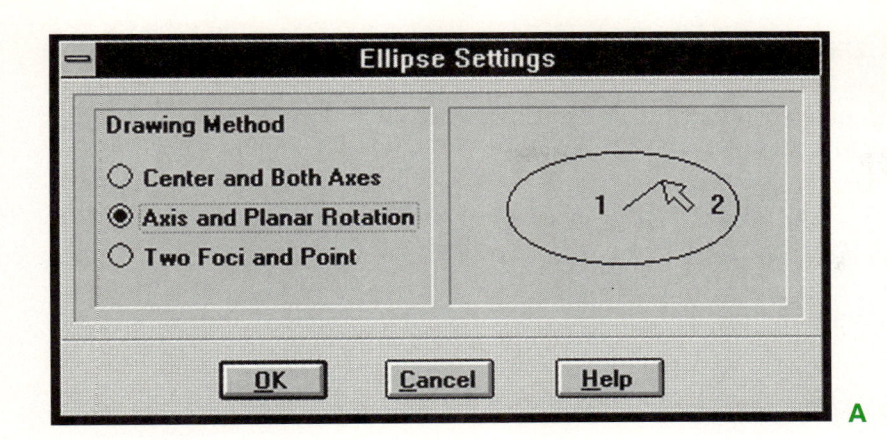

Figure 4-29.
Drawing an ellipse using the **Axis and Planar Rotation** option.

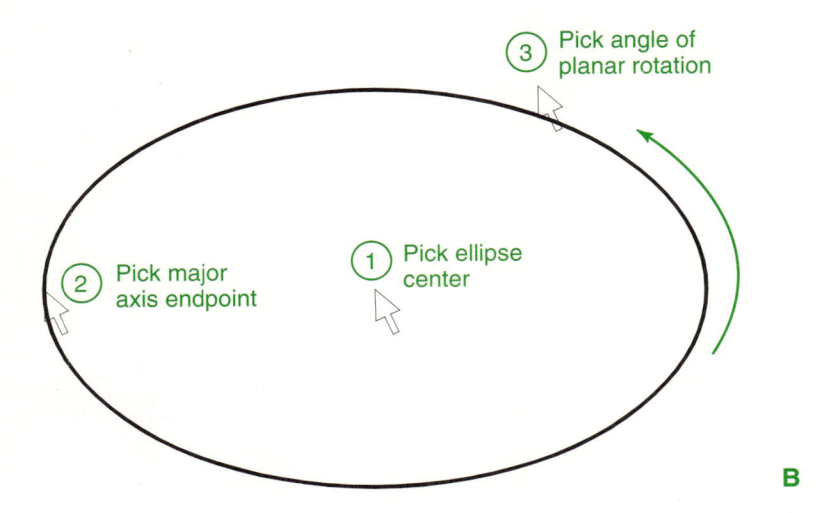

Pick where you want the center of ellipse to be. AutoSketch then asks you for the endpoint of the major axis. The major axis is the longest of the two axes.

<u>Ellipse</u> Axis endpoint:

Pick where you want the major axis to end. The angle between the center and this point determines the rotation angle of the ellipse. AutoSketch then asks for a rotation angle.

<u>Ellipse</u> Rotation around major axis:

As you move the cursor, the ellipse will change rotation. This rotation is called planar rotation. *Planar rotation* refers to the angle that the circle is viewed at. For example, imagine looking at a circle on a flat sheet of paper. Now imaging picking up the paper and turning it. Turning the paper makes the circle look like an ellipse. The more you turn the paper, the narrower it looks. When you've turned the paper 90 degrees, the ellipse becomes a straight line.

Thus, when drawing an ellipse using this method, the angle of your cursor from the major axis determines the planar rotation. When satisfied with size of the ellipse, pick again. The point you pick will not necessarily lie on the ellipse.

Drawing an Ellipse Using Two Foci and Point

To draw an ellipse by picking the two focal points and a point on the ellipse, open the **Ellipse Settings** dialog box by double-clicking on the **Ellipse** tool in the **Draw** menu. Choose **Two Foci and Point,** Figure 4-30. Then, pick **OK** to close the dialog box. The following prompt appears.

  Ellipse First focus of ellipse:

Pick where you want the first focus of the ellipse to be. AutoSketch then asks you for the second focus.

  Ellipse Second focus:

Pick where you want the second focus of the ellipse to be. The angle between the first focal point and the second point determines the rotation angle of the ellipse. AutoSketch then asks you for a point on the ellipse.

  Ellipse Point on ellipse:

As you move the cursor, a temporary ellipse will form around the two focal points and through this point. When satisfied with size of the ellipse, pick again.

Figure 4-30.
Drawing an ellipse using the **Two Foci and Point** option.

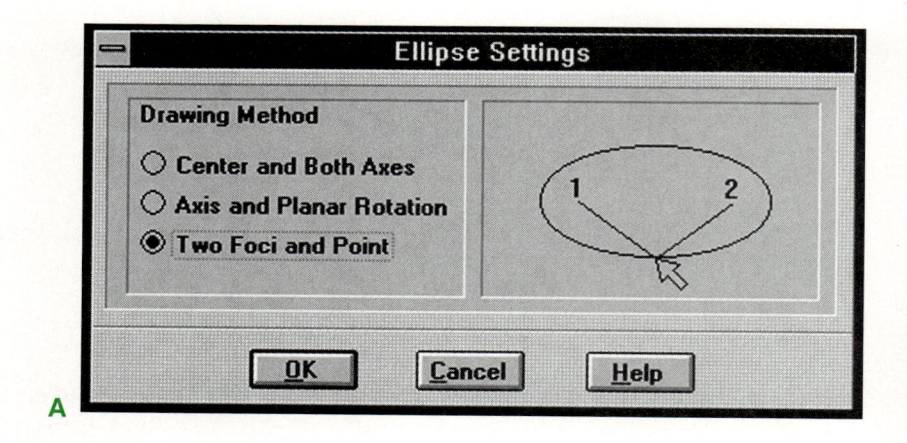

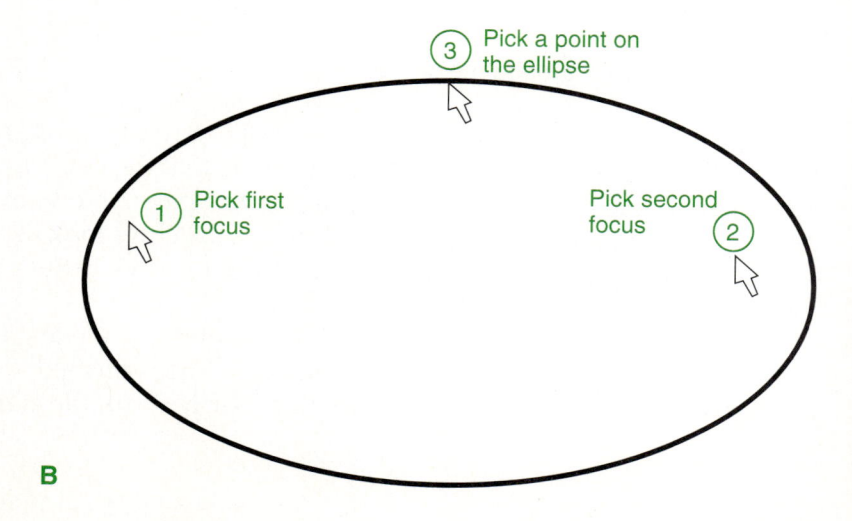

Exercise 4-7

1. Start a new drawing using the **Normal** template.
2. Use the **Center and Both Axes** option of the **Ellipse** tool to draw a ellipse.
3. Use the **Axis and Planar Rotation** option of the **Ellipse** tool to draw a ellipse.
4. Use the **Two Foci and Point** option of the **Ellipse** tool to draw an ellipse.
5. Save the drawing as C4E7 on your work diskette and then close the drawing.

Drawing Curves

A *curve* is a smooth-bending line. Curves are often drawn to show land contours in landscape architecture and map making. They are also used on special occasions in mechanical and architectural drafting. To draw a curve in AutoSketch, you pick the start point, points that control the shape of the curve, and an endpoint. The shape of the curve is guided by *control points* you pick between the curve's start and endpoints, Figure 4-31. Imaginary lines connecting the control points make up a *frame*. The frame is displayed only when you select a curve to edit. AutoSketch provides several methods to draw curves.

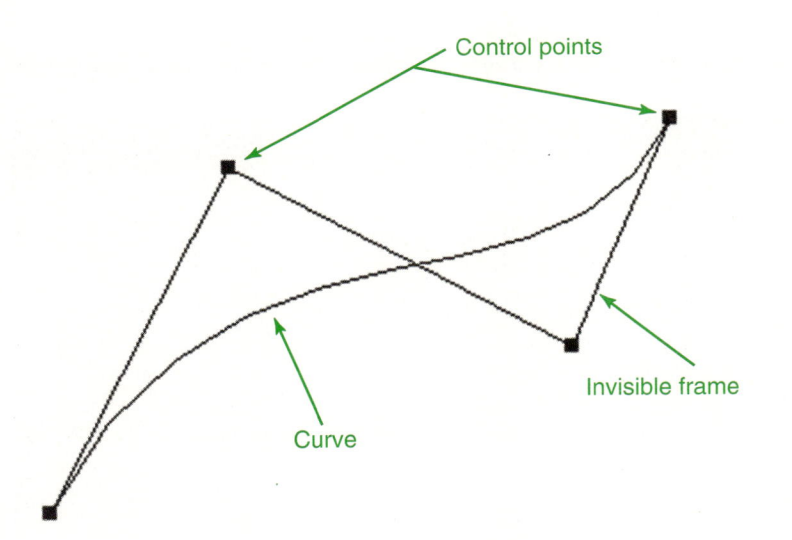

Figure 4-31.
A curve is defined by control points that you pick.

Setting Curve Settings

To access the **Curve Settings** dialog box, double-click on the **Curve** tool in the **Draw** menu. The sample in the dialog box adjusts to the setting you choose. If you change any setting while drawing a curve, the new settings won't take effect until you draw another curve. After selecting an option, pick **OK** to close the dialog box. The different options are covered in the next sections.

Curve type

The curve type refers to the mathematical equation that determines the shape of the curve. The curve type options in AutoSketch are **B-Spline** and **Bezier**. A *B-Spline* curve is very tightly controlled by the control points. A *Bezier* curve bends more freely around the control points. See the samples in the **Curve Settings** dialog boxes shown in Figure 4-32.

Figure 4-32.
Bezier and B-spline curves follow the control points differently. The preview in the **Curve Settings** dialog box shows this.

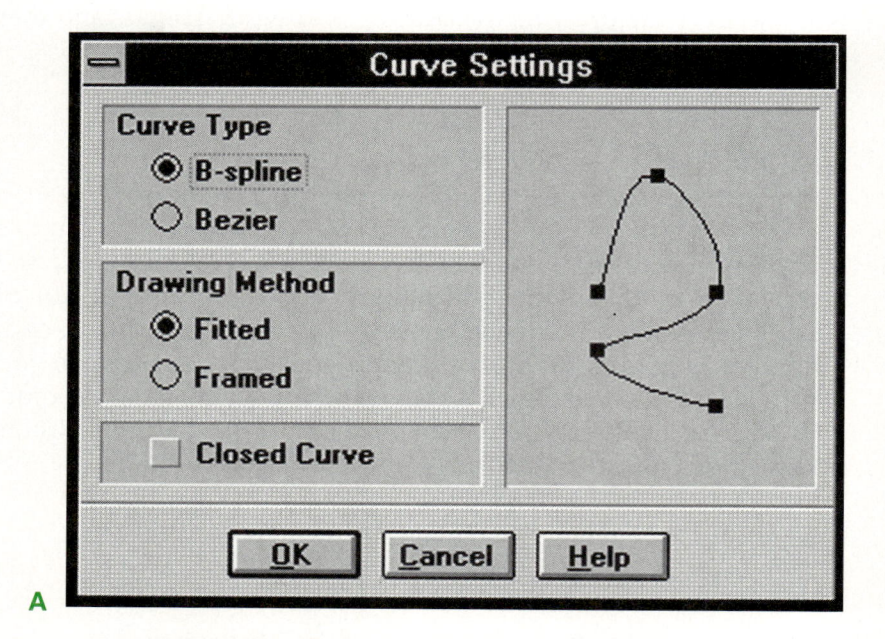

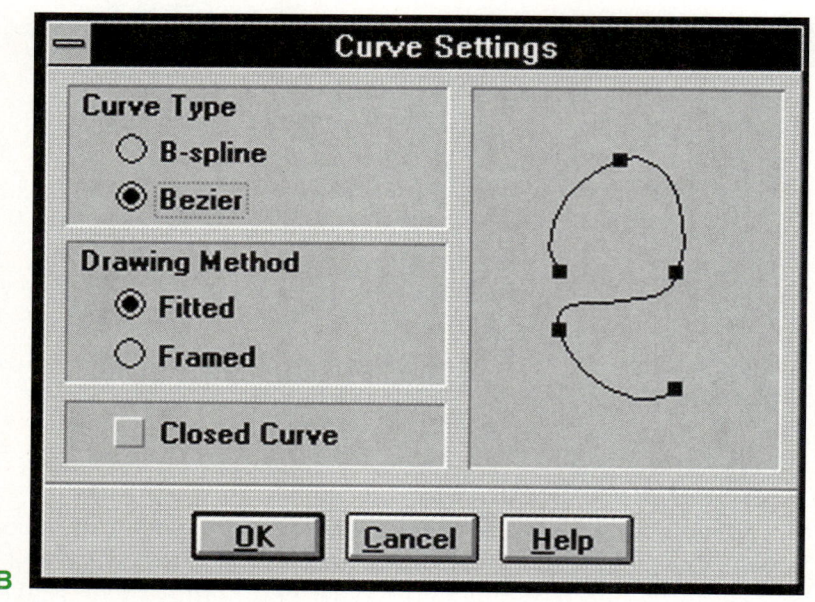

A

B

Drawing method

The way that the curve is drawn can also be selected. The curve can be drawn to pass through control points or inside of the control points. The **Fitted** option causes the curve to pass through each control point. The **Framed** option causes the curve to pull away from the control points.

With the **Framed** option, each control point pulls at the curve, causing it to bend. The frame will be displayed as you pick the control points. See the sample in the **Curve Settings** dialog box shown in Figure 4-33.

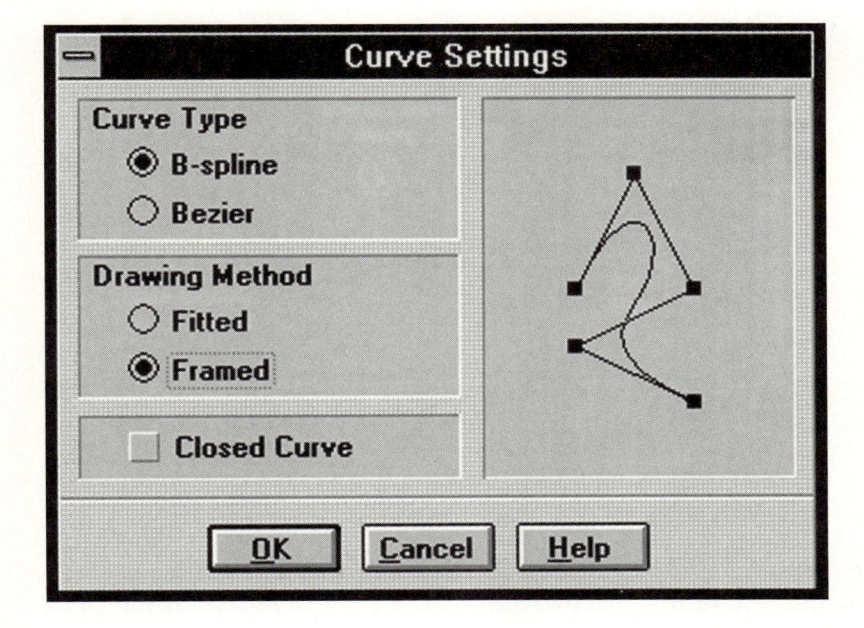

Figure 4-33.
The **Framed** option changes how the control points pull the curve.

Closed Curve

The **Closed Curve** option causes the curve to connect and bend between the first and last points chosen, Figure 4-34. The curve will close even if you don't pick the last control point on top of the first. When this option is not selected, the curve will not close itself. Even if you pick the last point on top of the first point, the curve will connect, but not bend at this intersection.

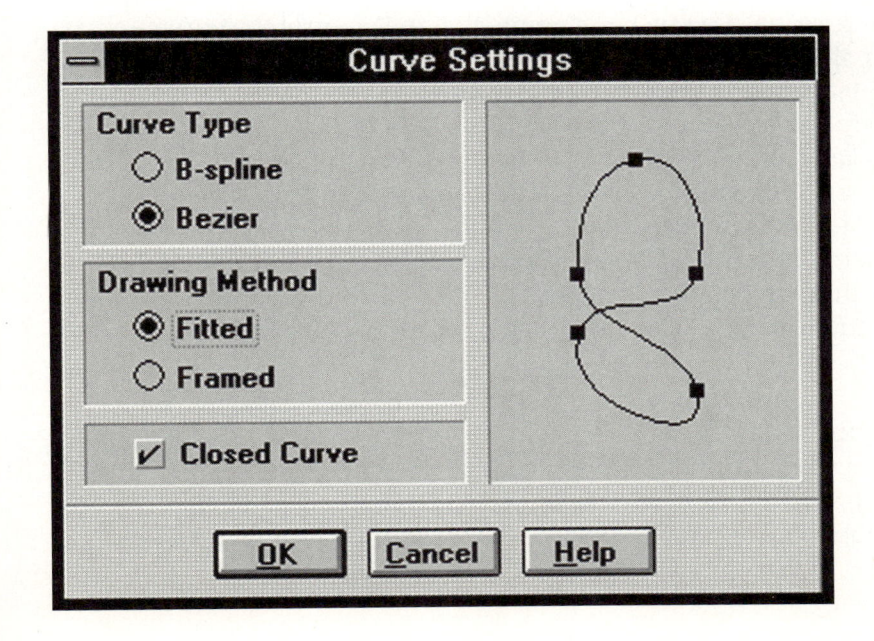

Figure 4-34.
The **Fitted** option makes the curve pass through the control points. The **Closed Curve** option causes the curve to connect the first and lasts control points picked.

Figure 4-35.
To draw a curve, pick the control points.

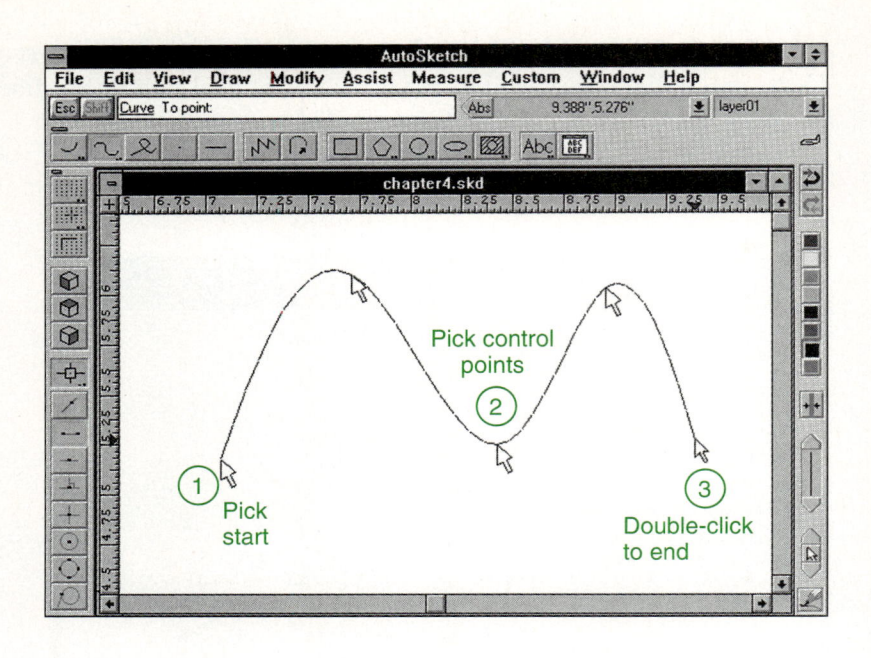

Drawing a Curve

To draw a curve using any of the curve settings, pick the **Curve** tool in the **Draw** menu, Figure 4-35. The following prompt appears.

<u>Curve</u> First point:

Pick where you want to start the curve. AutoSketch then asks you for the next point.

<u>Curve</u> To point:

This prompt will repeat as you pick the points that will control the curve's shape. The closer the points, the closer the curve will follow the frame. You can select up to 100 control points. A temporary curve will appear on-screen as you pick further control points. There are four ways to end the curve.
- Double-click the last control point
- Choose any tool or command
- Pick the first point of the curve again
- Enter /lpoint after you pick the last control point. (The /lpoint (last point) variable is discussed in more detail in Chapter 5.)

How the final curve appears depends on what options you set in the **Curve Settings** dialog box.

Drawing Freehand Curves

You can also draw curves without picking control points using the **Freehand** tool. This method of drawing is like using a pencil to sketch. To draw a freehand curve, pick the **Freehand** tool in the **Draw** menu, Figure 4-36. The following prompt appears.

<u>Freehand</u> Enter point:

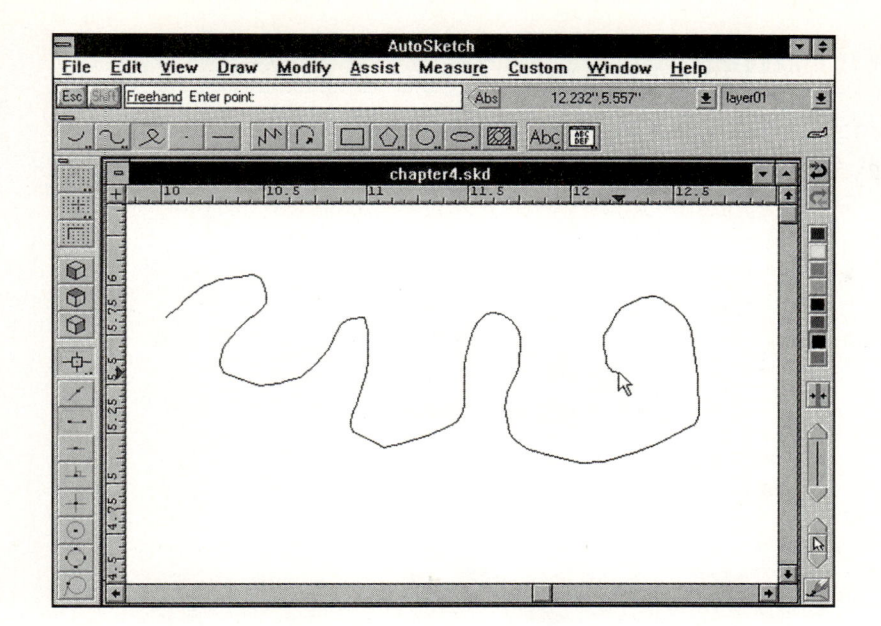

Figure 4-36.
To draw a freehand curve, hold down the pick button and draw. AutoSketch adds the proper control points to create the final curve.

Hold down the left mouse button and move the cursor around to create the curve. When finished drawing the curve, release the mouse button. A freehand curve is always a Bezier curve. Since this is a freehand tool, it is usually hard to get a smooth curve.

Exercise 4-8

1. Start a new drawing using the **Normal** template.
2. Draw four sets of the three lines as shown below.
3. Use the **B-Spline** and **Fitted** options to draw a curve by picking control points at the endpoints of one set of lines.
4. Use the **Bezier** and **Fitted** options to draw a curve by picking control points at the endpoints of one set of lines.
5. Use the **B-Spline** and **Framed** options to draw a curve by picking control points at the endpoints of one set of lines.
6. Use the **Bezier** and **Framed** options to draw a curve by picking control points at the endpoints of one set of lines.
7. Use the **Bezier** and **Fitted** options, plus the **Closed Curve** option, to draw a curve that appears similar to an egg.
9. Draw a figure eight using the **Freehand** tool.
9. Save the drawing as C4E8 on your work diskette and then close the drawing.

Drawing Polylines

A *polyline* is a single object consisting of connected lines and arcs. The main benefit of polylines is that AutoSketch miters the joints of each segment to create a smooth connection. Refer to Figure 4-37.

Figure 4-37.
The angle that segments of a wide polyline meet at will determine how AutoSketch draws the connection.

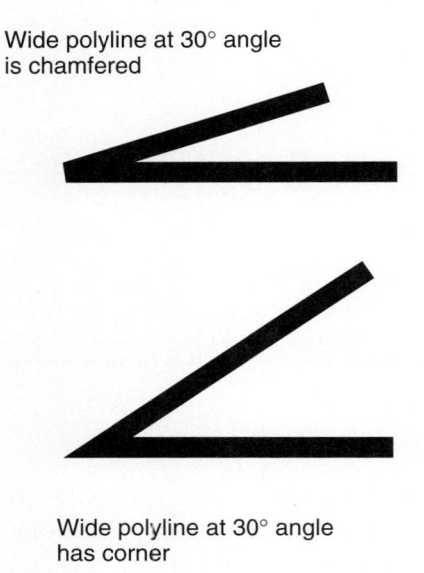

Wide polyline at 30° angle is chamfered

Wide polyline at 30° angle has corner

If two segments of a wide polyline connect at an angle more than 30 degrees, AutoSketch connects the outer edges with a zero radius fillet. When they connect at an angle less than 30 degrees, AutoSketch connects the segments with a chamfer that is the width of the polyline.

Any linetype can be used for a hairline polyline. However, only hairline polylines can have a linetype other than solid.

Drawing a Polyline with Only Line Segments

To draw a polyline, pick the **Polyline** tool in the **Draw** menu, Figure 4-38. The following prompt appears.

<u>Polyline</u> First point:

Pick where you want the polyline to start. AutoSketch then asks you for the next point.

<u>Polyline</u> To point:

This prompt will repeat as you pick more points along the polyline. When drawing thick polylines, an outline of the polyline follows your cursor. Completed segments of wide polylines will fill as you continue picking points. The fill always lags behind one point. This is because AutoSketch must know the angle of intersection to determine whether to fillet or chamfer the connection.

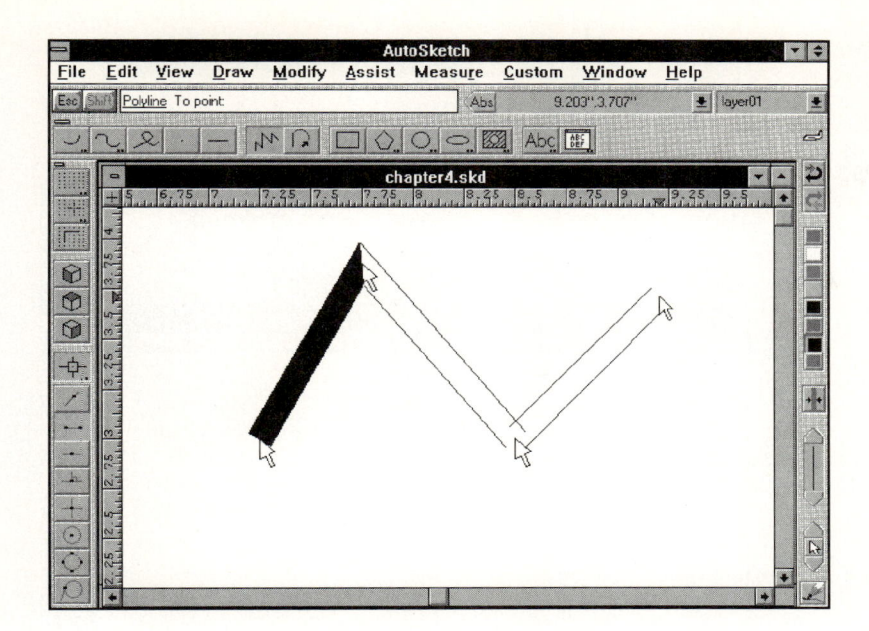

Figure 4-38.
Drawing a polyline with only line segments.

To end the polyline, either pick the first point again, pick the same point twice, or select another tool or command. Do not press the **Esc** button or key. This will cancel the polyline. If you pick the first point again, AutoSketch creates a closed polyline. The intersection of first and last points will be mitered. You must pick precisely on the first point again to create a closed polyline.

Drawing a Polyline with Line and Arc Segments

To draw a polyline with both line and arc segments, you use the **Arc Mode** tool. Line and arc segments of the polyline are interchanged by toggling in and out of **Arc Mode.** This allows you to chose whether each segment of a polyline is curved or straight. To toggle in and out of **Arc Mode,** click on the **Arc Mode** tool in the **Draw** menu.

First, decide if the segment will be an arc. If it will be an arc, pick the **Arc Mode** tool. When **Arc Mode** is on, the **Arc Mode** tool appears "pushed in." Note that when you are in **Arc Mode,** both the **Arc Mode** tool and the **Polyline** tool are active in the toolbox. All arc segments are created with the **Start, Point, End** method of drawing arcs. Changing the option in the **Arc Settings** dialog box will *not* change this.

Starting with a line segment

To draw a polyline starting with a line segment, first make sure that **Arc Mode** is off, Figure 4-39. Then pick the **Arc Mode** tool in the **Draw** menu. The following prompt appears.

Polyline First point:

Pick where you want the polyline to start. AutoSketch then asks you for the next point.

Polyline To point:

> **SHORTCUT**
> You can also toggle in and out of **Arc Mode** by holding the [Ctrl] key and pressing the function key [F1].

Figure 4-39.
Drawing a polyline starting with a line segment and having an arc segment.

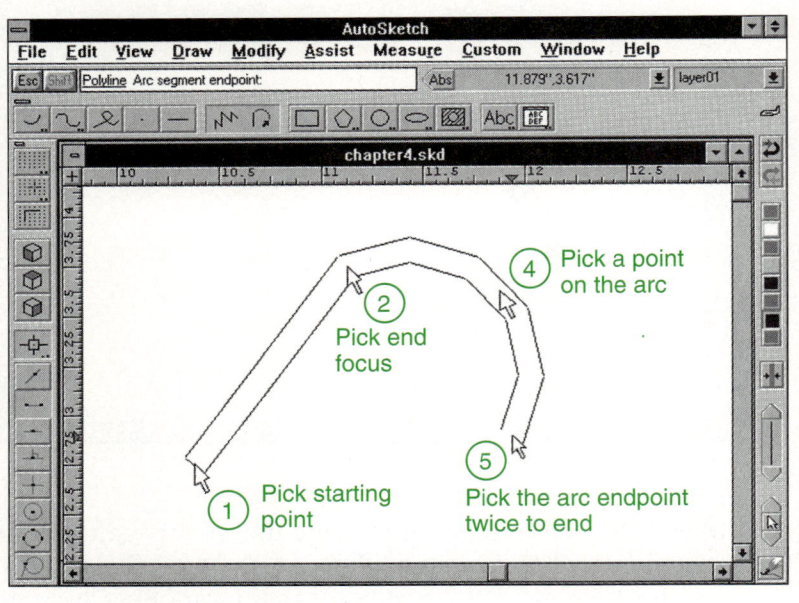

③ Pick arc mode tool

④ Pick a point on the arc

② Pick end focus

① Pick starting point

⑤ Pick the arc endpoint twice to end

Pick a second point to create the first line segment. Now, to make the second segment an arc, pick the **Arc Mode** tool. After you pick the tool, it will appear depressed. The prompt will also change.

<u>Polyline</u> Point on arc:

Pick a point or object where you want the arc to pass through. AutoSketch will then ask you for the endpoint of the arc.

<u>Polyline</u> Arc segment endpoint:

Locate the cursor to where you want the arc to end. As you move the cursor, a temporary polyline arc segment will enlarge, shrink, flip, and rotate. When satisfied with the size and rotation of the arc segment, pick again.

If the next segment should be an arc, continue on to pick a point on the next arc segment. If the next segment should be a line, pick the **Arc Mode** tool. The prompt will change back to the original prompt.

<u>Polyline</u> To point:

Starting with an arc segment

To draw a polyline starting with an arc segment, first make sure that **Arc Mode** is on. Then pick the **Polyline** tool in the **Draw** menu. The following prompt appears.

<u>Polyline</u> Arc segment start point:

Pick where you want the polyline to start. AutoSketch then asks you for a point on the arc.

Polyline Point on arc:

Pick a point or object where you want the arc segment to pass through. AutoSketch then asks you for the endpoint of the arc.

Polyline Arc segment endpoint:

Locate the cursor to where you want the arc segment to end. As you move the cursor, a temporary polyline arc segment will enlarge, shrink, flip, and rotate. When satisfied with the size and rotation of the arc segment, pick again.

Now, to make the second segment an line, pick the **Arc Mode** tool again. The button will no longer appear "pushed in." The prompt will also change.

Polyline To point:

Pick a point to create the line segment. If the next segment should be a line, continue on to pick the next endpoint of the line segment. If the next segment should be an arc, pick the **Arc Mode** tool. The prompt will change to the original prompt.

Exercise 4-9

1. Start a new drawing using the **Normal** template.
2. Draw a box using the **Polyline** tool
3. Using a single polyline, draw a slot with rounded ends, as shown below, using the **Polyline** tool.
4. Set the width 1/2".
5. Draw a three-segment polyline where one angle between segments is about 10 degrees, and the other angle is about 60 degrees.
6. Save the drawing as C4E9 on your work diskette and then close the drawing.

Erasing Objects

Not everything you draw will be right the first time. Sometimes, it is necessary to erase one or more objects. To do so, pick the **Erase** tool from the **Modify** menu. The cursor changes to a hand. The following prompt will also appear.

Erase Select object:

You can select objects one at a time. You can also select many objects at a time using AutoSketch's *crosses/window box* function.

> **SHORTCUT**
>
> You can access the **Erase** tool by pressing the function key [F3].

Erasing Single Objects

It is easy to erase one object at a time. After picking the **Erase** tool, locate the hand's finger over the object to erase. Press the pick button on your pointing device to select that object. The object will disappear from the screen.

This option is good when only one or two objects need to be erased. This option is also good when many objects are in a small area and only some of them need to be erased.

Erasing Multiple Objects with the Crosses/Window Box

You can select several objects at once using AutoSketch's *crosses/window box* function. This function lets you pick corners of a box that encloses all or some of the objects you want to erase. The term *crosses/window* refers to the two ways to select objects with the box.

This mode is automatic if you miss picking an object. Or, you can purposely pick a location that is not on an object.

By carefully choosing your first point, you can use either the window box or crosses box. These same selection techniques are used for other editing commands, such as **Move** and **Copy**. To cancel the crosses/window box without selecting objects, drag the box so that it does contain any objects and pick. You can also cancel the command by picking the **Esc** button or hitting the [Esc] key.

> **SHORTCUT**
>
> The crosses/window function of AutoSketch is a quick, easy way to select objects. You will likely use this function very often.

Erasing only objects within the box

Pick a point that is to the *left* of the objects you want to erase. Make sure that the point you pick is not an object. The following prompt appears.

Erase Crosses/window corner:

Move the cursor to the right of the first corner. A solid-line box appears. This is the *window box.* Only those objects entirely within the box will be erased. Refer to Figure 4-40.

Erasing objects within and crossing the box

Pick a point that is to the *right* of the objects you want to erase. Make sure that the point you pick is not an object. The following prompt appears.

Erase Crosses/window corner:

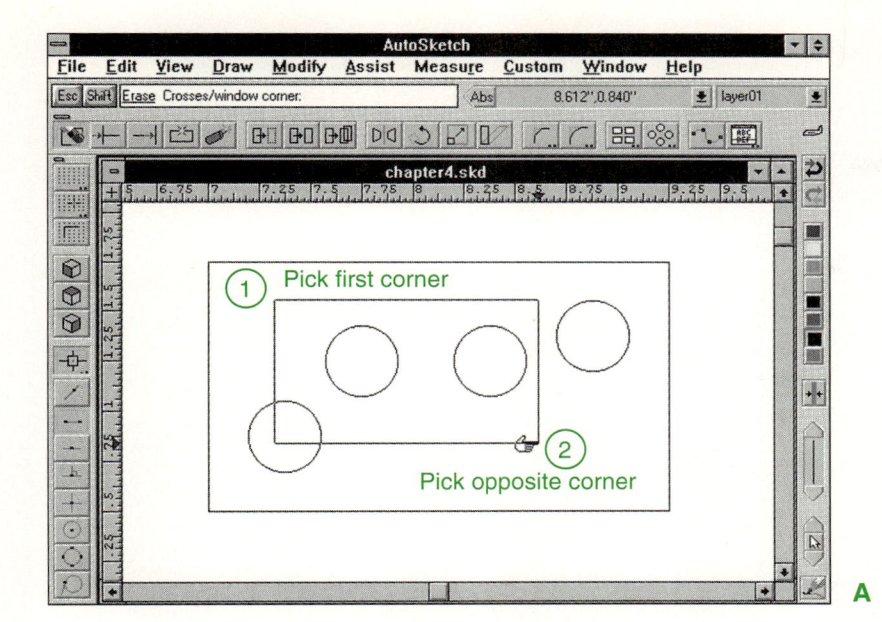

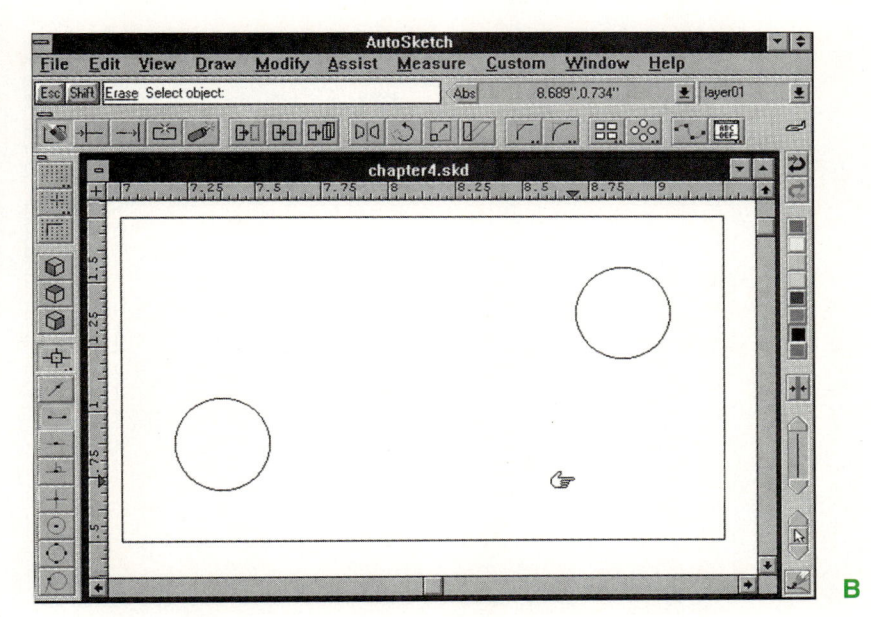

Figure 4-40
Erasing several objects with the window box. A—Pick opposite corners, from left to right, of a box around the objects. B—Only those objects entirely within the box are erased.

Move the pointer to the left of the first corner. A dotted-line box appears. This is the *crosses box*. All objects within the box *and* all objects crossing the box are erased. Refer to Figure 4-41.

Selecting Objects Using a Selection Set

There is another way to select objects. A selection set allows you to select multiple objects without having to select all objects within a crosses/window box. This technique is discussed in Chapter 6.

Figure 4-41.
Erasing several objects with the crosses box. A—Pick opposite corners, from right to left, of a box around the objects. B—Objects within and objects crossing the box are erased.

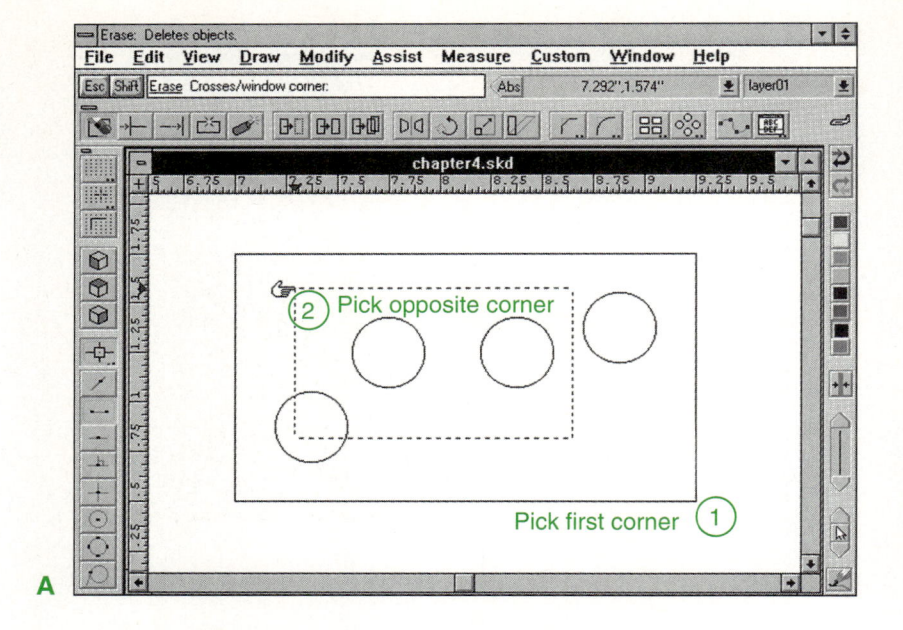

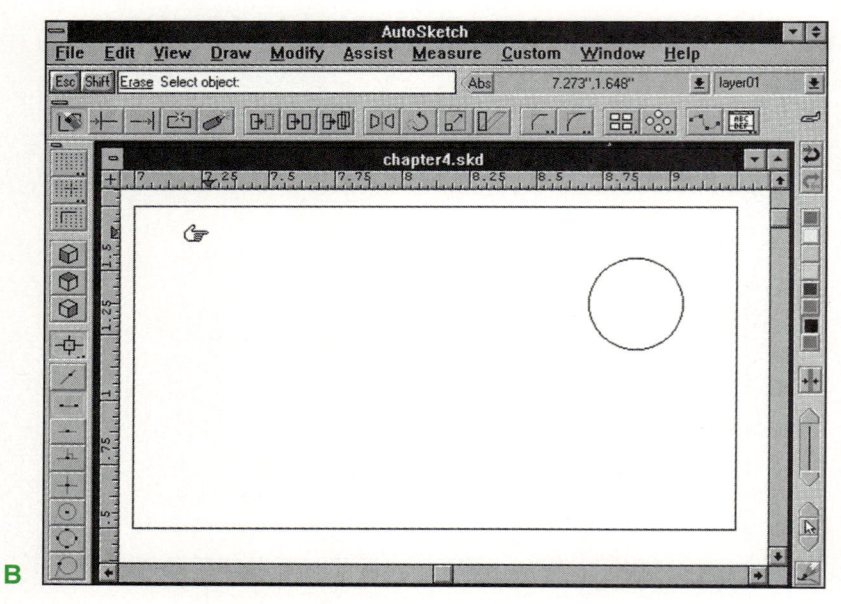

Exercise 4-10

1. Start a new drawing using the **Normal** template.
2. Draw one each of all of the objects offered by AutoSketch.
3. Erase just the circle.
4. Erase all of the remaining objects at one time using a crosses/window box.
5. Close the drawing without saving your changes.

Undoing and Redoing Your Commands

AutoSketch forgives your mistakes with the **Undo** command. You can reverse the effect of any previous drawing or editing function. For example, suppose you accidentally erased a line. To restore it, select the **Undo** button from the vertical status bar. You can also select the **Undo** tool from the **Edit** menu.

Select **Undo** once for every command you want to undo. This happens in reverse order. Continue picking **Undo** to cancel changes one at a time. If necessary, you can step back one command at a time all the way to when you opened or last saved your drawing. However, you cannot undo tasks done in a previous drawing session. If you backup to the beginning of your drawing session, the **Undo** button will be greyed out and you cannot select it.

How **Undo** works depends on how you chose the previous command. For example, suppose you just erased several objects within a window box. Using **Undo** once will replace all those objects. However, suppose you erased several objects one at a time. In this case, you would need to select **Undo** for each individual object that you erased to replace them on-screen.

If you accidentally undo too many steps, select the **Redo** button from the vertical status bar. You can also select the **Redo** tool from the **Edit** menu. This command reverses the effects of an undo. You can step back through all the **Undo** commands issued. If you have not used the **Undo** command, **Redo** will be greyed out.

> **SHORTCUT**
>
> You can quickly undo something by pressing the function key [F2].

> **SHORTCUT**
>
> You can quickly redo an undo by holding the [Alt] key and pressing the function key [F2].

Exercise 4-11

1. Start a new drawing using the **Normal** template.
2. Draw two lines.
3. Pick the **Undo** button to remove the second line.
4. Pick the **Redo** button to restore the second line.
5. Close the drawing without saving your changes.

Summary

AutoSketch allows you to create a wide variety of drawings. However, even the most complex drawing is simply an assortment of objects. AutoSketch objects include points, lines, boxes, polygons, polylines, circles, arcs, and curves. All of these items are found in the **Draw** menu. This chapter showed you how to add objects to the drawing. Chapter 5 will discuss how to draw objects in precise size and locations.

While creating drawings, you are likely to make at least a few mistakes. If the action you just completed did not turn out right, select **Undo**. If you need to remove objects from the drawing, use the **Erase** command. You can erase objects one at a time, You can also erase several objects at one time using the crosses/window box function.

Important Terms

Arcs	Diameter	Object properties
Bezier	Ellipse	Objects
B-spline	Frame	Planar rotation
Circles	Hairline	Point, Polygon
Circumference	Included angle	Polyline
Circumscribed	Inscribed	Radius
Control points	Line	Regular polygons
Crosses box	Linetype	Vertex
Crosses/window box	Line width	Width
Curve		Window box

New Commands and Tools

Arc Mode tool	**Customize**	**Polygon** tool
Arc tool	**Ellipse** tool	**Polyline** tool
Box tool	**Erase** tool	**Previous LineType** arrow
Change Properties tool	**Esc** button	**Redo** button
Circle tool	**Freehand** tool	**Redo** tool
Current LineType button	**Next LineType** arrow	**Undo** button
Curve tool	**Point** tool	**Undo** tool
		Width button

Review Questions

Give the best answer for each of the following questions.

1. Lines, boxes, circles, arcs, and other items you draw are called _____ in AutoSketch.
2. A(n) _____ marks an exact location on the drawing. It appears as a "dot" on-screen.
3. The color, width, and linetype of objects are called _____.
4. How many colors are displayed in the **Color Palette** in the vertical status bar? _____
5. You can use up to _____ colors in an AutoSketch drawing.
6. True or false. You can specify up to six custom line widths in the **Width** pop-up box.
7. How do you access the **Line Type Settings** dialog box? _____
8. True or false. AutoSketch automatically connects the endpoints of multiple lines drawn with the **Line** tool.
9. How many point locations are picked to draw a box? _____
10. How do you access the **Settings** dialog box for the **Circle, Arc, Polygon,** and **Curve** tools?
    A. **Circle** _____
    B. **Arc** _____
    C. **Polygon** _____
    D. **Curve** _____
11. How many options are available for drawing circles? _____

12. What feature does a **Settings** dialog box provide to help you determine how the points you pick will create the object. _____
13. The distance from the center to a place on the circumference on a circle is called the _____.
14. When drawing a two-point circle, the points you pick define the _____ of the circle.
15. When drawing a circle using the **Center, Radius** option, how can you select an exact radius without using the cursor? _____
16. How many options are available for drawing arcs?
17. What two arc drawing options allow you to enter an included angle? _____
18. What ellipse option allows you to enter the angle that the circle is viewed at? _____
19. The shape of the curve is guided by _____.
20. What type of curve, Bezier or B-spline, bends more freely around the control points? _____
21. What curve option, **Fitted** or **Framed**, causes the curve to pull away from the control points? _____
22. To form a closed curve shape, you must _____.
23. List two options to end a curve.
    A. _____
    B. _____
24. If two segments of a wide polyline connect at an angle less than 30 degrees, AutoSketch connects the segments with _____.
25. What tool is used to switch between a straight line and an arc segment of a polyline? _____
26. An arc segment of a polyline is automatically drawn with the _____ method.
27. True or false. You can draw a 1/2" wide polyline using the hidden linetype.
28. The button used to end most commands should *not* be pressed to end a polyline. What button is this? _____
29. True or false. To erase objects using a crosses/window box, you should use the **Box** tool.
30. How do you start the crosses/window box function to select objects to erase? _____
31. What determines whether the crosses box or window box is used to select objects? _____
32. True or false. Only those objects entirely within a window box are erased.
33. True or false. Only those objects entirely within a crosses box are erased.
34. What tool can be used to restore an object you have deleted? _____
35. How far can you step backwards through commands using the **Undo** tool? _____

Activities

1 - 4. Draw the illustrations shown below. Save your drawings using the filename given under each picture.

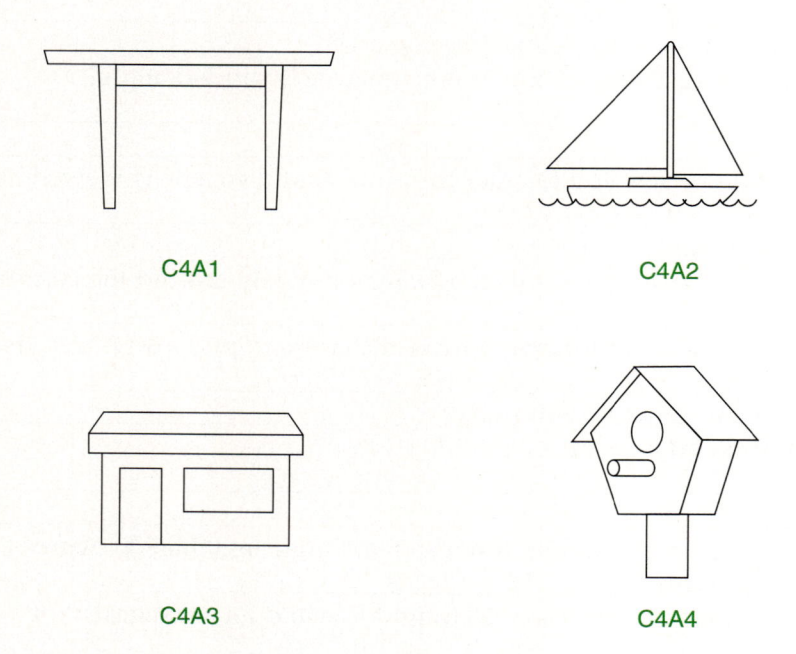

C4A1

C4A2

C4A3

C4A4

5. Think of a product you would like to invent or see built. Put some thought into the design. Then, use AutoSketch to make some rough sketches of your ideas. When finished, save the drawing as C4A5. Also, save the drawing under a special name you will remember easily.

CHAPTER 5

Drawing Precisely with Coordinates and Drawing Aids

Objectives

After studying this chapter, you will be able to:
* Explain how the Cartesian coordinate system is used to locate objects accurately.
* Type in absolute, relative, and polar coordinate values to place objects and enter distance.
* Identify AutoSketch functions that help you draw precisely with a mouse.
* Use the **Coordinate Display** to keep track of cursor position.
* Describe the purpose of a grid and set the grid spacing.
* Describe the purpose of snap and set the snap spacing.
* Use the orthogonal mode to draw perfect horizontal and vertical lines.
* Use attach modes to connect new objects to existing objects.

In Chapter 4, you learned how to place objects in a drawing. You placed these objects without regard to size or location. However, most designs require that you draw objects using exact measurements. This chapter shows you how to enter exact distance and size values using your keyboard and cursor.

Cartesian Coordinate System

Chapter 1 stated that computer-aided drafting systems are much more accurate than manual drawing. You might wonder how AutoSketch can make your drawings so accurate. The answer is that AutoSketch, and all CAD programs, use a standard method of object location called the Cartesian coordinate system.

Locating Points on the Cartesian Coordinate System

The *Cartesian coordinate system* consists of two axes at 90 degrees to each other, Figure 5-1. The horizontal axis is called the *X axis*. The vertical axis is called the *Y axis*. The intersection of the two axes is the *origin*. The location of any point or object can be found by measuring distance along the two axes. The distance may be measured in feet, inches, or millimeters.

Figure 5-1.
The Cartesian coordinate system has two axes perpendicular to each other.

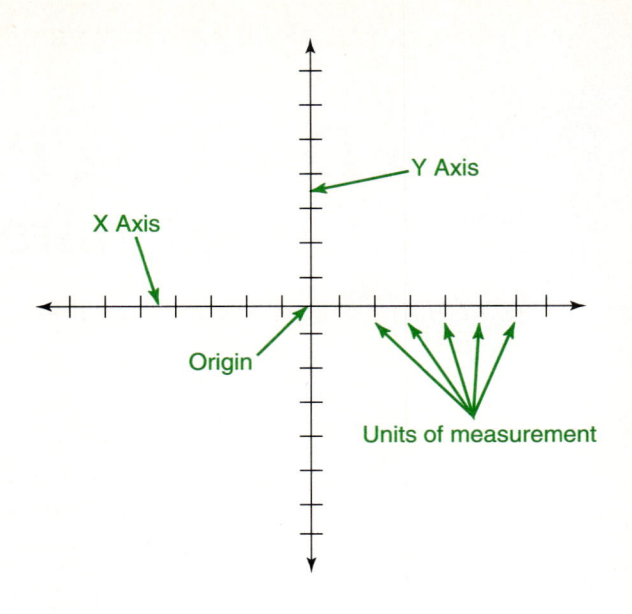

Typing in Coordinate Locations

Any time AutoSketch prompts for a location, you can pick the location with the cursor. You can also enter a coordinate with the keyboard. There are three ways to enter coordinate locations and distance using the keyboard. You can enter the location in absolute, relative, or polar coordinate values. Absolute coordinates measure distance from the origin. Relative and polar coordinates measure distance from the previous point.

Absolute Coordinates

Absolute coordinates mark an exact position measured from the origin of the Cartesian coordinate system. AutoSketch initially aligns the origin of the coordinate system with the lower-left corner of your screen, depending on the template chosen. An absolute coordinate is entered with a coordinate pair. A coordinate pair is entered as *X,Y*. The X value is the distance from the origin along the X axis. The Y value is the distance from the origin along the Y axis. The coordinate pair 0,0 is the location of the origin.

Drawing with absolute coordinates

An example of drawing a line using absolute coordinates is shown in Figure 5-2. After choosing the **Line** tool, you might enter the following values.

Line Enter point: **2,2**↵

Line To point: **10,7**↵

Figure 5-2. The endpoints of this line are located with absolute coordinates. Notice that both points are measured from the origin.

You must separate the X and Y coordinate values with a comma. Press the [Enter] key after typing in the values and the line appears. You can also enter negative absolute coordinates. A negative X value locates position to the left of the Y axis. A negative Y value locates position below the X axis.

Using unit notation

A distance is measured using the current unit of measurement. To enter distance using a different unit, place the unit notation behind the values. For example, if the current unit of measurement is inches, you can enter 10mm, 7mm as the coordinate pair.

Preparing for Exercises and Activities in This Chapter

Before starting the exercises or activities in this chapter, set your printer for landscape orientation. You can do this in the Windows Control panel. You can also use the following procedure.
1. Opening a new drawing with the **Quick Start** template.
2. Select the **File** menu and **Print Setup...** command.
3. Highlight the printer and select the **Setup...** button.
4. Select **Landscape** for the paper orientation and pick **OK**.
5. Pick **OK** to leave the **Print Setup** dialog box.
6. Close the drawing without saving your changes.

Doing these steps allows your drawing sheet to fit on a single A-size sheet of paper horizontally. You will only need to do this step once, unless you use another Windows application and change the orientation back to portrait.

Exercise 5-1

1. Start a new drawing using the **Normal** template.
2. Use the **Line** tool to draw four lines that make up a 3" square. Do not use your cursor. Start the left corner of the box at the origin.
3. Save the drawing as C5E1 on your work diskette and then close the drawing.

Relative Coordinates

Relative coordinates measure position from the previous point. The previous point may have been typed at the keyboard or picked using the cursor. A relative coordinate is entered in AutoSketch as R(X,Y). The X value is the distance from the previous point measured along the X axis. The Y value is the distance from the previous point measured along the Y axis. When entering relative coordinates, you must use the opening and closing parentheses and the letter R. The letter R can be either uppercase (R) or lowercase (r).

Drawing with relative coordinates

Following is an example of drawing a line using both absolute and relative coordinates, Figure 5-3. First select the **Line** tool from the **Draw** menu. Place the first endpoint of the line with an absolute coordinate.

Line Enter point: **2,2**↵

This places the first point two units to the right of the origin on the X axis and two units above the origin on the Y axis. At the next prompt, enter a relative coordinate. Refer to Figure 5-3 as you enter the following value.

Line To point: **R(9,5)**↵

Figure 5-3.
The second endpoint of this line is located with a relative coordinate. Notice that it is measured from the position of the first endpoint (2,2).

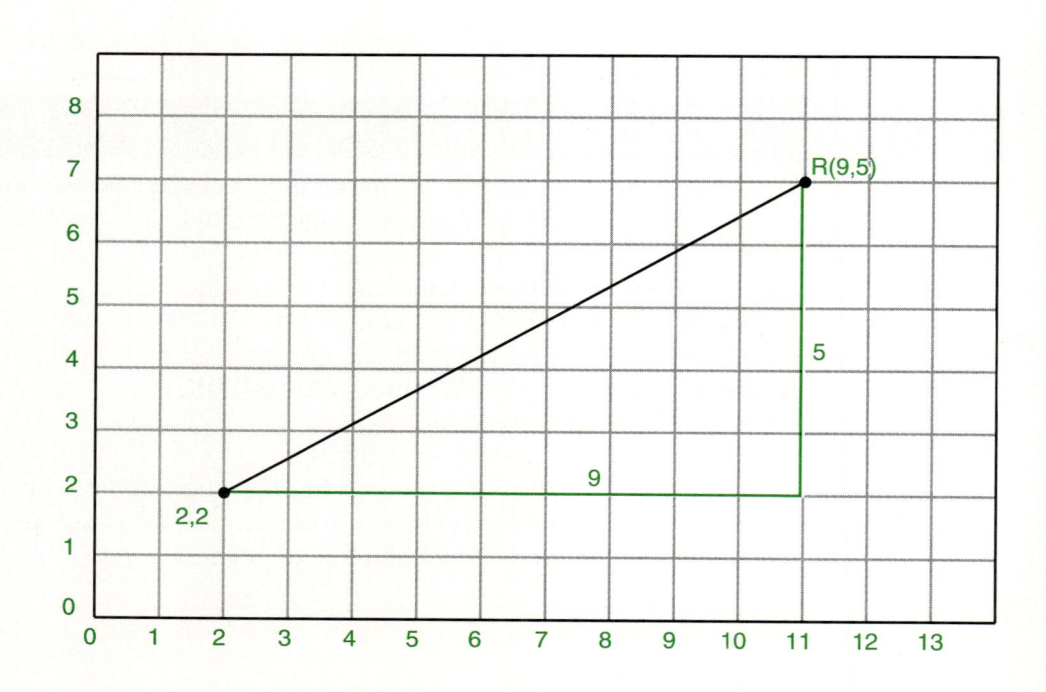

This value places the second endpoint nine units to the right and five units above the first endpoint of the line.

Simply stated, the absolute coordinate measures from the origin. The relative coordinate tells how many units to take in each direction from the point you are currently at. You can also enter negative coordinate values. For example, entering R(-9,-5) places the second endpoint nine units to the left and five units below the first endpoint of the line.

Exercise 5-2

1. Start a new drawing using the **Normal** template.
2. Use the **Box** tool to draw a 3" square. Do not use your cursor. Place the first corner of the box at the origin using an absolute coordinate. Place the opposite corner using a relative coordinate.
3. Save the drawing as **C5E2** on your work diskette and then close the drawing.

Polar Coordinates

Polar coordinates measure position from the previous point by distance and by the angle from the X axis. The previous point may have been typed at the keyboard or picked using the cursor. A polar coordinate is entered in AutoSketch as P(D,A). The D value is the distance from the previous point. The A value is the angle measured counterclockwise from the X axis, Figure 5-4. When entering polar coordinates, you must use the opening and closing parentheses and the letter P. The letter P can be either uppercase (P) or lowercase (p).

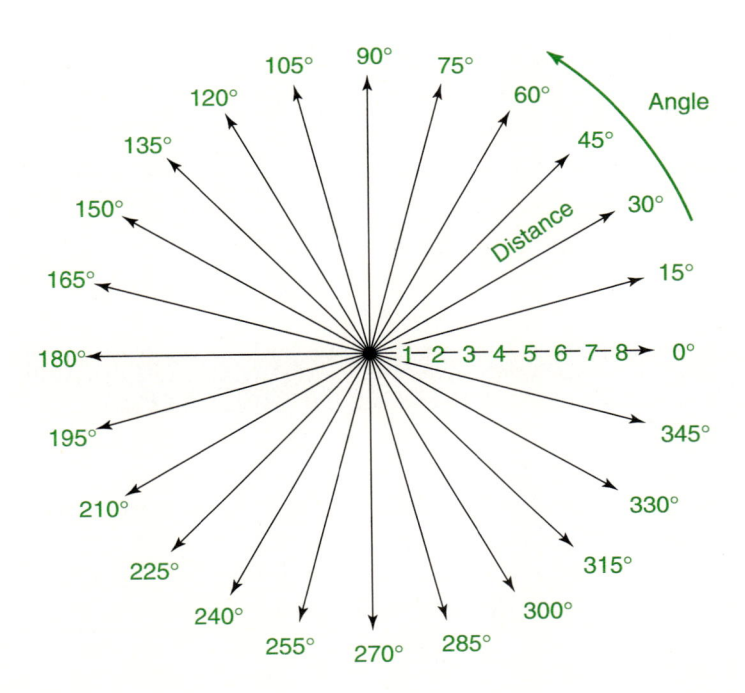

Figure 5-4.
The angle value (A) of a polar coordinate is the angle measured counterclockwise from the X axis. The distance value (D) is measured from the previous point.

Drawing with polar coordinates

Following is an example of drawing a line using absolute and polar coordinates. Refer to Figure 5-5. First, select the **Line** tool from the **Draw** menu. Place the first endpoint. For this example, an absolute coordinate is entered with the keyboard.

    <u>Line</u> Enter point: **2,2**↵

At the next prompt, enter a polar coordinate. Refer to Figure 5-5 as you enter the following value.
    <u>Line</u> To point: **P(8.5,35)**.↵

The polar coordinate value makes the line 8.5 units long at a 35 degree angle from the X axis. Simply stated, the absolute coordinate measures from the origin. The polar coordinate tells how far and in what direction to go from the point you are currently at.

Figure 5-5.
The second endpoint of this line is located with a polar coordinate. Notice that the distance (8.00) is measured from the previous point and the angle (35) is measured counterclockwise from the X axis.

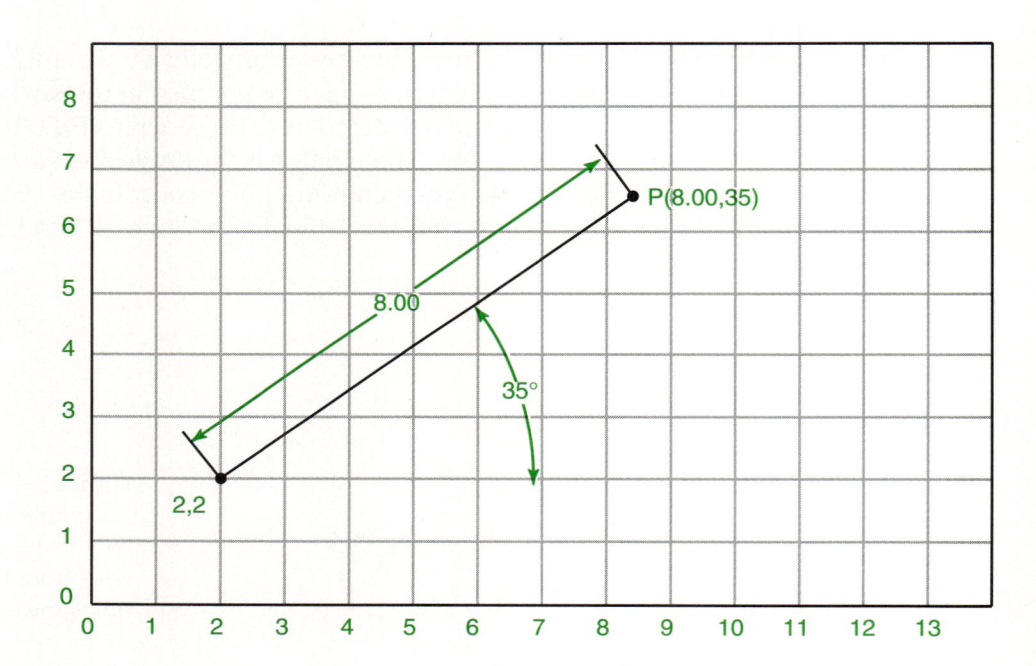

Exercise 5-3

1. Start a new drawing using the **Normal** template.
2. Use the **Polygon** tool to draw a six-sided circumscribed polygon. Place the center of the polygon in the middle of the screen with your cursor.
3. Use polar coordinates to locate the midpoint of one side of the polygon. Locate this point so that the polygon is 4 units wide between opposite sides (a radius of 2).
4. Save the drawing as C5E3 on your work diskette and then close the drawing.

Format for Entering Coordinate Values

When you type in coordinate values, the current unit of measurement is used. However, you can also enter distance using other units of measurement. For example, even though your unit of measurement might be inches, you can enter a metric value. To do so, you must add the unit notation behind the value you type in. The formats are shown in Figure 5-6.

Unit	Notation	Example
inches (decimal)	" or in	2" or 2in
inches and fraction	"	3 1/2"
feet (decimal)	'	1.5'
feet and fraction	'	2-1/2'
feet-inches (decimal)	' and "	3.5'-1.5"
feet-inches and fraction	' and "	1'-2 3/4"
millimeters	mm	5mm
centimeters	cm	2cm
meters	m	7m
kilometers	km	1km
miles	mi	4mi
yards	yd	2yd

Figure 5-6.
When entering coordinates in units of measure other than the current drawing units, you must use the proper unit notation.

In most cases, you will not mix customary (feet or inch) and metric values. Most mechanical drawings are either all customary or metric. However, in architectural drafting, you will often combine feet-inch and inch. Also, remember that when you enter a measurement using a different notation, AutoSketch converts the distance to the current unit of notation. For example, these values all convey the same measurement.
- 12'-1 3/4"
- 13 3/4"
- 13.75 in
- 1.146'
- 349.25mm
- 34.925cm

You can use up to six digits behind the decimal point (0.000000) for precision. This applies to all units of measure. Also, when entering architectural units, follow these guidelines.
- Separate feet and inch units with a dash.
- Separate fractions of an inch from whole inches with a space.
- Do not separate feet and inches with a space.
- Separate the numerator and denominator of a fraction with a front slash (/).
- The denominator of the fraction must be a power of 2, up to 64. In other words, the denominator must be 2, 4, 8, 16, 32, or 64.

Exercise 5-4

1. Start a new drawing using the **Normal** template. The default unit of measurement is fractional inches.
2. Use the **Line** tool to draw the lines shown below using the length given. You can place the first endpoint of each line with the cursor.
3. Save the drawing as **C5E4** on your work diskette and then close the drawing.

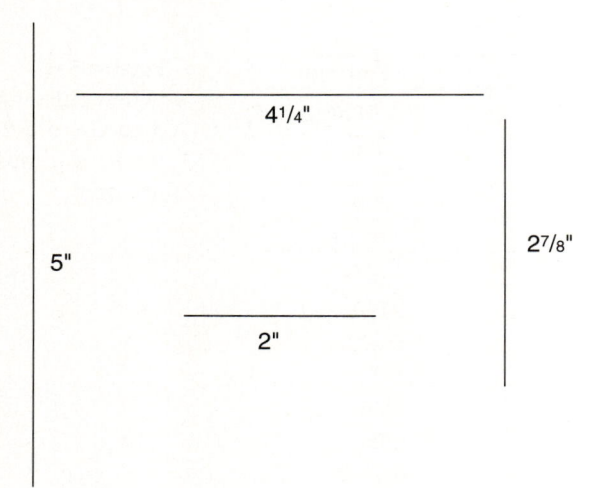

```
           4¼"
5"                      2⁷⁄₈"
           2"
```

Combining Absolute, Relative, and Polar Coordinates

You will probably use absolute, relative, *and* polar coordinates on any given drawing. In fact, it is best to combine the three methods along with using your cursor to pick locations. Type in coordinates when you know exact position, know how far you need to travel from your current position, or need to enter a distance or point location that would be impossible using the cursor (such as 2.006173). Use the cursor to locate a position on another object quickly. Later in this chapter, you will learn how to draw accurately using the cursor.

When to use relative coordinates

Relative coordinates are handy when you are drawing horizontal and vertical lines at a known distance. For example, suppose you are drawing a wall for a house plan. After selecting the **Polyline** tool, pick one corner of the house using the pointer. Then, to draw a horizontal wall 9' long, enter R(9'-0",0) at the first prompt.

Relative coordinates also let you enter negative values to move left or down. For example, enter R(-5,0) to locate a point five units to the left of the previous point. Enter R(-2,-5) to locate a point two units to the left and five units down from the previous point. Figure 5-7 shows how to draw a rectangle with the **Polyline** tool using one absolute and four relative coordinates. Note that both positive and negative values are entered.

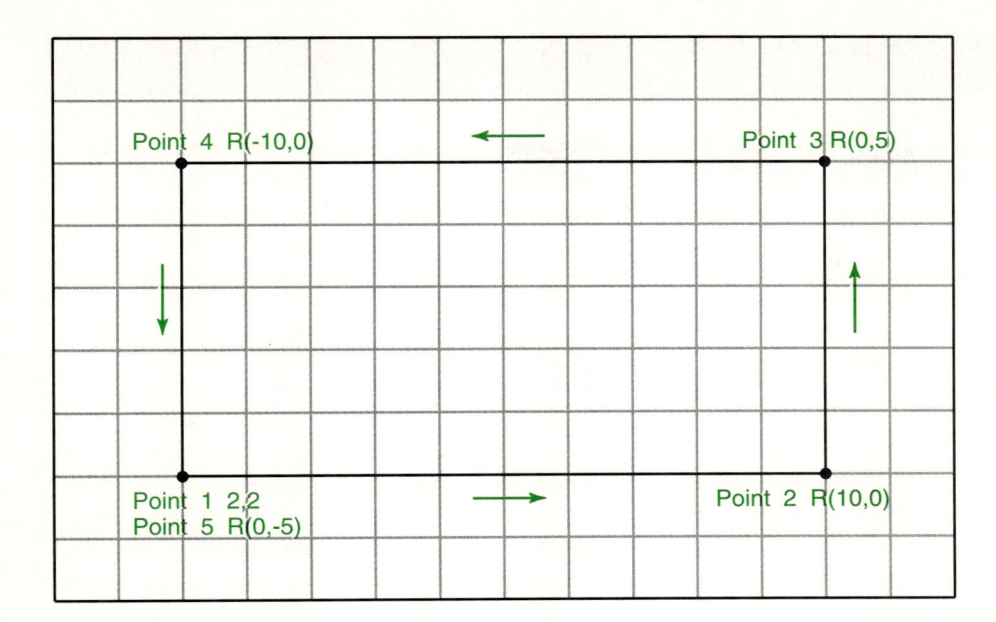

Figure 5-7.
The corners of this rectangle are located with relative coordinates. Notice that both positive and negative values are used.

When to use polar coordinates

You will probably use polar coordinates more than any other keyboard entry method. You can draw lines in any direction and at any distance. In addition, to draw horizontal and vertical lines, simply enter 0, 90, 180, or 270 degree angles.

Using /lpoint to Enter Location

Whenever you pick a location or enter a coordinate, AutoSketch stores the location of that place. You can recall that value by entering **/lpoint**. The **/lpoint** command means "last point." For example, suppose you just drew a line. Now, you need to connect the first endpoint of a new line to the second endpoint of the previous line. Instead of using an **Attach** mode, enter the following command.

<u>Line</u> Enter point: **/lpoint**↵

This method is easier to use than **Attach**. The **Attach** tools are covered later in this chapter. Also, there are times when the only method to find the previous point is with **/lpoint**. For example, suppose you just drew a circle by picking the center and a point on the circle. The only way to recall the point you placed on the circle is with **/lpoint**.

Exercise 5-5

1. Start a new drawing using the **Normal** template.
2. Use the **Line** tool to draw four lines that make up a 3" square. Pick only the first point with your cursor. Then, use **/lpoint** and relative coordinates to complete the square.
3. Save the drawing as C5E5 on your work diskette and then close the drawing.

Changing the Origin of the Drawing

When you open a new drawing, the (0,0) origin is at the lower-left corner. When the rulers are visible, you can easily change this origin. Pick the **Change Origin** button at the intersection of the rulers. This button is found near the upper-left corner of the drawing area, Figure 5-8. The cursor changes to the **Change Origin** cursor. Then, pick a point on your drawing where the origin should be. You can also enter a coordinate to relocate the origin.

Figure 5-8.
To change the origin of a drawing, pick the **Change Origin** button. With the **Change Origin** cursor, pick a new origin (0,0) for the drawing.

Exercise 5-6

1. Start a new drawing using the **Normal** template.
2. Change the origin of the drawing so that it is placed in the center of your drawing window.
3. Use the **Circle** tool to a 3" diameter circle where the center of the circle is at the origin.
4. Save the drawing as C5E6 on your work diskette and then close the drawing.

Drawing Precisely with the Cursor

In Chapter 4, you learned that it is easy to use the cursor to draw objects. You can see the cursor's position on-screen and quickly find where an object will be placed. You can also "drag" objects to size. However, your ability to draw accurately with the cursor depends on the resolution of your monitor and your skills in handling the mouse. If lines should meet, they should do so precisely. Even if two objects that should meet are .001" apart, this could be bad for the design.

AutoSketch provides *drawing aids* that help you use the cursor to locate position more accurately. There are five types of drawing aids: coordinate display, visible grid, invisible snap grid, attach modes, and orthogonal mode. These functions are *toggle modes*. Toggle modes mean that the functions are either on or off.

Coordinate Display

When using the cursor to draw, the **Coordinate Display** shows the current location of the cursor. The displayed values constantly change as you move the cursor. The display will read in the current unit of measurement. By default, the **Coordinate Display** shows the cursor position in absolute coordinates. However, you can have the **Coordinate Display** show all or just selected coordinate types.

Changing the Coordinate Type

To change the coordinate type displayed, pick the **Coordinate Type** button to the left of the **Coordinate Display**, Figure 5-9. The letters on the **Coordinate Type** button indicate the current type.
- Abs refers to absolute coordinates
- Rel refers to relative coordinates

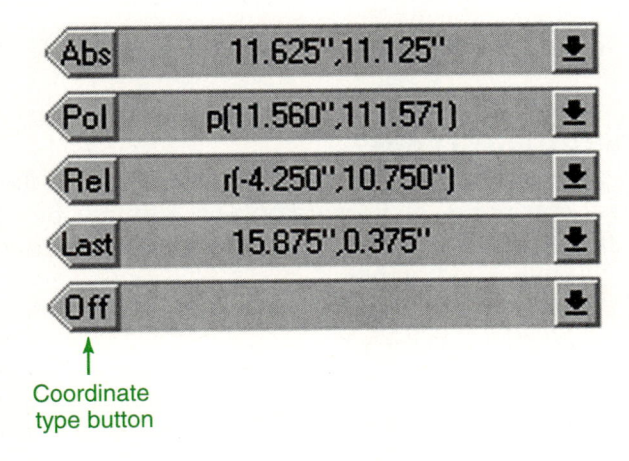

Figure 5-9.
To change the values shown in the **Coordinate Display**, pick the **Coordinate Type** button.

- Pol refers to polar coordinates
- Last refers to the last point entered (*lpoint* variable)
- Off means no coordinate value is displayed

You can rotate through the different coordinate types by picking this button. If you do not want the coordinates displayed, pick the button until Off is shown.

Displaying All Coordinate Types

To open a pop-up window with all the coordinate types displayed, pick the **Coordinate Window** button on the right side of the **Coordinate Display.** A pop-up display appears, Figure 5-10. Since this is a pop-up window, you can move it around the screen. To do so, place the cursor over the pop-up window. Hold down your left mouse button and drag the window to a new location. As you drag the window, an outline will appear to help locate a new position.

Figure 5-10.
Pick the pop-up arrow on the right side of the **Coordinate Display** to open a pop-up window with all the coordinate types displayed.

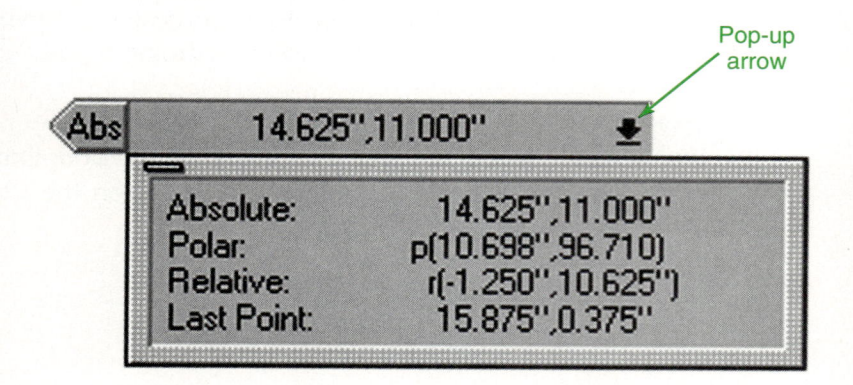

```
Abs       14.625",11.000"              ↓   ← Pop-up arrow

Absolute:         14.625",11.000"
Polar:            p(10.698",96.710)
Relative:         r(-1.250",10.625")
Last Point:       15.875",0.375"
```

Displaying Certain Coordinate Types

You may want to display only certain coordinate types. To do so, pick the control menu button at the upper-left corner of the coordinate pop-up window. By default, a check mark appears next to all four coordinate types. Pick the types you do not want shown, Figure 5-11.

Shrinking and Closing the Coordinate Display Pop-up Window

If the **Coordinate Display** pop-up window is too large, pick the control menu button and then pick **Shrink.** The width of the window will shrink. Notice that the abbreviations are no longer shown. To restore the full pop-up window, pick the **Control Menu** button and then **Expand.** To close the pop-up window, pick the **Control Menu** button and then **Close.**

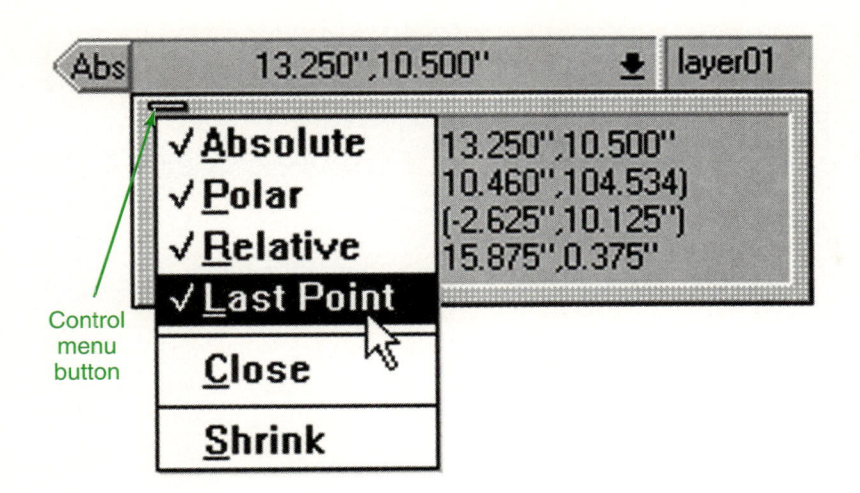

Figure 5-11.
Use the **Coordinate Display** control menu to set which values are shown in the pop-up window.

Exercise 5-7

1. Start a new drawing using the **Normal** template.
2. Change the **Coordinate Display** so that relative coordinates are shown.
3. Open the **Coordinate Display** pop-up window.
4. Change the **Coordinate Display** pop-up window so that only absolute and relative coordinate values are displayed.
5. Move the **Coordinate Display** pop-up window to the lower-right corner of your drawing.
6. Shrink the **Coordinate Display** pop-up window.
7. Close the **Coordinate Display** pop-up window.
8. Close the drawing without saving it.

Grid and Snap

A *grid* is a pattern of dots on-screen, Figure 5-12. These dots are used much like graph paper. The grid pattern lets you see distance between units of the Cartesian coordinate system. When adding an object, you can use the grid to locate position.

The grid does not control the movement of the cursor. It is only a visual reference. This is why AutoSketch offers an invisible *snap grid.* When **Snap** is turned on, a small crosshair, like a plus (+) mark, "jumps" along beside the default cursor (the arrow), Figure 5-13. If you change from the default cursor, the entire cursor moves with the snap grid. The cursor moves in precise increments measured by the snap spacing. Any location you pick is placed at the crosshair. This lets you place and size objects with precision.

Setting Grid and Snap Spacing

The spacing between grid dots and snap increments is set in the **Grid/Snap Settings** dialog box, Figure 5-14. To access it, double-click either the **Grid** tool or **Snap** tool in the **Assist** menu. By default, AutoSketch shows the **Assist** toolbox at the left side of the screen.

Figure 5-12.
A grid is a pattern of dots used much like graph paper.

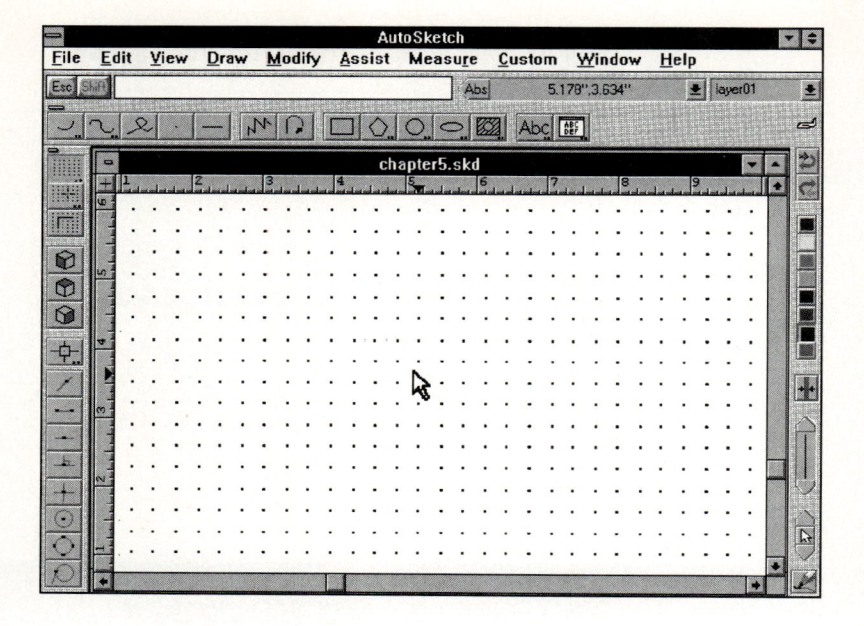

When finished making settings, pick **OK** at the bottom of the dialog box. You can reset the spacing, or turn off the grid and snap, at any time during your drawing session, even while using another command. The settings are covered in the next sections.

X: Value

The **X:** value is the horizontal grid and snap spacing. By default, the horizontal grid spacing equals the snap spacing. When you enter a value, it is placed under both **Snap** and **Grid** columns.

Y: Value

The **Y:** value is the vertical grid and snap spacing. By default, the vertical grid spacing equals the snap spacing. When you enter a value, it is placed under both **Snap** and **Grid** columns. When you set the **X:** spacing, the **Y:** spacing will change to the same value. However, you can enter a different **Y:** spacing value if necessary.

Figure 5-13.
When snap is on, a small plus mark jumps beside an arrow cursor. The plus mark will snap to the increments (in this case the grid). When using a crosshair cursor, the cursor itself jumps to snap increments.

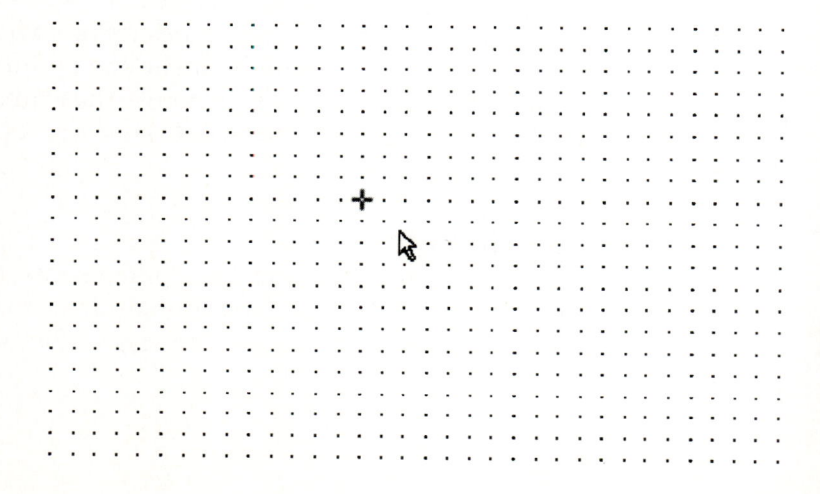

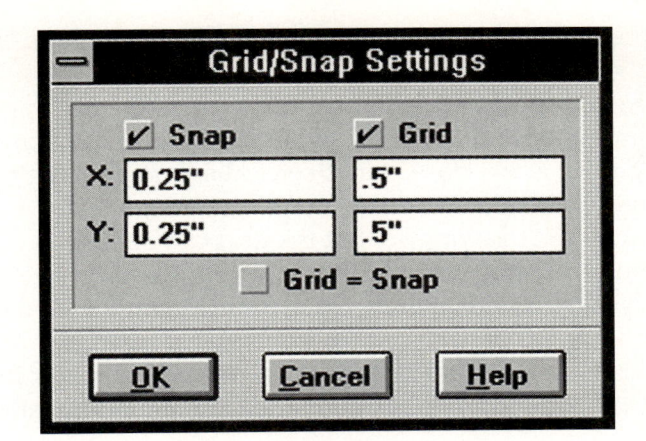

Figure 5-14.
Double-click on either the **Grid** or **Snap** tools to display the **Grid/Snap Settings** dialog box.

Grid = Snap

The **Grid = Snap** option forces the grid and snap to have the same X and Y values. This option is active when the box is checked. However, often you will want the grid spacing larger than the snap spacing. The grid serves as a general reference of your location on-screen. A small snap spacing helps you draw precisely. If the **Grid = Snap** option is not checked, you can enter different values for the grid and snap spacing.

Grid and Snap boxes

When checked, the **Grid** option causes the grid to be displayed. The **Snap** box, when checked, causes the cursor to move with the snap interval.

Turning the Grid and Snap on and off

The **Grid** and **Snap** tools are toggle commands. Grid and snap are either on or off. When you pick the **Grid** tool, the tool appears depressed and the grid is displayed. Click again to turn the grid off.

When you pick the **Snap** tool, the tool appears depressed and the cursor moves in snap intervals. Click again to turn snap off.

Guidelines for Using Grid

Here are some guidelines for using a grid. These should help you use the **Grid** function more effectively.
- The X and Y spacing is usually the same value.
- Grid spacing does not have to be in whole units. You may want grid dots spaced every .25 or .5 unit.
- The distance between grid dots can be large or small. The grid spacing depends on the size of the objects to be drawn. The distance between grid dots for a detailed drawing could be as small as .001 units. If the drawing is large or simple, space the grid dots further apart.

SHORTCUT
You can also toggle **Grid** on and off by holding the [Alt] key and pressing the function key [F6].

SHORTCUT
You can also toggle **Snap** on and off by holding the [Alt] key and press the function key [F7].

- If you turn on the grid and do not see grid dots on-screen, your grid spacing is too large in relation to your current view. For example, if you are looking at a 1" by 1" view of the drawing, a grid spacing of 2" will likely not appear on-screen.
- Setting the distance too small will blur the grid or prevent it from appearing at all. If the grid spacing is too close, AutoSketch displays a dialog box saying "Grid is too dense to display." Reenter the grid spacing using a larger value. You can also zoom-in on the drawing so that the grid can be displayed.
- The grid is for your reference. Set the spacing how it will help you best.
- Use the grid only when it is useful. Sometimes the grid dots make the drawing look cluttered.
- The visible grid is not a permanent part of the drawing. Even if the grid is on when you print the drawing to paper, the grid dots are not printed.

Guidelines for Using Snap

Here are some guidelines for using snap. These should help you use the **Snap** function more effectively.

- The snap spacing is sometimes the same as the grid spacing. However, more often the values are different.
- If the snap spacing is not the same value as the grid spacing, it is usually set to evenly divide the grid spacing. For example, if the grid spacing is 1 unit, the snap spacing might be .25 units. This means there are four snap locations for every grid point.
- Make the snap spacing useful. For example, many users set the snap spacing to the smallest distance needed to create the drawing. Suppose the size and spacing of most objects is .25, .5, and .75. Set the snap spacing to .25. Then, all distances you need to move or draw with the cursor are multiples of the snap spacing. The visible grid could be set at .5 to show half units, or 1 to show whole units of the coordinate system.
- The invisible snap grid can be turned on and off at any time. Turn the snap on only when it will help you draw. Remember, with snap on, you cannot freely pick locations other than the snap spacing. If you want to place a point freely, you must turn snap off. Also, snap may interfere with picking an object to erase or edit.

Exercise 5-8

1. Start a new drawing using the **Normal** template.
2. Set the grid spacing to .5 and turn it on.
3. Set the snap spacing to .25 and turn it on.
4. Use the **Polyline** and **Circle** tools to draw a 3.25" diameter circle and 3.75" square box.
5. Save the drawing as **C5E8** on your work diskette and then close the drawing.

Ortho Tool

Many lines you draw must be perfectly horizontally or vertically. However, it is difficult to draw lines or move objects perfectly horizontally or vertically. AutoSketch helps you with the **Ortho** tool. Ortho refers to *orthogonal*. This means perpendicular, or at right angles. Using the **Ortho** tool can be compared to using a T-square when drafting manually.

The **Ortho** tool is a toggle that you can turn on or off in the middle of another command. To turn it on, pick the **Ortho** tool from the **Assist** menu. Your drawing and editing tasks are then restricted to horizontal and vertical movement, Figure 5-15.

The **Ortho** tool acts differently when using the isometric drawing mode of AutoSketch. This function is discussed in Chapter 16.

SHORTCUT

You can also toggle **Ortho** on and off by holding the [Alt] key and pressing the function key [F5].

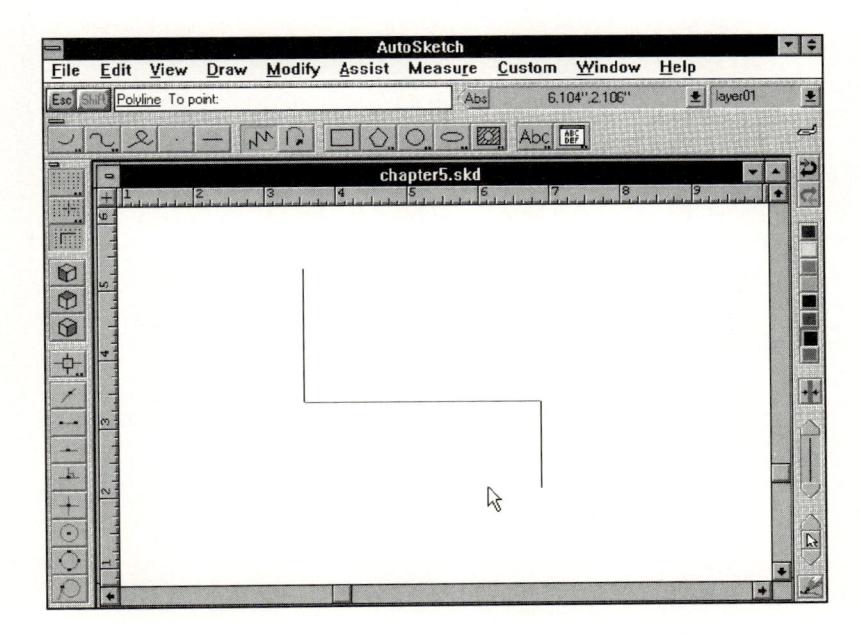

Figure 5-15.
You can draw perfectly horizontal and vertical lines with **Ortho** on.

Attaching to Objects

Attach is a drawing aid that helps you connect to precise places on existing objects. You might want to draw a line that meets precisely at the endpoint of another line, or the center of a circle. **Attach** is needed because objects tend to look jagged on the monitor. This makes it difficult to connect objects. When the **Attach** tool is active, you can pick anywhere on an object and AutoSketch locates the attach point for you.

You must pick very close to the object on-screen for **Attach** to work. Otherwise, the point picked will be placed freely and not be connected to the object. For example, suppose the **Center** tool is on and you are drawing a line to the center of a circle. Locate your cursor on the circle and pick it. The endpoint of the line will jump to the circle's center. If you try to pick in the center of the circle, the line is placed freely. This is because you did not pick close enough to the object for AutoSketch to decide what object to attach to.

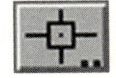

Attach Options

There are eight ways to attach new objects to existing objects. By default, the **Assist** menu is shown as a toolbox located vertically on the left side of the screen. You can use any one, or a combination, of these tools, Figure 5-16.

Center

The **Center** tool attaches to the center of a circle, arc, ellipse, or polygon. You must pick on, or near, the existing object to attach to the center point of that object.

To use the **Center** tool for circles and ellipses, you must turn off the **Quadrant** tool. For objects that have endpoints and midpoints (arcs and polygons), you must turn off the **Endpoint** and **Midpoint** tools. These tools have priority over the **Center** tool.

Endpoint

The **Endpoint** tool attaches to the nearest endpoint of an existing line, arc, polyline, text, or dimension. It will also attach to the nearest corner of an existing box, polyline segment, polygon, or pattern fill. This option ignores circles and ellipses since they do not have endpoints.

Figure 5-16.
Since **Attach** tools are used so much, the **Assist** toolbox is open on the left side of the screen when you start AutoSketch. When **Attach** is on, a point you pick close to an object will "attach" to one of eight places. The location depends on the object type and which **Attach** tools are selected.

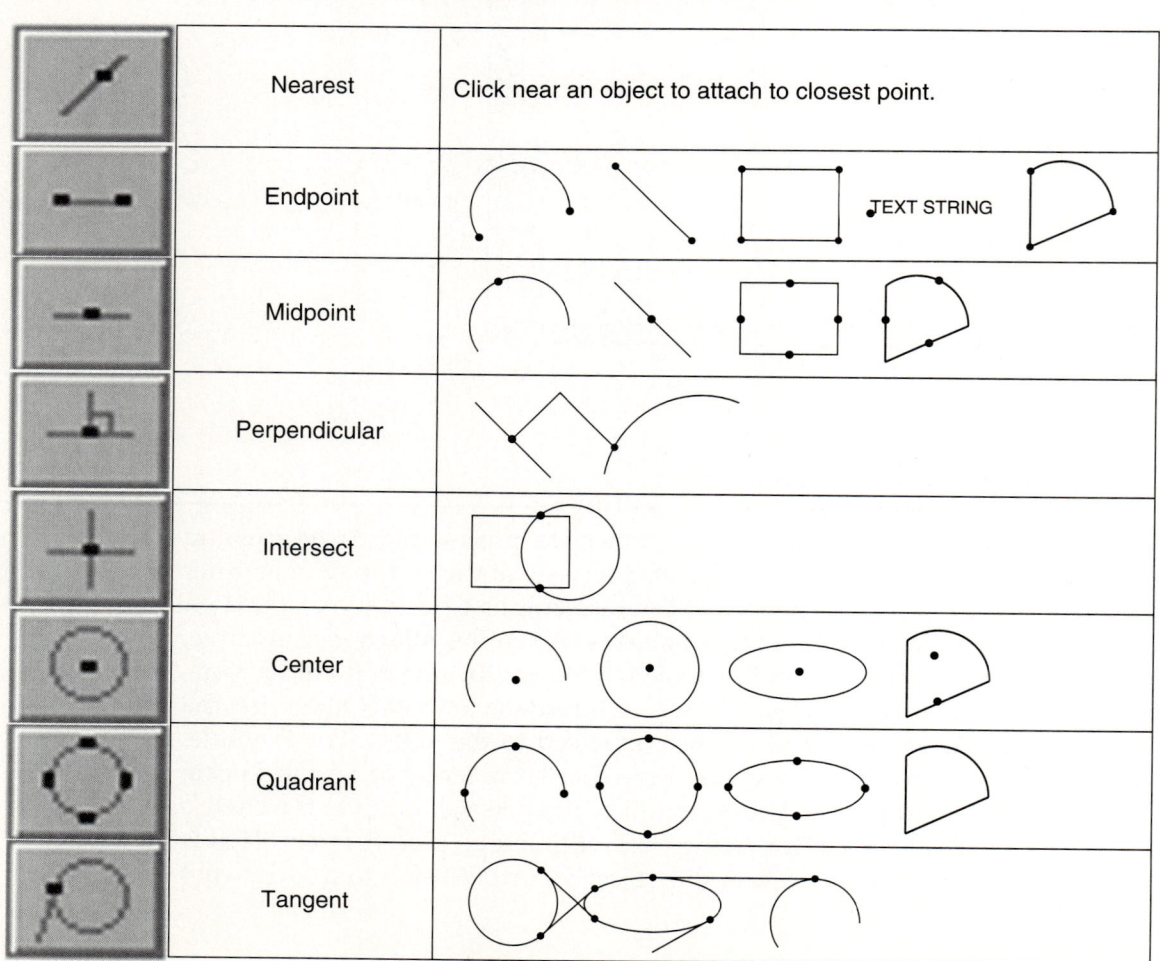

	Nearest	Click near an object to attach to closest point.
	Endpoint	
	Midpoint	
	Perpendicular	
	Intersect	
	Center	
	Quadrant	
	Tangent	

Intersect

The **Intersect** tool attaches to the place where two existing objects intersect.

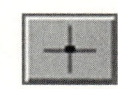

Midpoint

The **Midpoint** tool attaches to the midpoint of an existing line, box segment, arc, polyline segment, polygon segment, or pattern fill segment. This option ignores circles, ellipses, and text since they do not have midpoints.

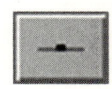

Nearest point

The **Nearest** tool will attach to the closest point on the existing object. Because this tool overrides all other options, it should be turned off when using any other **Attach** tool. This option does not attach to curves and freehand objects.

Perpendicular

The **Perpendicular** tool causes a segment you draw to attach perpendicular (90 degree angle) to the existing object you pick. This mode will work when attaching to an arc, circle, ellipse, box, polygon, line, polyline, and pattern fill. Since all tools except **Tangent** override **Perpendicular** they should be turned off when using the **Perpendicular** tool.

Quadrant

The **Quadrant** tool will attach to the quadrant of an existing circle, arc, or ellipse. The four quadrant points lie at the intersection of an object with horizontal and vertical lines drawn through its center. In other words, an arc that is less than 90 degrees in rotation has no quadrant points. For an ellipse, quadrant points are found at the endpoints of major and minor axes.

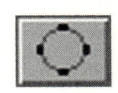

Tangent

The **Tangent** tool will attach to the tangent point of an existing circle, arc, ellipse, polyline arc segment, or fillet arc. A line that is *tangent* touches the circle, arc, or ellipse at only one point, but does not intersect. Two circles or arcs can also be tangent if they touch, but do not intersect. Since all other tools override **Tangent,** they should be turned off when using the **Tangent** tool.

Turning Attach Tools On and Off

You can turn on one or any combination of **Attach** options. First, pick the **Attach** tool. By default, all tools are active. Next, turn off the tools you don't want by clicking on the tool. The active tools will appear depressed. Double-clicking on an **Attach** tool will make it active, *and* turn off all other **Attach** tools. You can also double-click on the **Attach** tool to bring up the **Attachment Settings** dialog box, Figure 5-17. Here you can select attach modes by name.

When more than one **Attach** tool option is active, AutoSketch looks in the area where you pick until it finds the first match. For example, if both the **Endpoint** and **Midpoint** tools are active, where you pick on a line determines what option is used. Pick closer to an endpoint to use the **Endpoint** tool. Pick closer to the midpoint to use the **Midpoint** tool.

Figure 5-17.
Double-click on the **Attach** tool to display the **Attachment Settings** dialog box.

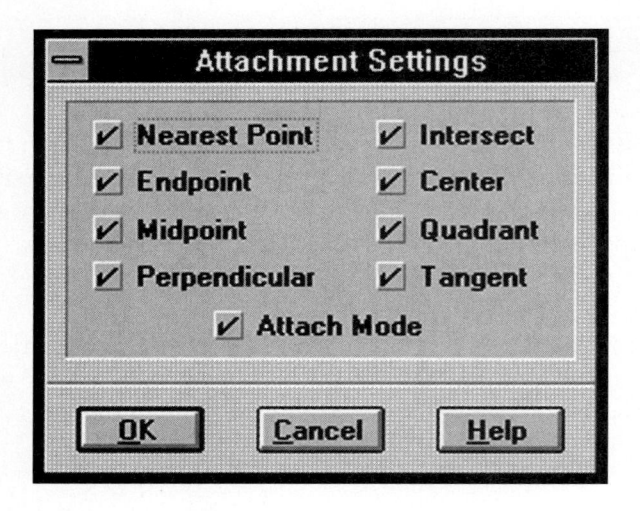

Remember that **Attach** is a toggle function. You can turn all options on or off by picking the **Attach** tool, even in the middle of a drawing command. Also, you can turn on and off individual attach options while using a drawing command.

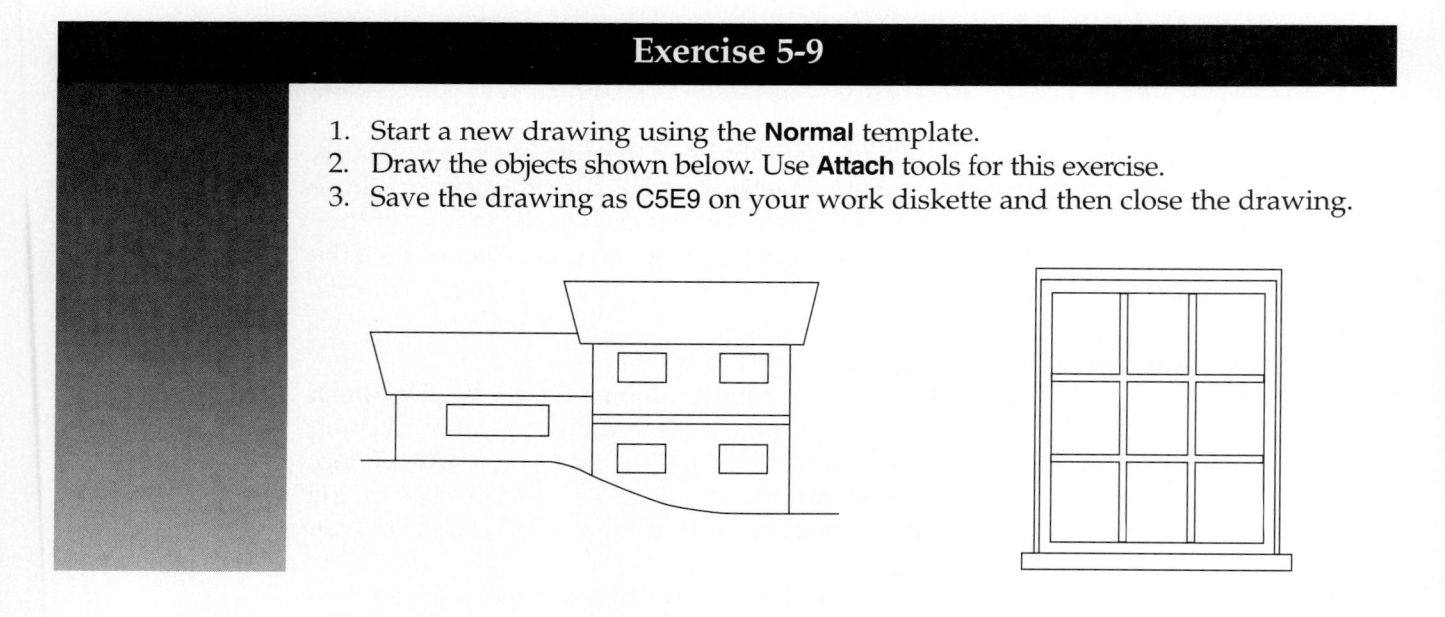

Exercise 5-9

1. Start a new drawing using the **Normal** template.
2. Draw the objects shown below. Use **Attach** tools for this exercise.
3. Save the drawing as C5E9 on your work diskette and then close the drawing.

Rulers

Rulers appear along the top and left sides of the drawing area. The rulers show the size of the current view using the current unit of measurement. If you change the view of your drawing, the rulers also change. If you do not want to use the rulers, you can turn them off. Select the **Preferences** tool from the **Custom** menu. Click on the box next to **Show Rulers,** Figure 5-18. The check mark disappears. Click **OK.** The rulers are no longer shown.

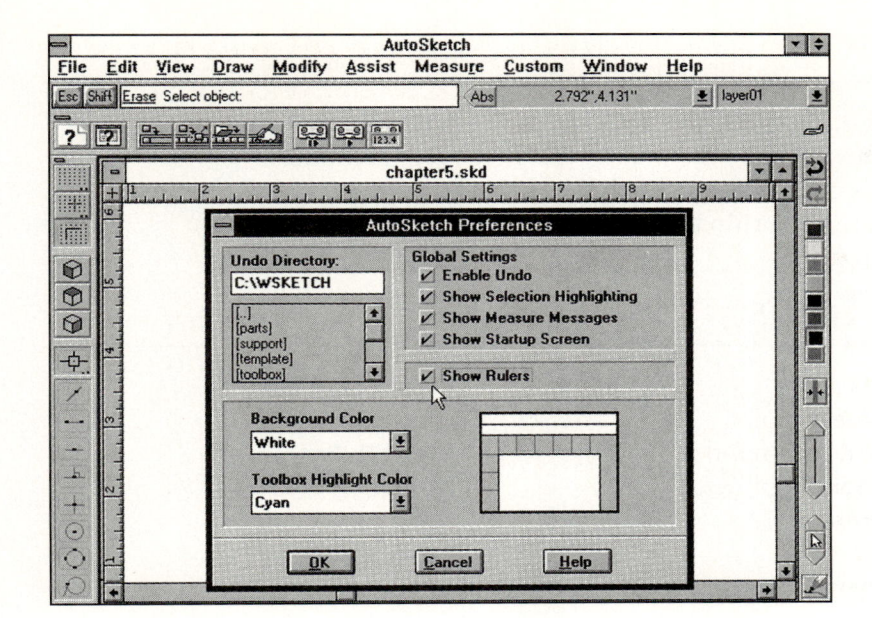

Figure 5-18.
If you do not use rulers, turn them off by clicking on the **Show Rulers** box. The check mark will disappear.

Summary

Most designs require that you draw objects at precise locations using exact dimensions. AutoSketch provides precision drawing aids that let you type in coordinate values and pick precise points with your mouse.

AutoSketch and all other CAD systems use a standard method of point location called the Cartesian coordinate system. You can type in absolute, relative, or polar coordinate values when AutoSketch asks for a point. Absolute coordinates measure distance from the origin. Relative and polar coordinates measure distance from the previous point.

AutoSketch provides several drawing aids to help you draw accurately with the cursor. Two aids that work well together are **Grid** and **Snap**. A grid is a pattern of dots on-screen. This is used much like graph paper. The grid alone does not control movement of the cursor. **Snap** is an invisible snap grid. This allows you to draw objects in precise increments. The snap spacing can be the same as the grid spacing, or it can be a different value.

Another useful drawing aid is **Attach,** This helps you connect to existing objects. When **Attach** is on, a point you pick on a new object will "attach" to one of eight places on an existing object. This can be a center, endpoint, intersection, midpoint, nearest point, perpendicular, quadrant, or tangent. You can make sure that objects meet precisely by using **Attach.**

Three other drawing aids are the **Coordinate Display, Ortho,** and **/lpoint** The **Coordinate Display** shows the coordinate location of the cursor. **Ortho** lets you draw only horizontally and vertically. A new object is connected to the previous point by entering **/lpoint** on the prompt line, rather than using the cursor or coordinates. The **/lpoint** option means last point.

Important Terms

Absolute coordinates
Attach
Cartesian coordinate system
Drawing aids
Grid
Origin
Orthogonal
Polar coordinates
Relative coordinates
Snap grid
Tangent
Toggle modes
X axis
Y axis

New Commands and Tools

Attach mode
Center tool
Change Origin button
Coordinate Type button
Coordinate Window button
Endpoint tool
Expand
Grid tool
Intersect tool
/lpoint command
Midpoint tool
Nearest tool
Ortho tool
Perpendicular tool
Preferences tool
Quadrant tool
Show Rulers button
Shrink,
Snap tool
Tangent tool

Review Questions

Give the best answer for each of the following questions.

1. AutoSketch and other CAD systems use a standard method of point location called the _____.
2. The X and Y axes meet at the _____.
3. When can you type in coordinate values? _____
4. In AutoSketch, an absolute coordinate value is entered as _____.
5. In AutoSketch, a relative coordinate value is entered as _____.
6. In AutoSketch, a polar coordinate value is entered as _____.
7. Relative and polar coordinates measure distance from _____.
8. Explain why you might use several point entry methods during a drawing session.
   _____
   _____
   _____
   _____
9. When entering coordinates, how do you enter units other than the current unit of measurement?
   _____
10. Entering **/lpoint** when asked for a point location will locate the point at _____.
11. Which button is used to change the origin of the drawing and where is this button found?
    _____
12. Functions that help you draw accurately with the cursor are called _____.
13. By default, the **Coordinate Display** shows the cursor location as a(n) _____ coordinate.
14. The **Ortho** mode restricts your drawing to what two directions? _____
15. True or false. A grid restricts cursor movement to precise increments.
16. Define X spacing and Y spacing as they apply to the **Grid** and **Snap** commands.
    _____
    _____
    _____
    _____

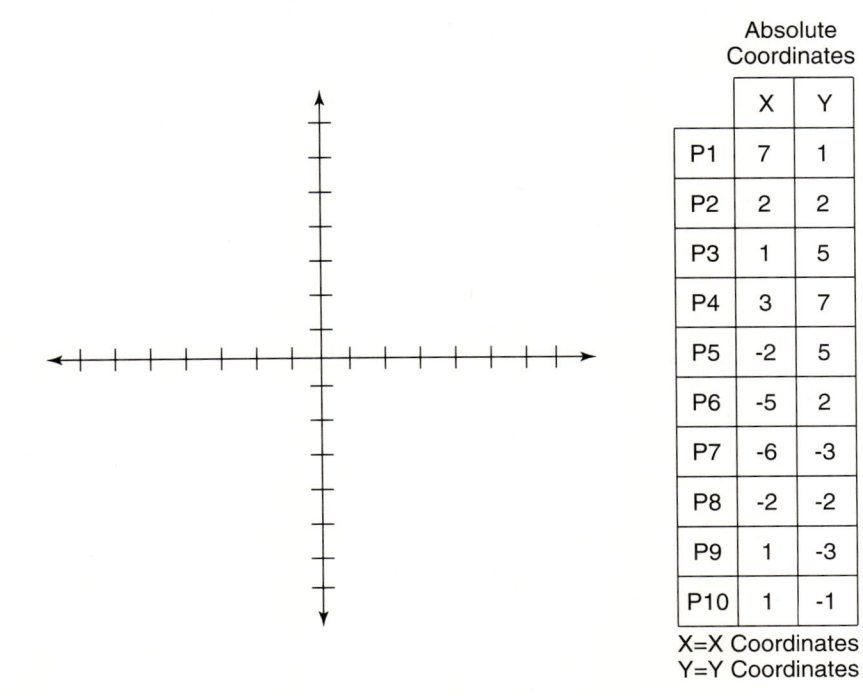

17. Describe two guidelines that determine what values you enter for the grid and snap spacing.
    _____
18. True or false. When **Snap** is on, a point you pick will be placed at the tip of the arrow cursor.
19. List the eight locations where **Attach** tools helps you to connect.
    A. _____
    B. _____
    C. _____
    D. _____
    E. _____
    F. _____
    G. _____
    H. _____
20. Why do some **Attach** modes need to be turned off so that others will work?
    _____
    _____
    _____

Activities

Here is a mixture of written and drawing activities to develop your skills using drawing aids. For the written activities, place your answers on a separate sheet of paper. For each activity using AutoSketch, select the **New...** command from the **File** menu. Save your AutoSketch drawings on your work diskette using the name given. After saving each activity, select the **Close** command from the **File** menu to clear the drawing from the screen.

1. Place a dot on the Cartesian coordinate axes shown below to locate points P1 through P10. The values are given as absolute Cartesian coordinates.

	Absolute Coordinates	
	X	Y
P1	7	1
P2	2	2
P3	1	5
P4	3	7
P5	-2	5
P6	-5	2
P7	-6	-3
P8	-2	-2
P9	1	-3
P10	1	-1

X=X Coordinates
Y=Y Coordinates

2. Record the absolute X and Y Cartesian coordinate values for the points labeled P1 through P10. Place the values in the chart below.

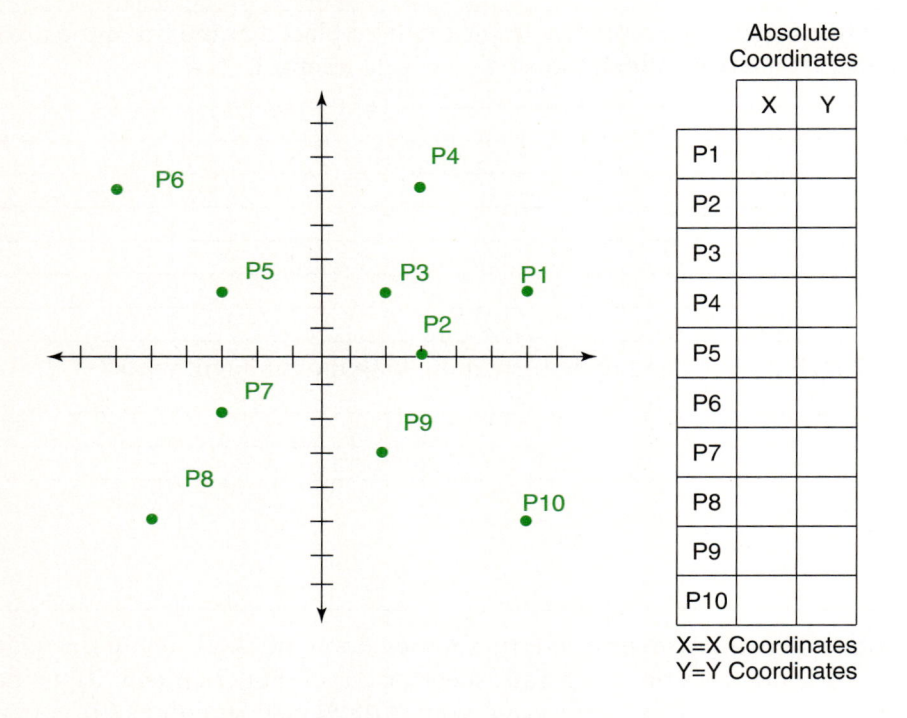

	Absolute Coordinates	
	X	Y
P1		
P2		
P3		
P4		
P5		
P6		
P7		
P8		
P9		
P10		

X=X Coordinates
Y=Y Coordinates

3. Starting at point P1 with an absolute coordinate of (2,2), list the relative X and Y coordinate values to travel from point to point. Go to each point in order, starting with point P2 through P10. Place your answers in the chart below. The gray lines show units of the Cartesian coordinate system.

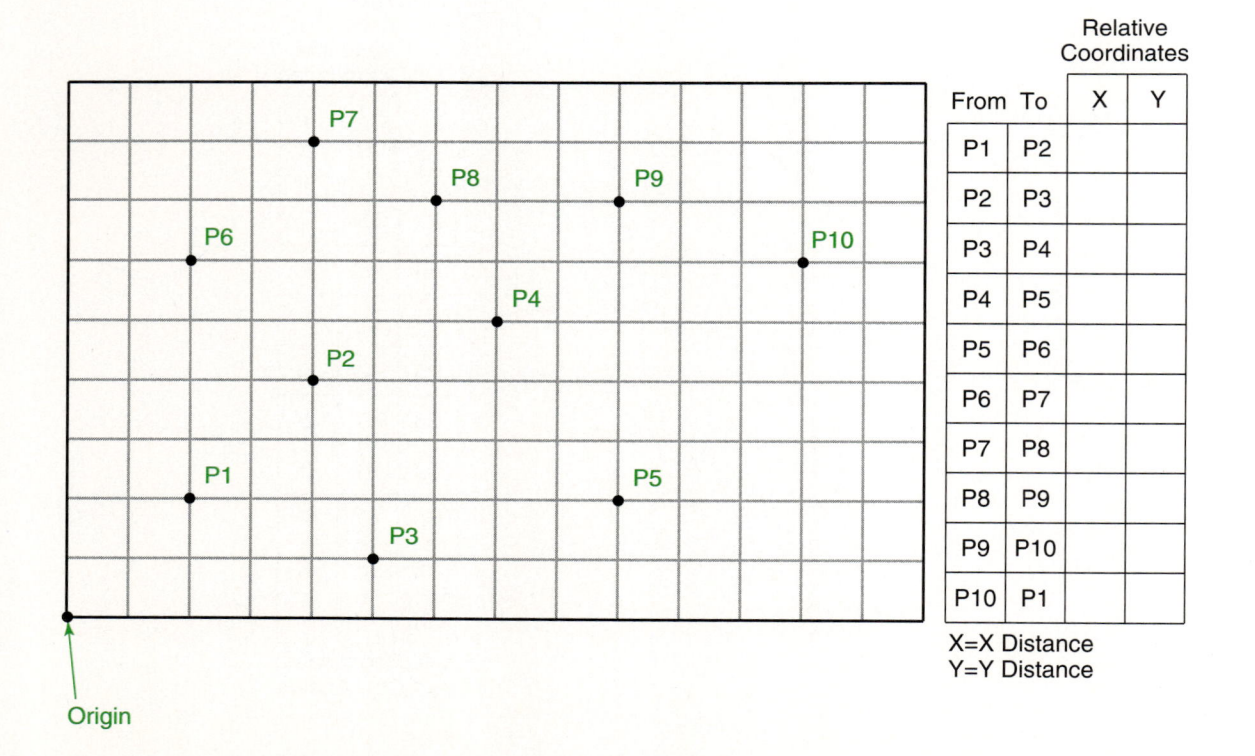

		Relative Coordinates	
From	To	X	Y
P1	P2		
P2	P3		
P3	P4		
P4	P5		
P5	P6		
P6	P7		
P7	P8		
P8	P9		
P9	P10		
P10	P1		

X=X Distance
Y=Y Distance

Origin

4. Place a dot to show angle and distance for each of the polar coordinates given in the chart below. Label each point with the proper name (such as P1). Measure all of the points from the center.

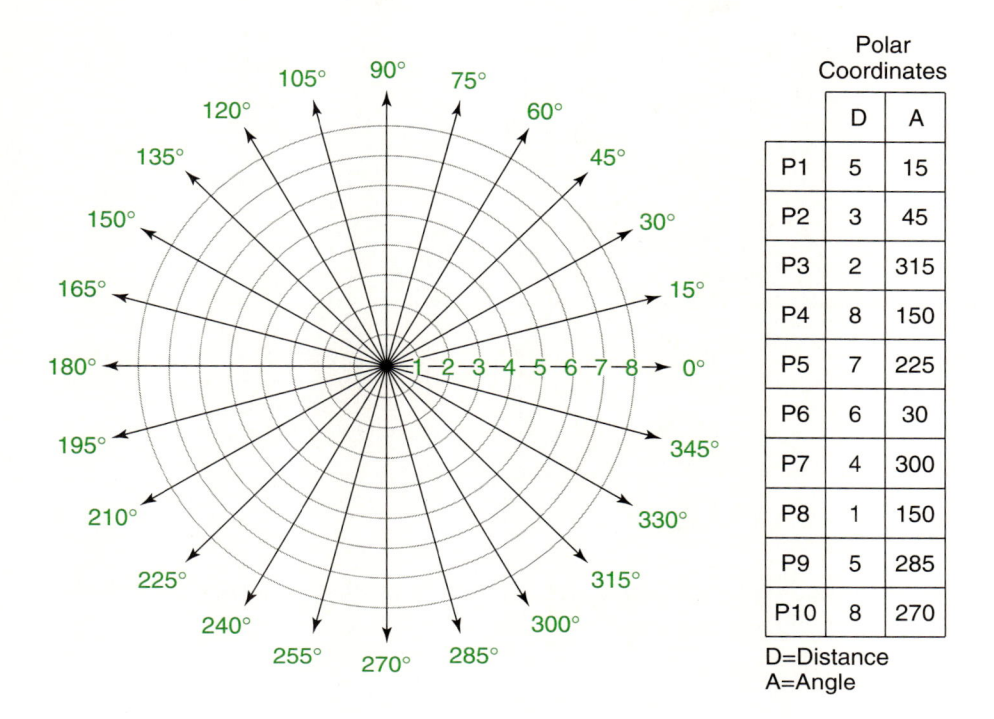

Polar Coordinates

	D	A
P1	5	15
P2	3	45
P3	2	315
P4	8	150
P5	7	225
P6	6	30
P7	4	300
P8	1	150
P9	5	285
P10	8	270

D=Distance
A=Angle

5. Starting at point P1 with an absolute coordinate of (0,0), list the polar coordinate values to travel from point to point. Go to each point in order, starting with point P2 through P4. Place your answers in the chart below.

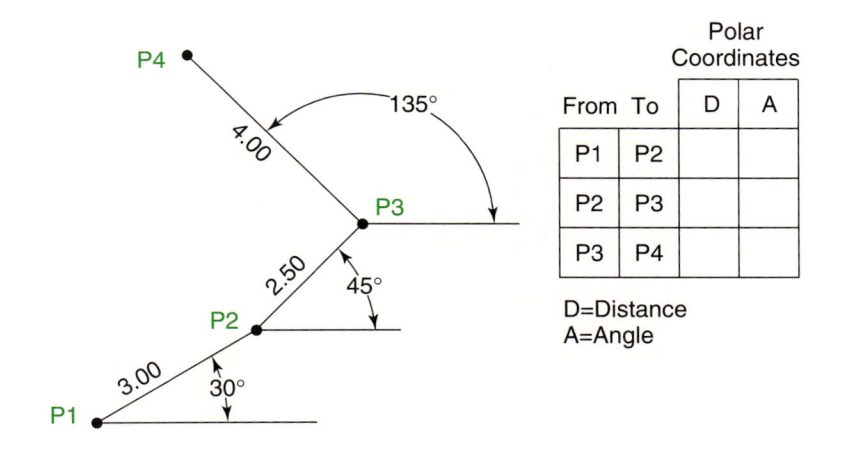

Polar Coordinates

From	To	D	A
P1	P2		
P2	P3		
P3	P4		

D=Distance
A=Angle

6. Given the dimensioned drawing below, identify the absolute coordinates of the points labeled P2 through P5. Place your answers in the chart below.

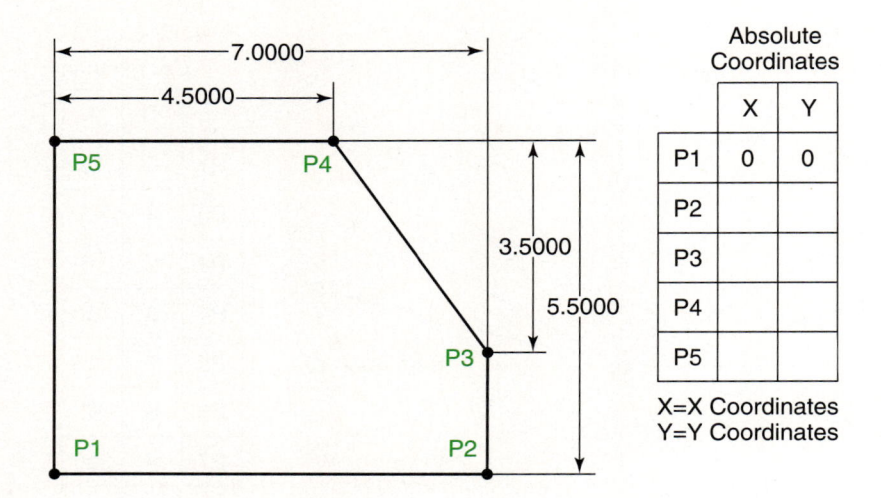

Absolute Coordinates

	X	Y
P1	0	0
P2	7	0
P3	7	2
P4	4.5	5.5
P5	0	5.5

X=X Coordinates
Y=Y Coordinates

7. Given the dimensioned drawing in Activity 6, list the relative X and Y coordinate values to travel point to point. Place your answers in the chart below.

Relative Coordinates

From	To	X	Y
P1	P2	7	0
P2	P3	0	2
P3	P4	-2.5	3.5
P4	P5	-4.5	0
P5	P1	0	-5.5

X=X Distance
Y=Y Distance

8. Given the dimensioned drawing below with P1 at (0,0), identify the coordinates of the points labeled P2 through P4. Use polar coordinate values. Place your answers in the chart below.

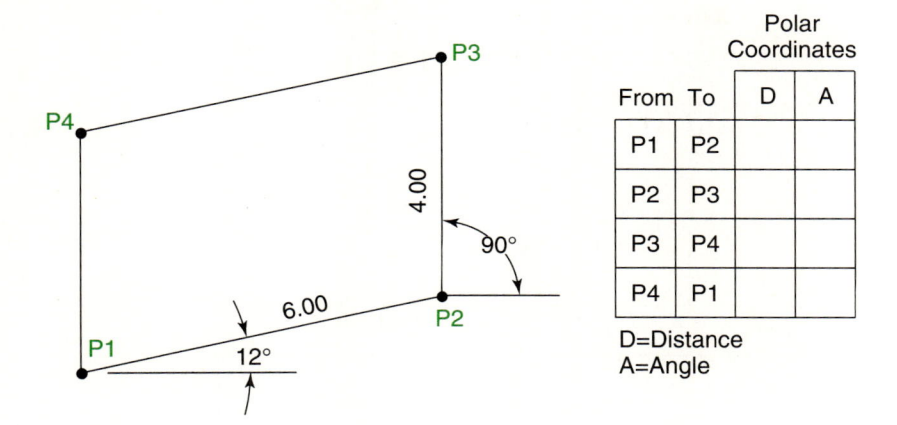

From	To	D	A
P1	P2		
P2	P3		
P3	P4		
P4	P1		

D=Distance
A=Angle

9. Start a new AutoSketch drawing. Select the **Quick Start** template. Select **Inches (decimal)** as the unit of measurement. Create the drawing shown below. Save the drawing as C5A9.

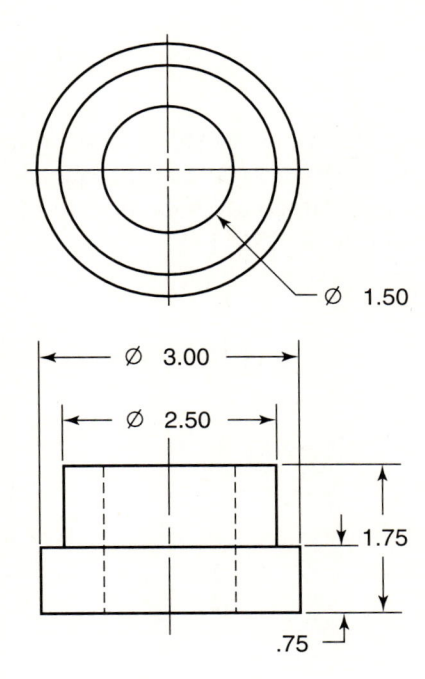

10. Start a new AutoSketch drawing. Select the **Quick Start** template. Select **Inches (decimal)** as the unit of measurement. Make sure that **Grid** and **Snap** are off. Draw the objects shown below in order, from screen A to E. Use **Attach** tools to connect the objects. Save the drawing as C5A10.

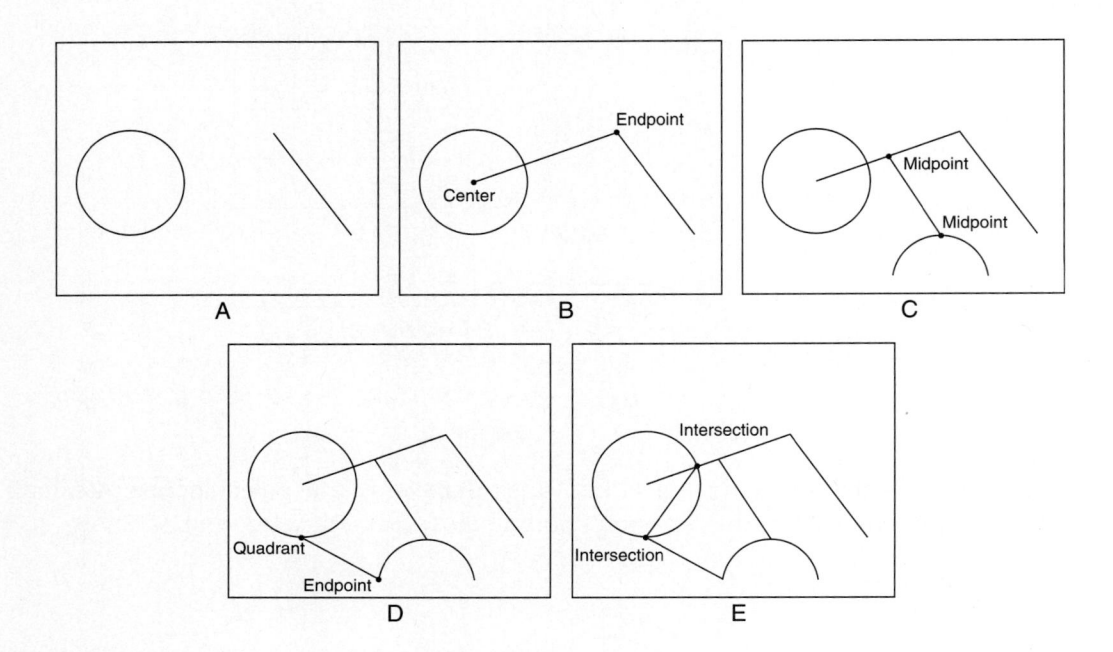

11. Start a new AutoSketch drawing. Select the **Quick Start** template. Select **Inches (decimal)** as the unit of measurement. Make sure that the **Grid** and **Snap** are off. Draw the objects shown below in order, from screen A to D. Use **Attach** tools to connect the objects. Draw the objects with the linetype shown. Save the drawing as C5A11.

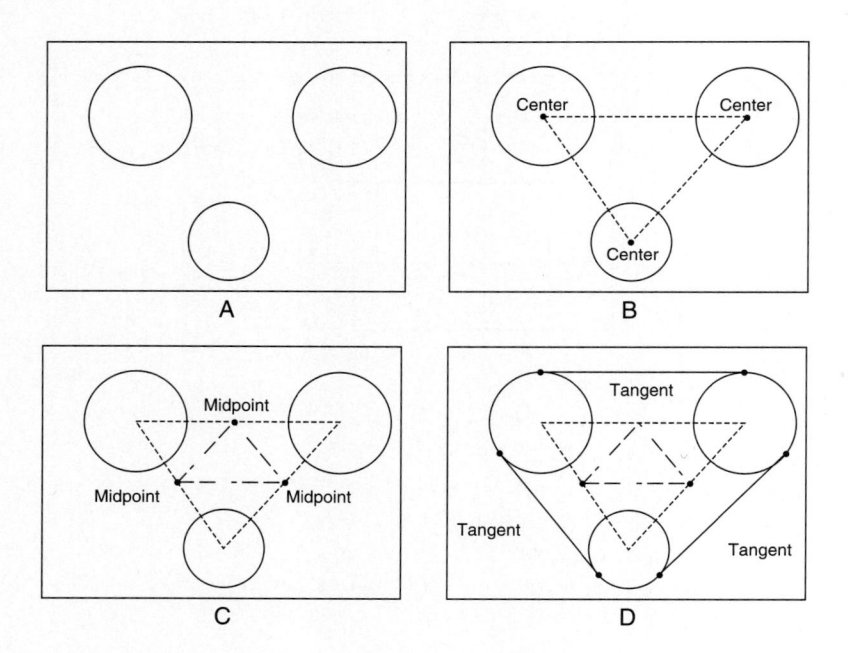

12. Start a new AutoSketch drawing. Select the **Quick Start** template. Select **Inches (decimal)** as the unit of measurement. Make sure that the **Grid** and **Snap** are off. Refer to the drawings below and perform the following procedures. Draw a line. Then, draw a circle using the **Center, Point** option. Connect a new line from where you picked the point on the circle to the endpoint of the line using **/lpoint.** Save the drawing as C5A12.

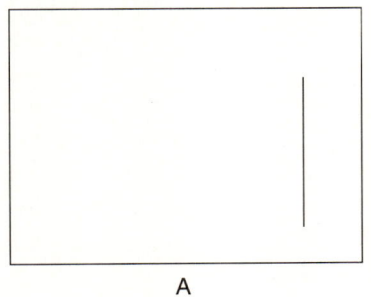

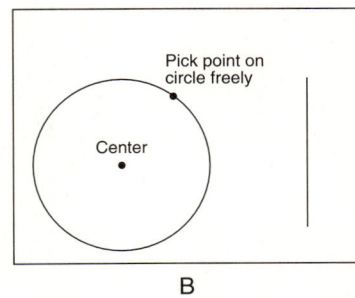

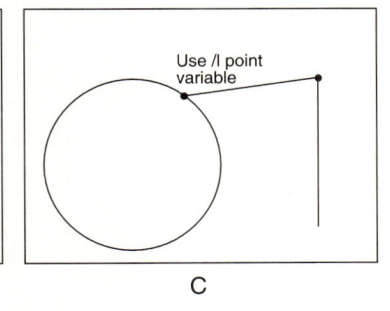

A

B — Pick point on circle freely; Center

C — Use /l point variable

13. Start a new AutoSketch drawing. Select the **Quick Start** template. Select **Inches (decimal)** as the unit of measurement. Make sure that **Grid** and **Snap** are off. Use **Ortho** to draw horizontal lines using all linetypes offered by AutoSketch. In addition, draw the lines using all eight colors in the **Color Palette,** even if you don't have a color monitor. Save the drawing as C5A13.
14. Start a new AutoSketch drawing. Select the **Quick Start** template. Select **Inches (decimal)** as the unit of measurement. Make sure that **Grid** and **Snap** are off. Refer to the drawing below as you perform the following procedures. Draw four concentric circles (circles that share the same center point). Make the smallest circle 1 unit in diameter. Increase the diameter 1 unit for each larger circle. Save the drawing as C5A14.

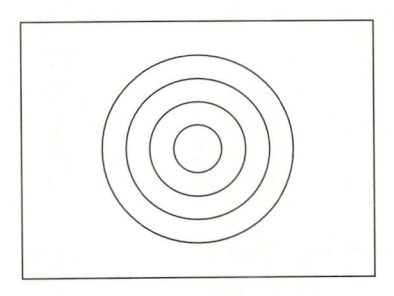

15. Start a new AutoSketch drawing. Select the **Quick Start** template. Select **Inches (decimal)** as the unit of measurement. Make sure that **Grid** and **Snap** are off. Refer to the drawing below as you perform the following procedures. Draw a 3 unit square box. Without typing coordinates or using **Grid** or **Snap,** draw a circle inside the square so that the circle touches all four sides. How did you do it? Save the drawing as C5A15. (Hint: You will have to draw one extra line and erase it later.)

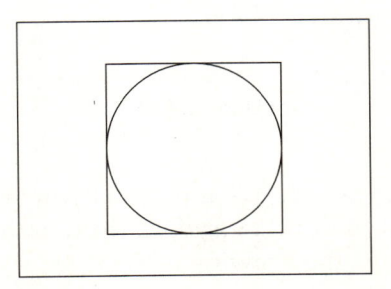

16 - 17. For each activity, start a new AutoSketch drawing and select the **Quick Start** template. Select **Inches (decimal)** as the unit of measurement. Draw the objects shown in Activities 6 and 8 by entering coordinates. Save the drawings as C5A16 and C5A17.

18 - 20. For each activity, draw the design or object shown below using drawing aids. **Grid** and **Snap** are especially useful. Draw the dimensioned drawings using exact measurements. Either type in coordinates or use your cursor with drawing aids. Save the drawings using the filenames given below.

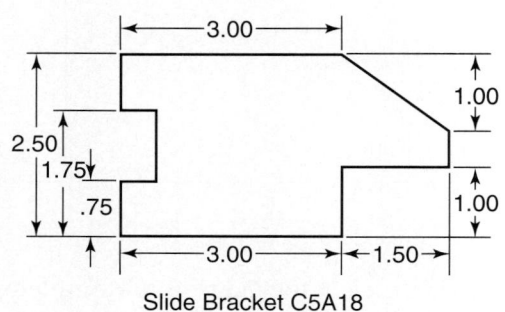

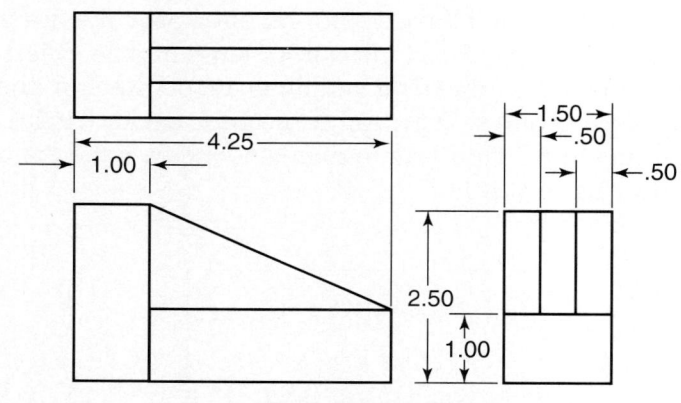

Slide Bracket C5A18

Clamp Spacer C5A19

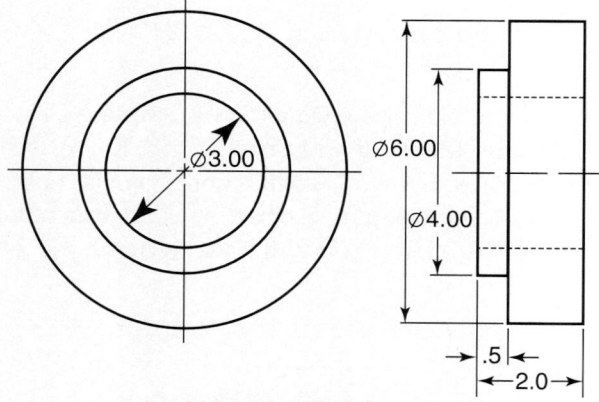

Motor Hub C5A20

21. In Chapter 4, you made rough sketches of a product you would like to invent or see built. Recall the drawing named C4A5 from your work disk. Refine your design using precise measurements. When finished, save the drawing as C5A21.

CHAPTER 6

Modifying and Editing Objects

Objectives
After studying this chapter, you will be able to:
- Choose objects to edit using selection sets and groups.
- Move objects to another location on the drawing.
- Make one or more copies of objects.
- Rotate objects.
- Make a mirror copy of objects.
- Scale objects.
- Change properties of existing objects.

Editing is the process of changing objects you have drawn. You will probably spend more time editing drawings than you will creating them. Drafting departments spend vast amounts of time editing. Most designs pass through many revisions before they are complete. As one drafter said, "There may be one way to create a line, but there are 100 ways to change it." One of the benefits of AutoSketch is that editing is quite easy.

AutoSketch provides a variety of editing functions to suit every situation. You can erase, move, copy, rotate, mirror, and scale one or several objects at once. You can also change an object's properties, such as color and linetype. The commands to do these tasks are found in the **Modify** and **Edit** pull-down menus and toolboxes, Figure 6-1. This chapter will cover basic editing commands.

Choosing Objects to Edit

In traditional drafting, you use an eraser to make changes. In AutoSketch, you select an editing command and the objects to edit. Chapter 4 showed you how to erase objects. You can erase objects one at a time, and with a crosses/window box. You will use these same selection techniques to select objects to edit. You can also use selection sets and groups to select objects.

Selecting Objects with Selection Sets

A *selection set* is a temporary group of objects. You can modify or edit everything in the group at the same time. Unlike the crosses/window box, a selection set does not have to contain all objects in a given area. You pick each object and "build" the selection set.

To create a selection set, first select a tool that requires you to select objects. When you select the tool (**Copy** in this case), a prompt will appear asking you to select an object.

Copy Select object:

Figure 6-1.
Commands to edit AutoSketch objects are found in both the **Modify** and **Edit** menus. You can choose to use a toolbox or pull-down menu.

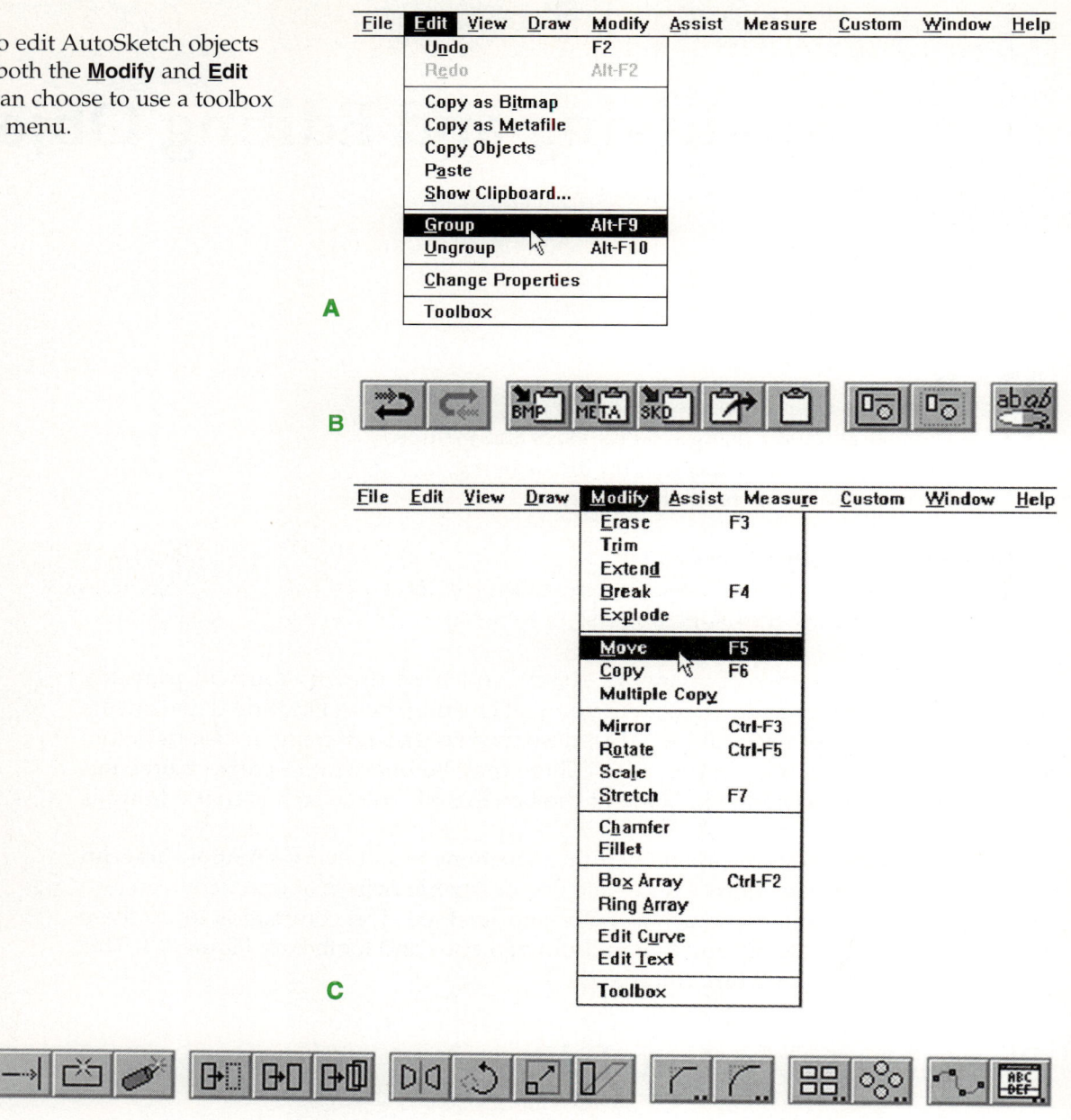

A

B

C

D

At this point, there are two ways to create a selection set.
- Hold down the [Shift] key as you pick objects
- Pick the **Shift** button on the horizontal status bar as you pick objects

AutoSketch will highlight the objects as you select them. You can also use the crosses/window box to select multiple objects to include in the selection set.

If you accidentally pick an object that should not be in the selection set, pick it again to remove it. All other objects in the selection set remain selected. When finished selecting objects, release the [Shift] key or pick the **Shift** button again. The objects stay selected, and you can continue your command. However, if you press the [Esc] key or choose another tool to end the command, the selection set is canceled.

Before starting the exercises or activities in this chapter, set your printer for

Chapter 6 Modifying and Editing Objects 165

Preparing for Exercises and Activities in This Chapter

landscape orientation. You can do this in the Windows **Control Panel**. You can also use the following procedure.
1. Opening a new drawing with the **Quick Start** template.
2. Select the **File** menu and **Print Setup** command.
3. Highlight the printer and select the **Setup** button.
4. Select **Landscape** for the paper orientation and pick **OK**.
5. Pick **OK** to leave the **Print Setup** dialog box.
6. Close the drawing without saving your changes.

Doing these steps allows your drawing sheet to fit on a single A-size sheet of paper horizontally. You will only need to do this step once, unless you use another Windows application and change the orientation back to portrait.

Exercise 6-1

1. Start a new drawing using the **Quick Start** template. Select **Inches (decimal)** as the unit of measurement.
2. Draw the objects shown below on the left. Create them in about the same size and location.
3. Select objects to erase using a selection set so that the drawing appears as shown on the right. Erase all objects as a group. Do *not* erase the objects one by one.
4. Save the drawing as C6E1 on your work diskette and then close the drawing.

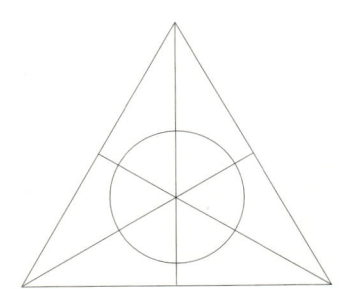

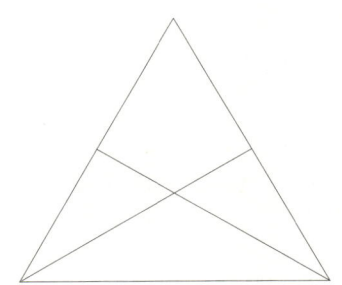

Selecting Objects with Groups

Grouping is another method to select more than one object at a time. A *group* is a collection of objects that is treated as one item, not as individual objects. If you copy the group, all objects that belong to the group are copied. Unlike a selection set, the objects remain together as a group until you tell AutoSketch to ungroup them. When you save a drawing, any groups are also saved. This means that the next time you open that drawing, the objects will still be grouped.

SHORTCUT
You can also access the **Group** tool by holding down the [Alt] key and pressing the function key [F9].

Building a group
To group objects, select the **Edit** menu and the **Group** tool. The following prompt appears.

    Group Select object:

Begin picking objects that make up the group. To pick objects one at a time, hold down the [Shift] key or click on the **Shift** button. You can also select several objects at once with the crosses/window box. Any object chosen to be in the group becomes highlighted on-screen. Picking an object a second time deselects it.

When you are finished picking objects, release the [Shift] key or click on the **Shift** button a second time. You can also choose another tool or simply click on the **Group** tool again. From now on, the group acts as a single object.

A group may contain up to 1000 objects. Each of these items can be an individual drawing object, such as a line, or another group that itself contains up to 1000 objects. Placing groups in other groups is called nesting.

Nesting groups
A *nested group* is a group that is part of another group. You can use nested groups to create more complex shapes and layouts. For example, look at the office plan in Figure 6-2. There are similar items found in each room. In addition, two rooms contain identical desk layouts. Groups, including nested groups, helped create this drawing quickly. The following steps were used to create this layout.
1. Draw the floor plan.
2. Draw the simple shapes such as the plants, telephones, chairs, and computers. Draw only one of each. Then, group each shape individually.
3. Next, draw an executive desk. Copy a telephone and place it on the desk. Also, copy chairs and place them next to the desk. Since these shapes are already grouped, you can pick any part of the group to select the entire shape.
4. Group together the entire executive desk layout. Be sure to include the telephone and chairs.
5. Once the groups are made, layout is easy. Copy the executive desk group to OFFICE 2 and OFFICE 3. In this example, you would need to draw the desk for OFFICE 1 separately because the room is too small for an executive desk and chairs. However, the computer, telephone, plant, and two chairs are easily copied because each is grouped.
6. Finally, copy the plant into the offices and add the utilities in the two restrooms.

Figure 6-2.
Groups and nested groups were used to create this office plan quickly.

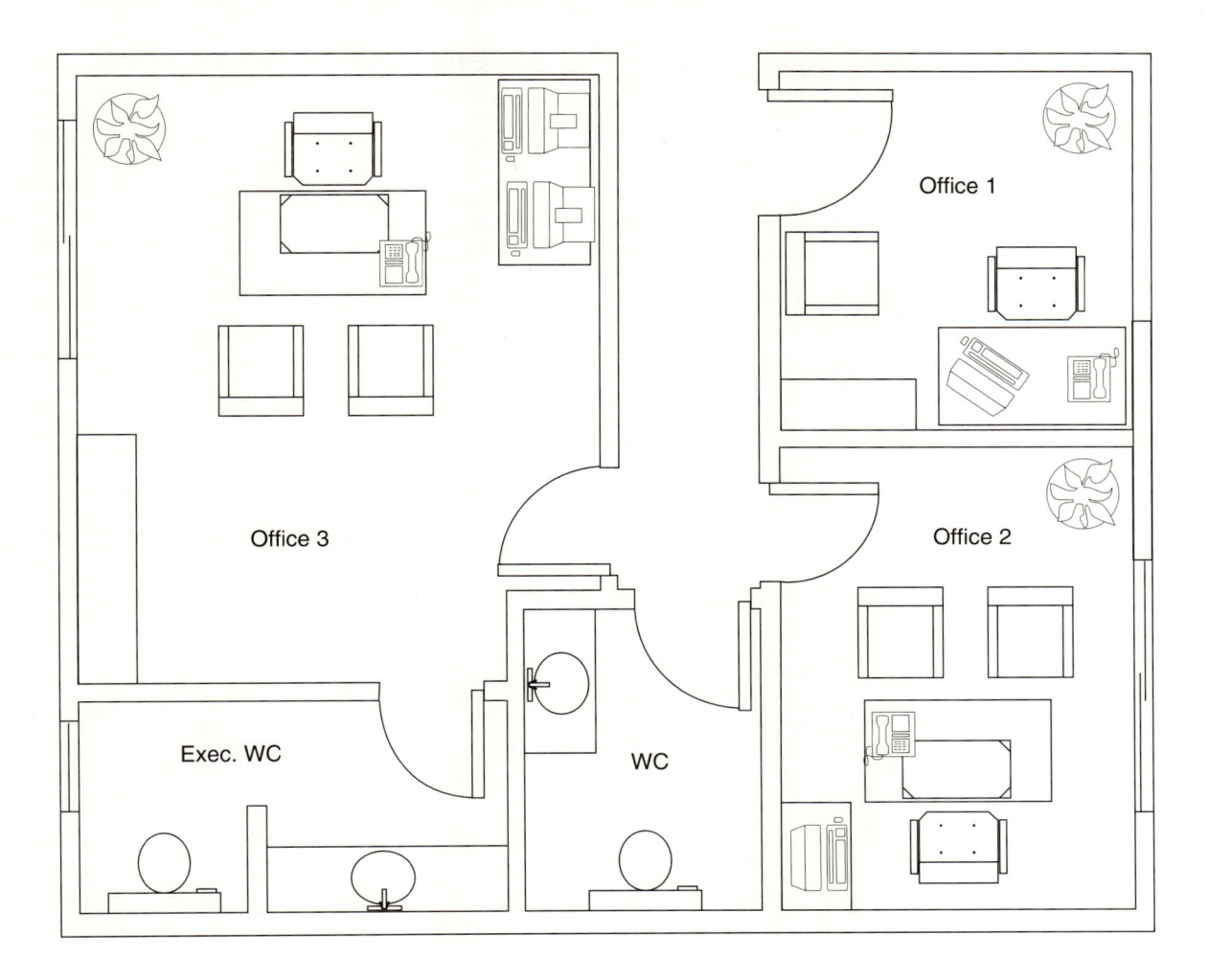

The executive desk group contains nested groups. The telephone and chairs are groups. When the executive desk group was created, the telephone and chair groups became nested.

You can nest groups to a maximum depth of eight. For example, Figure 6-3 shows that the office layout had a nesting depth of two.

Figure 6-3.
A group within group is called a nested group. You can nest groups to a depth of eight. The office plan in Figure 6-2 has two levels of groups.

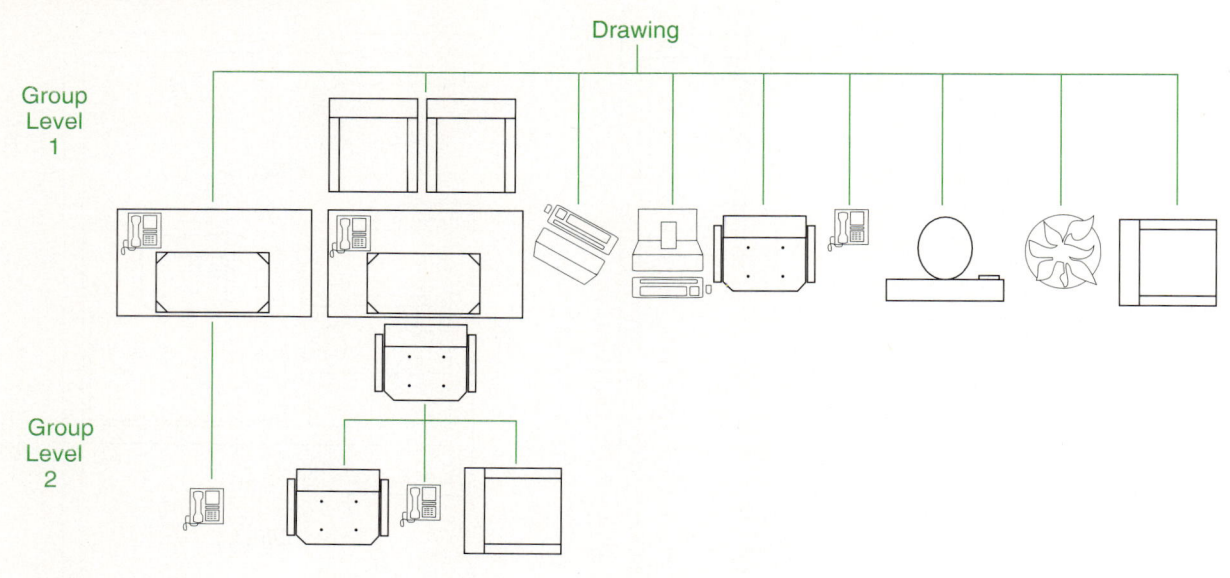

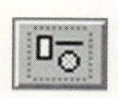

Ungrouping objects

At some point, a group may no longer be useful. For example, you may want to move an object within a group. You cannot work on individual objects until you ungroup them. If you try to select the object, the entire group is chosen instead. The **Ungroup** tool splits a group into its original objects or component groups. Select the **Ungroup** tool, found in the **Edit** menu. The following prompt appears.

Ungroup Select object:

Pick any object in the group. The entire group is then split into its component shapes.

Suppose you want to change the size of the executive desk shown in the previous example. You cannot select just the desk because it was grouped with the telephone and chairs. You must first use the **Ungroup** tool. With the tool active, pick any part of the executive desk layout to ungroup it. Suppose you then want to change the shape of the telephone. Since it was also grouped separately, you must ungroup it to make any changes. The **Ungroup** tool ungroups only one nesting level at a time, from the top down.

Using editing commands on groups

You can erase, move, copy, rotate, scale, and mirror groups just as you do single objects. However, the **Break** tool cannot be used to break an object that belongs to a group. This is because objects that belong to a group lose their identity as single objects. You also cannot chamfer or fillet objects within a group.

You can use the **Box Array, Ring Array** and **Stretch** tools on groups. These tools will be covered in Chapter 8. Also, **Attach** tools will connect to objects within a group.

SHORTCUT

You can also access the **Ungroup** tool by holding down the [Alt] key and pressing the function key [F10].

Why Use Selection Sets and Groups?

You might wonder why a drafter would take the time to group objects rather than use the crosses/window box. Suppose you want to make a copy of the output shaft shown in Figure 6-4. It would be impossible to enclose the objects that make up the output shaft inside a crosses/window box without also selecting parts of the key and spring. In this situation, you can use a selection set or the **Group** tool to select only those objects that make up the shaft. Then you can move, copy, or perform some other command on the group. If you only need to edit it once, use a selection set. If you want the output shaft to remain a single object for future use, group it.

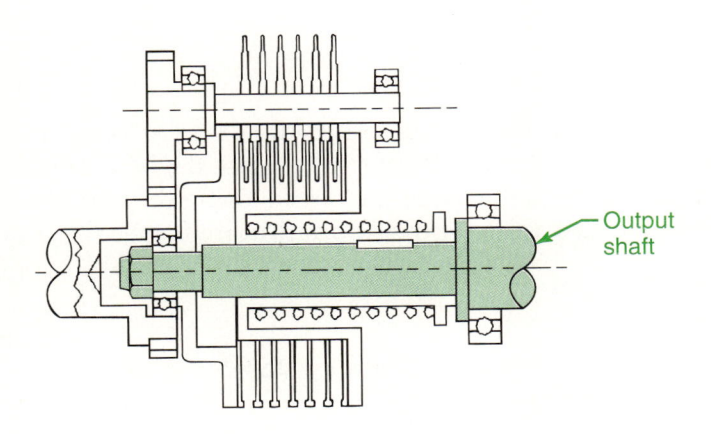

Figure 6-4.
It would be nearly impossible to copy just the output shaft selecting with the crosses/window box. Instead, group the objects or use a selection set. The output shaft can then be easily moved or copied.

Exercise 6-2

1. Start a new drawing using the **Quick Start** template. Select **Inches (decimal)** as the unit of measurement.
2. Draw the objects shown below on the left. Create them in about the same size and location.
3. Create a group of one circle and three vertical lines.
4. Erase the group with a single command.
4. Save the drawing as C6E2 on your work diskette and then close the drawing.

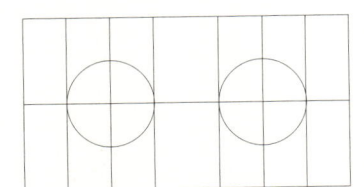

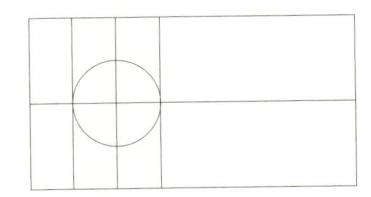

Moving Objects

The **Move** tool lets you move one or more objects to a new location on the drawing. To move objects, select the **Move** tool from the **Modify** menu, Figure 6-5. The following prompt appears.

Move Select object:

Pick the object or group to move. You can select several objects with the crosses/window box or a selection set. AutoSketch then asks you for a starting point for the move.

Move From point:

This first point you pick is a base point. The cursor will hold on to this "base" to drag the objects around on-screen. You can use an **Attach** tool to place the base point at a precise place. For example, you might want to drag the objects using the end of a line or center of a circle. AutoSketch then asks you for the stopping point for the move.

Move To point:

This second point tells where to move the objects. Simply stated, the first point you pick will be "moved" to the second point.

SHORTCUT

You can also access the **Move** tool by pressing the function key [F5].

Figure 6-5.
To move an object with the **Move** tool, first pick a base point. In this example, a corner of the box is used as the base point. As you drag the object to a new location, notice that the cursor drags the object by the base point.

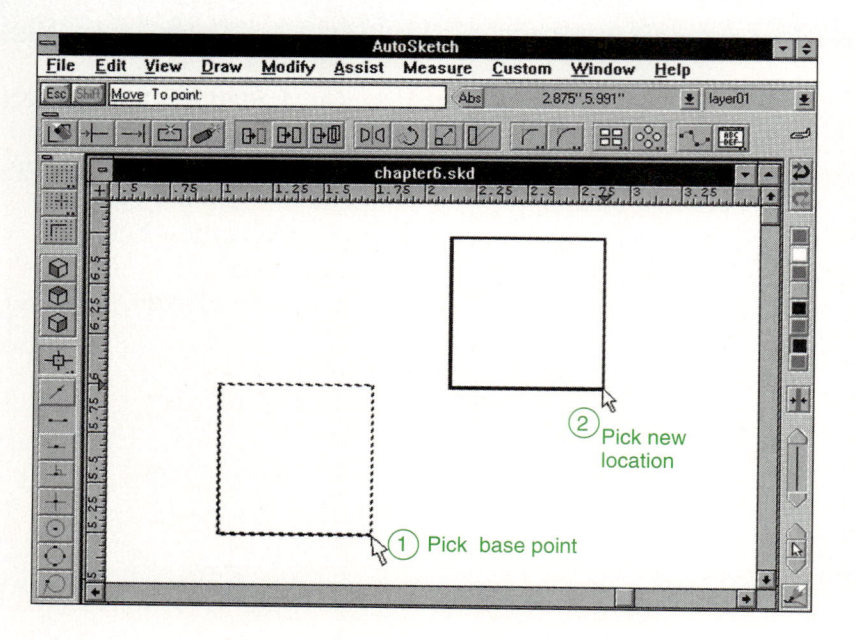

The easiest way to move objects is to drag them across the screen. As you move the cursor, an image of the objects moves along. The image helps you "pull" the objects to their new location. You can use snap or **Attach** tools to position the objects precisely. You can also refer to the **Coordinate Display.** This will show you where the objects are being moved to.

You do not have to drag the objects to their new location. If you know exact distance, enter coordinates of the second point, Figure 6-6. Here is an example of moving objects using a polar coordinate.

<u>Move</u> To point: **P(3,30)**

This tells AutoSketch to move the objects 3 units away from the current position at a 30 degree angle. Remember to press the [Enter] key after typing in coordinates.

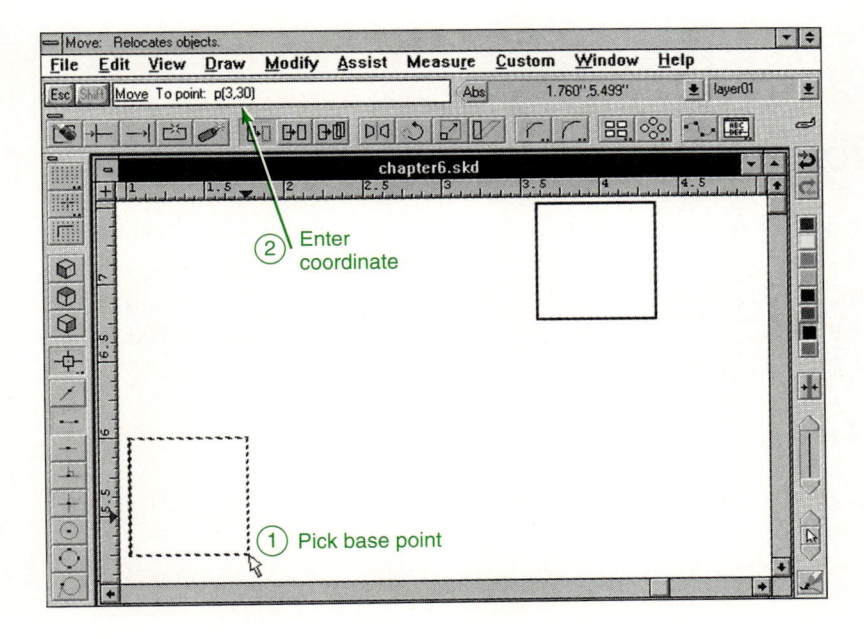

Figure 6-6.
When you know exact distance and direction to move objects, enter a relative or polar coordinate. In this example, a polar coordinate is used to locate the second point.

Exercise 6-3

1. Start a new drawing using the **Quick Start** template. Select **Inches (decimal)** as the unit of measurement.
2. Refer to the drawings shown below.
3. Draw a triangle using the **Line** tool and the **Endpoint** tool.
4. Move the left line by dragging it with the cursor.
5. Move the right line 3 units to the right and 3 units up by entering a relative coordinate.
6. Save the drawing as **C6E3** on your work diskette and then close the drawing.

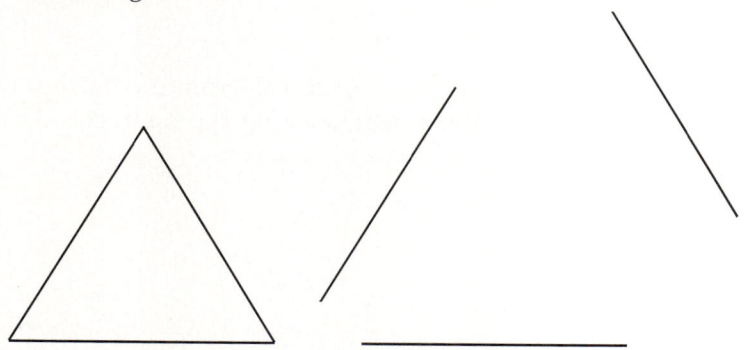

Copying Objects

The **Copy** tool is one of the most used functions of a CAD system. As stated in Chapter 1, you should never draw the same object twice when using CAD. Instead, copy the shape, Figure 6-7. The **Copy** tool works like the **Move** tool. However, the original objects are left unchanged. To copy objects, select the **Copy** tool from the **Modify** menu. The following prompt appears.

Copy Select object:

> **SHORTCUT**
>
> You can also access the **Copy** tool by pressing the function key [F6].

Pick the object or group to copy. You can also select several objects with the crosses/window box or a selection set. AutoSketch then asks for a first point.

Copy From point:

This first point you pick is a base point. The cursor will hold on to this "base" to drag the objects around on-screen. For accuracy, use an **Attach** tool or snap to place the base point. AutoSketch then asks for a second point.

Copy To point:

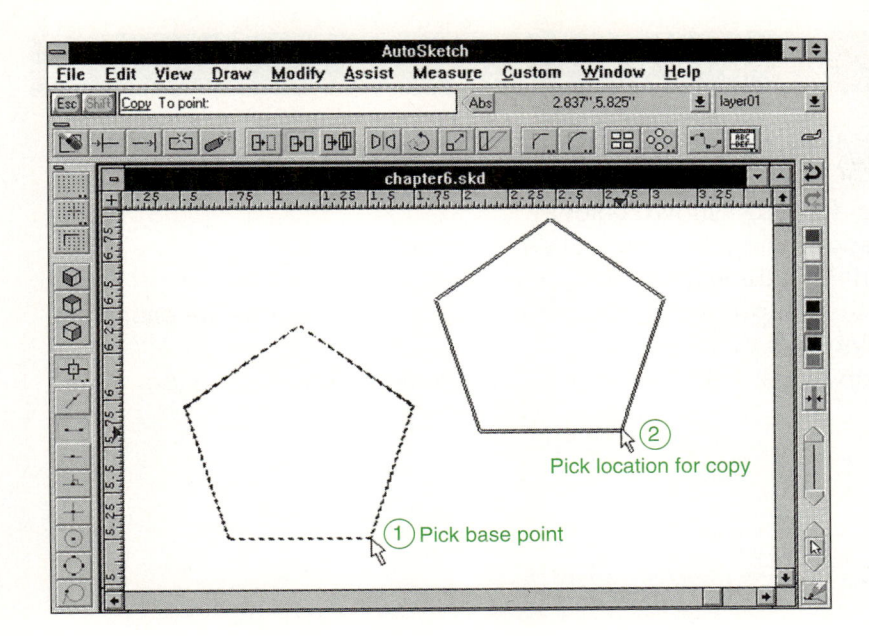

Figure 6-7.
To copy objects with the **Copy** tool, first pick a base point. Then, drag the copy to its location. Remember, you can also type in a coordinate to place the copy.

This second point tells where to copy the objects. Simply stated, all of the objects you selected will be duplicated at the second point. The relationship of the first point to the selected objects will be matched by the second point and the duplicated objects.

The easiest way to copy objects is to drag them across the screen. As you move the cursor, an image of the objects moves along. The image helps you "pull" the objects to their new location. You can use snap, **Attach** tools, or enter a coordinate to position objects precisely. You can also refer to the **Coordinate Display.** This will show you where the objects are being copied to.

The **Copy** tool is useful when you have many identical shapes in your drawing. Draw only one and copy it when needed. This will save a lot of time, especially on complex objects.

You can also use the **Copy** tool to experiment with designs. For example, suppose you want to alter some part of your drawing. Make a copy and work on the copy rather than the original. If you like the revised shape, erase the original. If you don't like your changes, simply erase the copy.

Exercise 6-4

1. Start a new drawing using the **Quick Start** template. Select **Inches (decimal)** as the unit of measurement.
2. Refer to the drawings shown below.
3. Draw a triangle using the **Line** tool and the **Endpoint** tool.
4. Copy the triangle three times as shown below. A corner of each copy should connect to the midpoint of a side of the original triangle. Use the **Midpoint** and **Endpoint** tools to do this.
5. Save the drawing as C6E4 on your work diskette and then close the drawing.

Making Multiple Copies

The **Copy** tool ends after you make one copy. If you want to make another copy, you must repeat the entire process, including reselecting the objects. If you need more than one copy of an object, use the **Multiple Copy** tool.

The **Multiple Copy** tool works like the **Copy** tool. However, after picking the second point you can continue to pick locations for more copies, Figure 6-8. If you enter relative or polar coordinates, the copies are placed relative to the previous copy. They are not placed relative to the original object(s) you selected. To end the command, you must press the [Esc] key or click on the **Esc** button.

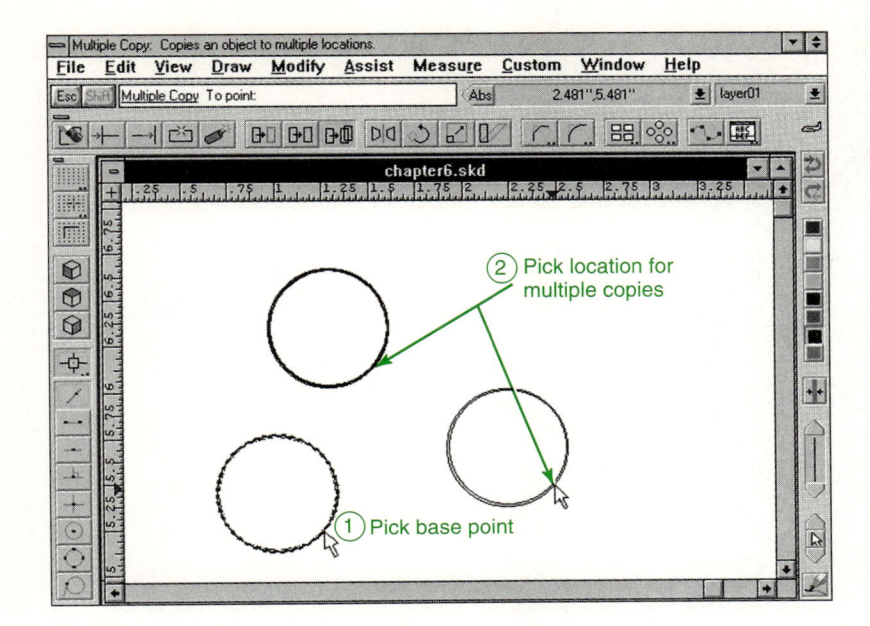

Figure 6-8.
To make multiple copies, use the **Multiple Copy** tool. First, pick a base point. Then, drag each copy to its location. Remember, you can also type in a coordinate to place copies relative to the previous one.

Rotating Objects

The **Rotate** tool is used to rotate one or more objects. To rotate objects, select the **Rotate** tool from the **Modify** menu. The following prompt appears.

Rotate Select object:

Pick the object or group to rotate. You can also select several objects with the crosses/window box or a selection set. AutoSketch then asks for the center of rotation.

Rotate Center of rotation:

Pick the point that the objects should rotate around. This is the *center of rotation.* The position of this point is important, Figure 6-9. If a shape must rotate around part of an existing object, use an **Attach** tool to pick the point. You may have to turn snap off to pick this point precisely. AutoSketch then asks for the angle of rotation.

Rotation angle:

SHORTCUT
You can also access the **Rotate** tool by holding down the [Ctrl] key and pressing the function key [F5].

Figure 6-9.
Objects rotate about a center point that you pick. Notice how the rectangle rotates depending on where the center point is chosen.

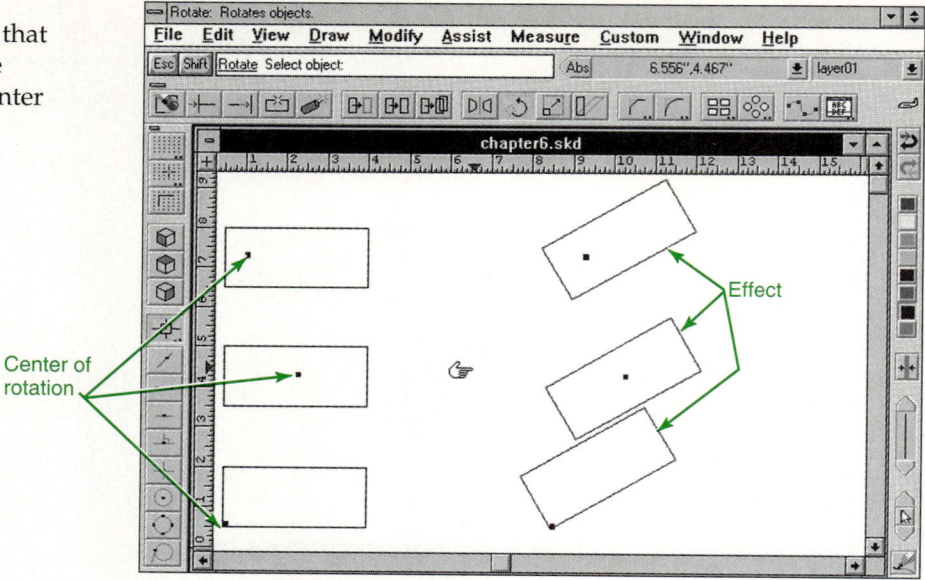

You can use the cursor to pick the angle. As you move the cursor, a temporary image will rotate around the center. Also, a readout in the prompt box shows the angle of rotation, Figure 6-10. If the **Ortho** tool is active, the cursor can only select angles of 0, 90, 180, and 270 degrees. Also, if **Snap** is on, the cursor will move only in snap increments. To enter an exact rotation, type an angle value in the prompt box.

Rotate Rotation angle: **30**

This entry will rotate the objects 30 degrees counterclockwise. Note that all objects selected rotate as a group around the pivot point.

Figure 6-10.
As you move the cursor to rotate selected objects, the angle of rotation appears in the prompt box.

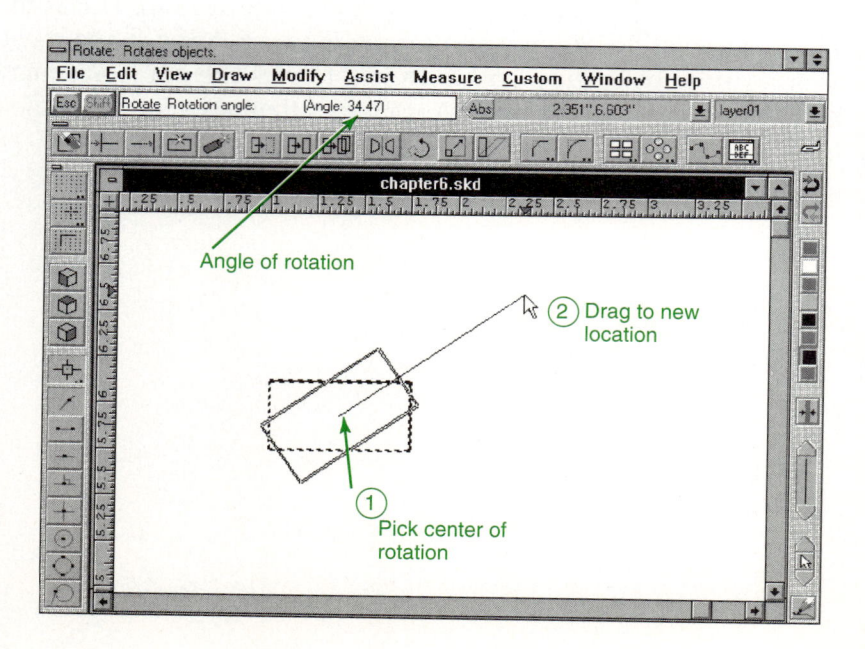

Exercise 6-5

1. Start a new drawing using the **Quick Start** template. Select **Inches (decimal)** as the unit of measurement.
2. Draw two boxes approximately the same size and place them end-to-end, as shown below.
3. Rotate the left box 90 degrees (one-quarter turn) by picking a corner as the center point and rotating the box with your cursor.
4. Rotate the right box by picking a center of rotation near the center of the box and entering a rotation angle.
5. Save the drawing as C6E5 on your work diskette and then close the drawing.

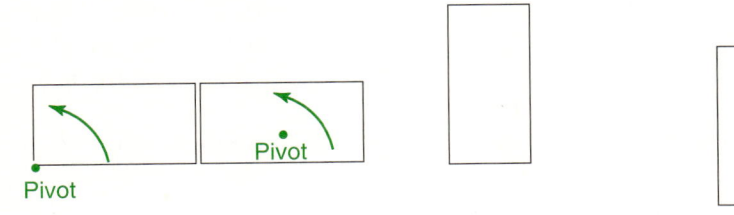

Mirroring Objects

The **Mirror** tool makes a mirror image copy of objects, Figure 6-11. The copied objects are placed on the opposite side of an invisible *mirror line.* The **Mirror** tool is often used for designs that are symmetrical. *Symmetrical* means one half is a reflection of the other half. To mirror objects, select the **Mirror** tool from the **Modify** menu. The following prompt appears.

Mirror Select object:

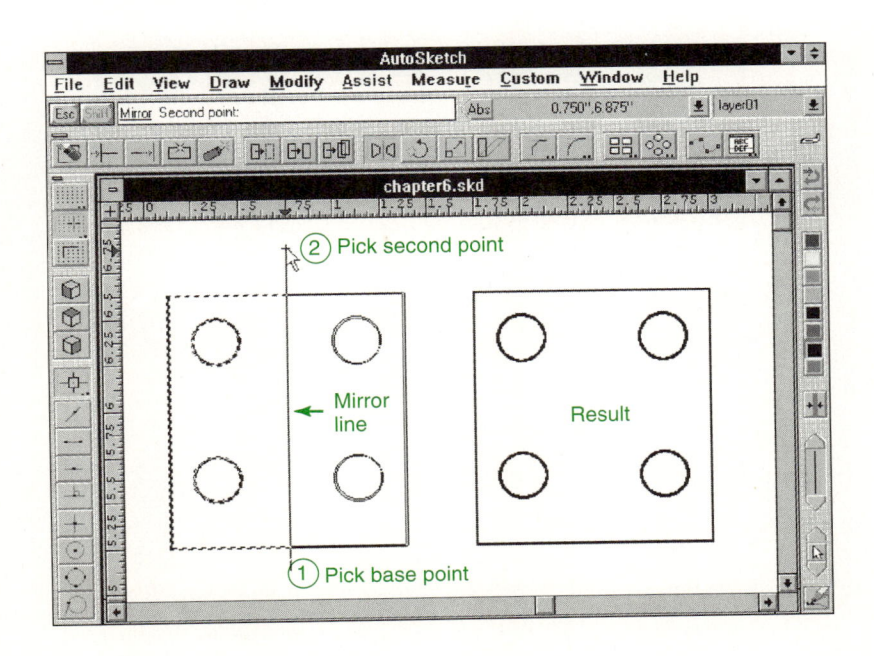

Figure 6-11.
To mirror objects using the **Mirror** tool, first select the objects to mirror. Then, pick two endpoints for the mirror line. The objects are reflected and copied about the mirror line.

Select the object to mirror. You can select multiple objects using the crosses/window box or a selection set. AutoSketch then asks you for a base point.

Mirror Base point:

This base point is the first end of the mirror line. You can use snap, an **Attach** tool, or a coordinate to place the base point at a precise location. AutoSketch then asks you for the second point of the mirror line.

Mirror Second point:

As you move the cursor, an image of the mirrored objects drags and rotates with it. The image helps you position the mirror image. You might use snap or an **Attach** tool to position the mirror line precisely. You can also enter coordinates for the base and second points of the mirror line. Once you select the second point, the mirrored objects are placed.

All objects can be mirrored. Text is mirrored differently than other objects. When you mirror text, it still reads from left to right. The letters are not mirrored. However, text that is left or right justified will switch justification. Text that is center justified remains the same.

> **SHORTCUT**
>
> You can also access the **Mirror** tool by holding down the [Ctrl] key and pressing the function key [F3].

Exercise 6-6

1. Start a new drawing using the **Quick Start** template. Select **Inches (decimal)** as the unit of measurement.
2. Create the drawing shown below on the left.
3. Use the **Mirror** tool twice to create the completed drawing shown below on the right. Do not draw any new objects. You may need to use **Attach** tools to place the mirror lines precisely.
4. Save the drawing as **C6E6** on your work diskette and then close the drawing.

Scaling Objects

The **Scale** tool lets you enlarge or reduce the size of existing objects, Figure 6-12. The size is scaled proportionally, meaning that the object is not distorted. The **Stretch** tool is used to resize an object in one direction only. This tool is covered in Chapter 8.

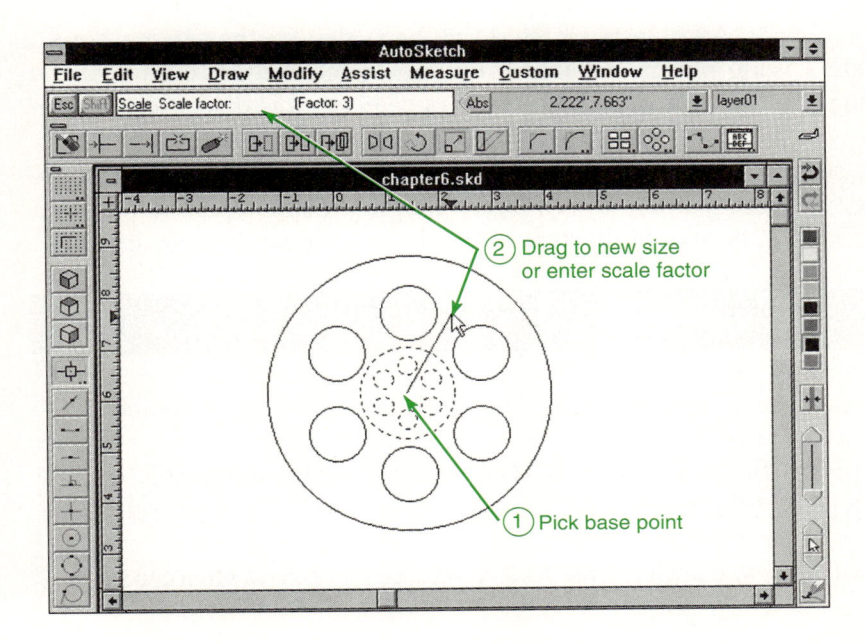

Figure 6-12.
To scale objects using the **Scale** tool, first select the objects to scale. Then, select a base point. Next, move your cursor in or out to change the size. The base point remains fixed. The scale factor appears in the prompt box.

To scale an object, select the **Scale** tool from the **Modify** menu. The following prompt appears.

<u>Scale</u> Select object:

Select the object or group. You can also select multiple objects using the crosses/window box or a selection set. AutoSketch then asks you for the base point.

<u>Scale</u> Base point:

Select a base point. The base point remains fixed while the objects are enlarged or reduced around it. Select your base point carefully. Usually, there is one point on the shape that should remain in the same location. AutoSketch then asks for the scale factor.

<u>Scale</u> Scale factor:

The second point you pick determines the scale factor. Move your cursor to increase or decrease the size in 1/10 increments. As you do this, an image of the objects expands and reduces in size. Also, the scale factor appears in the prompt box. Press your mouse button when you reach the desired scale factor. To enter an exact scale, type a value in the prompt box. You can also enter the coordinates of the second point.

<u>Scale</u> Scale factor: **2**↵

This entry would make the object(s) twice as big. A scale factor less than 1 would reduce the size of the objects. All objects selected will be scaled by the same amount. However, if you scale a wide object, the width remains unchanged. To change the width, use the **Change Properties** tool in the **Edit** menu. This tool is discussed later in this chapter.

When you scale several objects, AutoSketch also scales the distance between the objects. For example, suppose you have two boxes that are 1 unit apart. Entering a scale factor of 2 will double the size of the boxes *and* place the boxes 2 units apart.

Exercise 6-7

1. Start a new drawing using the **Quick Start** template. Select **Inches (decimal)** as the unit of measurement.
2. Draw the objects shown below on the left. Do not add dimensions.
3. Make two copies of the objects.
4. On one copy, use the **Scale** tool to double the size of the objects.
5. One the second copy, use the **Scale** tool to reduce the size of the objects by one-half.
6. Save the drawing as C6E7 on your work diskette and then close the drawing.

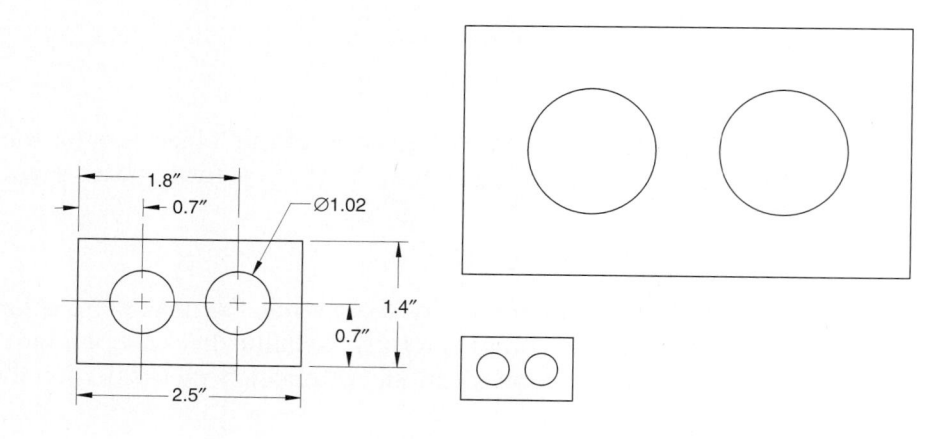

Modifying Object Properties

There are times when you need to change an object's color, linetype, or layer. Layers are discussed in Chapter 10. For example, you might draw several shapes in black using a solid wide line. However, later you may want the shapes in red with a hairline hidden linetype. This is easy to do with AutoSketch.

Showing Object Properties

The **Show Properties** tool lists an object's type, layer, color, width, and linetype. In addition, there may be other information displayed about the object you select. For example, if you select a curve, AutoSketch will display the method (Bezier or B-spline) used to create the curve. For a line, AutoSketch will show the location of the endpoints. The **Show Properties** tool is in the **Measure** toolbox.

The **Show Properties** tool is helpful for several purposes.
- A monochrome monitor does not show different colors. Remember from Chapter 4 that to plot drawings with different pens, you must draw objects in different colors. Select the object with **Show Properties** to find out an object's color.
- A rectangle shown on-screen may have been created as a box, polygon, polyline, or four separate lines. The difference can be important if you plan to edit the shape. The **Show Properties** tool lists the type of object you have chosen.
- You can use the **Show Properties** tool to find out what layer an object is on.

To view object properties, select the **Show Properties** tool from the **Measure** menu. The following prompt appears.

Show Properties Select object:

After you pick an object, a dialog box appears listing various values, depending on the object that was selected. The values for a polyline are shown in Figure 6-13. Pick **OK** to continue with your drawing session.

If you pick a text object, information on text features will be given, Figure 6-14. The linetype is not given since text always has a solid linetype.

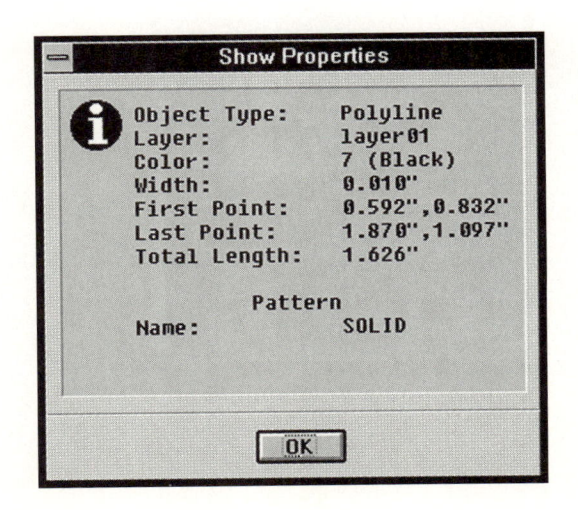

Figure 6-13.
The **Show Properties** tool gives you detailed information about an object. In this example, the properties of a polyline are given.

Figure 6-14.
The properties for text will be different from other objects.

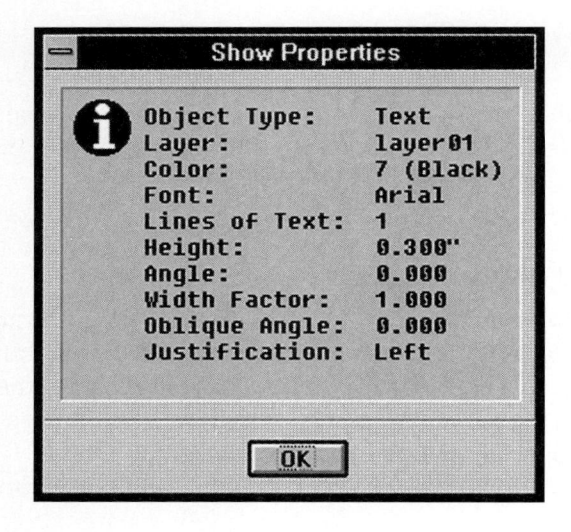

```
Show Properties

Object Type:    Text
Layer:          layer01
Color:          7 (Black)
Font:           Arial
Lines of Text:  1
Height:         0.300"
Angle:          0.000
Width Factor:   1.000
Oblique Angle:  0.000
Justification:  Left

        OK
```

Changing Object Properties

The **Change Properties** tool allows you to change the properties of objects to the current values. You can select certain values to change. This is done in the **Property Settings** dialog box, Figure 6-15.

Figure 6-15.
Double-click on the **Change Properties** tool to display the **Property Settings** dialog box.

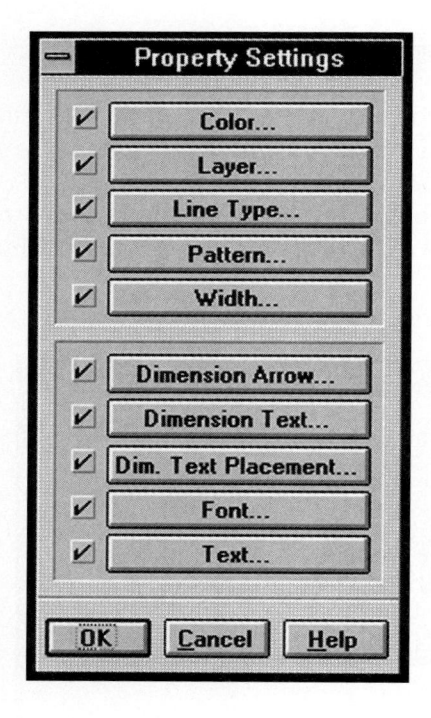

```
Property Settings

✔   Color...
✔   Layer...
✔   Line Type...
✔   Pattern...
✔   Width...

✔   Dimension Arrow...
✔   Dimension Text...
✔   Dim. Text Placement...
✔   Font...
✔   Text...

   OK    Cancel    Help
```

To open the **Property Settings** dialog box, select the **Edit** menu and double-click on the **Change Properties** tool. Pick the box next to the properties you want to change. A check mark means that property will be changed. If you pick on the property button, the settings dialog box for that property will appear, Figure 6-16.

Figure 6-16.
Pick a property button to change the current settings for that property. In this example, the linetype is being changed to hidden.

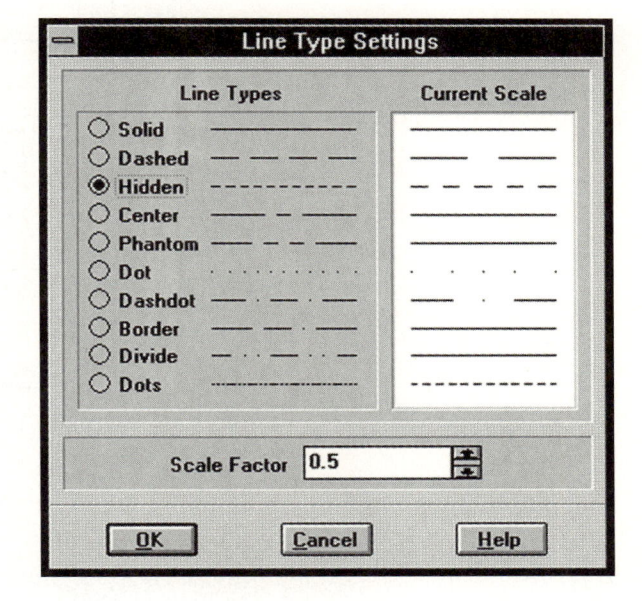

Pick **OK** to close the dialog box. The following prompt appears.

<u>Change Properties</u> Select object:

Select the object or group. You can also use the crosses/window box or a selection set to select several objects. As you select objects, they change to the current values, Figure 6-17. Remember, only the items checked in the **Property Settings** dialog box will be changed.

If a selected object does not have one of the properties checked in the **Properties Settings** dialog box, it will not be affected. For example, text height affects text objects only.

Figure 6-17.
When using the **Change Properties** tool, the objects selected are changed to the current settings. In this example, the bottom three lines are being changed to a hidden line type.

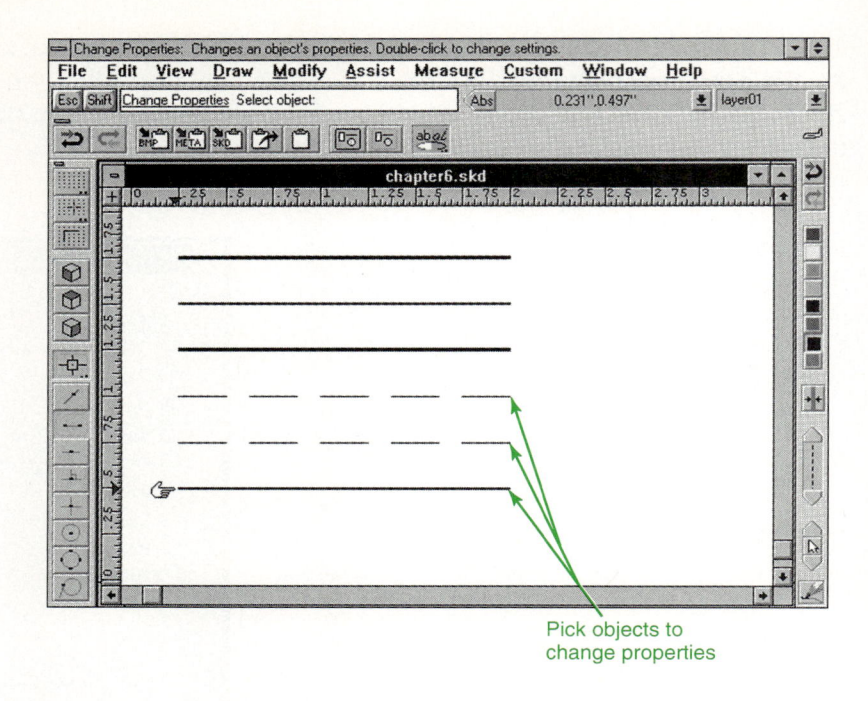

Pick objects to change properties

Exercise 6-8

1. Start a new drawing using the **Quick Start** template. Select **Inches (decimal)** as the unit of measurement.
2. Draw one of each object in the **Draw** menu.
3. Use the **Show Properties** tool and pick each of the objects. Note what information is provided.
4. Draw six horizontal lines, all in black, using a solid linetype.
5. Use the **Change Properties** tool and change each line to a different color and linetype.
6. Save the drawing as **C6E8** on your work diskette and then close the drawing.

Summary

This chapter has covered basic methods to select, modify, and edit objects. You can select objects individually or with the crosses/window box. You can also group objects or use a selection set. Grouped objects act like a single item. They can be copied, moved, or otherwise edited with one pick. These items remain as a group until ungrouped. A group can contain single objects or other groups, called nested groups. A selection set is a temporary group.

Editing is the process of changing objects you have drawn. You will probably spend more time editing drawings than you will creating them. AutoSketch provides a variety of editing functions to suit every situation. Some of the simplest tools are **Move, Copy, Multiple Copy, Mirror,** and **Rotate.** These tools work on every object, set of objects, or group. The **Scale** tool allows you to modify the size of an object proportionally. Finally, you may also want to edit the properties of objects, such as their color or linetype. The **Change Properties** tool is used for this. The **Show Properties** tool can be used to find out the current properties of an object.

Important Terms

Center of rotation
Editing
Group

Mirror line
Nested group

Selection set
Symmetrical

New Commands and Tools

Change Properties tool
Copy tool
Group tool
Landscape

Mirror tool
Move tool
Multiple Copy tool
Rotate tool

Scale tool
Shift button
Show Properties tool
Stretch tool
Ungroup tool

Review Questions

Give the best answer for each of the following questions.

1. AutoSketch's editing tools are found in what menu(s)? _____
2. How can you select multiple objects without using a crosses/window box? _____
3. What keyboard key can you hold down to select multiple objects? _____
4. True or false. Grouped objects remain as a group until you select another drawing tool.
5. A group that is chosen to be a part of another group is called a(n) _____.
6. Describe how you can place more than 1000 objects in a group. _____

7. Explain how the **Ungroup** tool affects a nested group. _____

8. When using the **Move** or **Copy** tools, what two methods can be used to specify the distance to place the objects?
   A. _____
   B. _____
9. After selecting objects to rotate, you must pick a(n) _____ that the objects will rotate around.

10. When rotating objects, what two methods can you use to determine the angle of rotation?
    A. _____
    B. _____
11. After selecting objects to mirror, you must pick endpoints of a(n) _____.
12. When using the **Scale** tool, entering a scale factor less than 1 will _____ the size of the objects.
13. List three situations when you might use the **Show Properties** tool.
    A. _____
    B. _____
    C. _____
14. The **Change Properties** tool allows you to change the properties of objects to _____.
15. The properties that are changed by the **Change Properties** tool depend on _____.

Activities

Here is a mixture of drawing activities to develop your skills using editing tools. Make sure your printer orientation is set on landscape. For each activity, select the **New...** command from the **File** menu. Select the **Quick Start** template and select **Inches (decimal)** as the unit of measurement. Save your AutoSketch drawings on your work diskette using the name given. After saving each activity, select the **Close** command from the **File** menu to clear the drawing from the screen.

1-5. For each activity, complete the part or shape shown on the left. Use a grid, snap, or other drawing aids to help you create the drawing. You do not need to add dimensions. Then, make a copy to the right of the original. Use only editing tools to change the copy so that it looks like the drawing shown on the right. Do not draw any new objects. Use the filename given under the illustration.

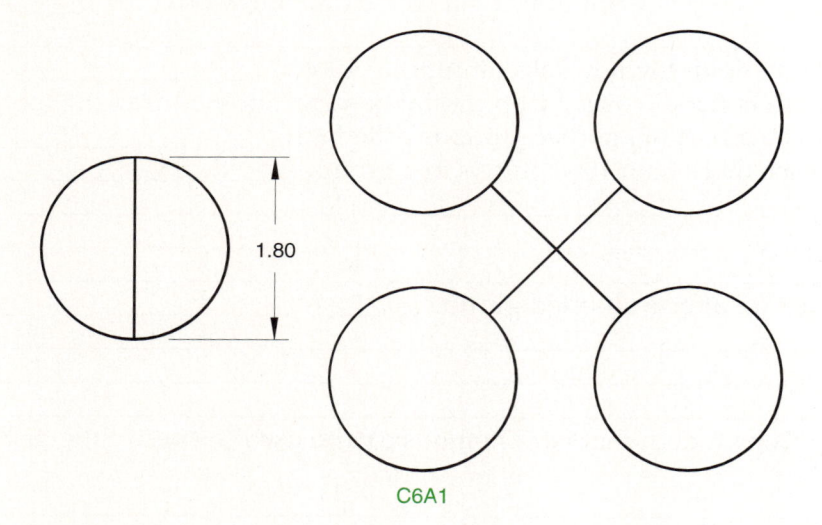

C6A1

Chapter 6 Modifying and Editing Objects

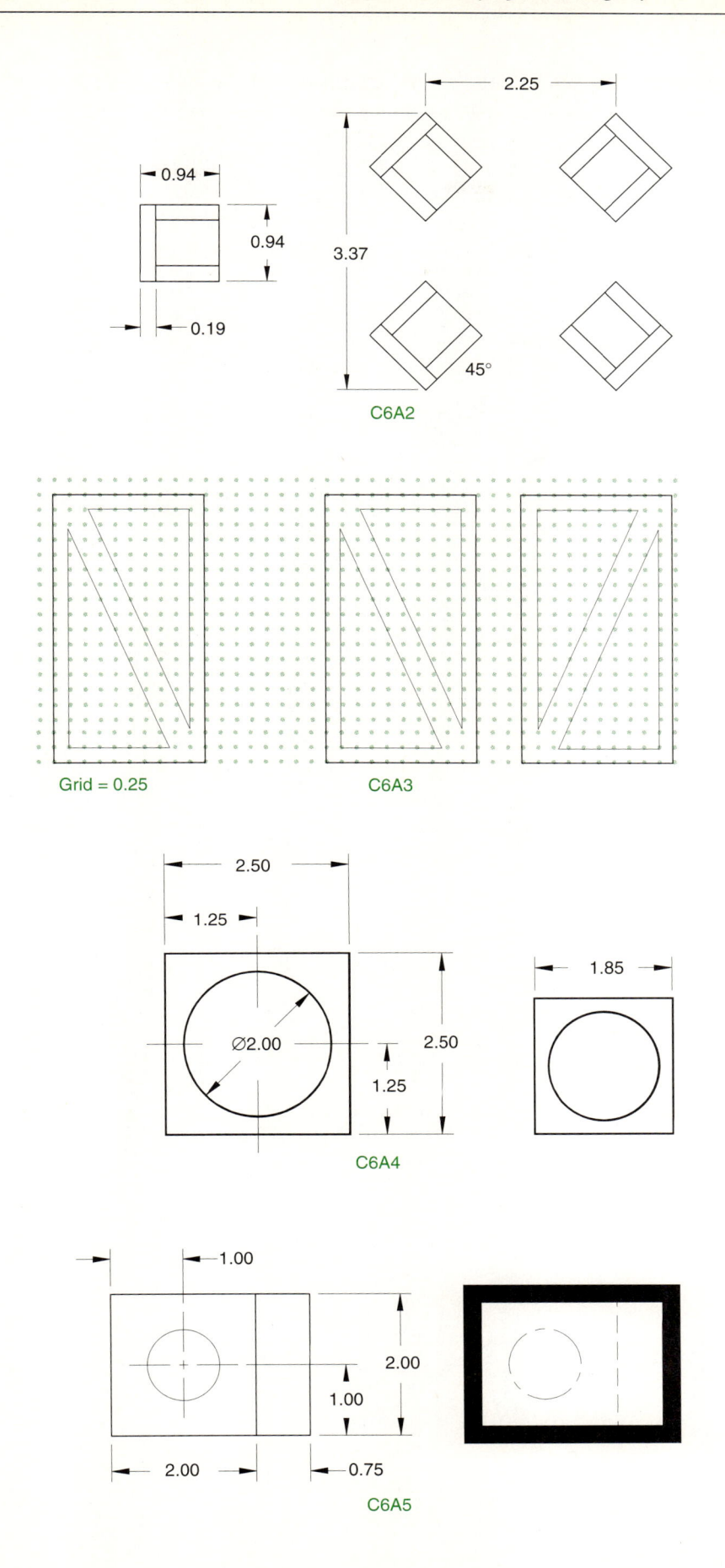

C6A2

Grid = 0.25   C6A3

C6A4

C6A5

Different parts of this drawing were "zoomed" to help precisely locate objects. (Autodesk, Inc.)

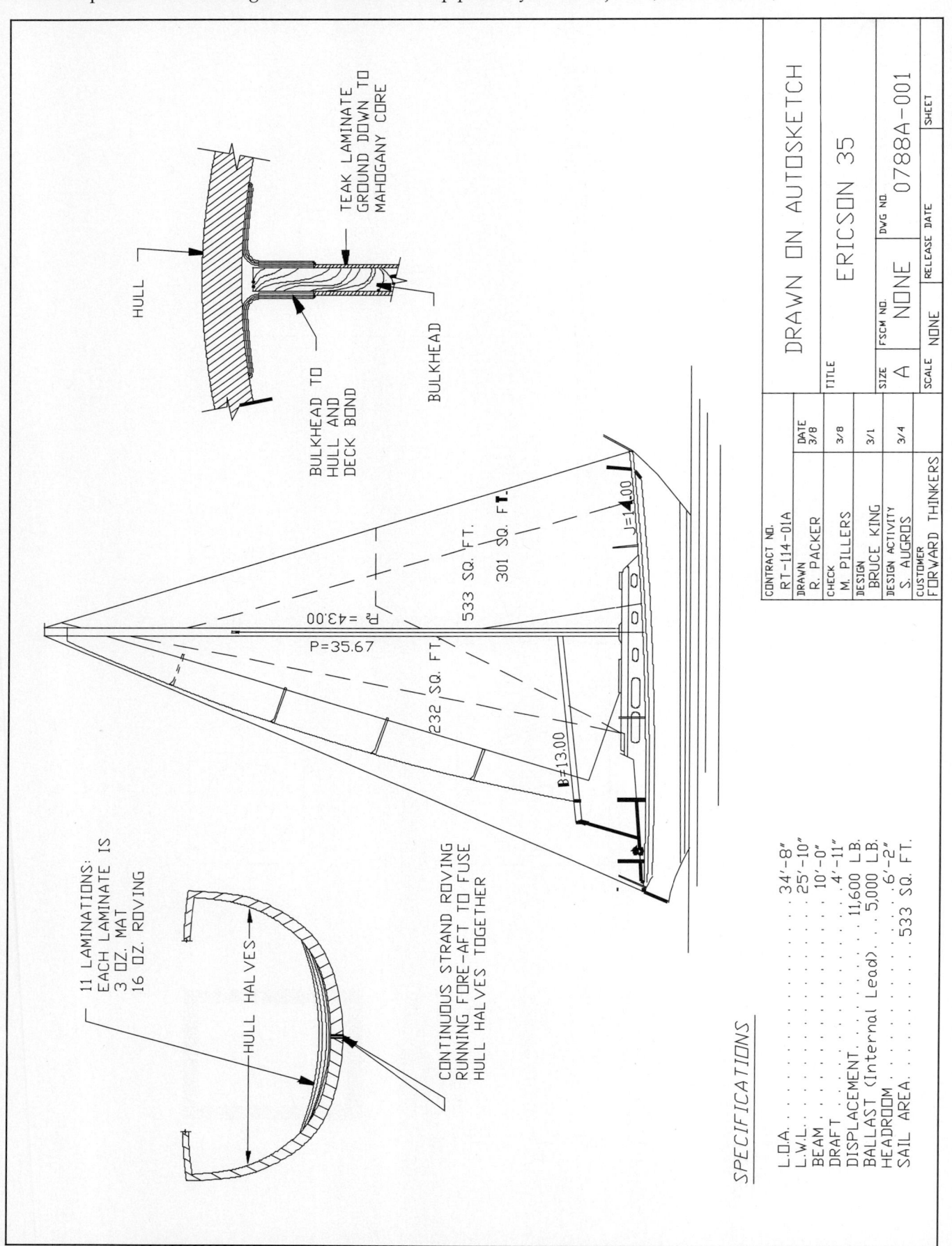

CHAPTER 7

Changing Views of Your Drawing

Objectives

After studying this chapter, you will be able to:
- Enlarge or reduce the current view using different AutoSketch options.
- Change the view so that the entire drawing fits on-screen.
- Show the portion of the drawing that will be printed.
- Move across an enlarged view of the drawing.
- Recall the previous displayed view.
- Display an aerial view of your drawing.
- Redraw the screen to clean it of unwanted clutter.
- Set view preferences.

Your monitor is the "window" where you see and work on your drawing. Since your drawing is held as data in computer memory, AutoSketch can show you as much or as little as you want. **View** commands determine how much and what parts of your drawing appear on-screen. With **View** tools, you might magnify a complicated part of your drawing to do detail work. Then, you could move across the magnified view. Finally, you might reduce the view to see entire drawing again. **View** tools are found in the **View** menu and toolbox, Figure 7-1.

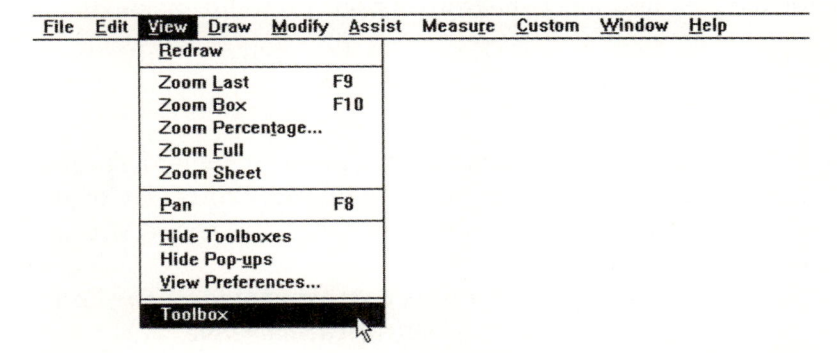

Figure 7-1.
The **View** menu contains tools that determine how much and what parts of your drawing appear on-screen. You can use either a pull-down menu or a toolbox.

189

Zoom Tools

Zoom tools magnify or reduce your view of a drawing just like a zoom lens on a camera. You can use these tools to change the view of your drawing even while using another command. Since most drawings are larger than the size of your monitor, you will likely use zoom tools often in AutoSketch.

Zoom Box

Zoom Box tool enlarges the area within a box you pick, Figure 7-2. The area inside the box expands to fill the drawing window. This command is probably the most used view tool. To change the view, select the **View** menu and the **Zoom Box** tool. The following prompt appears.

Zoom Box First corner:

Pick the first corner of an imaginary box that surrounds that area you want to enlarge.

Zoom Box Second corner:

Pick the opposite corner of the box. As you locate the second corner, your cursor drags the outline of the zoom box. Once you pick the second corner, the portion of the drawing inside the box enlarges to fill the screen.

SHORTCUT

You can also press the function key [F10] to access the **Zoom Box** tool.

Exercise 7-1

1. Start a new drawing using the **Quick Start** template. Select **Inches (decimal)** as the unit of measurement.
2. In the middle of the screen, draw a small circle.
3. Use the **Zoom Box** tool so that the circle fills the drawing window.
4. Save the drawing as C7E1 and leave the drawing on-screen for the next exercise.

Zoom Percent

The **Zoom Percent** tool lets you enlarge or reduce the current view by a specific magnification factor. When the view expands or reduces in size, the center of the drawing remains centered on-screen. This can be helpful when the shape you are working on needs to remain centered on-screen.

To change the view, select the **View** menu and the **Zoom Percent** tool. The **Zoom Percentage** dialog box appears, Figure 7-3. Enter a *magnification value.* This value determines if the view will be enlarged or reduced, and by how much. A factor less than 100 reduces the size of the view. A factor more than 100 enlarges the view. For example, a magnification factor of 200 would make objects look twice as big. A magnification factor of 50 would make objects look one-half as big. This would allow twice as much of the drawing to fit on-screen. After entering a value, click on **OK** and the view changes. *Note:* After you click **OK**, the value in the magnification box becomes the default value. This is important since other AutoSketch tools use this value.

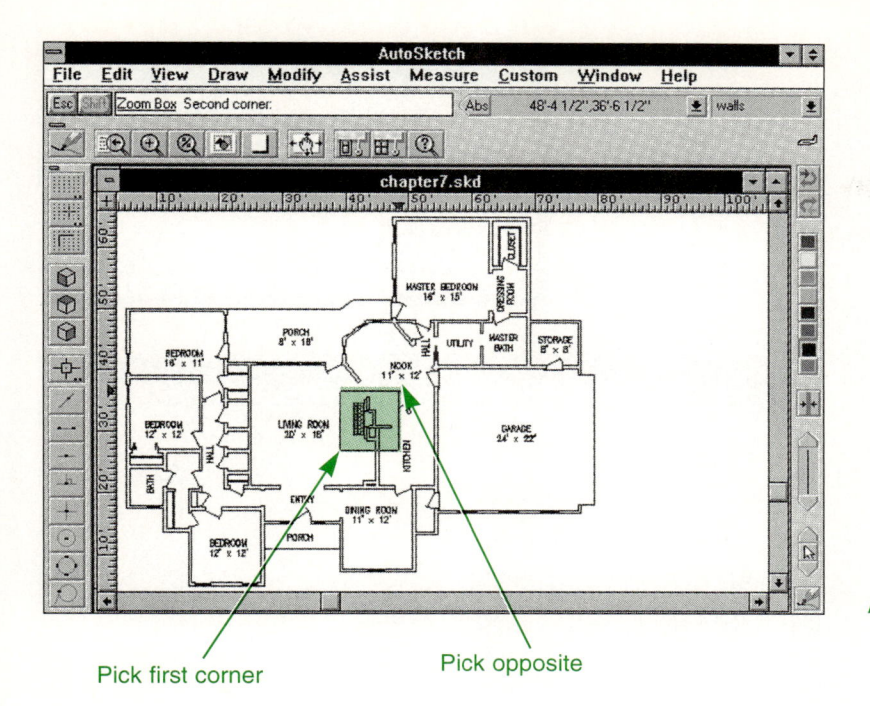

Figure 7-2.
Using the **Zoom Box** tool. A—Pick opposite corners around the portion of the drawing to view. This is shown here with a tint. B—The selected part of the drawing expands to fill the screen.

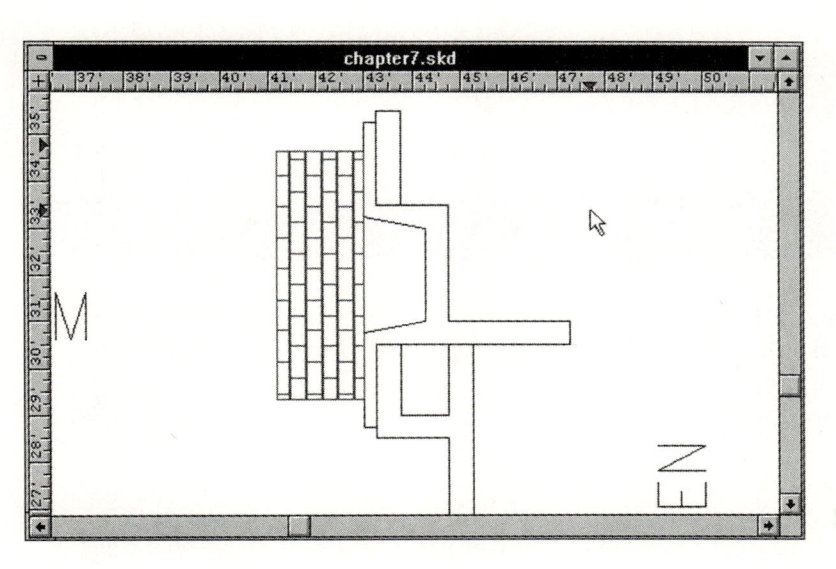

Pick first corner

Pick opposite

Figure 7-3.
Using the **Zoom Percent** tool. A—After selecting the tool, enter a magnification factor in the **Zoom Percentage** dialog box. B—The view reduces (or enlarges), but remains centered on-screen. This view was reduced 50 percent from Figure 7-2B.

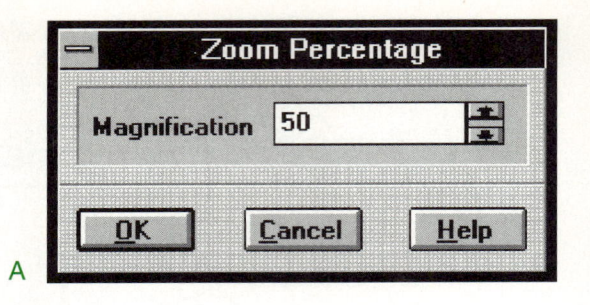

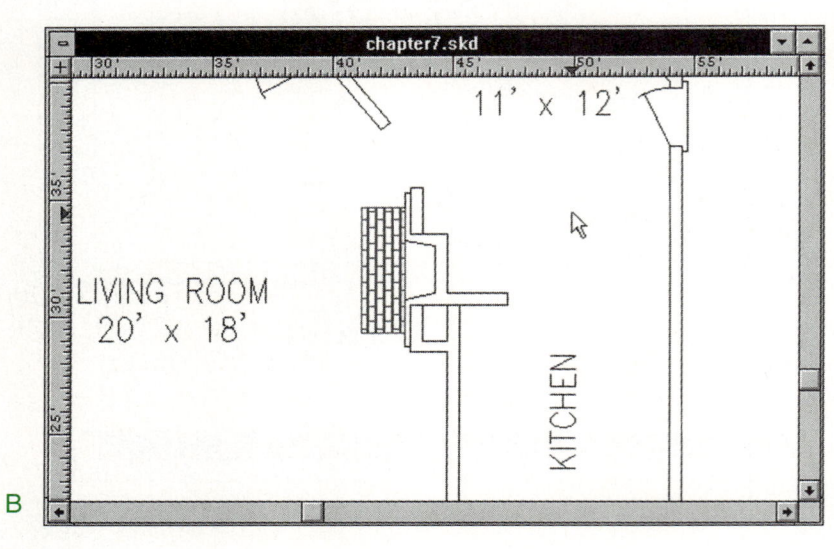

Exercise 7-2

1. Start with the drawing from Exercise 7-1 on-screen (filename C7E1).
2. Use the **Zoom Percent** tool and enter a magnification value of 50. The circle should appear half as large in the drawing window.
3. Draw a box on the left side of the circle, and a polygon on the right side of the circle.
4. Save the drawing as **C7E2** and leave the drawing on-screen for the next exercise.

Zoom Full

The **Zoom Full** tool changes the view so that all the objects in your drawing fill the drawing window, Figure 7-4. This command is good if you get "lost" using other zoom tools. This command will always show the entire drawing. It does not matter how large or small your drawing is. To zoom the entire drawing, select the **View** menu and the **Zoom Full** tool. There are no prompts with this command.

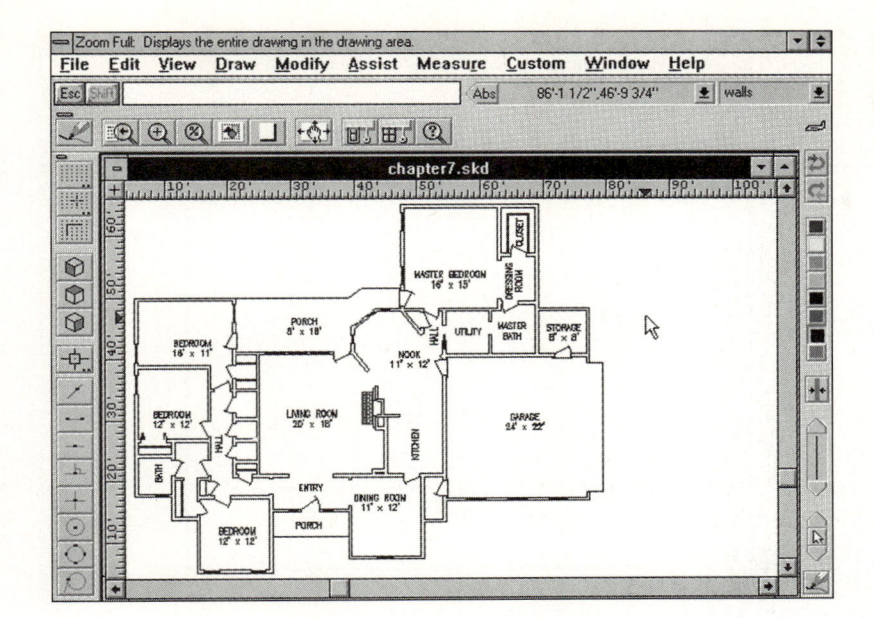

Figure 7-4.
The **Zoom Full** tool always changes the view so that the entire drawing fills the screen.

Exercise 7-3

1. Start with the drawing from Exercise 7-2 on-screen (filename C7E2).
2. Use the **Zoom Full** tool so that the objects you have drawn fill the drawing window.
3. Save the drawing as C7E3 and leave the drawing on-screen for the next exercise.

Zoom Sheet

The **Zoom Sheet** tool displays the portion of the drawing you wish to print. Setting-up drawing sheets is discussed in Chapter 10 and Chapter 16. You may have noticed from previous chapters that using the **Quick Start** template places both solid and dotted boxes in the drawing window. These represent the drawing sheet. The solid line represents the paper size. The dotted line shows the margins. These boxes are not actually part of your drawing.

To view what will be printed, select the **View** menu and the **Zoom Sheet** tool. What happens next depends on the following conditions.

- If a sheet has not been defined for the drawing, a message appears telling you that no sheet was found, Figure 7-5. If you start a new drawing with the normal template and try to use the **Zoom Sheet** tool, this message will be displayed.

Figure 7-5.
This message appears if you use the **Zoom Sheet** tool and no sheet has been defined for the drawing.

Figure 7-6.
The **Zoom Sheet** tool shows the portion of the drawing that will be printed. Only objects inside of the solid black line of the sheet will be printed.

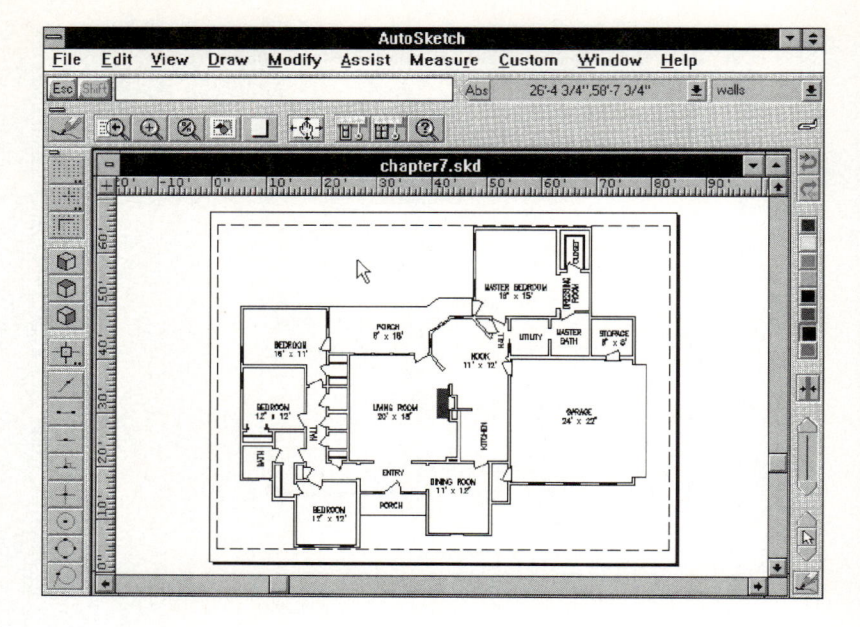

- If the drawing has only one sheet, AutoSketch displays the area of the drawing it contains, Figure 7-6. It does not matter whether the sheet is hidden or visible. (You will learn how to hide sheets later in this chapter.)
- Drawings may also have more than one sheet. One sheet is current, while other noncurrent sheets may be visible or hidden. The sheets are used for printing. If the drawing has more than one sheet, the **Zoom Sheet** dialog box appears. The **Current Sheet** box shows the name of the current sheet. Pick the sheet you want to view and then pick the **Zoom** button. You can view either the current sheet or a non-current sheet. Viewing a non-current sheet makes it current. (Note: Sheets placed on invisible layers are not shown or listed.)

Exercise 7-4

1. Start with the drawing from Exercise 7-3 on-screen (filename C7E3).
2. Use the **Zoom Sheet** tool so that the sheet and the objects you have drawn fill the drawing window.
3. Save the drawing as C7E4 and leave the drawing on-screen for the next exercise.

Zoom Last

The **Zoom Last** tool returns to the previous view. For example, suppose you enlarged a portion of a drawing to do detail work. When you are finished, select the **Zoom Last** tool to return to the previous view.

You can pick **Zoom Last** repeatedly to toggle between the current and previous views. This tool is also used when you make a mistake using one of the other view commands. Suppose you have zoomed in on a portion of the drawing. However, you realize that you did not quite get the view you wanted. Select the **Zoom Last** tool to return to the previous view. You can then redo your view command. There are no prompts with this command.

SHORTCUT
You can also press the function key [F9] to return to the last view

Exercise 7-5

1. Start with the drawing from Exercise 7-4 on-screen (filename C7E4). The view should be of the entire drawing sheet. If not, use the **Zoom Sheet** tool.
2. Draw two small boxes, one in the upper-left corner of the drawing windows, and another in the lower-right corner.
3. Use the **Zoom Box** tool to magnify the box in the upper-left corner.
4. Start a line from one corner of the box.
5. Use the **Zoom Last** tool to return to view of the drawing sheet.
6. Use the **Zoom Box** tool to magnify the box in the lower-right corner.
7. Complete the line by connecting it to one corner of the box.
8. Use the **Zoom Last** tool to return to view of the drawing sheet. Note how your line connects the two boxes.
9. Save the drawing as C7E5 and close the drawing.

Panning and Scrolling

After using the **Zoom Box** tool or the **Zoom Percent** tool to enlarge a view, you may no longer see the entire drawing. Sometimes it is necessary to see an object that is just off the screen. *Panning* and *scrolling* allow you to move around the magnified view. Imagine you are using a magnifying glass to look closely at a drawing. As you move the glass, your view slides over the drawing. This is panning and scrolling.

Using the Pan Tool

You can use the **Pan** tool to move across the drawing view, Figure 7-7. Select the **View** menu and the **Pan** tool. The following prompt appears.

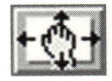

Pan Pan reference:

You must either pick a point or enter a coordinate. The *pan reference* point marks a "handle" you hold to "pull" the drawing into your viewing window. Pick a point to be the pan reference point. AutoSketch then asks you for a second point.

Pan Pan destination:

The *pan destination* point marks the distance and direction you want to move the view. Move your cursor and pick a point, or enter a coordinate. The current view then moves across the drawing in that distance and direction. For example, suppose at the second prompt, you enter the following.

Pan Pan destination: **P(6,30)**↵

This entry tells AutoSketch to shift the view 6 units at a 30 degree angle from the pan reference point.

Using Scroll Bars to Pan

Scroll bars along the bottom and right edges of the drawing window let you move horizontally and vertically across your drawing. Each scroll bar has a scroll arrow at each end and a scroll box, Figure 7-8. The scroll bars allow you to pan across the drawing three ways.

- Pick a scroll arrow to move across the drawing 1/16 screen. The new view will be in the direction of the arrow.

> **SHORTCUT**
>
> You can also press the function key [F8] to access the **Pan** tool.

Figure 7-7.
Using the **Pan** tool. A—Pick pan reference and pan destination points. B—The view moves that distance in the direction of the pan destination point.

A

Pick Pan reference point

Pick Pan distinction point

B

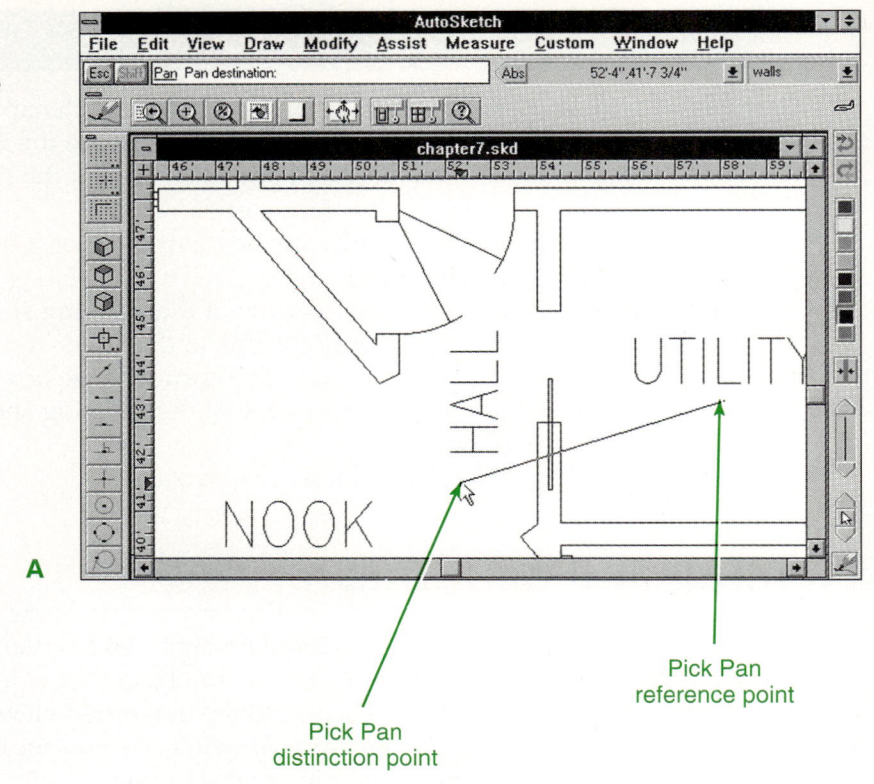

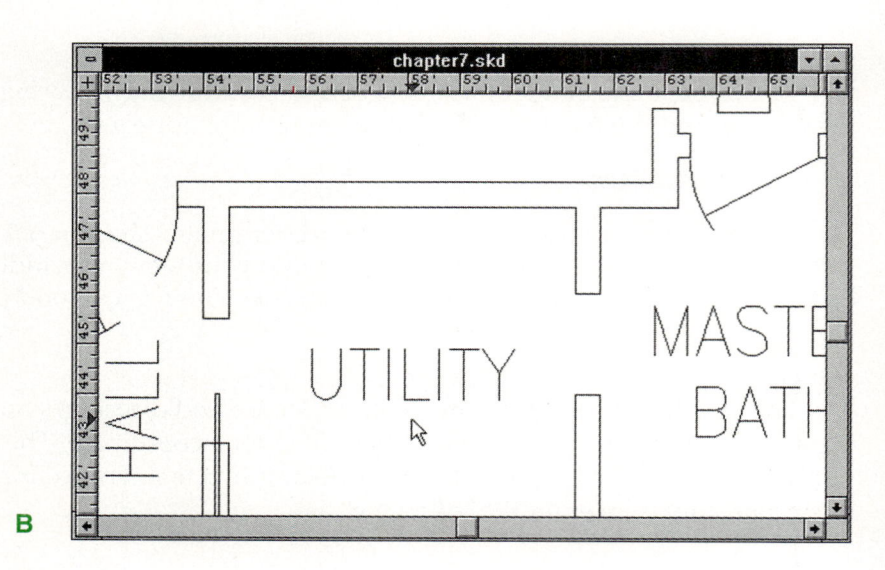

Figure 7-8.
Scroll bars allow you to pan across the drawing.

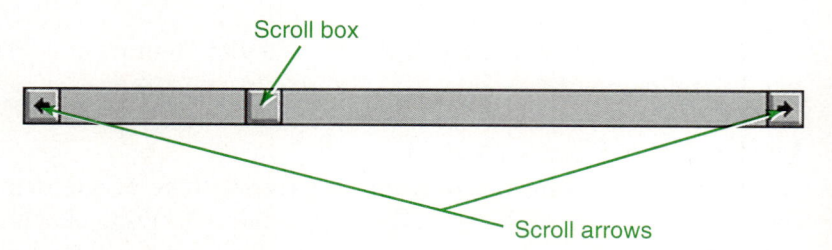

Scroll box

Scroll arrows

- Pick in the rectangular region between the scroll arrow and scroll box to pan across the drawing one full screen. The new view will be in the direction of the arrow on the side you pick.
- Pick and drag the scroll box to the position you want. The scroll box shows the approximate position of your view relative to the entire drawing. For example, dragging the scroll box to the middle of the scroll bar positions your view near the midpoint of your drawing.

Exercise 7-6

1. Open the drawing from Exercise 7-2 (filename C7E2). The view should show a box, circle, and polygon that fill the drawing window. If not, select the **Zoom Full** tool.
2. Use the **Zoom Box** tool so that the circle fills the drawing window.
3. Use the **Pan** tool to move your view so that the box is shown in the drawing window.
4. Use the horizontal scroll bar so that the polygon is shown in the drawing window.
5. Save the drawing as **C7E6** and close the drawing.

Aerial View

The *Aerial View pop-up window* is a helpful tool to navigate around large drawings. This feature displays your current view relative to the entire drawing. To display the **Aerial View** pop-up window click on the **Aerial View** button. This button appears as a small airplane on the top of the vertical status bar, Figure 7-9.

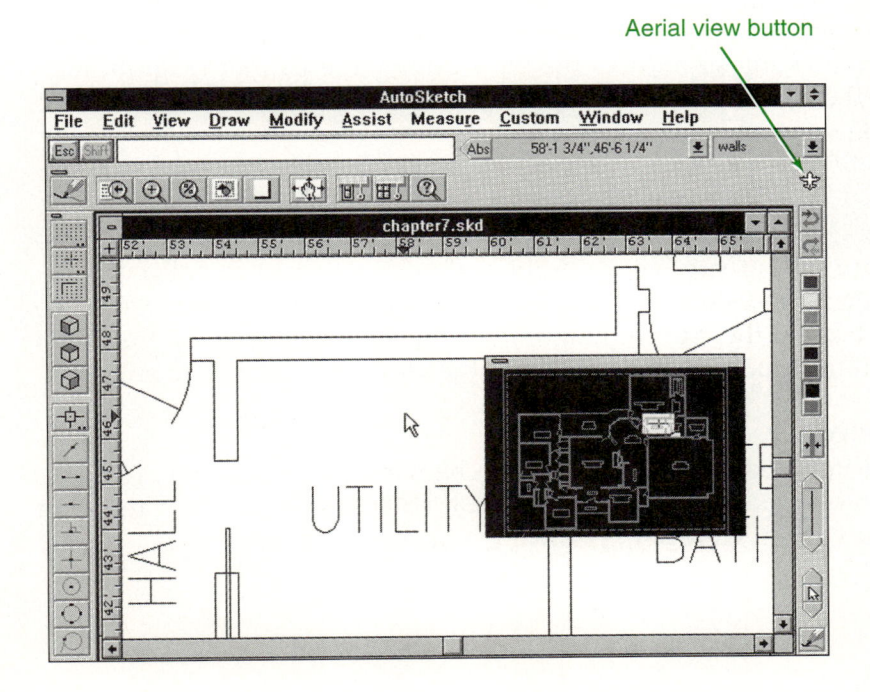

Figure 7-9.
Pick the **Aerial View** button to display the **Aerial View** pop-up window. When the pop-up window is displayed, the button will appear as the top view of an airplane. When the pop-up window is not displayed, the button will appear as the side view of an airplane.

When you click on the **Aerial View** button, the drawing appears in the pop-up window. This view will be of the whole drawing, as if you used the **Zoom Full** tool. In this pop-up window, your current view of the drawing is shown as a smaller box. This is called the *view box*. The part of the drawing inside the view box appears in the same colors as your drawing. The rest of the drawing in the pop-up window appears in reverse video. Text, patterns, and bitmaps appear as outlines since these items increase the redraw time.

You can leave the **Aerial View** pop-up window on-screen while you continue to edit your drawing. Each time you use a **View** tool, the view box in the **Aerial View** window also changes to reflect the new view. To close the **Aerial View** window, click on the **Aerial View** button again. With the **Aerial View** window open, the button appears as the top view of an airplane.

Moving the Aerial View Window

You can leave the **Aerial View** pop-up window displayed while you work on your drawing. However, the window might cover a portion of your drawing you need to edit. In this case, you can move the **Aerial View** window. To do so, place the cursor over the top border of the window, hold down your left mouse button, and drag the **Aerial View** window to a new location.

Using the Aerial View as a View Tool

The **Aerial View** window also serves as a method to zoom and pan the drawing. You can resize and move the view box, or create a new view. Each time you move or resize the view box, AutoSketch redraws the main drawing window to reflect the new view.

Panning

To pan using the view box, drag the center of the crosshairs within the view box to another location in the **Aerial View** window.

Zoom in

To zoom in, drag the square handle at the lower-right corner of the view box to make it smaller. You can also double-click anywhere in the **Aerial View** window using the left mouse button. Doing so reduces the view box by the current magnification percentage used by the **Zoom Percent** tool. Note: If this number is greater than 100, the view box will still be reduced.

Zoom out

To zoom out, drag the square handle at the lower-right corner of the view box to make it larger, Figure 7-10. If you have a multi-button mouse, you can also double-click anywhere in the **Aerial View** window using the right mouse button. Doing so increases the view box by the current magnification percentage used by the **Zoom Percent** tool. Note: If this number is less than 100, the view box will still be enlarged.

Creating a new view box

You can also create a new view box instead of changing the current one. To do so, hold down the left mouse button anywhere within the **Aerial View** window and drag a new view box. The current view of the drawing will change to match.

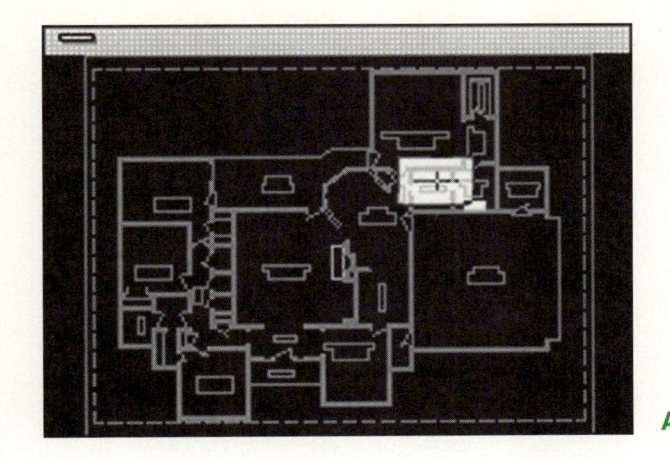

Figure 7-10.
Drag the square handle of the view box to zoom in or out. Here, the view box was dragged larger to show more of the drawing. The view in the drawing window will change to match.

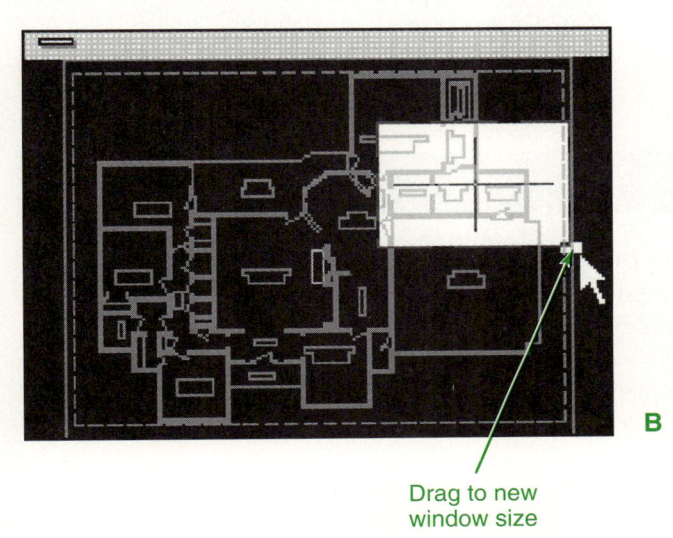

A

B

Drag to new window size

Using the Aerial View Control Menu

The **Aerial View** window also has a *control menu*. The view commands in this menu affect the **Aerial View** window only. When you first open the **Aerial View** window in a drawing, the window includes the entire drawing. You may want this window to show only a portion of the drawing. Click on the control bar in the upper-left corner of the **Aerial View** window, Figure 7-11. The commands are as follows.

- **Zoom Box** expands the view box to fill the entire **Aerial View** window. It does *not* change the view shown in the drawing window. You can then create a view box to move around within the **Aerial View** window. Once you create the view box, the drawing window changes to match.
- **Zoom Full** displays the entire drawing in the **Aerial View** window. Again, you must then create a view box to move around within the **Aerial View** window so the drawing window will change.
- **Zoom Sheet** displays the area of the drawing within the current sheet. This option will be "greyed" out if no sheet is defined.
- **Close** closes the **Aerial View** window. This can also be done by picking the **Aerial View** button as discussed earlier.

Figure 7-11.
The commands in the **Aerial View** control menu will affect only the **Aerial View** pop-up menu.

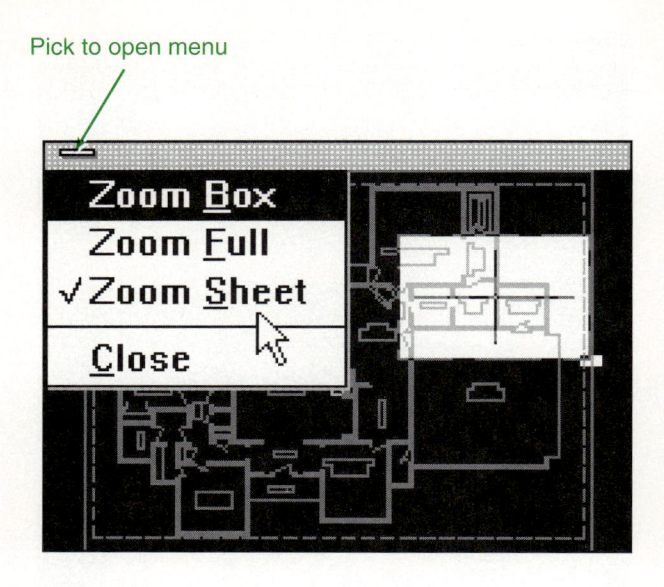

Pick to open menu

Exercise 7-7

1. Start a new drawing using the **Quick Start** template. Select **Inches (decimal)** as the unit of measurement.
2. Draw a box that nearly fills the entire sheet (sides are just inside the dotted-line margins of the sheet).
3. Draw several circles, polygons, and lines in different areas inside the box you drew.
4. Display the **Aerial View** pop-up window.
5. Select the **Aerial View** control menu and select **Zoom Full**.
6. Create a new view within the **Aerial View** window. Zoom in on one circle you drew.
7. Drag the view in the **Aerial View** window to pan over one of the polygons.
8. Zoom out by resizing the view in the **Aerial View** window.
9. Close the **Aerial View** pop-up window.
10. Save the drawing as C7E7 and close the drawing.

Redrawing the View

Selecting the **Redraw** tool or button "cleans" the drawing. The **Redraw** tool is found in the **View** menu. The **Redraw** button is found in the **Vertical Status Bar**. Redrawing removes "marks" and "holes" left by some drawing and editing commands. The **Redraw** command clears the current display and redraws the view. There are no prompts with this command.

> **Exercise 7-8**
>
> 1. Start a new drawing using the **Quick Start** template. Select **Inches (decimal)** as the unit of measurement.
> 2. Draw two circles the same size in exactly the same location. Use two different colors.
> 3. Use the **Erase** tool and pick on the circle. Do this only once. Notice it appears that no circles remain on-screen.
> 4. Pick the **Redraw** tool. Notice how the circle that was not erased reappears on-screen.
> 5. Save the drawing as **C7E8** and close the drawing.

Using View Tools within Other Commands

You can select a **View** tool, or any of the other commands mentioned in this chapter, at any time. You can select these even during a drawing or editing operation. Once you change the view, the drawing or editing command resumes.

For example, suppose you need to draw a long line between two small objects that are far apart. You can zoom in on one object and place the first endpoint. Next, you can select the **Pan** tool to move across to the other object. Your line command will still be active. Then, you can place the second endpoint of the line.

Changing What is Seen with View Preferences

AutoSketch allows you to hide certain parts of your drawing and AutoSketch screen that are not always needed. For example, you can hide pattern fills and text to speed up redraw time. These two elements are slow to redraw. You can also hide certain features associated with printing your drawing. Select the **View Prefs** (Preferences) tool to display the **View Preferences** dialog box, Figure 7-12.

Hiding Features that Slow Performance

Certain features will slow down the time it takes for AutoSketch to redraw a view. You can turn these off when you don't need them to speed up the redraw time.

Hide Fill

When selected in the **View Preferences** dialog box, **Hide Fill** makes all pattern fills invisible. Before printing, you need to turn off this feature so that pattern fills are printed. When selected, you only see the boundaries of wide objects and pattern objects.

Figure 7-12.
You can choose to show or hide various items in the **View Preferences** dialog box.

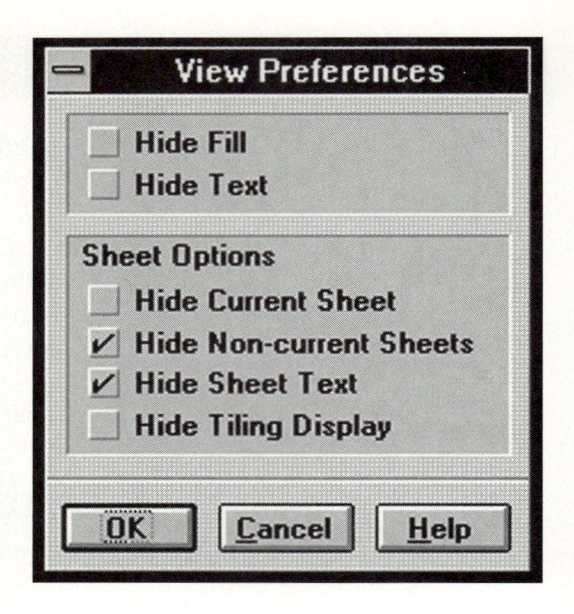

Hide Text

When selected in the **View Preferences** dialog box, **Hide Text** makes all text objects appears as box outlines where text is, Figure 7-13. If you add new text, it is visible until you redraw the drawing. Before printing, you need to turn off this feature so that text is printed. Otherwise, boxes are printed where text is.

Hiding Sheet Options

Sheet Options are features that are used when printing your drawing. Sheets appear as an outline with an inner dotted line to represent the paper margins. These items are covered in much greater detail in Chapter 16.

Figure 7-13.
The **Hide Text** option of the **View Preferences** dialog box makes all text objects appears as box outlines to speed redraw time. Here, the shaded boxes represent where text is.

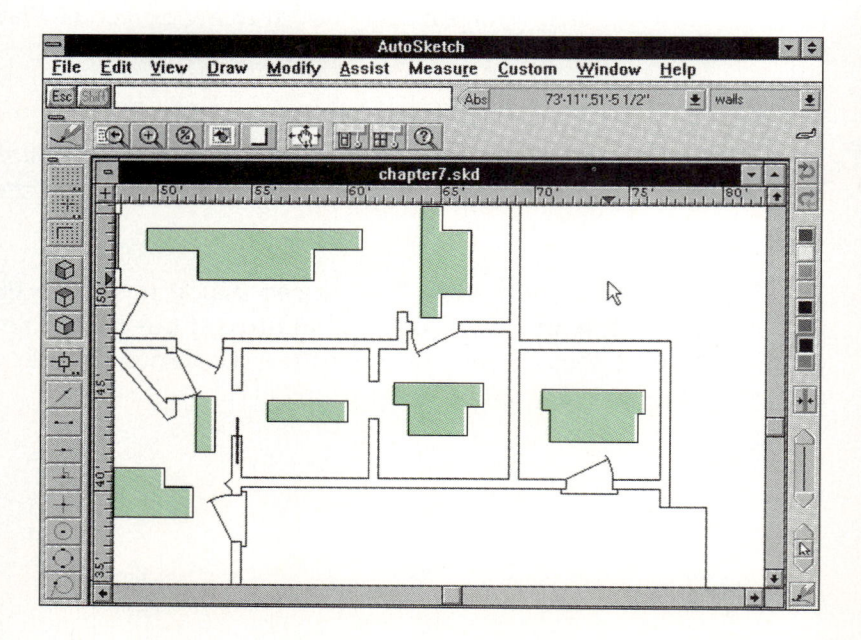

Hiding the current sheet

When selected in the **View Preferences** dialog box, the **Hide Current Sheet** option hides the lines that represent the current sheet. The current sheet represents the portion of the drawing printed when you select the **Print** command from the **File** menu.

Hiding non-current sheets

AutoSketch allows you to have more than one sheet. When selected in the **View Preferences** dialog box, the **Hide Non-current Sheets** option hides the lines that represent sheets other than the current sheet. This option is selected by default.

Hiding sheet text

When selected in the **View Preferences** dialog box, the **Hide Sheet Text** option hides the size and scale information on visible sheets. This option is selected by default.

Hiding the tiling display

If your drawing is bigger than the paper in your printer, AutoSketch will print the drawing across multiple sheets of paper. This is called *tiling*. To help you see what parts of your drawing are on which sheets, AutoSketch faintly shows the position of the paper on-screen. When selected in the **View Preferences** dialog box, the **Hide Tiling Display** option hides this on-screen tiling.

Hiding Other Features

You may want to hide other features of AutoSketch. If you don't use toolboxes, pop-up windows or rulers, you may wish to turn them off.

Hiding toolboxes

When you need more space in your drawing area, you can easily hide all open toolboxes, Figure 7-14. Simply click on the **Hide/Show Toolboxes** tool in the **View** menu. The toolboxes will return when you select a menu. If you are using pull-down menus, the **Hide Toolboxes** command changes to **Show Toolboxes** in the **View** menu. To show the toolboxes, you can either select this command, or select another menu.

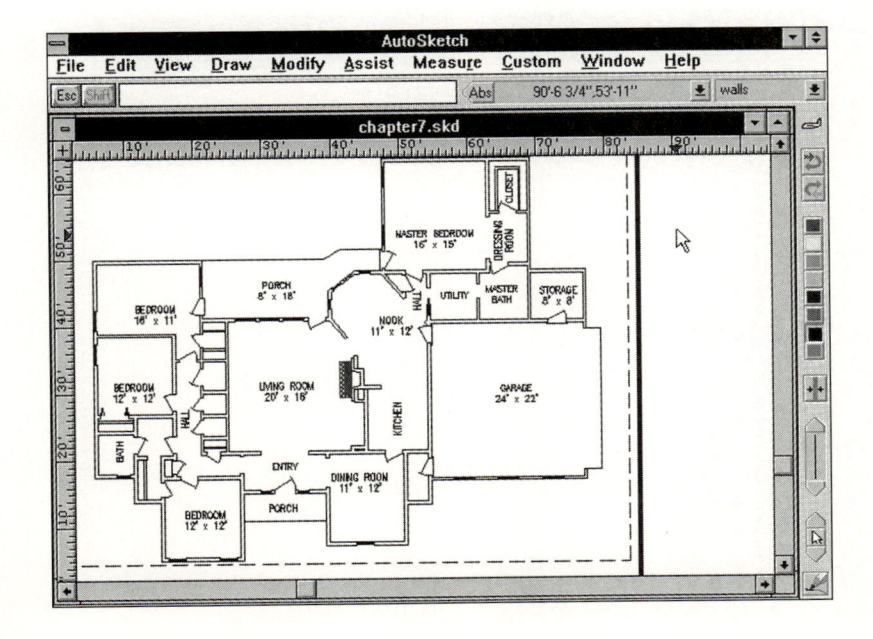

Figure 7-14.
The **Hide Toolboxes** tool hides the toolboxes to give you more room to draw.

Hiding pop-up windows

You can also quickly hide all floating pop-up windows when you need more space in your drawing area. *Pop-up windows* are small windows that remain on-screen. You can drag these to different locations in or out of the drawing area. They include the **Aerial View** pop-up, **Coordinate Display** pop-up, and **Width** pop-up. Toolboxes can float on-screen if you select **Copy to Float** from the toolbox control menu. These can also be moved around like pop-up windows. See Figure 7-15.

To hide pop-up windows and floating toolboxes, click on the **Hide/Show Pop-ups** tool or **Hide/Show Toolboxes** tool in the **View** toolbox. You must click on this tool again to show floating toolboxes and pop-up windows. This tool only affects *floating* toolboxes and pop-up windows.

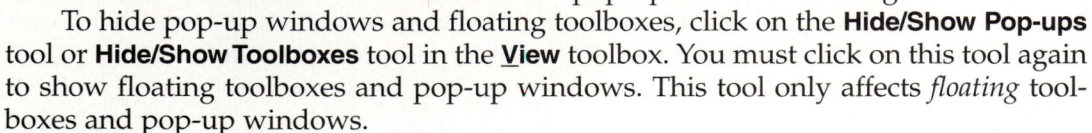

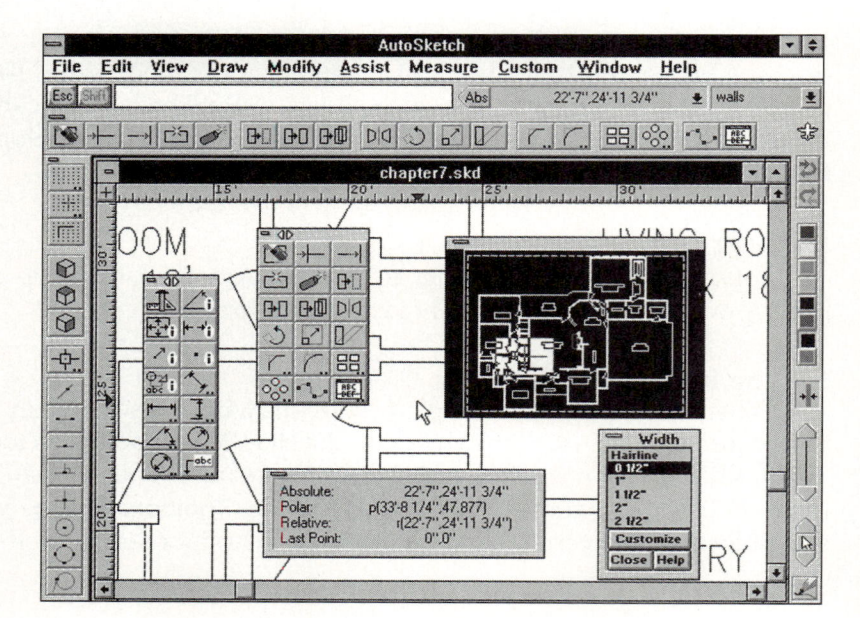

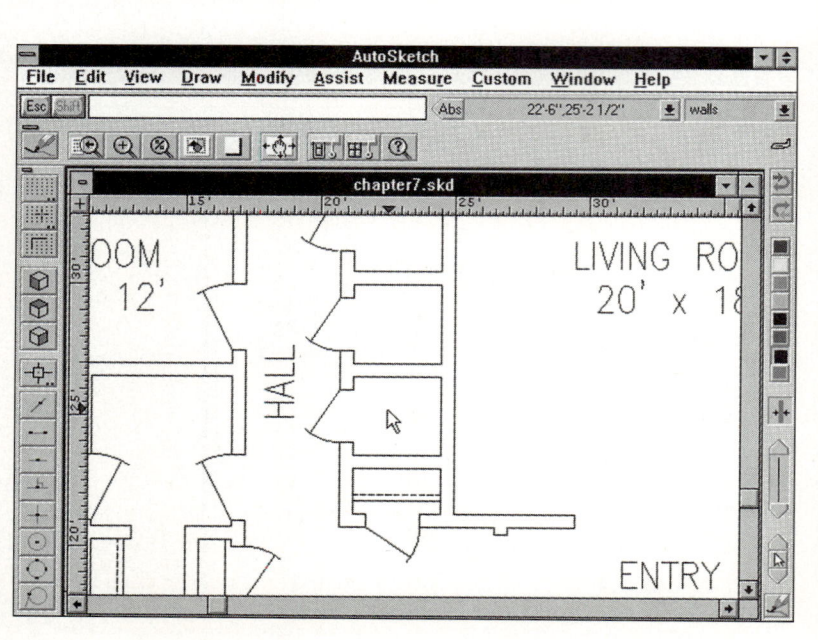

Figure 7-15.
Floating toolboxes and pop-up windows can take too much space. Use the **Hide Pop-ups** tool to temporarily turn them off without closing them.

Hiding rulers

Rulers are found along the left and top sides of your drawing window. These are used to show location. Small arrows in the ruler mark the cursor position. Rulers are not always needed since AutoSketch displays coordinate location. To hide the rulers, select the **Preferences** tool in the **Custom** menu. The AutoSketch **Preferences** dialog box appears, Figure 7-16. Deselect the **Show Rulers** option and pick **OK**. The rulers will no longer be shown. If you want them to show, select the **Show Rulers** option and pick **OK**.

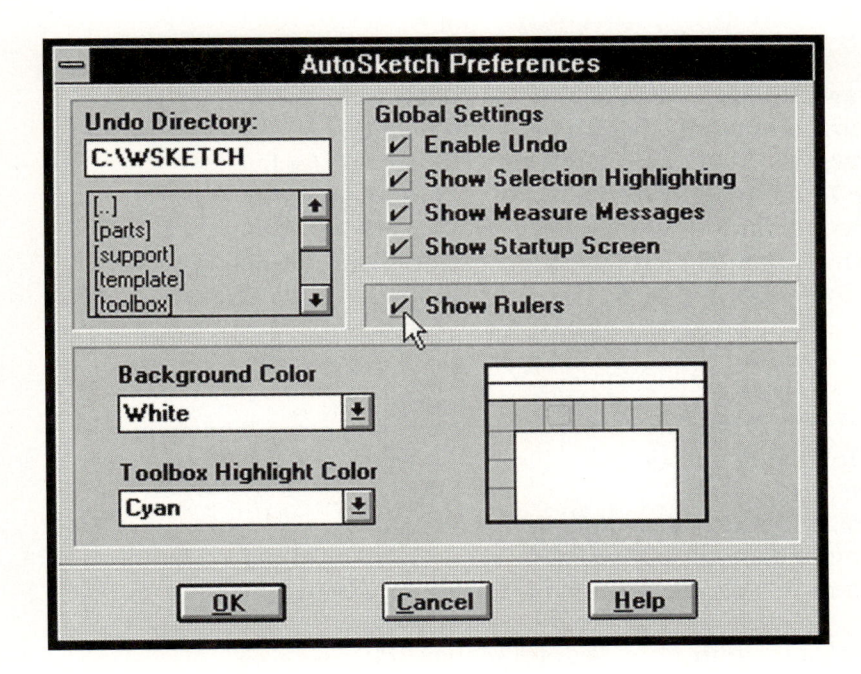

Figure 7-16.
Deselect the **Show Rulers** option in the AutoSketch **Preferences** dialog box to hide rulers. To deselect this option, click on the check mark next to the option. The check mark will disappear. Then pick **OK**.

Summary

View tools are tools that determine how much and what parts of your drawing appear on-screen. AutoSketch can show you as much or as little of the drawing as you want to see. There are many **View** tools offered by AutoSketch. All can be used from within another drawing or editing task.

The **Zoom Box** tool enlarges the portion of the drawing within a box you pick. The **Zoom Percent** tool lets you enlarge or reduce the current view by a specific magnification factor. The **Zoom Full** tool changes the view so that your entire drawing fits on the screen. The **Zoom Sheet** tool shows the entire drawing, including the drawing sheet. The **Zoom Last** tool returns the previous view. The **Pan** tool and scroll bars allow you to move around a magnified view. The **Redraw** tool cleans off the drawing, removing marks and holes left by drawing and editing objects.

The **Aerial View** pop-up window provides a "birds-eye" view of your drawing, and allows you to change the view. You can zoom and pan, as well as a create an entirely new view of the drawing. The **Aerial View** control menu determines what portion of the drawing is shown in the **Aerial View** pop-up window.

View preferences allow you to set which display features of AutoSketch are active. You can hide fills and text to speed the redraw time. You can also hide current and non-current drawing sheets. In addition, AutoSketch provides tools that let you hide toolboxes and pop-up windows.

Important Terms

Aerial View pop-up window
Control menu
Magnification value
Pan destination

Pan reference
Panning
Pop-up windows
Scroll bars

Scrolling
Tiling
View box
Zoom tools

New Commands and Tools

Aerial View button
Custom menu
Hide Toolboxes command
Hide/Show Pop-ups tool
Hide/Show Toolboxes tool
Pan tool
Preferences tool

Redraw button
Redraw command
Redraw tool
Show Toolboxes command
View commands
View menu
View Prefs tool

View tools
Zoom Box tool
Zoom Full tool
Zoom Last tool
Zoom Percent tool
Zoom Sheet tool

Review Questions

Give the best answer for each of the following questions.

1. After choosing the **Zoom Box** tool, you must pick opposite corners of a _____ around the portion of the drawing to display.
2. Which zoom tool allows you to pick exactly the portion of the drawing to magnify? _____
3. When using the **Zoom Percent** tool, a magnification factor less than 100 _____ the size of the view.
4. Which zoom tool keeps the center of the drawing centered on-screen when you zoom in or out? _____
5. The view tool that will always show all of the objects in your drawing is the _____ tool.
6. To move your view to see an object that is only partly shown on-screen, you usually select the _____ tool.
7. How are the scroll bars different from the **Pan** tool? _____
8. For what view tool can you also enter coordinates? _____ Explain how.
9. In what two situations is the **Zoom Last** tool often used?
   A. _____
   B. _____
10. The _____ tool renews the current view of the drawing to remove "marks" and "holes" left by drawing and editing tools.
11. What factors determine what portion of the drawing the **Zoom Sheet** tool shows?
    A. _____
    B. _____
    C. _____

12. To pan using the view box in the Aerial View pop-up window, you must _____.
13. What happens when you double-click within the **Aerial View** window using the left mouse button? _____
14. True or false. The commands in the **Aerial View** control menu determine the portion of the drawing shown in the drawing window.
15. What two view preferences can speed the redraw time of your drawing?
    A. _____
    B. _____

Activities

Here is a mixture of drawing activities to develop your skills using view tools. If you have not already changed your printer orientation to landscape as discussed in Chapter 6, do so now. For each activity, select the **New...** command from the **File** menu. Select the **Quick Start** template and select **Inches (decimal)** as the unit of measurement. Complete each of the drawings below. Use view options when appropriate to help you see details. Use the drawing aids you have learned about to make drawing easier. Save your AutoSketch drawings on your work diskette using the name given. After saving each activity, select the **Close** command from the **File** menu to clear the drawing from the screen.

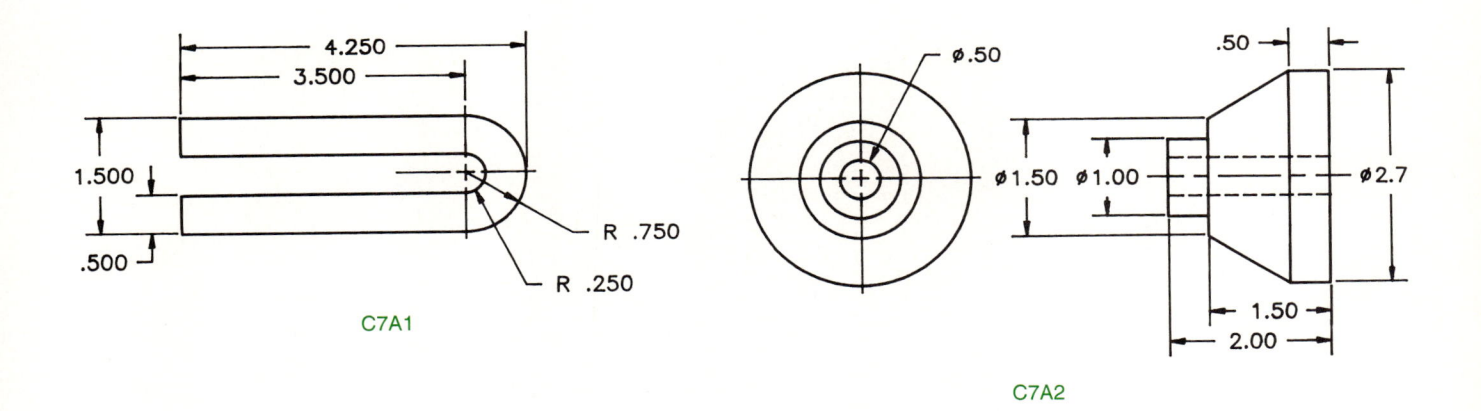

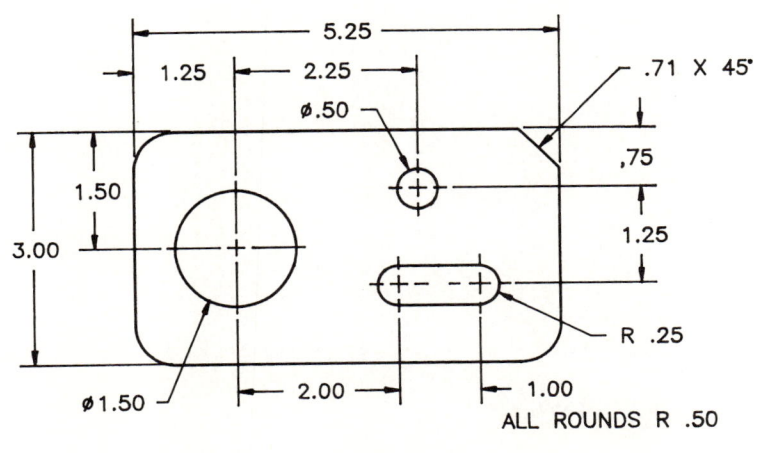

C7A1

C7A2

C7A3

ANGLE PLATE

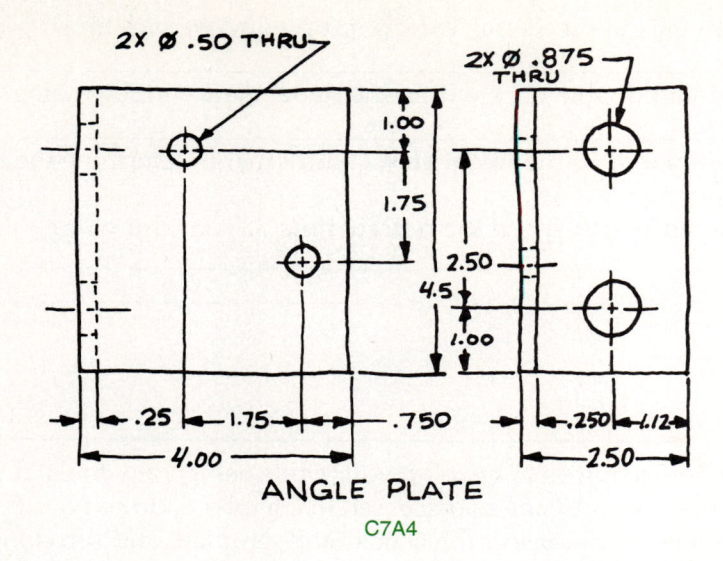

C7A4

CHAPTER 8

Advanced Editing and Measuring Tools

Objectives
After studying this chapter, you will be able to:
- Break and explode objects.
- Trim and extend objects.
- Draw fillets and chamfers.
- Stretch objects.
- Create multiple copies in a box or ring array pattern.
- Edit curves you have drawn.
- Measure distance, angle, area, and bearing.

Breaking Objects

The **Break** tool lets you remove a portion of an object, truncate an object, or create two separate objects. (Truncate means "cut off.") For example, you can remove a portion of a box to create a polyline. You can also use the **Break** tool to remove a portion of a line to make a shorter line. Finally, you can break a circle and make an arc.

You can break all closed objects, except pattern fills. Breaking a box or polygon turns it into a polyline. You can also remove a section from the middle of an open or closed object with the **Break** tool. For example, you can remove a section of a circle to create an arc. You cannot break text, points, dimensions, curves, or grouped objects.

To break an object, select the **Modify** menu and the **Break** tool. The following prompt appears.

Break Select object:

You can only break one object at a time. Notice that after you pick an object, it becomes highlighted on-screen. Now, you must pick two *break points*. AutoSketch asks you for the first break point.

Break First break point:

Select the first point where the object will break. If you are trimming an object, this point should be the point where you want to trim.

Break Second break point:

SHORTCUT

You can also access the **Break** tool by pressing the function key [F4].

209

Pick the second point that defines the portion to remove. If you are trimming the object, you can pick the second point beyond the object.

You can pick the same point twice to break an object, but leave it the same size and shape. For example, if you pick the same break point twice on a box, it creates a polyline in the shape of the box. Although no visible change occurs, AutoSketch breaks the object at the point you picked.

See Figure 8-1 for several examples of using the **Break** tool. The illustration also shows how the **Break** tool affects different objects. Note: When breaking a circle, arc, or box, the portion removed is counterclockwise from the first break point. If you select them in the wrong order, you can get some odd results.

Figure 8-1. Where and in what order you select points will affect how different objects break.

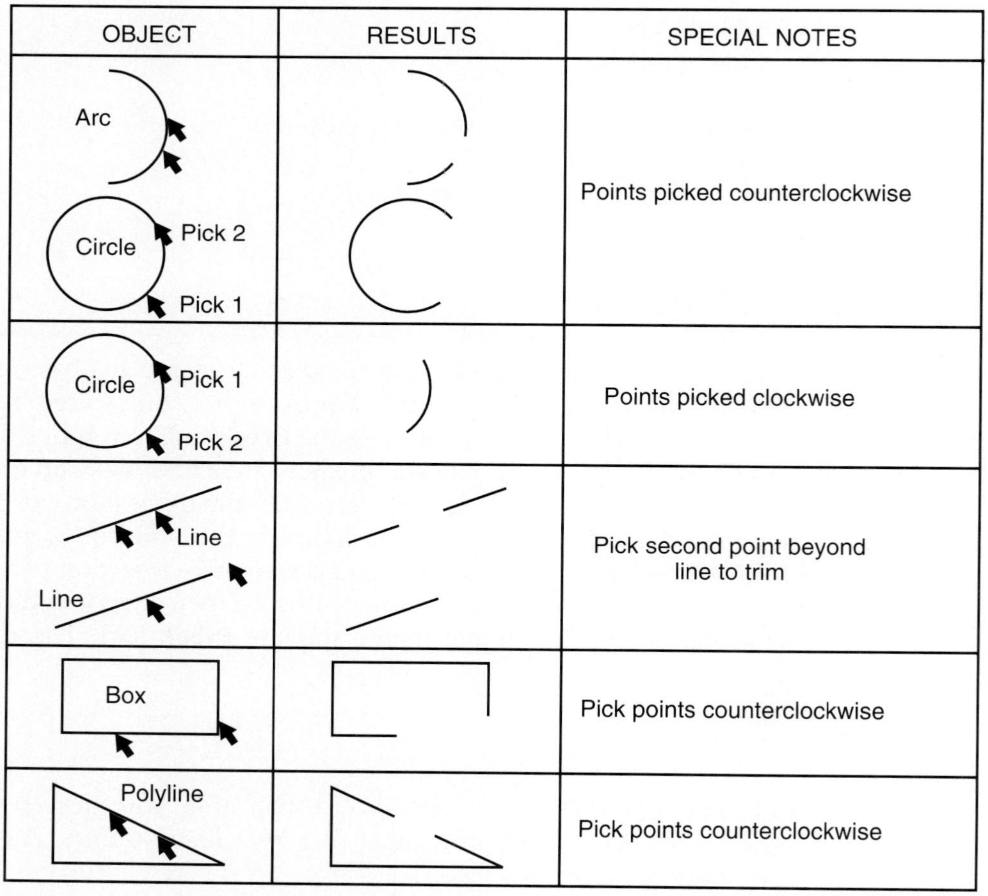

OBJECT	RESULTS	SPECIAL NOTES
Arc		Points picked counterclockwise
Circle (Pick 2, Pick 1)		
Circle (Pick 1, Pick 2)		Points picked clockwise
Line / Line		Pick second point beyond line to trim
Box		Pick points counterclockwise
Polyline		Pick points counterclockwise

Drawing aids, such as **Attach** tools, can affect where the break points are placed. If they cause the object to be broken incorrectly, select the **Undo** button. Then, turn the tools off and try again.

Exercise 8-1

1. Start a new drawing using the **Quick Start** template. Select **Inches (decimal)** as the unit of measurement.
2. Draw two intersecting roads using horizontal and vertical lines. Do not use a grid or snap, but do use the **Ortho** tool for this activity.
3. Draw a circle around the intersection.
4. Break the lines and circle to complete the intersection as shown below.
5. Save the drawing as C8E1 on your work diskette and then close the drawing.

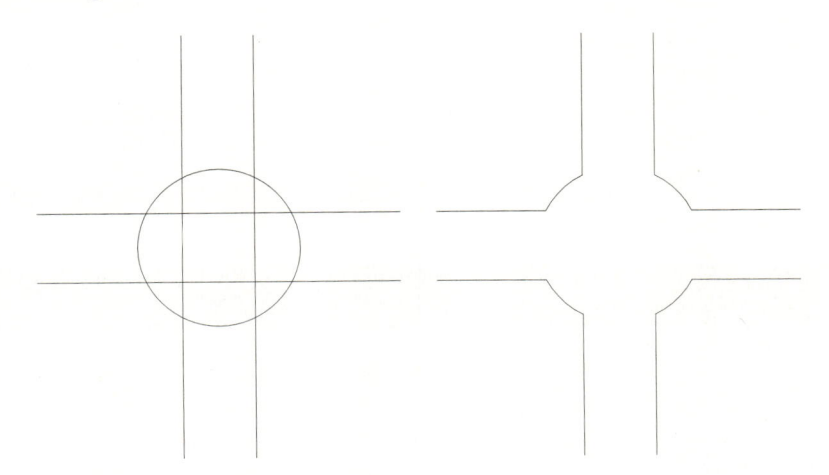

Exploding Objects

The **Explode** tool breaks an object into its component parts. For example, exploding a five-sided polygon breaks it into five separate lines. The **Explode** tool affects objects differently. You can explode boxes, polylines, polygons, dimensions, leaders, and multiline text. You can also explode objects that are drawn with a wide line width. Wide objects explode into components that include edges, endpoint edges, and pattern fill.

Although you can explode wide arcs, lines, circles, and ellipses, you cannot explode the hairline (zero-width) versions of these objects. You cannot use the **Explode** tool with curves, points, pattern fills, or single-line text. Also, grouped objects must be ungrouped before they can be exploded.

To explode an object, select the **Modify** menu and the **Explode** tool. The following prompt appears.

<u>Explode</u> Select object:

Pick the object to explode. You can also select several objects at once with the crosses/window box or a selection set. AutoSketch breaks the object into its parts so you can edit them separately. When you explode an object, the change will not be visible on-screen, Figure 8-2.

Figure 8-2.
Exploding an object does not show a visible change. However, you can select an individual segment of the object to edit. Here, a polygon was exploded to remove a side.

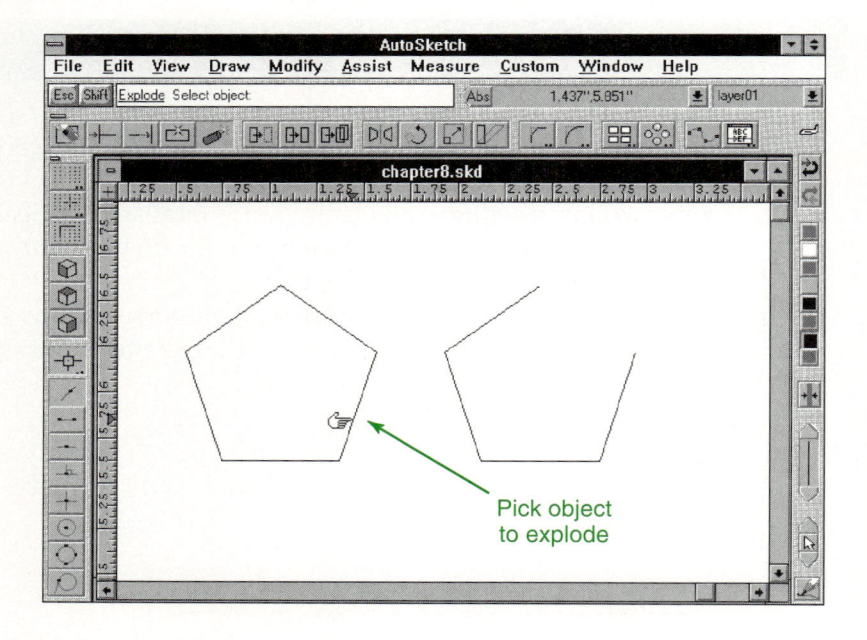

Exercise 8-2

1. Start a new drawing using the **Quick Start** template. Select **Inches (decimal)** as the unit of measurement.
2. Draw a six sided polygon.
3. Use the **Explode** tool to break the polygon into its component parts.
4. Erase opposite sides of the polygon. Note that you can now erase individual line segments of the polygon.
5. Save the drawing as **C8E2** on your work diskette and then close the drawing.

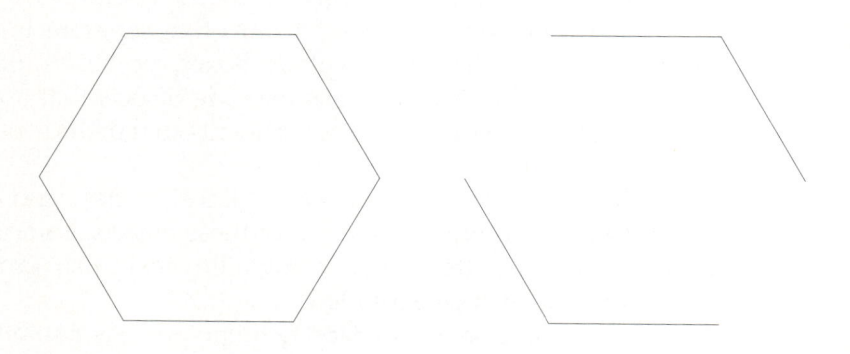

Trimming Objects

The **Trim** tool lets you clip crossing objects so they meet precisely. For example, you can trim a line that crosses a circle so that the endpoint of the line ends exactly on the circle, Figure 8-3. To trim an object, you need to select the object and a boundary for the trim. A *boundary* is an object that defines where to cut the object.

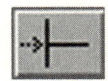

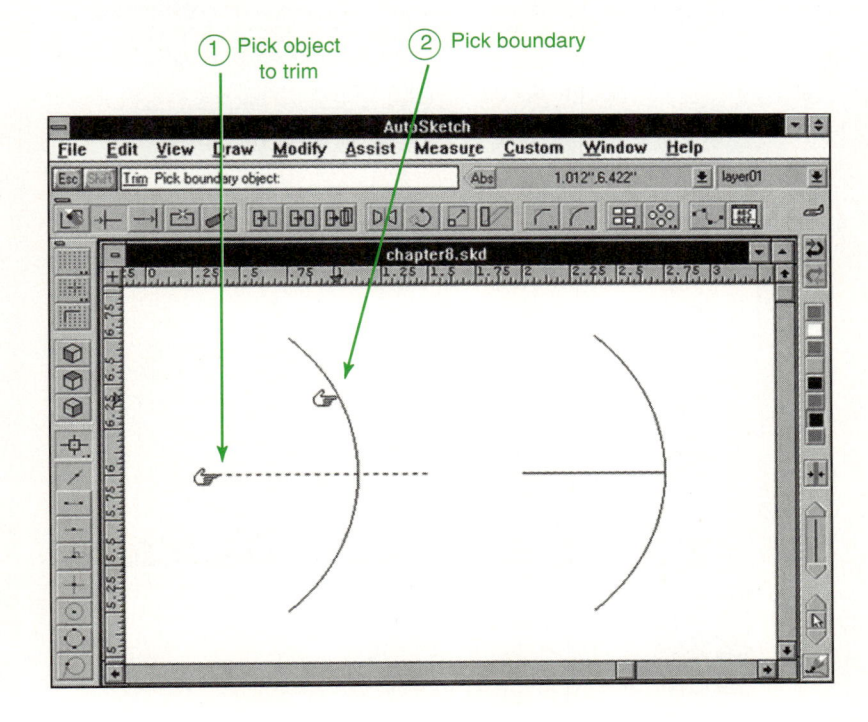

Figure 8-3.
To trim an object, select the object to trim. Pick the object on the side of the boundary that should remain. Then, select the boundary object.

You can trim lines, arcs, circles, and ellipses. Objects with widths other than hairline cannot be trimmed. For items that cannot be trimmed, such as a box or polygon, use the **Break** tool.

To trim an object, select the **Modify** menu and the **Trim** tool. The following prompt appears.

Trim Select object:

Select the object you want to trim. Where you select the object is important. The part of the object that you pick will remain. Everything on the other side of the boundary will be trimmed. AutoSketch then asks you for the boundary.

Trim Pick boundary object:

Select the boundary object. This defines the ending point of the trimmed object. Almost any object can be a boundary, except text. If the boundary is a wide object, pick the side that you want to trim to. Also, the object to trim may cross the boundary more

SHORTCUT
To trim an object with a width other than hairline, first change to a hairline width. Then trim the object, and change back to the needed line width.

than once. The object is trimmed where each endpoint touches the boundary. However, if a line or arc crosses a boundary, only one end of the object is trimmed at a time.

Figure 8-4.
Use a selection set or crosses/window box to select multiple items to trim to a common boundary.

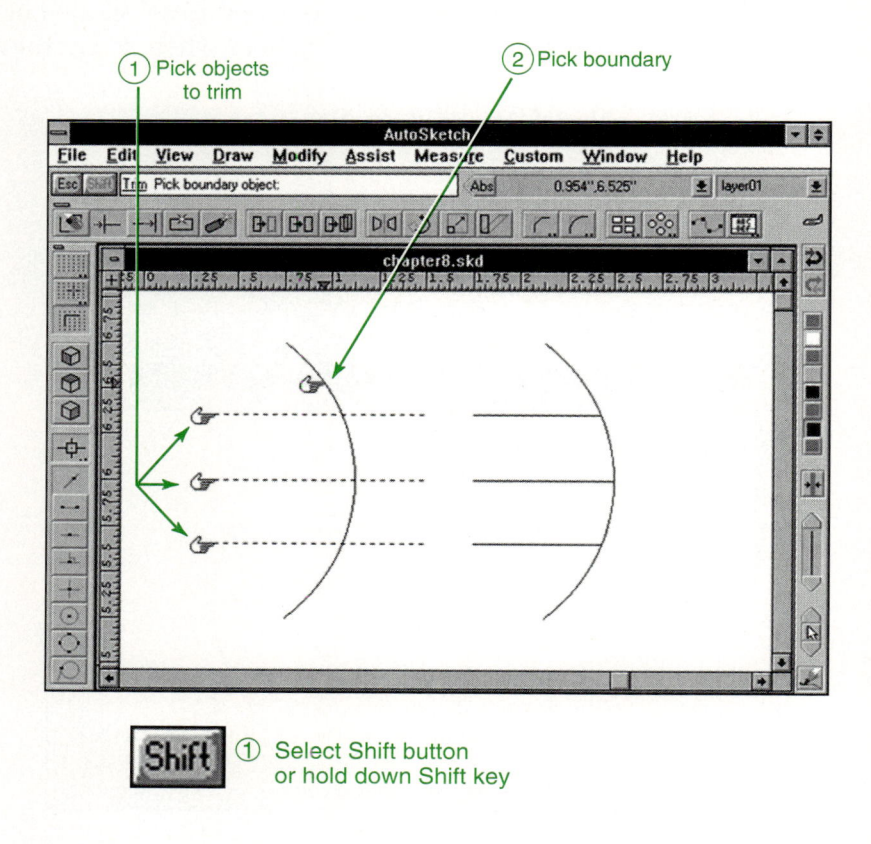

You can trim more than one object at a time, Figure 8-4. For example, you can select several lines, circles, and arcs that cross a common boundary to trim to that boundary. Remember, the side of each object you pick will remain intact after the trim. However, if you use a crosses/window box to select multiple objects, AutoSketch does not know what side you want to trim. In this case, AutoSketch prompts you to choose which side of the objects to trim.

Trim Pick a side of the boundary:

Move your cursor on each side of the boundary. AutoSketch highlights the trimmed objects. Pick the side you want to remain intact. The opposite portion of the objects disappears.

Exercise 8-3

1. Start a new drawing using the **Quick Start** template. Select **Inches (decimal)** as the unit of measurement.
2. Draw a circle and three lines through the circle as shown below on the left.
3. Use the **Trim** tool and select the lines to trim using a selection set. Remember to pick the lines in the right places so that they trim properly.
4. Pick the circle as the boundary object.
5. Save the drawing as **C8E3** on your work diskette and then close the drawing.

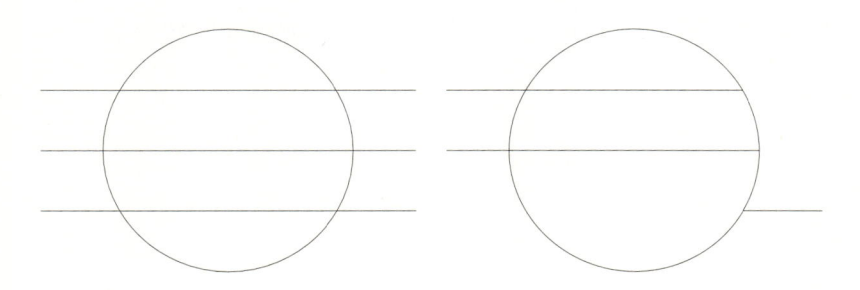

Extending Objects

The **Extend** tool lets you lengthen a line, arc, or broken ellipse to another object so that it joins precisely. For example, you can extend a line to a circle so that the endpoint of the line lies exactly on the circle, Figure 8-5. To extend an object, you need to select the object and a boundary.

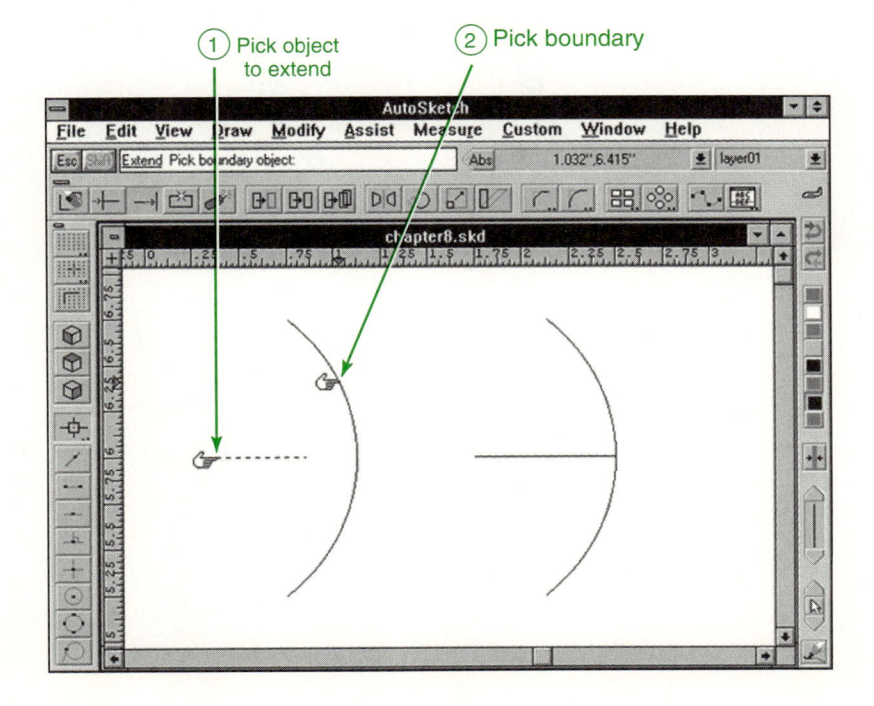

Figure 8-5.
To extend an object, first select the object. Then select a boundary object to extend to.

Almost any object can be a boundary, except text. Objects with widths other than hairline cannot be extended. You must change the width to Hairline, extend the object, and then change the width back.

To extend an object, select the **Modify** menu and the **Extend** tool. The following prompt appears.

<u>Extend</u> Select object to extend:

Select the object to extend. AutoSketch then asks you for the boundary.

<u>Extend</u> Pick boundary object:

Select the boundary object. This location defines the ending point of the extended object. Where you select the boundary is important. Pick the boundary object at a point close to the desired intersection. For example, if you plan to extend a line to a circle, the line could extend to the closest side or to the farthest side. Although the object to extend cannot be a wide object, the boundary can be. If the boundary is a wide object, pick the side that you want to extend to. See Figure 8-6.

Figure 8-6.
Extending to a wide object is similar to extending to a closed object. Pick the side of the boundary that you want the object extended to. Here, the line is extended to the far side of the wide arc, though on-screen this is not seen.

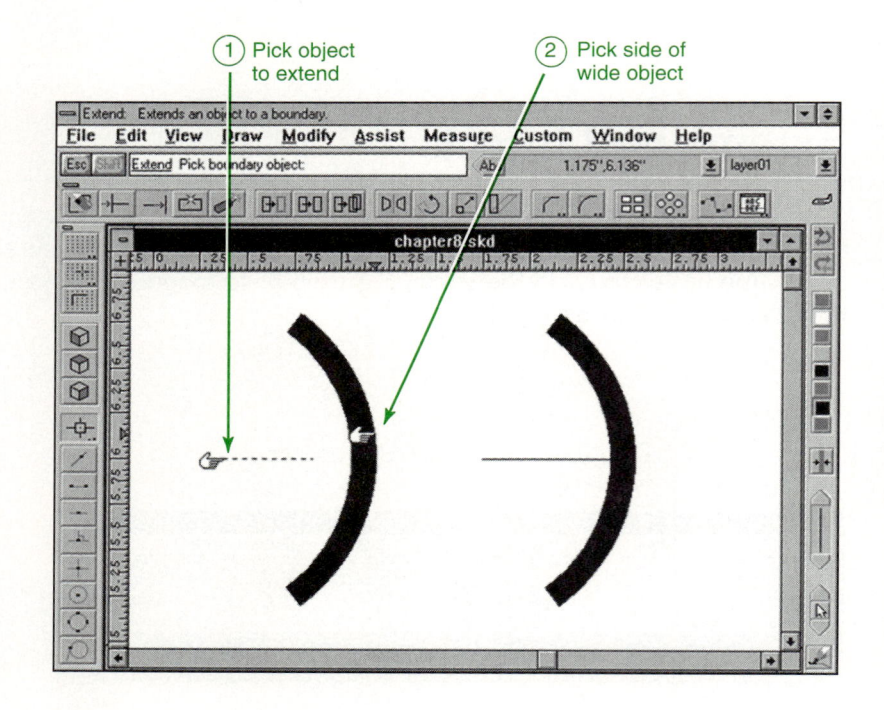

You can extend more than one object at a time. For example, you can select several lines or arcs to extend to a common boundary.

Exercise 8-4

1. Start a new drawing using the **Quick Start** template. Select **Inches (decimal)** as the unit of measurement.
2. Draw a wide horizontal line, an arc, and a vertical line, as shown below on the left.
3. Use the **Extend** tool and select the arc and vertical line to extend using a crosses box.
4. Pick the horizontal line as the boundary object.
5. Save the drawing as C8E4 on your work diskette and then close the drawing.

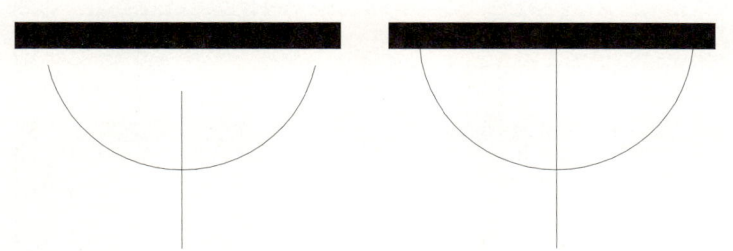

Drawing Fillets and Chamfers

Fillets, rounds, and chamfers refer to how inside and outside corners are treated on machined and welded parts. A *fillet* is a rounding of an inside corner to relieve the stress of machining. *Rounding* and *chamfering* (beveling) also relieve stress, but both are done on outside corners. Although the **Fillet** tool and **Chamfer** tool are named for mechanical drafting applications, they have many uses in all fields of drafting.

Drawing Fillets

The **Fillet** tool creates an arc between two non-parallel objects. To set the fillet radius, double-click on the **Fillet** tool in the **Modify** menu. The **Fillet Setting** dialog box will appear, Figure 8-7. Enter the radius of fillet you want to create. Then, pick **OK** to close the dialog box. The following prompt appears.

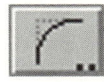

<u>Fillet</u> Select object:

Select the first object. AutoSketch then asks for the second object to be filleted.

<u>Fillet</u> Select second object:

Once both objects are selected, the fillet is drawn, Figure 8-8. You can also select both objects at the same time with a crosses/window box or a selection set.

Figure 8-7.
Double-click on the **Fillet** tool to display the **Fillet Settings** dialog box.

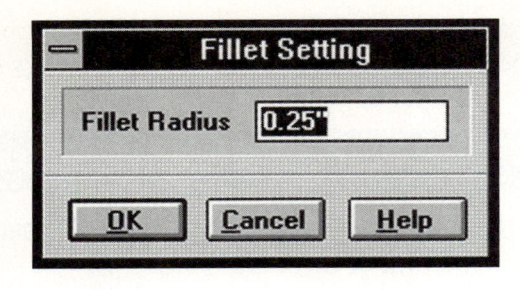

Figure 8-8.
With the fillet radius set, pick the objects to fillet. Once the second object is selected, the fillet is drawn.

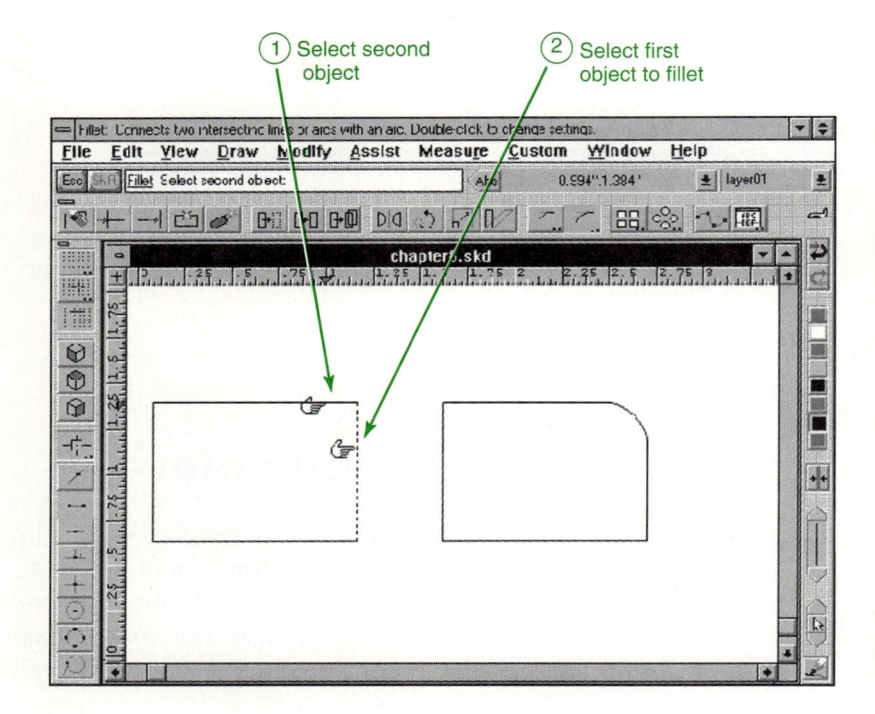

Effect of Fillet on Objects

Figure 8-9 shows the effects of applying a fillet to different objects. This figure also shows how where you pick an object to fillet determines the placement of the fillet. The following is a list of other important things to remember about fillets.

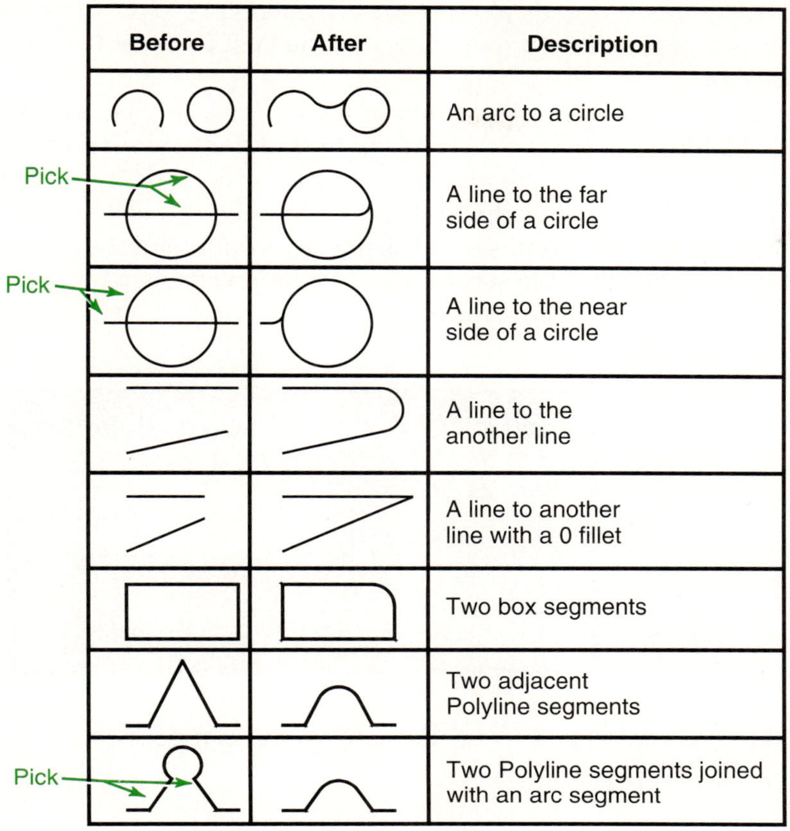

Before	After	Description
⌒ ○ (Pick)	⌒○	An arc to a circle
⊖ (Pick)	⊖	A line to the far side of a circle
⊖ (Pick)	⊖	A line to the near side of a circle
/ /	∫	A line to the another line
/ /	/ /	A line to another line with a 0 fillet
▭	▭	Two box segments
⋀	⌒	Two adjacent Polyline segments
⋀○ (Pick)	⌒	Two Polyline segments joined with an arc segment

Figure 8-9.
The **Fillet** command can affect objects differently.

- If the two objects cross, the objects are trimmed and then filleted. If the objects do not join, the fillet is drawn and the objects are extended to meet the fillet.
- A useful application of the **Fillet** tool is to extend or trim two nonparallel lines until they intersect. Enter a fillet radius of 0. This will extend the two lines until they meet, or trim two lines to their intersection.
- You cannot fillet parallel lines.
- You cannot fillet wide lines, wide arcs, wide circles, wide ellipses, or pattern fills. You must first change the width to "hairline," complete the fillet, and then change the objects back to their original width.
- You can fillet adjacent segments of boxes, polylines, or polygons by selecting each segment separately. Do not use a crosses/window box or you will get unexpected results.
- If you fillet a wide box, polyline or polygon, the fillet radius must be at least half the width of the object (line width).
- If you fillet an object containing a pattern fill, the object will fillet, but the pattern fill will not. You must remove the pattern fill, fillet the object, and redraw the pattern fill.
- To fillet objects in a group, you must first ungroup them.

Drawing Chamfers

A chamfer is a beveled edge drawn between two lines or line segments of a box, polygon, or polyline. The **Chamfer** tool trims or extends the two lines, and connects the ends with a chamfer line. You can also chamfer two polyline segments separated by an arc segment. AutoSketch deletes the arc and replaces it with a chamfer line. You cannot chamfer wide lines, arcs, circles, ellipses, or pattern fills.

To set the chamfer distances, double-click on the **Chamfer** tool. The **Chamfer Settings** dialog box will appear, Figure 8-10. Enter values for the **First Chamfer Distance** and **Second Chamfer Distance.**

The first distance is how far from the intersection of the two objects that the first object will be chamfered. The second distance is how far from the intersection of the two objects that the second object will be chamfered. In other words, you are setting how much of the object will be trimmed or extended.

To select different values for the first and second distances, you must set the **First Chamfer Distance** value first. Notice that the second distance value changes to match the first. Then, you can select a different value for the second distance.

Figure 8-10.
Double-click on the **Chamfer** tool to display the **Chamfer Settings** dialog box.

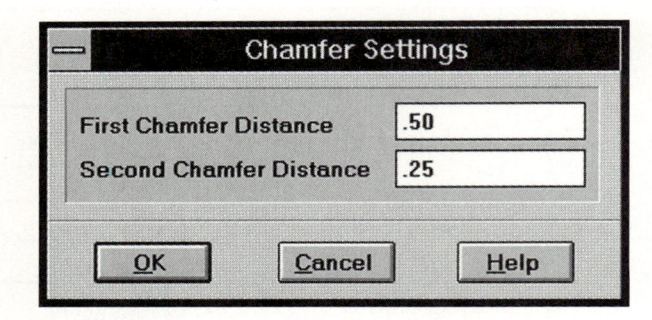

Finally, click **OK** to close the dialog box. The following prompts appears.

Chamfer Select object:

Pick the object that you want the **First Chamfer Distance** applied to. AutoSketch then asks for the second object.

Chamfer Select second object:

Pick the object that you want the **Second Chamfer Distance** applied to. Once you pick the second object, the chamfer is drawn, Figure 8-11. You can also select both objects at the same time with a crosses/window box or selection set. However, if the chamfer distances are not the same, pick each object separately.

Effect of Chamfer on Objects

Figure 8-12 shows the effects of chamfering different objects. The following is a list of other important things to remember about chamfers.
- If the two objects cross, the objects are trimmed and then chamfered. If the objects do not join, the chamfer is drawn and the objects are extended to meet the chamfer.
- If there are more than two objects selected, AutoSketch selects the objects in the order that they were drawn to create the chamfer.
- A chamfer between two lines is simply another line. You can erase it, or change its properties. If the two objects selected for a chamfer have the same property setting (for example, the same linetype), the chamfer will have that setting. If settings differ, the chamfer will have the current property settings.
- AutoSketch converts a chamfered box or polygon into a polyline. The chamfer line becomes a part of the polyline object. This means it will have the same properties, such as layer and color.

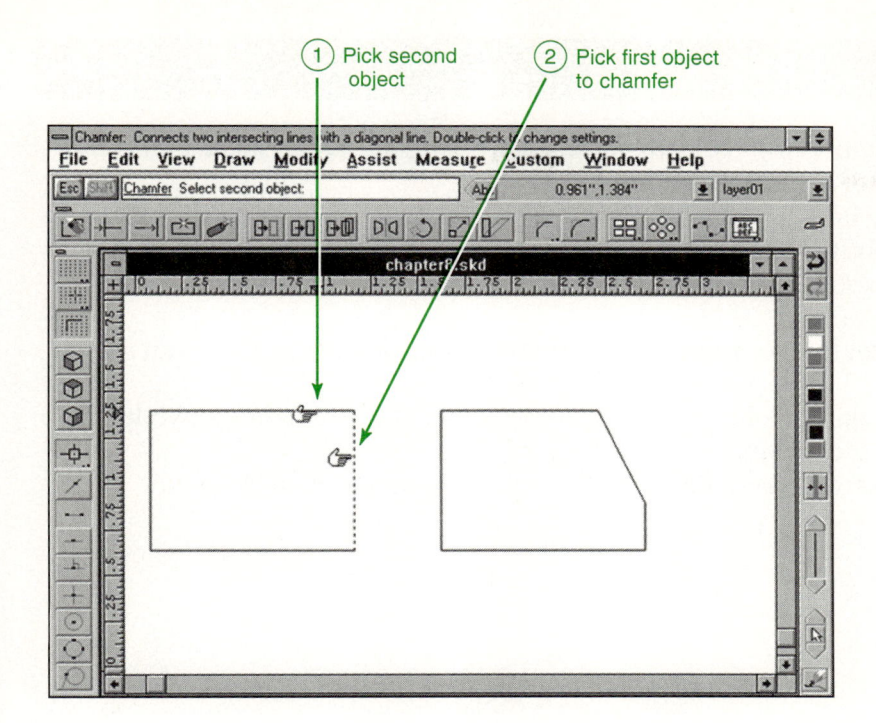

Figure 8-11.
To draw a chamfer, set the first and second chamfer distances in the **Chamfer Settings** dialog box. Then use the **Chamfer** tool and pick the objects. The order that you pick the objects determines their chamfer distance.

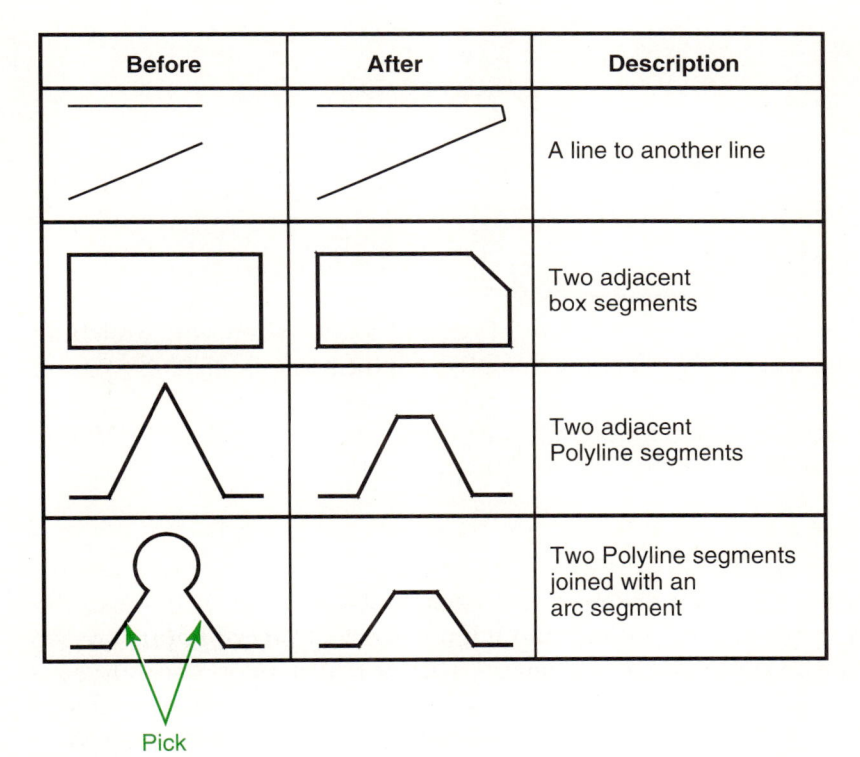

Figure 8-12.
The **Chamfer** command can affect objects differently.

Before	After	Description
		A line to another line
		Two adjacent box segments
		Two adjacent Polyline segments
		Two Polyline segments joined with an arc segment

- You cannot chamfer one closed object to another. For example, you cannot chamfer a box to another box. You also cannot chamfer one polyline to another.
- You cannot chamfer wide lines, wide arcs, wide circles, wide ellipses, or pattern fills. You must first change the width to "hairline," complete the chamfer, and then change the objects back to their original width.

Exercise 8-5

1. Start a new drawing using the **Quick Start** template. Select **Inches (decimal)** as the unit of measurement.
2. Draw three lines as shown below on the left. These lines should nearly form a triangle, but the endpoints should not connect.
3. Use **Fillet** tool with a zero radius fillet to connect the endpoints of two lines precisely.
4. Use the **Fillet** tool to connect the endpoints of another two lines with a small fillet arc.
5. Use the **Chamfer** tool to connect the endpoints of the remaining two lines with a small chamfer.
6. Save the drawing as C8E5 on your work diskette and then close the drawing.

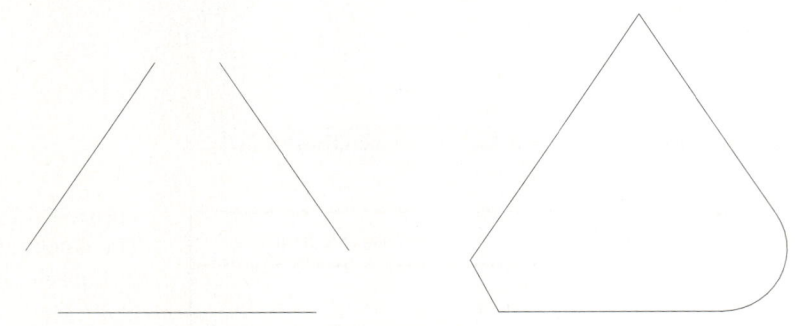

Stretching Objects

The **Stretch** tool is a unique combination of several tools. When you stretch an object, you move, scale, and extend pieces of that object at the same time. It allows you to stretch a shape as if it were made of rubber. You can distort the size, yet leave all connections between objects intact, Figure 8-13. To stretch an object, select the **Modify** menu and the **Stretch** tool. The following prompt appears.

Stretch First corner:

Notice that AutoSketch is not asking you to select objects. This is because the **Stretch** tool looks for control points rather than whole objects. *Control points* are the connections that define an object's size and shape. The four corners of a box are control points. The corners of a polygon are control points. When you stretch an object, you are really moving the control points. All components of an object connected to those control points are shortened or lengthened by the stretch.

SHORTCUT

You can also access the **Stretch** tool by pressing the function key [F7].

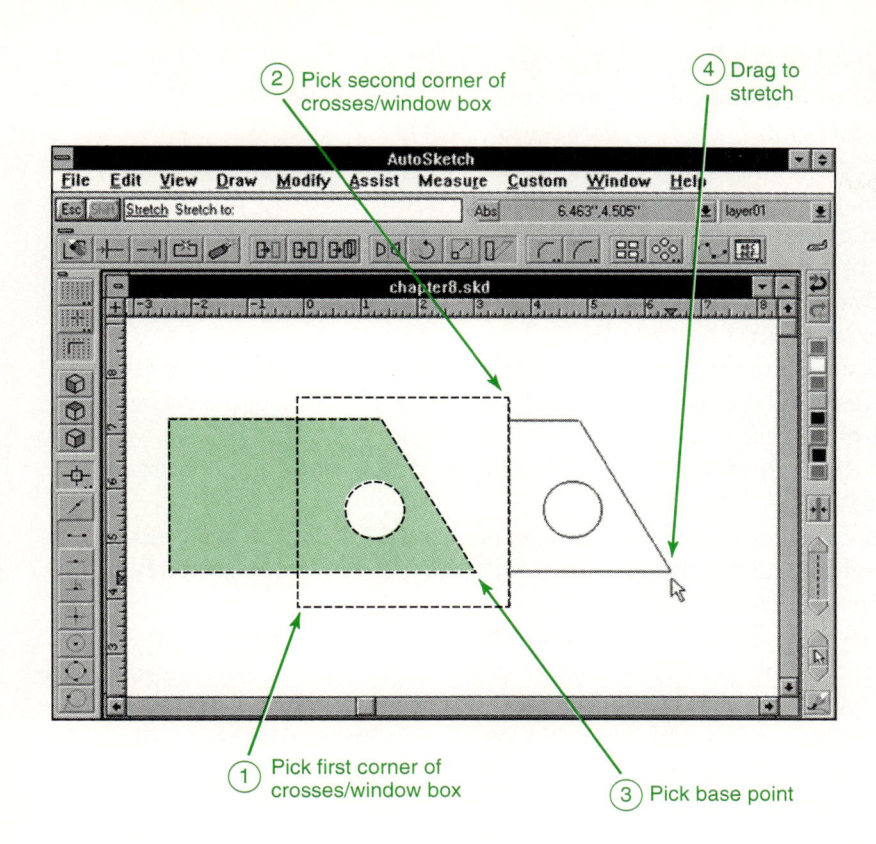

Figure 8-13.
To stretch an object, first select objects with a crosses box. Then, pick a stretch base point and drag the base point to stretch the objects. In this example, the original object is shown with a tint. The right side of the object has been stretched to a new location.

① Pick first corner of crosses/window box
② Pick second corner of crosses/window box
③ Pick base point
④ Drag to stretch

You must select control points using a crosses/window box. Pick the first corner of the crosses/window box. AutoSketch then asks for the other corner of the box. Notice as you move your cursor that the box is a crosses box, no matter what direction you move the cursor.

Stretch Crosses/window corner:

Pick the second corner of the box around the control points to move. Make sure the box crosses only the objects you want to stretch. AutoSketch then asks you for a base.

Stretch Stretch base:

The *stretch base* point is the place the cursor will hold to stretch the object(s). You can use an **Attach** tool to place the base point at a precise place, such as the corner of a box. However, this point doesn't have to be on the object. It is only a reference point for the cursor. AutoSketch then asks where you want to stretch the object to.

Stretch Stretch to:

At this prompt, move your cursor to stretch the shape to its new length or shape. The cursor holds on to the stretch base point. An image of the modified objects appears as you move the cursor. You can also type in coordinates or use the **Coordinate Display** to stretch the object. In many cases, you will stretch the shape to meet another object.

How Objects Are Affected by Stretch

The **Stretch** tool affects objects different ways. Keep the following list in mind when using the **Stretch** tool.

- All objects with endpoints and corners can be stretched. This includes lines, arcs, boxes, polylines, polygons, and pattern fills. Stretching a box or polygon turns the object into a closed polyline.
- Curves cannot be stretched. Instead, use the **Edit Curve** tool in the **Modify** menu to change the shape of the curve.
- Ellipses and circles cannot be stretched.
- Text and dimensions are affected by the **Stretch** tool. If you stretch one end of a text string, the text will compress or expand horizontally. Text can also be rotated using the **Stretch** tool.
- If you stretch a dimension along with objects, the dimension text changes to show the new distance, Figure 8-14.
- The **Stretch** tool can be used to stretch objects within a group. If grouped objects are within the selection box, the objects stretch only if they can be stretched normally. Otherwise, the group is unaffected.
- If you completely enclose an object in the crosses/window selection box, the **Stretch** command simply moves the object (including a group). The **Stretch** command will also move a circle or ellipse if you include at least two quadrant points or the center and one quadrant. However, remember that you will not be able to *stretch* a circle or ellipse.

Start a new drawing using the **Quick Start** template. Select **Inches (decimal)** as the unit of measurement.

Figure 8-14.
If you stretch a dimensioned object, the dimension will be updated. Here, a door symbol is stretched to a new location. Note how the dimensions also change to reflect the new position.

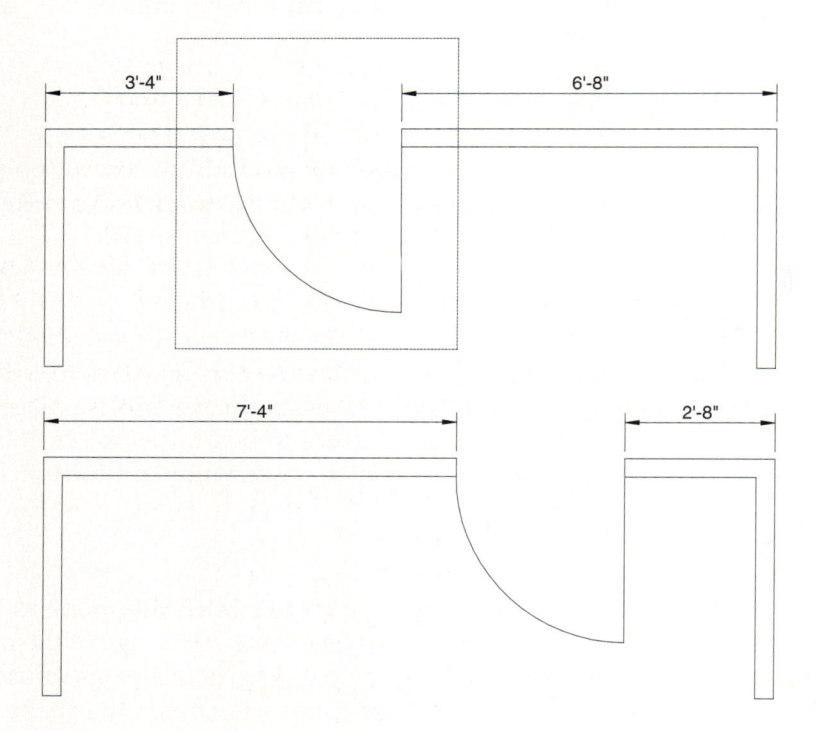

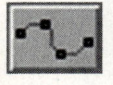

Exercise 8-6

1. Create the shaft and bearing shown below on the left. Use a grid, **Attach** tools, and/or coordinates to draw the objects to the size given. Do not dimension the drawing.
2. Use the **Stretch** tool to lengthen the shaft to the right of the bearing by 2".
3. Use the **Stretch** tool to increase the diameter of both the shaft and bearing. Increase each diameter by 1".
4. Save the drawing as **C8E6** on your work diskette and then close the drawing.

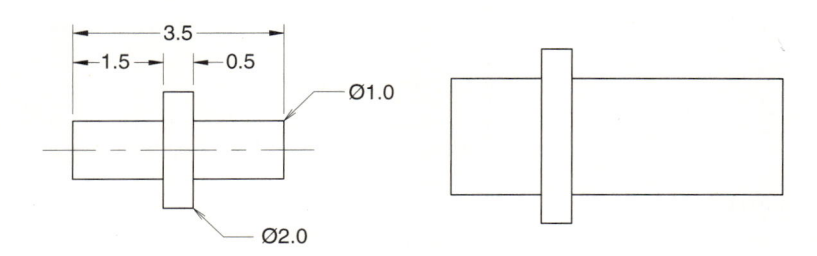

Drawing a Box Array

The **Box Array** tool makes multiple copies of an object or group in a rectangular pattern. This is easier than using the **Multiple Copy** tool in situations where you need a number of copies in an exact pattern. You can enter the distance between copies, or pick the distance with your cursor. You can also have the array of copies placed at an angle. Before actually drawing the array, you need to determine the box array settings.

Setting Box Array Settings

To access the **Box Array Settings** dialog box, double-click on the **Box Array** tool in the **Modify** menu, Figure 8-15. The different options are discussed in the next sections.

Point

This section allows you to determine the distance to space copies. This value represents the distance between *edges* or *features* of the objects, *not* the space between objects. For example, if you will be copying a square that is 2 units, a value of 1 for the distance will cause copies to overlap. However, a distance of 3 will result in 1 unit of space between copies.

If either of the boxes in this section is checked, AutoSketch will prompt you to use the cursor to determine the distance. If neither box is checked, AutoSketch uses the values listed to determine the distance. Enter the vertical distance between copies next to **Row Distance**. Enter the horizontal distance between copies next to **Column Distance.**

> **SHORTCUT**
> To access the **Box Array** tool at another time, you can also hold down the [Ctrl] key and press the function key [F2].

Figure 8-15.
Double-click on the **Box Array** tool to display the **Box Array Settings** dialog box.

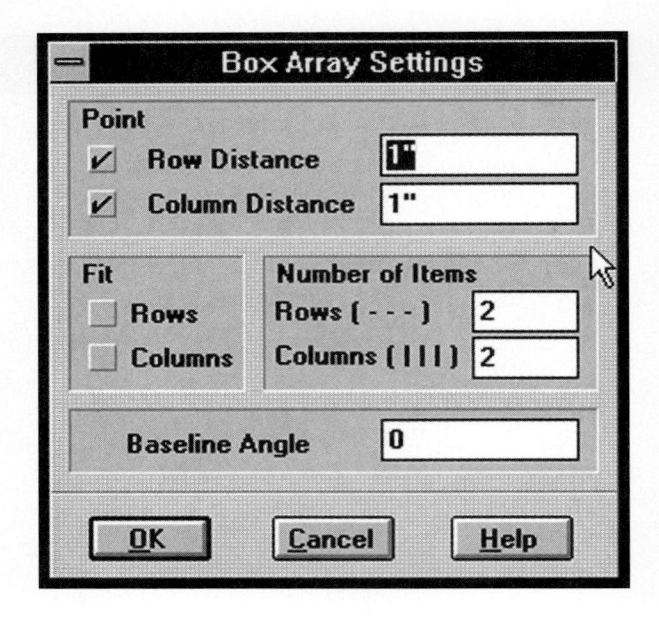

Number of items
This section allows you to enter the number of copies that will be made. Enter the number of copies you want vertically next to **Rows.** Enter the number of copies you want horizontally next to **Columns.** For example, if you enter 3 rows and 4 columns, there will be a total of 11 copies made. There will be 12 total objects, including the original.

Fit
This section allows you to define the total height and width of the array. All copied objects are fit within the locations you pick. The distances entered in the **Point** section are ignored.

Baseline Angle
The **Baseline Angle** determines what angle the copies are placed at. The copies themselves are not rotated. In other words, you are setting how much the array is "slanted."

Drawing a Simple Box Array
First, draw a small object such as a circle. Open the **Box Array Settings** dialog box by double-clicking on the **Box Array** tool in the **Modify** menu. In this example, the boxes next to **Point** and **Fit** are not checked. Instead, enter a value such as 2 next to both of these options. Leave the **Baseline Angle** at the default 0. Also, enter a value of 2 for the **Rows** and **Columns.**

Once the box array settings are made, pick **OK** to close the dialog box. The following prompt appears.

> Box Array Select object:

Select the object you want to be part of the array. You can select a single object, groups, or multiple objects. To select multiple objects, use the crosses/window box or selection set. Once you select the object(s), the copies are instantly drawn at the distance set in the **Box Array Settings** dialog box, Figure 8-16.

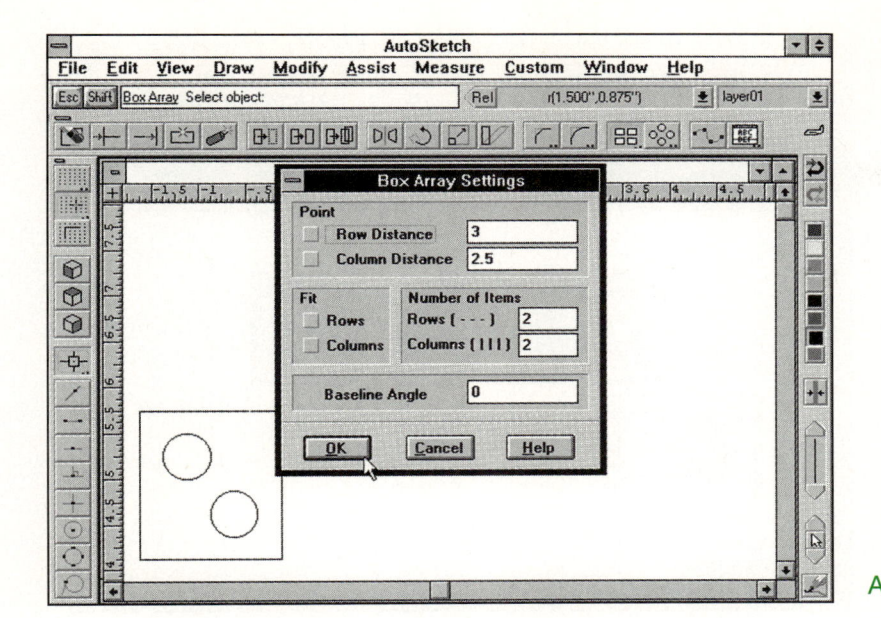

Figure 8-16.
The box array settings on the left result in the completed array on the right.

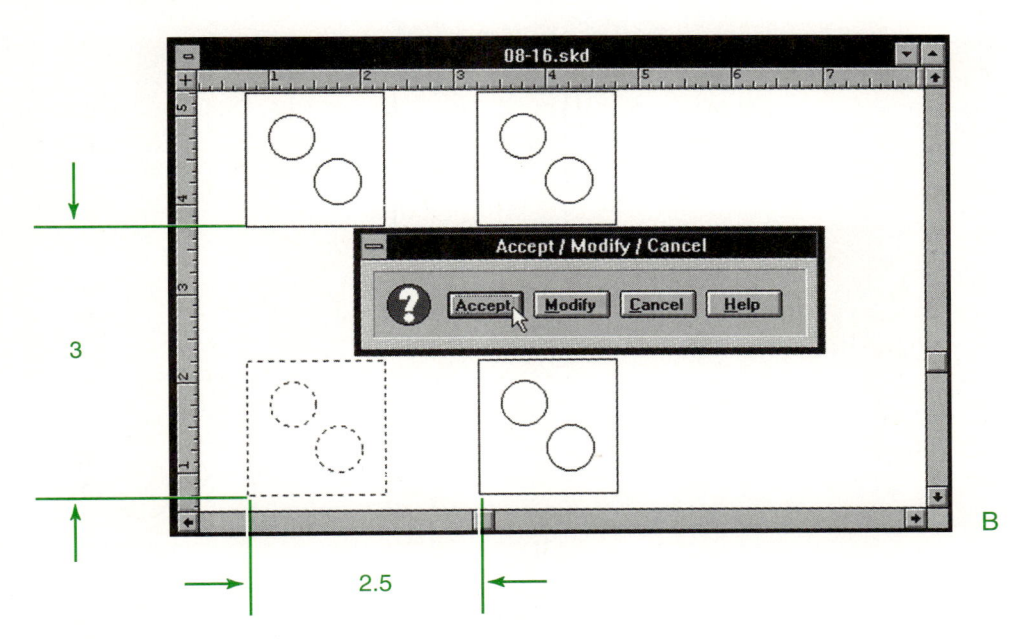

If for some reason your settings were not correct, AutoSketch allows you to make changes. After the copies are placed, the dialog box in Figure 8-17 appears. You can either **Accept** the array, select **Modify,** or select **Cancel**. If you pick **Cancel,** the **Box Array** command will end without making any copies. If you pick **Modify,** the **Box Array Settings** dialog box reappears. You can enter different values and try again. Once **OK** is picked in the dialog box, the array is recreated automatically.

Drawing a Box Array by Picking Distances

When the **Point** box is checked in the **Box Array Settings** dialog box, any settings next to **Row Distance** and **Column Distance** are ignored. After closing the dialog box and selecting an object, AutoSketch prompts you for points. These points define the distance between copies, Figure 8-18. After selecting the **Box Array** tool or closing the **Box Array Settings** dialog box, the following prompt appears.

Figure 8-17.
After creating an array, AutoSketch asks whether you want to accept or modify it. In this case, the array did not turn out as expected and you would select **Modify**.

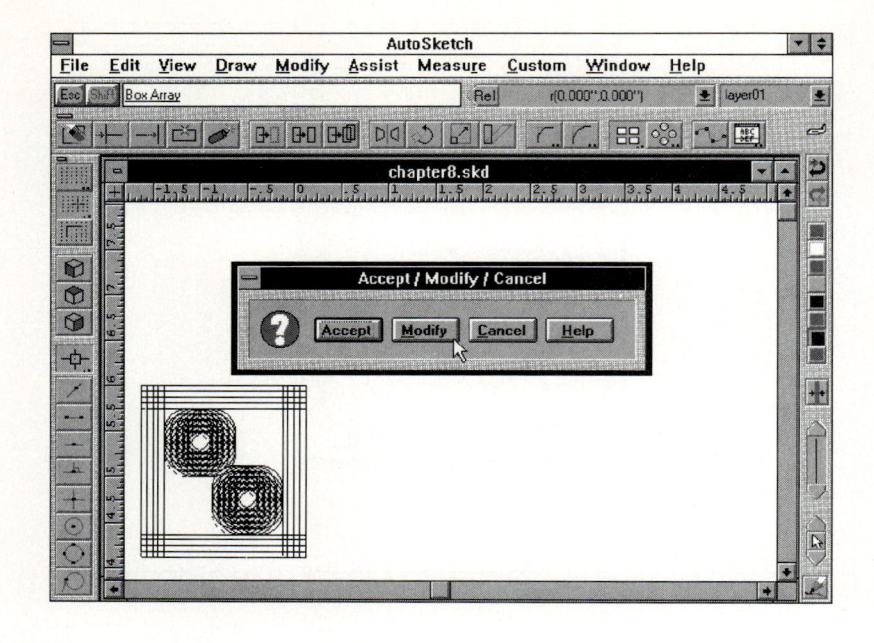

Figure 8-18.
When **Point** is checked in the **Box Array Settings** dialog box, you must manually define the distance between copies.

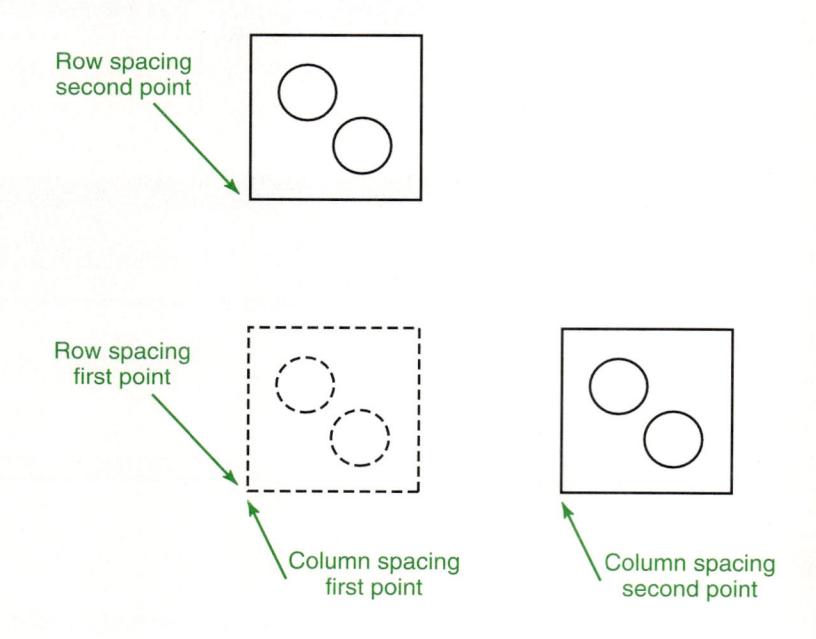

Box Array Select object:

    Select the object to copy. AutoSketch then asks you for the first of two points to define the horizontal distance between copies.

Box Array Column spacing First point:

    Pick a base point for column spacing. This does not need to be on the object. The points you choose are just reference points. AutoSketch then asks you for a second point to define the horizontal distance.

Box Array To point:

Drag the object left or right to specify the direction and horizontal spacing of the box array. Notice that the object will only move horizontally, even if you do not have **Ortho** on. Pick this distance. AutoSketch then asks for the first of two points to define the vertical spacing.

<u>Box Array</u> Row spacing First point:

Pick a base point for row spacing. This point is a reference point and does not need to be on the object. AutoSketch then asks for a second point to define the vertical spacing.

<u>Box Array</u> To point:

Drag the object up or down to specify the direction and vertical spacing of the box array. Notice that the object only moves vertically, even if you do not have **Ortho** on. Pick this distance. The box array is then drawn.

AutoSketch places the values for the distances you picked in the **Box Array Settings** dialog box. If you want to draw another box array with the same values, open the dialog box. Pick the boxes in the **Point** section so that the check marks disappear. AutoSketch will then use the values for the next array.

Drawing Box Arrays Using Fit and Baseline Angle

The **Fit** box in the **Box Array Settings** dialog box changes the meanings of **Row Distance** and **Column Distance** settings. When this box is checked, these values mean the total distance between farthest copies, *not* between *each* copy, Figure 8-19. This tool works well if you know the number of objects that must fill a specific area.

Another option in the **Box Array Settings** dialog box is **Baseline Angle.** This setting allows you to set the angle that copies are placed at. The copies are not rotated, only the "box" that they are placed in. See Figure 8-20.

Figure 8-19.
Fitting a box array. Checking the boxes in the **Fit** section of the **Box Array Settings** dialog box fits all copies within the points you pick. Notice in this example that the points picked define the *total* distances. The array is fit within these distances.

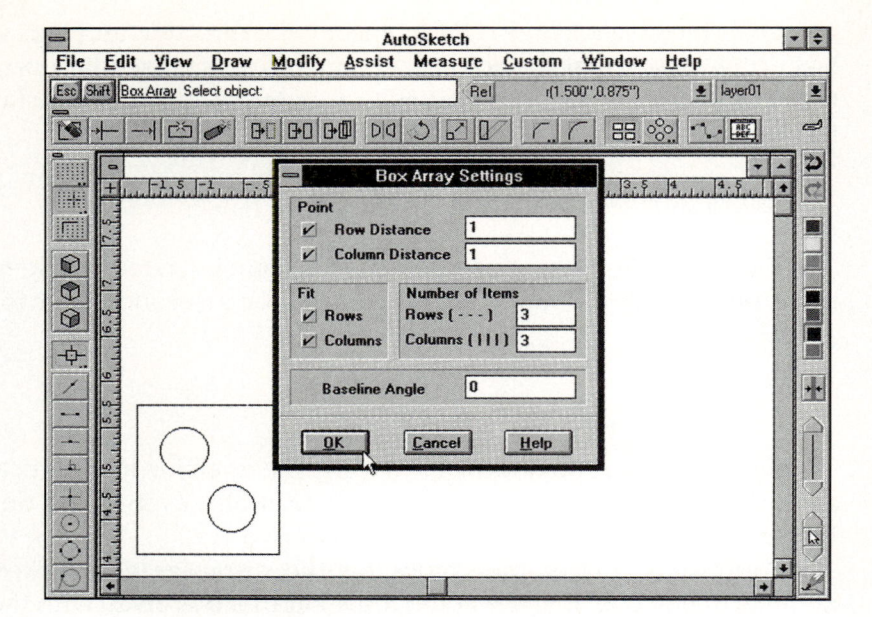

A

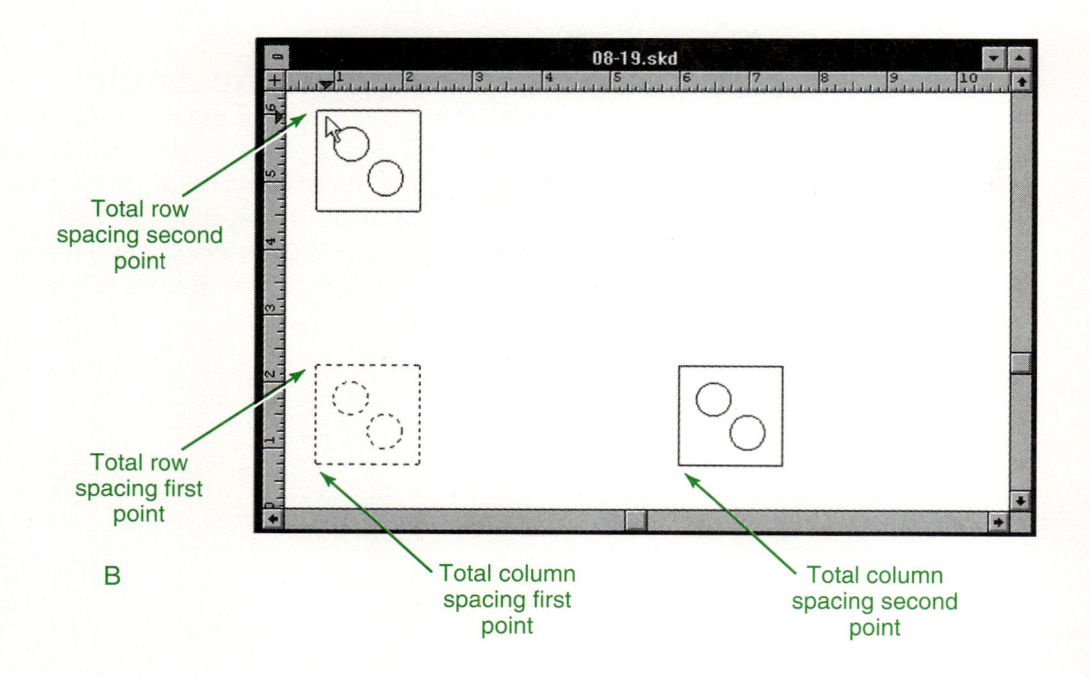

Total row spacing second point

Total row spacing first point

B

Total column spacing first point

Total column spacing second point

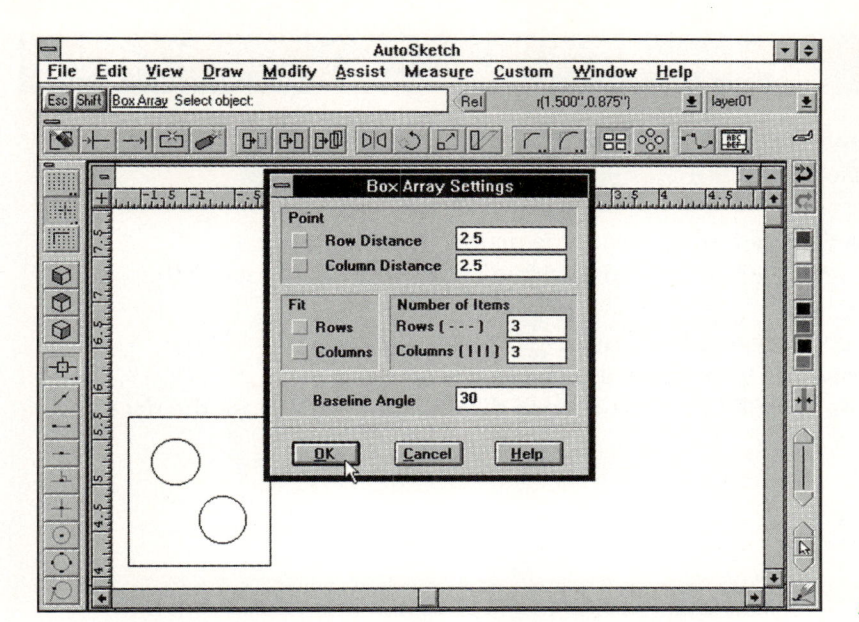

Figure 8-20.
To place a box array at an angle, enter an angle value opposite the **Baseline Angle** box. The copied objects are not rotated. However, the "box" that defines the array is rotated.

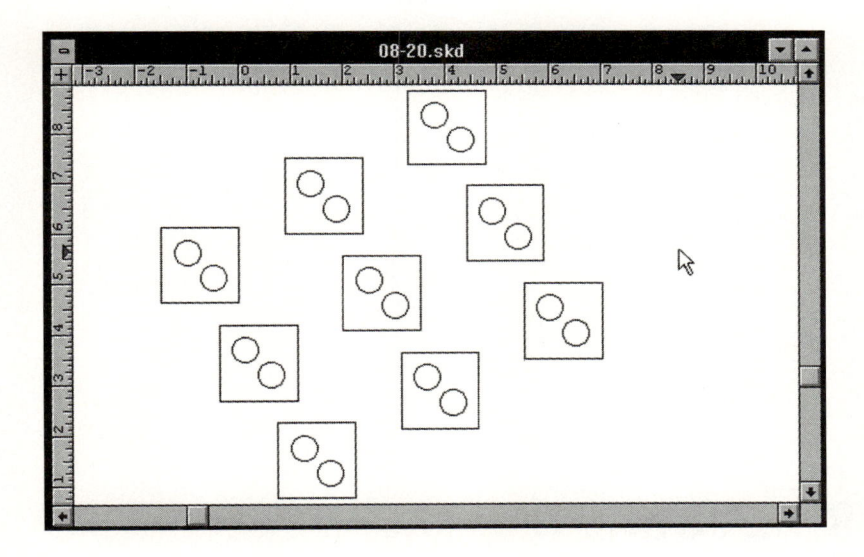

A

B

Exercise 8-7

1. Start a new drawing using the **Quick Start** template. Select **Inches (decimal)** as the unit of measurement.
2. Draw a 1" square.
3. Use the **Multiple Copies** tool to create the array shown below.
4. Save the drawing as C8E7A on your work diskette, but do not close the drawing.
5. Erase all of the copies except for the original.
6. Use the **Box Array** tool to recreate the array.
7. Save the drawing as C8E7B on your work diskette and then close the drawing.

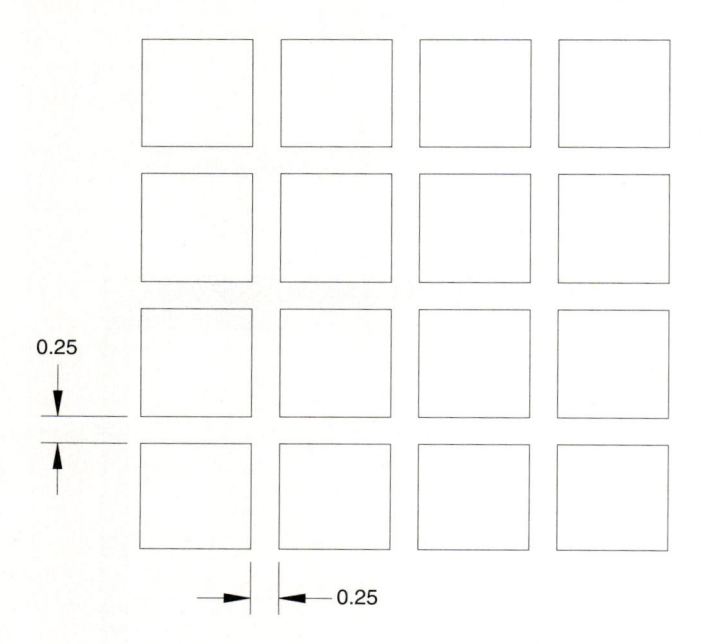

Drawing a Ring Array

The **Ring Array** tool is much like the **Box Array** tool. However, the copies are placed in a circle around a center point. Before actually drawing the array, you need to determine the ring array settings.

Setting Ring Array Settings

To access the **Ring Array Settings** dialog box, double-click on the **Ring Array** tool in the **Modify** menu, Figure 8-21. **Ring Array Settings** set the number of items in a ring array, the rotation angle of the items, the center point of the array, and the pivot point of the items in the array. The different options are discussed in the next sections.

Number of Items

This number is the total number of items. This number includes the original, so the number of *copies* will always be one less than this number.

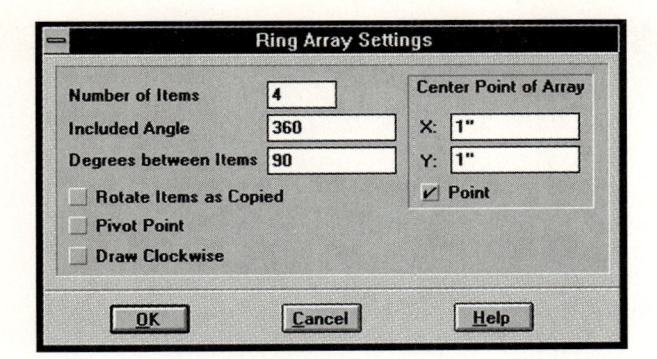

Figure 8-21.
Double-click on the **Ring Array** tool to display the **Ring Array Settings** dialog box.

Included Angle
This is the total angle included from the first copy to the last copy in the array. This value cannot exceed 360 degrees (a full circle). As you change this value, the **Degrees between Items** value will change to evenly space the copies.

Degrees between Items
This is the angle between copied objects. As you change this value, the **Included Angle** value will change to account for the total angle from the first copy to the last copy.

Rotate items as Copied
If this box is checked, the items are individually rotated as the array is created around the center point.

Pivot Point
If **Pivot Point** is checked, you can pick a new pivot point other than the default. The default is the center of the object(s) being copied. If this option is checked, you will be prompted for the pivot point while using the tool.

Draw Clockwise
By default, copies are made counterclockwise. If you check the **Draw Clockwise** option, AutoSketch will create the array clockwise from the original object.

Center Point of Array
The values in this section allow you to determine the center point that the objects are rotated around. If the **Point** box is checked, AutoSketch will prompt you for the center point of the array. If the box is not checked, AutoSketch uses the absolute coordinate values for **X:** and **Y:** to determine the center.

Drawing a Simple Ring Array

For this example, the default settings are used. First draw a small object, such as a rectangle. Then, select the **Modify** menu and the **Ring Array** tool. The following prompt appears.

Ring Array Select object:

Select the object you want to be part of the array. You can select single objects, groups, or multiple objects. To select multiple objects, use the crosses/window box or selection set. Once you select the object, the following prompt appears. (Note: this only appears if the **Point** box is checked in the settings.)

Ring Array Center point of array:

Once you pick the center point of the array, the copies are instantly drawn at the proper angles. If for some reason your settings were not correct, AutoSketch allows you to make changes. After the array is created, the dialog box in Figure 8-22 appears. You can either **Accept** the results of your array, select **Modify**, or select **Cancel**. If you pick **Cancel**, the **Ring Array** command will end without making any copies. If you pick **Modify**, the **Ring Array Settings** dialog box reappears. You can enter new values and try again. Once **OK** is picked in the **Ring Array** dialog box, the array is recreated automatically.

Figure 8-22.
After creating an array, AutoSketch asks whether you want to accept or modify it.

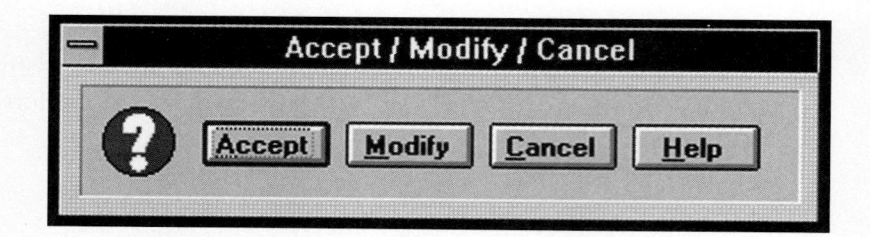

Figure 8-23 shows a bicycle pedal assembly. A ring array was used to reflect the position of the pedal around the crankshaft. The array was made using the settings shown at the left. Figure 8-24 through Figure 8-27 show how the different options can change the array.

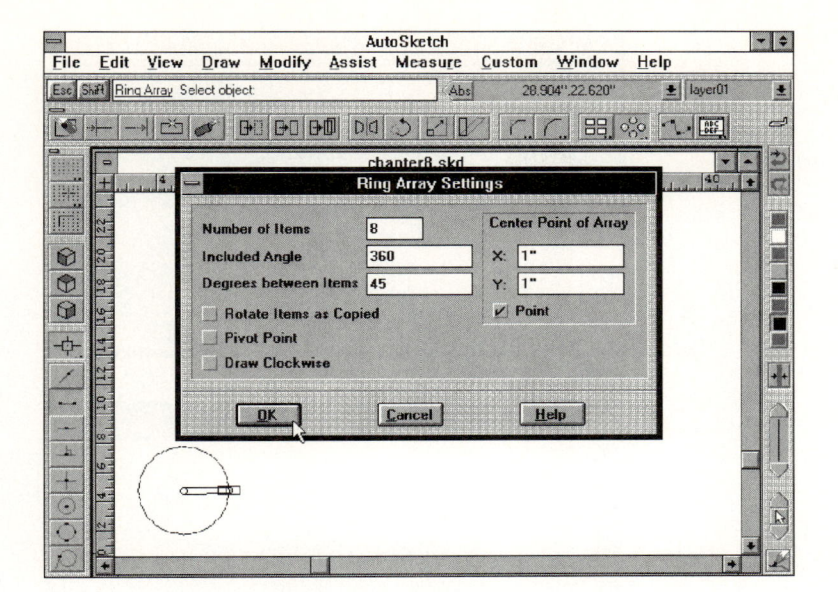

Figure 8-23.
This array of a bicycle pedal was created using the values shown on the left.

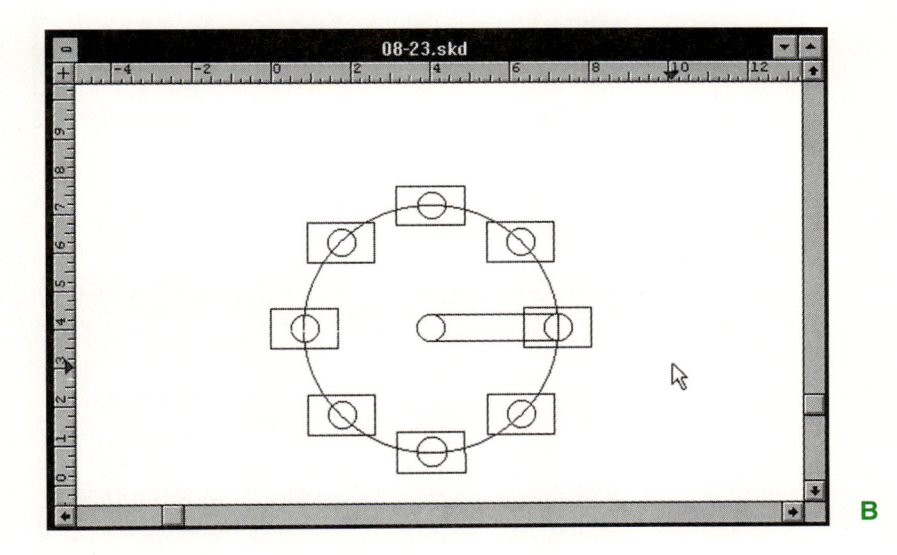

Figure 8-24.
This ring array has five copies that extend 180 degrees (half circle).

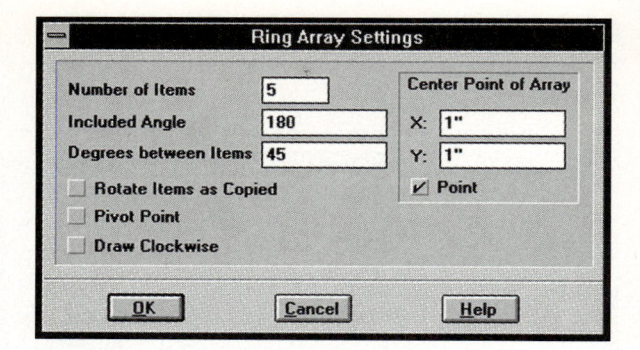

A

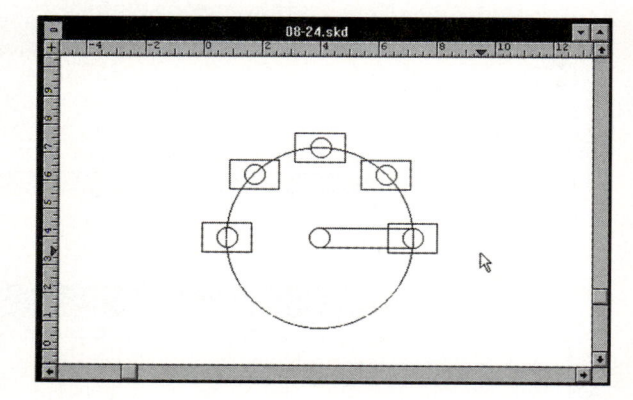

B

Figure 8-25.
This ring array has five copies that extend 180 degrees (half circle). However, the **Draw Clockwise** box was checked. Notice how this differs from the array in Figure 8-24

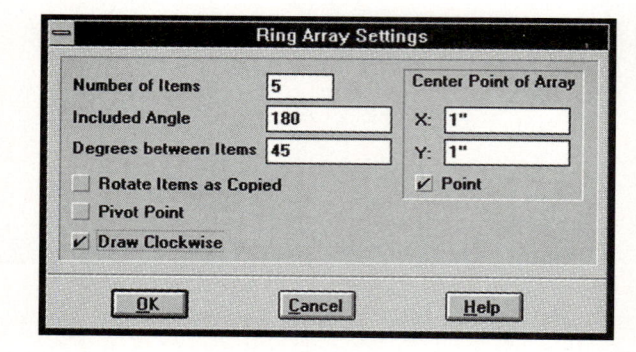

A

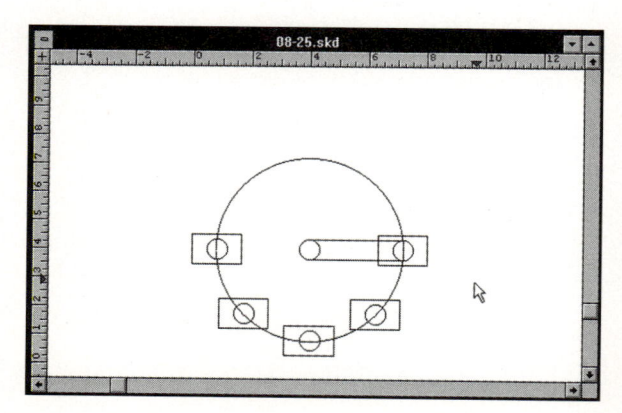

B

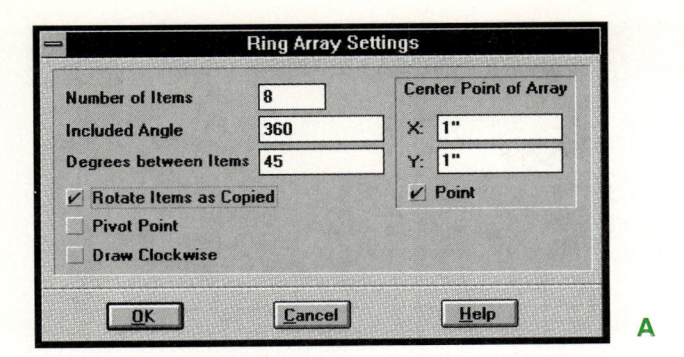

Figure 8-26.
In this ring array example, the **Rotate Items As Copied** box was checked.

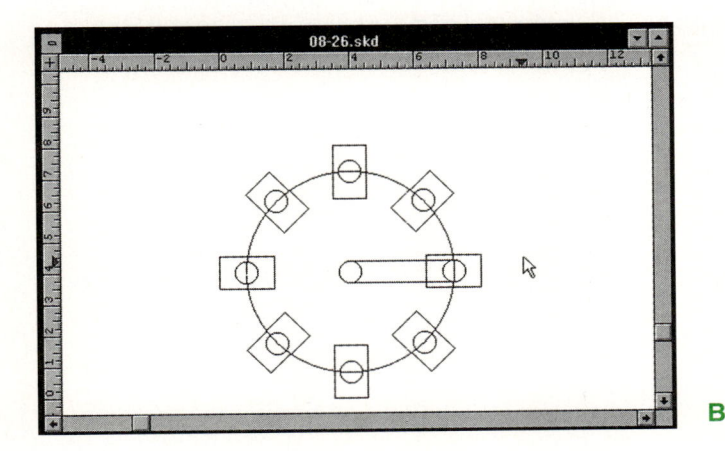

Figure 8-27.
In this ring array example, the **Pivot Point** box was checked, and the user picked a different pivot spot on the pedal. Notice how the pedal is offset.

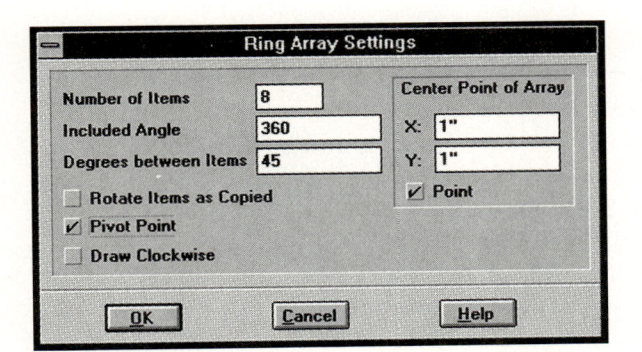

Exercise 8-8

1. Start a new drawing using the **Quick Start** template. Select **Inches (decimal)** as the unit of measurement.
2. Draw a triangle using the **Polygon** tool.
3. Use the **Ring Array** tool to create the array shown below on the left.
4. Save the drawing as C8E8A on your work diskette, but do not close the drawing.
5. Erase all of the copies except for the original.
6. Use the **Ring Array** tool to create the array shown below on the right. (Hint: You will need to draw an extra line and then erase it to complete this array.)
7. Save the drawing as C8E8B on your work diskette and then close the drawing.

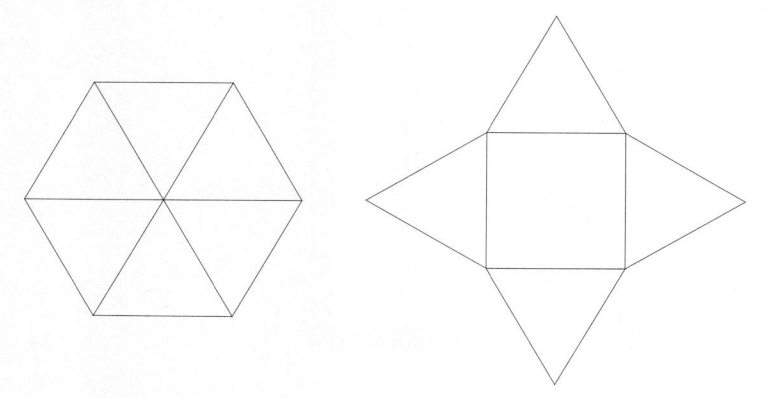

Editing Curves

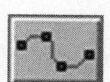

Curves and freehand objects fit into an invisible frame that controls their shape. The curve settings you chose when drawing the curve determines how the curve fits into the frame. There are several control points on both the curve and the frame. These points are invisible until you select the **Edit Curve** tool. This tool allows you to change the shape of a curve by moving, deleting, or inserting control points. You can make changes on the curve itself or on the frame that defines it.

To edit a curve or freehand drawing, select the **Modify** menu and the **Edit Curve** tool. The following prompt appears.

Edit Curve Select object:

Pick a curve or freehand object. If you select a Bezier curve or freehand object, the **Edit Bezier** dialog box appears, Figure 8-28. If you select a B-spline curve, the **Edit B-spline** dialog box appears. Control points appear along the curve. You can delete, move, and insert control points. When finished editing the curve, click on **OK**. The control points will disappear.

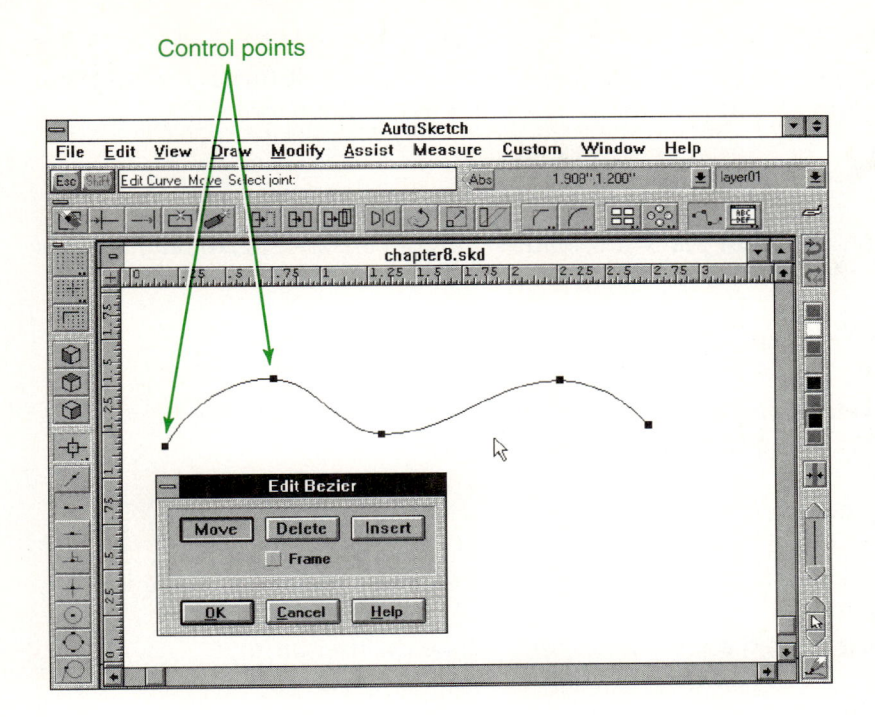

Figure 8-28.
Use the **Edit Curve** tool to select a curve to edit. Once the curve is selected, control points appear on the curve.

Displaying the frame

When you first select a curve to edit, only the control points on the curve appear. To see the control points that define the frame, click on the **Frame** box in the **Edit Bezier** (or **Edit B-spline**) dialog box. A check mark will appear, and the frame will appear with its control points, Figure 8-29. You can then move, delete, or insert control points on the frame.

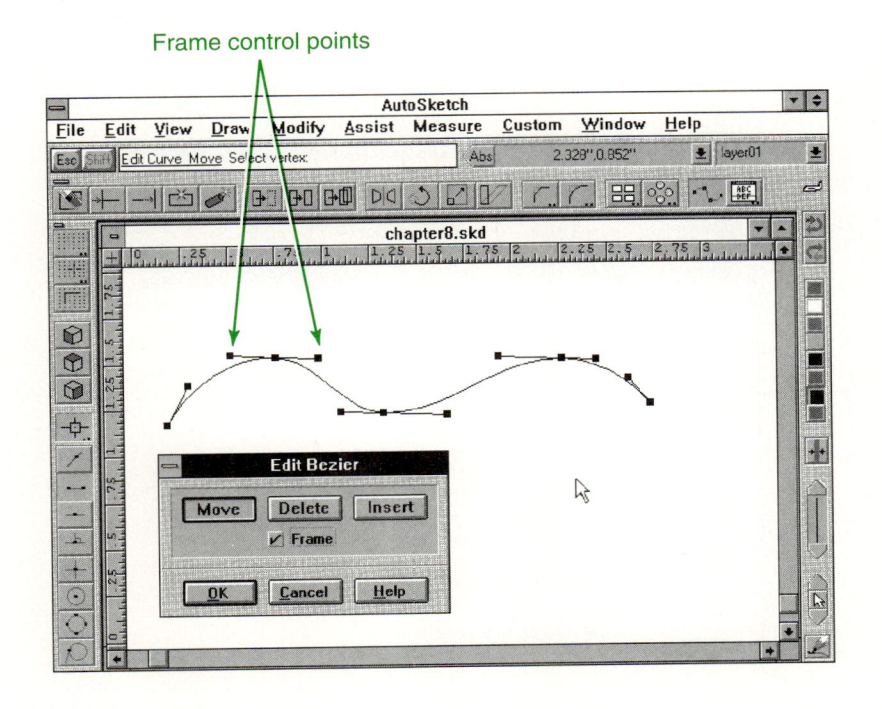

Figure 8-29.
Checking the **Frame** box displays the frame of the curve. The control points of the frame can then be edited.

Moving a control point

To move a control point, click on the **Move** button in the **Edit Bezier** (or **B-spline**) dialog box. The prompts that appear are very similar to the **Move** tool.

<u>Edit Curve Move</u> Select joint:

Select the control point that you want to move, Figure 8-30. AutoSketch then asks for the start point of the move.

<u>Edit Curve Move</u> From point:

This point is a reference point. It can be on the curve, but does not have to be. AutoSketch then asks for the ending point for the move.

<u>Edit Curve Move</u> To point:

As you move the cursor, the curve will move on-screen. When the shape is right, click to finish the command.

When the frame is turned on, the following will be the first prompt.

<u>Edit Curve Move</u> Select vertex:

The other prompts are the same as if moving a point on the curve.

Figure 8-30.
Moving a control point alters the shape of the curve.

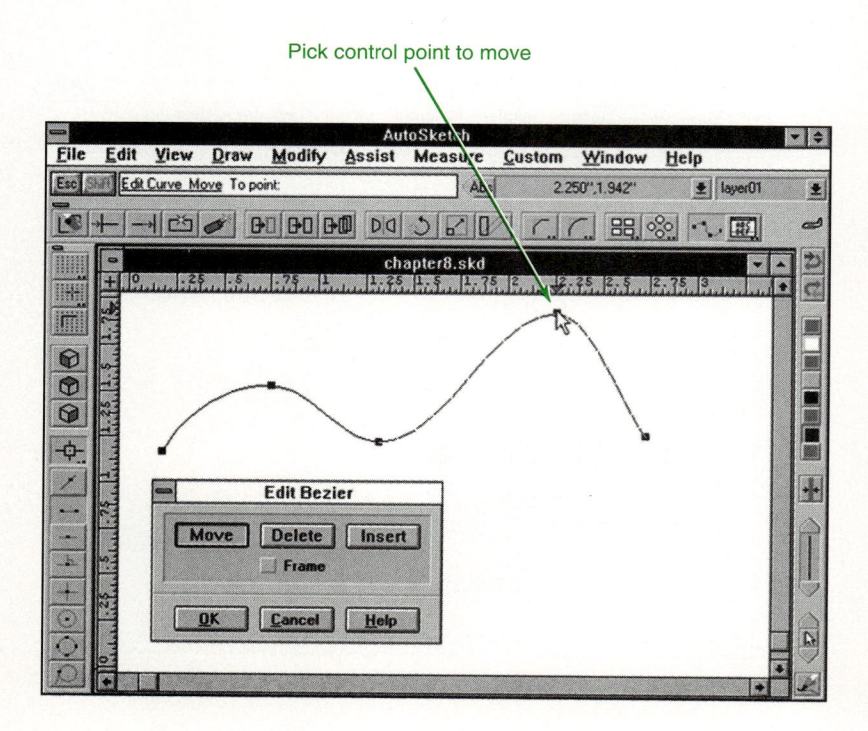

Deleting a control point

To delete a control point, choose the **Delete** button in the **Edit Bezier** (or **Edit B-spline**) dialog box. Then, pick the control point you want to delete. The point is deleted and the curve realigns with the remaining control points, Figure 8-31. Deleting a control point will lessen the curvature of a curve.

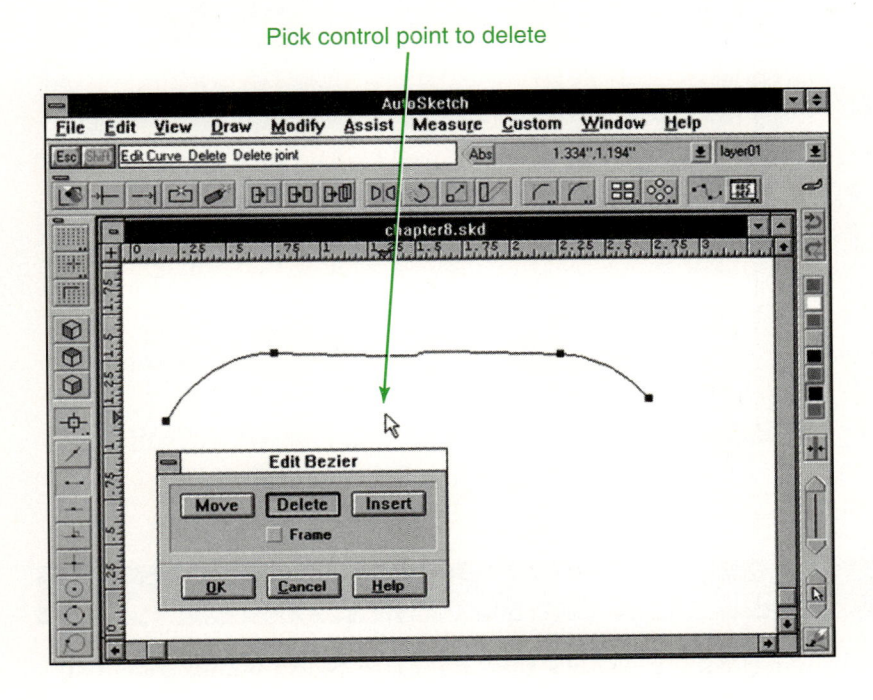

Figure 8-31.
Deleting a control point "flattens" the shape of the curve.

You can only delete points on the curve. You cannot delete control points on the frame. However, you can delete points on the curve even when the **Frame** box is checked.

Inserting a control point

To insert a control point, choose the **Insert** button in the **Edit Bezier** (or **Edit B-spline**) dialog box. Then, pick on the curve where you want the new point, Figure 8-32. Inserting a control point adds curvature to your curve.

You can only add points on the curve. You cannot add control points to the frame. However, you can add points to the curve even when the **Frame** box is checked. If this box is checked, the frame adjusts when the new point on the curve is added.

Figure 8-32.
Control points can be added to a curve. These new points can then be moved to

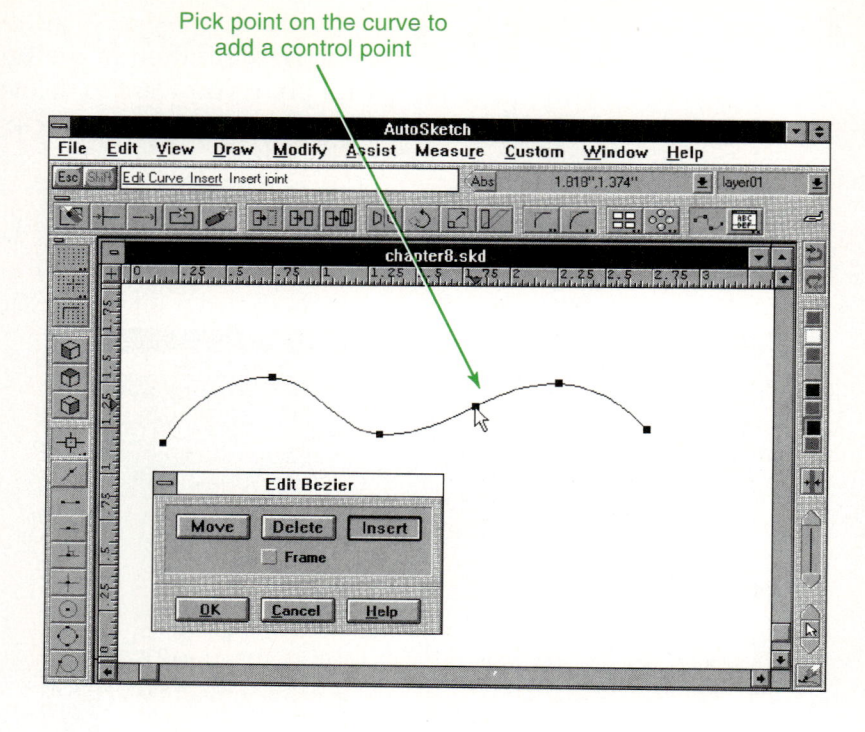

Pick point on the curve to add a control point

Exercise 8-9

1. Start a new drawing using the **Quick Start** template. Select **Inches (decimal)** as the unit of measurement.
2. Draw a simple Bezier curve with a start point, two control points, and an endpoint. The curve should be similar to the one shown below on the left. The curve should bend very little.
3. Use the **Edit Curve** tool so that the bends are more extreme. Also, add one extra bend in the curve. The curve should now look similar to the one shown below on the right.
4. Save the drawing as C8E9 on your work diskette and then close the drawing.

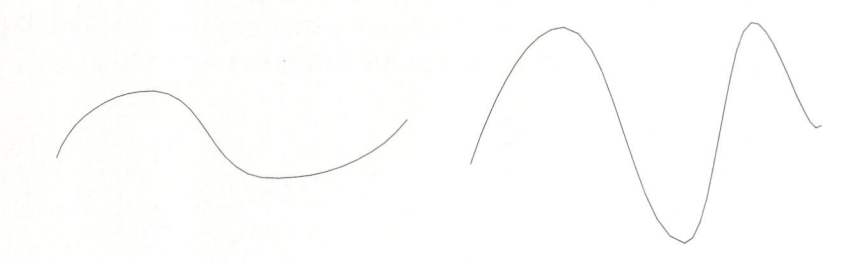

Measuring Distance, Angle, and Area

When editing, you may find it necessary to measure a distance or angle without placing a dimension on the drawing. You might want to know the distance, perimeter, area, or other data about the design. The tools in the **Measure** menu can be used for this.

In manual drafting, distance is measured with a ruler, and angles are measured with a protractor. The area inside a closed shape is calculated by multiplying measurements. These same functions can be done in AutoSketch.

Measuring Distance

Distance is measured between two locations you pick on the drawing. It is important to remember that distance is a straight line. For example, you may want to know the distance between the endpoint of a line and the center of a circle. Select the **Measure** menu and the **Measure Distance** tool. The following prompt appears.

<u>D</u>istance From point:

Select the first point of the distance you want to know. *Important:* It is possible to pick points that are not part of an object. Use drawing aids to precisely pick the locations you want to measure. After you select the first point, AutoSketch asks for the second point of the distance.

<u>D</u>istance To point:

Select the second point of the distance you want to measure. After you pick the second point, a dialog box appears telling the distance, Figure 8-33.

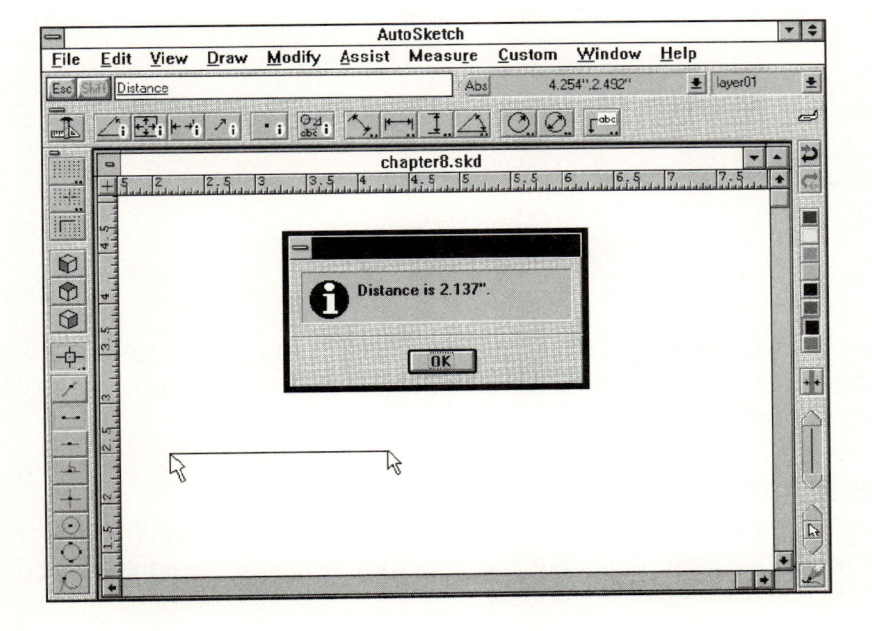

Figure 8-33.
To measure distance, pick two points that define the distance.

Measuring an Angle

AutoSketch lets you measure the angle formed by three points, Figure 8-34. Select the **Measure** menu and the **Measure Angle** tool. The following prompt appears.

Angle Base point:

Figure 8-34.
To measure an angle, pick a vertex and two locations that define the angle.

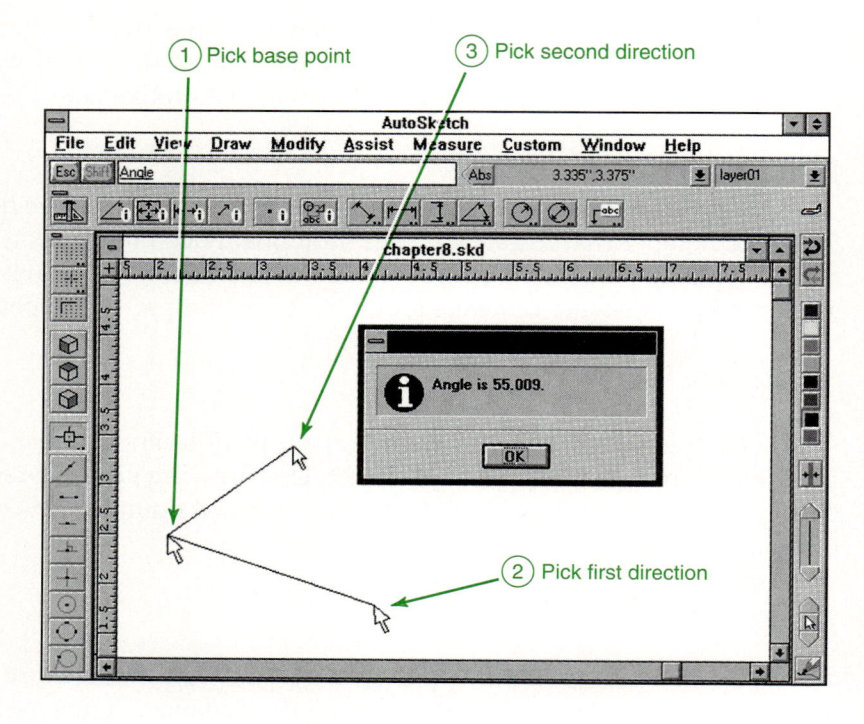

This first point to pick is the base point, or vertex of the angle. *Important:* It is possible to pick points that are not part of objects. Use drawing aids to precisely pick the points. After you select the base point, AutoSketch asks for the first direction of the angle.

Angle First direction:

Select a point on the first "leg" of the angle. AutoSketch then asks for the second direction of the angle.

Angle Second direction:

Select a point on the second "leg" of the angle. As you pick a point in each direction, a rubber band line extends from the base point. Once you pick the second direction, a dialog box appears telling the angle.

Measuring Perimeter and Area

Perimeter is the distance traveled around a shape. *Area* is the surface enclosed by a shape. Area is expressed in square units. An architect may measure the square feet enclosed by a house. This is the area of the house.

To measure perimeter and area, you must pick the corners of the shape, Figure 8-35. Select the **Measure** menu and the **Measure Area** tool. The following prompt appears.

Area First perimeter point:

Select the first point of the shape. An X will appear to mark the location. You do not have to pick on an object. You can measure the invisible area of an empty section of the drawing. Therefore, it is important to use drawing aids to precisely pick points on objects. After you select the first point, AutoSketch then asks for the next point on the perimeter.

Area Next point:

Pick the second, third, and other corners of the shape. As you do, an X appears to mark each location. These will appear smaller than the first X so that you can easily see the first point. Continue picking corners until you pick the first point again to define a closed shape. A dialog box then appears to display both the area and perimeter.

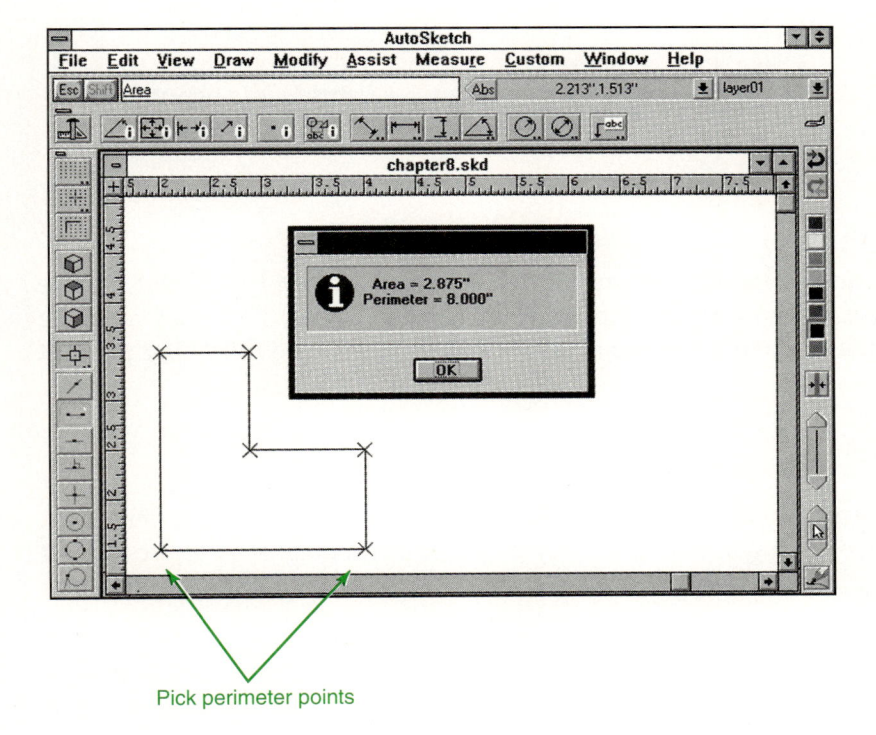

Figure 8-35.
To measure perimeter and area, pick the corners of the enclosed area.

Displaying Point Coordinates

The **Measure Point** tool lets you find the exact Cartesian coordinate values of any location on-screen. Select the **Measure** menu, and the **Measure Point** tool. The following prompt appears.

Point Point coordinates:

As you move the pointer, its location on the Cartesian coordinate system appears in the prompt box. The location is displayed in the current units of measurement.

When you pick a location, a dialog box appears giving the exact coordinate values. See Figure 8-36.

Figure 8-36.
To show the exact location of a point, use the **Measure Point** tool. Then pick the point to show the coordinates.

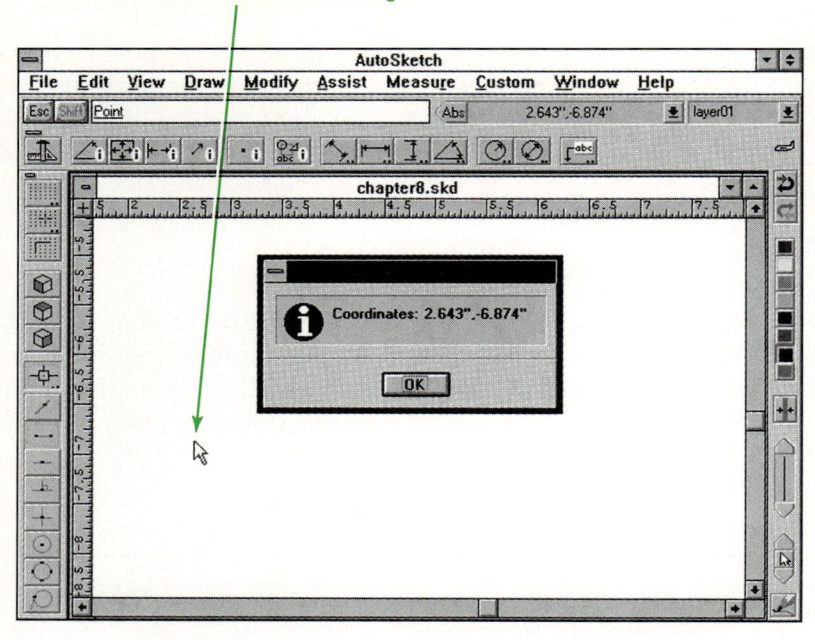

Measuring Bearing

The **Measure Bearing** tool lets you measure a direction from a base point, Figure 8-37. The angle measurement is given in degrees measured counterclockwise from the X axis. Select the **Measure** menu and the **Measure Bearing** tool. The following prompt appears.

Bearing Base point:

Pick the base point. You can select a point that is not part of an object. Therefore, it is important to use drawing aids to precisely pick points on objects. After you select the base point, AutoSketch then asks for the second point of a line defining the bearing.

Bearing Enter point:

As you move the pointer, the bearing angle appears in the prompt box. Be sure to use drawing aids to accurately pick points on objects. Once you select the second point, a dialog box appears giving the exact angle.

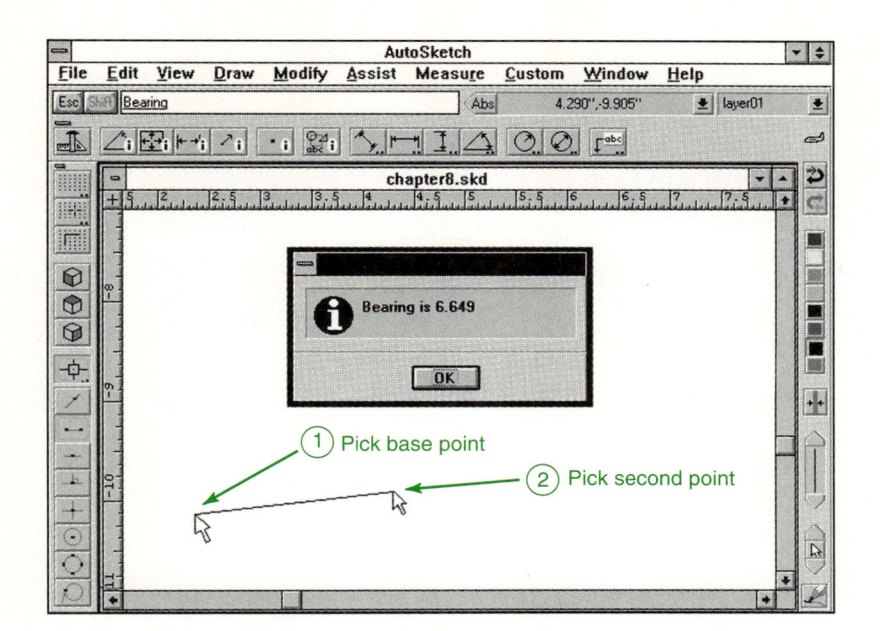

Figure 8-37.
To measure a bearing, pick a base point and the direction you want to measure.

Exercise 8-10

1. Draw an arc and a circle. Measure the distance between their centers. Hint: Make sure the **Attach** tool **Center** is on and **Quadrant** is off. Erase the objects and redraw the screen.

    Distance between centers: _____

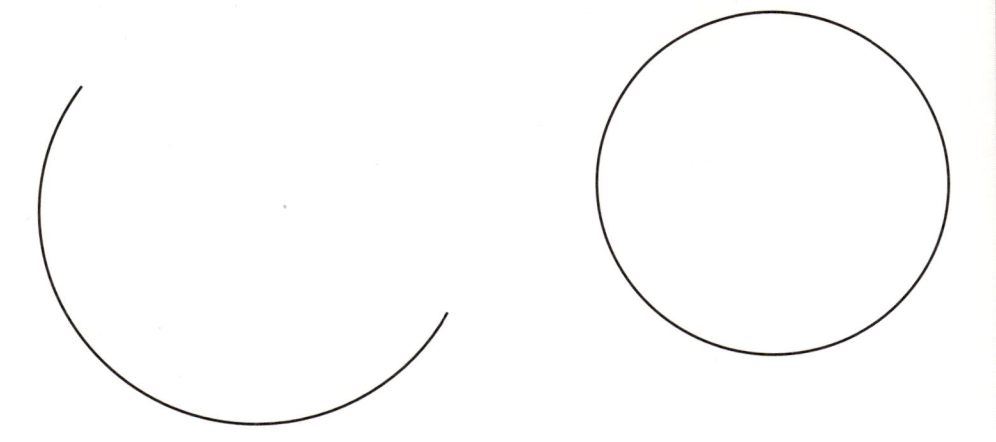

A

2. Draw three single lines that form a triangle. Use the **Attach** tool **Endpoint** to make sure the lines connect precisely. Label the lines A, B, and C. Measure the angle between each pair of lines.

   Angle between lines A and B: _____
   Angle between lines B and C: _____
   Angle between lines C and A: _____

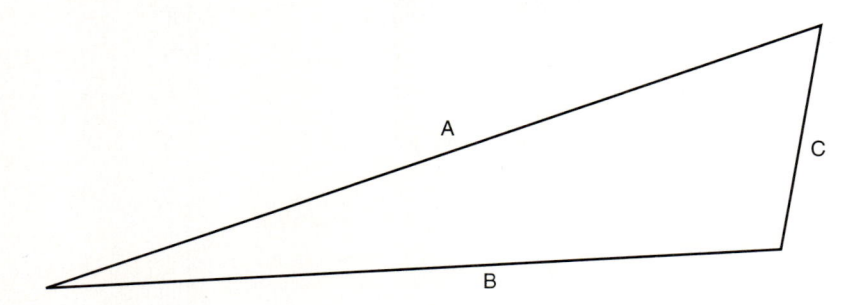

3. Draw a series of four connected lines that form a closed shape. Measure the perimeter and the area. Measure the coordinate location of the four corners of this shape.

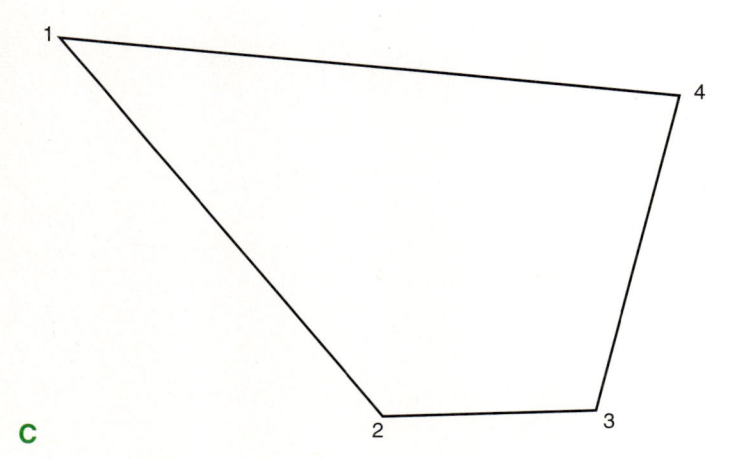

   Perimeter: _____
   Area: _____
   Location of first corner: _____
   Location of second corner: _____
   Location of third corner: _____
   Location of fourth corner: _____

4. Draw a series of five connected lines using the **Line** tool and the **Attach** tool **Endpoint.** Label the lines A, B, C, D, and E. Measure the bearing of each line.
   Bearing of line A: _____
   Bearing of line B: _____
   Bearing of line C: _____
   Bearing of line D: _____
   Bearing of line E: _____

D

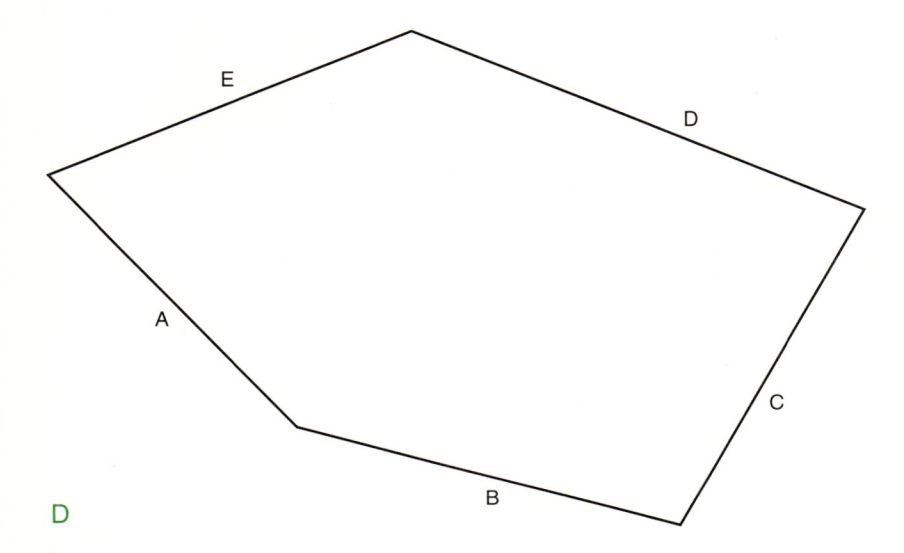

5. Save the drawing as C8E10 on your work diskette and then close the drawing.

Summary

This chapter presented a mixture of drawing techniques used in mechanical drafting, architectural drafting, and technical illustration. Some of the techniques include advanced methods of editing, constructing complex shapes, and ways to manipulate curves.

The **Trim, Extend,** and **Break** tools allow you to modify the length of objects, or cut pieces out of an object. The **Explode** tool separates an object into its component parts, such as the four lines that make up a box. The **Stretch** tool allows you to modify the shape of an object, either proportionally or unevenly.

The **Box Array** tool allows you to make multiple copies in a rectangular pattern. The **Ring Array** tool also allows you to make multiple copies, but in a circular pattern.

The **Fillet** tool modifies the corners of an object by creating an arc. The **Chamfer** tool also modifies the corners of an object, but creates a straight line.

Curves can be edited with the **Edit Curve** tool. Curves are made up of control points. Control points can be moved, deleted, or additional control points added.

When editing, you may find it necessary to measure distance or angle without placing a dimension on the drawing. Tools to do this are found in the **Measure** menu. These tools let you measure distance, angle, area, bearing, and point location. These tools can be used while using another tool.

Important Terms

Area
Boundary
Break points
Chamfering
Control points
Fillet
Perimeter
Rounding
Stretch base

New Commands and Tools

Box Array tool
Break tool
Chamfer tool
Edit Curve tool
Explode tool
Extend tool
Fillet tool
Measure Angle tool
Measure Area tool
Measure Bearing tool
Measure Distance tool
Measure Point tool
Ring Array tool
Stretch tool
Trim tool

Review Questions

Give the best answer for each of the following questions.

1. After picking an object to break, you must pick two _____.
2. True or false. When breaking a circle, the portion removed is clockwise from the first to second break point.
3. The _____ tool splits an object into its component parts.
4. The _____ tool clips a line or arc until it meets precisely with a boundary object.
5. What determines to which side of a closed object (such as a box or circle) the **Extend** tool will extend a line? _____
6. What method of selection is used to stretch a shape? _____
7. Explain how the **Stretch** tool affects circles, boxes, curves, text, and dimensions.
   _____
   _____
8. Which options of a box array must be selected so that you can use your cursor to specify the row distance and column distance values? _____
9. True or false. Copies made using the **Box Array** tool can be placed at an angle other than horizontal.
10. Suppose you want to make a ring array that has 9 copies and spans three-quarters of a full circle. What would be the settings for **Included Angle** and **Degrees Between Items?**
    _____
11. True or false. Parallel lines can be chamfered.
12. What fillet radius can be entered so that the **Fillet** tool extends the endpoints of two lines to meet precisely? _____
13. If the chamfer distances are not the same, which selection method should you use?
    _____

14. When editing a curve, what method(s) can you use to add more "bend" to the curve?

15. What locations must you pick to measure the area of a shape? _____

Activities

Here is a mixture of drawing activities to develop your skills using editing tools. If you have not already changed your printer orientation to landscape as discussed in Chapter 6, do so now.

For each activity, select the **New...** command from the **File** menu. Select the **Quick Start** template and select **Inches (decimal)** as the unit of measurement. Save your AutoSketch drawings on your work diskette using the name given. After saving each activity, select the **Close** command from the **File** menu to clear the drawing from the screen.

1 - 10   For each activity, complete the part or shape shown on the left. Use a grid, snap, or other drawing aids to help you create the drawing. Do not dimension the drawings. Then, make a copy to the right of the original. Use editing tools *only* to change the copy so that it looks like the drawing on the right. Do not draw any new objects. Save each drawing using the filename given under the illustration.

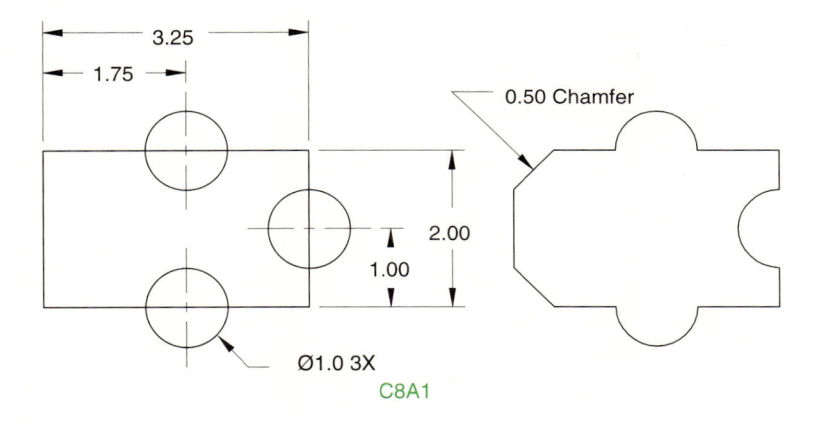

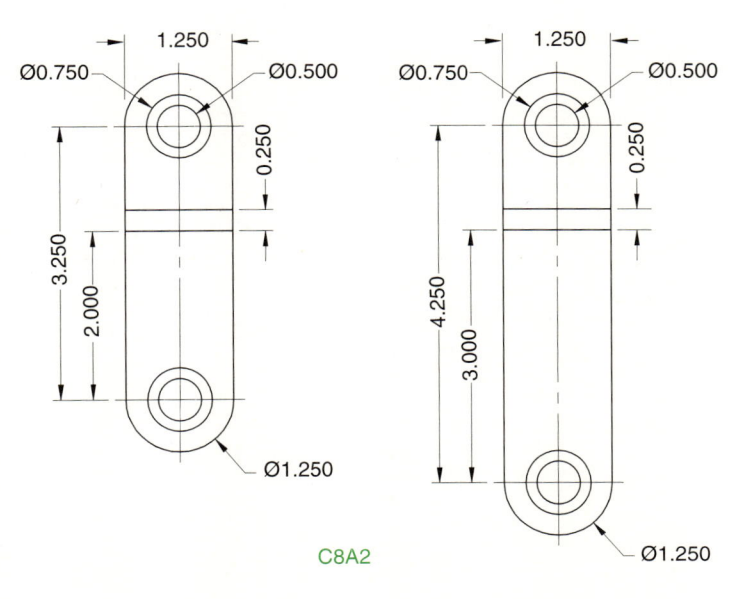

C8A1

C8A2

C8A3

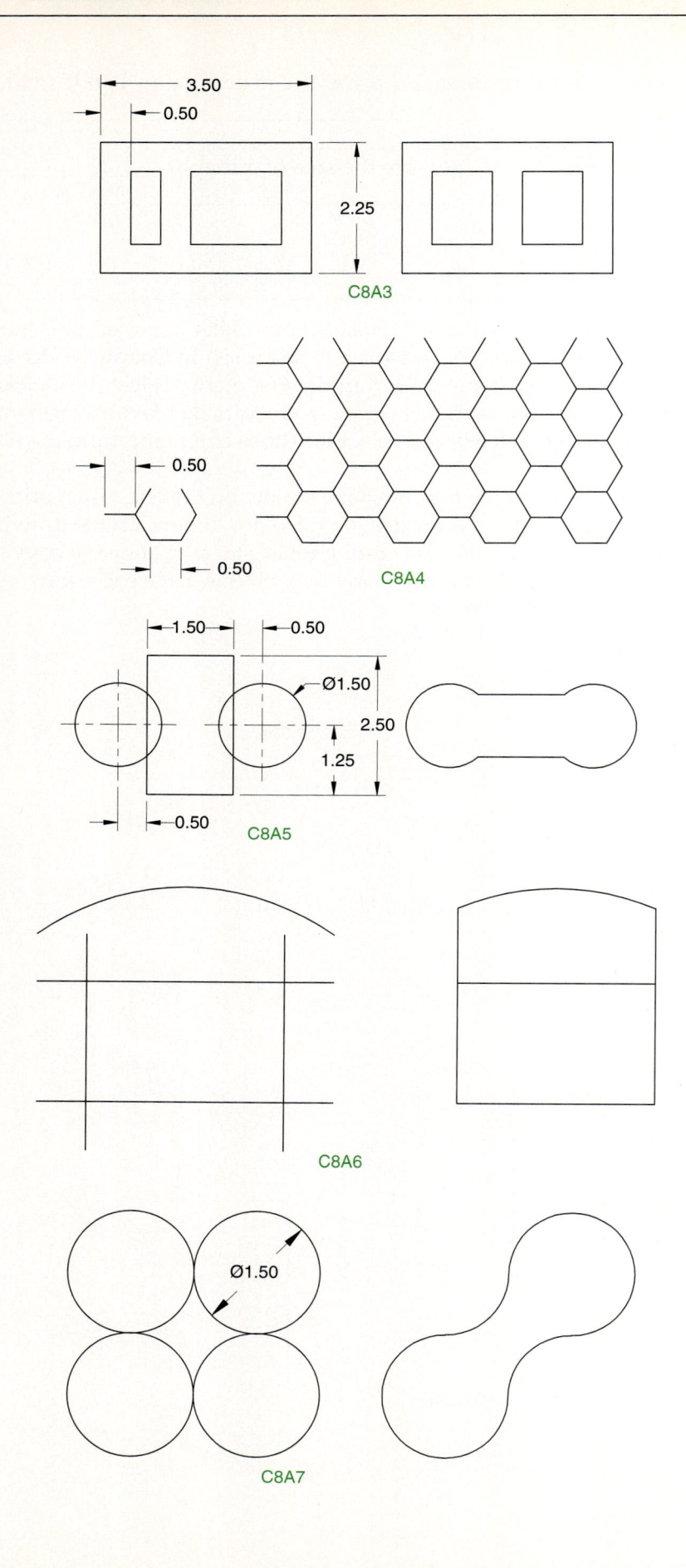

C8A4

C8A5

C8A6

C8A7

Closed Bezier Curve

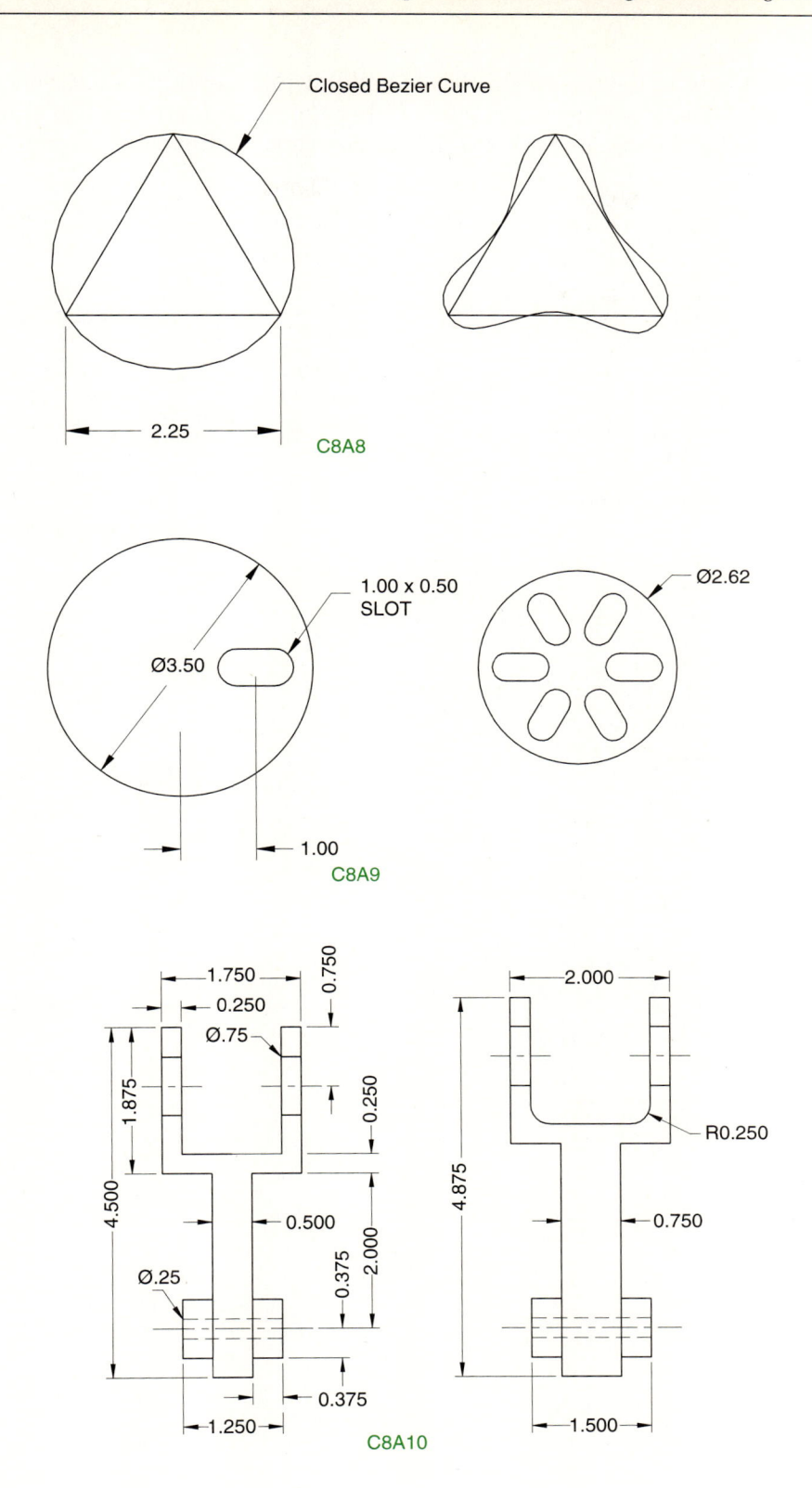

2.25

C8A8

1.00 x 0.50 SLOT

Ø3.50

Ø2.62

1.00

C8A9

1.750
0.750
0.250
Ø.75
1.875
0.250
4.500
0.500
0.375
2.000
Ø.25
0.375
1.250

2.000
4.875
R0.250
0.750
1.500

C8A10

11 - 15. For each activity, draw the design or object shown using drawing aids. **Grid** and **Snap** are especially useful. Use exact measurements. Either type coordinates or use your cursor with drawing aids to locate objects. Save the drawings using the filename given under the illustration.

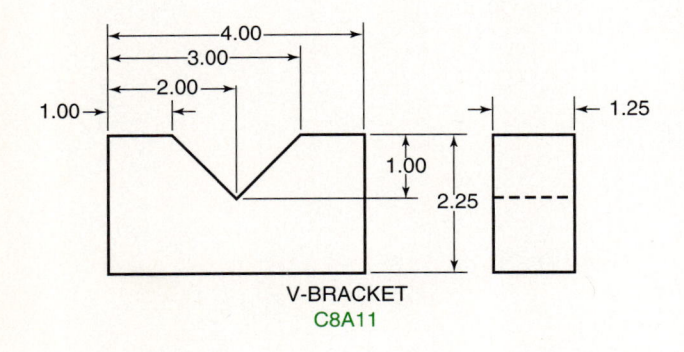

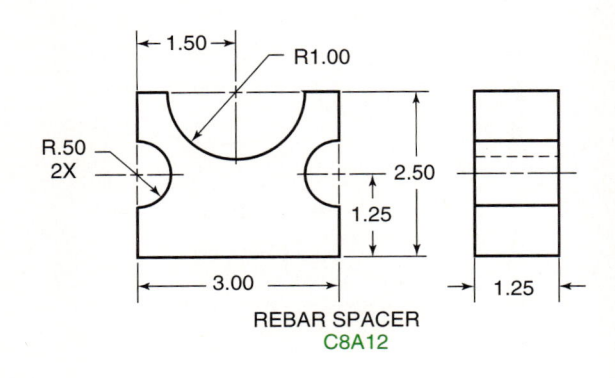

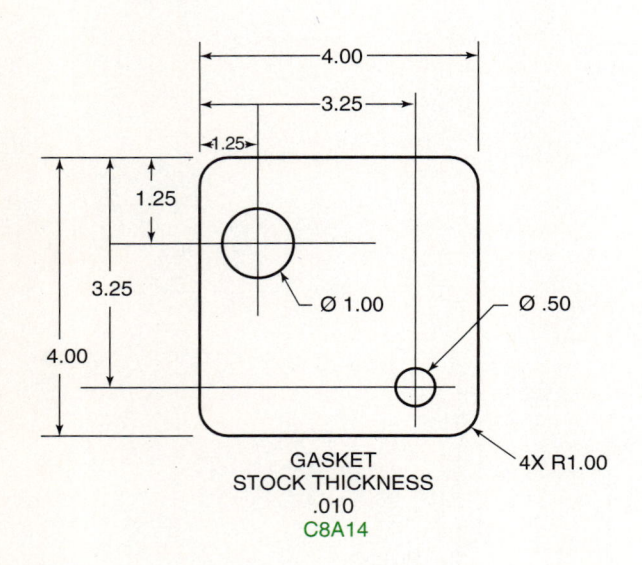

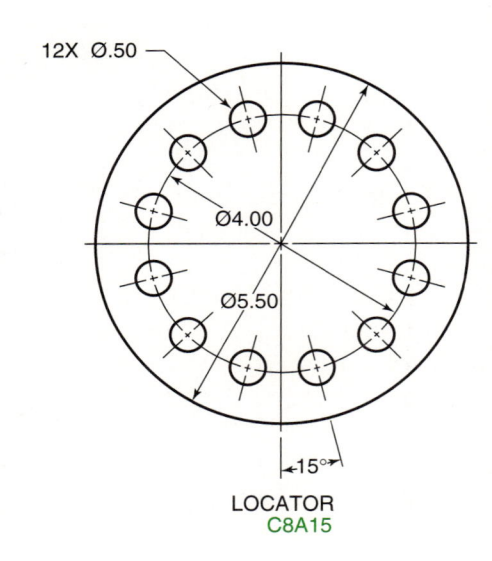

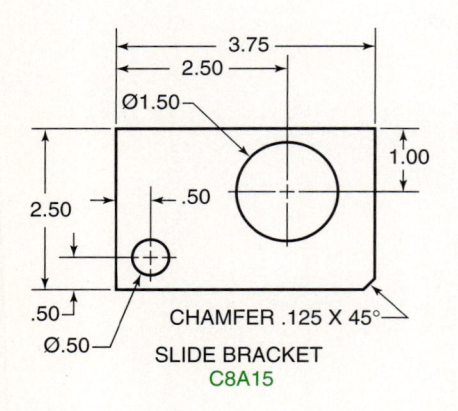

V-BRACKET
C8A11

REBAR SPACER
C8A12

**GASKET
STOCK THICKNESS
.010**
C8A14

LOCATOR
C8A15

SLIDE BRACKET
C8A15

CHAPTER 9

Adding and Editing Text

Objectives

After studying this chapter, you will be able to:
- Describe the benefits of CAD text over manual lettering.
- Set text features such as font, height, width factor, and angle.
- Explain the difference between SHX and TrueType fonts.
- Set the justification of text.
- Place text on the drawing using the **Quick Text** tool.
- Place text on the drawing using the **Text Editor** tool.
- Import text from an ASCII file.
- Draw special characters
- Edit words and letters in an existing text string.
- Change the text features of an existing text string.
- Explain the effects of editing tools such as **Move, Stretch,** and **Mirror** on text.

Almost all technical drawings, graphs, and illustrations contain information given as text. *Text* refers to letters, numbers, words, and sentences that are added to explain features that cannot be described with graphics, Figure 9-1. Text is needed because a drawing by itself often cannot convey all the features of a design idea. Text can specify several things.
- Text can be used to specify the material used to make the product.
- Special instructions for manufacturing or constructing the product can be given as text.
- The drawing name, part name, revision number, and other necessary information to document the project are given as text.
- Dimensions and tolerances are given as text.

Benefits of CAD Text

There are many benefits to adding text with a CAD program compared to lettering in manual drafting. First, text is easy to place on the drawing. There are only three steps to place text.
1. Select text features. Features refer to the font, height, width, and angle of the text.
2. Pick a location for the text on the drawing.
3. Type in the text using the keyboard, or import a text file.

A second benefit is neatness. Since a technical drawing must inform the user, text must be readable. Some drafters have sloppy lettering habits. However, text drawn with AutoSketch and plotted on a printer or plotter is clear and precise.

Figure 9-1.
This chart would be meaningless without text. (Autodesk)

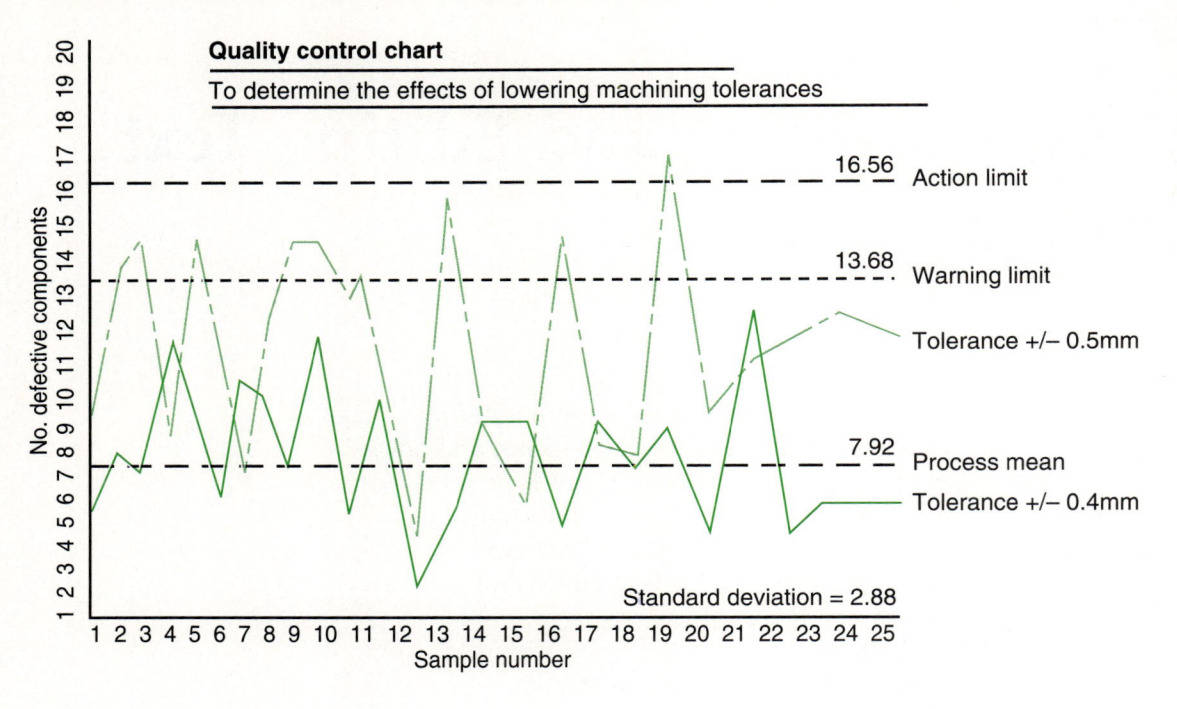

A third and very important benefit of CAD text is standardization. Companies can specify standard text features that should be used for all drawings. The text font, height, width, and angle are set as company policy. Any drawing created will follow that standard, no matter who the drafter is. All lettering will look the same. Also, if more than one person works on a drawing, the text style remains the same. This makes drawings easier to understand.

Selecting Text Features

Text features must be set before placing text on a drawing. *Text features* determine the appearance of the text. Features include font, height, width, oblique angle, and angle. To access the **Text Settings** dialog box, double-click on either the **Quick Text** tool or **Text Editor** tool in the **Draw** menu, Figure 9-2. The sample in the dialog box will adjust to reflect the settings you choose. The options are discussed in the next sections.

Font

The *font* is the style of the lettering. The characters may look quite fancy, or may be made of simple lines to look like printed letters. Fonts that are bold are easily seen. These fonts are used for titles and labels. The font may also consist of complex shapes, or even symbols. All of the characters in a font follow the same general style.

AutoSketch can use TrueType and SHX fonts. To select one, pick the arrow of the **Font Type** list box and then pick the font.

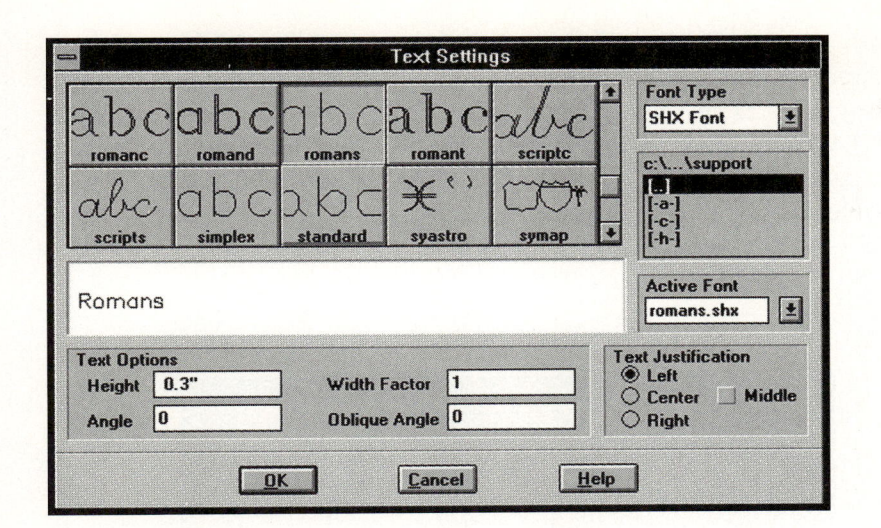

Figure 9-2.
Text features are set in the **Text Settings** dialog box.

SHX fonts

SHX fonts are fonts from AutoSketch and from AutoCAD. The SHX font set included with AutoSketch for Windows has more than 20 alphanumeric fonts containing letters and numbers. There are also five symbolic fonts containing special symbols for mapping, mathematics, music, astronomy, and meteorology. See Figure 9-3. When you use symbolic fonts, pressing a specific key on the keyboard produces a symbol. SHX fonts take longer to redraw, but you can place them at precise italic angles.

Figure 9-3.
These are just a few of the many SHX fonts offered by AutoSketch. The ROMANS is the SHX font often used for text. Note that in symbol fonts, capital and lower case letters can represent different symbols.

TrueType fonts

TrueType fonts are provided by Windows. They are the same fonts used by most word processing programs. They look better than SHX fonts. When you select TrueType fonts, the **TrueType Options** box appears in the **Text Settings** dialog box, Figure 9-4. Here you can make TrueType fonts bold, italic, underlined, and strike out. These options are not available with SHX fonts.

Figure 9-4. When you select TrueType for the **Font Type** the dialog box changes and includes **TrueType Options.** This area of the dialog box is identified here with a tinted background.

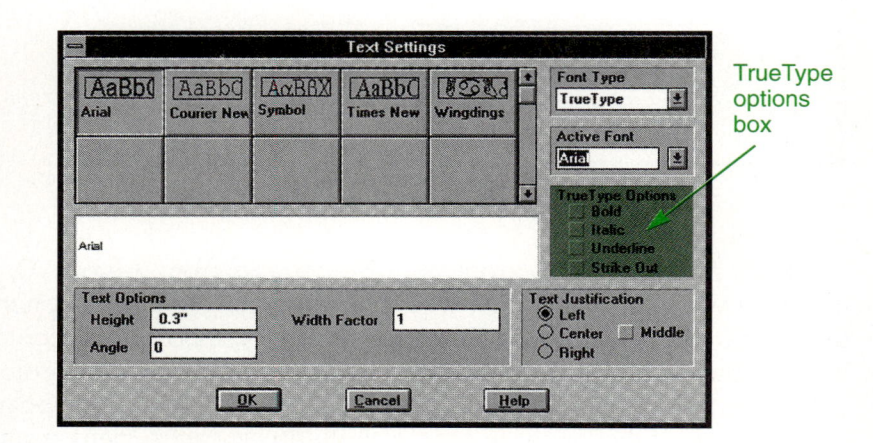

Selecting a font

After you select the font type (SHX or TrueType), a display of available styles is shown. Pick on the font style you want to use. If you selected a SHX font, the **Active Font** box shows the name of the font you chose. A sample of the font with the settings you choose appears. You can also select a font by name. To select one, pick the arrow of the **Active Font** list box and then pick the font name. Samples of the different fonts and the keys for symbol characters are found in the appendix. For most technical drawings, the SHX fonts ROMANS and SIMPLEX, and the TrueType font Arial look best.

Height

The **Height** option in the **Text Options** area determines how tall the text is going to appear. The height is based on the height of capital letters, Figure 9-5. Lowercase letters will obviously be smaller. Most text you add to a technical drawing should be capital letters. Occasionally, capital and lowercase letters will be used for notes.

To change the height, place the cursor in the box opposite **Height.** Enter the new text height. A standard text height for printed technical drawings is .125" to .25". The value you enter is measured using the current unit of measurement. However, you can also enter a height in other units of measurement by adding the unit notation behind the value. After entering a new height, press the [Tab] key or click in another box so that the sample shows the text with the new height.

The height of the text is affected by the sheet scale. Remember that you draw using real units. However, to print the drawing on a smaller sheet of paper, you must scale your drawing. For example if the scale of your drawing is 1/4"=1", a text height

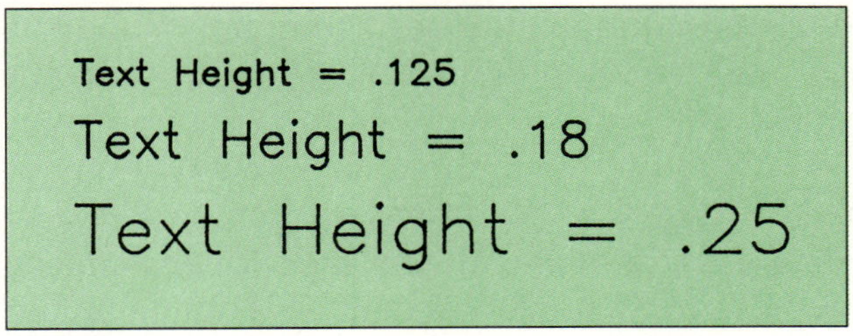

Text Height = .125
Text Height = .18
Text Height = .25

Figure 9-5.
On most drawings, the text height looks best between .125" and .25". Titles and labels may be larger.

of .125 is printed at only .03125" high. This is too small to be seen clearly. Keep the sheet scale in mind when setting the text height. If the sheet scale is 1/4"=1" and you want text to appear .125" high on the printed drawing, enter a text height of .5 in the **Height** box. Sheet scale is discussed in Chapter 10.

All new text entered will assume the current height setting. All text previously entered will remain at the height that it was entered at.

Width Factor

The **Width Factor** option in the **Text Options** area determines the width of each character. Changing the width squeezes and expands the text string. It does not change the height. The default factor is 1. This is normal text width. Entering a factor greater than 1 expands the text. For example, entering 2 makes the text twice as wide. Entering a factor less than 1 compresses the text. For example, entering .5 makes the text half as wide. See Figure 9-6.

To change the width, place the cursor in the box opposite **Width Factor** Enter any positive whole value, decimal value, or fraction for the new text width. For most technical drawings, use the default value. Entering another factor might cause the text to look awkward and not conform to standards. After entering a new width factor, press the [Tab] key or pick in another box so that the sample shows the text with the new width factor.

Angle

The **Angle** determines the rotation for the entire text object. Rotation is measured from horizontal in a counterclockwise direction, Figure 9-7.

To change the rotation, place the cursor in the box opposite **Angle.** The default rotation is 0 degrees (horizontal). If you enter a negative rotation angle, the text string is rotated clockwise. All text is placed at the current rotation until the angle value is reset. After entering a new angle, the sample will *not* show text with the new angle.

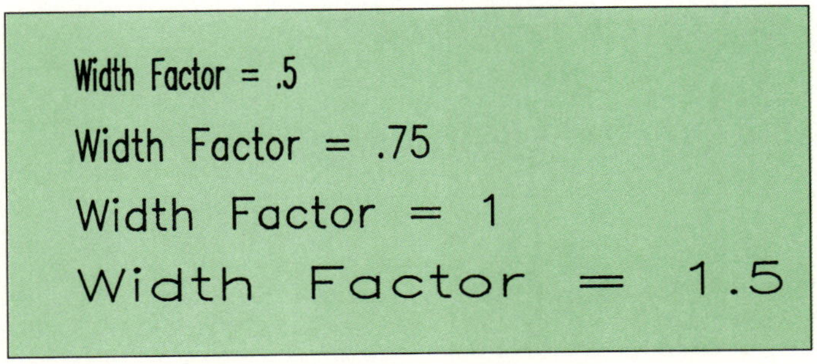

Width Factor = .5
Width Factor = .75
Width Factor = 1
Width Factor = 1.5

Figure 9-6.
A **Width Factor** less than 1 squeezes text. A factor greater than 1 expands text. The height does not change.

An angle of 90 degrees makes the text read from bottom to top. An angle of 270 degrees makes the text read from top to bottom. An angle of 180 degrees places text horizontally upside down.

Oblique Angle

Oblique Angle is the slant, or angle, of each individual character in the text. This box will only appear when an SHX font is selected. This option is used to make an SHX font look italic. The oblique angle is also used when drawing an isometric drawing, as you will learn in Chapter 15.

The oblique angle is measured clockwise from vertical. To change the angle, place the cursor in the box opposite **Oblique Angle.** When the slant is 0 degrees, text is upright. This is the default value. A positive value slants the text clockwise (to the right). A negative value slants the text counterclockwise (to the left). A +15 degree slant is frequently used on technical drawings. See Figure 9-8.

Figure 9-7.
The angle of a text object is measured counterclockwise from horizontal.

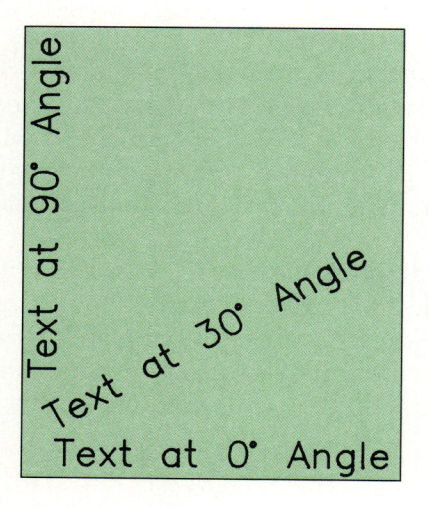

Figure 9-8.
The oblique angle is the slant of text characters for SHX fonts. You can also have negative oblique angles for text that slants to the left.

All new text entered will be at the new oblique angle. Previously drawn text will not be affected. The recommended range of angle values is between -30 and +30 degrees. After entering a new oblique angle, press the [Tab] key or click in another box so that the sample shows the text with the new angle.

TrueType Options

When you select a TrueType font type, four additional options are shown under **TrueType Options** These options are **Bold, Italic, Underline,** and **Strike Out**

You can use bold lettering for titles. Italic and underline text might be useful to emphasize information. Strike out text has a horizontal line running through it. This text might be used to represent revisions you have made to text in a drawing.

Text Justification

Text justification determines the alignment of the text in relation to the point you pick to place the text, Figure 9-9. The three types of justification in AutoSketch are **Left, Right,** and **Center**. For example, if you select **Center** each line of text is centered with your insertion point. If you select **Right,** all of the text will be to the left of the insertion point (right justified).

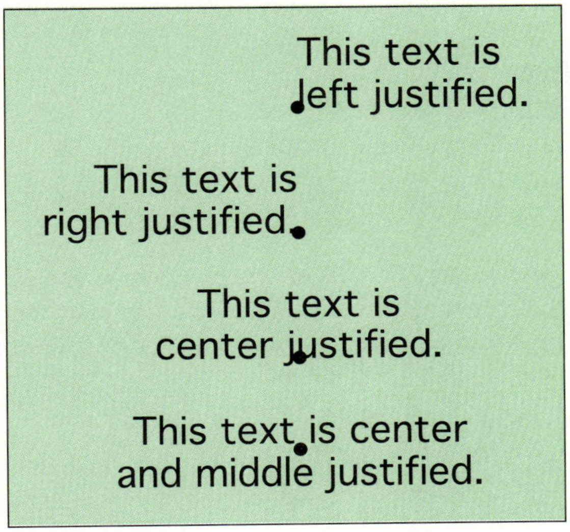

Figure 9-9.
Text justification affects the placement of the text object relative to the point you pick to place the text.

The **Middle** box is for alignment above and below the insertion point. This option makes the text object centered vertically over the point you pick to place the text. For example, if you enter three lines of text with the **Text Editor,** the entire text object is centered over the insertion point. See Figure 9-9. Note: This option calculates the centering based on the height of a capital letter, even if there are none in the text object.

Color

A setting for text color is not found in the **Text Settings** dialog box. However, because text is an object, you can set the color with the **Color Palette**.

There are two reasons to make text a different color. First, special notes and specifications should stand out from shapes and dimensions. Second, AutoSketch uses the color to determine what pen to pick when plotting. This allows you have a different plotted color or pen width for text.

Exercise 9-1

1. Start a new drawing using the **Quick Start** template. Select **Inches (decimal)** as the unit of measurement.
2. Determine the default text features of this template by double-clicking on **Quick Text** tool and recording the following information.
   Font Type _____
   Active Font _____
   Height _____
   Width Factor _____
   Oblique Angle *(if SHX font)* _____
   Angle _____
   Text Justification _____
3. Close the drawing without saving it.

Placing Text on the Drawing

Once you have set the text features, you are ready to add text to the drawing. AutoSketch allows you to place text using either the **Quick Text** tool or **Text Editor** tool.

Placing Text Using Quick Text

Quick Text is a fast way to add single lines of text to your drawing. Pick the **Quick Text** tool found in the **Draw** menu. The following prompt appears.

Quick Text Enter point:

Pick the insertion point of the text string. The *insertion point* marks the placement of the text on the drawing. AutoSketch places text according to the justification you set. Once you pick a point, a small underline appears on the drawing. AutoSketch then asks you to enter the text.

Quick Text Enter text:

Type in the text at this prompt. Use the keyboard to enter characters, letters, numbers, words, and sentences. As you type, the text appears on the drawing as well as in the prompt box, Figure 9-10. As the text appears on-screen, it will have the features that you set.

If you make a mistake, press the [Backspace] key to back up. When you finish typing in the line of text, press [Enter]. This command repeats, so the first prompt will appear again in the prompt box.

More lines of Quick Text

You may want to type in more than one line of text at a time. This can be done using the **Text Editor** tool. However, for just a couple lines of text, it is easier to use the **Quick Text** tool.

After typing the first line of text, do not press the [Enter] key. Instead, place the cursor in the drawing window and press the pick button on your mouse. This causes the cursor to begin another line of text. Continue to enter a new line of text. When finished, press [Enter]. Note: Each line is treated as a separate object. Also, once you press the mouse button, you cannot backspace to a previous line.

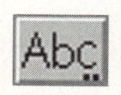

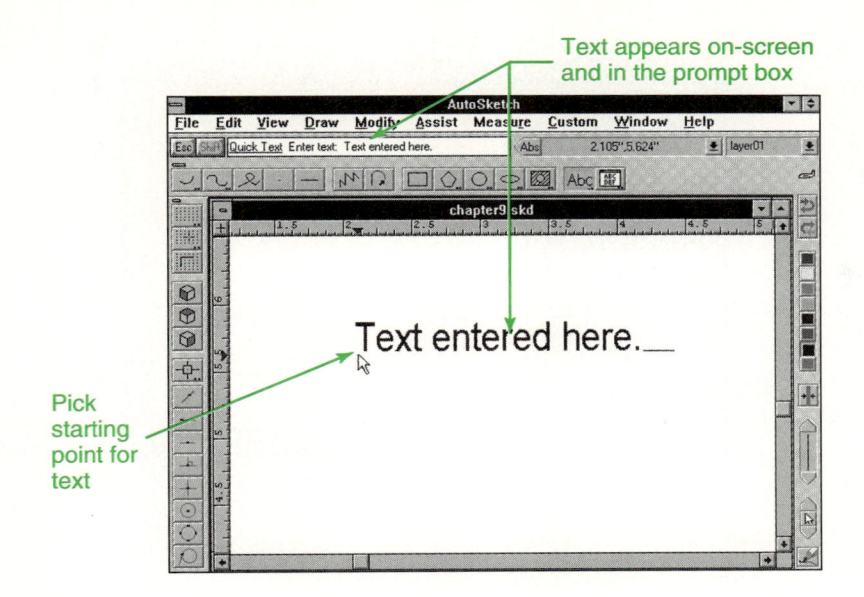

Text appears on-screen and in the prompt box

Pick starting point for text

Figure 9-10.
After picking the **Quick Text** tool and a text location, type in the text. The text appears both on the drawing and in the prompt box as you type.

Exercise 9-2

1. Start a new drawing using the **Quick Start** template. Select **Inches (decimal)** as the unit of measurement.
2. Draw a 3" square box.
3. Enter the paragraph shown below using the **Quick Text** tool. Read through the paragraph first to know which text features to set.
4. Save the drawing as C9E2 on your work diskette and then close the drawing.

```
This is .125 high text
in the ROMANS font.
To end each line, press
the pick button on your
mouse. This jumps the
cursor to the next
evenly spaced line.
When you are finished
entering text, press
the [Return] key.
```

Placing Text Using the Text Editor

AutoSketch includes a powerful way to enter big blocks of text. This is the **Text Editor** tool. The text editor allows you to add entire paragraphs of text to the drawing. The paragraph of text is treated as a single object. Select the **Draw** menu, and the **Text Editor** tool. The following prompt appears.

Text Editor Enter point:

Pick the start point of the text string. Once you pick the start point, the **Text Editor** dialog box appears, Figure 9-11. The options found in the **Text Editor** dialog box are discussed in the next sections.

Figure 9-11.
The **Text Editor** dialog box provides many functions for working with text.

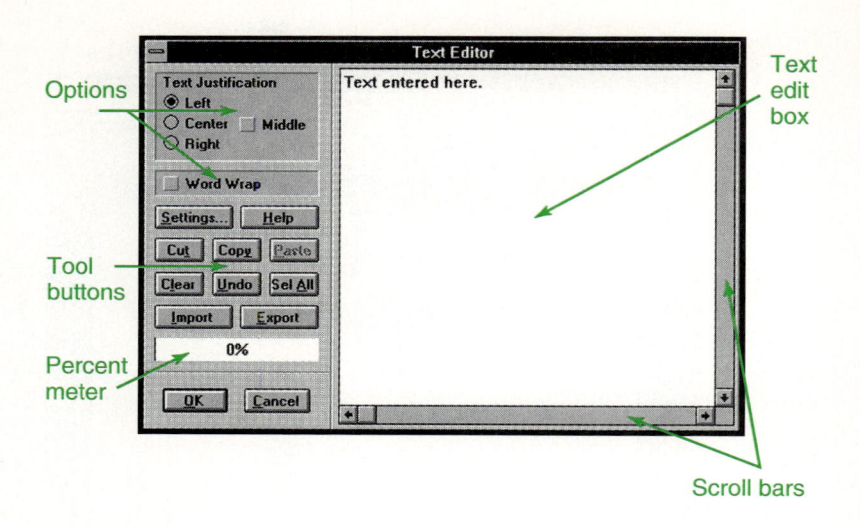

Text edit box

The *text edit box* is the open box on the right side of the dialog box. This is where you type in the text. The box is limited to 2048 characters. Scroll bars along the top and side allow you to move left, right, up, and down across large amounts of text.

Text Justification

Text Justification in the **Text Editor** dialog box options affects only the current text object you are entering. It does not change the settings in the **Text Settings** dialog box.

Word Wrap

Word Wrap creates a new line when your typing reaches the edge of the text edit box. When selected, a check mark appears in the box next to it. You do not have to press the [Enter] key at the end of each line. If you deselect the **Word Wrap** option, the text goes back to a single line.

Settings

The **Settings** button brings up the **Text Settings** dialog box. Again, changes you make here do not change the settings for future objects.

Cut

The **Cut** button removes the selected text and stores it in case you want to paste it in elsewhere. To select text, pick in the text edit box, hold the mouse button down, and drag the cursor to highlight the text to cut. You can move and copy words and lines of text in the edit box by cutting and pasting text. To move text, simply cut it and paste it where you want it.

Copy

The **Copy** button copies the selected text and stores it in case you want to paste it in elsewhere. To select text, pick in the text editing box, hold the mouse button down, and drag the cursor to highlight the text to copy.

Paste

The **Paste** button inserts the text that you last cut or copied. The text is placed at your cursor position in the text edit box. In addition, you can use the **Cut** or **Copy** commands in another Windows application, and then use the **Paste** button here to insert that text. Consult your Windows documentation for more information about cutting, copying, and pasting text between two applications.

Clear

The **Clear** button removes the selected text. However, unlike the **Cut** button, it does not store it for future use. This button is helpful if you want to clear a lot of text, but do not want the contents of the last copy or cut overwritten by your action.

> **SHORTCUT**
>
> You can quickly clear selected text by pressing the [Delete] key.

Undo

The **Undo** button reverses the effect of the last selected button. For example, if you cleared the text, but want it back, select **Undo**.

Sel All

The **Sel All** button selects all of the text in the text edit box. This is easier than dragging your cursor over all the text to select it.

Import

The **Import** button brings in a text file that you created outside of AutoSketch. This file must an ASCII file. An ASCII file can be created with the Windows Notepad program or DOS EDIT program. These files do not contain formatting, such as bold, italic, or underline. In addition, many common word processing programs can save a text file as an ASCII file.

Since AutoSketch does not support text blocks larger than 2048 characters, any characters over this limit are not imported. The file will simply be cut short when the 2048 limit is reached.

When you pick the **Import** button, the **Import Text Files** dialog box appears. Select the drive and directory where the text file is stored. By default, AutoSketch looks for files with a .TXT file extension. To look for any file, enter *.* for the **Filename.**

Export

The **Export** button saves the text in the edit box to an ASCII file. When you pick the **Export** button, the **Export Text File** dialog box appears. Select the drive and directory where the text file should be stored, and name the file. By default, AutoSketch saves files with a .TXT file extension unless you also type a different extension.

> **SHORTCUT**
>
> By creating text in a word processor, you can utilize tools provided with that program such as spell check.

OK

The **OK** button accepts the entered text and places it on the drawing using the current text features.

Cancel

The **Cancel** button closes the **Text Editor** dialog box without placing text on the drawing. If you have text entered and select **Cancel** a dialog box will appear asking you if you want to save the changes.

Percentage meter

The *percentage meter* shows you how much of the 2048 character limit you have used up.

Exercise 9-3

1. Start a new drawing using the **Quick Start** template. Select **Inches (decimal)** as the unit of measurement.
2. Select the **Text Editor** tool and select a point at the top of your drawing.
3. Enter the two paragraphs shown below using the default text features. However, do not click **OK** yet to place the text on the drawing.
   > This is the first paragraph of text. It will be moved below the second paragraph of text. The text features will also be changed.
   > This is the second paragraph of text. It will be moved above the first paragraph of text. The text features will also be changed.
4. Select the **Word Wrap** option.
5. Pick the **Settings** button and change the settings to the SHX font ROMANS with a text height of .20. Also change the text justification to **Center.**
6. Select the **Export** button and enter the filename JUNK.TXT in the **Export Text File** dialog box. Place this file in the directory path C:\.
7. Highlight the bottom paragraph and use the **Cut** and **Paste** buttons to move the bottom paragraph to the top.
8. Move your cursor below the second paragraph. Select the **Import** button and enter the filename JUNK.TXT in the **Import Text File** dialog box. This file will be in the directory path C:\.
9. Pick **OK** to place the text on the drawing.
10. Save the drawing as C9E3 on your work diskette and then close the drawing.

Drawing Special Characters

AutoSketch allows you to draw several special characters without having to use a symbol font. This is done by typing two percent signs (%%) followed by a letter. Using this method, you can overscore or underscore text, or place degrees, plus/minus tolerance, and diameter symbols. (An overscore is a straight, horizontal line above a letter. An underscore is a straight, horizontal line below a letter.)

You can enter the special character code anywhere in your text string. Figure 9-12 shows the letters and symbols to enter to obtain the special characters. You will not see the overscore, underscore, or special character in the drawing until you pick **OK** or press [Enter] to place the text on the drawing.

Figure 9-12.
Using two percent signs and a letter, you can enter commonly used symbols and effects.

Typed	Entered on-screen
%%oToggle overscore%%o	T̄oggle overscorē
%%uToggle underscore%%u	T̲oggle underscore̲
%%uUnder %%oand%%u overscored%%o	Under a̲n̲d̲ overscored̄
%%dDraw degrees symbol	°Draw Degrees Symbol
%%pDraw plus/minus symbol	±Draw plus/minus symbol
%%cDraw diameter symbol	ØDraw diameter symbol
%%%Draw percent sign	%Draw percent sign

Exercise 9-4

1. Start a new drawing using the **Quick Start** template. Select **Inches (decimal)** as the unit of measurement.
2. Place the line of text shown below on your drawing. Use the proper codes to enter each of the special characters.
   <u>The temperature can vary ± 5°in a ⌀5.25" can.</u>
3. Save the drawing as C9E4 on your work diskette and then close the drawing.

Editing Text

You can perform the same editing functions on text as you can other AutoSketch objects. In addition, you can also change the words in the text object and change text features.

Changing Words and Letters

You can edit the characters of a single line of text, and text object, using the **Edit Text** tool found in the **Modify** menu. When you select this tool, the following prompt appears.

<u>Text Editor</u> Select object:

Pick the text object you want to edit. It does not matter whether the text was created with the **Quick Text** or **Text Editor** tool. Once you pick the text, it appears in the **Text Editor** dialog box. Make any changes and then pick **OK.**

Changing Text Features

If you pick the **Setting** button while in the **Text Editor** and change the values, those settings will take effect only on the text you are editing. The new values do not apply to future text. To change the settings for all future text, reset the text settings by double-clicking on the **Quick Text** or **Text Editor** tools in the **Draw** menu.

Exercise 9-5

1. Open the drawing from Exercise 9-3 (filename C9E3).
2. Select the **Edit Text** tool and pick the text object to edit.
3. Pick the **Settings** button and change the settings as follows.
   Font Type TrueType
   Active Font Courier New
   Height .18
   Width .75
   TrueType options: Italic
   Justification Left
4. Save the drawing as C9E5 on your work diskette and then close the drawing.

Changing Text with Other Editing Commands

Text can be moved, copied, scaled, or rotated in a similar manner to lines, circles and other objects. These commands see text as any other object. When you edit text, the cursor drags a box that represents the "area" of the text, rather than the text itself, Figure 9-13.

Figure 9-13.
When using many editing tools, text will appear as a box that represents the "area" of the text. Here, a text string is being copied. The box representing the copied text is shown with a tinted background.

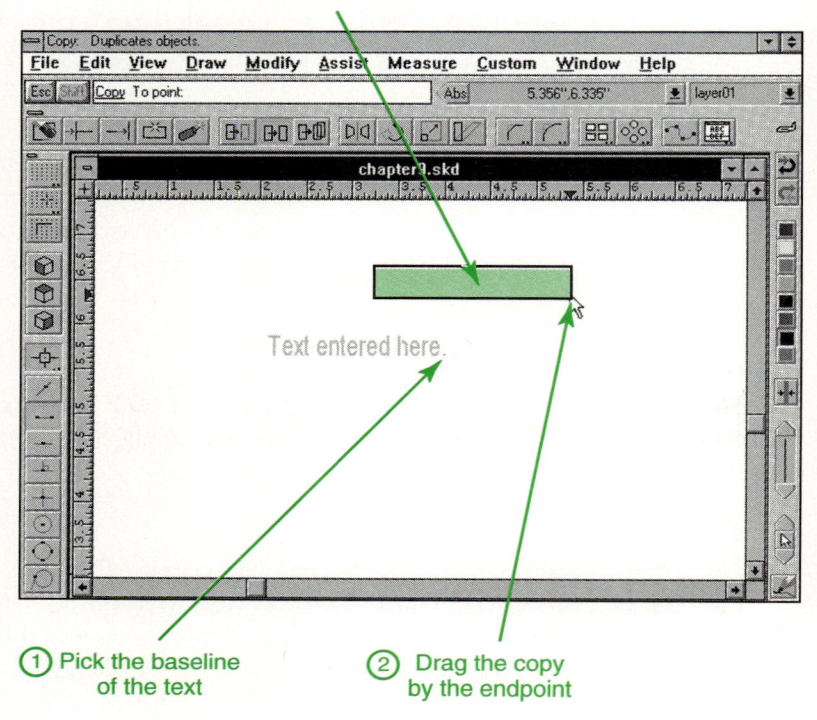

Box representing the text

① Pick the baseline of the text

② Drag the copy by the endpoint

Some commands have special effects on text, Figure 9-14. If you stretch one end of a text string, the text will compress or expand lengthwise, and also rotate with your cursor. If you mirror text, the text location mirrors, but the characters continue to read from left to right. You cannot break, trim, or extend a text string. You can explode a multiline text object into single, individual lines of text. However, you cannot explode a single line of text into individual characters.

You can connect to text using **Attach** tools. Text has two points that define each end of the baseline. You can use the **Attach** tool **Endpoint** to grab either of these points.

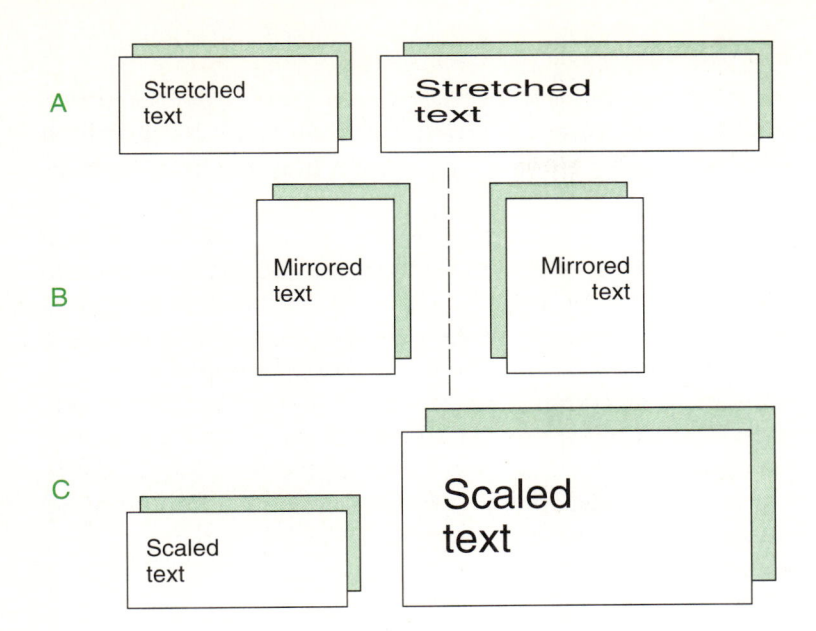

Figure 9-14.
Some editing commands affect text strangely. A—Stretched text changes in width factor. If stretched diagonally, the oblique angle will also change. B—Individual characters of mirrored text are not reflected, but the position of the text string is. C—Scaled text changes in height and width.

Exercise 9-6

1. Start a new drawing using the **Quick Start** template. Select **Inches (decimal)** as the unit of measurement.
2. Draw a triangle and place the following text within the drawing. Make a copy below the original. Refer to the drawing shown below.
    Text Triangle
3. Mirror the triangles and text horizontally as shown.
4. Stretch one of the triangles and text horizontally as shown.
5. Save the drawing as **C9E6** on your work diskette and then close the drawing.

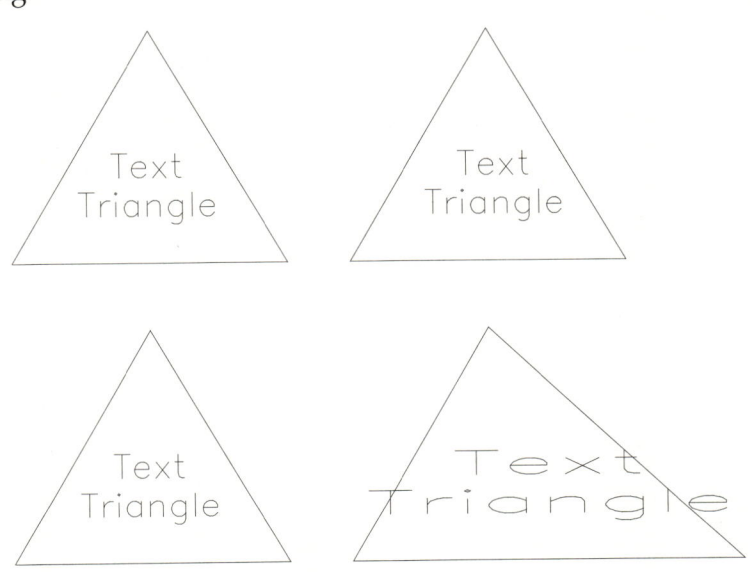

Summary

Technical drawings contain text to convey information that cannot be described with graphics. Text can take on many appearances, depending the text features selected.

The font refers to the style of the text. The height determines how tall the text will be shown. AutoSketch uses a width factor to determine the width of text in relation to the height. The oblique angle is the slant, or angle, of each individual character in an SHX font text string. The angle determines the rotation for a text object. Justification determines the alignment of the text in relation to the point you pick to place the text.

To place text on a drawing, pick the **Draw** menu and **Quick Text** or **Text Editor** tool. **Quick Text** is a fast way to add single lines of text to your drawing. **Quick Text** is also an easy way to add a couple of lines of text. The **Text Editor** allows you to add entire paragraphs of text to the drawing. For either tool, you must pick an insertion point for the text after picking the text tool. Finally, the text is entered. Special characters can be typed in using a control codes.

Important Terms

Font
Insertion point
Percentage meter

SHX fonts
Text
Text edit box

Text features
Text justification
TrueType fonts

New Commands and Tools

Active Font
Angle
Bold
Cancel button
Center
Clear button
Copy button
Cut button
Edit Text tool
Export Text File dialog box
Export button

Font Type
Height
Import Text File dialog box
Import button
Italic
Left
Middle
Oblique Angle
Paste button
Quick Text tool
Right

Sel All button
Settings button
Strike Out
Text Editor tool
Text Justification
Text Options
Text Settings dialog box
TrueType Options
Underline
Width Factor
Word Wrap

Review Questions

Give the best answer for each of the following questions.

1. List three benefits of CAD text over manual lettering.
    A. _____
    B. _____
    C. _____
2. In what dialog box do you set text features, and how do you access this dialog box?

3. What advantage do SHX fonts have over TrueType fonts? _____
4. What advantage do TrueType fonts have over SHX fonts? _____
5. Which fonts are most appropriate for a technical drawing? _____
6. True or false. In symbol fonts, capital letters mean the same symbol as lower case letters.
7. Which text features are available for SHX fonts, but not for TrueType fonts? _____
8. Which text features are available for TrueType fonts, but not for SHX fonts? _____
9. The **Height** option is based on the height of _____.
10. Suppose you are using a drawing scale of 1/4"=1'-0". For text to be printed 1/8" high, what value should you enter for text height? _____
11. True or false. The **Width Factor** value you set refers to the actual width of individual characters.
12. Explain the difference between **Oblique Angle** and **Angle** when setting text features for SHX fonts. _____
13. An **Oblique Angle** of _____ degrees is typically used on technical drawings to make text appear italic.
14. Suppose you select both **Center** and **Middle** options for text justification. When you draw a text object with three lines, where is the text placed relative to the point you pick on the drawing?
15. The _____ marks the placement of the text on the drawing.
16. When using the **Quick Text** tool, how can you place more lines of text without having to select a new insertion point? _____
17. What type of file can be imported into the **Text Editor** dialog box? What is the maximum number of characters that can be imported? _____
18. What code would be entered to draw a degree symbol? _____
19. True or false. Text is considered to be an object, just like a line, circle, or arc.
20. How do the following tools affect text?
    A. **Copy** _____
    B. **Move** _____
    C. **Break** _____
    D. **Explode** _____
    E. **Trim** _____
    F. **Extend** _____
    G. **Scale** _____
    H. **Stretch** _____
    I. **Mirror** _____
    J. **Rotate** _____

Activities

Here is a mixture of drawing activities to develop your skills drawing and editing text. If you have not already changed your printer orientation to landscape as discussed in Chapter 6, do so now. Then, for each activity using AutoSketch, select the **New** command from the **File** menu. Select the **Quick Start** template and select **Inches (decimal)** as the unit of measurement. Save your AutoSketch drawings on your work diskette using the name given. After saving each activity, select the **Close** command from the **File** menu to clear the drawing from the screen.

1. Draw a four-sided irregular polygon. Find the bearing of two sides, and place text at the proper angle along the two sides. See the illustration below. Save the drawing as C9A1.

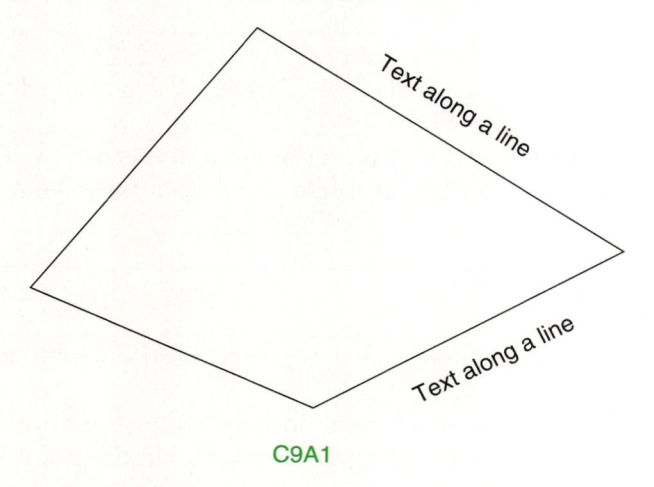

C9A1

2 - 4. For each activity, draw the design or object shown using the proper text features. Save the drawings using the filename given underneath each.

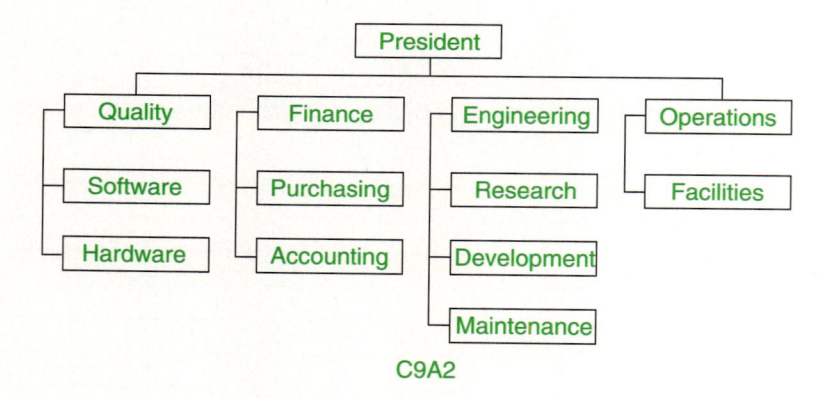

C9A2

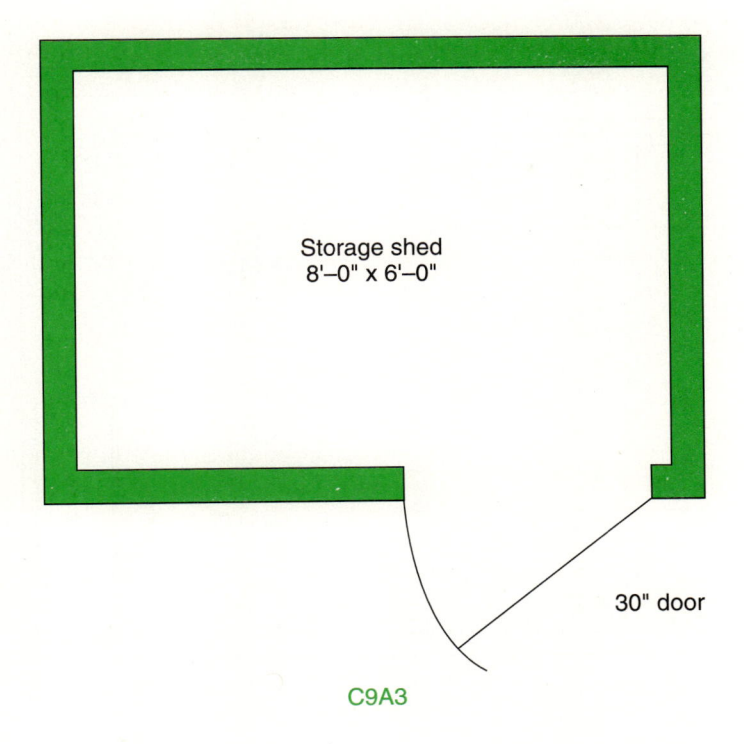

Chapter 9 Adding and Editing Text   273

Storage shed
8'–0" x 6'–0"

30" door

C9A3

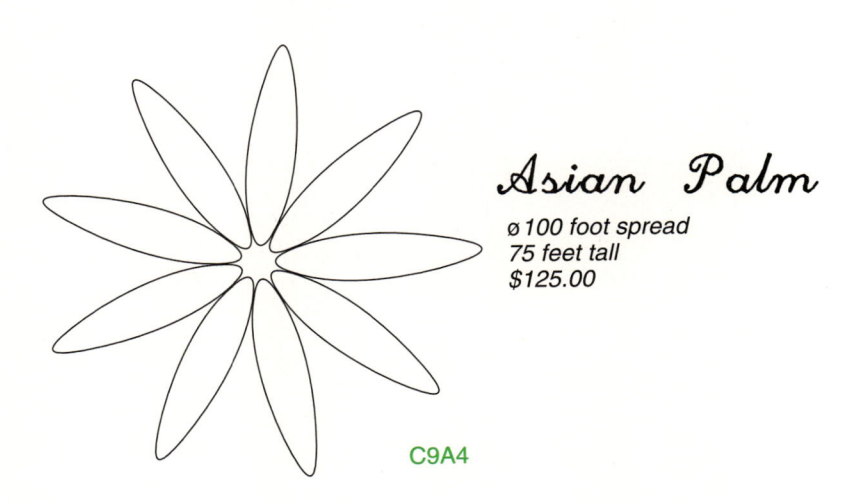

Asian Palm
ø 100 foot spread
75 feet tall
$125.00

C9A4

5. Draw a simple floor plan of a living room, bedroom, music studio, or other area of a house. Add furniture to the room and label each piece. An example is shown below. Save the drawing as C9A5.

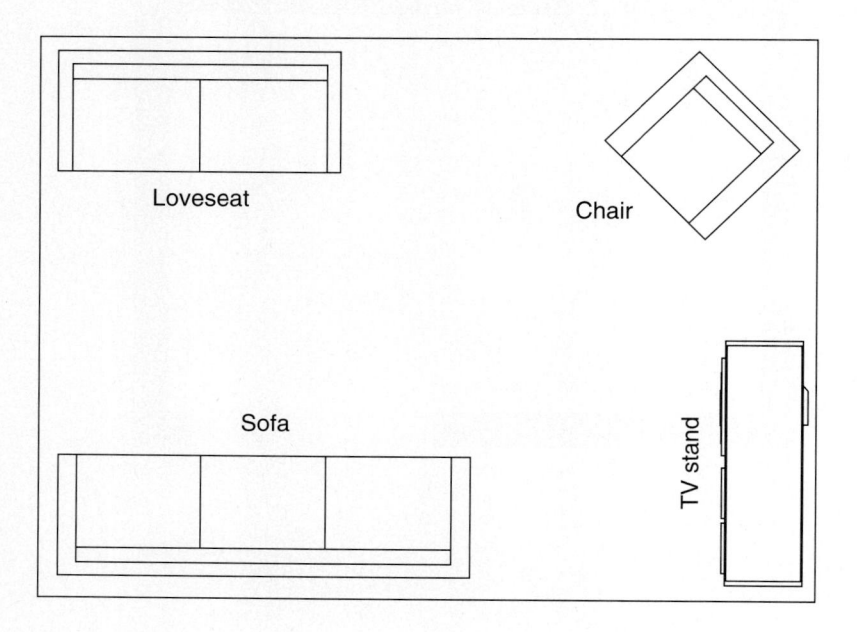

CHAPTER 10

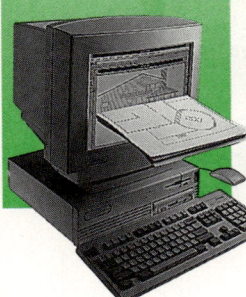

Developing New Drawings and Templates

Objectives
After studying this chapter, you will be able to:
- Explain the steps of the design process.
- Create a drawing plan and sketch a design idea.
- Complete the steps needed to set up a drawing.
- Set the units of measurement for a drawing.
- Set up a drawing sheet, complete with name, size, and scale.
- Use layers to sort information in a drawing.
- Add a border and title block to the drawing.
- Create, save, and use AutoSketch templates.
- Identify useful tips for developing drawings.

The chapters up to now have shown how to draw and edit single objects and basic shapes. This chapter pulls together your skills and shows you how to set up a working drawing.

You will first learn some guidelines that affect how you approach a design problem. Then, you will go through the steps taken to set up and construct a drawing template, complete with a border and title block. Finally, you will explore tips and techniques to help you construct drawings quickly.

The Design Process

Technical drawings are made to document a design idea. Designs do not pop out of thin air. Rather, they evolve over time. There are six steps that lead to solving a design problem. These steps are known as the *design process.*
1. Identify the problem or need.
2. Set guidelines.
3. Gather information.
4. Develop possible design solutions.
5. Select the final design solution.
6. Evaluate the solution.

In this chapter, you will apply the design process to solving a problem. For example, a design sketch might describe the size and shape of a computer table. A drafter then develops the working drawings for that design using AutoSketch, or another CAD program.

The design process applies to every drawing. Drawings of manufactured products, buildings, and circuit boards are all developed using the design process.

Identify the Problem or Need

New designs are the result of a problem, need, or market opportunity. Designs can be as complex as a fuel-efficient car to meet environmental standards. The problem can also be as simple as needing a shelf system to hold compact discs (CDs) for your multimedia computer.

When you start the design process, recognize what is needed and what is not. For example, solving the need for a CD holder means that you design a new computer desk. Focus your efforts on solving the problem at hand. It helps to write out a description. The following statement might be a description of the problem.

"Computer CDs are stacking up on my computer desk. A method to organize the CDs without taking up valuable desk space is needed."

Write a clear statement that does not imply any one solution. Keep it general. The following statement is too specific.

"A shelf is needed above my computer to hold CDs."

This statement limits the possible solutions. A well-written statement gives the other design steps direction.

Set Guidelines

Guidelines provide a focus to your problem solving. Guidelines are also called *design criteria*. The design criteria set the limits that you must work in. Your criteria can vary, depending on the problem or need, but typically include the following.
- Who will use the product?
- When will they use it?
- Where and how will they use it?
- Must the product follow a particular style?
- What is the maximum or minimum size?
- What is the maximum or minimum weight?
- What is the maximum or minimum strength?
- What is the cost range?

For example, guidelines for solving the "organizing CDs" problem might be the following.
- Standard CD cases are 5 5/8"x 5"x 3/8" thick.
- The title on the edge of the CD case must be visible.
- The CDs must be within easy reach from the keyboard.
- The case must hold at least ten CDs, preferably more.
- The case must have a lock to prevent access by young children.
- The case must not interfere with use of the computer.
- The case must be affordable.

Make sure that you do not limit the number of potential solutions with your guidelines. For example, suppose you are still setting guidelines for a CD organizer. By writing "must fit under the keyboard," you exclude any ideas for organizers that sit on top of the monitor, or under the computer. Be careful when setting guidelines.

Gather Information

The research stage of the design process is often the most neglected step. Be informed. Gather all the information needed to develop potential solutions to the

problem or need. The ideas you can generate are only limited by your knowledge of the subject. The following information should be included for most projects.
- Similar products and competing products.
- Feedback received on similar products.
- Market research.
- Customer opinion.
- Available materials to make the product.
- Existing parts and supplies that could reduce cost and labor.
- Recent technical information from trade publications.
- Supplier catalogs.

Check every detail, even ones that may seem worthless at first. You might find an item, perhaps buried deep in some trade publication, that helps solve your problem. For example, you might find information on a product that stores audio CDs in a floor-standing tower. This could apply to your product.

Develop Possible Solutions

After identifying the problem and gathering information, begin to think of possible solutions. Write down any idea that is related to the design. Do not dismiss any idea at first. Record any thought that comes to your mind or is offered by others in your group. This process is called **brainstorming.** At this stage, do not ignore even "far-out" ideas. Look for *quantity* of ideas, not quality.

Once you have exhausted ideas, begin to review them. Look for good and bad points in each. See which ideas begin to solve the problem. Sketch those ideas on paper. These first drawings are only rough sketches because the solutions are not yet well thought out. Many times, you will combine good points from two or more ideas. This is all part of refining your design.

Sometimes, you can build on previous designs. By making small changes to an existing product, you may answer the design problem. The data you gather may identify a product similar to what you need. For example, remember the floor-standing tower product to hold CDs? That model may be bigger than what you need, or not easily accessible from the keyboard. However, a smaller version of that type of organizer could be one possible solution.

As you refine possible solutions, prepare detail sketches, models, and renderings. When possible, test models using actual conditions the product must endure. Look for function, safety, appearance, and durability as they relate to your design. Your tests may lead to changes.

Select the Final Solution

The final design comes after you refine the best ideas to form one effective solution. Weed out weak ideas and combine the good points of all practical options.

Sometimes the best solution of those presented is not acceptable. It may not meet all the guidelines you set earlier in the design process. For example, you might develop a CD holder that fits under the keyboard. Yet, it raises the keyboard to an uncomfortable height. No one would buy it. If you look back at your guidelines, one of the items was "Must not interfere with use of the computer." Always check the solutions to see that they fall within the guidelines.

Often, the final design is a compromise between the most ideal and the most practical solutions. Take into account any factor that relates to your design. This might include the cost to build, the purchase price, service and repair, the market, the environment, social aspects, looks, and durability. There will also be many factors unique to your design that you must consider.

In most companies, the final design is presented to management before it is manufactured or built. This involves presentation drawings, written reports, cost estimates, and market analysis. Many times, you must prove that the design will be a profitable venture.

Once the final design has been selected, the work of the drafter begins. The design ideas, sketches, and specifications must be converted into working drawings. This includes detail drawings, assembly drawings, bills of materials, parts lists, and other data. Making working drawings often requires cooperation among many people and departments.

Evaluate the Solution

The design process does not end with the final design solution. Next comes evaluation, or *feedback*. Check to see if the designed product or method has fulfilled its purpose. Talk with users of the product. A customer might have found a flaw that you did not foresee. Sometimes, the compromises you make when choosing a final design solution affect the quality of the product.

Feedback is important because the design process is not a one-time routine. The process cycles so that feedback helps refine the product (and future products), Figure 10-1. The feedback you receive on one product is also important data for future products.

Figure 10-1.
The design process is a cycle, where feedback refines the design of a product, making it "new and improved."

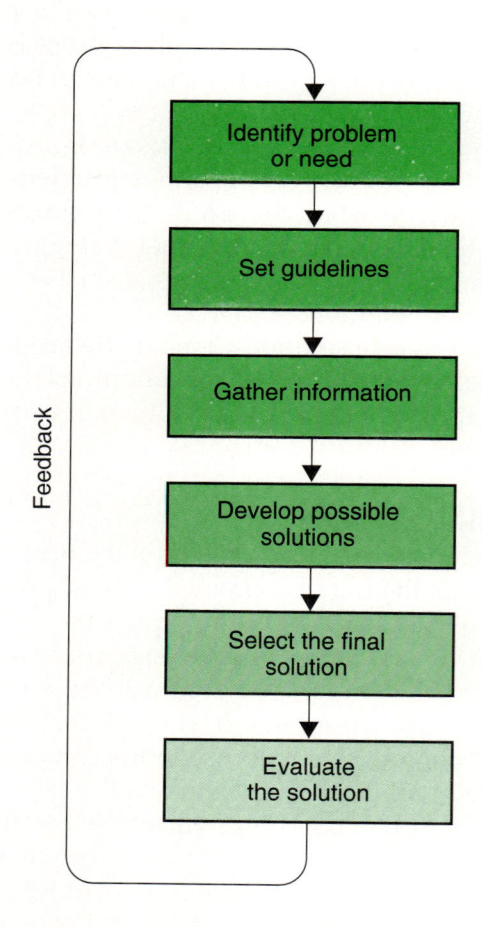

Exercise 10-1

1. Decide on a product or structure that you would like to design. Write out a clear description.
2. Set guidelines for the design, such as who will use it and how they will use it.
3. Gather information about the product you intend to design. Use the library, or possibly visit a company in your city to find this information.
4. Develop at least three possible designs that meet the guidelines for your design problem. All of the designs should be viable solutions to the design problem.
5. Select a final design solution.

Sketching Designs

Throughout the design process you will need to make notes of your ideas. Many popular products begin as a pencil sketch on a piece of scratch paper. During the initial stages, sketch your rough ideas. A sketch often communicates better than words.

Documenting an idea, using manual drafting tools or CAD, can take time. Yet, with a sketch you can quickly jot down dozens of ideas. The design evolves as sketched concepts are reviewed. When the best ideas are combined, refined sketches can be drawn. Engineering sketches are very refined sketches that are given to a drafter to convert into drawings.

When you sketch ideas, there are three suggestions to remember.
1. Proportion is important. A circle twice the size of another circle should look twice the size. Two lines that form a 30 degree angle should appear to be at 30°, Figure 10-2.

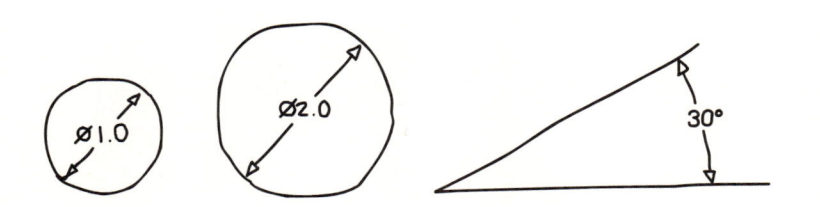

Figure 10-2.
Your sketches should be proportional. Shapes should appear to be the proper size, location, angle, and shape.

2. Add dimensions. Show the size and location of features so that your sketches are not misunderstood, Figure 10-3. Do not worry about proper dimensioning methods at this time. You will learn them in Chapter 13.
3. Use graph paper when available. It helps you draw more accurately, Figure 10-4.

Figure 10-3.
Add dimensions to sketches. You may draw well, but your drawing most likely will not be perfectly accurate.

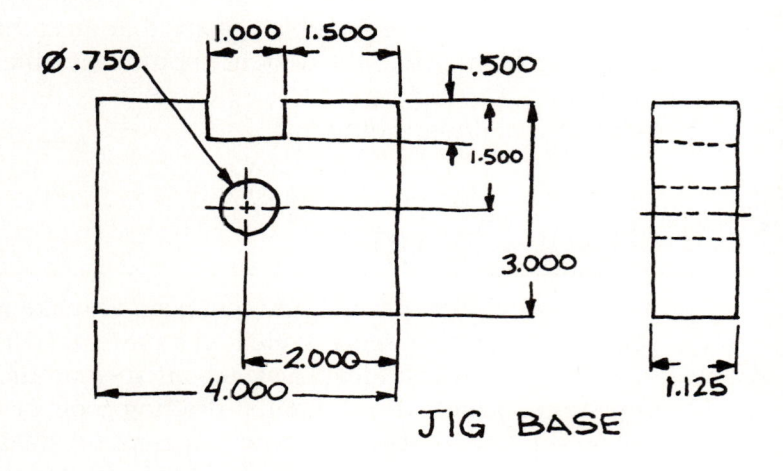

Figure 10-4.
Use graph paper to help your sketches be more accurate. Assign each square of the grid some measurement, such as .25" or 10mm.

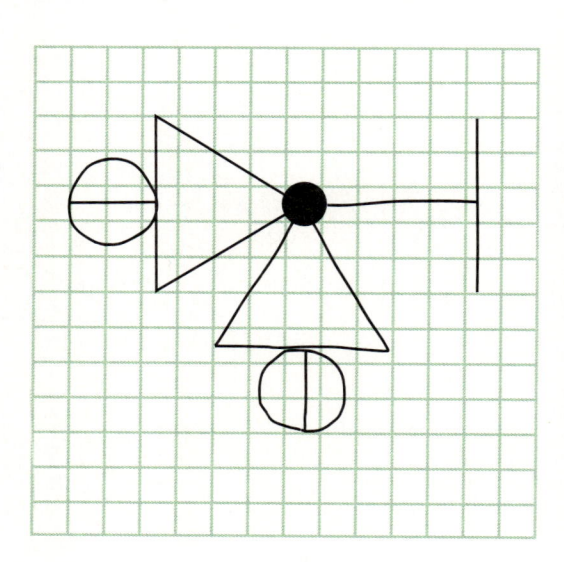

Drawing Plan

When you are sketching a product design, it is a good idea to have a plan on paper. This can be done on a standard form. This form is often called a *drawing plan*. Some companies have formal documents that track the progress of a project. The plan shown in Figure 10-5 is one that you might use in the classroom. The main parts of the drawing plan are the sketch and the documentation that identifies the part name, drafter, drawing setup, and any drafting standards to be applied.

Figure 10-5.
This is a typical drawing plan sheet that you can use for your sketches.

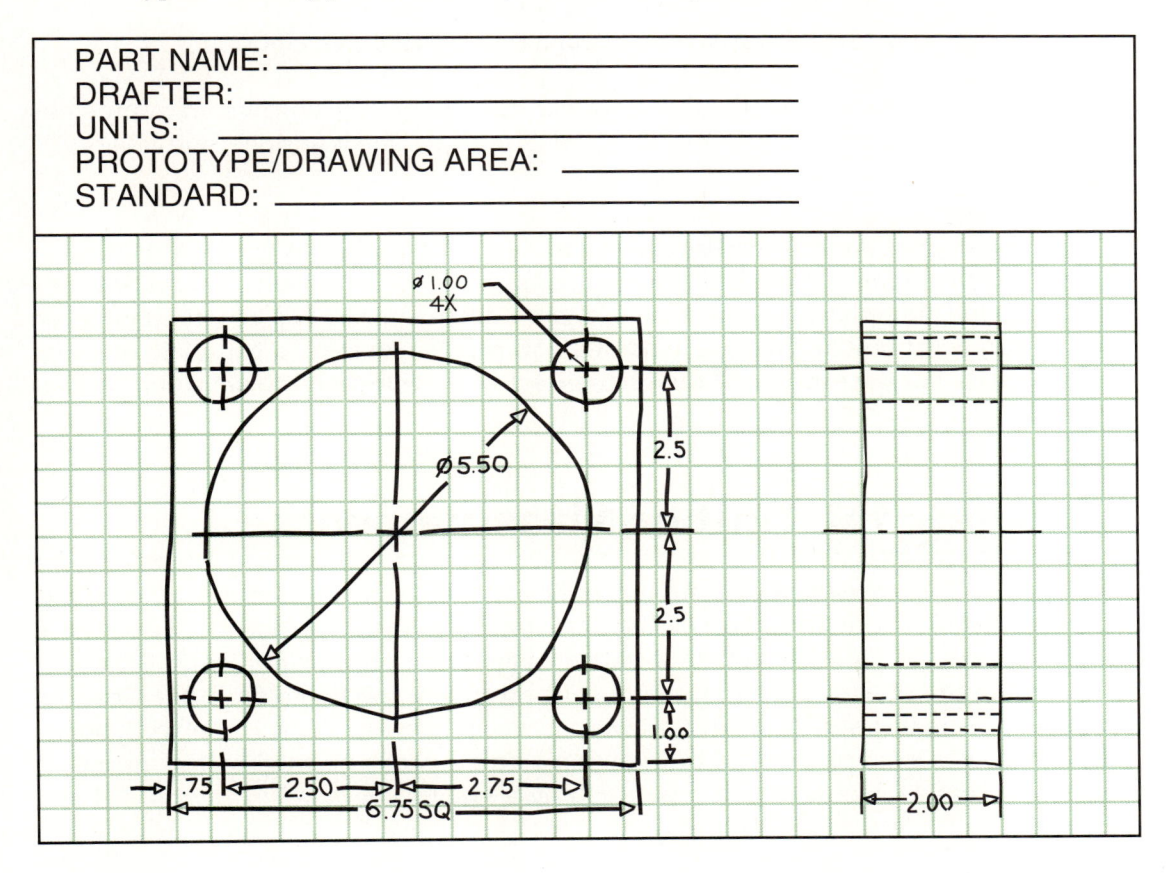

Exercise 10-2

1. Design a drawing plan.
2. Create a sketch of your design in Exercise 10-1. Make sure that your sketch is accurate and that you include measurements.

Setting Up a New Drawing

With a final sketch or drawing plan sheet in hand, it is time to start AutoSketch. Your first steps after starting the program are known as drawing setup. *Drawing setup* is the actions you take to prepare the drawing file before adding objects. Mostly, this includes entering values for units of measurements, sheet size, drawing aids, linetype, width, and color. In addition, you might draw a border and title block. It will take a little time to set all the following values.

- Units of measurement
- Sheet size and scale
- Layers
- Object properties (linetype, width, and color)
- Drawing aids (**Grid, Snap**, rulers, **Attach** modes, **Orthogonal** mode)
- Text font and features

Object properties, drawing aids, and text features have been covered in previous chapters. This chapter will cover units of measurement, sheet settings, and layers.

Templates

When starting AutoSketch, it asks you to choose a template. A *template* is a drawing file that contains a complete drawing setup, usually including a border and title block as well. Templates are discussed in more detail later in this chapter. However, before you learn how to create a template, you must first set up a complete drawing from scratch. For the purpose of this chapter, select the **Normal** template when starting AutoSketch.

Defining the Unit of Measurement

The *unit of measurement* determines how AutoSketch measures size, distance, and location of objects you place on the drawing. You can select a standard unit (such as inches, feet, and millimeters) or specify a custom unit of measurement.

Setting the default unit of measurement should be the first step in the drawing process. This makes sure that your drawing correctly represents the real object you are designing. For example, if you are drawing a small part, you might draw in inches or millimeters. However, if you are designing a house, you would draw in feet and inches.

To set the unit of measurement, access the **Units of Measurement** dialog box in one of the following ways.

- From the **File** menu, select the **Sheet Setup...** command and then the **Units of Measurement** button.
- Double-click on the **Coordinate Display** box found on the horizontal status bar.
- Double-click on the **Coordinate** pop-up window (if it is open)
- From the **Measure** menu, select the **Units** tool
- From the **Custom** menu, select the **Drawing Settings** tool. Then select the **Units...** button.

Figure 10-6.
Settings for the unit of measurement and precision are found in the **Units of Measurement** dialog box.

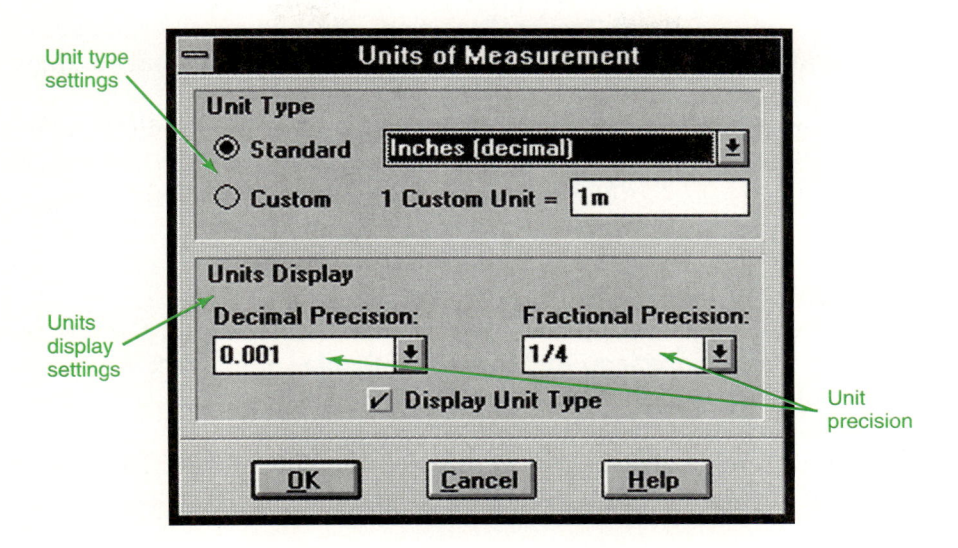

Unit type settings → Unit Type: Standard [Inches (decimal)], Custom 1 Custom Unit = 1m

Units display settings → Units Display: Decimal Precision: 0.001, Fractional Precision: 1/4, Display Unit Type ✓

Unit precision

Units of Measurement Settings

The settings discussed in the following sections are available in the **Units of Measurement** dialog box, Figure 10-6.

Unit Type

The **Unit Type** is the format of the measurement, such as inches, millimeters, or feet. All objects you add are measured and displayed using the current unit type. You can choose from 12 standard unit formats. Pick the arrow next to **Unit Type** to display a drop-down list.

Some units have both decimal and fractional settings. Decimal formats are chosen for most mechanical drawings. These are typically measured in whole units and parts divisible by 10 (such as tenths, hundredths, and thousandths).

Use fractional formats for architectural drawings. On these drawings, you usually want dimensions placed in feet, inches, and fractions of an inch (12'-4 1/2" for example).

The **1 Custom Unit =** option is used when you want to specify a special unit of measurement. For example, you may want each unit on the drawing to represent 5 millimeters. You would enter 5mm in the box next to **1 Custom Unit =** and click on the **Custom** button.

Units Display

The **Units display** setting determines the precision that you want units displayed to. Pick the arrow under either **Decimal Precision:** or **Fractional Precision:** to display a drop-down list. You can select the precision from this list.

For units that use **Decimal Precision**, select the number of decimal places (from .1 to .000001). The precision for mechanical drawings is usually 3 digits behind the decimal point (0.001).

For units that use **Fractional Precision**, select the smallest fraction that you want units displayed to (from 1/2 to 1/64). Some architectural drawings may be measured to the nearest 1/8". Although you can enter coordinates with greater precision (1/128 for example), AutoSketch rounds off the measurement.

Display Unit Type

The **Display Unit Type** box determines whether or not the unit of measurement is shown in edit boxes and the **Coordinate Display.** If this box is not checked, a measurement of 3mm appears as 3. It is best to have this box checked to remind you which unit of measurement is being used.

This setting does not affect dimensions. They have their own setting for the display of units within dimensions.

Setting Up a Drawing Sheet

A *drawing sheet* is used to tell AutoSketch what part of your drawing to print. After you set up a drawing sheet, solid-line and dotted-line boxes appear in the drawing window. The solid-line box represents the paper size. The dotted-line box shows the margins. These boxes are not actually part of your drawing.

You do not need to set up a drawing sheet until you are ready to print. However, it is best to set up a sheet during drawing set up for the following reasons.

- Defining your sheet in the beginning lets you draw a border and title block in the right places.
- Setting-up the right size sheet allows you to create a drawing that will print or plot on a single piece of paper.

To set up a sheet, select the **File** menu and **Sheet Setup...** command. The **Sheet Setup** dialog box shown in Figure 10-7 will appear. The order that you should set up your sheet is as follows.

1. Name the sheet. AutoSketch allows you to have more than one sheet, so each one needs a different name.
2. Select the sheet size. This should match the paper size that you will print or plot your drawing on.
3. Select a drawing scale (sheet scale) if your school or company standardizes on a scale. You can also enter the total size of your drawing (world size) and let AutoSketch determine the scale so that the drawing fits on the paper size.

Figure 10-7.
Settings for sheet name, size, and scale are found in the **Sheet Setup** dialog box.

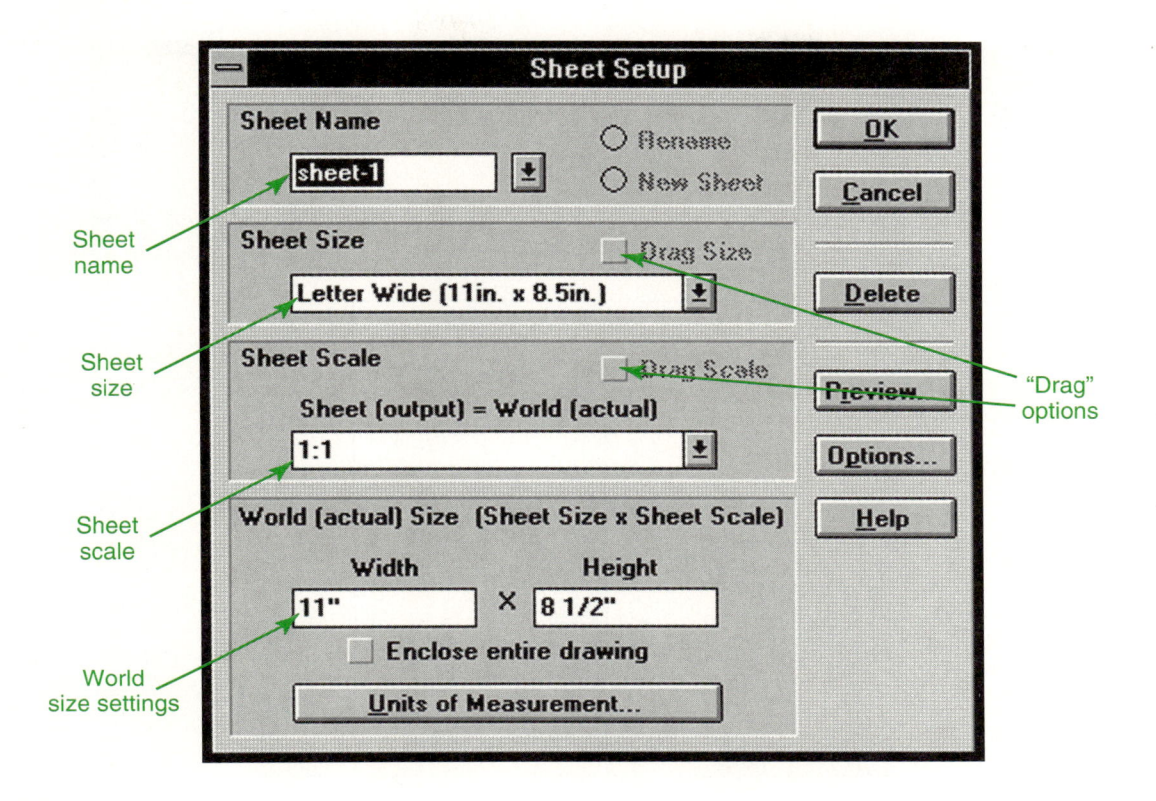

Sheet Name

You can name the sheet of paper that you will draw on. The default name is sheet-1. You can keep this name, change the default name, or add more sheets. Additional sheets are used when printing your drawing. These sheets may be different sizes so that you can print just a portion of your drawing.

To change the default sheet name, pick in the **Sheet Name** box in the **Sheet Setup** dialog box. Type the new name. Then, select **Rename** to change the default sheet name to the new name. When setting up a drawing, it is only necessary to define your default sheet name.

To add another sheet, pick in the **Sheet Name** box. Type the new name. Then, select **New Sheet** to create a new sheet.

Sheet Size

The **Sheet Size** setting determines the size of the invisible paper that you will draw on. This size should be the same size as the paper that you will print or plot your final drawing on. Most technical industrial drawings are created on C-size and D-size paper. However, for most of the activities in this book, you will use A-size paper. You can also create a custom sheet size if the size of paper you need is not listed. Also note that standard sizes of printer paper are listed. Since you are using this textbook for drafting, select a standard drafting sheet size, Figure 10-8.

Figure 10-8.
These are the standard sheet sizes for mechanical and architectural drawings.

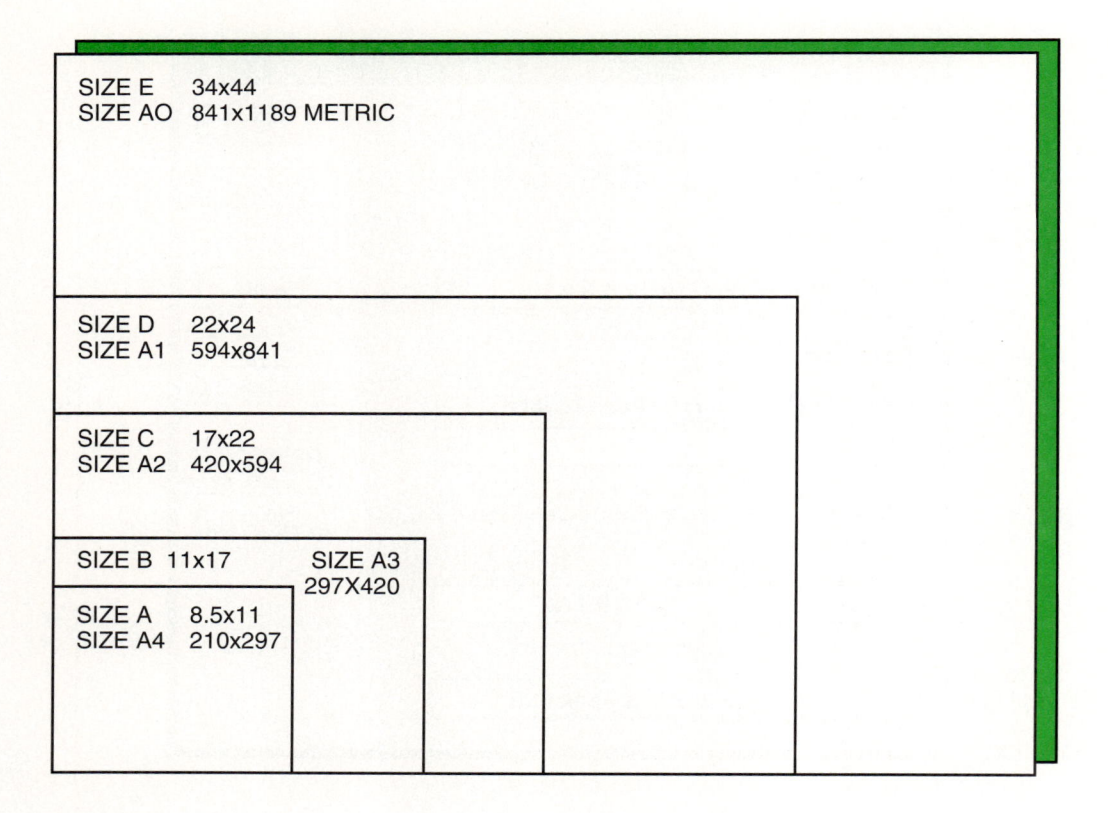

```
SIZE E     34x44
SIZE AO    841x1189 METRIC

SIZE D     22x24
SIZE A1    594x841

SIZE C     17x22
SIZE A2    420x594

SIZE B  11x17        SIZE A3
                     297X420
SIZE A     8.5x11
SIZE A4    210x297
```

To set the sheet size, pick the arrow next to the sheet size in the **Sheet Setup** dialog box. Note that both "tall" and "wide" options are listed. This is the orientation of the paper. For this book, pick the **Letter Wide** or **A Wide** option for the sheet size.

The **Drag Size** option is used when creating additional sheets other than the default. This allows you to drag a box around the portion of the drawing area you want included in the sheet. This option is covered in Chapter 16.

Sheet Scale and World Size

The **World Size** and **Sheet Scale** settings work together. The **World Size** setting refers to the entire area that your drawing will require. Remember that with CAD you draw in real units. That means when you draw a line a mile long, AutoSketch thinks of it as a mile.

However, you cannot print a line a mile long on a piece of paper. You need to scale the drawing down so it fits on the paper. The **Sheet Scale** setting determines how much to scale down your drawing so that it will print on paper.

There are two ways to set these values.
- Set the sheet scale to the standard scale of your school or company. Most drafters are accustomed to a standard scale because they have been using it for years with manual drafting. For example, floor plans for a house are almost always at a 1/4"=1'-0" scale. This makes sure that a floor plan will fit on a C-size or D-size piece of paper. At this scale, you can draw a floor plan as big as 88 feet by 68 feet in size on a C-size sheet of paper.
- There are times when a standard scale just doesn't work. For example, suppose you are designing a row of apartment buildings that is 200 feet long. What scale would you choose? Rather than guess at the scale, simply set the **World Size** to 250 feet wide (to allow for notes). AutoSketch adjusts the height and sets the scale to 1"=11'–4 23/64".

More on World Size

There are times when you may not know exactly how large your drawing is, or how large it will be when finished. It may be easier to set the world size by picking **Enclose entire drawing.** Thus, as you create your drawing, the world size setting adjusts accordingly. However, when using this option, do not add your border or title block until you complete the drawing.

More on Sheet Scale

When you edit the world size, the **Sheet Scale** setting will automatically change so that your drawing will fit on the sheet size chosen. In most cases, you will not edit the sheet scale, since you always want your drawing to fit on the sheet.

Combinations of Sheet Size, Drawing Scale, and World Size

Figure 10-9 shows some standard combinations of sheet size, world size, and scale.

Figure 10-9.
This chart shows the available world size for standard sheet sizes and drafting scales.

	Selected sheet size	Desired scale Sheet=World	World size
Mechanical	11 x 8 1/2 A	2" = 1" 3/4" = 1" 1/2" = 1" 1/4" = 1"	5.5, 4.25 14.6, 11.3 22, 17 44, 34
	17 x 11 B	2" = 1" 3/4" = 1" 1/2" = 1" 1/4" = 1"	8.5, 5.5 22.6, 14.6 34, 22 68, 44
	22 x 17 C	2" = 1" 3/4" = 1" 1/2" = 1" 1/4" = 1"	11, 8.5 29.3, 22.6 44, 34 88, 68
	34 x 22 D	2" = 1" 3/4" = 1" 1/2" = 1" 1/4" = 1"	17, 11 45.3, 29.3 68, 44 136, 88
	44 x 34 E	2" = 1" 3/4" = 1" 1/2" = 1" 1/4" = 1"	22, 17 45.3, 29.3 68, 44 136, 88
Metric	297mm x 210mm A4	1mm = 5mm 1mm = 10mm 1mm = 20mm 10mm = 1mm	1485, 1050 2970, 2100 5940, 4200 29.7, 21.00
	420mm x 297mm A3	1mm = 5mm 1mm = 10mm 1mm = 20mm 10mm = 1mm	2100, 1485 4200, 2970 8400, 5940 42.0, 29.7
	594mm x 420mm A2	1mm = 5mm 1mm = 10mm 1mm = 20mm 10mm = 1mm	2970, 2100 5940, 4200 11880, 8400 59.4, 42.0
	841mm x 594mm A1	1mm = 5mm 1mm = 10mm 1mm = 20mm 10mm = 1mm	4205, 2970 8410, 5940 16820, 11880 84.1, 59.4
	1189mm x 841mm A0	1mm = 5mm 1mm = 10mm 1mm = 20mm 10mm = 1mm	5945, 4205 11890, 8410 23780, 16820 118.9, 84.1

Exercise 10-3

1. In this exercise you will complete the drawing set up steps that would be needed to design the deck shown below.
2. Start a new drawing using the **Normal** template.
3. Set the unit of measurement as **Feet, Inches and Fraction** and the precision as 1/2.
4. Name the sheet DECK-1.
5. Select a wide, B-size sheet.
6. Set the **Sheet Scale** to 1/2"=1'. Make sure that the **World Size** will enclose your entire deck design.
7. Draw a box on your drawing that is the outside dimensions of the finished deck. Use drawing aids as needed. It is not necessary to draw all the details of the deck, unless you want to test your skills.
8. Save the drawing as C10E3 on your work diskette and then close the drawing.

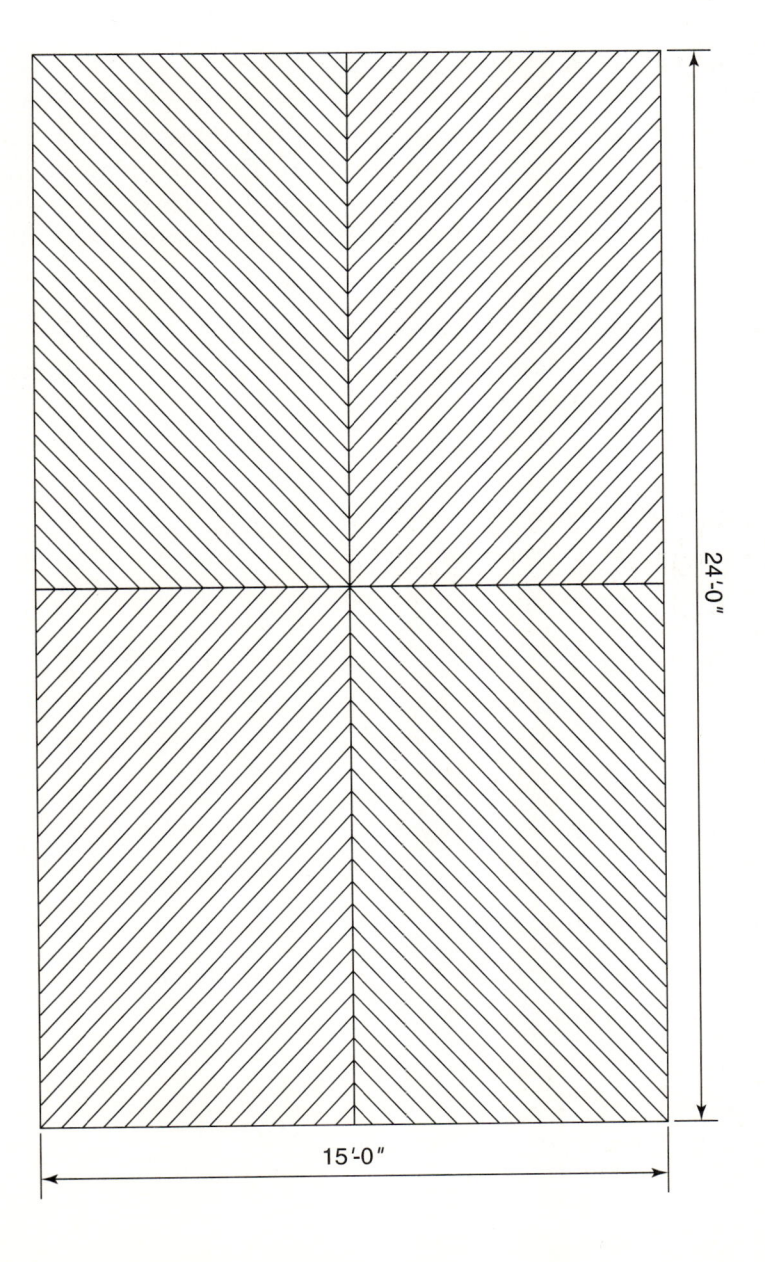

Exercise 10-4

1. In this exercise you will complete the drawing set up steps that would be needed for your design from Exercise 10-2.
2. Start a new drawing using the **Normal** template.
3. Select the appropriate unit of measurement and precision.
4. Name the sheet EXER3.
5. Select sheet size for the printer or plotter you are using.
6. Set the **World Size** so that the drawing sheet will enclose your entire design.
7. Edit the **Sheet Scale** so that the scale is easily read, and so that the **World Size** still is large enough to enclose the design. This may require that you enter a custom scale.
8. Draw your design.
9. Save the drawing as C10E4 on your work diskette and then close the drawing.

Sorting Information with Layers

In manual drafting, different parts of a design are often drawn on separate sheets of paper. For example, suppose you are drawing a house plan. The floor plan might be drawn on one sheet, the foundation plan on another, and the electrical plan on still another. Placing similar details on the same sheet of paper helps organize your drawing.

AutoSketch provides up to 256 layers that work much the same way. Each *layer* is like a sheet of paper that you can draw on. However, unlike manual drafting, these sheets can be shown together in any combination, Figure 10-10. Each sheet is transparent, so objects on the ones behind it show through.

Figure 10-10.
Each layer is like a clear sheet of plastic that you draw on. Layers are always in perfect alignment.

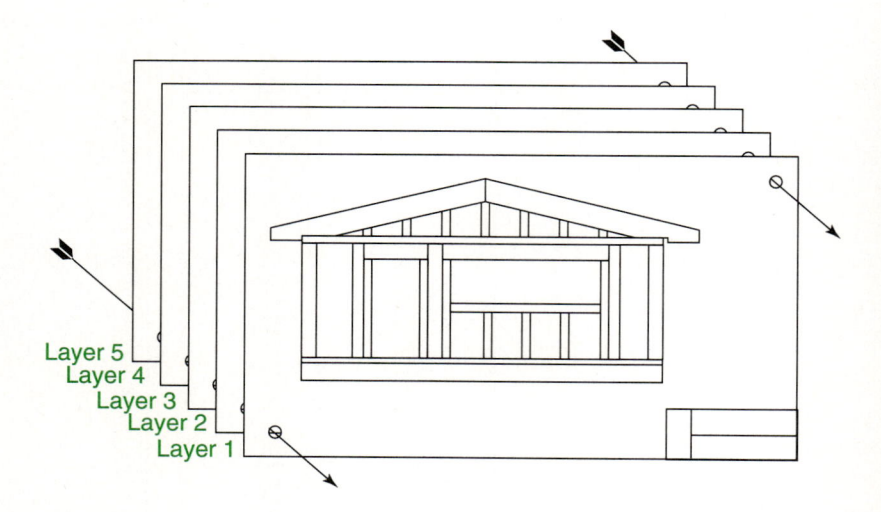

Objects you add to the drawing are assigned to the *current layer,* shown in the **Current Layer** box. You can change the current layer so that new objects are placed on a different layer. Pick the **Current Layer** button (arrow) and select the new layer from the list box. See Figure 10-11. The first ten layers are shown in alphabetical order. Although you can draw on only one layer at a time, there is no limit to the number of objects that can be included on any one layer.

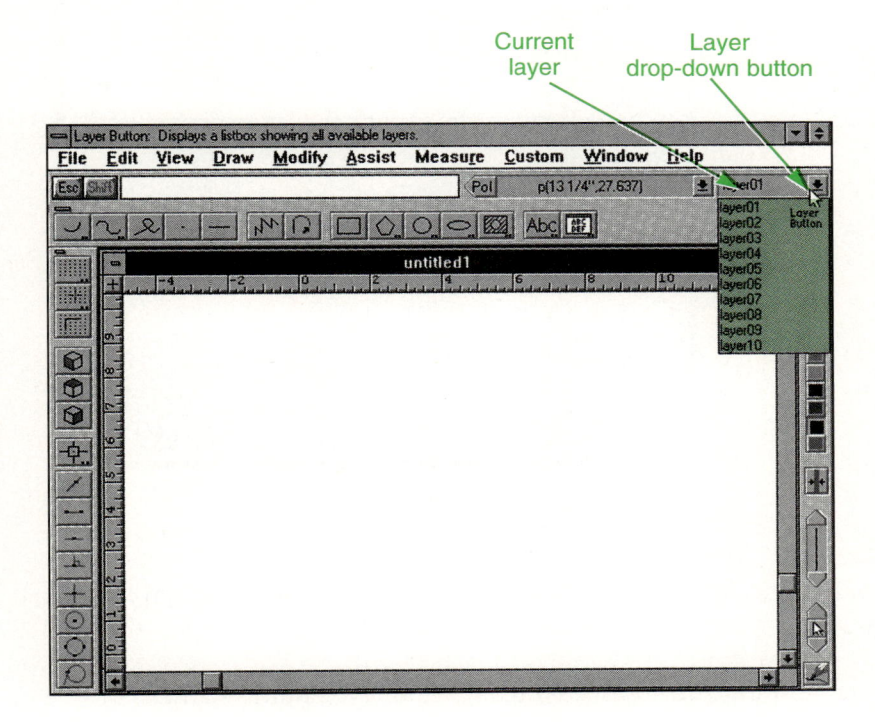

Figure 10-11.
The first ten layers are shown in alphabetical order in the layer list box. Simply pick a layer name to make it current.

You can display or print any one, all, or any combination of the layers at the same time. Turn off layers that contain unneeded information. For example, suppose you were drawing a house plan. The floor plan might be drawn on one layer, the foundation plan on another, and the electrical plan on another. You might turn off the layer having the foundation plan to see only the floor plan and electrical plan together. Only visible layers are printed.

Adding, Renaming, and Deleting Named Layers

AutoSketch allows you to name your layers. For example, the layer with the floor plan might be named "floor." The layer with dimensions might be named "dimen." All layer settings are made in the **Layer Settings** dialog box, Figure 10-12. You access this dialog box by double-clicking on the **Current Layer** box in the horizontal status bar.

By default, AutoSketch names the first ten layers "layer01" through "layer10." You can rename these layers, delete one or all of them, or add new named layers.

Figure 10-12.
Settings for layer names and their visibility are in the **Layer Settings** dialog box.

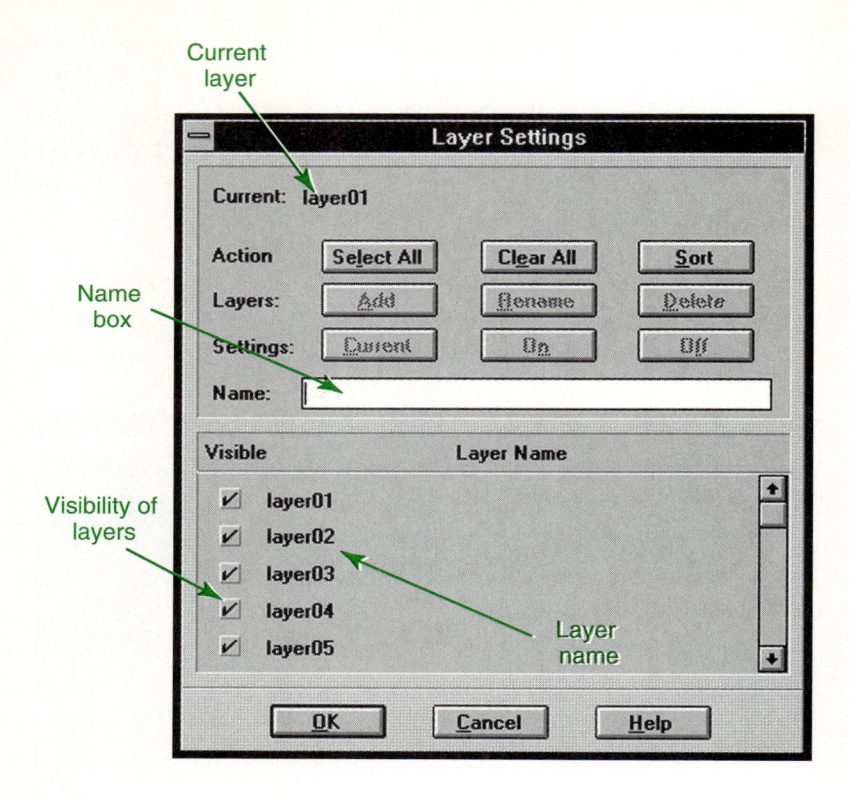

Adding a New Layer

To add a new layer, first open the **Layer Settings** dialog box. Then, pick in the box next to **Name:**. Type in the name of the layer you want to add. Then pick the **Add** button.

The layer name can contain up to 31 characters. The layer name will always be lowercase. Once you pick the **Add** button, AutoSketch adds the layer to the list and resorts the list in alphabetic order.

Renaming a Layer

To rename one of the default layers, or rename one of your own layers, first open the **Layer Settings** dialog box. Then, pick the layer so that it is highlighted. Its name will appear next to **Name:** when you pick it. Pick in the **Name:** box and type in a new name. Then pick the **Rename** button. The list may not be in alphabetic order until you choose the **Sort** button.

Deleting a Layer

To delete a layer first pick that layer. Its name will appear next to **Name:**. Then, pick the **Delete** button. You cannot delete the current layer and layers that contain objects. To delete all unused layers, select the layers by picking on the **Select All** button. Then, pick **Delete.** The current layer and any layers that contain objects will not be deleted, even if they were selected.

Making a Layer Current

When you draw an object, it is assigned to the current layer. This layer's name appears in the **Current Layer** box. To draw objects on a different layer, pick the drop-down button next to the **Current Layer** box. Pick the layer that you want to make current from the list. You can edit objects on any visible layer, regardless of which layer is current.

Making Layers Visible and Invisible

You can modify objects only on visible layers. Even though you draw on only one layer at a time, any one or all of the layers may be displayed. Objects on a layer not displayed are still there, just hidden. These objects are not affected by editing commands, nor are they printed. On the other hand, sometimes it may be hard to select an object within a complex drawing. It may be easier to edit the drawing if you hide unneeded layers. Then, make them visible again as needed. You can hide and show one layer, or a number of layers, as follows.

Single layer

To hide or show a single layer, pick the box next to the layer name. A check mark appears when the layer is visible.

All layers

To hide or show all layers, pick **Select All** and then the **Off** button or **On** button.

Selected layers

To hide or show selected layers, hold the [Ctrl] key and select several layer names. Then, select the **Off** button or **On** button.

Moving Objects to a New Layer

Objects drawn on one layer can be moved to another layer. For example, suppose you are making an architectural drawing and have added several electrical outlets to the floor plan layer. However, you should have placed the outlets on the electrical plan layer. Instead of erasing each outlet and drawing them on another layer, you can change their layer.

First, set the current layer to the electrical plan layer. Then, double-click the **Change Properties** tool from the **Edit** menu. Pick each option so only the **Layer** option has a check mark beside it, and pick **OK**. Finally, pick the outlets to assign them the new layer.

Layering Schemes

It is wise to establish standards for using layers. Some companies specify a *layering scheme* that determines what details are to be placed on what layer. A layering scheme helps you keep track of where different information is.

In a mechanical drawing, dimensions, object lines, title block and border, hidden lines and centerlines, and specifications typically all have their own layer. Architectural drawings have the different plans on different layers. Other designers then know which layer to select to review or edit those details. AutoSketch makes it easy to name the layers to represent their content.

Exercise 10-5

1. Start a new drawing using the **Normal** template.
2. Rename the following default layers as follows.

   layer01 floor
   layer02 foundation

3. Add a layer named ELECTRIC.
4. Draw a circle on the floor layer, a box on the foundation layer, and a line on the electric layer.
5. Hide the floor and foundation layers.
6. Use the **Select All** and **Delete** buttons to delete all layers that do not have objects on them.
7. Save the drawing as C10E5 on your work diskette and then close the drawing.

Setting Object Properties, Drawing Aids, and Text

Object properties, drawing aids, and text font and features should also be chosen during drawing setup. These topics were covered in previous chapters. Refer to those chapters for specific information.

Adding a Border and Title Block

A *border* is a solid, wide line that surrounds the drawing sheet. It provides a margin between the edge of the paper and the drawings, usually .5".

The *title block* is an area on the drawing that gives important information about the product and progress of the project. Companies using CAD systems frequently have a standard title block saved as part of a template. The title block contains labels like "Drawn By:" and "Drawing Name:," but leaves spaces to add the information.

Drawings created with CAD software usually have the words "CAD Drawing" on them. This alerts drafters that the drawing should be edited using CAD, not manual techniques, so that all edits are incorporated into the drawing file.

A typical title block is shown in Figure 10-13. Any title block might include the following information.

- Part name or drawing title
- Drawing number
- Scale of the drawing
- Material specifications, manufacturing methods, and special treatments
- Company name and address
- Date of original completion
- Revision dates
- General tolerances
- Names of the drafter, checker, and persons responsible for engineering, materials, and production approvals

Figure 10-13.
This is a typical industry title block.

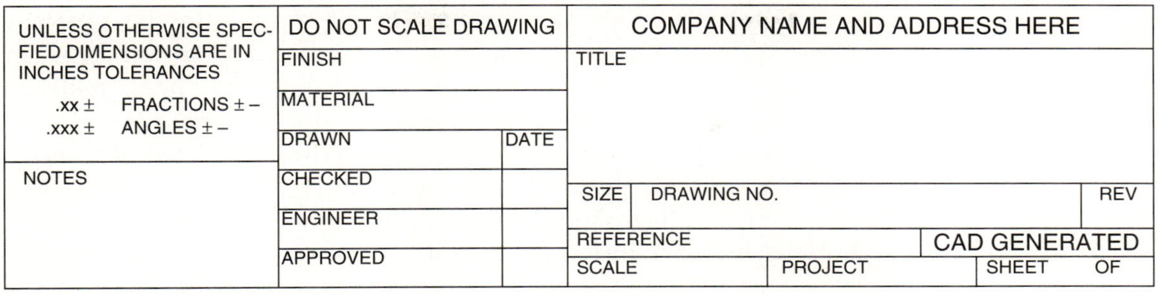

UNLESS OTHERWISE SPECIFIED DIMENSIONS ARE IN INCHES TOLERANCES .xx ± FRACTIONS ± – .xxx ± ANGLES ± – NOTES	DO NOT SCALE DRAWING		COMPANY NAME AND ADDRESS HERE			
	FINISH		TITLE			
	MATERIAL					
	DRAWN	DATE				
	CHECKED		SIZE	DRAWING NO.		REV
	ENGINEER		REFERENCE		CAD GENERATED	
	APPROVED		SCALE	PROJECT	SHEET	OF

Every drawing, from your first AutoSketch problem to a complex industrial print, should have a title block. The style of the title block varies between companies. Generally, it is found in the lower-right corner or along the bottom of the print. A standard style and size of title block for a practice drawing is shown in Figure 10-14.

Figure 10-14.
This title block can be used for practice drawings. This is shown full size so that you can measure it.

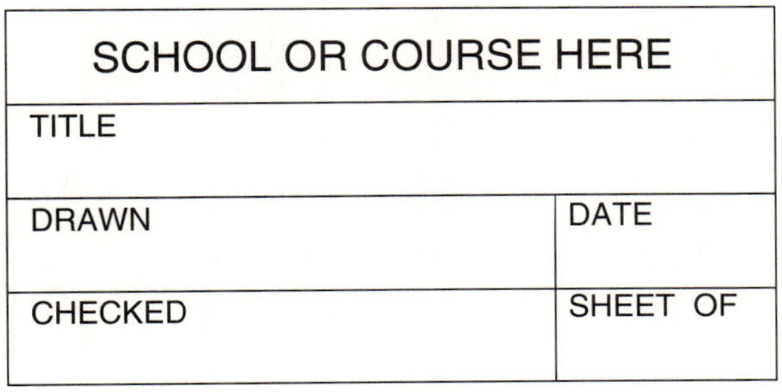

SCHOOL OR COURSE HERE	
TITLE	
DRAWN	DATE
CHECKED	SHEET OF

Exercise 10-6

1. Start a new drawing using the **Normal** template.
2. Complete the drawing setup steps needed for a wide, A-size drawing sheet. Use **Inches (decimal)** as the units of measurement and a 1:1 scale.
3. Draw a border .5″ from the outer edges. Use a wide linetype.
4. Add a title block to the lower, right-hand corner using the sample shown in Figure 10-14.
5. Save the drawing as C10E6 on your work diskette and then close the drawing.

Using Templates

A manual drafter begins a drawing with a sheet of paper, vellum, or film. The media is typically preprinted with a border, title block, and may be labeled with the company name, address, and other headings. The drafter lays down the sheet, fills-in the title block, and begins the drawing. The printed sheet of paper is really a starting point, or template, that the drafter adds objects to.

Templates can be developed in computer-aided drafting. A *template* is a drawing file with setup steps already performed. The drawing might also contain a border and title block. The drafter calls up the template drawing file, adds some information, and saves the drawing under a drawing filename. The template file remains unchanged.

Templates save much time since the drafter does not have to proceed through drawing setup. A number of templates can be set up for each drafting situation and every paper size. Also, having all drafters use the same templates means that drawings conform to specific standards.

Figure 10-15.
To save your drawing setup as a template, select the **Template Files (*.skt)** option of the **Save As...** command.

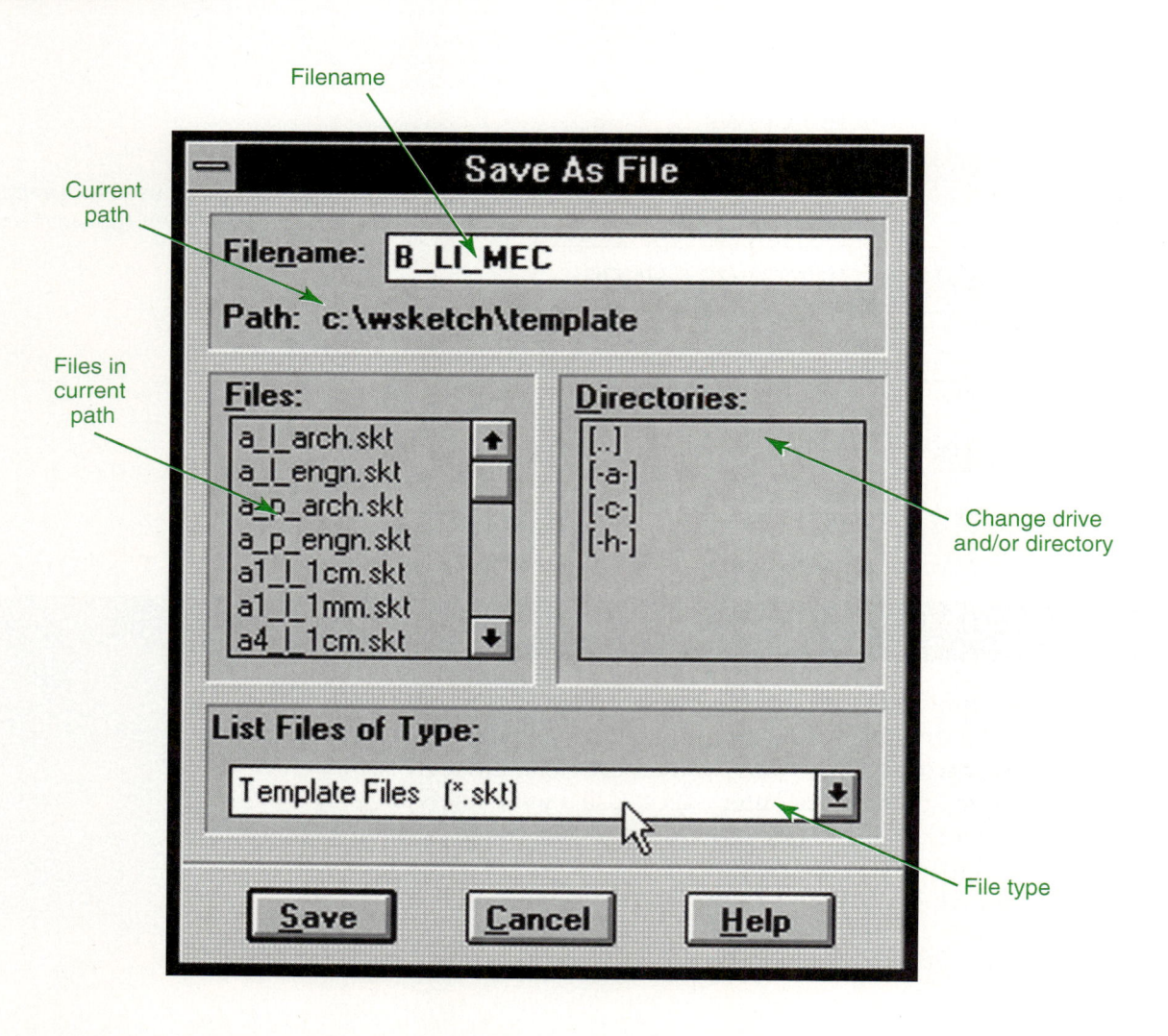

Saving a Template

If you set up a drawing and want to save it as a template, select the **File** menu and the **Save As...** command. Pick the arrow next to **List File of Type:** and pick the **Template Files (*.skt)** option. Then pick the **Filename:** box and enter a name. When you pick the **Save** button, the **Template Description** dialog box appears. Type in a description for the template and pick **OK**

Name the template according to its setup, Figure 10-15. For example, a B-size mechanical template in landscape format might named B_LI_MEC, where "B" stands for a B-size sheet, "L" stands for landscape, "I" stands for inches, and "MEC" stands for a mechanical template.

Starting a Drawing with a Template

When you start a new drawing in AutoSketch, the **Select Template** dialog box appears asking you to select a template. As you select each template, the description also appears, Figure 10-16. Pick **OK** to start a new drawing using the highlighted template.

Figure 10-16.
The template description gives important information about the template file.

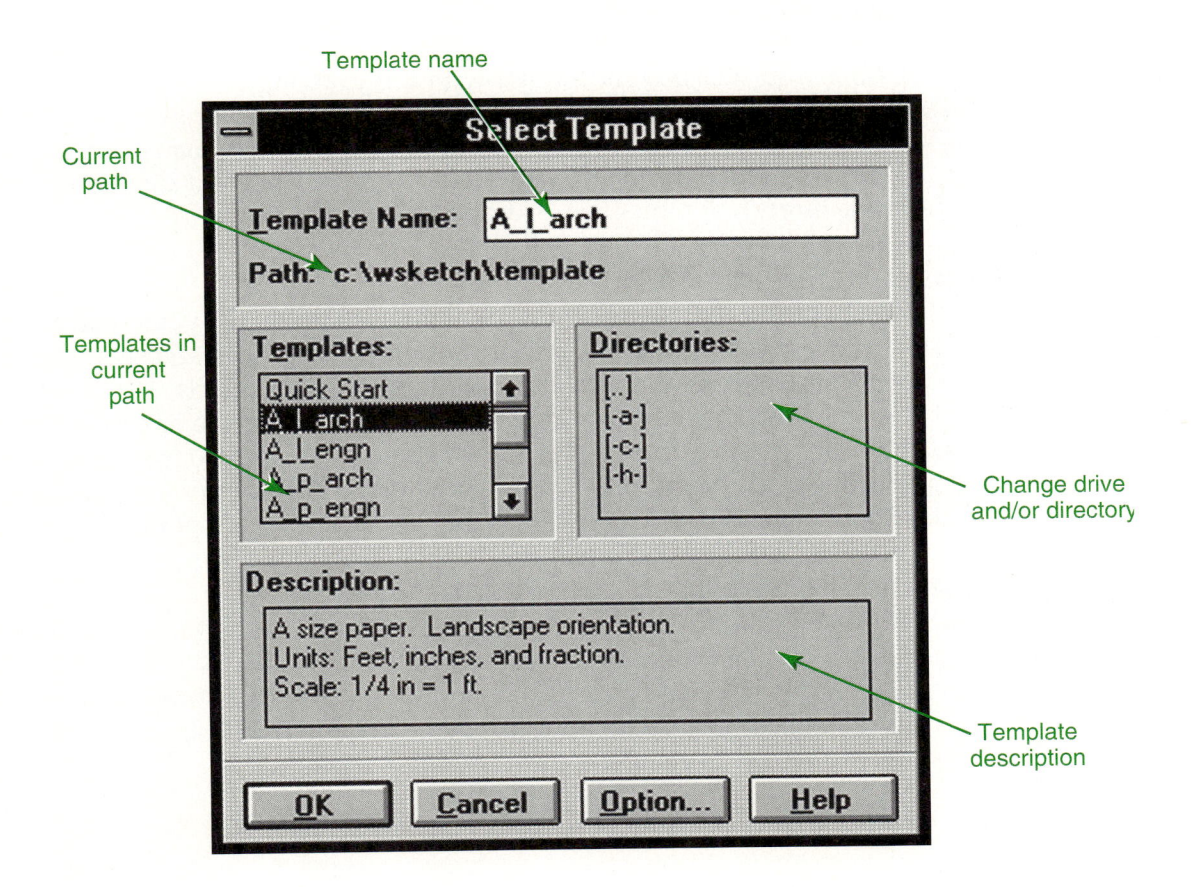

If you mostly use just one template, you can turn off the display of the **Select Template** dialog box. The next time you start AutoSketch, an untitled document using the last template you chose opens directly.

To turn off the **Select Template** dialog box, pick the **Template Option...** button in the dialog box. The **Template Option** dialog box appears. Pick the box next to **Show template dialog on startup** to deselect it.

If you want to use another template later, you can choose **New...** from the **File** menu. The **Select Template** dialog box will appear for you to select a different template.

Exercise 10-7

1. Open the drawing C10E6 from this chapter.
2. Save the drawing as a template on your work diskette using the filename A_LI_MEC. Then close the drawing.
3. Start a new drawing using the template you just created.
4. Draw a simple design.
5. Save the drawing as C10E7 on your work diskette and then close the drawing.

Tips for Developing Drawings

Over time, you will become skilled with the drawing and editing functions of AutoSketch. You will learn that there are often several ways to draw a complex shape. When starting a new drawing, don't jump right into it. Instead, plan a strategy to construct the drawing efficiently. Although each design will present different problems, consider these important topics before drawing the first line.

Begin at Reference Points

It is best to begin a drawing at one key reference point. On your sketch, look for points that other objects are measured or located from. Usually, these points are the corners of rectangular objects and the centers of circular objects, Figure 10-17. On mechanical drawings, the reference might be a machined surface or corner. On architectural drawings, features such as doors and windows are located from exterior walls. You may wish to reset the origin of the drawing to that reference point. This allows the **Absolute Coordinate Display** to show location from the reference point.

Construct the Minimum Number of Objects

Rarely do you need to construct every object in the drawing. Many drawings have similar features that can be copied. Symmetrical designs can be mirrored. A design with rectangular or circular patterns can be made using **Box Array** and **Ring Array** tools. Thus, you need only construct a few basic shapes. Then use editing techniques as the primary method to complete the drawing.

Figure 10-17.
In most cases, start your drawing at a reference point, usually a corner or center.

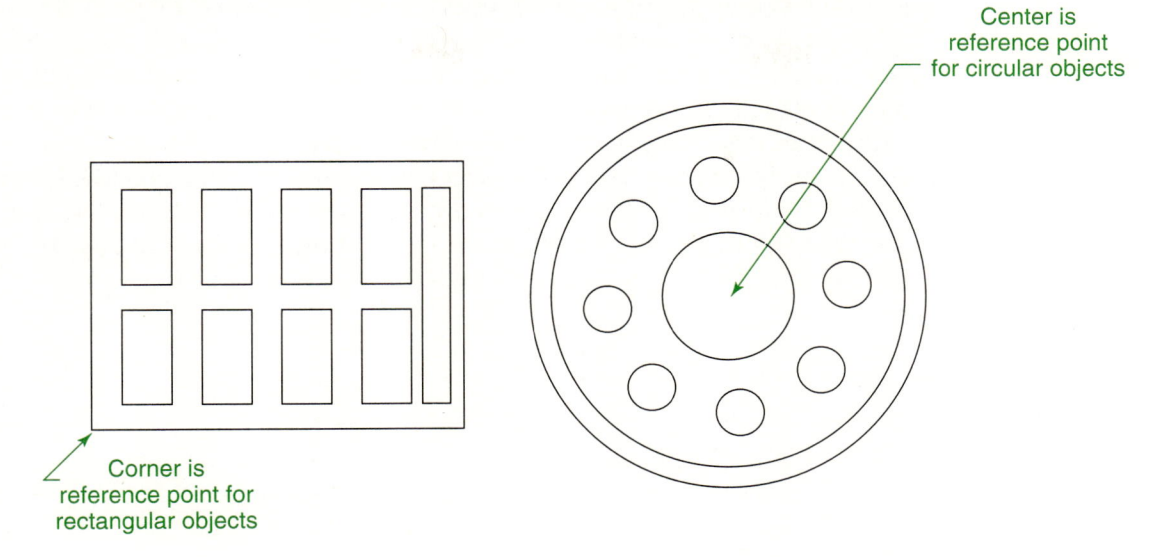

Center is reference point for circular objects

Corner is reference point for rectangular objects

Plan Your Commands

You are not limited to using just a few commands. A drawing might require that you copy, mirror, *and* rotate one or several shapes to form a complex object. Before you begin, look at the design sketches. Determine which commands will work best to complete the drawing. Some CAD systems have more features than others. If you later work on a system other than AutoSketch, be aware that it may have more or less functions. This can affect how you approach a drawing.

Use Hairline Width

Construct drawings using the hairline width. Most editing commands will not work with wide objects. When finished drawing and editing, change to wider lines where needed.

Avoid Switching Properties Often

A drawing may contain many different linetypes and colors. Avoid constantly switching values for these properties as you develop the drawing. This consumes precious time. It is better to construct all shapes with solid, black lines. Then, use the **Change Properties** tool to change the objects to the needed style.

Summary

Most technical drawings are made to record a design. You might create drawings for products you invent or build. The final drawing is started after the designer works through a series of steps, called the design process. These steps include identifying the need, setting guidelines, gathering information, developing possible solutions, selecting a solution, and evaluating the solution. Whether you realize it or not, these steps are done each time you make a decision.

With a design in mind, you will probably sketch it first. Sketching is the quickest and easiest way to jot down your ideas. As you prepare to use AutoSketch, transfer your sketches to a drawing plan. Consider how you will construct the drawing with AutoSketch.

Once you have a refined sketch, start AutoSketch and proceed to set up the drawing file. This includes setting the unit of measurement, sheet settings, layers, color, linetype, grid spacing, snap spacing, attach modes, and text features. Of these options, the layers and sheet settings were first introduced in this chapter. Sheet settings define the size of your "paper" and the scale of the drawing. Layers allow you to separate objects on different levels to help organize your drawing. All setup settings can be stored in a template drawing. If so, simply call up the template and begin.

As you draw, remember the tips to help you create drawings quickly. Remember, with a computer you should never have to draw the same shape twice. Also, add a title block to the drawing to give important information about you and the design. The title block can be placed in the template drawing.

Important Terms

Border
Brainstorming
Current layer
Design criteria
Design process

Drawing plan
Drawing setup
Drawing sheet
Feedback
Layer

Layering scheme
Template
Title block
Unit of measurement

New Commands and Tools

1 Custom Unit
Add button
Current Layer box
Current Layer button
Custom button
Decimal Precision
Display Unit Type
Drag Size
Drawing Settings tool
Enclose entire drawing
Fractional Precision
Layer Settings dialog box
Letter Wide

List Files of Type:
Name:
New Sheet
Off button
On button
Option... button
Rename
Rename button
Select All button
Select Template dialog box
Sheet Name
Sheet Scale
Sheet Setup...

Sheet Setup dialog box
Sheet Size
Show template dialog on startup
Sort button
Template Description dialog box
Template Files (*.skt)
Template Option dialog box
Unit Type
Units Display
Units of Measurement button
Units of Measurement dialog box
Units tool
World Size

Review Questions

Give the best answer for each of the following questions.

1. Explain how you might use each step of the design process to design a chair.

2. List four sources other than those given in the text that you might reference for design information.
   A. _____
   B. _____
   C. _____
   D. _____
3. At what stage of the design process do you usually begin making sketches? _____
4. Many times making minor changes to a(n) _____ will solve your design problem.
5. Why might the best functional solution to your design problem *not* be acceptable?
   _____
   _____
6. Often, the final solution is a(n) _____ between the most ideal and the most practical ideas.
7. The design process cycles so that _____ helps refine the product.
8. The two main parts of a drawing plan are the _____ and _____.
9. What options or values should be set during drawing setup?
   _____
10. You can choose both _____ and _____ units of measurement.
11. What are the range of values for both decimal and fractional unit precision?
    A. _____
    B. _____
12. What four options are set when preparing a drawing sheet using the **Sheet Setup** command?
    A. _____
    B. _____
    C. _____
    D. _____
13. What are two reasons why you should set up a drawing sheet before you begin drawing, rather than waiting until you print?
    A. _____
    B. _____
14. What sheet setup option should you choose when you may not know exactly how large your drawing is, or how large it will be when finished? _____
15. What would be the maximum world size for a drawing at half scale (.5=1) on a wide B-size sheet? Use AutoSketch to determine the answer. _____
16. Suppose you have a boat design that is 80 feet long and 60 feet tall. What scale would allow this design to fit on a C-size mechanical sheet? _____
17. Objects you add to the drawing are assigned to the _____ layer.
18. True or false. Only the current layer can be visible.
19. True or false. You can erase objects on layers that are not visible.
20. Describe three ways AutoSketch can save time when constructing drawings.
    A. _____
    B. _____
    C. _____

Activities

1 - 2. Gather pictures (magazine clippings, photographs, etc.) of products designed in industry. For two of the products, write down the problem or need you think each solved. Also, write a brief list of information you feel may have been needed to develop the design.
3. Use AutoSketch to develop a drawing plan sheet like that shown in this chapter. It should fit an 8 1/2"x11" sheet of paper. Print the sheet and make ten photocopies to have handy when you need them.
4. Create three templates in addition to the A_LI_MEC template completed in Exercise 10-7. All of the templates should have a border and a title block. Use the settings shown below. Use the default AutoSketch setting for any setting not shown below.

Settings	Template Name			
	A_LI_MEC	B_LI_MEC	A_LI_ARC	B_LI_ARC
Units	Inches (decimal)	Inches (decimal)	Feet, Inches, and Fraction	Feet, Inches, and Fraction
Precision	.001	.001	1/4	1/4
Sheet Name	A_LI_MEC	B_LI_MEC	A_LI_ARC	B_LI_ARC
Sheet Size	A Wide	B Wide	A Wide	B Wide
Drawing Scale	1:1	1:1	1/4"=1'	1/4"=1'
Layers	object hidden center dimension title	object hidden center dimension title	foundation floor electric heating dimensions	foundation floor electric heating dimensions
Grid	.5	.5	6"	1'-0"
Snap	.25	.25	6"	6"
Attach Modes	none active	none active	none active	none active
Orthogonal Mode	off	off	off	off
Coordinate Display	pop-up window	pop-up window	pop-up window	pop-up window
Rulers	not shown	not shown	not shown	not shown
Color	black	black	black	black
Linetype	solid	solid	solid	solid
Width	hairline	hairline	hairline	hairline
Text Features	.18 ROMANS	.18 ROMANS	6" ROMANS	6" ROMANS

5 - 10. For each activity below, fill out a drawing plan sheet. Include a sketch so that you can practice your sketching skills. Then, select one of the templates from Activity 4 and complete the drawing. Save the drawing using the filename given beneath the illustration.

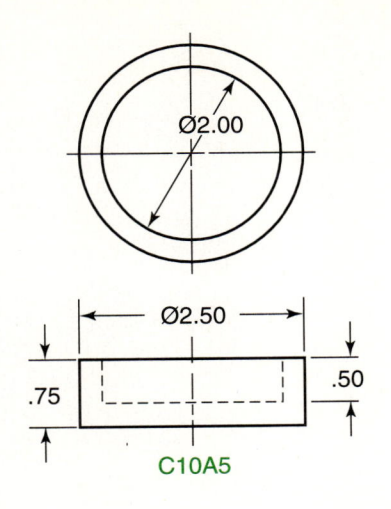

Chapter 10 Developing New Drawings and Templates

C10A5

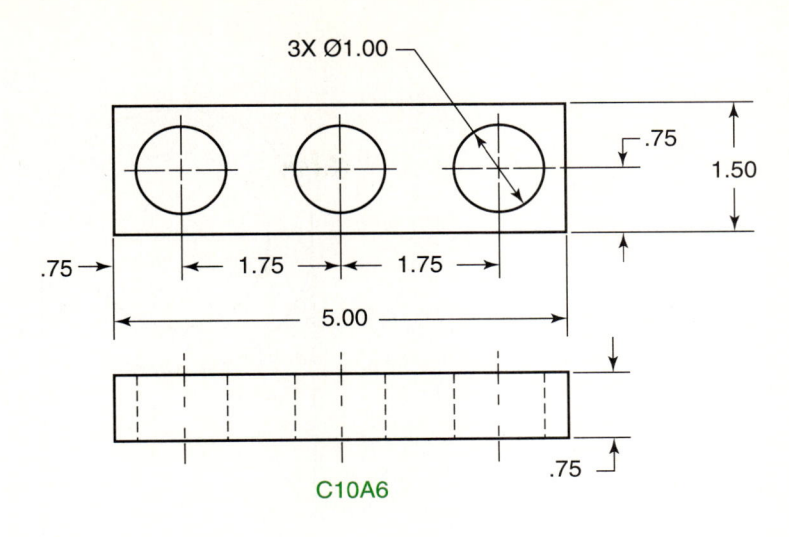

C10A6

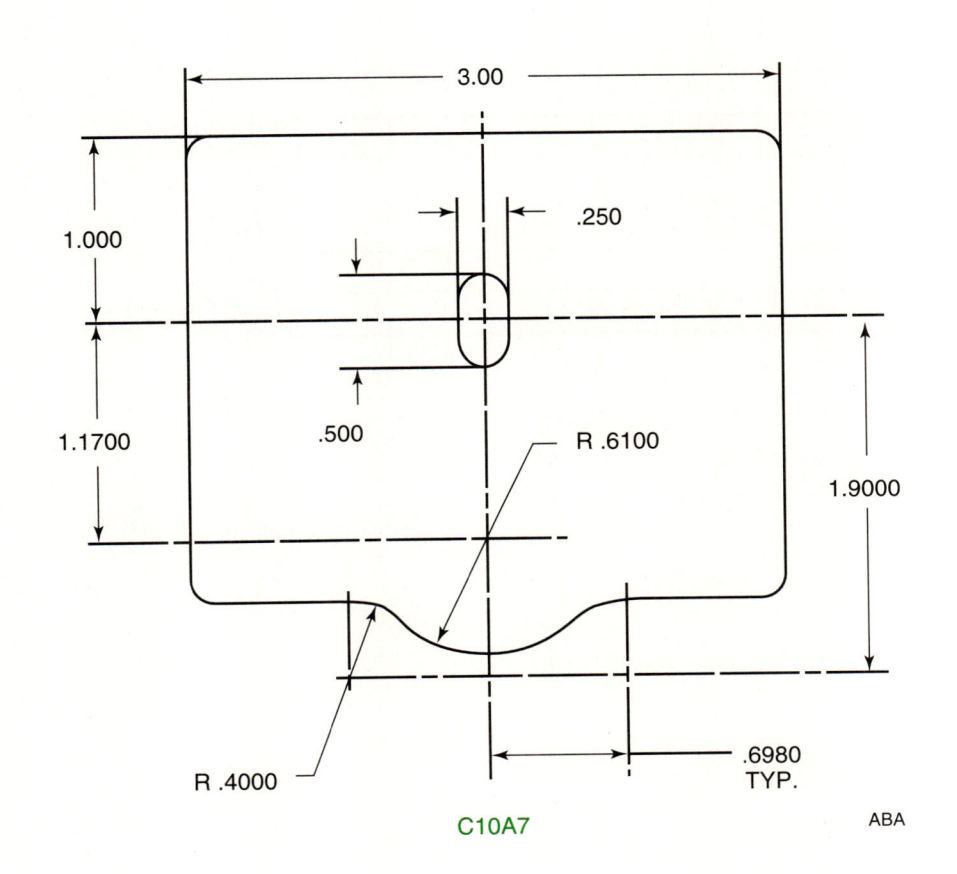

C10A7 ABA

Ø 2.00
Ø 3.00
6.00
2X R .25
2X R 1.50
2X R 1.00
2X R .50
6.00

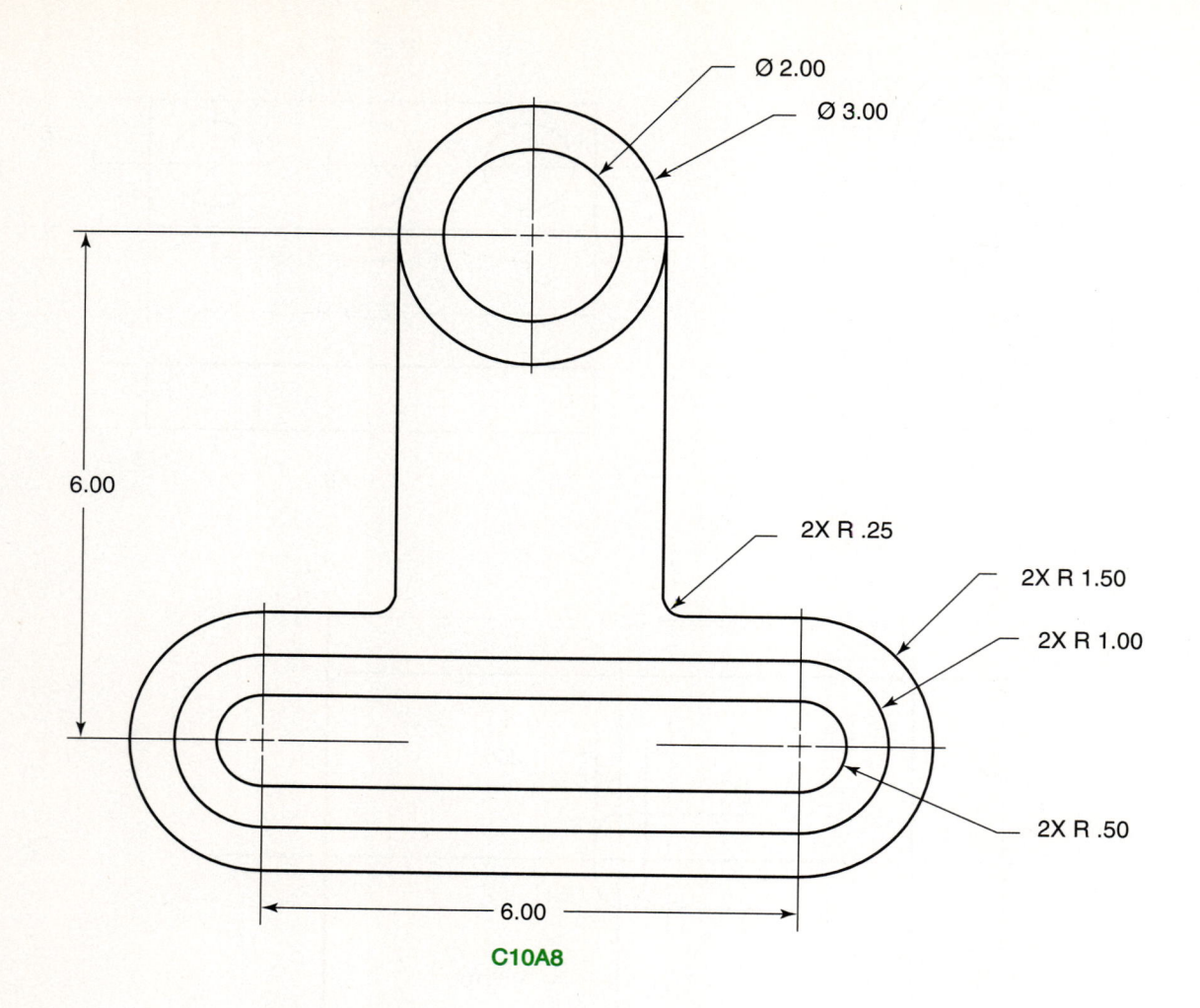

C10A8

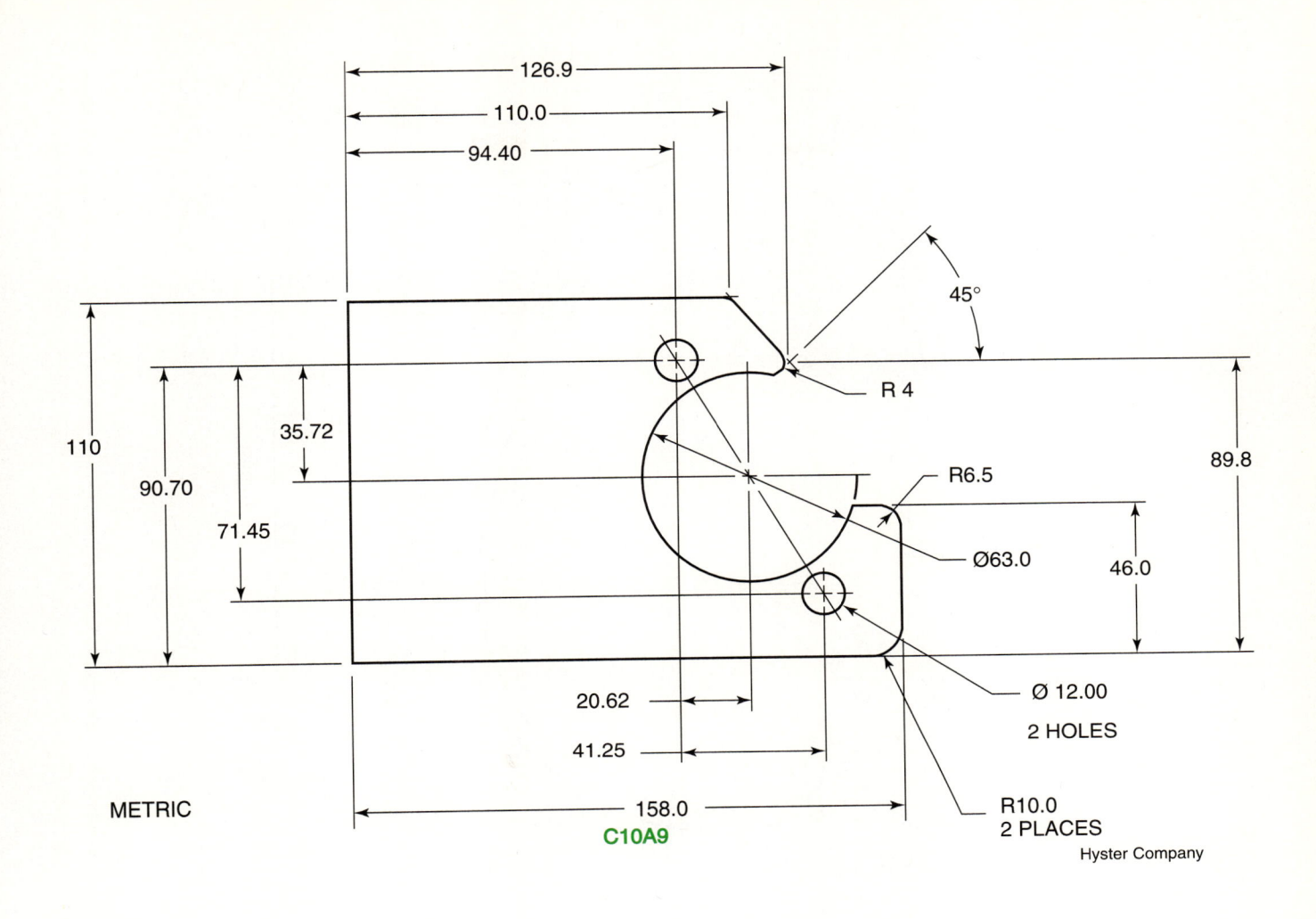

Chapter 10 Developing New Drawings and Templates

METRIC

C10A9

Hyster Company

Dimensions shown: 126.9, 110.0, 94.40, 45°, R 4, 110, 90.70, 71.45, 35.72, 89.8, R6.5, Ø63.0, 46.0, Ø 12.00, 2 HOLES, 20.62, 41.25, 158.0, R10.0, 2 PLACES

306 AutoSketch for Windows

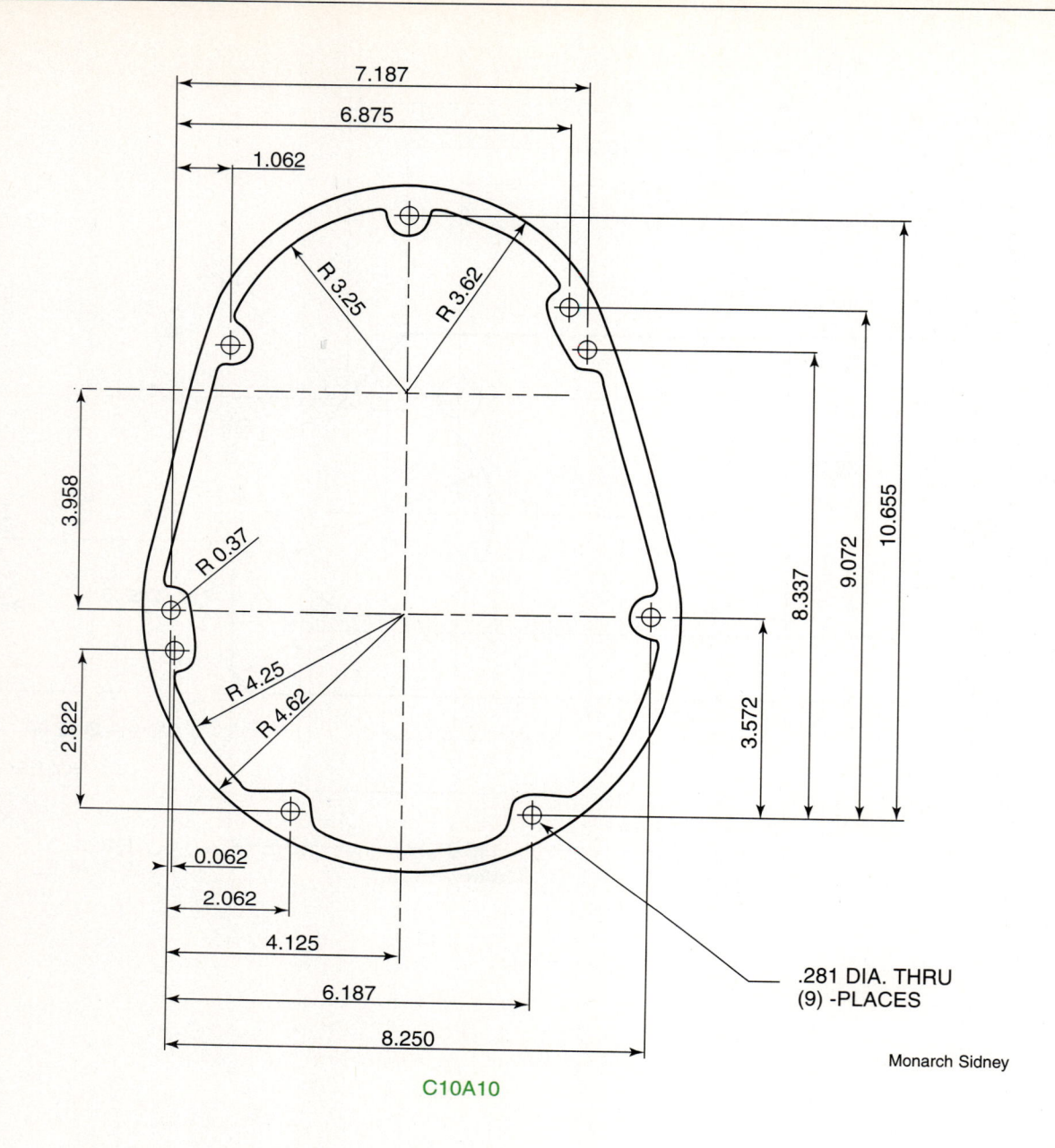

.281 DIA. THRU
(9) -PLACES

Monarch Sidney

C10A10

CHAPTER 11

Applying AutoSketch

Objectives

After studying this chapter, you will be able to:
- Explain how multiview drawings are used to document an idea in mechanical drafting.
- Identify how plans are used to document structures in architectural drafting.
- Identify how different linetypes and widths are used to indicate features in a drawing.
- Determine the sheet size and scale required to develop a drawing.
- List the drawings that make up a set of architectural plans.
- Draw a space diagram, floor plan, and elevation plan.
- Identify how AutoSketch can be used with desktop publishing applications.

Nearly ever industry creates drawings to explain and document ideas. In fact, can you think of an industry that does not use drawings? This chapter covers making drawings that help describe products to be manufactured or constructed. *Mechanical drafting* is the process of creating drawings for items to be manufactured (such as a car). *Architectural drafting* is the process of creating drawings for items to be constructed (such as a house). Many times, these drawings are incorporated into desktop publishing layouts.

The drawings you develop will convey the size, shape, material, and other features of a design idea. When complete, drawings are printed and later reproduced for distribution to people who will manufacture or construct the product. In addition, the data may be sent to a computer-aided machining system.

The graphics used to describe an object with computer-aided drafting are no different than those developed with manual drafting tools. You must follow accepted drafting practices no matter which tools you choose. However, as you learned in previous chapters, the *method* of creating the drawing is different using AutoSketch. With AutoSketch, you rely on drawing commands and special techniques that help simplify the drawing process.

Mechanical Drafting

Every field of drafting has its own method to describe the product to be manufactured or constructed. In mechanical drafting, the main presentation style used to convey the design idea is called a *multiview drawing.*

Multiview Drawings

When sketching an idea, most people make it look three dimensional. This type of drawing is called a *pictorial drawing,* because it looks like a picture, Figure 11-1. Unfortunately, pictorial drawings do not show the true size, shape, and relationship of features. Holes and curves appear oval, and edges may not be drawn true size. Although the drawing looks nice, it rarely gives enough information to create the product.

Figure 11-1.
A pictorial drawing gives a three-dimensional view of an object. However, the drawing does not provide true size and shape of features.

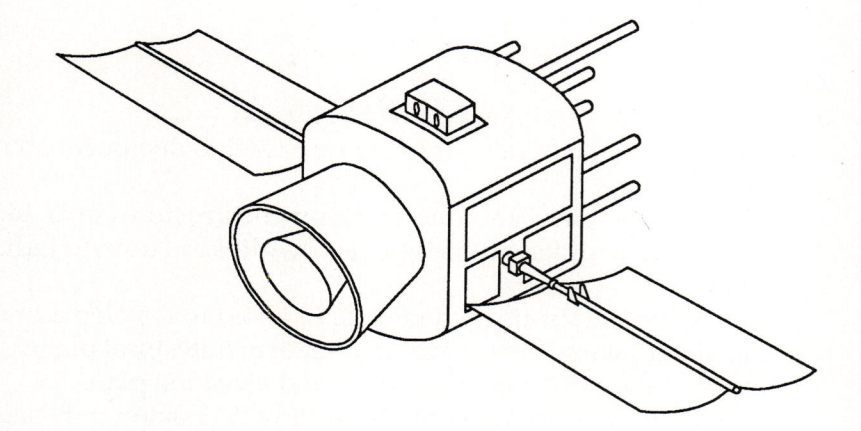

A multiview drawing gives true shape and size by showing the object as viewed straight-on from different positions, Figure 11-2. In these views, each face of the object appears true size and shape. The views are developed through a process known as *orthographic projection. Orthographic* refers to right (90°) angles. *Projection* means to project images outward. Thus, orthographic projection means to project views of an object onto planes placed at 90° angles.

This may all sound somewhat confusing. There is an easy way to learn orthographic projection. It begins by imagining the object as if it is inside a glass box.

Figure 11-2.
This multiview drawing gives the three views needed to describe the object.

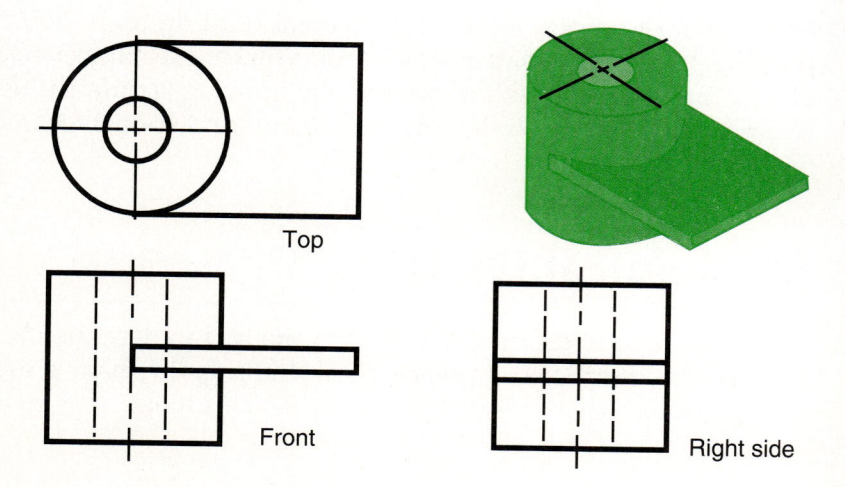

The Glass Box

The *glass box* is a technique many drafters use to develop multiview drawings. Imagine an object as if it is placed inside a glass box, Figure 11-3. When you look directly through the front of the box, you see the front view of the object. Looking through the side of the box, you see the object's side. In the same manner, by looking through the top of the box, you see the top of the object. You might also look through the back, bottom, and other side to see these views.

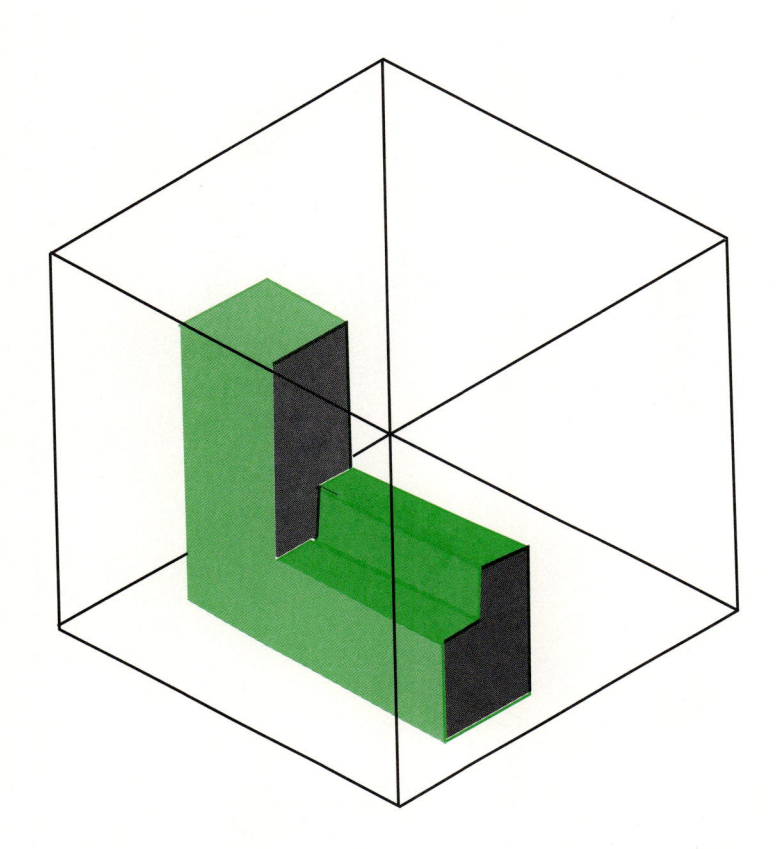

Figure 11-3.
To help visualize the views of a multiview drawing, imagine that the object is placed inside a glass box.

Now, imagine that you projected, or painted, each of the views seen onto the surface of the glass, Figure 11-4. Each side of the box becomes a *projection plane*, because you are projecting an image onto it. Looking at the six planes, you see views showing the front, top, sides, bottom, and back of the object. By unfolding the glass box (the projection planes), the views fold out next to each other in an orderly manner, Figure 11-5. You have just created a multiview drawing.

Most often, three views are sufficient to describe an object. These views are the front, top, and side. Sometimes only two views are needed. However, more views are needed if you find that three views fail to totally describe the object.

Sketching Your Ideas

Obviously, you cannot place a glass box around a product idea you might have. Thus, it may be difficult to think of each of the views. To help, prepare a sketch of your idea. Sketch the approximate size and shape of the object to be drawn. Show the outlines and estimated dimensions of the object. If you are a drafter, this type of sketch may be given to you by an engineer. You can then sketch a multiview drawing of the design.

Figure 11-4.
By painting the image seen on each face of the glass box, you "project" the views.

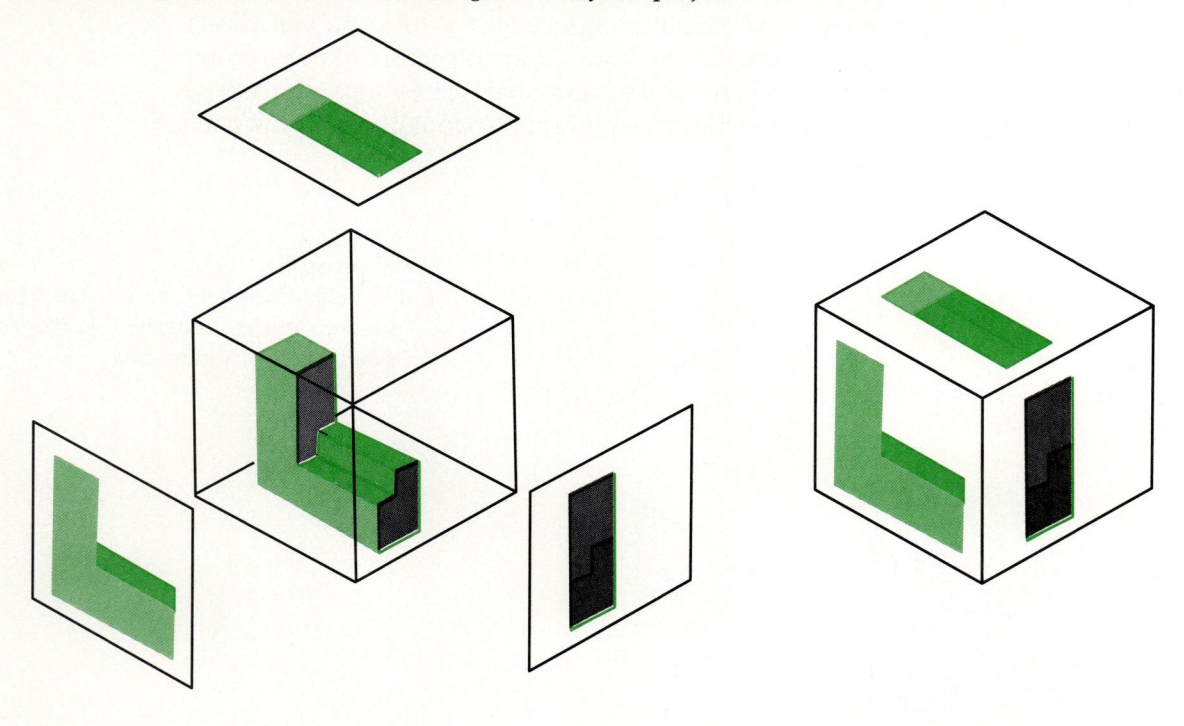

Figure 11-5.
After you project the views, "unfold" the glass box to position the views into a multiview drawing.

Some drafters are able to visualize (picture) an object and do not use a sketch. However, a sketch is a wise choice in both manual and computer-aided drafting. The sketch becomes a useful document for showing your idea to other designers for review. Then, use the sketch to select how to lay out the multiview drawing.

Relationship of Views

The purpose of a drawing is to communicate information. To do so, you must place the views in an orderly fashion. Each view should align with adjacent views. This makes it easier to look from one view to the next and "see" the shape. Also, the relationship of views is very important. Each view must be in a position that can be understood by another person reading the drawing. The relationship of the views is different between the United States and Europe.

Third-angle projection

In the United States, *third-angle projection* is used. The relationship of the views for a third angle projection is shown in Figure 11-6. The front view is the central view. The left side view is on the left of the front view, and the right side view is on the right. The top view is above and the bottom view is below the front view. The rear view, if needed, is shown beside the left side view. Third-angle projection is used in this text. The International Standards Organization (ISO) symbol for third-angle projection is also shown in Figure 11-6. This symbol should appear in the title block of any drawing that uses third-angle projection.

First-angle projection

In Europe, *first-angle projection* is used. The relationship of the views for a first-angle projection is shown in Figure 11-7. The front view is once again the central view. However, the projection technique for the remaining views creates "shadow" images. The left side view is on the right. The right side view is on the left. In a similar manner, the top view is below and the bottom view is above the front. The rear view is shown beside the right side view. The ISO symbol for first-angle projection is shown in Figure 11-7. This symbol should appear in the title block of any drawing that uses first-angle projection.

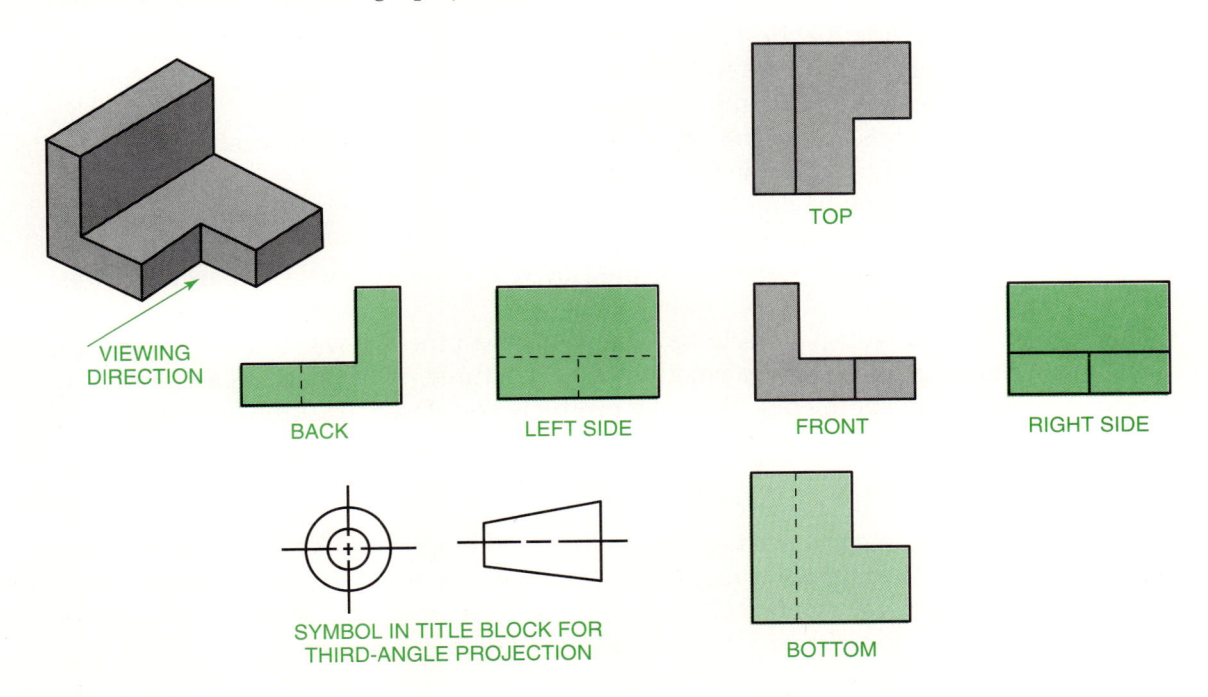

Figure 11-6.
Location of views for third-angle projection. Note the symbol that should appear in the title block to indicate that this is a third-angle projection.

Figure 11-7.
Location of views for first-angle projection. Also note the symbol that should appear in the title block to indicate that this is a first-angle projection.

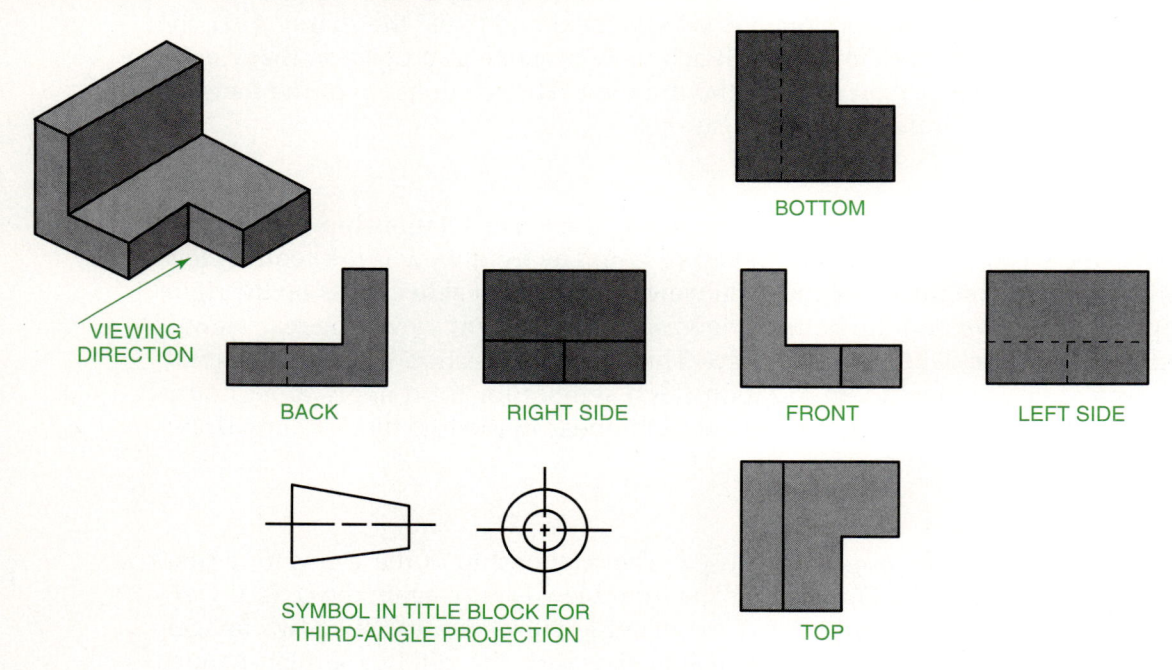

Selecting Views to Draw

In all cases, you should choose the fewest number of views that will still fully describe the object. For many parts, you will need three views. They describe the part's shape and features, and also provide the width, height, and depth. Too many views are unnecessary and can clutter the drawing.

Use your pictorial sketch to decide which side of your object should be the front view. There are several key rules to consider. The front view normally should be one or more of the following.
- The side that best shows the overall shape of the object.
- The side of the object with the most detail.
- The part's normal position in use.
- The position of the object you are most likely to see in real life.
- The view with the fewest hidden edges.
- The view having the longest dimension.
- The most stable position of the product.
- Also, when possible, use the side with the most surfaces parallel to the projection planes.

Once you have selected the front and other views to draw, sketch a multiview drawing. Check to see whether one, two, three, or more views are needed to describe the object completely. For most products you design at home or in school, three views will be required.

Two-view drawings

Simple parts, such as basic cylindrical objects, can be drawn with just two views. Notice the diameter dimension given to show the size of the cylinder in Figure 11-8. The diameter symbol (Ø) appears before diameter dimensions. A top view for this item would only repeat what was drawn in the front view.

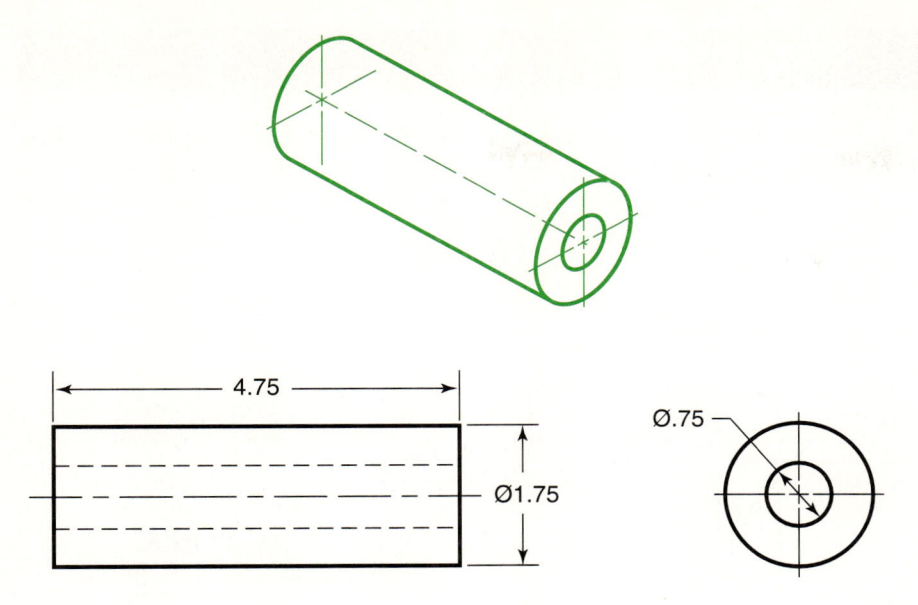

Figure 11-8.
Simple parts can often be described with just two views. This cylinder needs only two views. A top view would merely repeat the front view.

One-view drawings

At times, an object can be fully described using one view. A product with little depth, such as a gasket or flat sheet metal design, requires only one view. The thickness is given as a note, Figure 11-9. This is called a one-view drawing, rather than a multiview.

Figure 11-9.
This gasket can be described with one view and a note giving the thickness.

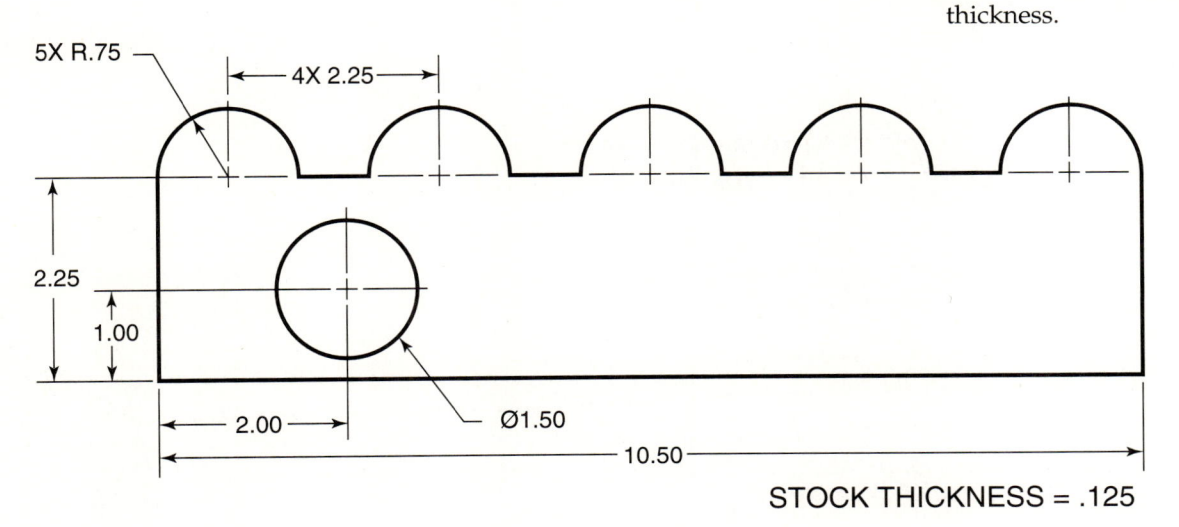

STOCK THICKNESS = .125

Exercise 11-1

For each of the following pictorial drawings, identify how many views would be required to fully describe the object. All holes are drilled completely through the object.

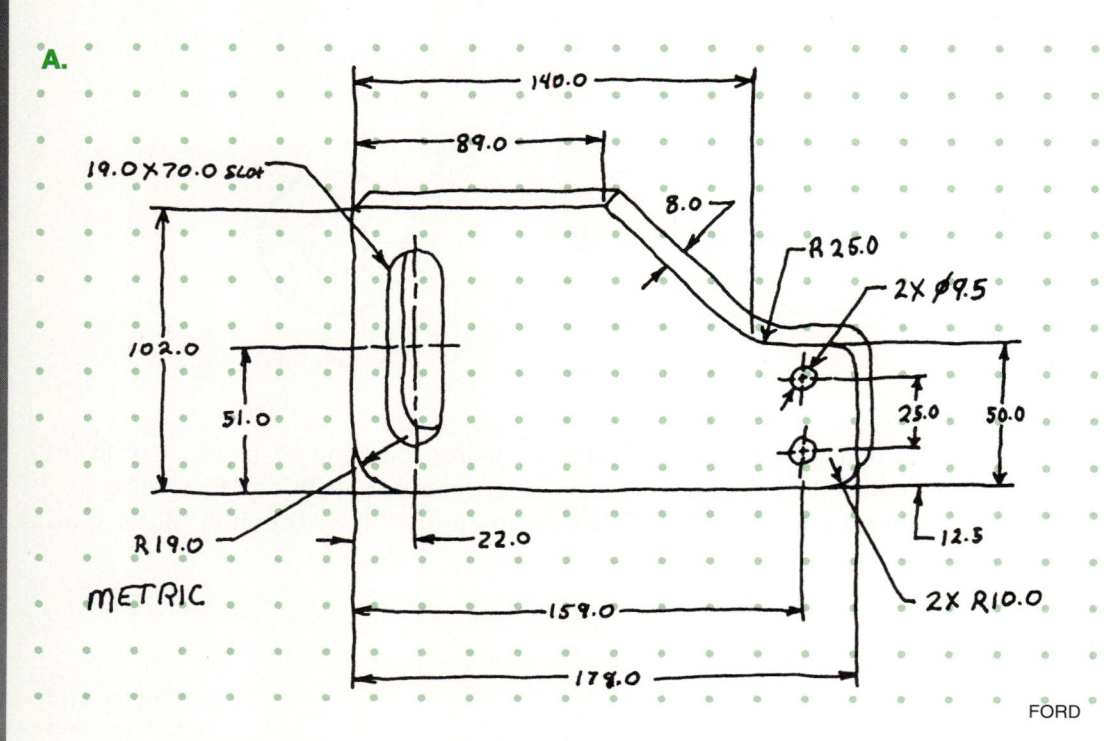

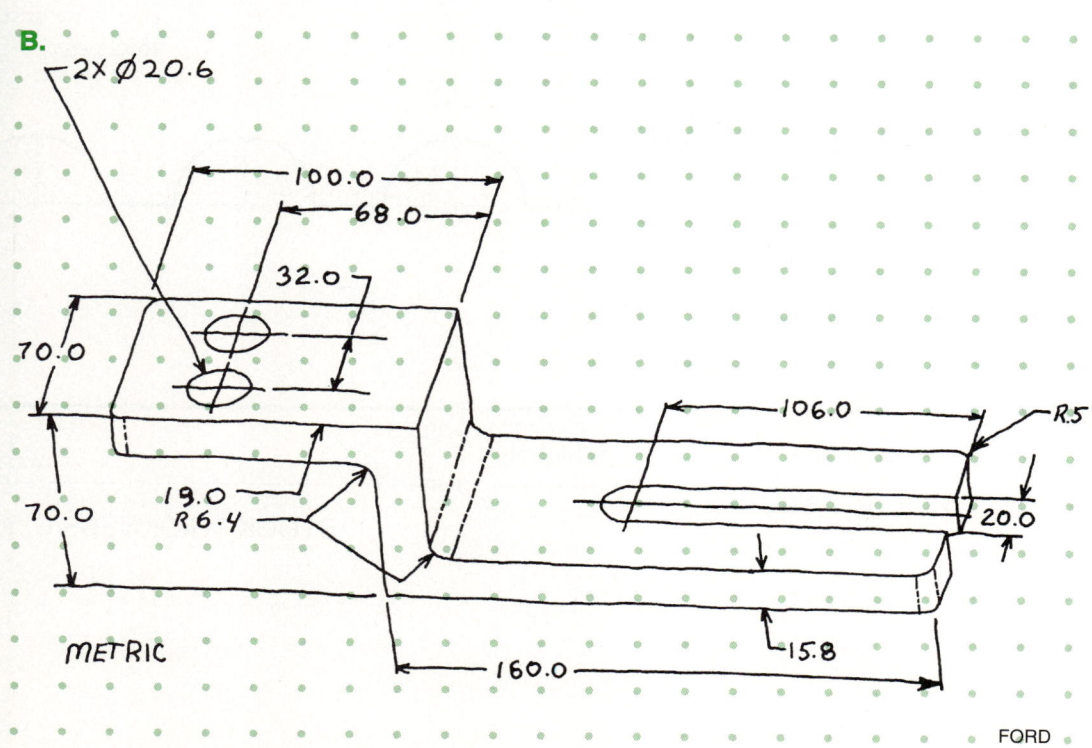

A.

METRIC

FORD

B.

METRIC

FORD

Showing Features in Mechanical Drawings

In any view, there can be both visible and hidden edges of the object. A complex object may have many hidden edges. For example, Figure 11-10 shows an object with several holes. In each view, you can see one hole, but the other two cannot be seen. Thus, hidden lines (dashed lines) are placed in the view to indicate the location of the hidden holes.

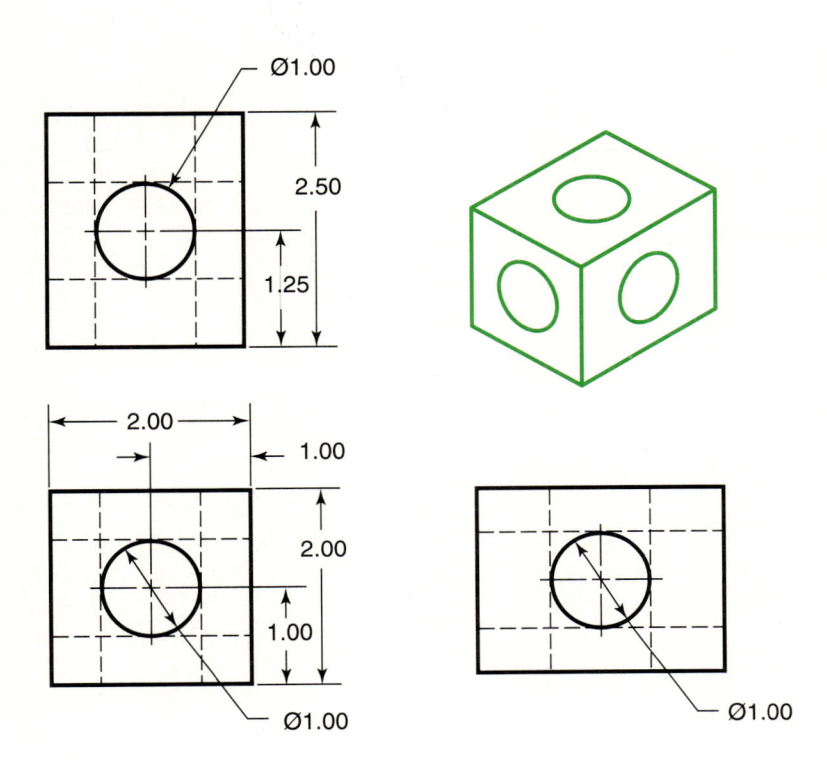

Figure 11-10.
This multiview drawing describes an object with three holes. In each of the views, the holes appear true size and shape. Also, in each view the edges of the holes are shown with hidden lines. Centerlines show the centers of the holes.

A drawing may contain many types of lines to represent centers, breaks, cutaways, and other details. To help show features, each line on a technical drawing has a certain meaning. A thick, solid line usually defines the visible edge of a shape. A thin, dashed line generally depicts a hidden edge of an object. A pattern of short and long dashes indicates the center of a hole.

The American National Standards Institute (ANSI) has set guidelines that assign meanings to various line patterns and widths. These guidelines are called the *alphabet of lines*. See Figure 11-11. With this standard, each drafter can "read," or interpret, the drawing the same way. Thick lines are drawn .030" (.7mm) wide. Thin lines are drawn .015" to .020" (.5mm to .7mm) wide. Thin lines can be drawn using the hairline setting in AutoSketch.

As stated in Chapter 10, it is often best to draw all lines first with a hairline width. This is because certain commands do not work on wide lines. Then, before printing the final drawing, change lines to their proper width.

You can also change the colors of lines if you are using a plotter. Since a pen plotter selects the pen based on the object's color, you can use different pen widths in the plotter. When plotting the drawing, select pens that follow the standard line widths prescribed by ANSI.

Figure 11-11.
The alphabet of lines gives meaning to various combinations of linetypes and line widths.

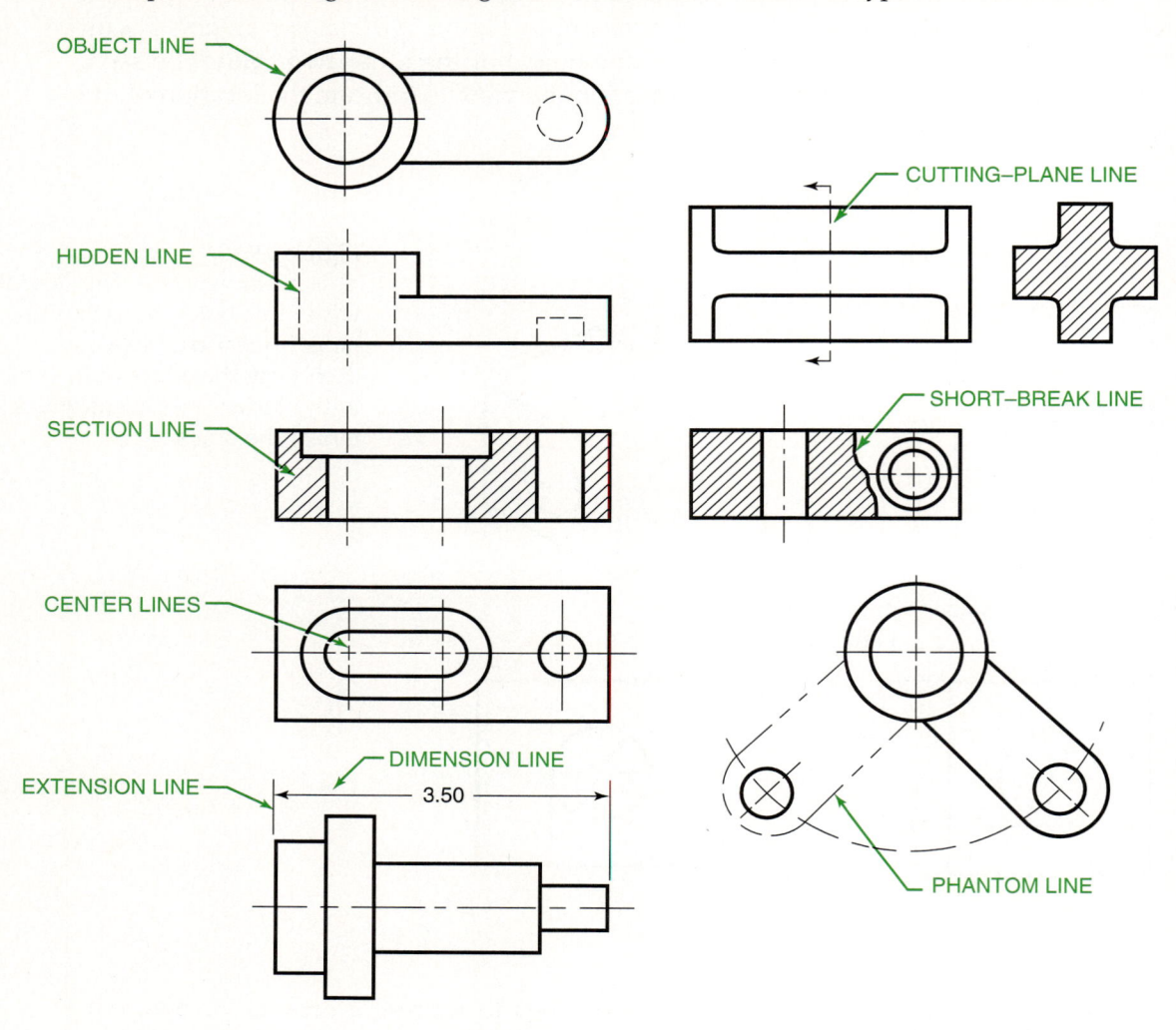

OBJECT LINE

CUTTING–PLANE LINE

HIDDEN LINE

SHORT–BREAK LINE

SECTION LINE

CENTER LINES

DIMENSION LINE

EXTENSION LINE — 3.50

PHANTOM LINE

Showing visible features in a view

Object lines show edges and contours of the object that are visible in a view. Object lines are drawn solid and thick so that they stand out. They should clearly contrast with other lines on the drawing. Make all object lines the same color. For the activities in this text, use black for object lines.

Showing hidden features in a view

Hidden lines show edges of the object that are not visible in a view. Hidden lines are drawn with a dashed, thin linetype, Figure 11-12.

With AutoSketch, select the **Hidden** linetype in the **Current Line** setting on the vertical status bar. You should also make hidden lines a different color. For the activities in this text, use yellow for hidden lines.

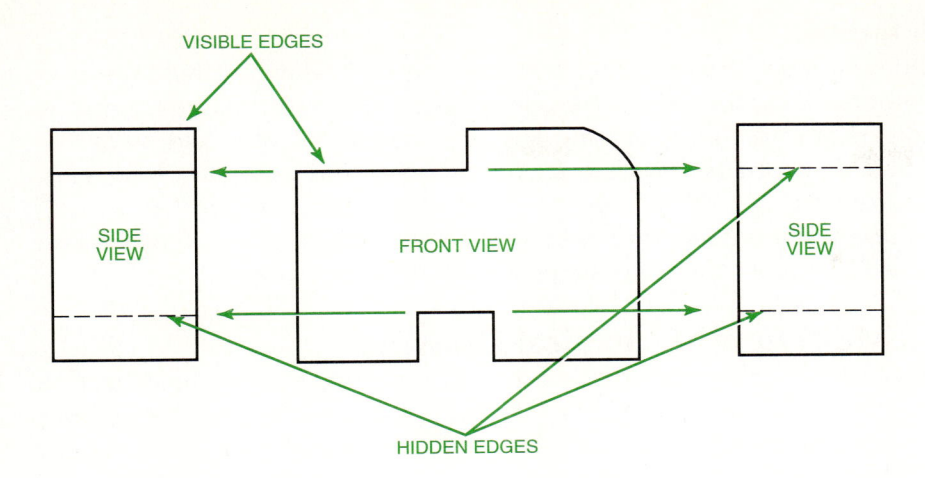

Figure 11-12. Hidden lines indicate edges that are hidden in the view. These lines may, or may not, be visible in other views.

Showing centers and symmetry

Centerlines show the centers of holes, cylindrical objects, and symmetrical objects. Along an object's length, a centerline notes the axis extending lengthwise through the object's center. In an end view, centerlines cross to show the center, Figure 11-13. Centerlines are thin, like hidden lines.

Select the **Center** linetype in the **Current Line Type** setting on the vertical status bar. Also select a different color from the **Drawing Color** dialog box. For the activities in this texts, use cyan (light blue) for centerlines.

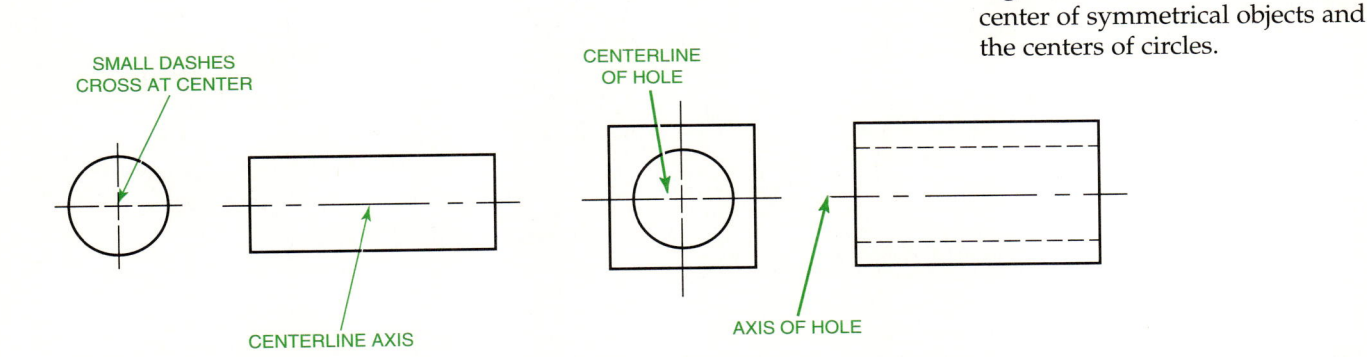

Figure 11-13. Centerlines mark the center of symmetrical objects and the centers of circles.

Precedence of lines

There are times when two or more linetypes coincide on a drawing. The rule that decides whether to draw an object line, hidden line, or centerline is called the *precedence of lines.* The order is as follows.
1. Object lines
2. Hidden lines
3. Centerlines

For example, if a hidden line coincides with an object line, draw the object line. If a centerline coincides with a hidden line, draw the hidden line.

Showing other features

There are different lines to show other features of a part. Many of these are covered in later chapters. A *cutting-plane line* designates where an imaginary cut is. *Section lines* indicate surfaces in a section view where material was "cut away" to show interior detail. *Long-break lines* are used to shorten the view of a long part. *Short-break lines* are used in a single view to show where material was "cut away." An assortment of *dimensioning lines* are used to show size and location of features on an object. *Phantom lines* show the alternate positions of moving parts.

Developing a Multiview Mechanical Drawing

Once you have chosen the views to draw, sketch a multiview drawing. Check to see whether one, two, three, or more views are needed to describe the object completely. Also, add in hidden and centerlines on your sketch. Finally, add dimensions. See Figure 11-14. This planning step is important. The sketch and dimensions help you set up the drawing or select a template. Also, your drawing session is more productive because the sketched views serve as a reference.

Figure 11-14.
After sketching a multiview drawing, add dimensions. Then, use the sketch to create a CAD drawing.

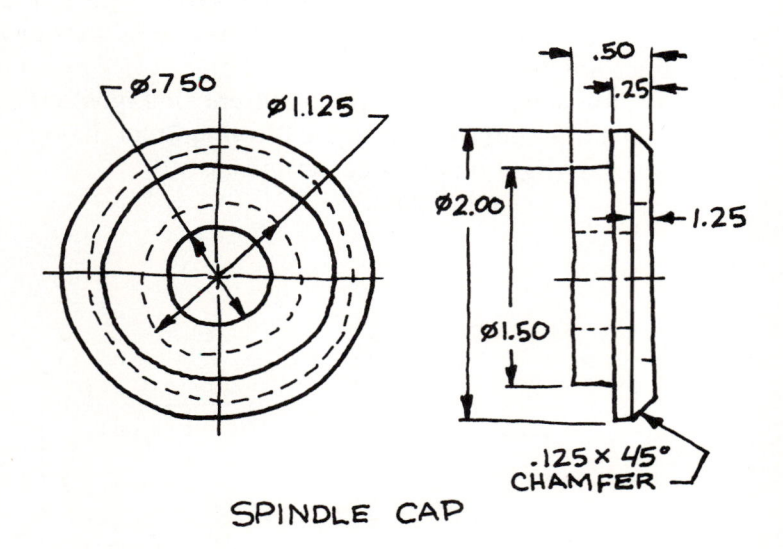

Exercise 11-2

For each of the pictorial drawings in Exercise 11-1, sketch a multiview drawing for the part. Each grid dot represents 1/2".

Determine the drawing area required

Chapter 10 discussed how to set the size of your drawing area using the **Sheet Size, World Size,** and **Sheet Scale** settings. To determine the area you need, calculate the space required by the drawing. Consider the drawing itself plus dimensions, notes, a border, and title block. It is important that you allow ample space for these items. Of course, you can always reset the world size or sheet scale. However, if you start a template with a predrawn border and title block, resetting the sheet scale can causes problems. It is better to get it right the first time.

The space for the views in a simple three-view drawing is calculated using the height, width, and depth of the object. Also, leave space between and around the views. A general rule for simple drawings is to allow 1″ to 2″ between and around views. However, more complex drawings require space for details, auxiliary views, section views, and specifications. It is better to have too much room rather than not enough.

Figure 11-15 shows how to calculate the horizontal and vertical drawing area needed for a simple three-view drawing. The horizontal distance is the sum of the object's width and depth, plus space between views. The vertical distance is the sum of the height and depth, plus space between views. With these measurements, you can select a template or proceed with drawing setup. Remember that adding a revisions schedule, bill of materials, dimensioning table, or other information later will further increase the drawing area needed.

Figure 11-15. When calculating the required drawing area, take into account the height, width, and depth of the object, and leave 1″ to 2″ between views.

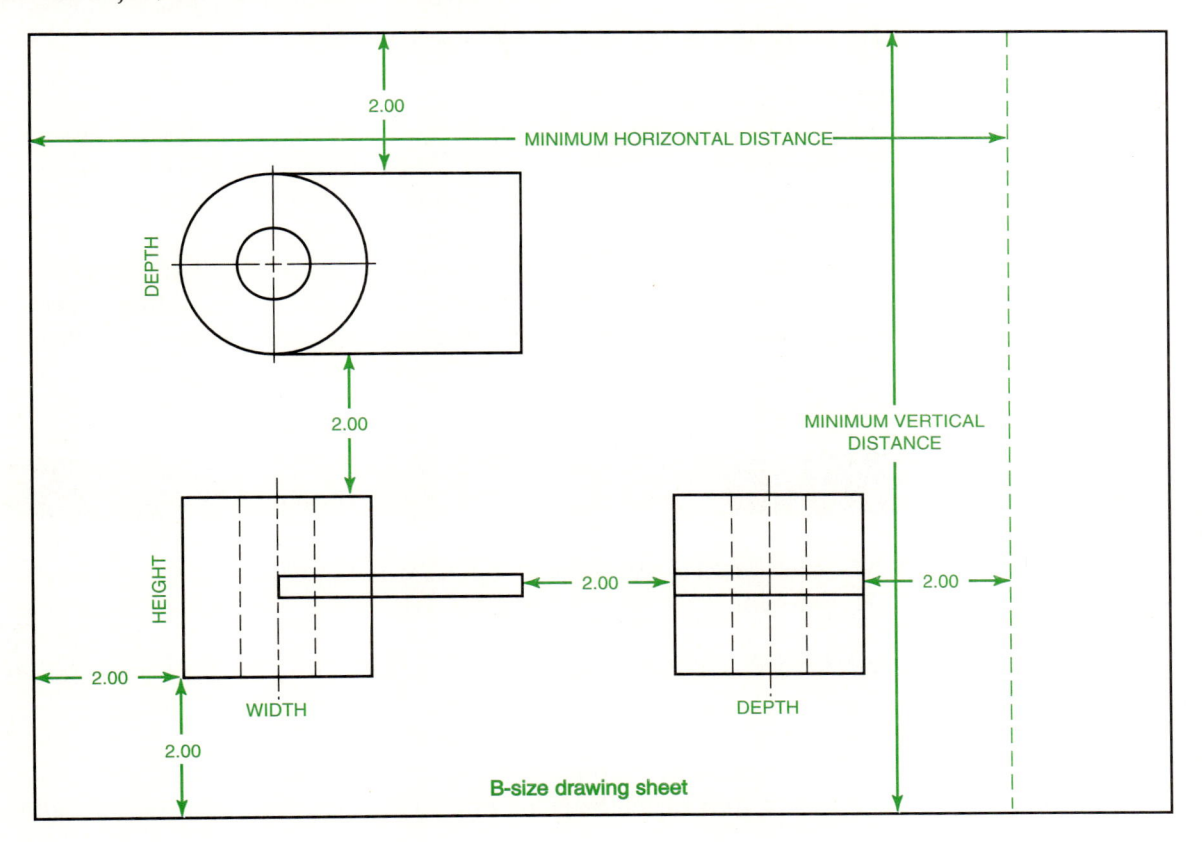

Set up the drawing or select a template

Once the planning is complete, start AutoSketch. Load the proper size template, if one is available, using the **New...** command in the **File** menu. Some templates come with AutoSketch and others were developed in activities of Chapter 10. The drawing in Figure 11-15 requires a template with a B-size sheet at a 1:1 sheet scale. Of course, using a 1:2 sheet scale only requires an A-size sheet.

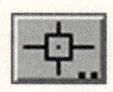

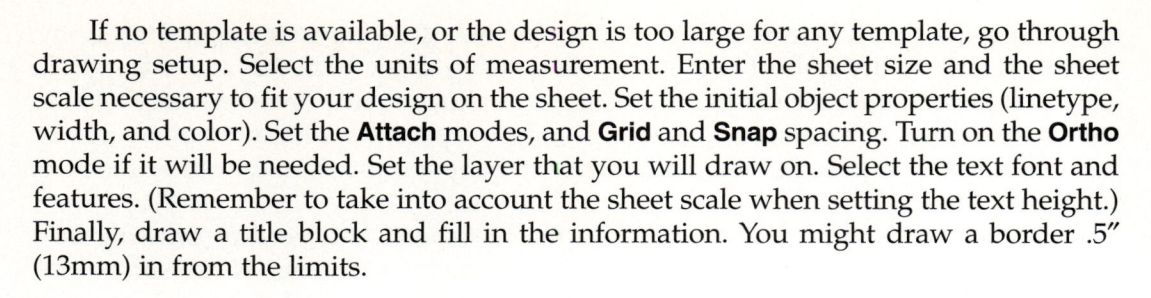

If no template is available, or the design is too large for any template, go through drawing setup. Select the units of measurement. Enter the sheet size and the sheet scale necessary to fit your design on the sheet. Set the initial object properties (linetype, width, and color). Set the **Attach** modes, and **Grid** and **Snap** spacing. Turn on the **Ortho** mode if it will be needed. Set the layer that you will draw on. Select the text font and features. (Remember to take into account the sheet scale when setting the text height.) Finally, draw a title block and fill in the information. You might draw a border .5" (13mm) in from the limits.

Laying out views

When finished with drawing setup, you are ready to add objects to the drawing. Because of the planning steps taken thus far, the task of laying out the views should be easy. Use drawing aids when possible to locate objects precisely. Also, avoid switching between linetypes and colors while drawing one view. You can draw all items for each linetype and color at one time. Or, draw the views using one linetype and color. Then, change the colors and linetypes later.

Begin with the front view, about 2" to the right of, and above, the lower-left corner of your border. Draw object lines in the front view. Then, add the top and side views. Add hidden lines. Finally, in the example of Figure 11-15, you need to add centerlines. These locate the center of the hole in both views.

Spacing and centering views

There are two basic ways to space views. The first way is to calculate the exact distance between views based on the number of views and size of the drawing area. The other method is to space the views approximately. The second method is preferred by most CAD drafters.

First, use a grid to make sure the views align. However, if the drawing looks crowded, space the views further apart using the **Move** tool. You might need more space between views for dimensions or other information. Make sure the moved views still align.

Exercise 11-3

For each of the sketches from Exercise 11-2, use AutoSketch to complete a multi-view drawing. Select an appropriate template from those you created in Chapter 10.

Save the drawing

Don't forget to save your drawing. Whether you set up the drawing, or recalled a template, your drawing remains "Untitled" until you save it. After saving the drawing, remember to select **Close** from the **File** menu so that the drawing is removed from memory. Then, select **New** from the **File** menu to begin a new drawing, or **Exit** to end your drawing session.

Architectural Drafting

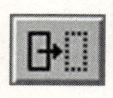

Architectural drafting involves preparing drawings that describe the structure and materials for residential and commercial buildings. This section shows how to apply AutoSketch to several basic architectural drawings. For a comprehensive look at architectural conventions, advanced architectural drawings, and the business of architecture, refer to *Architecture: Residential Drawing and Design* published by The Goodheart-Willcox Company.

Architects play a unique role because they must both understand how people live as well as construction techniques. A significant amount of time is devoted to analyzing the needs of the client. Only after these needs are established can the drafter begin developing working drawings. One of the first steps is space planning.

Space Planning

Space planning is working with the "space" of a building to meet the needs of the client. The planning process begins by establishing the client's needs. Suppose you are an architect designing a residential structure (house). You need to consider the lifestyle of the family plus special requirements. The following items need to be looked at.

- What activities take place in the home? For example, does the family have children? Are they hobbyists? Do they entertain guests often?
- What rooms or other areas of activity are needed, and are any divisions of those areas needed? Examples of rooms include kitchen, dining, living, den, bedrooms, bathrooms, storage, office, and utility.
- What are the general space requirements (in square footage) of each area? For example, a woodworking hobbyist might require more space than a person whose hobby is collecting stamps.
- What areas need to be next to each other? For example, the kitchen and dining room might be adjacent. The activities in these two rooms (cooking and eating) are related to each other. Placing these rooms next to each other also supports the natural flow of traffic.
- Look at the future expansion potential and the internal flexibility of the space. For example, do the rooms allow for rearrangement of furniture or other equipment? Is there a natural way to extend the area in the future with an add-on structure?
- What are the traffic patterns? Are there areas that could become bottlenecks? Consider both the natural flow of people, plus any special equipment, vehicles, or materials.
- What is the financial situation of the client? A millionaire may request a different style of structure than most middle-class families.

Residential areas

There are several different areas of a residential structure. These are outlined below.

- Entry—A separate area with an outside door and a closet, with easy access to other areas of the home.
- Living—An area for conversation, television, music, reading, and entertaining, perhaps with a fireplace.
- Family—A more informal room for play, hobbies, and television.
- Dining—An area usually separate, but near the kitchen.
- Kitchen—An efficient area for food preparation. A typical rule for efficiency is that the total distance between the refrigerator, sink/counter, and cooking appliances be at least 12′ and no more than 21′. This is called the ***work triangle.*** There are also different styles of kitchens, including the "U," "L," two wall, one wall, island, and peninsula styles.

Figure 11-16.
A bubble diagram is a simple diagram used for space planning.

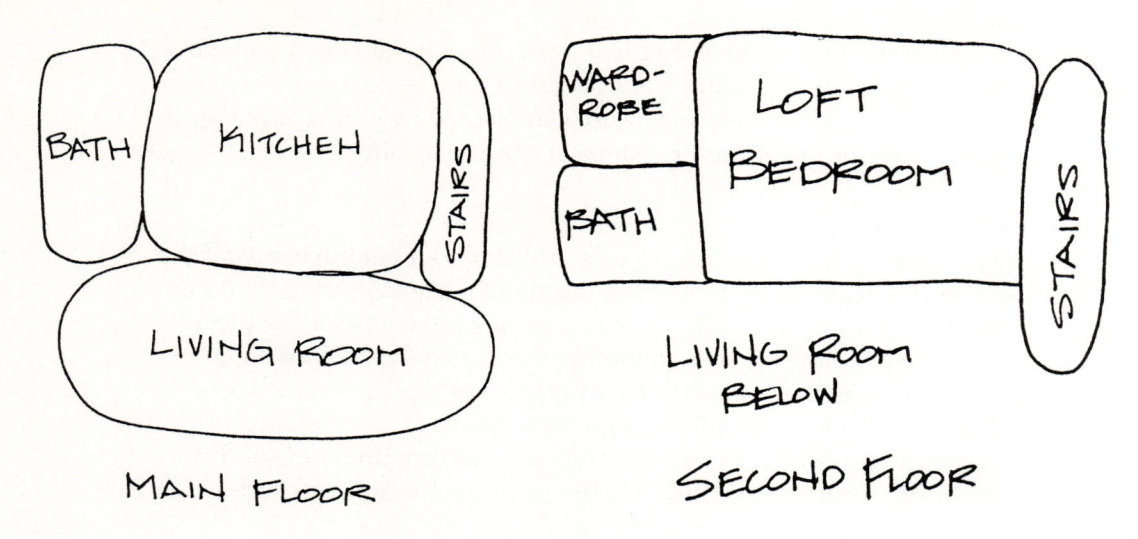

Figure 11-17.
A single-line space diagram is better proportioned than a bubble diagram.

- Utility—Room for laundry, clean up, and usually also containing major home systems such as the water heater and furnace.
- Sleeping—Bedrooms for sleeping, dressing, reading, and rest. The size of these rooms depends on the number of people and privacy wanted.
- Bathroom—At least one must be accessible from all bedrooms. A master bedroom usually has a separate bathroom. A half-bath (no tub/shower) is usually close to living and family areas. Consider typical designs, fixture sizes, and space requirements as given in an architectural reference book. You must also consider fixture styles and sizes, lighting, ventilation, heating, and efficient layout of plumbing.
- Garage—For automobiles, workshop, storage, and outdoor activity equipment. Sometimes a freezer and laundry facilities are located here. A garage can be either attached or detached, and should follow the style of the house.

Developing a space diagram

Typically a *space diagram* is made to chart all of the above information. This diagram is not a formal sketch or furniture layout. It is a freehand sketch showing the proposed areas to meet all the criteria listed above, Figure 11-16. A space diagram may start out as a bubble diagram with bubbles, squares, or other symbols are used to represent the key areas. Sometimes a sketch can bring out problems that cannot be discovered in discussions with the client. You can use AutoSketch to create a single-line space diagram, Figure 11-17.

After a space diagram is made, scaled sketches are created. These allow the architect to see how the sketch converts into a real structure. The actual size of areas is considered, Figure 11-18. The width of doors, windows, and other access areas are all represented to scale. If all the areas seem to work, the single-line space diagram can be converted into a *wall diagram,* Figure 11-19. This is the basic design of the floor plan.

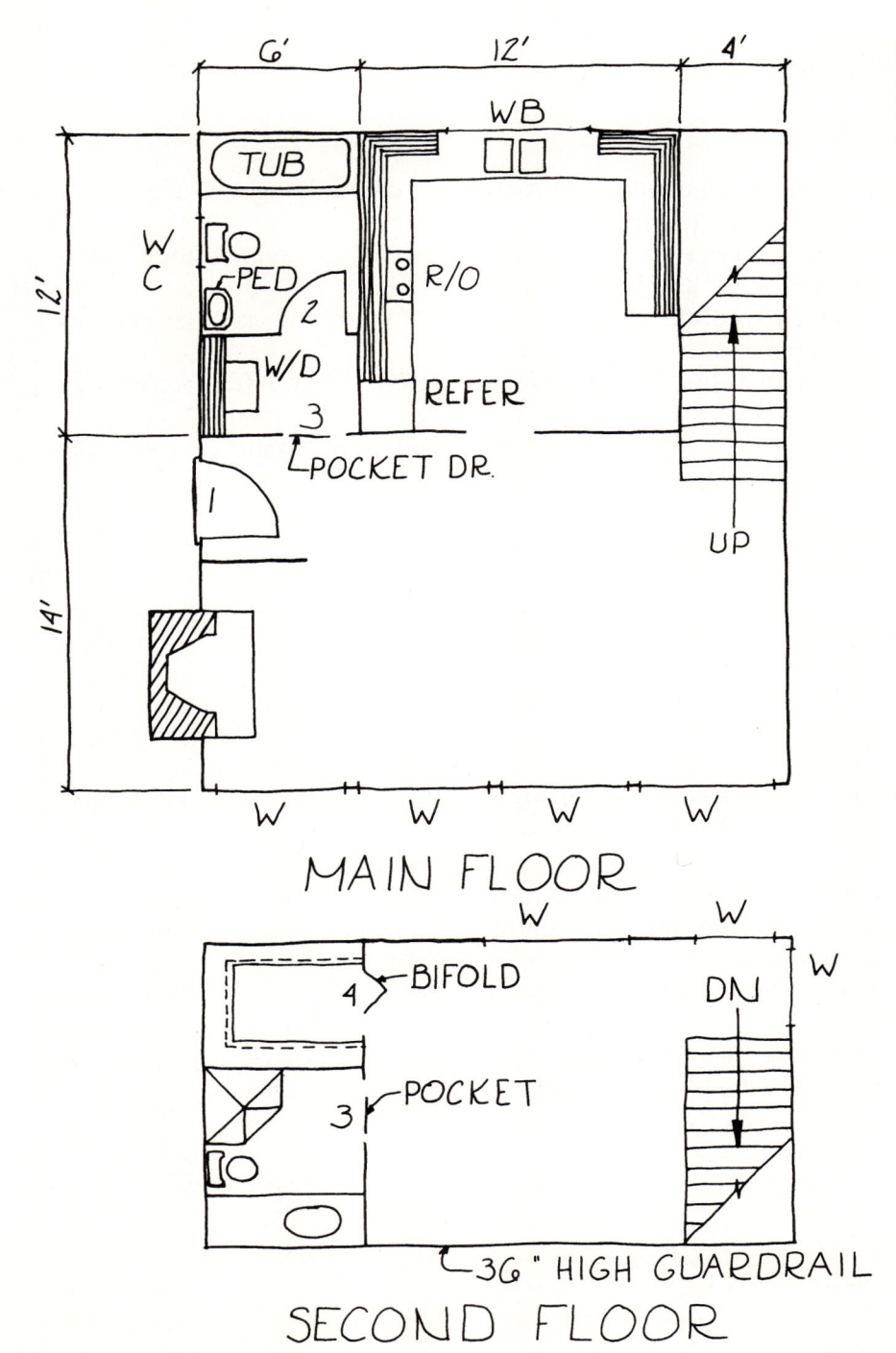

Figure 11-18.
A sketch derived from a single-line space diagram. Note the detail added.

Figure 11-19.
A wall diagram created from a single-line space diagram.

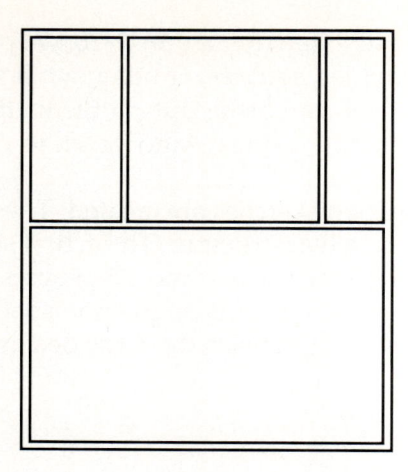

Plot Plan

The *plot plan* is the placement of a structure on the building site, Figure 11-20. There are a number of factors to consider, including the following.
- Legal restrictions, such as setback requirements, utility easements, zoning ordinances, deed restrictions, and building permits.
- Physical characteristics of the property, such as location of trees, other buildings, terrain of the property, and street access.
- Site orientation, such as the way the sun passes over the structure, preferred views, and sound.
- In a commercial site, sidewalks, curbs, gutters, landscape, drainage, and parking spaces must also be looked at.

Developing the plot plan

Some sites are rectangular, while many sites are not. Use polar coordinates to draw a nonrectangular property boundary. The lines that make up the boundary are called the *property line.* The sheet scale is typically 1/8″ = 1′-0″.

Once you establish the property line, locate the necessary property line setbacks for reference purposes. The example shown in Figure 11-21 shows a 35′ front setback, 10′ side setbacks, and 20′ rear setback. Once they are established, insert the space diagram into the plot plan to verify that it fits. Use the **Part Insert** command. Move the space diagram into the most desirable orientation. Once the space diagram is moved into its final position, the setback lines can be erased or moved to a separate layer.

You can also try several different arrangements by placing each insertion of the space diagram on a different layer. Remember to dimension the distance from the house to two adjacent property lines. You might also add the overall length and width of the house, and the location of any landscaping.

This example was very simple. Some plot plans are much more complex. The elevation of uneven property is noted at the corners, and contour lines are placed within the plan to show the slope. The plot might also be noted by the legal description. This is found on the survey and in the legal deed. A legal description is typically noted by lot and block in residential areas.

Figure 11-20.
A plot plan shows details about the property and how the buildings sit on the property.
(E. Henry Fitzgibon, Architect)

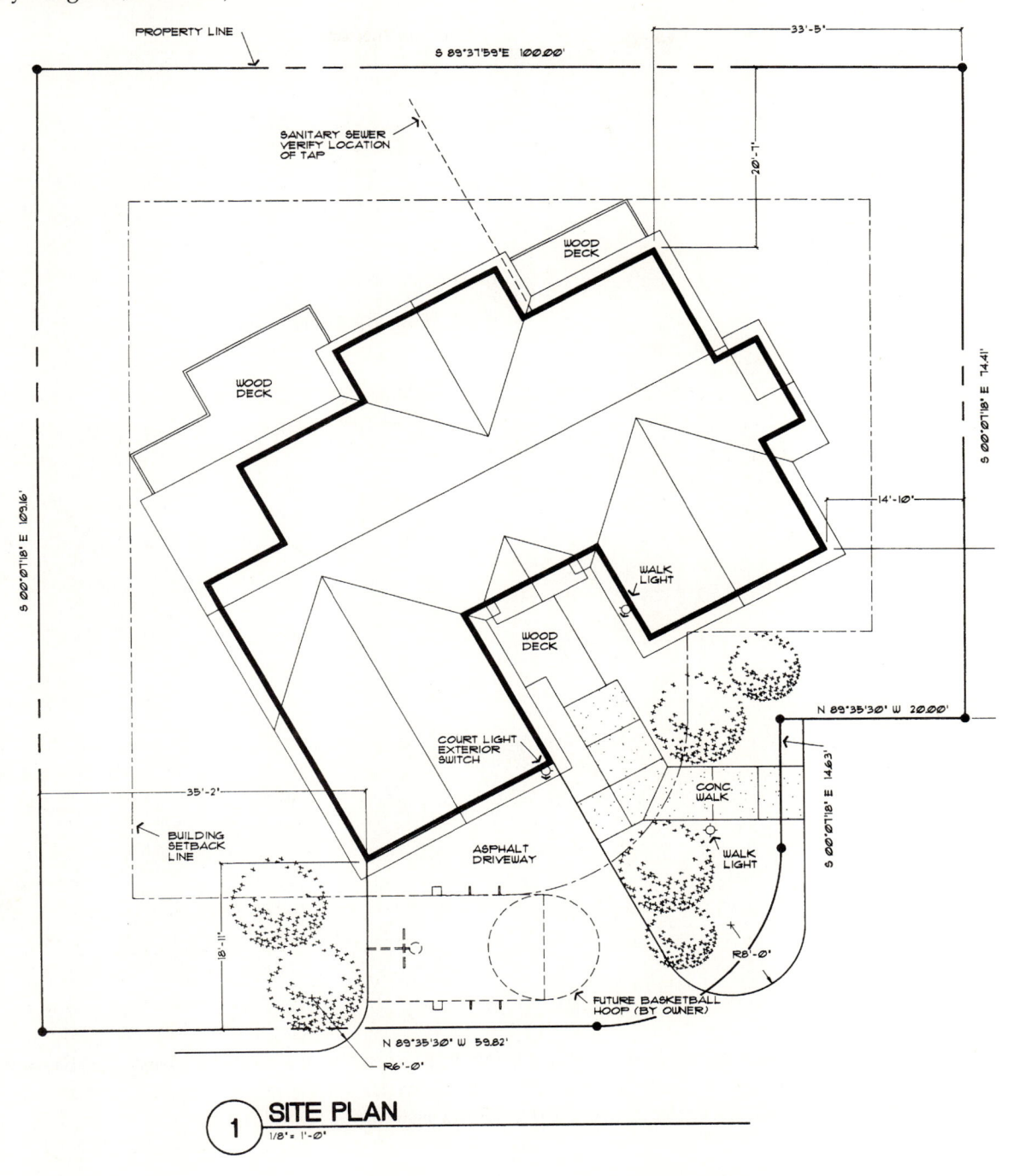

Floor Plans

The *floor plan* begins with the wall diagram. Then, doors, windows, cabinets, appliances, and some accessories are added. A floor plan uses symbols extensively. Symbols are the shorthand of architectural drawing to reduce drawing time. A completed floor plan is shown in Figure 11-22. Your AutoSketch drawing will likely

Figure 11-21.
This is a plot plan in development. Setback lines are included. The space planning diagram is used to locate the structure.

include electrical and possibly heating or plumbing symbols. However, these are on a separate layer and printed separate from the floor plan. Dimensions are measured from exterior walls to the center of interior walls.

Developing a floor plan

Earlier in this chapter, you created a space diagram and a wall diagram. These are preliminary floor plans. A floor plan begins to take shape when the space diagram is turned into a wall diagram, Figure 11-23. The wall diagram formally divides the space into exact measurements.

The actual thickness of the interior wall is 4 1/2" (3 1/2"wall stud plus drywall on either side), but is generally drawn as 4" on the floor plan. Exterior walls are typically

Figure 11-22.
A completed two-story floor plan. (E. Henry Fitzgibon, Architect)

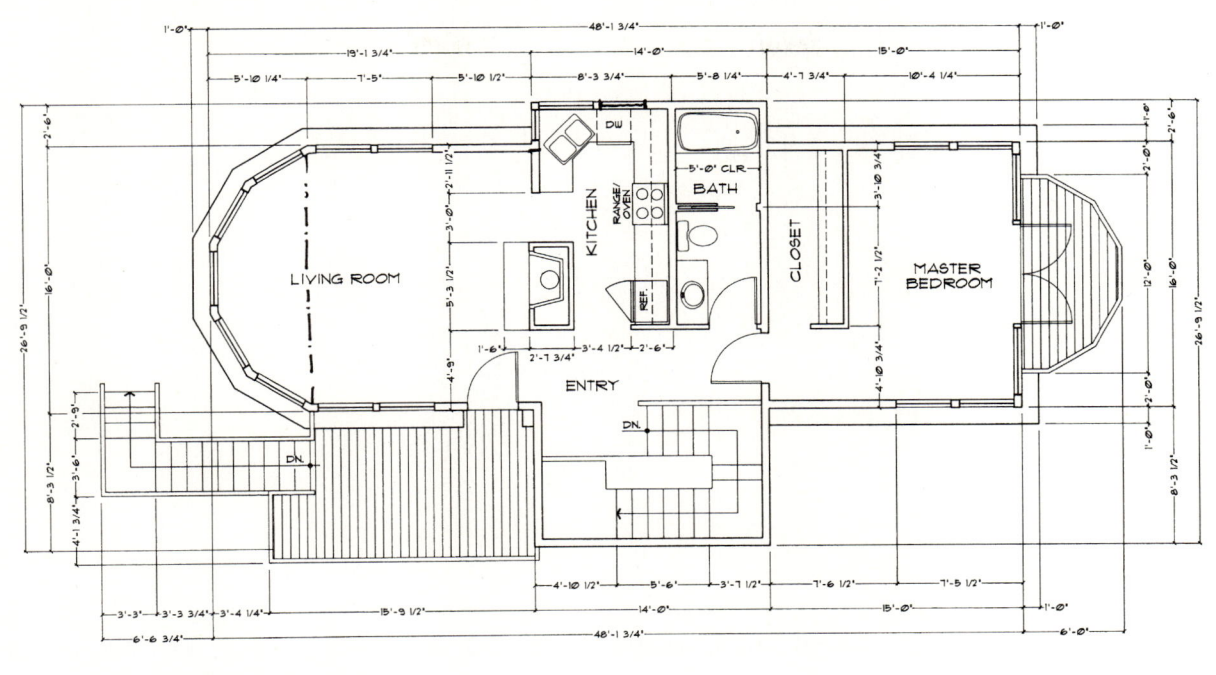

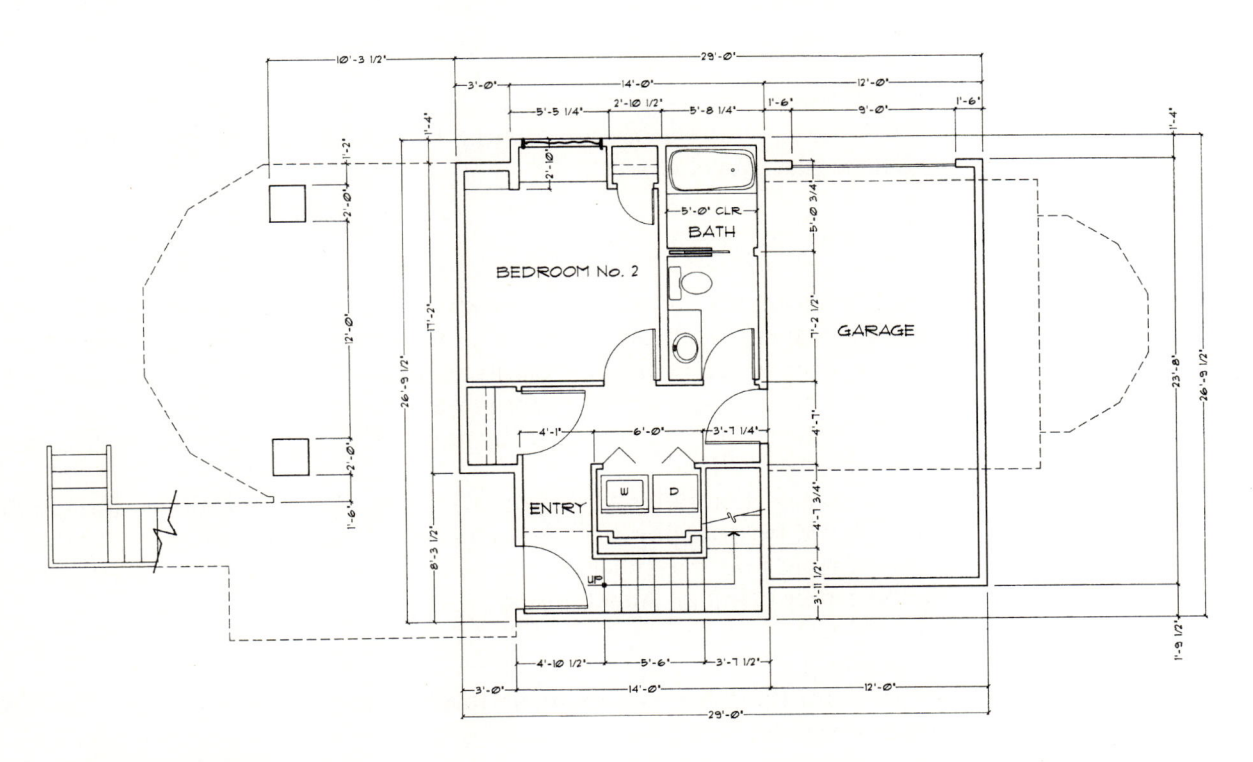

Figure 11-23.
A wall diagram. Make sure that all overlapping lines are trimmed, as shown here.

6 1/2" (3 1/2" wall stud plus interior drywall, exterior insulation, and siding) but generally drawn as 6" on the floor plan.

Lines that intersect at corners, crossing walls, and T intersections should be trimmed so that the lines do not overlap. To draw the walls, use a 4" snap and 1'-0' grid for the rough layout. To measure interior walls on either side of a centerline, use a 2" snap.

Note: AutoSketch framing symbols are 5 1/2" for external walls and 3 1/2" for internal walls. These symbols are for creating framing plans and should not be used for general floor plan development.

Adding doors and windows

Once the walls are laid out, doors and windows are added to the floor plan, Figure 11-24. Some CAD systems automatically remove wall lines when a symbol is placed. These systems usually add headers and sills as needed. AutoSketch does not do this.

Some door and window symbols come with AutoSketch. The symbols for floor plans have the letters "PL" at the end of the name. The symbols without "PL" are for elevation views.

Consider the insertion point when you place a symbol. While doors and windows are typically placed by dragging a corner of a symbol, they are measured by their centers. For this reason, you may wish to adjust the symbol once it is placed. To do so, drag it by the center using an **Attach** mode.

Drawing stairs

Stairs are complex. However, there are three basic parts of any stairs. These parts are the tread, risers, and stringer. The *tread* is the part of the stairs that you walk on. The *riser* is the vertical part between the treads. The *stringer* is the "side" of the stairs that the treads and risers are fixed to.

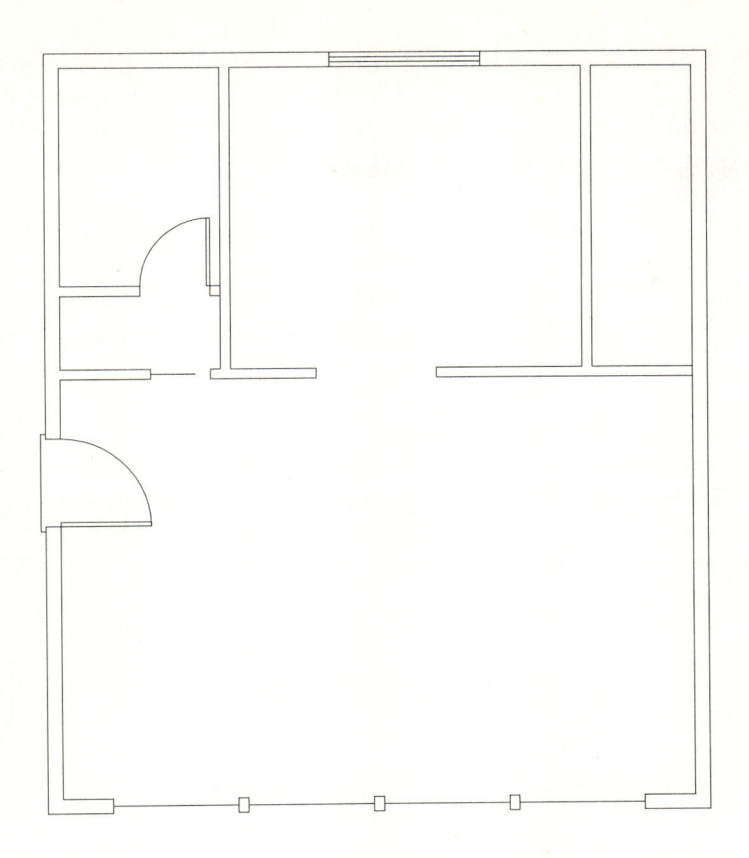

Figure 11-24.
Doors and windows have been added to the wall diagram shown in Figure 11-23. This is now a partial floor plan.

Figure 11-25.
Stairs have been added to the partial floor plan shown in Figure 11-24.

Figure 11-25 shows straight stairs added to the floor plan. A break line is used to indicate that the stairs enter the next floor. There are also L-shaped and U-shaped stairs. When drawing stairs, you must consider the direction (noted with an arrow), width of tread, depth of tread, and floor-to-floor height. Make sure that you include the necessary handrails.

Adding fireplace

Add a fireplace according to proper design techniques. Use the appropriate pattern fill symbol. You may want to place the fireplace, or any other specific details, on a separate layer.

Locating cabinets, appliances, and fixtures

Use the **Line** and **Box** tools to create the cabinets. Remember to note upper cabinets with a dotted line. Appliances and fixtures are noted with symbols. A symbol for most every appliance needed in residential and commercial buildings comes with AutoSketch. Symbols are located in the \WSKETCH\PARTS\HOME\FIX_APPL directory. See Figure 11-26.

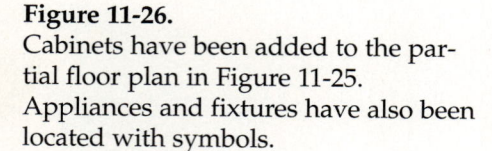

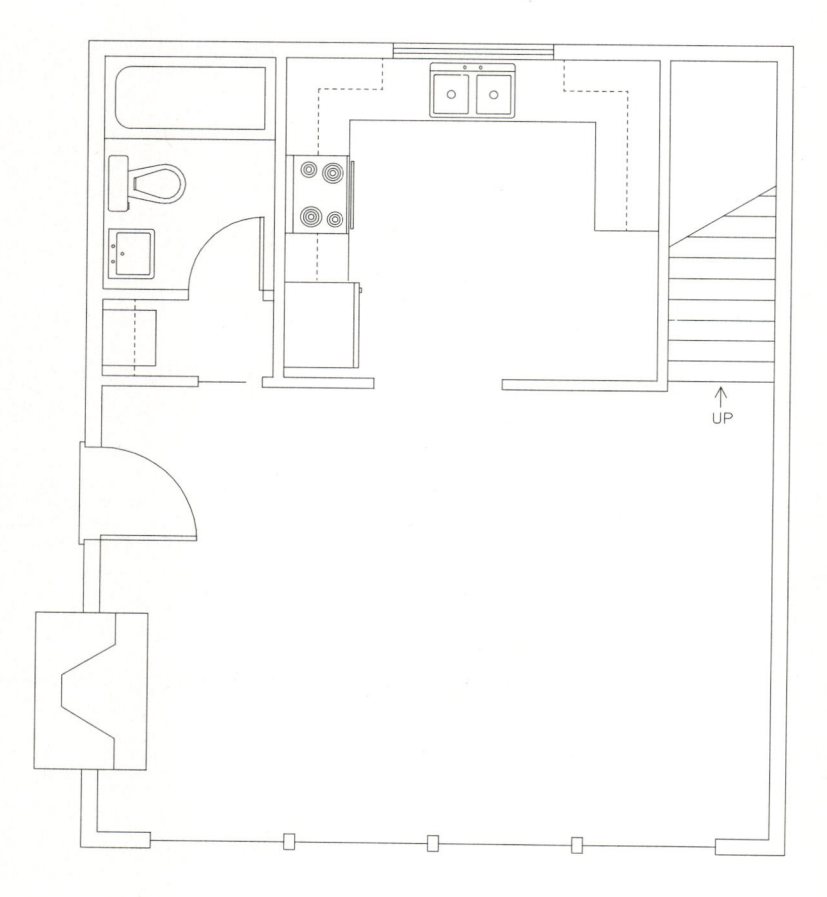

Figure 11-26.
Cabinets have been added to the partial floor plan in Figure 11-25. Appliances and fixtures have also been located with symbols.

> **SHORTCUT**
> Remember, symbols to be used in the floor plan have the letters "PL" at the end of the name. Those without "PL" are elevation views.

Dimensioning

All construction features of the floor plan should be dimensioned unless the location or size is obvious. For example, a door placed at a standard 4" offset from an adjacent wall need not be dimensioned. However, if there is any doubt, dimension the feature. Dimensions extend from exterior walls to the center of interior walls. See Figure 11-27.

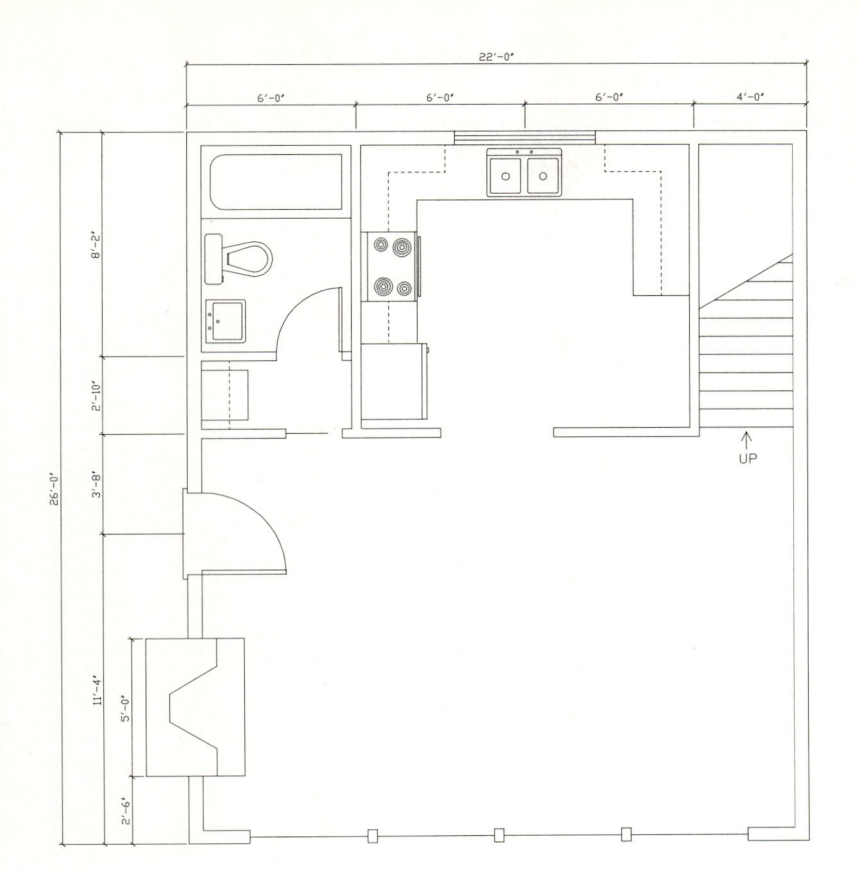

Figure 11-27.
Dimensions have been added to the partial floor plan in Figure 11-26 to locate walls, doors, windows, and other features. This is now a completed floor plan.

Add room names, notes, and materials

Once the dimensions are complete, add text to note the room names and sizes. Doors and windows are typically noted with a letter in a bubble at the end of a centerline. This letter is then referenced in a door and window schedule. Also, make sure that the title block is completed.

Drawing the electrical plan

The electrical plan may be drawn directly on the floor plan. However, many times it is drawn on a separate layer. The *electrical plan* shows the location of switches, lights, and receptacles. Dashed lines curve from the switches to the lights or receptacles they control. See Figure 11-28.

Commercial applications may be drawn differently. For example, a store might have a ceiling light grid versus single fixtures. AutoSketch comes with most of the electrical symbols that you will need. These symbols are in the \WSKETCH\PARTS\HOME\ELECTRCL directory. All electrical symbols for residential and commercial construction are located in the \WSKETCH\PARTS\AEC_ELEC directory.

Background Drawings

Architects rely on experts in the mechanical engineering, plumbing, and structural engineering fields when designing a building. The details of framing, heating/ventilation/air conditioning (HVAC), and plumbing may not appear in the plans for simple residential drawings. More complex structures require these details.

Figure 11-28.
The location of electrical switches, fixtures, and outlets can be added to the floor plan. Also shown here is the location of telephone jacks, cable TV connectors, and other electrical devices.

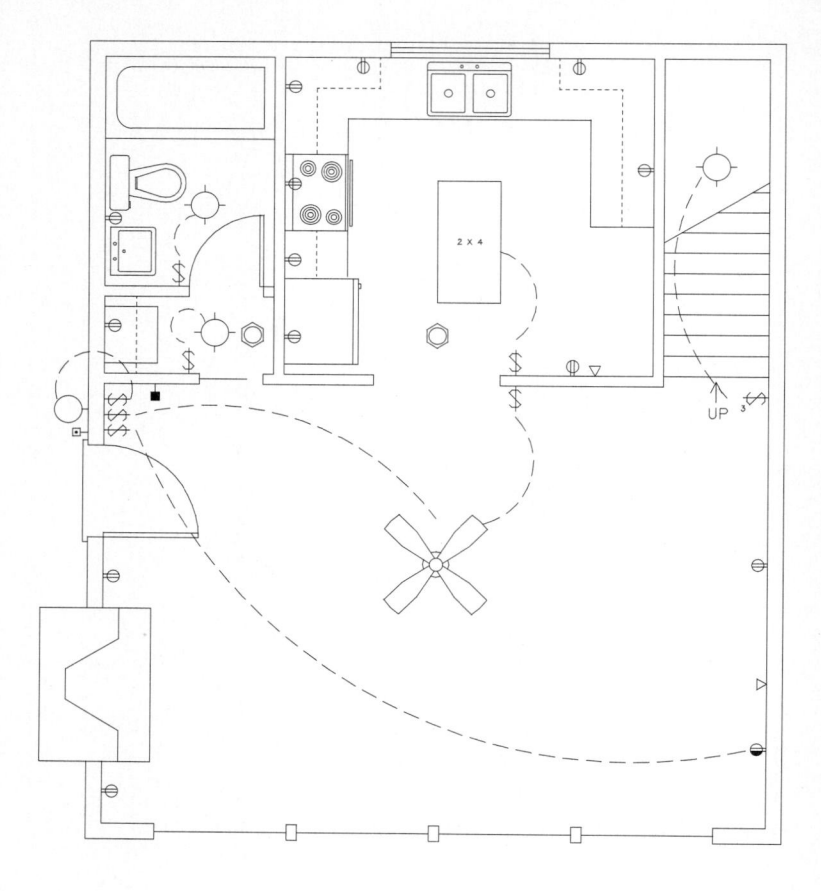

An architect must coordinate the work of professionals. All of these professionals need accurate and timely information from the architect. In the practice of *overlay drafting,* the architect provides consultants with background drawings. *Background drawings* are a subset of the information contained in the architect's floor plan or elevations. However, not all consultants require the same set of information. For example, the plumbing and mechanical engineers require background drawings with sinks, toilets, and showers. The electrician does not need this information.

When creating background drawings, avoid complex layering schemes. Remember that the purpose of the drawing is to reduce confusion and increase drawing efficiency. In addition, the drawing should be simple so that it is easy for the architect to modify the drawing.

Elevations

Elevation plans show each side of the building as it will appear from the outside. Sometimes you may draw interior elevations, but these are considered detail drawings.

An elevation shows the finished appearance. It includes brick or siding, windows, doors, roof features, grade line, and vertical heights. A typical elevation is shown in Figure 11-29. A good knowledge of construction techniques will help you determine the type of roof the structure requires.

Figure 11-29.
An elevation plan shows how the side of a building will look once it is constructed. (E. Henry Fitzgibon, Architect)

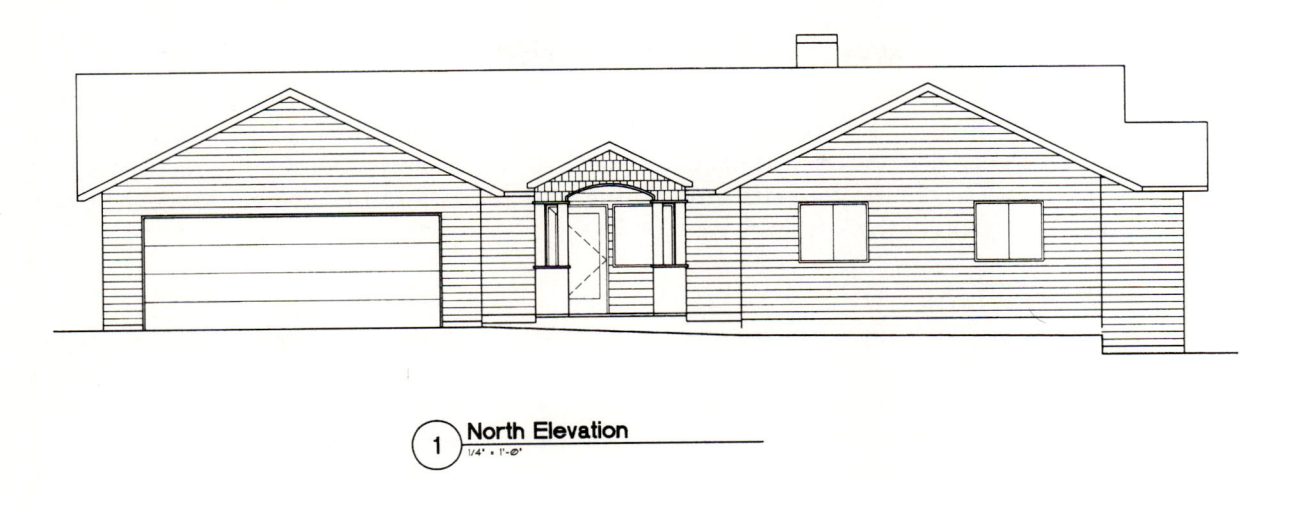

Developing an elevation plan

There are two items required to develop an elevation plan. First, you must know the height of the walls and how they will be constructed. This is shown with a section view, Figure 11-30. For a comprehensive look at section views, refer to *Architecture: Residential Drawing and Design* published by The Goodheart-Willcox Company.

Figure 11-30.
An architectural section view can show the details of how a wall is constructed.

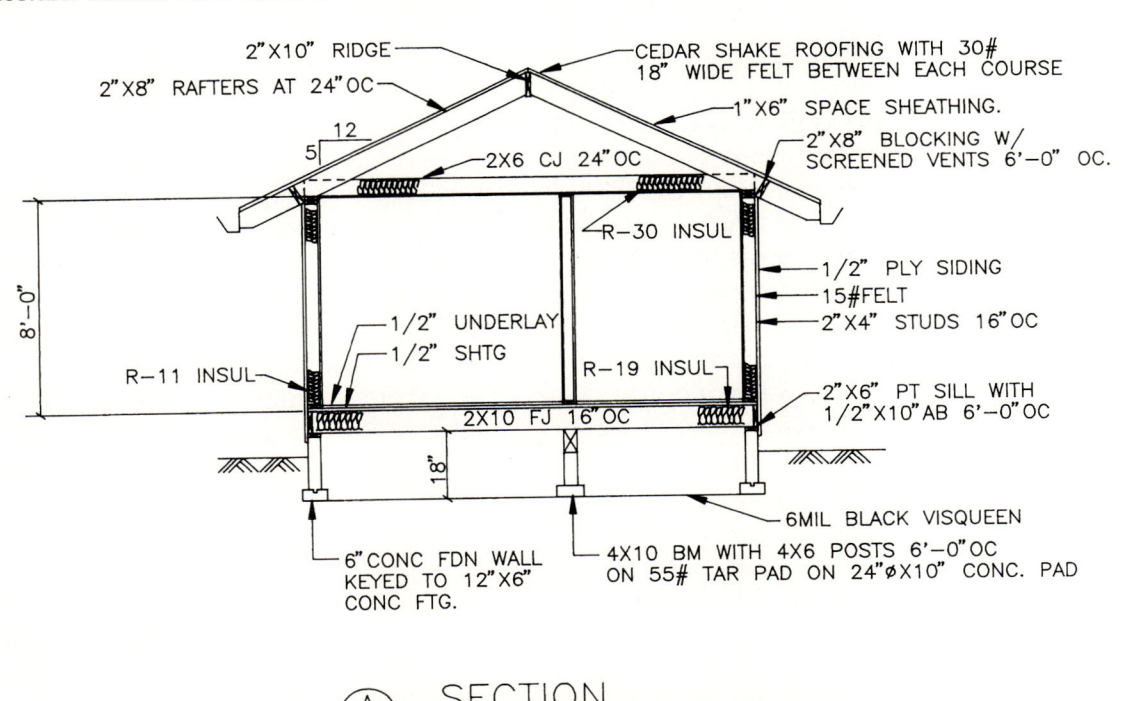

Once the information about the walls is known, the floor plan then provides the width and depth of the structure. The floor plan is placed on the drawing above the elevation. The elevation is projected from the floor plan. Features (such as doors, windows, and corners) are placed at a height that matches the section view. Use construction lines placed on a separate layer to help you.

Once you have located all features, proceed to draw them on the elevation layer. Add details such as window features, overhangs, and roof details not seen in the floor plan. Finally, use pattern fills to create bricks, shingles, and other repetitive features. Be careful not to clutter your drawing with too many pattern fills. If you do so, the overall impression of the structure may be lost.

Other Plans

Other plans found in architectural drafting include foundation, climate control, plumbing, framing, and detail drawings. Whether or not these drawings are required depends on the expertise of your building contractor and local codes.

For example, a foundation plan is necessary to determine what support the structure will have. Yet, an experienced builder does not need a framing plan unless local building codes requires it. Framing plans should be drawn to scale. Structural sizes, spans, and spacing symbols should be obtained from engineering tables or codes.

An experienced plumber will not need a detailed plumbing plan to install plumbing in a family dwelling. The plumber only needs to know the types of plumbing fixtures and their locations. However, a separate detailed plumbing plan will help an inexperienced person purchase the plumbing lines and fixtures. Commercial buildings require a separate plumbing plan.

Also included with architectural drawings are specifications. *Specifications* are a typed list with the descriptions of construction details, fixtures, appliances, and finish of a construction job. Architectural drawings serve as part of legal contract. Therefore, all information about the structure must be on the drawings or in the specifications. No detail should be left uncovered. Fixtures, appliances, and finishes not in a schedule or on a drawing should be placed in the specifications.

Desktop Publishing

Desktop publishing (DTP) is the process of creating a document using electronic means. In other words, a person involved in desktop publishing uses a computer to create books, magazines, flyers, and other printed material. Creating graphics for desktop publishing is another application of AutoSketch.

There are several different ways that a drawing or illustration created in a CAD program can be placed into a desktop publishing layout. With AutoSketch, there are three primary ways of doing this. A drawing or illustration can be exported as a DXF file, printed to a PostScript file, or plotted to a file. Exporting a DXF file is covered in Chapter 17. Printing and plotting to files are covered in Chapter 16. However, the applications of placing these types of files in DTP layouts are covered in the next sections.

Overview of DTP Programs

There are several popular desktop publishing programs. Most of the popular programs have basic features that operate in the same way. However, all of these programs have features and functions that are different from the others.

With most programs, you first need to set up a document. This is similar to setting up a drawing in AutoSketch. You need to define a page size and margins. You also need to select the orientation of the page (portrait or landscape), just as in AutoSketch. Finally, you need to set up and select the correct printer driver. Refer to the manuals that came with your desktop publishing program for the specific setup steps that you need to perform.

Once the setup is complete, the document can be created. A *document* is usually made up of a combination of text and art. This book is an example of a document that uses both text and art. Text and art are usually created in a program separate from the DTP program. These other programs are usually more specialized in creating text or art. AutoSketch can be used to create art. Word processors can be used to create the text. Once text and art are created, they must be *imported* into the DTP program.

Importing text and art into a DTP program is similar to importing a drawing into AutoSketch. However, there is one big difference. DTP programs often require that you draw a frame first for the text or art to be placed in. A *frame* is part of the page that is defined to have text or art, Figure 11-31. Frames can be resized or moved around as needed to change the layout of the page.

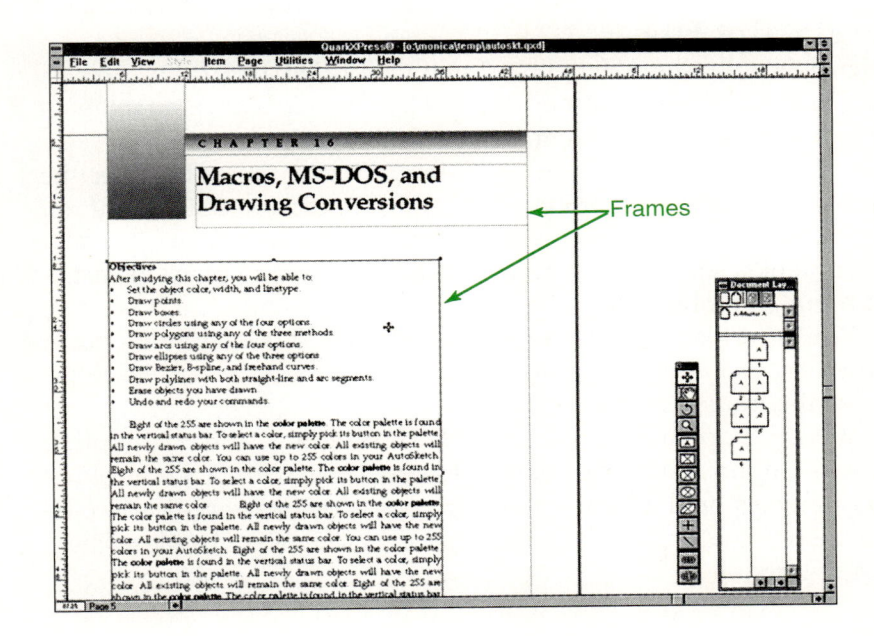

Figure 11-31.
Frames in a desktop publishing program are used to place text and art into a document.

After the document is completed, it must be saved or printed. Simple black and white documents can be printed on a laser printer. However, larger documents often require printing processes not available to the average desktop publisher. These documents are saved to a file. That file can then be given to a *film house* that produces film of the document. These pieces of film are then used to "burn" plates for printing presses.

Working with Drawing Exchange Format (DXF) Files

You can export your drawing from AutoSketch using the *Drawing Exchange Format (DXF)*. This format saves your drawing in a standard "language" that many programs can use. For more information on exporting a DXF file from AutoSketch, refer to Chapter 17.

SHORTCUT
Always do a couple of tests using your DTP program and the type of file you want to use *before* you get too far along in your project.

Specific steps for using a DXF file in a DTP program may vary. However, the following are basic steps. The DTP program you are using should have similar, if not the same, steps.
1. After the document is set up and you have decided where to place the art, draw an art frame (if required).
2. Import the DXF file into the frame. (This command is usually found in the **File** menu.)
3. Move or resize the frame as necessary.

DXF files can be used by most DTP programs, and many other programs as well. This type of file is widely used in the publishing field. However, DXF files may not be suitable for all applications. In applications where maintaining proper proportion in line weight, a DXF file may not hold these properties accurately. In other certain instances, a DXF file may also not be suitable.

If you plan on using several DXF files in your document, be sure to test a few files. This will make sure that every item you are concerned with (such as line weight) will be acceptable. If the DXF files will not work for you, it is better to find this out before you have exported all of your AutoSketch drawings to this format.

Working with PostScript Files

Like DXF, *PostScript* is a standard "language" that many DTP programs can use. To create a PostScript file in AutoSketch, you must print to file. This is covered in detail in Chapter 16. Refer to that chapter for more information.

Specific steps for using a PostScript file in a DTP program may vary. However, the following are basic steps. The DTP program you are using should have similar, if not the same, steps.
1. After the document is set up and you have decided where to place the art, draw an art frame (if required).
2. Import the PostScript file into the frame. (This command is usually found in the **File** menu.)
3. Move or resize the frame as necessary.

PostScript files are very popular. Almost every DTP program, and many other types of programs, can import PostScript files. However, as with a DXF file, a PostScript file may not always be the best option for your particular application. Be sure to do a few tests before you print all of your drawings to PostScript files. These files also tend to be very large.

Working with Plot Files

Hewlett-Packard developed a language for their plotters to use. This is called the *Hewlett-Packard Graphics Language (HPGL).* Many desktop publishing programs can use HPGL files. These files are similar to PostScript files. However, the big difference is that you have to "map" plotter pens.

An HPGL file does *not* contain information that makes up the image (as a PostScript file does). Instead, the file contains information that tells a plotter how to move the pens to make the image. Line weights and colors are not stored in the file. These properties are determined by the color and width of the pen used. However, the type of line (such as hidden or centerline) is stored in the file.

When an HPGL file is imported into a DTP program, the color and width of the line must be set. Some programs allow you to change these properties. However, some do not. Be sure to check the program you are using to see what properties, if any, you can change. This will directly affect whether or not an HPGL file will work for your needs.

> **SHORTCUT**
>
> **Important:** Be sure to keep a record of the color and linetype assigned to a pen. This information is needed by the desktop publisher to accurately use an HPGL file.

Importing HPGL files

Specific steps for using an HPGL file in a DTP program may vary. However, the following are basic steps. The DTP program you are using should have similar, if not the same, steps. Note that this procedure is slightly different from that used for DXF and PostScript files.

1. After the document is set up and you have decided where to place the art, draw an art frame (if required).
2. Import the HPGL file into the frame. (This command is usually found in the **File** menu.)
3. If the program you are using allows you to change the color and linetype, a dialog box will appear after the file is selected (before it is actually imported into the document). Refer to the pen assignments that were written when the file was created. Make the necessary changes to the current pen settings.
4. Move or resize the frame as necessary.

Advantages of HPGL files

Using an HPGL file is a bit more complex than DXF or PostScript files. However, there are several advantages. First, these types of files are usually much smaller than either DXF or PostScript files. When working with large documents (such as this book), the file size is an important factor.

A second advantage is the flexibility that HPGL files offer. Since the pen color and width can be changed as the file is imported, certain problems can be corrected. If the drawing is created in color, but the final document will only have one or two colors, HPGL files allow you to adjust the colors on the drawing to match your needs. DXF and PostScript files do not allow you to do this. Also, when the image is resized in the DTP program, the lines to not become "blocky."

An advantage related to drafting publications is that the line weight can be controlled. There are specific line weight requirements for mechanical drawings. HPGL files allow you to precisely control these line weights.

Disadvantages of HPGL files

The biggest disadvantage of HPGL files is that they are not as easy to work with as DXF and PostScript files. If you do not have specific color and line weight requirements, consider using a DXF or PostScript file instead of an HPGL file.

A second disadvantage is that HPGL files do not reproduce text very well in most cases. If you have a graphic (drawing) with large blocks of text, it would probably be best to use a DXF or PostScript file.

A final disadvantage of HPGL files is that not all DTP programs can use them. Also, some programs that *can* use HPGL files do not allow you to change the color or pen width. If you are using one of these programs, you may need to import the HPGL into a different program first, convert it into a format that your DTP program can use, then import it into your DTP layout.

Other Ways of Using CAD with DTP

Since AutoSketch for Windows is a Windows based program, there is another way of placing your drawings into a DTP layout. This is with the "cutting and pasting" function of Windows. However, this method of placing art is not used much, since the art is usually created well before the document is laid out. This means that the art has to be saved to a file.

To use the cut and paste funtions, simply select the **Copy as Metafile** tool in the **Edit** menu of AutoSketch. Then select the objects you want to place in the layout. Finally, switch to the DTP program and paste the object into the layout.

Summary

Each field of drafting uses slightly different presentation styles for their drawings. Most styles give views of the building or product showing the different sides. In mechanical drafting, these views form a multiview drawing. Each view shows one face of the object in true size and shape. In architectural drafting, the floor plans show the "top" view of the structure while section and elevation plans show the side views.

A multiview drawing is made by projecting each view onto a projection plane. The planes can be thought of as the sides of a glass box. Once the views are projected, the "glass box" is unfolded to position the views. This text uses third-angle projection. Arrange your projection so that the front view meets one or more of the guidelines given in this chapter. Usually, three views are needed to fully describe the object.

An architectural drawing begins by sketching the space needed to satisfy the needs of the client. The space diagram becomes a wall diagram. This in turn becomes the floor plan. Additional details are added to the floor plan for doors, windows, appliances, and other features. An elevation view is then created to show the appearance.

When drawing views, use different linetypes and colors for unique features. For example, use thick, solid, black lines to show outlines and visible edges. Use thin, dashed, colored lines to show hidden features. This will help other drafters "read" your drawing.

Before you start any drawing, make a sketch. This will help you lay out the drawing area needed. A sketch also helps you develop the views. You might choose to do this on a drawing plan sheet like that shown in Chapter 10.

Begin laying out the front view in the lower-left corner. As a general rule, draw the object lines of the views, and then add hidden lines, centerlines, and other lines. You can also draw all lines in all views using one linetype and color and then change them to the proper linetype when complete. This prevents having to change linetypes often. Make sure the views align. If the drawing seems crowded, move the views apart.

If you are planning to use you drawing or illustration in a desktop publishing (DTP) program, you have several options. You must first look at the DTP program that you are using to see what kinds of files it can use. Then, you must look at your needs.

If you have specific color or line weight needs, consider using an HPGL file. An HPGL file may also be best for you if you are working with large documents, since these files are usually small. One thing to keep in mind about HPGL files is that they do not reproduce large amounts of text well.

If you do not have specific color or line weight needs, consider using a DXF or PostScript file. These files are easier to work with. Also, if you have large amounts of text on your drawing, these files may be best. However, these files do tend to be much larger than HPGL files.

The Windows cut and paste function can also be used to place art into a DTP program. However, most often the art is created well before a document is laid out. This means that the art must be saved to a file. For this reason, cutting and pasting is not used much to place CAD created art into a DTP layout.

Important Terms

Alphabet of lines
Architectural drafting
Background drawings
Centerlines
Cutting-plane line
Desktop publishing (DTP)
Dimensioning lines
Document
Drawing Exchange Format (DXF)
Electrical plan
Elevation plans
Film house
First-angle projection
Floor plan
Frame
Glass box
Hewlett-Packard Graphics Language (HPGL)
Hidden lines
Imported
Long-break lines
Mechanical drafting
Multiview drawing
Object lines
Orthographic
Orthographic projection
Overlay drafting
Phantom lines
Pictorial drawing
Plot plan
PostScript
Precedence of lines
Projection
Projection plane
Property line
Riser
Section lines
Short-break lines
Space diagram
Space planning
Specifications
Stringer
Third-angle projection
Tread
Wall diagram
Work triangle

Review Questions

Give the best answer for each of the following questions.

1. In mechanical drafting, the main presentation style for making drawings is called a(n) _____.
2. True or false. The graphics used to describe an object with CAD are different than those developed with manual drafting tools.
3. A sketch that is three dimensional is called a(n) _____ drawing.
4. A multiview drawing gives true _____ and _____ of the item being designed.
5. How does the "glass box" technique of developing a multiview drawing help you visualize the six possible views?
6. Why is it a good idea to make a sketch before starting a CAD drawing?
7. If the left side view is on the left of the front view, and the right side view is on the right, what projection method is being used? _____
8. For most products, the multiview drawing will require at least _____ views, and possibly more auxiliary views.
9. List the factors to consider when choosing which side of the object will be placed as the front view.
10. When might you need only one view to describe a product? _____
11. _____ is the process of placing a file (text or art) into a desktop publishing layout.

12. What linetype and thickness is given to lines that show edges of the object that are not visible in a view?
13. When a centerline and object line coincide, a(n) _____ should be drawn.
14. What sheet size would you set for a three-view drawing of a rectangular object that is 5" high, 12" wide, and 2" deep? Leave 2" between and around views. Assume that your scale is 1:1.

15. What can you do to save time when your drawing contains a mix of object lines, hidden lines, and centerlines?

16. True or false. The first architectural drawing created is the elevation. This is so that clients can see what their house will look like.
17. List four items that should be discussed with the client when examining space requirements.
    A. _____
    B. _____
    C. _____
    D. _____
18. What are the three file types most commonly used to share CAD drawings with DTP programs?

19. List four residential areas covered in this chapter. Identify two that were not.
    A. _____
    B. _____
    C. _____
    D. _____
    E. _____
    F. _____
20. _____ coordinates are typically used to draw the plot plan.
21. The thickness of interior walls as drawn on the floor plan is _____. Exterior walls are drawn _____ thick.
22. Define "desktop publishing."

23. On the electrical plan, how do you connect a light to the switch that controls it?
24. What two drawings are necessary to construct a proper elevation plan?

25. Why are some architectural drawings required, while others are not?

26. A(n) _____ is used in many DTP programs to define an area on the page for art or text.
27. When using HPGL files, you need to keep track of the _____ _____, since this information is needed when importing into a DTP program.
28. What does HPGL stand for? What does DXF stand for?

29. True or false. AutoSketch removes wall lines when a door or window symbol is inserted.
30. The standard that assigns meaning to different styles and thicknesses of lines is called the _____.

Activities

Here is a mixture of drawing activities to develop your skills. Begin each drawing by selecting one of the templates from Activity 4 of Chapter 10 and complete the drawing. Save the drawing on your work diskette using the filename given beneath the illustration. After saving each activity, select the **Close** command from the **File** menu to clear the drawing from the screen.

1 - 5. Develop each multiview drawing. Use the .25" grid spacing for size reference. Do not add dimensions to your drawing. Place different linetypes on different layers. Also, use the colors recommended in this chapter for the linetypes.

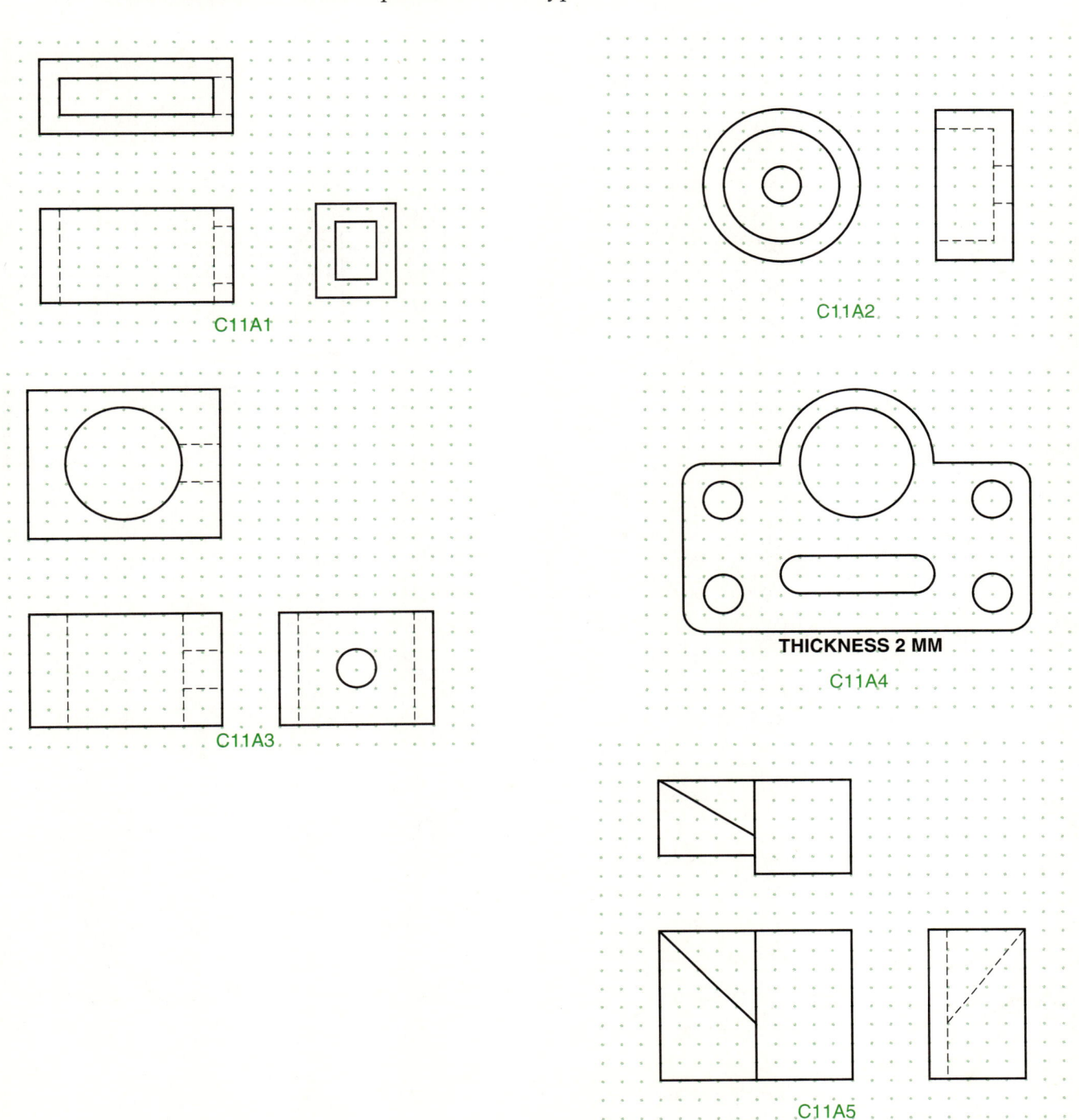

C11A1

C11A2

C11A3

THICKNESS 2 MM

C11A4

C11A5

6 - 10. Using the pictorial sketch, develop a multiview drawing. Draw only the number of views needed to describe the object.

11 - 12. Choose two simple objects around you and develop multiview drawings to describe them. Make sure that your AutoSketch drawings conform to good drafting practice. Use the proper layers, colors, and linetypes. Save the drawings as C11A11 and C11A12.

13 - 15. Develop a floor plan for the following three structures. Before starting each drawing, complete all the questions and sketches necessary for space planning. If you have access to an architectural drafting text such as *Architecture: Residential Drawing and Design,* attempt to also draw a foundation plan, section view, and elevation.
    A. A screened-in porch added onto a house
    B. A one-room cabin similar to the example provided in the book
    C. A three- or four-bedroom home

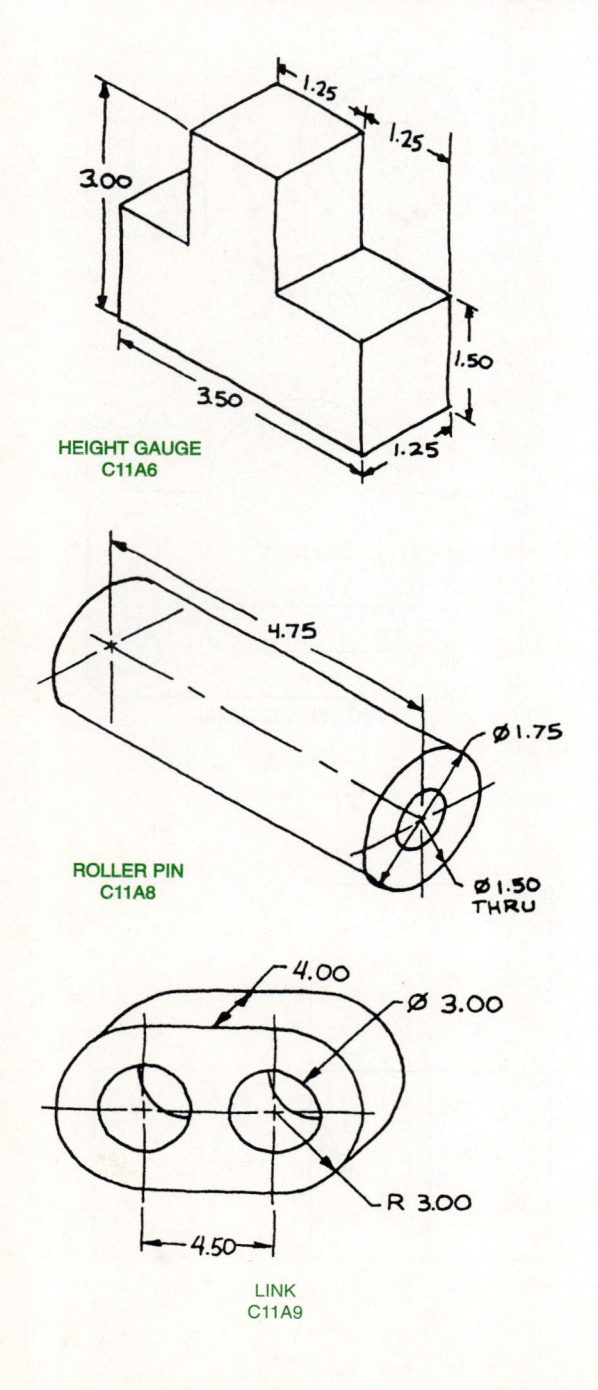

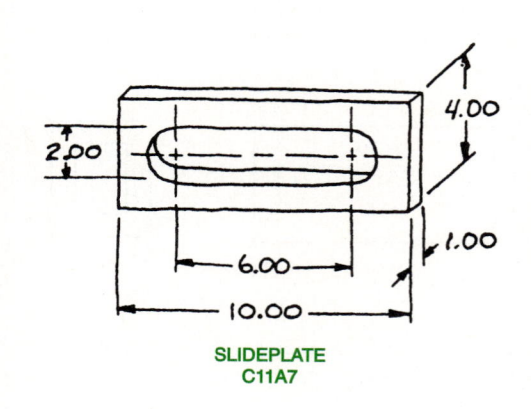

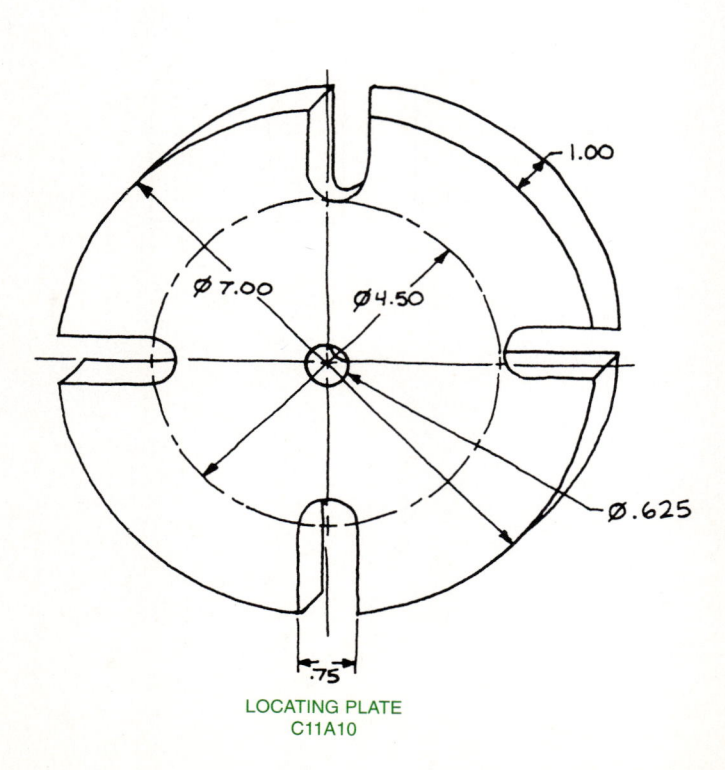

HEIGHT GAUGE
C11A6

SLIDEPLATE
C11A7

ROLLER PIN
C11A8

LINK
C11A9

LOCATING PLATE
C11A10

CHAPTER 12

Drawing Pattern Fills and Section Views

Objectives

After studying this chapter, you will be able to:
- Identify the elements of a section view.
- Draw a basic section view with AutoSketch.
- Choose pattern settings.
- Describe the effect of **Edit** and **Modify** tools on patterns.
- Change the features of existing patterns.
- Use **View** tools to alter the way patterns are displayed.
- Draw various types of section views.
- List various uses for pattern fills.

AutoSketch allows you to fill areas of your drawing with a pattern. The most common use of a pattern fill is for section views. A *section view* reveals interior features by showing the object with a portion cut away. The surface that is cut is represented with a pattern fill. In mechanical drafting, for simple products the interior features can be represented by hidden (dashed) lines. However, if the product is complex, hidden lines make the drawing look confusing. Section views are often used to describe a product with a complex interior. This also applies to architectural drafting where a section view of a wall reveals construction techniques.

Another use of patterns is for elevation views in architectural drawings. An elevation view is an exterior view of a building. A common example is a house, where a pattern fill might show brick or roofing. Although an exercise is provided to let you experiment with an architectural elevation view, this chapter applies pattern fills to the more complicated topic of section views in mechanical drawings.

Elements of a Section View

The elements of a section view are shown in Figure 12-1. The imaginary knife that cuts the object is known as the *cutting plane*. The location of the cutting plane is shown in one view as a *cutting-plane line*. This line is thick with long dashes separated by two smaller dashes. A cutting-plane line can also be represented by a series of medium-size dashes. Short perpendicular lines with arrows are placed at each end of the cutting-plane line. The arrows indicate the direction that the section is viewed at. Section views are placed on the drawing at a position opposite the direction that the cutting-plane line arrows point. It is best to project the section view directly behind the cutting-plane line, Figure 12-2.

Figure 12-1.
A section view shows the object as if a portion is "cut away."

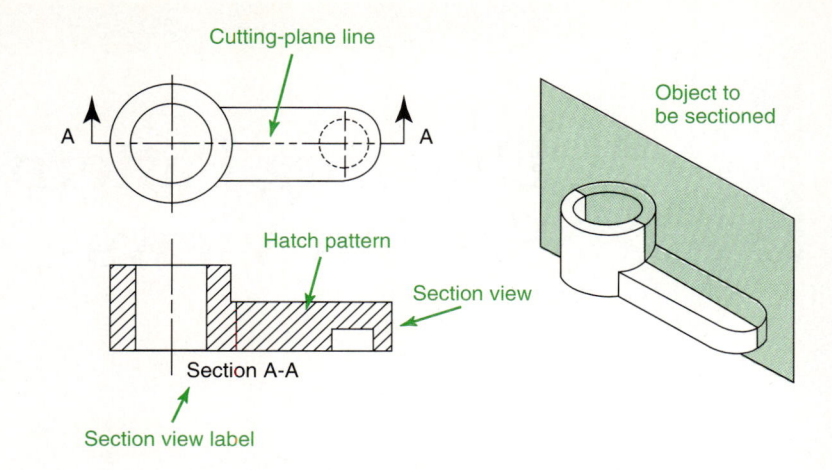

Figure 12-2.
Place the section view behind the cutting-plane line. The two examples shown on the right are correct. Do not use either of the two examples shown on the left. These are incorrect.

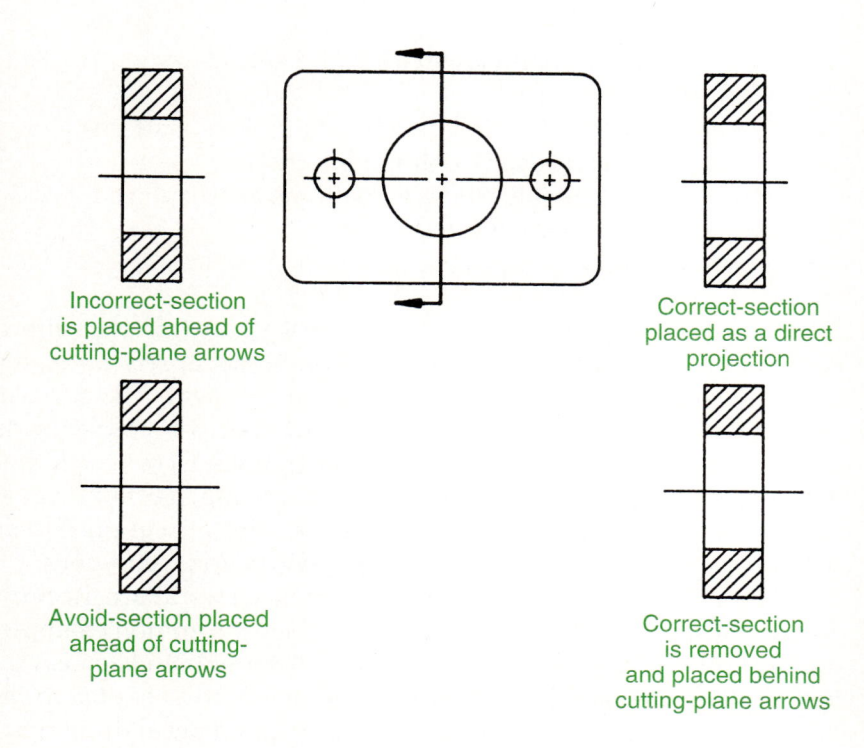

In the section view, the surface that is cut is represented with a ***hatch pattern***. See Figure 12-1. AutoSketch calls a hatch pattern a **Pattern Fill**. The linetype, spacing, and angle of lines in the hatch pattern indicates the material the object is made. Some of the patterns recommended by the American National Standards Institute (ANSI) are given in Figure 12-3. All of the patterns shown in Figure 12-3 come with AutoSketch.

Since a section view is a cut-away of an object, avoid using hidden lines in a section. They make the view look confusing. Hidden lines should be included *only* if needed for dimensioning. Centerlines are included on the section view and are typically added after the hatch pattern.

The cutting-plane line should be clearly labeled if more than one section view is included on the working drawing. Place bold, uppercase letters at each corner of the line. The section view is then labeled according to these letters. Refer to Figure 12-1. Place a note below the section view to indicate the cutting-plane line that is referenced. For example, the section view might be labeled as SECTION B-B.

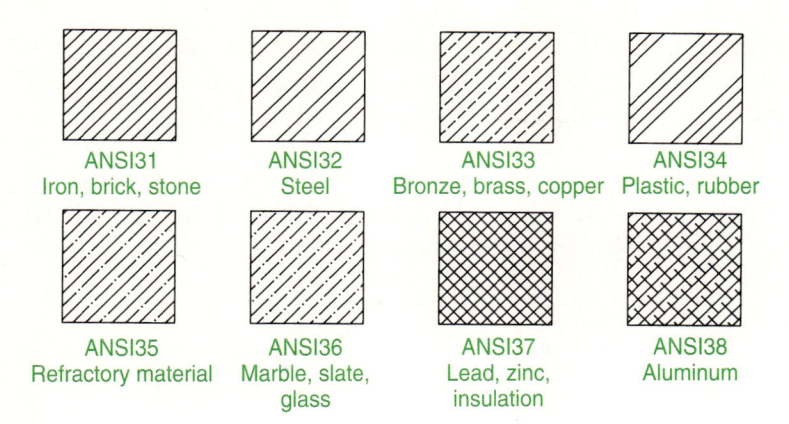

Figure 12-3.
Hatch pattern can indicate the material that the object is made from. Follow ANSI standards for the materials you are using.

ANSI31 — Iron, brick, stone
ANSI32 — Steel
ANSI33 — Bronze, brass, copper
ANSI34 — Plastic, rubber
ANSI35 — Refractory material
ANSI36 — Marble, slate, glass
ANSI37 — Lead, zinc, insulation
ANSI38 — Aluminum

Drawing a Basic Section View with AutoSketch

Most CAD programs let you draw section views quickly because they automatically add the hatch pattern within a specified area. AutoSketch provides this ability using the **Pattern Fill** tool. The basic steps when drawing a section view with AutoSketch are as follows.

1. Draw the section view outline. The section view may be one view of a multiview drawing or a stand-alone view. There are several types of section views (discussed later in the chapter).
2. Select the hatch pattern by double-clicking on the **Pattern Fill** tool in the **Draw** menu. This displays the **Pattern Settings** dialog box. The various pattern settings are discussed in the next section.
3. Finally, select the object that you want the pattern placed inside of. You might select a single object to fill, such as a circle. Or, pick a point inside a closed area to fill. Once placed, a pattern fill is an object. It can be given color or assigned to a layer.

Choosing Pattern Settings

Pattern settings are chosen in the **Pattern Settings** dialog box, Figure 12-4. Access this dialog box by double-clicking on the **Pattern Fill** tool in the **Draw** menu. AutoSketch comes with more than 200 patterns to choose from.

Choosing a pattern

Patterns can be chosen from the icons or from the **Active Pattern** box. Use the scroll bar to move through the available displayed patterns, or click on the scroll arrow of the **Active Pattern** box to see a list of available patterns. Click on the pattern icon or the pattern name you want.

Note: During installation, AutoSketch creates a \WSKETCH\SUPPORT subdirectory to store patterns. If pattern icons do not appear, change to this directory.

Pattern type

There are hatch and bitmap types of patterns. A *hatch pattern* is a pattern provided by AutoSketch. Hatch patterns are the default pattern type. A *bitmap pattern* is a pattern provided by Windows. When this pattern type is selected, the dialog box changes, Figure 12-5.

It is best to select a hatch pattern provided by AutoSketch. Bitmap patterns do not allow you to change some important settings, such as the hatch angle and scale. To select a **Pattern Type,** click on the drop-down button (arrow) and select the pattern type.

Figure 12-4.
Double-click on the **Pattern Fill** tool to access the **Pattern Settings** dialog box. The dialog box shown here will appear when a hatch pattern is selected for the pattern type.

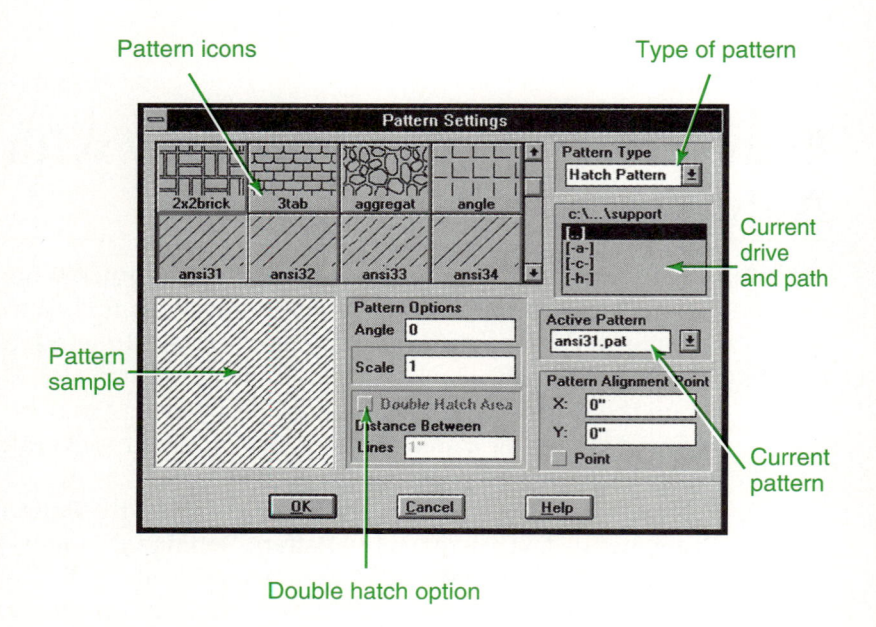

Figure 12-5.
When you select a bitmap pattern type, the dialog box and available options change.

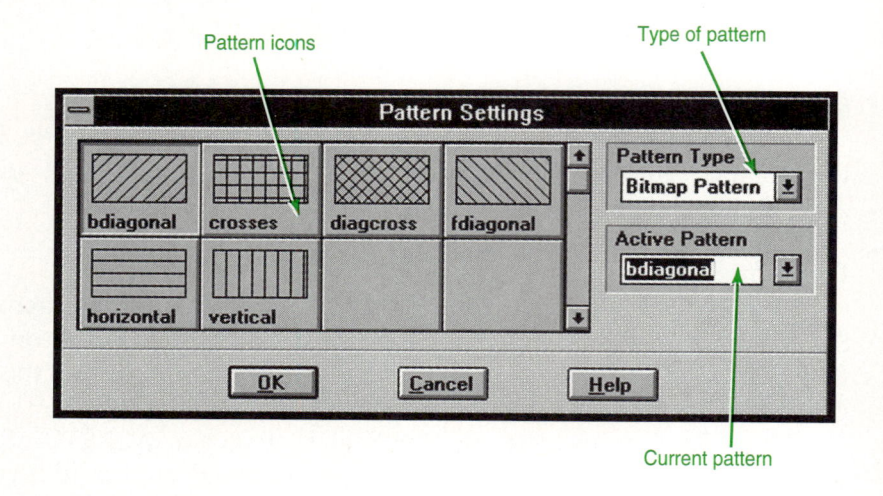

Pattern Options

Pattern Options allow you to change the angle and scale of the pattern. The **Angle** setting determines the angle the pattern is drawn, Figure 12-6. For some patterns, changing the angle can distort the pattern. The **Scale** setting determines the scale the pattern is drawn. When filling a smaller area, a smaller scale may look better, Figure 12-7. After changing these settings, click on the pattern again or press the [Tab] key so that the sample reflects the new settings.

When using the crsshtch.pat (crosshatch) pattern, you can also choose whether you want the pattern double-hatched. **Double Hatch** causes the crosshatch (crsshtch.pat) pattern to include vertical as well as horizontal lines, Figure 12-8. You can also specify the distance between lines.

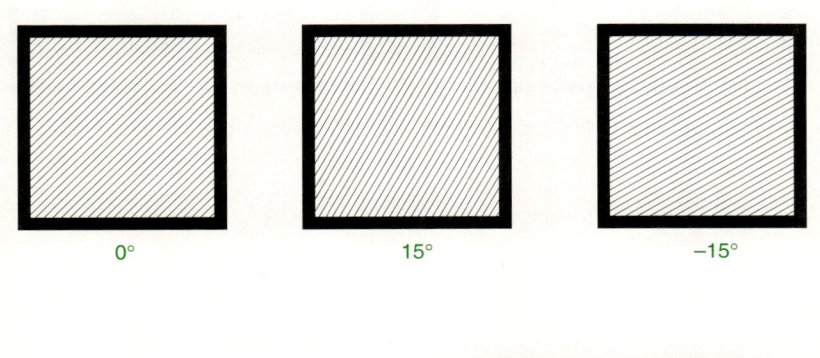

Figure 12-6.
The **Angle** in the **Pattern Settings** dialog box refers to the angle of the pattern. Lines in the ANSI31 pattern are drawn at a 45° angle. This is shown here on the left. Changing the **Angle** setting from 0 alters the pattern angle accordingly.

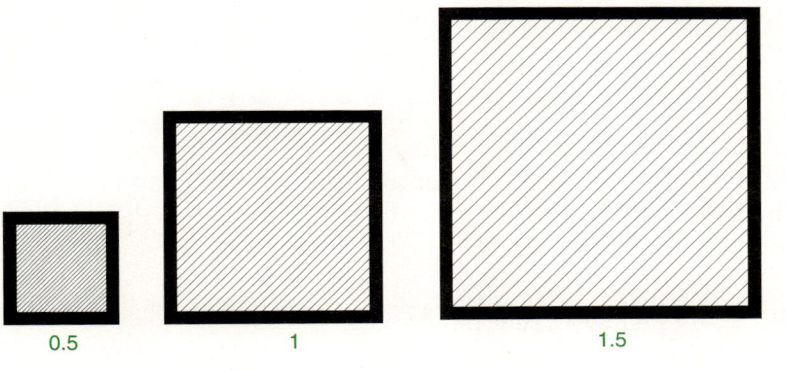

Figure 12-7.
Change the pattern **Scale** so that the pattern looks appropriate for the size of the object. Smaller objects usually need the pattern scale decreased. Larger objects usually need the scale increased.

Figure 12-8.
The crosshatch pattern allows you to set the distance between lines, and whether or not the pattern is double-hatched.

Pattern Alignment Point

The **Pattern Alignment Point** allows you to shift the pattern within the area being filled. This might be needed when a pattern appears odd, such as when it does not align well with the object, Figure 12-9. You can select the X and Y coordinates of the pattern origin. For example, entering .25 for both fields shifts the pattern one-quarter unit up and to the right. If you select **Point,** AutoSketch prompts you for the alignment point when you place the pattern.

Figure 12-9.
Use the **Point** option when the pattern should align with a certain point on an object. Here, the brick pattern needed to start in the lower-left corner of the wall.

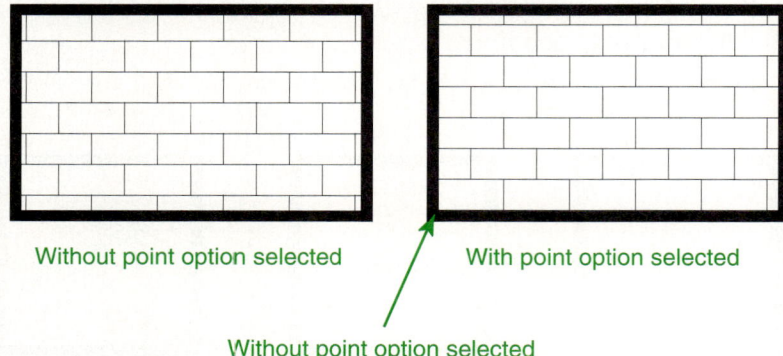

Without point option selected

With point option selected

Without point option selected

Placing the Pattern

To place a pattern fill, select the **Draw** menu and the **Pattern Fill** tool. The following prompt appears.

Pattern Fill Select closed object or area:

Select the objects you want the pattern placed inside, or click within a closed area to fill. You can also leave "islands" unhatched (discussed later in this chapter). When finished, pick the **Esc** button.

> **SHORTCUT**
> You can quickly end the **Pattern Fill** command by pressing the [Esc] key.

Closed objects

If you select an object to fill, the object must be a closed object. Examples of closed objects are circles, boxes, polygons, ellipses, and closed polylines. The pattern will fill the entire object, regardless of other objects that cross or connect with it, Figure 12-10.

Areas

To fill an area bordered by intersecting lines or objects, pick inside of the area to be filled. The area must have no gaps for the pattern to "leak". The pattern will fill the entire area. (The pattern will also overlap objects that do not form the boundary that the pattern is inside.)

Irregular shapes that are closed can be easily filled with a pattern fill. Simply click inside the closed shape and AutoSketch will do the rest, Figure 12-11.

If you get an error message when filling an object that contains other objects, pick a point between the right boundary of the area you want to fill and the objects it contains.

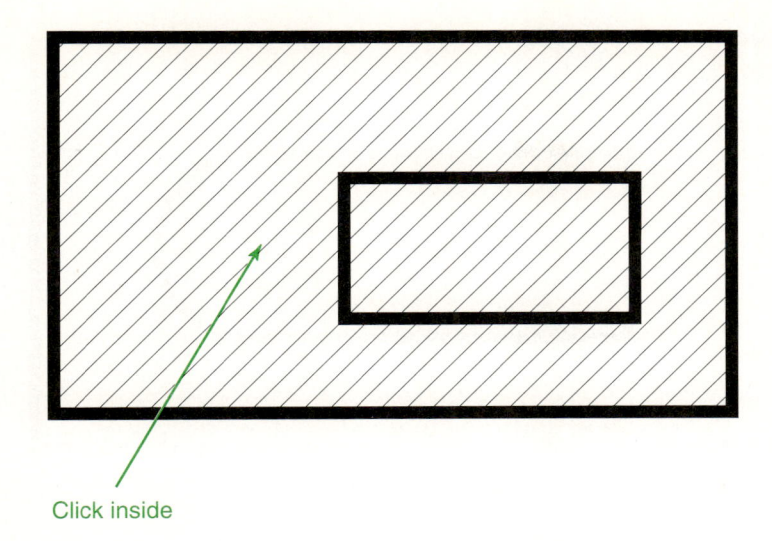

Figure 12-10.
When you select a closed object to fill, the entire area is filled with a pattern, including other objects within the closed object.

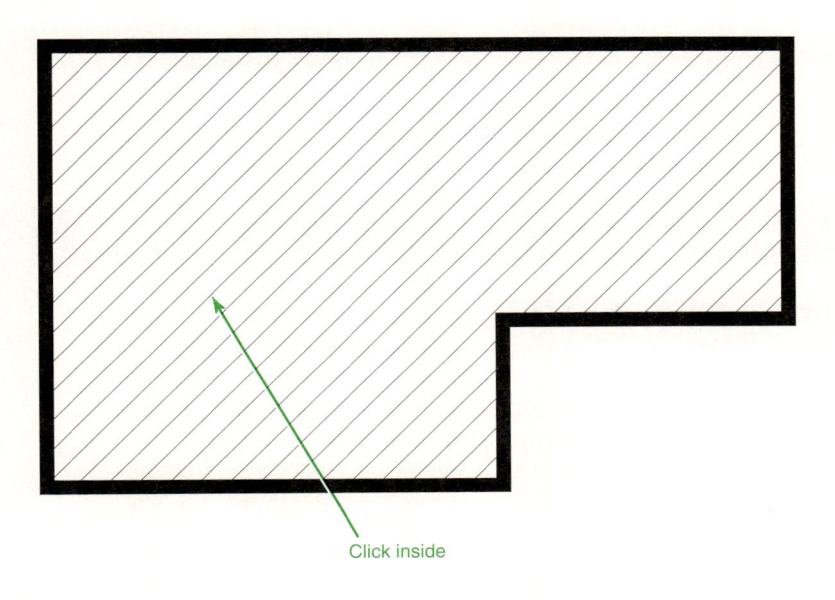

Click inside

Figure 12-11.
An irregular area can be filled simply by picking within the area. AutoSketch automatically calculates the pattern fill.

Click inside

Islands

Sometimes a section view requires more than one pattern fill. For example, the drawing shown in Figure 12-12 is difficult to hatch with one pattern fill command because it has islands. *Islands* are areas within a hatch pattern that are to remain open, or to be filled with a different hatch pattern. An island must be an object or area contained in, but not touching, a larger area being filled.

To leave an island unhatched, click on the **Selection Lock** button (the **Shift** button) on the horizontal status bar. When you click on **Selection Lock** again, the pattern is placed leaving selected islands unhatched. To place a different pattern fill in an island, use the **Pattern Fill** tool and select only the objects that make up the island.

Objects that cannot be filled

Curves, freehand objects, dimensions, text, points, leaders, and other pattern fills cannot be used for pattern fill boundaries.

SHORTCUT

Instead of using the **Selection Lock**, hold down the [Shift] key when selecting the objects. When you release the [Shift] key, the pattern is placed.

Figure 12-12.
When an area has islands that should not be hatched, use the [Shift] key or the **Selection Lock** button to select the objects. This causes the islands to remain unfilled. In this example, if the circle had not been selected, it too would remain unfilled.

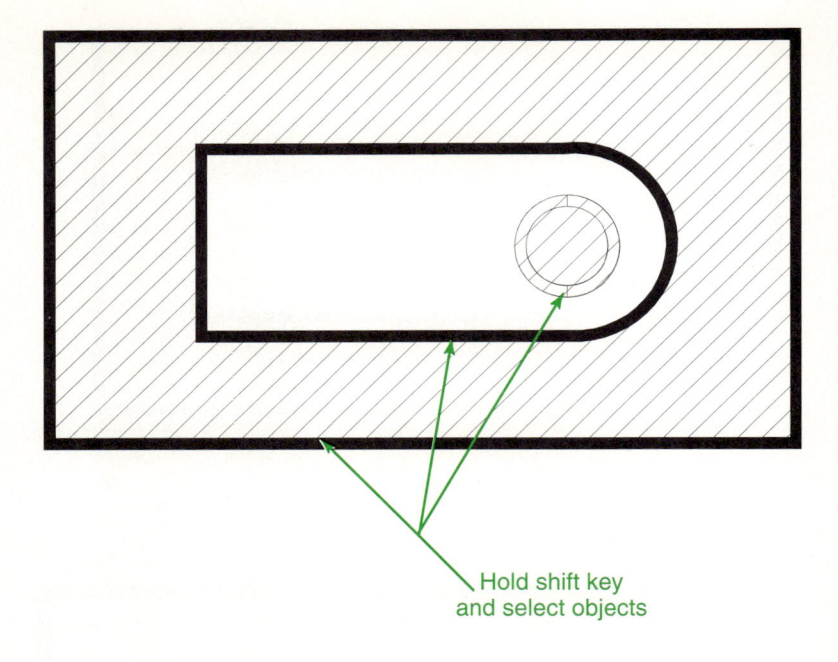

Hold shift key and select objects

Exercise 12-1

1. Start a new drawing using the **A_LI_MEC** template created in Chapter 10.
2. Draw the following shapes in the approximate positions. Size does not matter.
3. Add the ANSI31 pattern to the outer shape and the circle at the same time.
4. Change the pattern settings and use the ANSI31 pattern again to fill the area between the circle and the rectangle.
5. Save the drawing as C12E2 on your work diskette and then close the drawing.

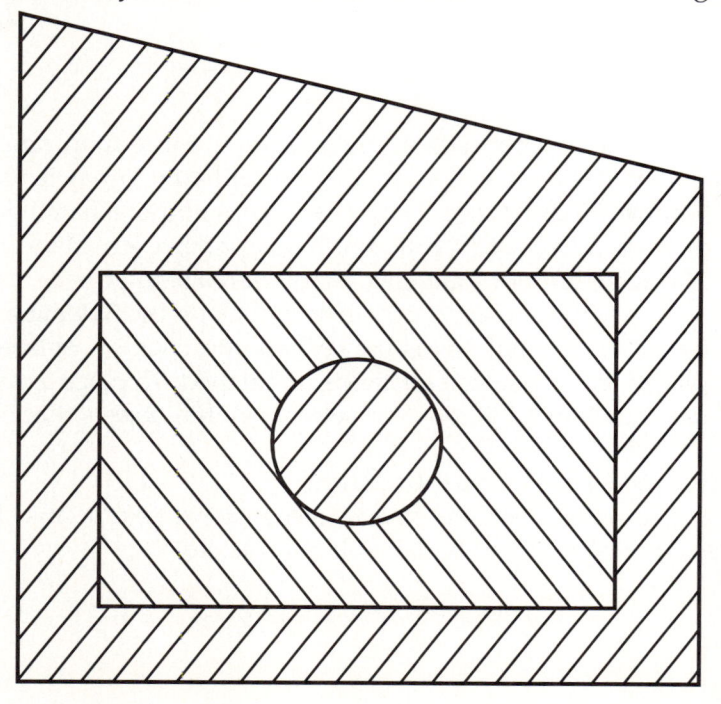

Editing Pattern Fills

Pattern fills can be edited and modified. However, certain tools may function differently with pattern fills, and some tools may not function at all. In addition, various properties of pattern fills can be changed. You can also determine whether drawn fills are displayed.

Edit and Modify Tools

A limited number of **Edit** and **Modify** tools apply to pattern fills. The **Break, Explode, Trim, Extend, Chamfer,** and **Fill** commands do not work with pattern fills. **Move, Copy,** and **Box Array** tools work with patterns just like they do with other objects. The **Erase, Scale, Stretch, Mirror, Rotate,** and **Ring Array** tools have some features that should be noted.

Erase tool

To erase a pattern fill, click on one of the lines that make up the pattern fill. To erase only the object(s) that the pattern is inside, carefully pick only the objects. Using a crosses/window box will likely result in the pattern fill also being erased.

Scale tool

The **Scale** tool will enlarge or reduce the pattern fill, but will not change the size or spacing of the lines that make up the pattern fill, Figure 12-13. The scale of the pattern must be set in the **Pattern Settings** dialog box.

Stretch tool

The **Stretch** tool will enlarge or reduce the pattern fill, but will not change the size, spacing, or angle of the lines that make up the pattern fill, Figure 12-14. The scale and angle of the pattern must be set in the **Pattern Settings** dialog box.

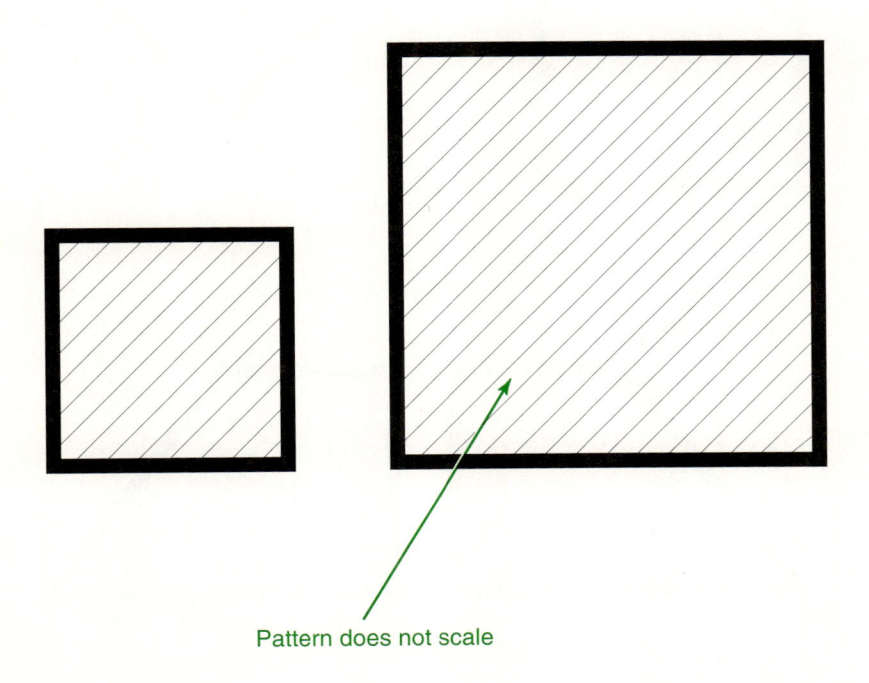

Figure 12-13.
When using the **Scale** tool on objects with pattern fills, the object will be scaled and the pattern fill will fill the new object size. However, the pattern itself remains unchanged.

Pattern does not scale

Figure 12-14.
When using the **Stretch** tool on objects with pattern fills, the object will be stretched and the pattern fill will fill the new object size. However, the pattern itself remains unchanged.

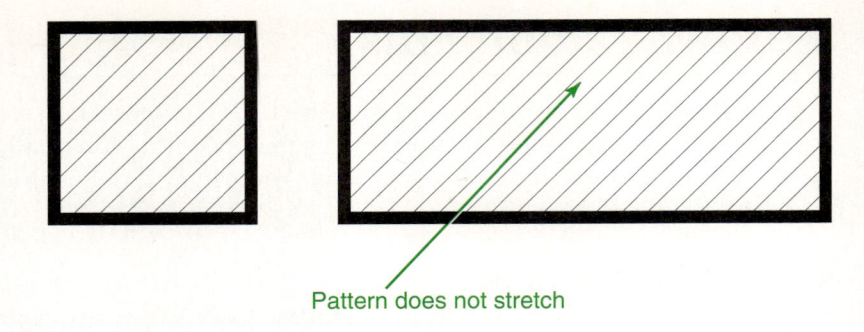

Pattern does not stretch

Rotate

The **Rotate** tool will change the angle of the entire pattern fill, but not the angle of the lines that make up the pattern fill, Figure 12-15. The angle of the lines that make up the pattern must be set in the **Pattern Settings** dialog box.

Mirror tool

The **Mirror** tool will mirror the entire pattern fill, but not the angle of the lines that make up the pattern fill, Figure 12-16. The angle of the lines that make up the pattern must be set in the **Pattern Settings** dialog box.

Figure 12-15.
The **Rotate** tool will rotate objects with pattern fills. The pattern fill will fill the object at its new position. However, the pattern itself will not be rotated.

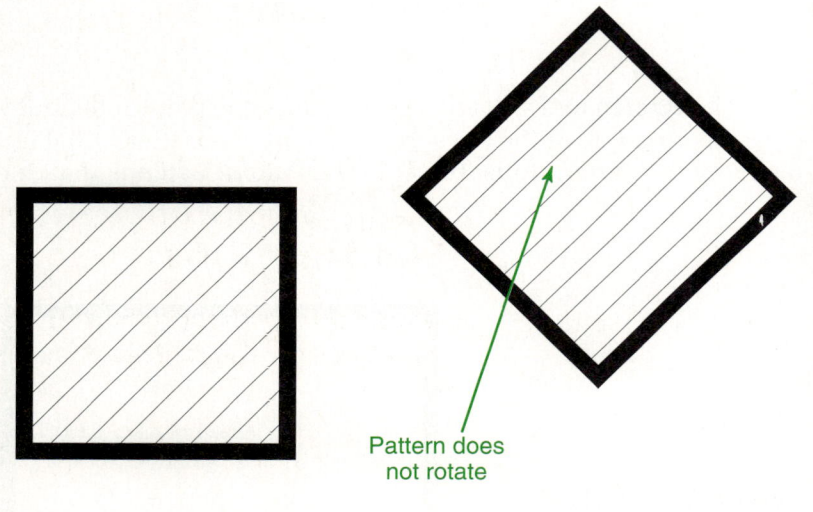

Pattern does not rotate

Figure 12-16.
You can use the **Mirror** tool on objects with pattern fills. However, note here how the pattern itself is not mirrored. The pattern simply fills the new object.

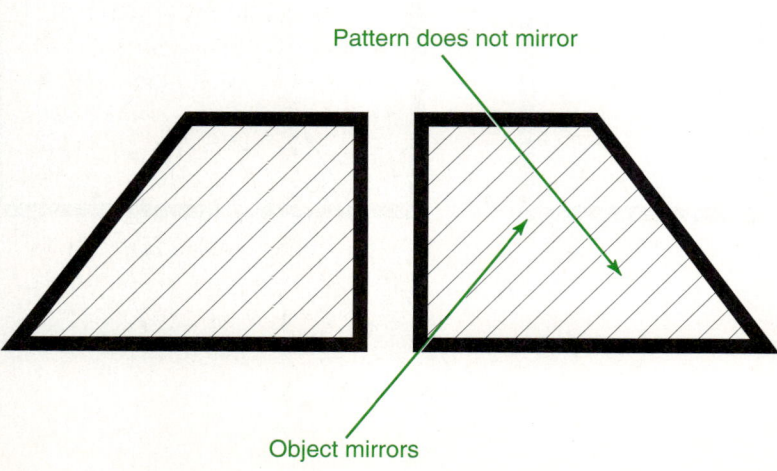

Pattern does not mirror

Object mirrors

Ring Array tool

Like the **Rotate** tool, the **Ring Array** tool will copy and change the angle of the entire pattern fill, but not the angle of the lines that make up the pattern fill, Figure 12-17. The angle of the lines that make up the pattern must be set in the **Pattern Settings** dialog box.

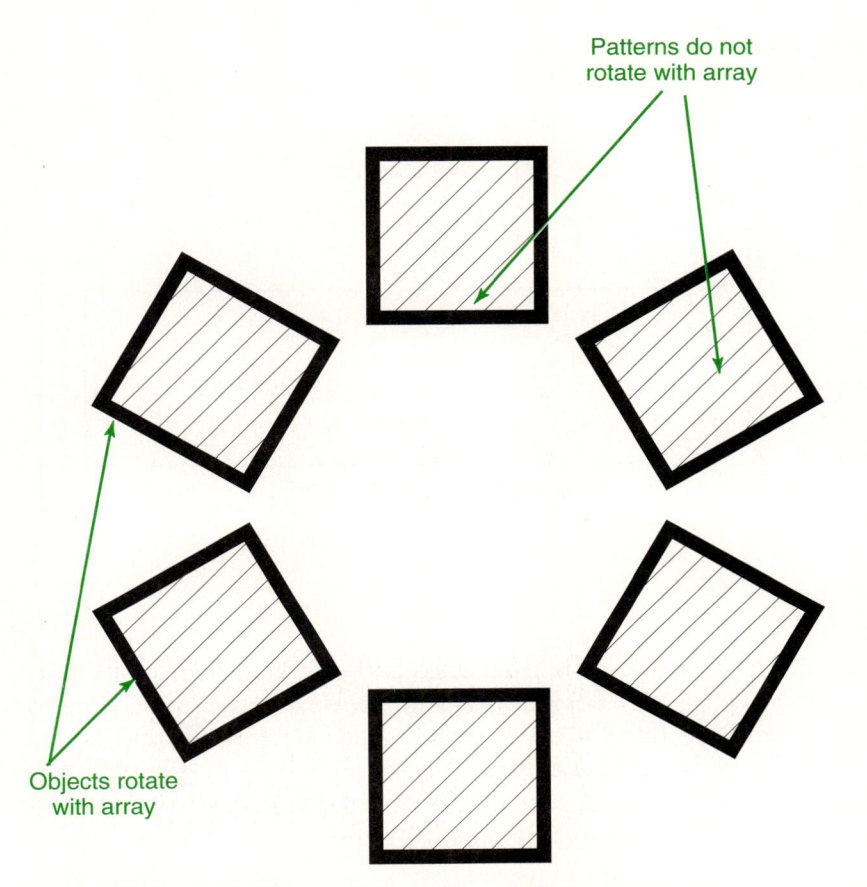

Figure 12-17.
Objects with pattern fills can be used to create ring arrays with the **Ring Array** tool. However, as with the **Rotate** tool, the pattern itself will not rotate. The pattern will simply fill all of the new objects in the array.

Changing Scale and Angle of Patterns

Pattern fills made with one pattern style can be changed. For example, suppose you just completed a section view for a product made of steel. However, a design change now dictates that the product be made of aluminum. Instead of erasing each pattern and redrawing them, change the options in the **Pattern Settings** dialog box. Then, double-click the **Change Properties** tool from the **Edit** menu. From the **Property Settings** dialog box, pick each option so only the **Pattern** option has a check mark beside it, and pick **OK**. Finally, pick the **Change Properties** tool again and pick the pattern fills to assign them the new pattern.

Exercise 12-2

1. Open C12E1 from Exercise 12-1.
2. Stretch the shape so that it is twice as wide. Notice that the pattern is not affected.
3. Change the pattern between the circle and rectangle to ANSI32 using the **Change Properties** tool.
4. Save the drawing as C12E2 on your work diskette and then close the drawing.

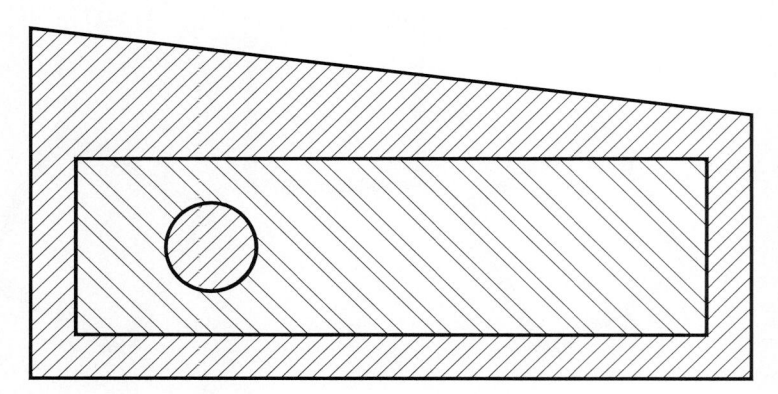

View Tools

A pattern fill is made up of many individual lines and objects. When many pattern fills are used in a drawing, using **Zoom** and **Redraw** tools can take much time to redraw the patterns. You can hide, but not erase, pattern fills to quicken your display speed. Select the **View** menu and **Preferences** tool. Pick the box next to **Hide Fill.** You can also quicken display speed by placing pattern fills on a separate layer. Then, simply make that layer invisible. Remember to redisplay the patterns before printing your drawing.

Creating Custom Patterns

Although AutoSketch provides most every pattern you will need, you can create custom patterns. Refer to the *AutoSketch for Windows User Guide* for more information on creating custom patterns.

Types of Sections

The cutting plane for a section view does not have to pass straight through the entire object. An object may be cut in other directions to show special detail. Types of sections include full, half, offset, revolved, broken-out, removed, thin, conventional breaks, and aligned.

Full Section

In a *full section*, the cutting plane passes straight through the entire object. The section view replaces an exterior view to show some interior feature, Figure 12-18. The cutting-plane line and section label may be omitted since the section view is one of the multiviews.

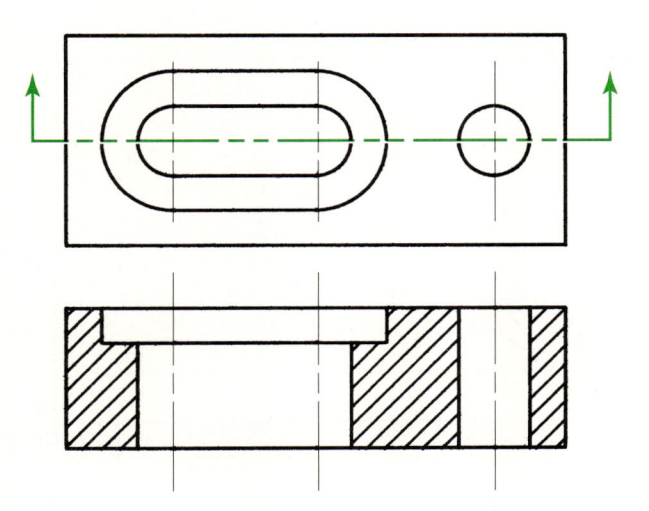

Figure 12-18.
In a full section view, the cutting plane passes straight through the object.

Half Section

In a *half section*, the cutting plane cuts away one-quarter of the object. Two cutting planes intersect at a 90° angle. Both internal and external features are shown in the same view, Figure 12-19. Half sections are used when the object is symmetrical. A symmetrical object is when one-half of the object is the mirror image of the other. Cutting-plane lines and section labels are omitted. Place a centerline where the exterior and half section views meet. Only one arrow is given on the cutting-plane line.

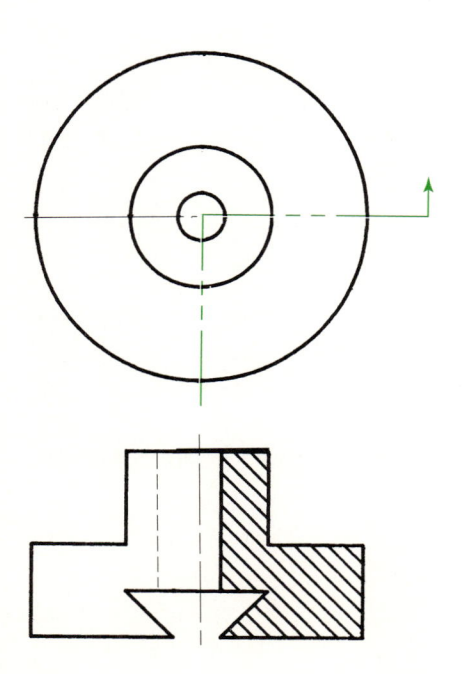

Figure 12-19.
A quarter of the object is cut away in a half section.

Offset Section

In an *offset section,* the cutting plane is not continuous. The plane is stepped, or offset, at one or more places to show or avoid a certain detail, Figure 12-20. An offset section is not clearly indicated on the section view. Thus the path of the offset cutting plane is indicated by the cutting-plane line in one of the other views.

Figure 12-20.
The cutting plane in an offset section may be "stepped" several times to pass through important features.

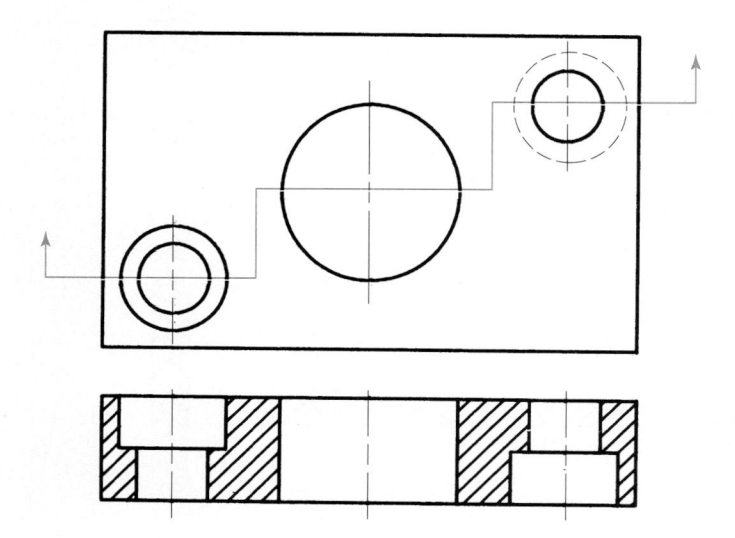

Broken-Out Section

A *broken-out section* appears as if a portion of the object was broken off to reveal the interior, Figure 12-21. A broken-out section reveals a small portion of interior detail. No cutting-plane line is indicated. A curved break line is placed between the section and exterior views.

Revolved Section

A *revolved section* takes a slice of the object, and rotates it 90° to show a cross-section, Figure 12-22. This section is commonly used for shafts, webs, spokes, and flanges.

Figure 12-21.
A broken-out section looks as if a portion of the object is broken away.

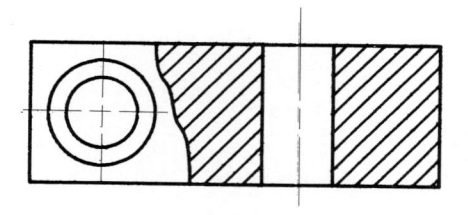

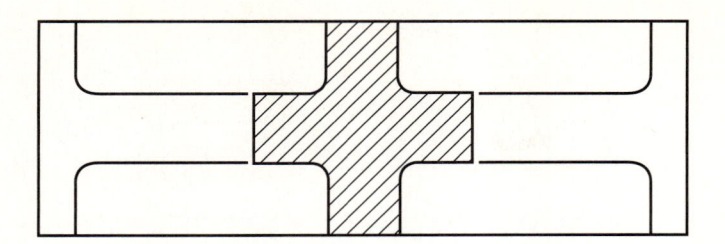

Figure 12-22.
A "slice" of the object is revolved in place for a revolved section view.

Conventional Breaks

A *conventional break* is used to shorten extremely long products, such as shafts and tubes, Figure 12-23. It would be impractical to show the entire length of these objects. If scaled to fit, details might be too small. A revolved section may be placed within the break. The overall distance is still given, even though the entire object is not drawn. The measurement should reflect the actual total length of the part.

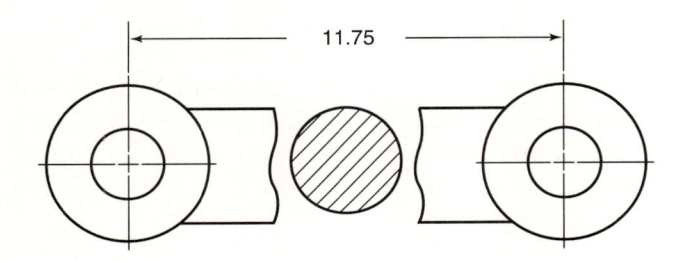

Figure 12-23.
A conventional break line is used for long objects that will not fit on paper. A revolved section view may be included (as shown here).

Removed Section

A *removed section* is formed when a section view is taken from its normal projected place on the drawing and moved elsewhere. The cutting-plane line and section view must be clearly labeled.

Thin Section

Section views of sheet metal, gaskets, or other thin products are often too narrow to add hatching. The hatch pattern would serve no practical purpose. In a ***thin section,*** the hatch pattern is replaced with a solid color, Figure 12-24. Use the solid pattern fill to place this color.

Figure 12-24.
Thin sections, such as this gasket, are given solid color rather than a hatch pattern.

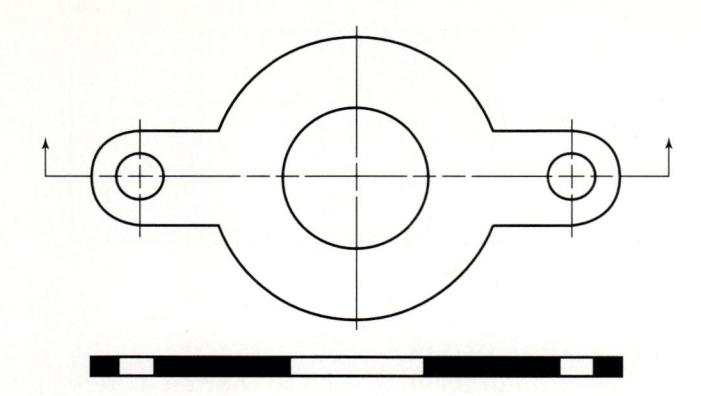

Aligned Section

It is not a good practice to make a full section of a symmetrical object that has an odd number of holes. In an *aligned section,* the cutting plane is rotated so that the section view shows two holes at their proper distance from the center. The actual projection may be misleading. See Figure 12-25.

Sectioning Assemblies

An *assembly view* shows connected parts. When sectioning assemblies, rotate the hatch pattern to a different angle on each piece, Figure 12-26. Shafts, set screws, bolts, pins, rivets, nuts, or ball bearings in an assembly are not sectioned when the cutting plane passes through them lengthwise. However, they are sectioned when the cutting plane passes through them across their axis.

Figure 12-25.
A—Some section views may be confusing if drawn at a true projection.
B—An aligned section view can help clarify the drawing.

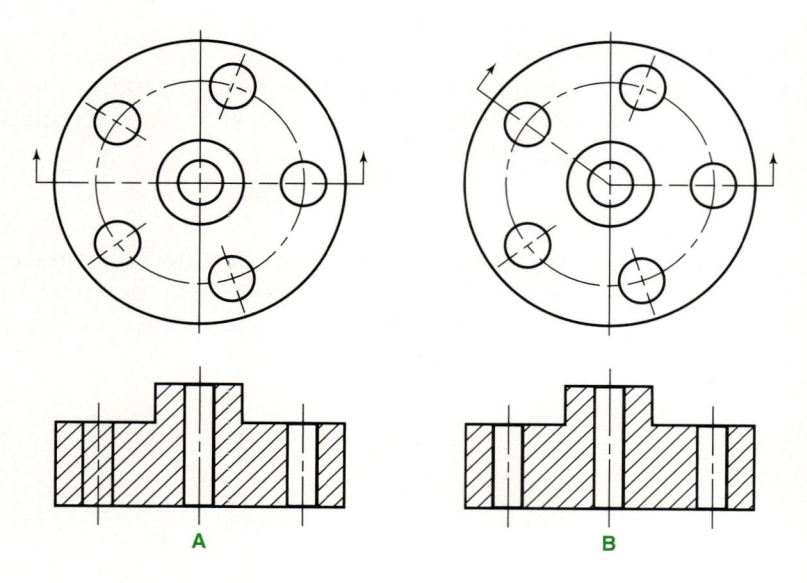

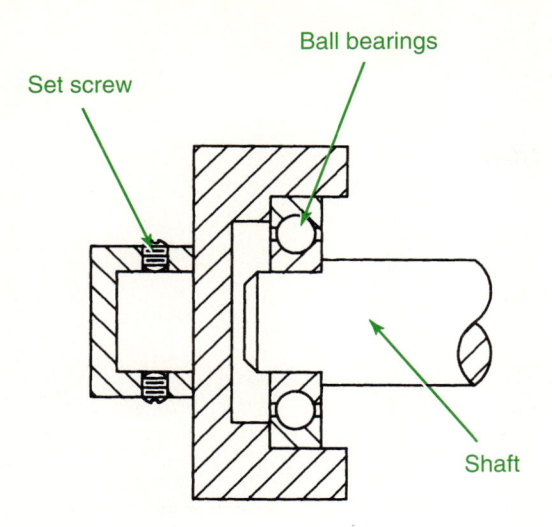

Figure 12-26.
Each part in a sectioned assembly has a hatch pattern at a different angle. Shafts, set screws, and ball bearings are not sectioned along their length.

Other Uses for Pattern Fills

AutoSketch provides many patterns for drafting tasks other than mechanical drawing. Architectural drawings use pattern fills for section views and elevation views. See Figure 12-27 and Figure 12-28. An elevation view shows the exterior of a building. The pattern fill might show brick, roofing material, or other features. In addition, pattern fills can be used for charts and graphs, Figure 12-29.

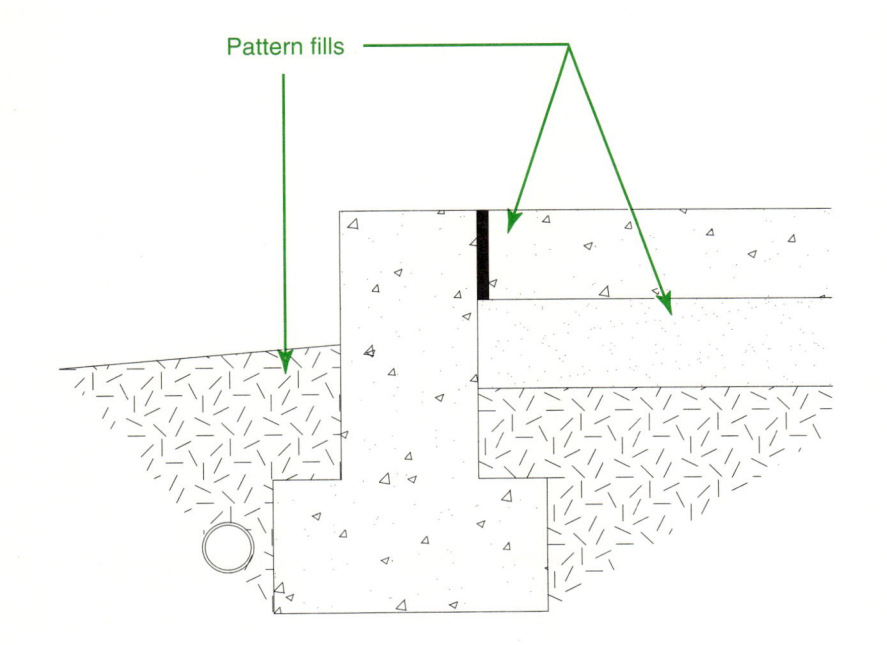

Figure 12-27.
A section view of a foundation wall may use different pattern fills to represent concrete block, a poured concrete slab, and earth fill.

Figure 12-28.
Architects may use many different pattern fills in the elevation of a house to show bricks, siding, roofing materials, and other features.

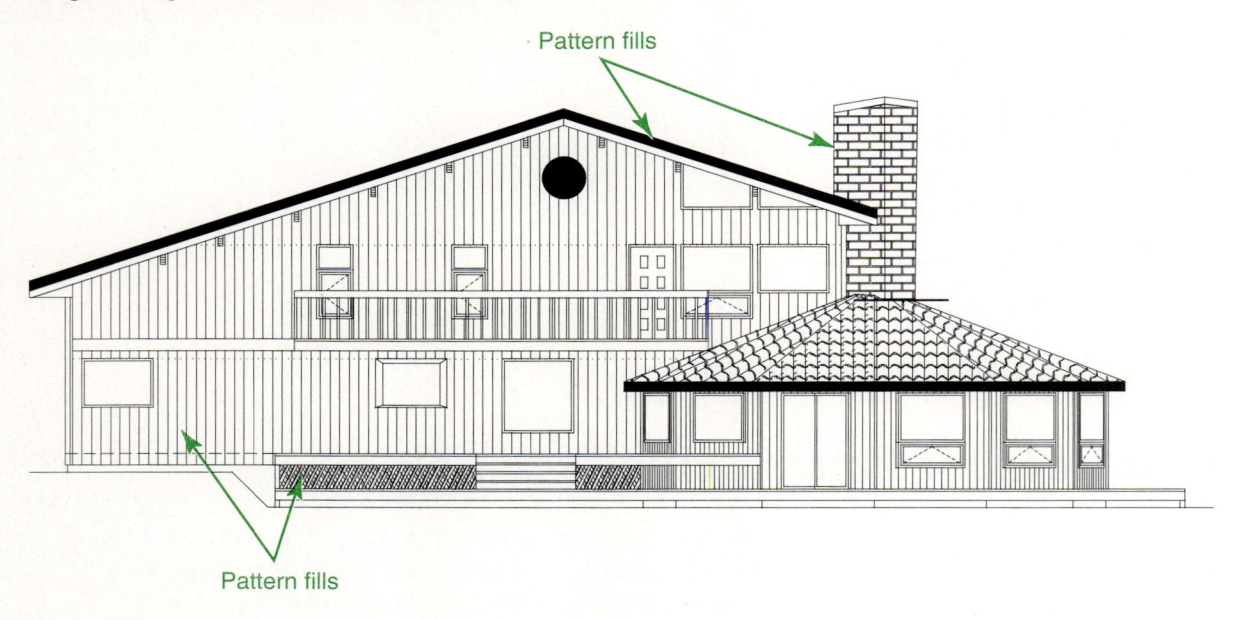

Figure 12-29.
Pattern fills can be used to enhance charts and graphs

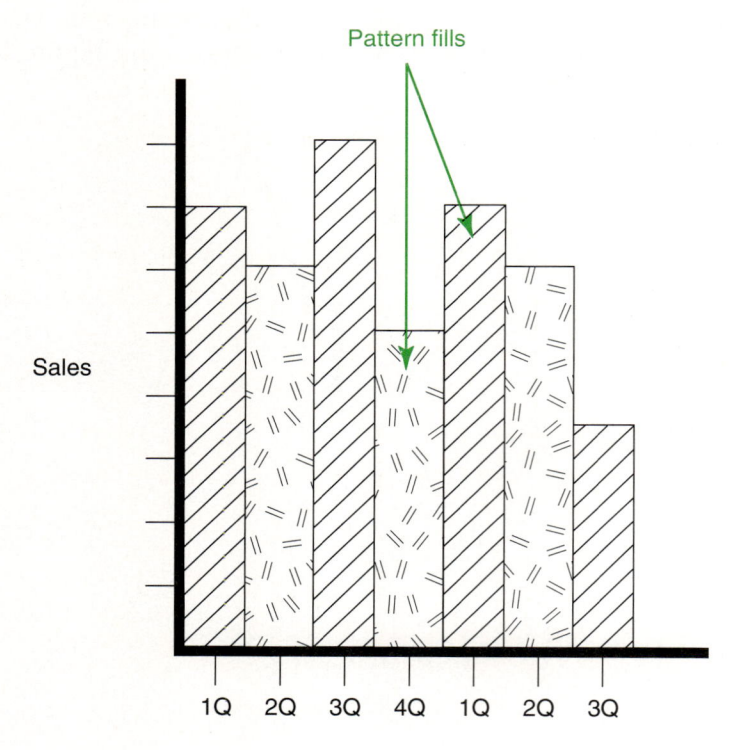

Exercise 12-3

1. Start a new drawing using the **A_I_arch** template found in the \WSKETCH\TEMPLATE subdirectory.
2. Draw the front elevation shown at A below. Approximate sizes and locations. Do not dimension your drawing.
3. Add pattern fills so that your drawing looks like the one shown at B. You may need to adjust the scale of the patterns.
4. Save the drawing as C12E3 on your work diskette and then close the drawing.

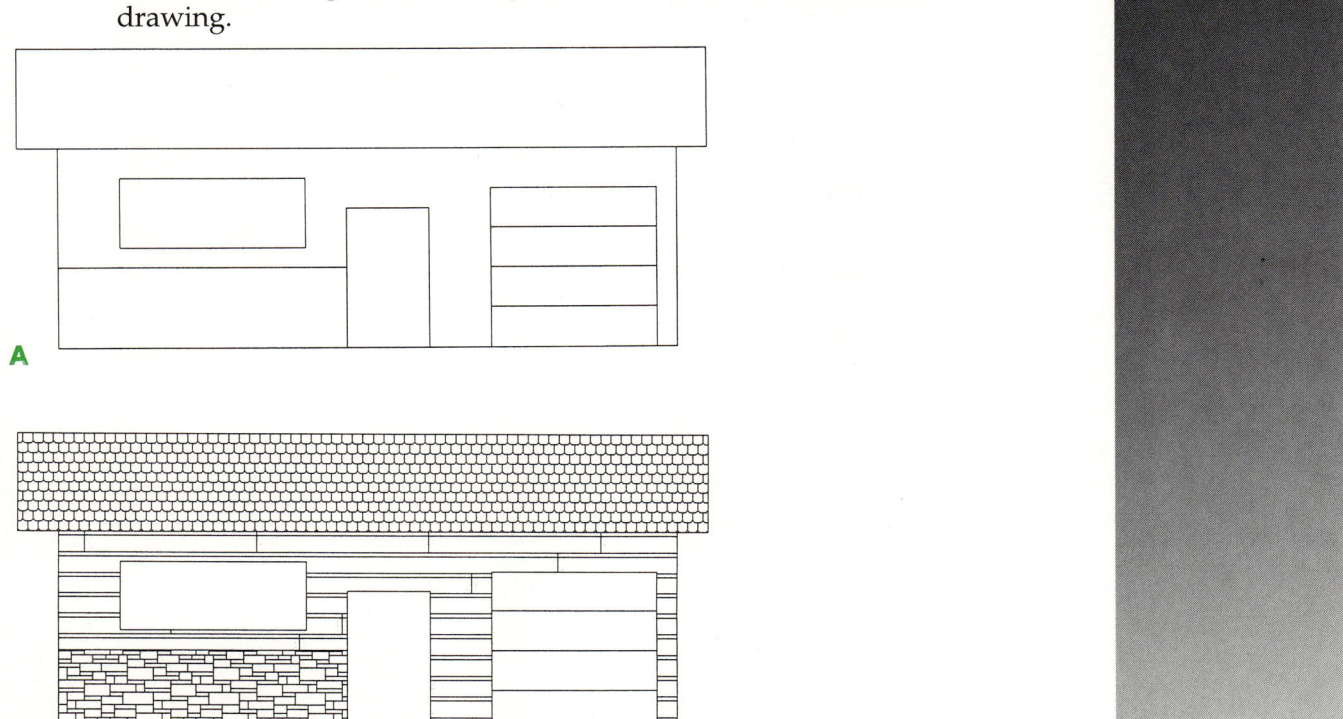

A

B

Summary

The **Pattern Fill** tool allows you to place patterns on your drawing. Patterns are used for mechanical section views, architectural elevations, and for other illustration purposes. In a section view, the cutting-plane line shows where the section view is cut from. The section view can be part of a view, or a separate view. The pattern indicates the material that the product or structure is made from. There are many types of section views for both mechanical and architectural drawings.

To place an AutoSketch pattern fill requires that you choose pattern settings and select the objects to fill. Pattern settings include bitmap and hatch patterns. You can use options to scale the size of the pattern or place it at an angle. When selecting the objects to fill, you can select any closed object or an area enclosed by several objects. If the area requires more than one pattern, you must select the interior areas, called islands, so that they are left unfilled. Areas bounded by curves, freehand objects, dimensions, text, points, leaders, and other pattern fills cannot be used for pattern fill boundaries.

You cannot edit the pattern of a pattern fill, however, you can copy, move, or array a pattern fill. Commands that rotate, stretch, or scale an object will not affect a pattern fill. You must go back to the pattern settings and change the properties.

Important Terms

Aligned section
Assembly view
Bitmap pattern
Broken-out section
Conventional break
Cutting plane
Cutting-plane line
Full section
Half section
Hatch pattern
Islands
Offset section
Removed section
Revolved section
Section view
Thin section

New Tools and Commands

Active Pattern box
Angle
Double Hatch
Hide Fill
Pattern Alignment Point
Pattern Fill
Pattern Options
Pattern Settings dialog box
Pattern Type
Point
Scale

Review Questions

Give the best answer for each of the following questions.

1. Why are section views required in mechanical and architectural drafting?
2. The location of a section view's cutting plane is shown in one view as a(n) _____.
3. Where should you place section views in relation to the direction that the cutting-plane line arrows point?
4. Hidden lines may be included in a section view only if needed for _____.
5. What is the advantage of using hatch pattern types rather than bitmaps? _____
6. Why would you want to change the scale of a pattern? _____
7. When might you select the **Point** option when placing a pattern? _____
8. True or false. To place a pattern in an area, you must pick each of the objects that make up the boundary.
9. To place a pattern in an area that contains an island, you must hold down the _____ key or the _____ button.
10. True or false. The **Scale** and **Stretch** tools will change the size and spacing of a pattern.
11. True or false. The **Rotate** tool will change the angle of a pattern.
12. What option should be selected to speed the redraw time of a drawing that contains many patterns? _____

13. In what type of section view does the cutting plane cut away one-quarter of the object? _____

14. In what type of section view is the cutting plane stepped at one or more places to show or avoid a certain detail? _____

15. What are two other common uses for patterns other than section views?
    A. _____
    B. _____

Activities

Here is a mixture of drawing activities to develop your skills with section views and pattern fills. Begin each drawing by selecting one of the templates from Activity 4 of Chapter 10 and complete the drawing. After saving each activity, select the **Close** command from the **File** menu to clear the drawing from the screen.

1. Draw four 1" square boxes and fill each of them with a different ANSI pattern provided with AutoSketch.

2 - 5. For each multiview drawing below, change one view into a section view as indicated. Use the .25" grid as a reference, and place your drawings on one of the prototypes developed in Chapter 10. Save the drawing on your work diskette using the filename given beneath each illustration.

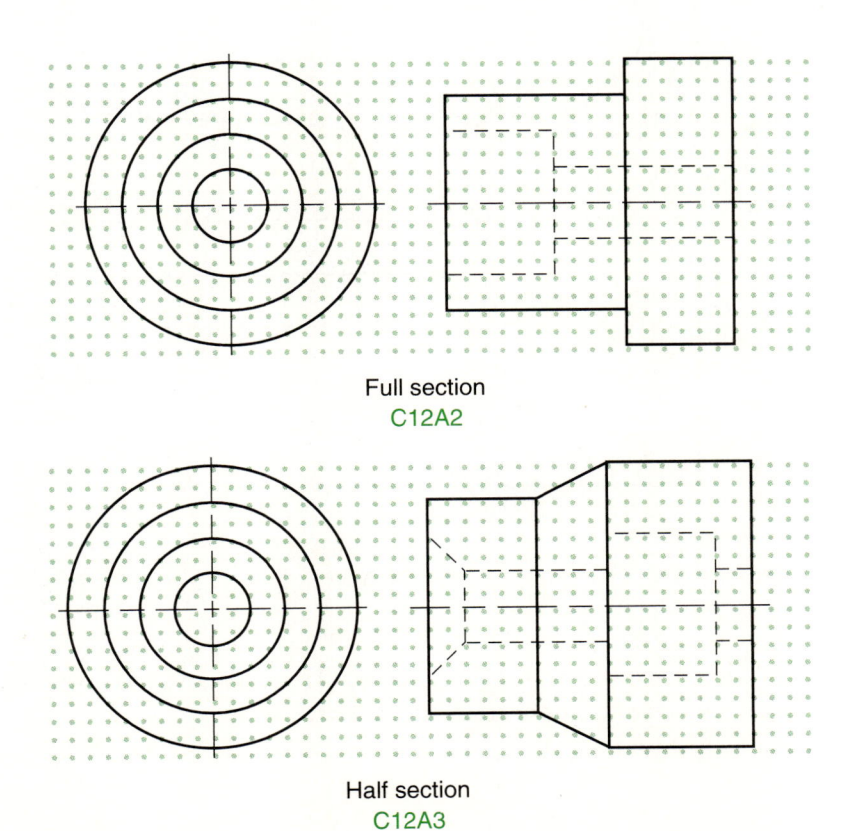

Full section
C12A2

Half section
C12A3

Offset section
C12A4

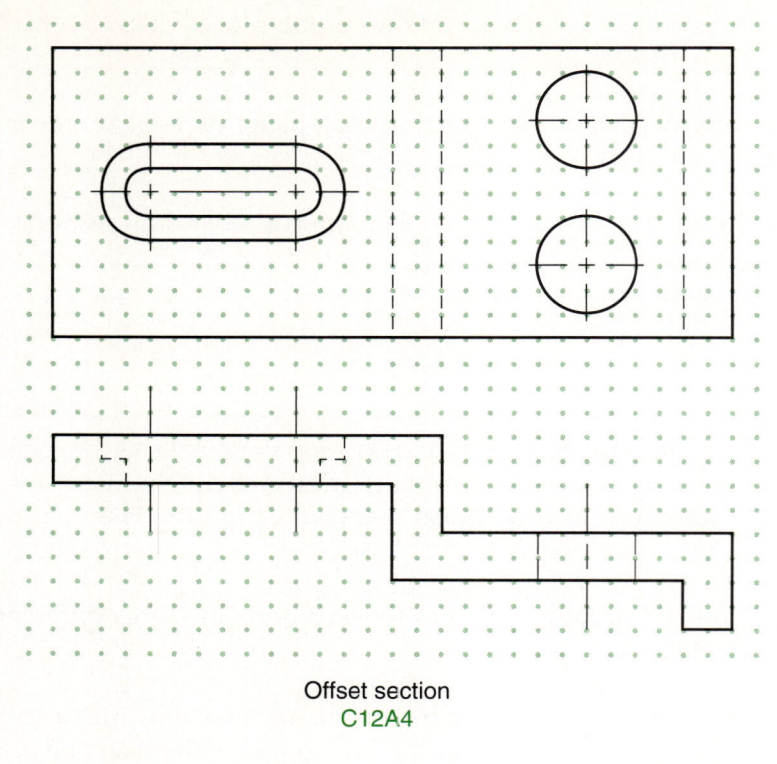

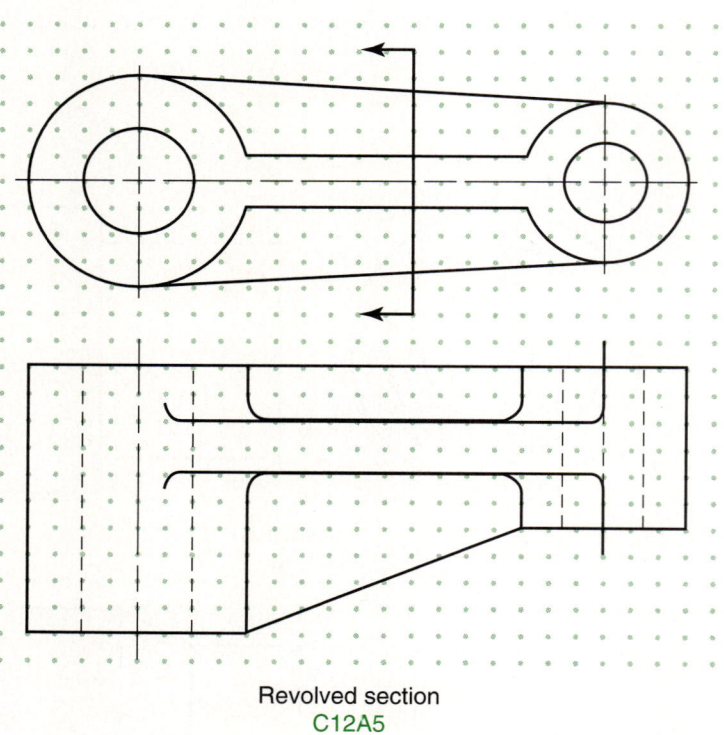

Revolved section
C12A5

6 - 10. For the objects in Activities 6 through 10 of Chapter 11, make an appropriate section view drawing. Save the drawings as C12A6 through C12A10.

CHAPTER 13

Dimensioning Drawings

Objectives

After studying this chapter, you will be able to:
- Define the difference between size and location dimensions.
- Identify the lines, symbols, and other elements used when placing dimensions.
- Follow recommended dimensioning standards.
- Select the unit of measurement and dimension settings for placing dimensions.
- Place horizontal, vertical, and aligned linear dimensions.
- Place angle dimensions.
- Place diameter and radius dimensions.
- Add leaders to a drawing.
- Explain the principle of associativity.
- Edit dimension text and dimension settings.
- Follow the general guidelines for dimensioning common shapes.

Two important aspects of product design are shape and size descriptions. The previous chapters discussed adding objects to create shapes. *Dimensions* are added to show the size and location of those shapes. Dimensions may consist of numbers, lines, symbols, and notes, Figure 13-1. Each drafting field dimensions drawings differently. For example, dimensioning for mechanical applications follows the American National Standards Institute (ANSI) document *Dimensioning and Tolerancing*. However, the basic techniques common to all fields are discussed here.

AutoSketch makes dimensioning a quick process. In traditional drafting, dimensions are drawn by hand and measurements are made using a scale. With AutoSketch, you only have to pick where the dimension begins and ends, and where to place it. AutoSketch adds the needed lines and arrows, and inserts the measurement.

Size and Location Dimensions

Generally, dimensions describe either the size or location of a feature, Figure 13-2. The term *feature* refers to any distinct part of a product. Features on a manufactured product can include a surface, edge, hole, or slot. Features on an architectural drawing can include walls, windows, doors, appliances, and cabinets. Even a complex product, such as a house, can be described by the size and location of the many simpler parts that it is made up of.

Size dimensions give the width, length, height, diameter, or radius of a feature. For example, size dimensions might show the thickness of a piece of metal or the width of a window. There can be many size dimensions on a drawing. Each one describes the measurements of a specific part of the product.

365

Figure 13-1.
Dimensions can consist of numbers, lines, symbols, and notes. (Autodesk, Inc.)

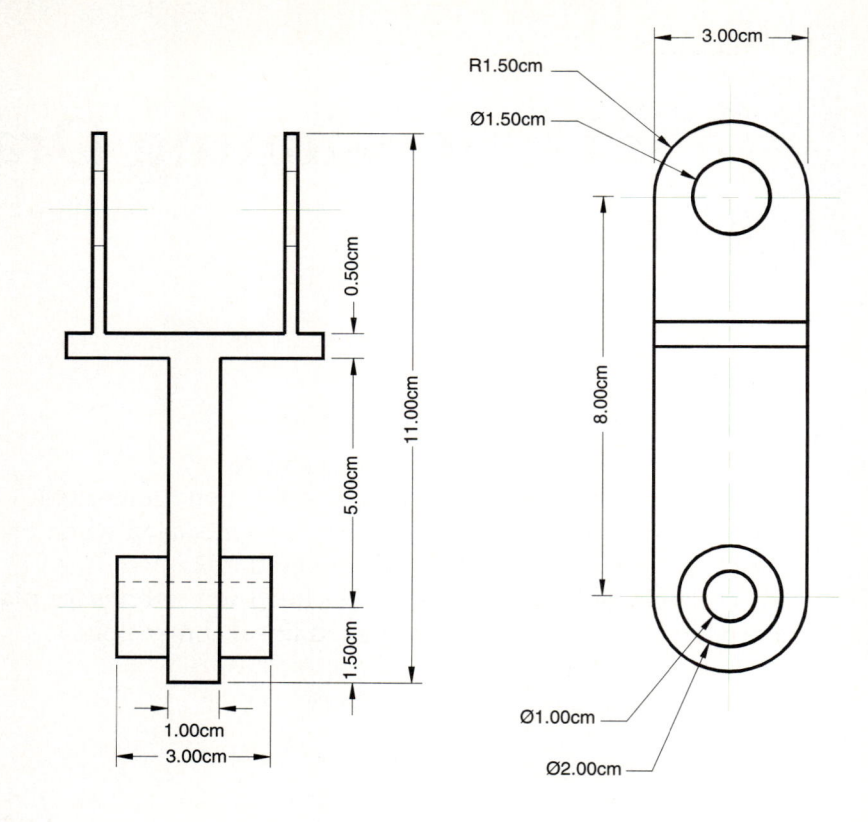

Figure 13-2.
In general, dimensions describe size and location.

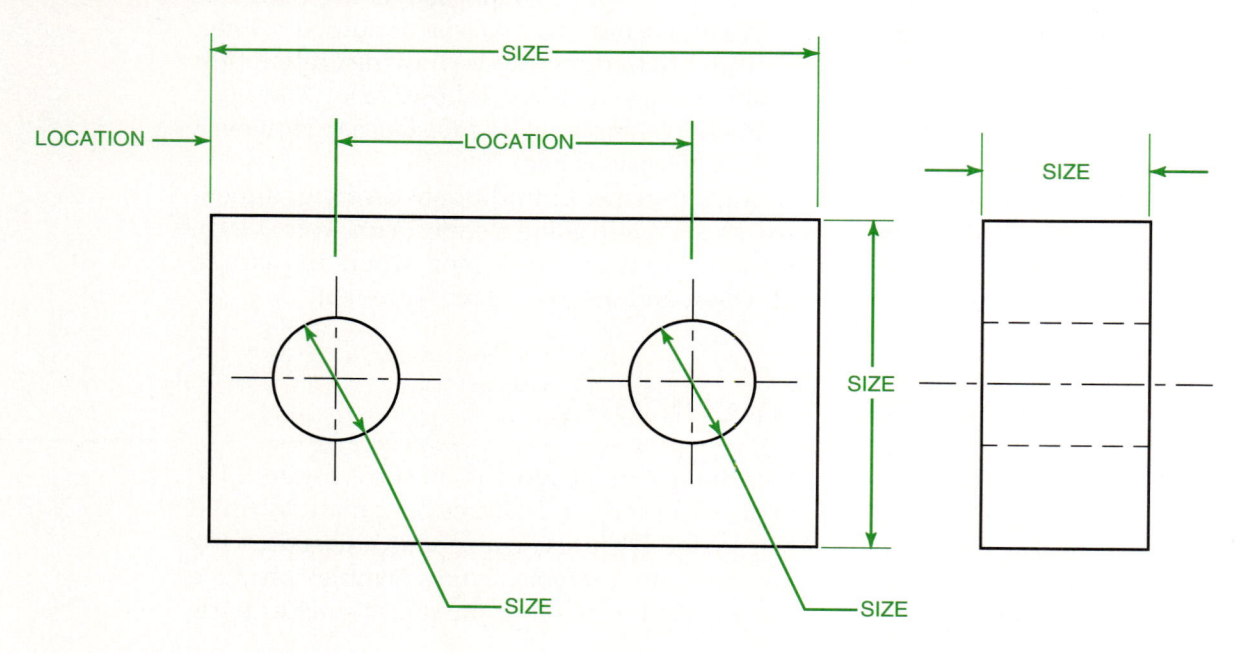

Location dimensions give the position of a feature or shape. The position is measured from an edge, center, surface, or another feature. For example, a location dimension might locate the center of a hole from a finished surface or from the center of another hole. In an architectural drawing, the center of a window might be measured from an exterior wall.

Elements in Dimensioning

A variety of lines, symbols, and notes are used when dimensioning a technical drawing, Figure 13-3. You should be familiar with the following terms.

Figure 13-3.
Many different lines, symbols, and notes can be used to dimension a technical drawing.

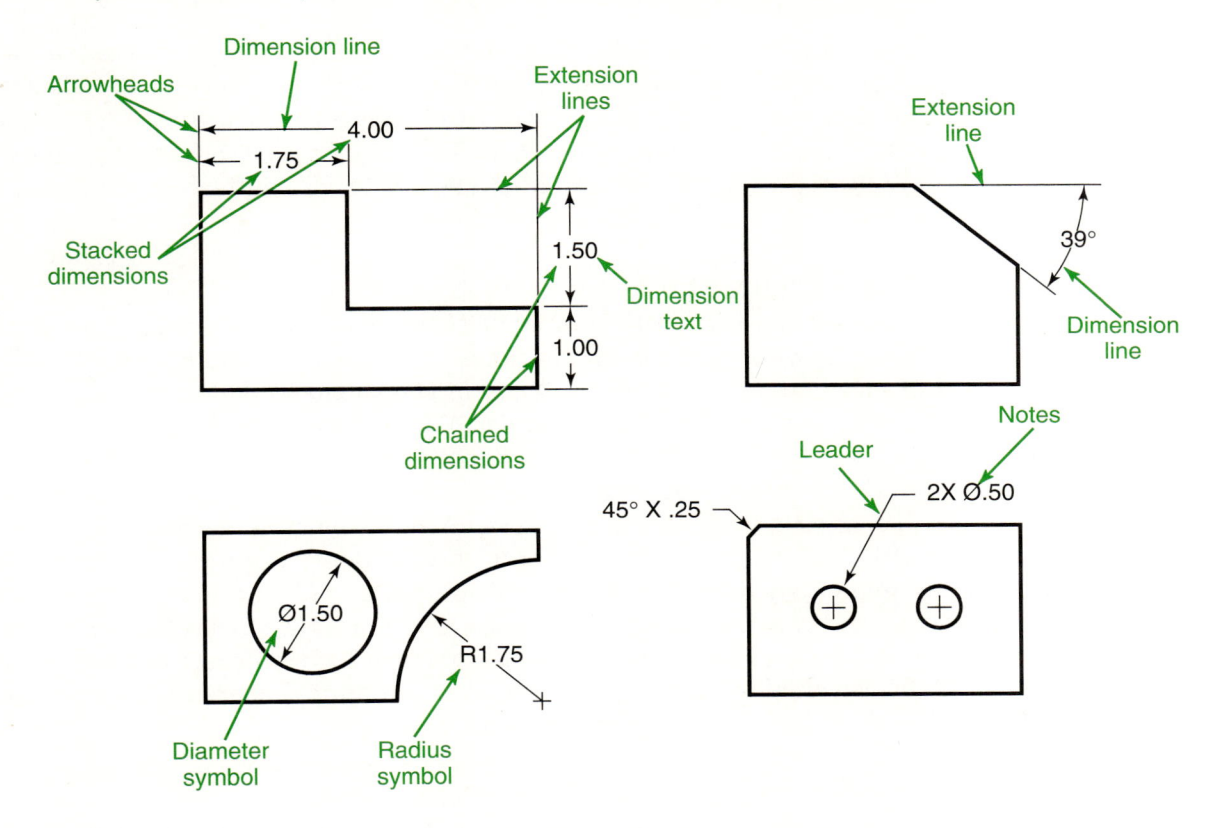

Extension Lines

Extension lines, also known as witness lines, mark the beginning and end of a dimension. The lines generally begin .06" away from the shape being dimensioned. This distance is called the *extension line offset*. Extension lines typically extend .12" beyond the dimension line. Extension lines are not used when the dimension is located within a shape, such as a circle.

Dimension Lines

Dimension lines show the direction and extent of a dimension. The line is straight for a linear dimension and curved for an angular dimension. The dimension line is typically broken near the middle for placement of dimension text. However, in architectural and structural drawings, the dimension text is often placed on top of the line.

A dimension line should be at least .375" to 1" away from the surface it measures. This depends on the space available and the complexity of the drawing. Stacked dimension lines should be spaced at least .375" to .5" apart. Where there is limited space, the dimension line and dimension text are placed outside the extension lines.

Arrows

In mechanical drafting, dimension lines end with *arrows*. In architectural drafting, dimension lines may end with arrows, tick marks, or dots. AutoSketch for Windows has five types of arrows. They are the standard arrow, solid arrow, tick, dot, and no arrow. If you create your own arrows, the recommended size when printed is .18" long and one-half as wide.

Dimension Text

Dimension text is the numerical value of the measurement plus any needed symbols or notes. AutoSketch calculates the distance picked and inserts the measurement automatically. AutoSketch draws dimension text using the height set in the **Dimension Settings** dialog box and the current font set for regular text. The dimension text height should be .125" to .18". Use larger text if the drawing is to be reduced in size when printed or plotted. Titles and subtitles should be .18" to .25" high when printed so that they stand out.

Unit of Measurement

The *unit of measurement* is the format of the numerical value used to place dimension text. Your drawing might be dimensioned in inches, feet, millimeters, or meters. The unit of measurement is set in the **Units of Measurement** dialog box. This can be accessed several ways. Refer back to Chapter 10 for a discussion on accessing this dialog box.

The standard unit of measurement on mechanical drawings is typically either decimal-inches or metric (stated in millimeters). A zero precedes metric dimensions for measurements less than 1, such as 0.03mm. A zero does *not* precede inch measurements less than 1, such as .03". Fractional dimensions may be used on mechanical drawings to indicate standard stock sizes, such as 1/2" COLD-ROLLED STEEL. Architectural drawings are dimensioned in fractional inches, or feet and inches written as 12'-0". When all dimensions are stated in one unit of measure, a general note should appear on the drawing indicating this.

Unidirectional and Aligned Dimension Text

Unidirectional and aligned systems refer to how dimension text is placed in relation to the dimension line, Figure 13-4. This is set in the **Dimension Settings** dialog box. Unidirectional and aligned dimensions should not both be used on the same drawing.

Unidirectional dimensions are placed horizontally. This means any dimension can be read from the bottom of the drawing. The manufacturing industry follows the unidirectional system for most mechanical drawings.

Aligned dimensions are placed parallel to the dimension lines. They read from the bottom or right side of the drawing. Most architectural drawings use aligned dimensions.

Position of Dimension Text

Dimension text on mechanical drawings is usually placed within a break in the dimension line. However, sometimes there is limited space for dimensions. In these cases, the dimension text is placed outside of the dimension line. Refer to Figure 13-4. The text placement is set in the **Dimension Settings** dialog box.

Architectural and structural drawings commonly have the dimension text placed above or below the dimension line, with no break in the dimension line. This works best with aligned dimensions.

Solid and broken dimension lines should not both be used on the same drawing.

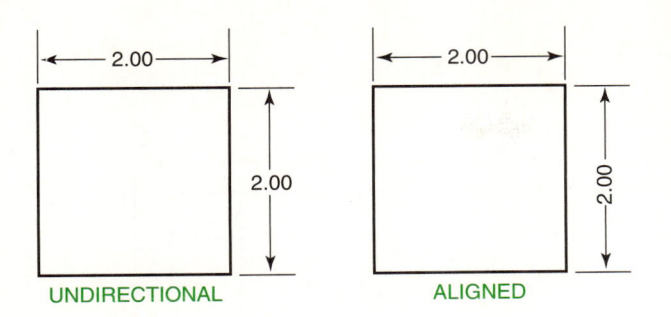

Figure 13-4.
Unidirectional dimensions are placed horizontally with a break in the dimension line for text. Aligned dimensions are placed parallel with the dimension line and text reading from the bottom or right.

Dual Dimensioning

Dual dimensioning uses both customary (inch/feet) and metric measurements. See Figure 13-5. The placement of the two dimensions varies. In the United States, the customary measurement appears above the metric. In other countries, the metric measurement appears above the customary. Most CAD systems do not support dual dimensions. However, you can place dual dimensions with AutoSketch. This must be done by manually adding the second measurement separated by a slash or brackets.

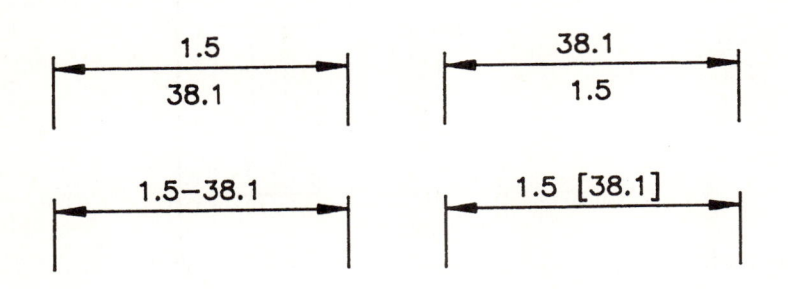

Figure 13-5.
Dual dimensions contain both conventional and metric measurements. There are several different ways that these dimensions can appear on a drawing.

Leader

A *leader* is a thin line leading from a note or dimension to where it applies. Refer to Figure 13-3. Leaders allow you to place dimensions away from the view. This way, the dimension does not interfere with the view. A leader consists of a short (.125") horizontal line and a second leader line that extends at a 15° to 75° angle to the place where it applies. The angle is usually either 45° or 60°. The leader usually ends with an arrow. However, a dot is used when the leader note applies to a surface.

Leaders placed close together should be parallel. However, leaders should *not* be placed parallel to extension or dimension lines. A leader that points to a circle or arc should be aligned with the center of that feature. Dimensioning circles and arcs with leaders is discussed later in this chapter.

Notes and Symbols

Notes and symbols appear frequently on technical drawings. Notes may be general or specific. *General notes* apply to the entire drawing. For example, a general note might read: HEAT TREAT AND TEMPER RC 44-48. General notes appear in the lower-left or upper-left corner of the drawing area, or next to the title block. Specific notes refer to specific features on a drawing. They are attached to the feature described or dimensioned by a leader line. See Figure 13-6. Notes are always placed horizontally, even if dimension text follows the aligned system.

Figure 13-6.
General notes appear next to the title block or a corner. Specific notes refer to the features indicated with a leader. (Autodesk, Inc.)

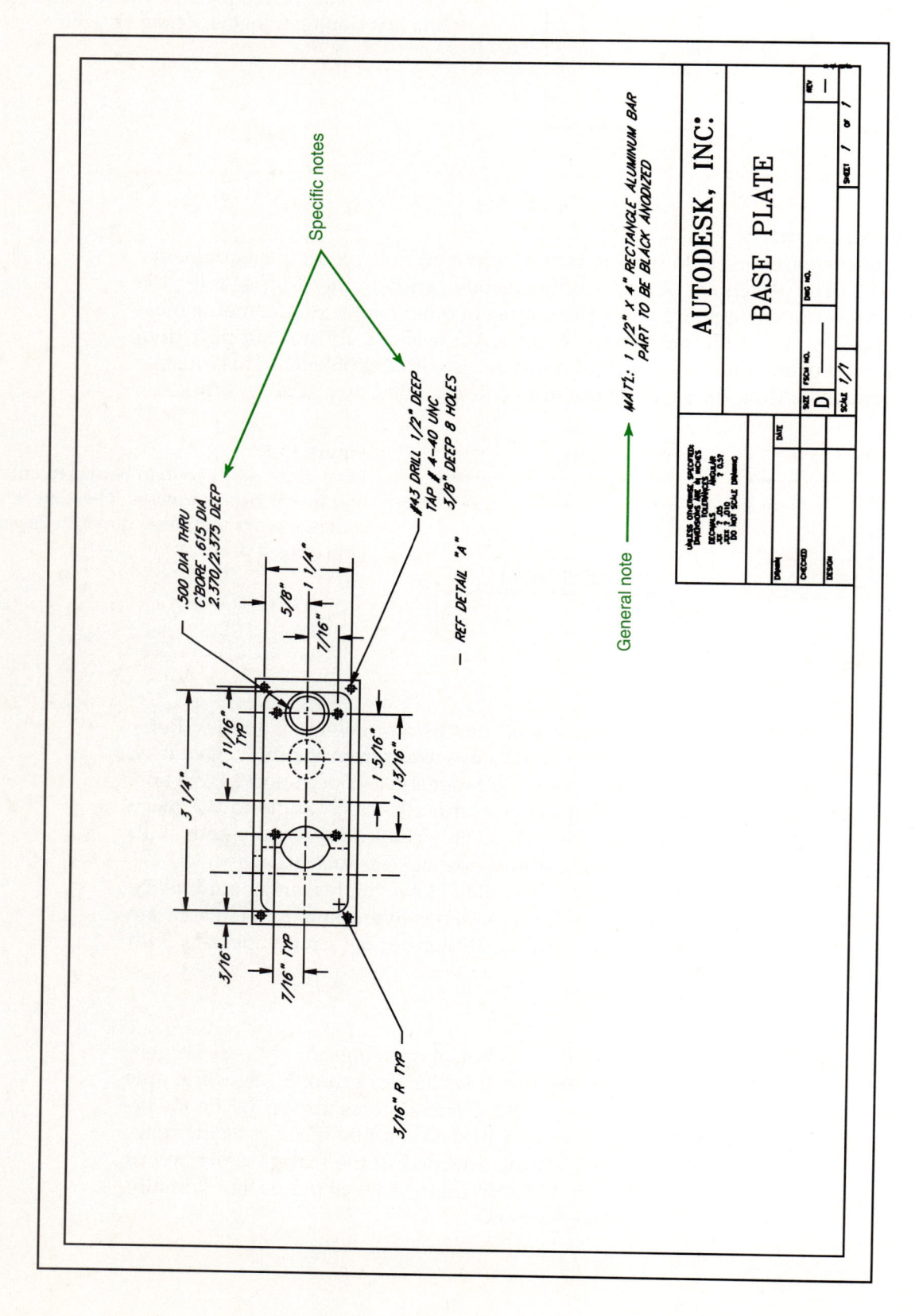

Symbols refer to certain standards. Some of the more common symbols that appear in mechanical drawings are shown in Figure 13-7. In addition, there is a specialized segment of dimensioning called *geometric dimensioning and tolerancing.* This advanced dimensioning technique uses many other symbols that describe the form, position, texture, or tolerance of features on a product.

Figure 13-7.
Several different symbols can be found on mechanical drawings.

MEANING	ASME Y14.5M–1994 SYMBOL
Dimension origin	 ⌖→
Conical taper	▷
Slope	◁
Counterbore/spotface	⌴
Countersink	⌵
Depth/deep	⤓
Square	□
Dimension not to scale	<u>15</u>
Number of times/places	8X
Arc length	⌒105
Radius	R
Spherical radius	SR
Spherical diameter	SØ
Diameter	Ø

Dimensioning with AutoSketch

AutoSketch supports most, but not all, of the dimensioning standards just discussed. For example, AutoSketch does not support automatic dual dimensioning. AutoSketch also does not support the addition of many special symbols. If your drawing requires strict adherence to dimension standards, you may have to place some dimensions manually.

Setting the Unit of Measurement

AutoSketch allows you to set the unit of measurement used to create and dimension your drawing. This task is typically done when you set up your drawing. However, you can change the unit of measurement in the middle of a drawing to place several dimensions with a different unit. For most drawings, it is not necessary to alter the unit of measurement when dimensioning the drawing.

Setting Dimension Settings

Once you have determined the unit of measurements, you should set your dimension settings. To change the dimension settings, access the **Dimension Settings** dialog box. See Figure 13-8. Access the dialog box in one of the following ways.
- Select the **Measure** menu and double-click on any dimension tool.
- Select the **Custom** menu, pick the **Drawing Settings** tool, and then pick the **Dimension** button.

Figure 13-8. Double-clicking on any dimension tool displays the **Dimension Settings** dialog box. Here you set the features for arrows, dimension text, and other features.

SHORTCUT

As you make changes in the **Dimension Settings** dialog box, the sample dimension changes to reflect your settings. This allows you to quickly see if your changes are best for the application.

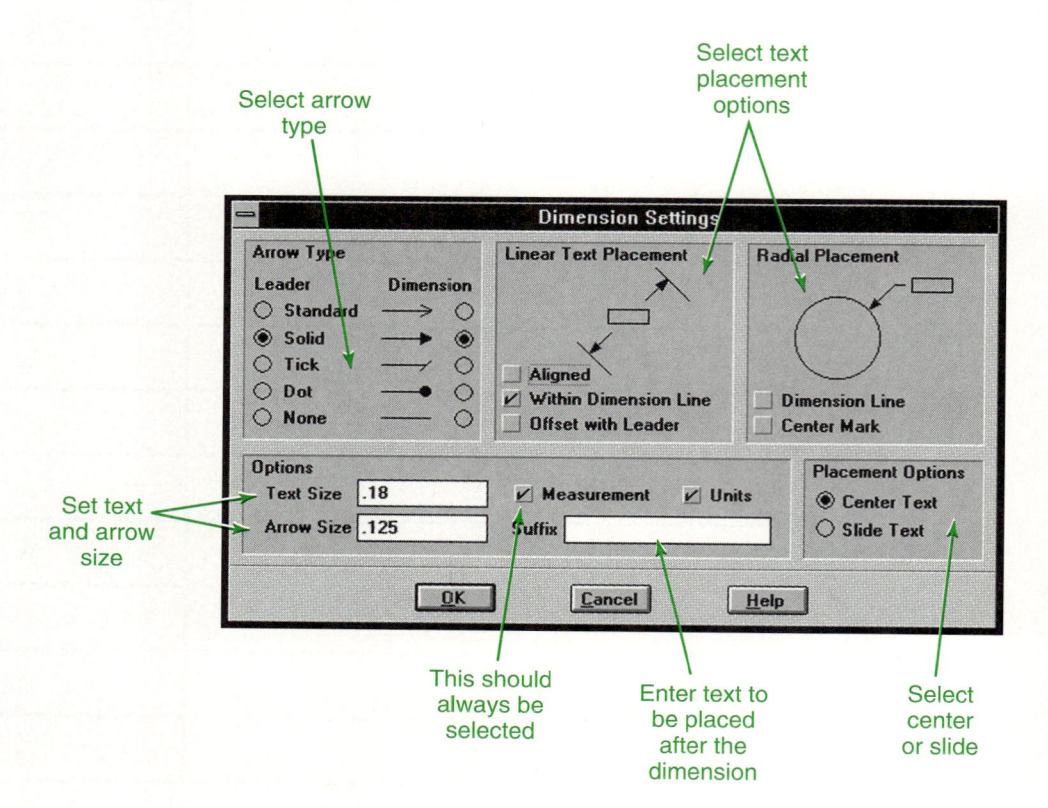

Arrow Type

The **Arrow Type** settings determine what style of arrow is used for dimensions and leaders. Simply pick the circle next to the settings you prefer. The size of the arrow is set with the **Arrow Size** option in the **Options** section. When you change the text size, AutoSketch automatically adjusts the arrow size to 80 percent of the text size. However, after changing the text size you can also change the arrow size.

Linear Text Placement

In this section you set the dimension text as aligned or unidirectional. You also set if the dimension text is placed within the dimension line or above it. When **Aligned** is on, dimension text is aligned with the dimension line. When off, text is placed horizontal.

When **Within Dimension Line** is on, dimension text is placed within a break in the dimension line. When off, dimension text is placed outside the dimension line.

When **Offset with Leader** is on, AutoSketch places linear dimensions with a leader. Simply pick the box next to the settings you prefer.

Radial Placement

In this section you set whether or not dimension lines and center marks are inserted for radial measurements (dimensions for curved objects). When **Center Marks** is on, a small crosshairs is placed at the center point of diameter and radius dimensions. The mark size is twice the arrow size. When **Dimension Line** is on, a dimension line will appear in the circle or arc you dimension with a leader.

Options

In the **Options** section, you set the size of the arrow and dimension text. **Text Size** is the height of the dimension text. This setting does *not* change other text placed on the drawing. **Arrow Size** is the length of the arrow. This is initially set to 80 percent of text size, but you can change it. **Measurement** determines whether or not the dimension text appears. This should *always* be turned on. **Units** sets if the unit of measurement symbol (such as " or mm) is added to the dimension text. Most often, this option is turned on for architectural drawings and off for mechanical drawings. **Suffix** allows you to add text behind the dimension, such as 0.75 SQUARE. You can also add special characters here, such as the tolerance and degree symbols.

Note: Once you add a suffix, it will be added to every dimension until you remove it from the **Dimension Settings** dialog box.

Placement Options

Here you control if dimension text is centered automatically or if can be moved as it is created. Choosing **Center Text** always centers the text within the dimension line. Choosing **Slide Text** allows you to move the dimension text within the dimension line. This gives the most flexibility in placing dimension text. Pick the box next to the setting you prefer.

Exercise 13-1

1. Start a new drawing using the **Normal** template.
2. Double-click on any dimensioning tool in the **Measure** menu to open the **Dimension Settings** dialog box.
3. Select each of the arrow types for leaders and dimensions and see the changes in the samples shown in the dialog box.
4. Select each of the options for **Linear Text Placement** and see the changes in the samples shown in the dialog box. List the settings that can be on together or separately. List the settings that override the others.
5. Select each of the options for **Radial Placement** and see the changes in the samples shown in the dialog box. Determine whether both settings can be on together, in addition to individually.
6. Close the drawing without saving it.

Exercise 13-2

For each of your drawing templates completed in Chapter 10, edit the dimension settings as shown below. Open each template using the **New...** command. When finished editing the template, select the **Save As...** command and save the modified drawing as a new template.

Settings	Template Name			
	A_LI_MEC	B_LI_MEC	A_LI_ARC	B_LI_ARC
Arrow	Solid	Solid	Dot	Dot
Linear Text Placement	Within Dimension Line, Not Aligned	Within Dimension Line, Not Aligned	Aligned, Not Within Dimension Line	Aligned, Not Within Dimension Line
Radial Placement	No Dimension Line No Center Mark	No Dimension Line No Center Mark	No Dimension Line No Center Mark	No Center Mark No Dimension Line
Options				
Text Size	.18	.18	6"	6"
Arrow Size	.125	.125	4"	4"
Measurement	yes	yes	yes	yes
Units	no	no	yes	yes
Placement Options	Slide Text	Slide Text	Slide Text	Slide Text

Dimensioning Linear Distance

Linear dimensions measure straight distances. The dimension can be placed vertically, horizontally, or aligned (parallel) to an object at an angle. Linear dimensions that measure size might show the height, width, or length of a product. Linear dimensions that measure location might show the position of the center of a hole measured from a finished edge.

The general method for linear dimensioning is to select the **Measure** menu and pick a dimensioning tool. Next, pick the two points to measure. Then pick the position of the dimension line and placement of the dimension text.

Note: AutoSketch often cannot accurately dimension objects created with a wide line width. This is because AutoSketch calculates distance to either the inner or outer edge of the wide line. It is best to create and dimension drawings using the hairline line width. This will make sure your drawing is accurately dimensioned. You can then later change the line width using the **Change Properties** tool.

Placing Horizontal Linear Dimensions

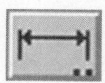

Horizontal dimensions measure a horizontal linear distance between two points. To place a horizontal dimension, select **Measure** and then the **Horizontal Dimension** tool. The following prompt appears. Refer to Figure 13-9.

Horizontal Dimension From point:

Select the first point. Use **Attach** tools to accurately select the point.

Horizontal Dimension To point:

Select the second point of the distance to measure. AutoSketch then asks you to select a location for the dimension line.

Horizontal Dimension Dimension line location:

Once you pick the dimension line location, the dimension is placed with the proper measurement. If you have the **Slide Text** option on, AutoSketch asks you for the location of the dimension text with the following prompt.

Horizontal Dimension Text location:

AutoSketch then draws in the extension lines, dimension lines, and arrows, and places the measurement.

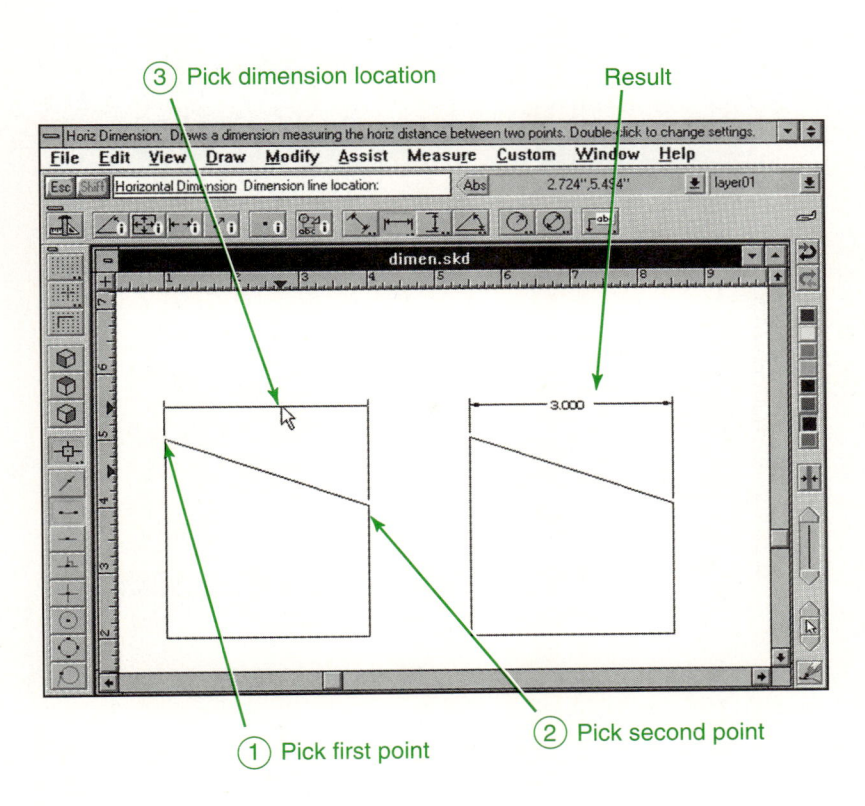

Figure 13-9.
To place a horizontal dimension, pick two dimension points and the location of the dimension line.

Placing Vertical Linear Dimensions

Vertical dimensions measure vertical linear distance between two points. To place a vertical dimension, select the **Measure** menu and then the **Vertical Dimension** tool. The following prompt appears. Refer to Figure 13-10.

Vertical Dimension From point:

Select the first point. Use **Attach** tools to help locate this point precisely.

Vertical Dimension To point:

Select the second point of the distance to be measured. AutoSketch then asks for the location of the dimension line.

<u>Vertical Dimension</u> Dimension line location:

Make sure the line is far enough out so that the measurement text does not overlap the object. Once you pick the dimension line location, the dimension is placed with the proper measurement. If you have the **Slide Text** option on, AutoSketch asks you for the location of the dimension text with the following prompt.

<u>Vertical Dimension</u> Text location:

AutoSketch then draws in the extension lines, dimension lines, and arrows, and places the measurement.

Figure 13-10.
To place a vertical dimension, pick two dimension points and the location of the dimension line.

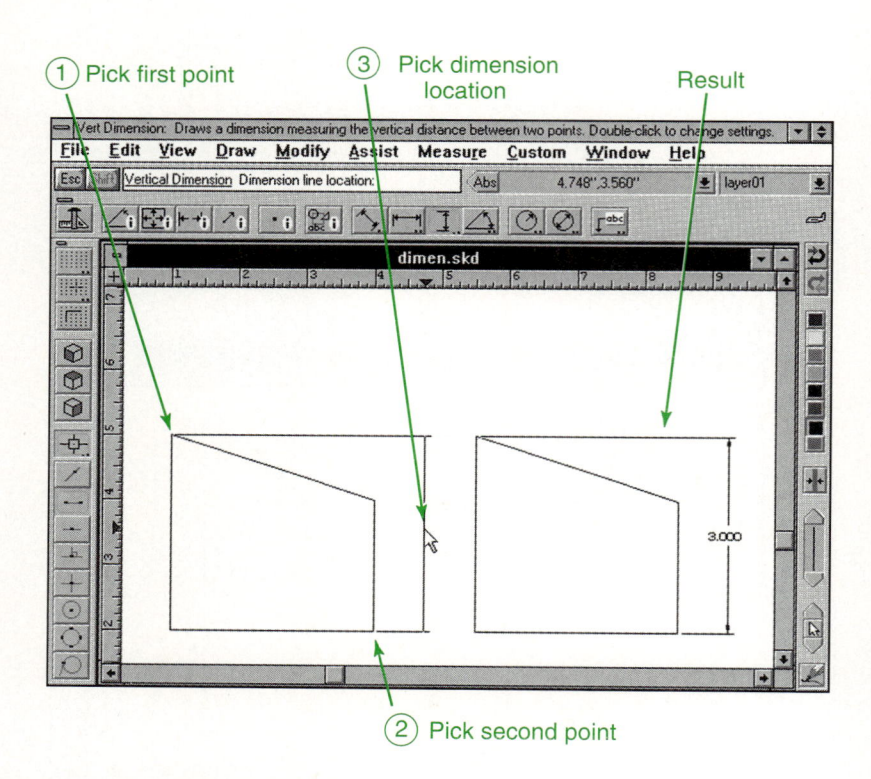

① Pick first point ③ Pick dimension location Result
② Pick second point

Placing Aligned Linear Dimensions

Aligned dimensions measure the true length of a surface drawn at an angle. The dimension line is parallel to the angled surface. To place an aligned dimension, select the **Measu<u>r</u>e** menu and then the **Aligned Dimension** tool. The following prompt appears. Refer to Figure 13-11.

<u>Aligned Dimension</u> From point:

Select the first point. Use **Attach** tools to locate the point precisely.

<u>Aligned Dimension</u> To point:

Select the second point. AutoSketch then asks you for the location of the dimension line.

<u>Aligned Dimension</u> Dimension line location:

Make sure the line is far enough out so that the measurement text does not overlap the object. Once you pick the dimension line location, the proper dimension is placed. If the **Slide Text** option is on, AutoSketch then asks you for the location of the dimension text.

<u>Aligned Dimension</u> Text location:

AutoSketch then draws in the extension lines, dimension lines, and arrows, and places the measurement at the proper angle.

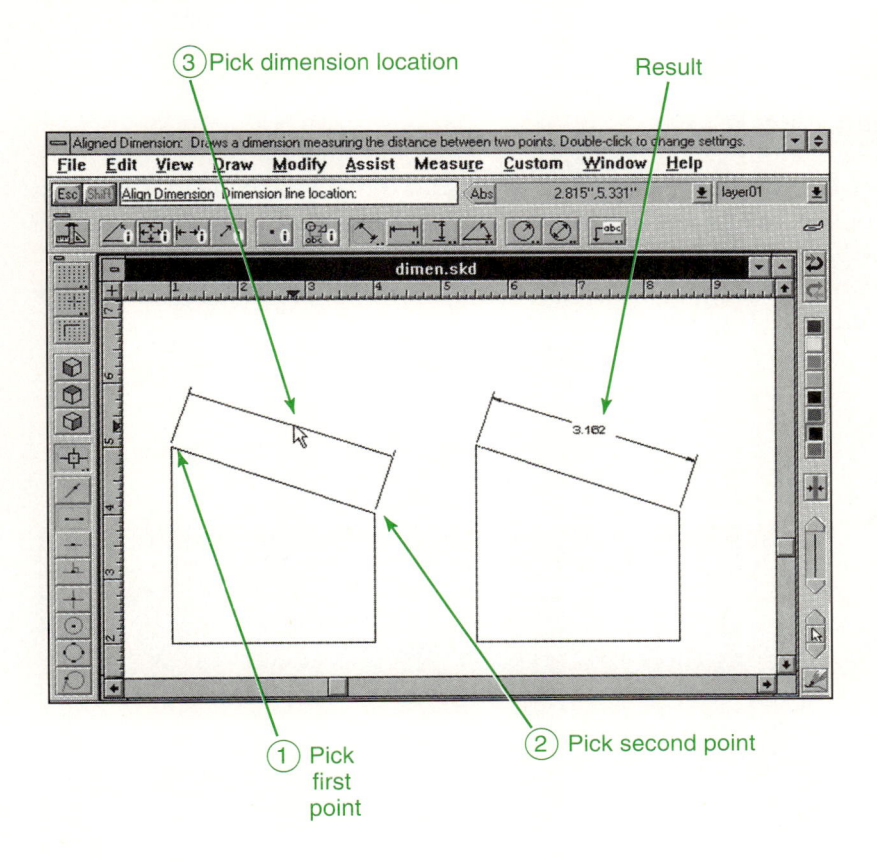

Figure 13-11.
To place an aligned dimension, pick two dimension points and the location of the dimension line.

Chained and Stacked Linear Dimensions

There are two basic methods to place linear dimensions that measure multiple features. These two methods are chained and stacked. The combination you choose depends on the intended accuracy of the drawing, and the drafting field.

Chained dimensions, or point-to-point dimensions, continue a linear dimension from the second extension line of the previous dimension. This breaks one long dimension into shorter segments that add up to the total distance. Chained dimensions are common in architectural drafting. They can also be found in mechanical designs requiring less precision. Chained dimensions are less accurate since each dimension depends on others in the chain.

Stacked dimensions, also called datum dimensions, continue linear dimensions from a common edge or surface. This edge or surface is called a ***datum.*** Stacked dimensions are often found when very precise locations and sizes are required. Dimension lines of stacked dimensions are spaced .375" to .5" apart.

Linear Dimensions with Leaders

When **Offset with Leader** is selected in the **Dimension Settings** dialog box, the linear dimensioning commands work a bit differently, Figure 13-12. First you must pick the two points to measure and the location of the dimension line. This is the same as dimensioning without a leader. However, after you pick the location of the dimension line, the following prompt appears (for a horizontal dimension).

Horizontal Dimension To point:

Move your cursor away from the dimension line to create a leader line. Pick when the leader is placed at an approximate 45° angle. AutoSketch then asks for the location of the dimension text.

Horizontal Dimension Text location:

Move the cursor to drag the dimension text so that the tail of the leader is about .125" long. The dimension text is placed at the point you select. An arrow is also placed on the leader. You will notice when you select the text location that you can only draw a horizontal line, even if **Ortho** is turned off. This is because all leader notes should be placed so that they are read horizontally.

Figure 13-12.
With the **Offset with Leader** option selected, you must pick points for the leader and text location.

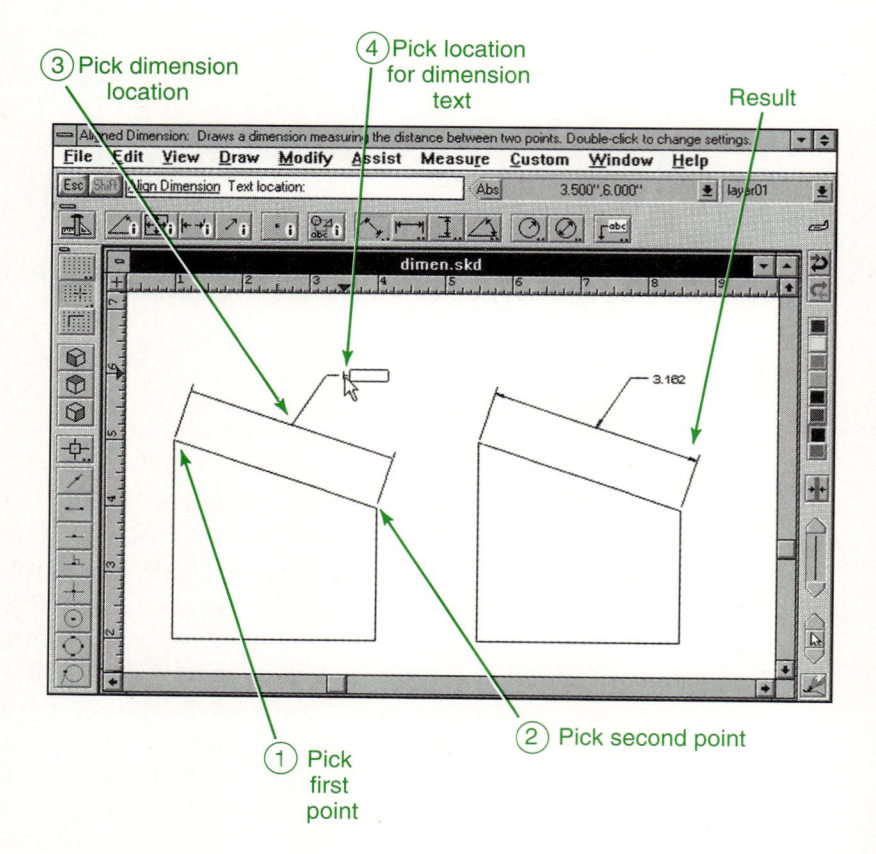

Exercise 13-3

1. Start a new drawing using the **A_LI_MEC** template.
2. Develop and dimension the following drawing as shown.
3. Save the drawing as C13E3 on your work diskette and then close the drawing.

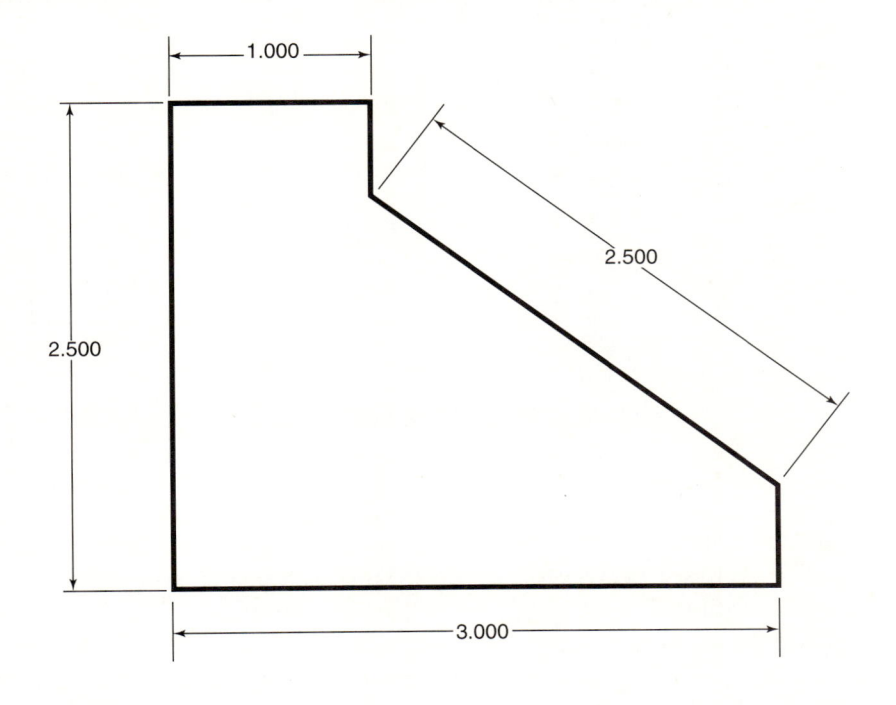

Dimensioning Angles

Angle dimensions measure the angle formed by nonparallel surfaces. In the past, angles were measured in degrees, minutes, and seconds (such as 45°40′33″). However, more industries now use decimal degrees (such as 45.676°). This is how AutoSketch measures angles. The precision (how many decimal places) is set in the **Units of Measurement** dialog box. In addition, AutoSketch adds the degree symbol (°) behind angular dimensions.

To place an angle dimension, select the **Measure** menu and the **Angle Dimension** tool. The following prompt appears. Refer to Figure 13-13.

Angle Dimension Select first line:

Pick the first of the two lines that form the angle. AutoSketch then asks for the second of the two lines.

Angle Dimension Select second line:
Select the other leg of the angle. AutoSketch then asks for the location of the dimension arc.

Angle Dimension Dimension line arc location:

Figure 13-13.
To place an angle dimension, pick the two lines forming the angle and the location of the dimension line.

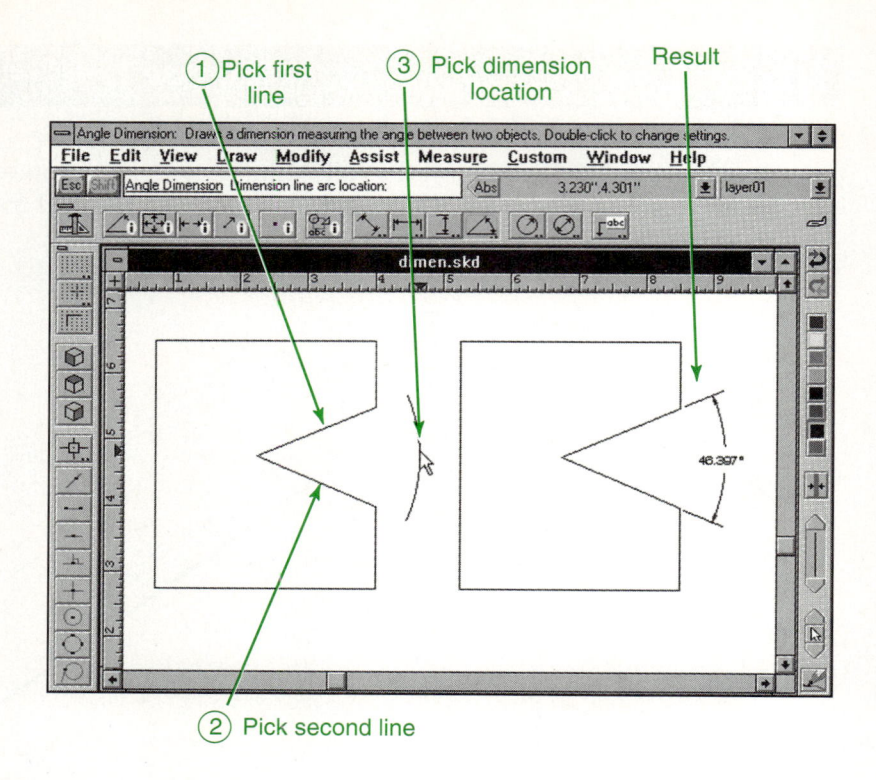

AutoSketch needs to know where to place the dimension line and measurement. If you move the cursor between the lines, AutoSketch dimensions the interior angle and places the dimension line and text within the two lines. If you move the cursor outside of the lines, AutoSketch dimensions the exterior angle formed between the two lines, Figure 13-14. If you dimension lines that form a small angle and the text will not fit between the lines, a dialog box appears asking you to reposition the text at another location. See Figure 13-15.

Figure 13-14.
Picking the location of the dimension line outside of the two lines causes AutoSketch to dimension the exterior angle formed by the two lines.

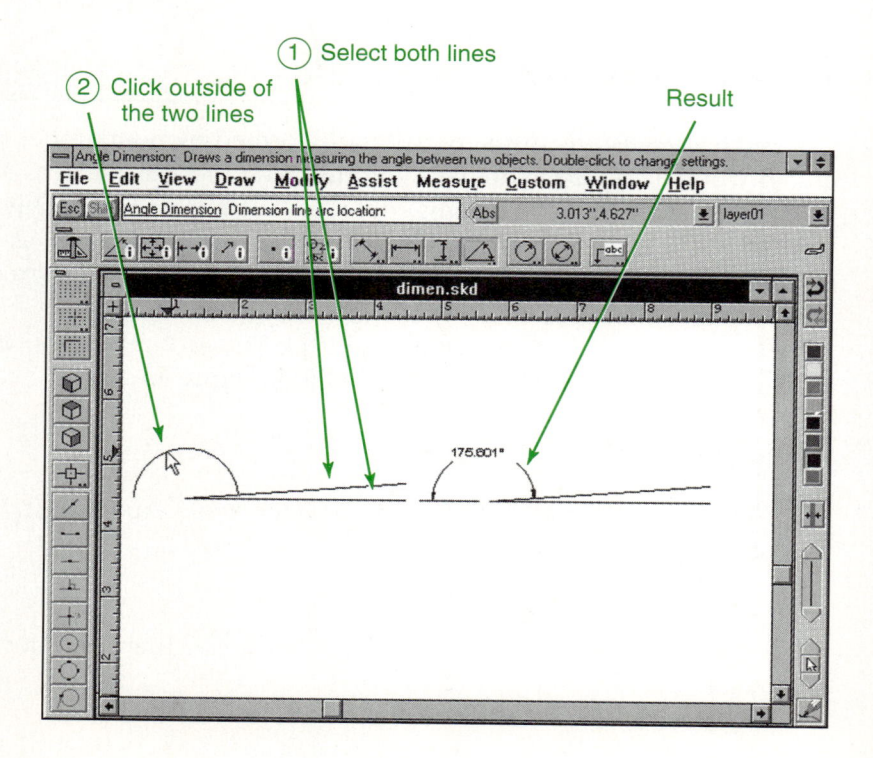

Exercise 13-4

1. Start a new drawing using the **A_LI_ARC** template.
2. Develop and dimension the following drawing as shown.
3. Save the drawing as C13E4 on your work diskette and then close the drawing.

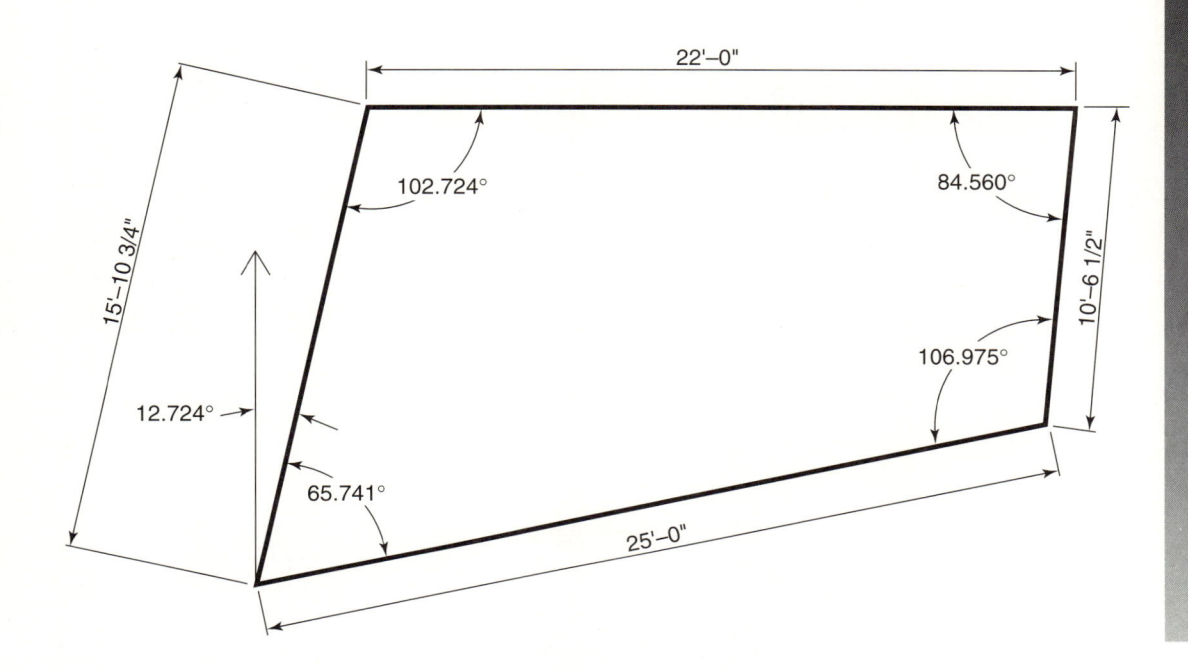

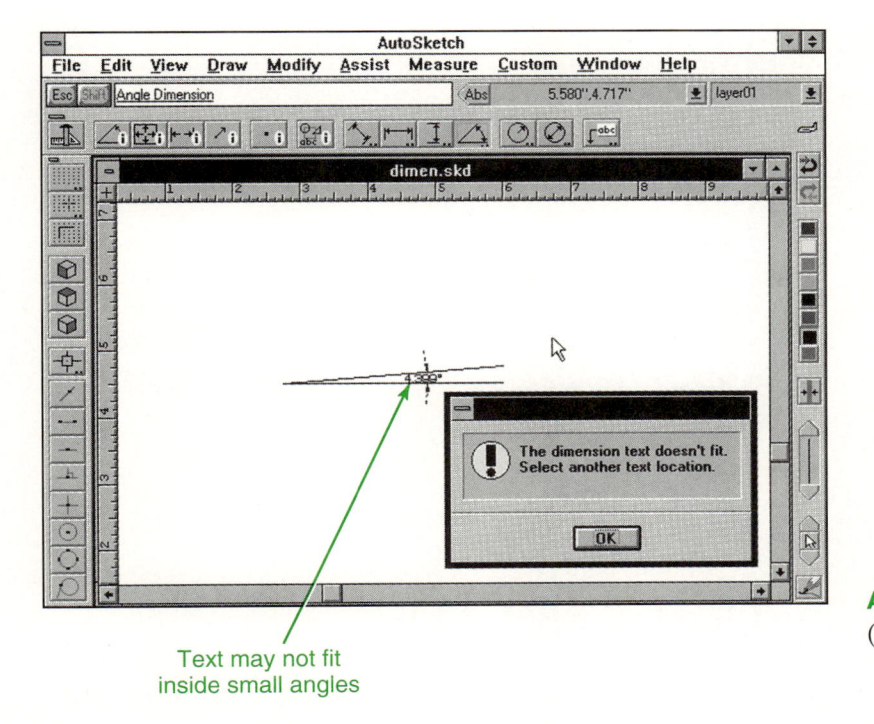

Figure 13-15.
If the dimension text won't fit between the two lines, AutoSketch lets you place it elsewhere.

Text may not fit inside small angles

A
(continued)

Figure 13-15.
(continued)

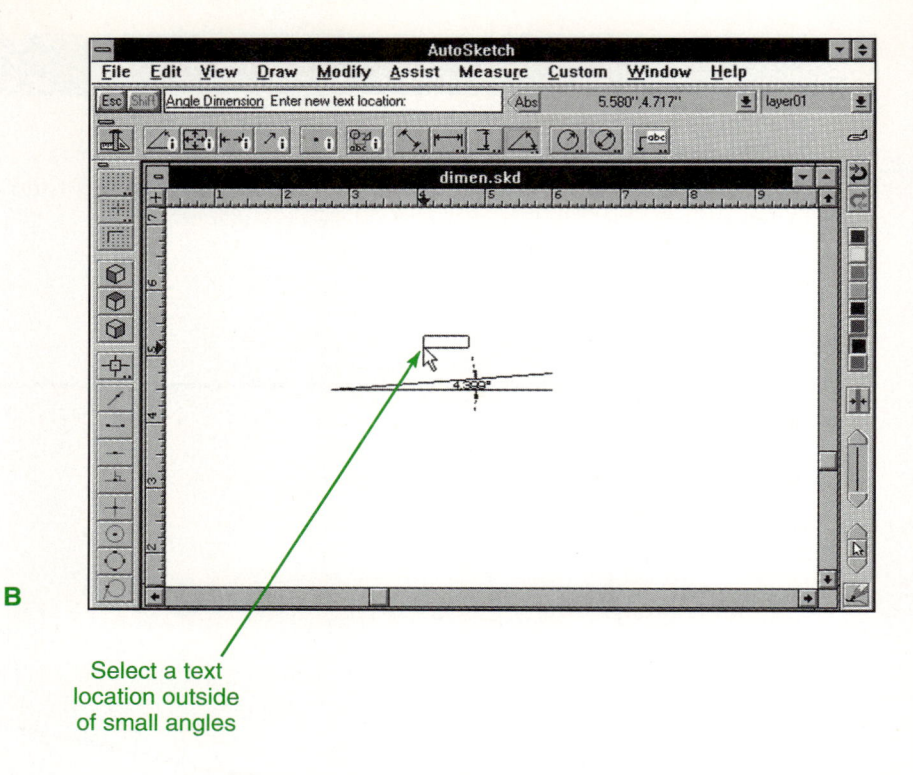

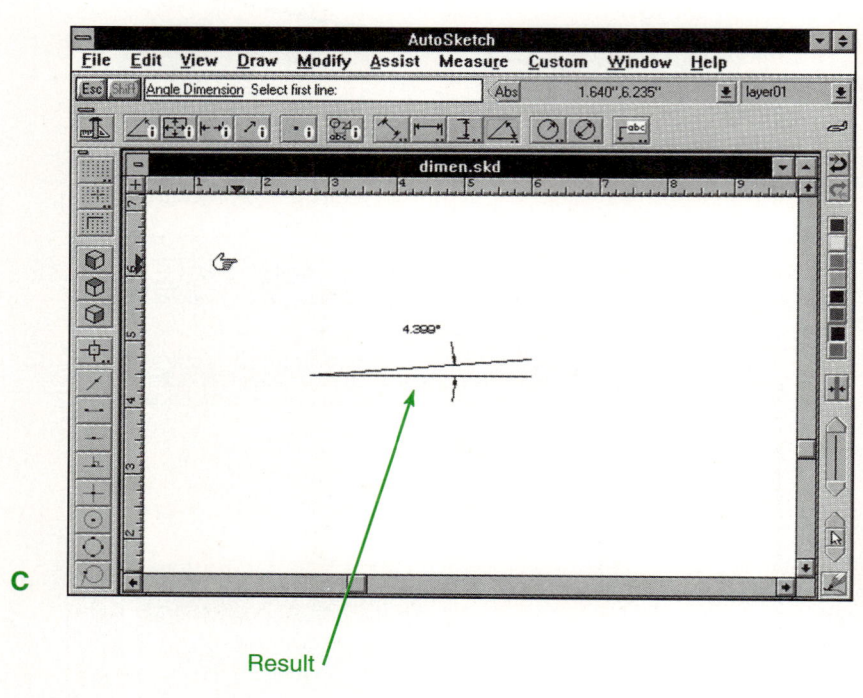

B

Select a text location outside of small angles

C

Result

Dimensioning Circles and Arcs

Circles and arcs are dimensioned using diameter and radius dimensions. Remember to set the radial placement settings **Center Marks** and **Dimension Lines** in the **Dimension Settings** dialog box before dimensioning circles and arcs.

Placing Diameter Dimensions

Diameter dimensions indicate the size of circles, cylinders, holes, and other circular shapes. AutoSketch also allows you to place diameter dimensions for arcs. Diameter dimensioning can be done with or without a leader. Without a leader, the measurement is placed within the circle or arc. Do this for larger circles. With a leader, the diameter is placed outside and the leader points toward the circle or arc's center. Leaders are used for smaller circles and arcs where the measurement will not fit within them. Leaders are also used when the measurement will clutter the view.

Dimensioning within circles

To place a diameter dimension within a circle, select the **Measure** menu and then the **Diameter Dimension** tool. The following prompt appears. Refer to Figure 13-16.

Diameter Dimension Select circle or arc:

Pick the circle to be dimensioned. AutoSketch then asks for the location of the dimension line.

Diameter Dimension Dimension line location:

Move your cursor inside the circle. The dimension line will move with the cursor. Pick when the dimension line is at an approximate 45° angle. (Be sure to pick *inside* the circle.) The dimension text is placed centered within the dimension line. If you have **Slide Text** turned on, drag the text where you want it placed and pick. AutoSketch places the diameter symbol before the dimension text.

Do not use the **Center Text** and **Center Marks** options together when placing a diameter dimension within a circle. They will overlap.

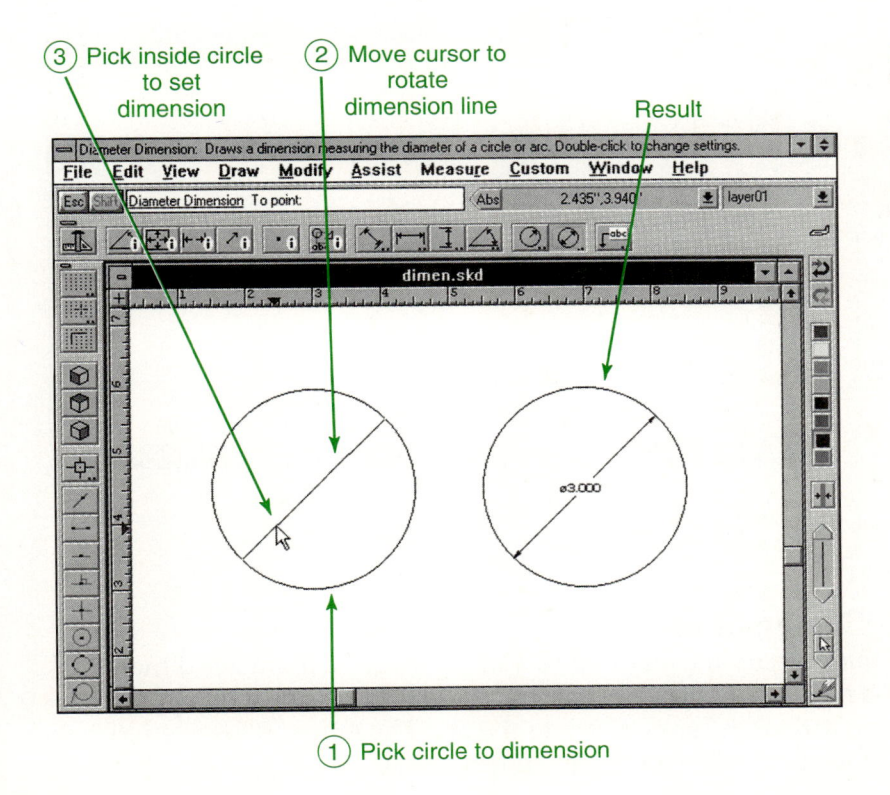

Figure 13-16.
To place a diameter dimension, pick the circle and then position your cursor so that the dimension line is at the proper angle. Pick inside the circle to place the dimension text.

Dimensioning a diameter with a leader

To place a diameter dimension with a leader, select the **Measure** menu and the **Diameter Dimension** tool. The following prompt appears. Refer to Figure 13-17.

<u>Diameter Dimension</u> Select circle or arc:

Pick the circle to dimension. AutoSketch then asks for the location of the dimension line.

<u>Diameter Dimension</u> Dimension line location:

Move your cursor outside the circle and pick the angle of the leader line. Pick when the leader is placed at an approximate 45° angle and the leader is long enough. AutoSketch then asks for the location of the dimension text.

<u>Diameter Dimension</u> Text location:

Move the cursor so that the tail of the leader is about .125" long. Pick at that point to place the text. The leader is drawn with the current arrow type. You can also place a dimension line and center mark within the circle when placing a leader diameter dimension. Set these options in the **Dimension Settings** dialog box.

Figure 13-17.
To place a diameter dimension using a leader, pick the circle or arc, position the cursor outside the object, and pick the length of the leader and position of the text.

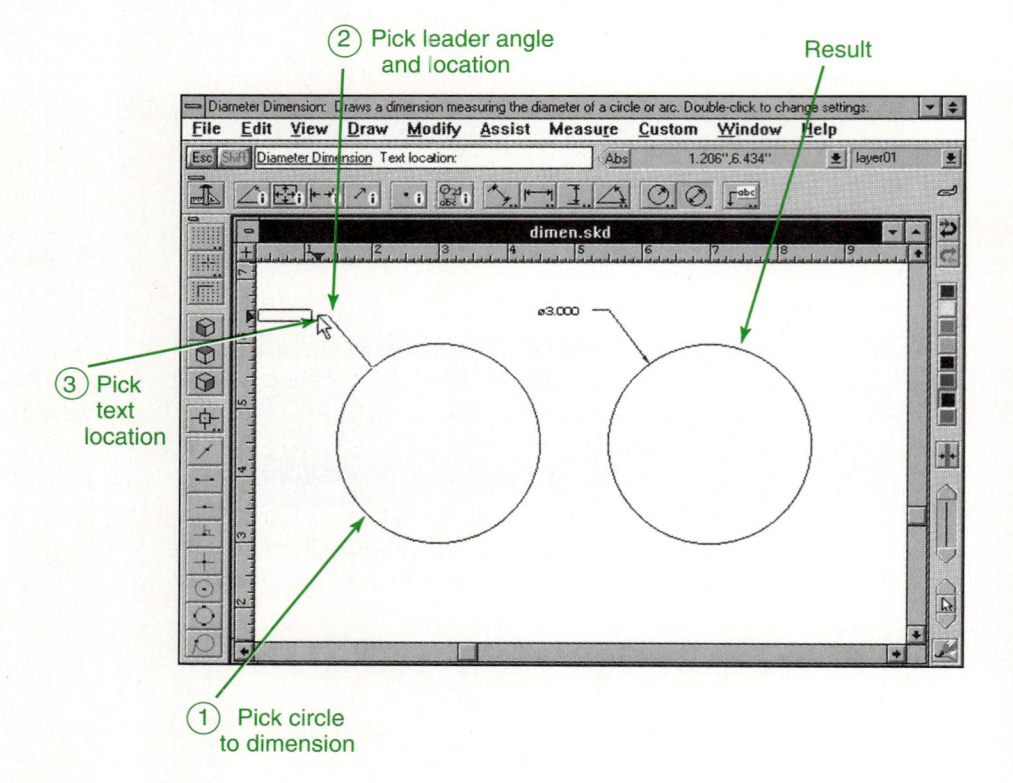

Placing Radius Dimensions

Radius dimensions specify the size of arcs and circles. Like diameter dimensions, the radius dimension measurement can be placed inside the arc. However, most radius dimensioning is done with a leader pointing to the arc. The letter "R" is used to indicate radius. The letter comes before all radius dimensions.

Dimensioning within arcs

To place a radius dimension within an arc, select the **Measure** menu and the **Radius Dimension** tool. The following prompt appears. Refer to Figure 13-18.

    Radius Dimension Select circle or arc:

Pick the circle to dimension. AutoSketch then asks for the location of the dimension line.

    Radius Dimension Dimension line location:

Move your cursor *inside* the arc. The dimension line will move with the cursor. Pick when the dimension line is placed at an approximate 45° angle. The dimension text is placed centered within the dimension line. If **Slide Text** is turned on, drag the text where you want it placed and pick.

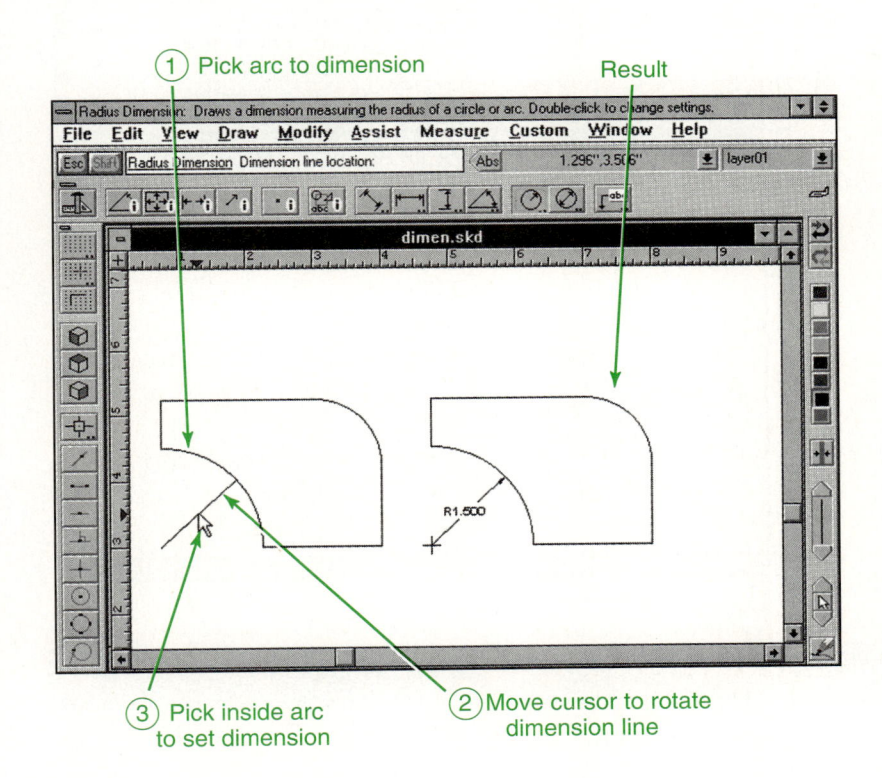

Figure 13-18.
To place a radius dimension, pick the circle or arc, and then position your cursor so that the dimension line is at the proper angle. Here, the **Center Marks** option was selected in the **Dimension Settings** dialog box.

Dimensioning an arc with a leader

To place a radius dimension with a leader, select the **Measure** menu and the **Radius Dimension** tool. The following prompt appears. Refer to Figure 13-19.

    Radius Dimension Select circle or arc:

Pick the arc to dimension. AutoSketch then asks for the location of the dimension line.

    Radius Dimension Dimension line location:

Move your cursor *outside* the arc and pick the angle of the leader line. Pick when the leader is placed at an approximate 45° angle. AutoSketch then needs to know where to place the dimension text.

Radius Dimension Text location:

Move the cursor until the tail of the leader is about .125" long. The dimension text is placed where you pick. The leader is drawn with the current arrow type.

You can also place a dimension line and center mark within the arc when placing a radius dimension with a leader. Set these options in the **Dimension Settings** dialog box.

Figure 13-19.
To place a radius dimension using a leader, pick the circle or arc, position the cursor outside the object, and pick the length of the leader and position of the text.

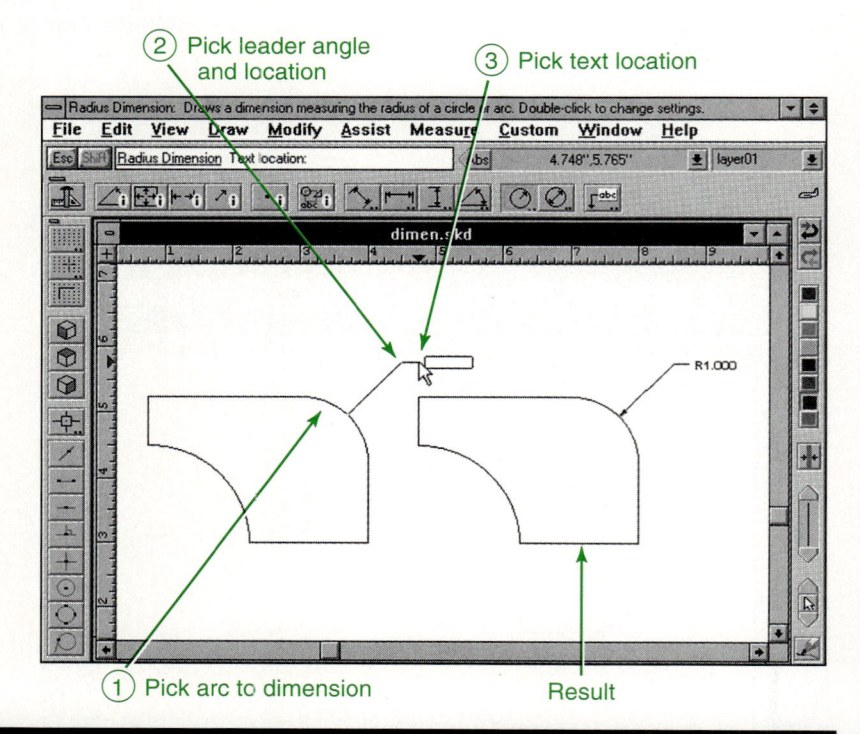

② Pick leader angle and location
③ Pick text location
① Pick arc to dimension
Result

Exercise 13-5

1. Start a new drawing using the **A_LI_MEC** template.
2. Develop and dimension the following drawing as shown.
3. Save the drawing as **C13E5** on your work diskette and then close the drawing.

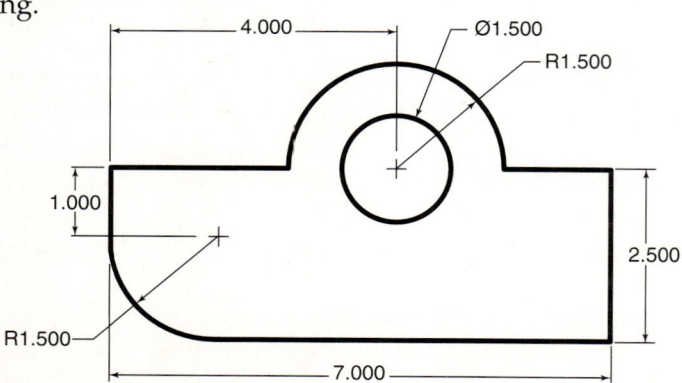

Leaders

Leaders are used to point out dimensions and notes. A leader consists of an optional arrow, a leader line, a tail, and a note. To place a leader, select the **Measure** and the **Leader** tool. The following prompt appears. Refer to Figure 13-20.

Leader Start point:

Pick the start of the leader. This is the end of the leader that will have the arrowhead (if one is currently selected). AutoSketch then asks for the second point of the leader.

Leader To point:

Move your cursor so that the leader line is at an approximate 45° angle and pick again.

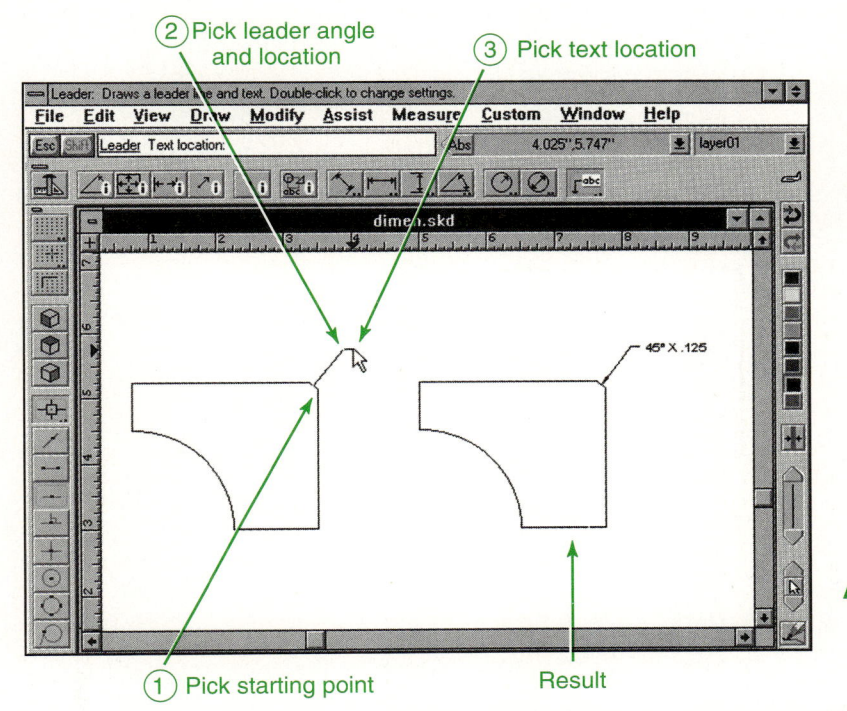

Figure 13-20.
To place a leader, pick the start point, length of the leader, and text position. The **Text Editor** dialog box will appear for you to enter text. Here, a special character sequence (%%d) was added to create the degree symbol.

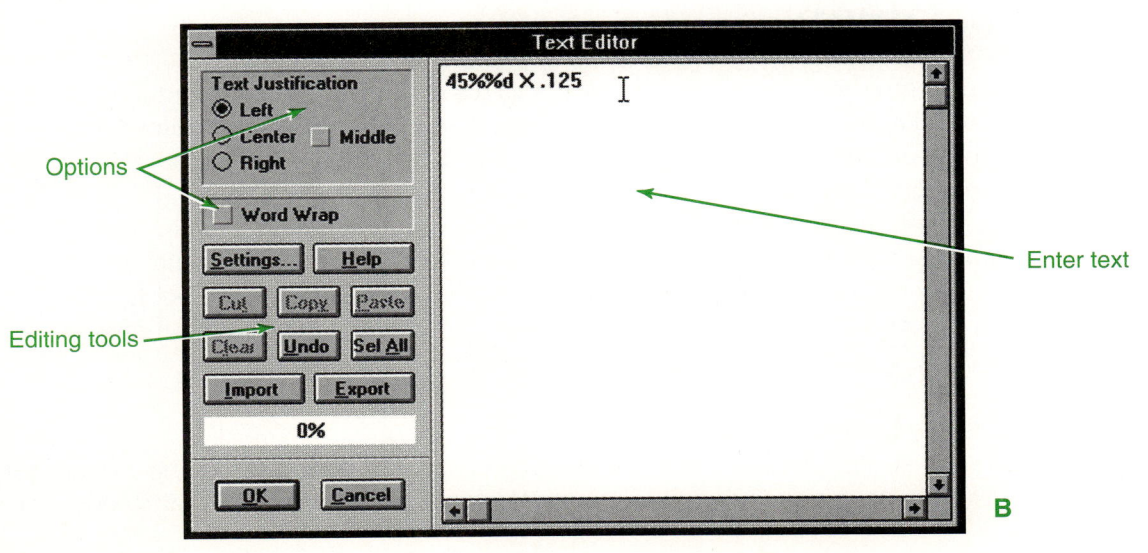

If the leader is drawn horizontally, the **Text Editor** dialog box appears after you select the second point of the leader. If the leader is not horizontal, the following prompt appears.

Leader Text location:

Move your cursor to create a .125" leader tail and pick again. The **Text Editor** dialog box appears.

Enter the text you want for the leader. Pick **OK** to close the dialog box and place the text. The settings for the **Text Editor** dialog box are the same as when adding text to a drawing.

Exercise 13-6

1. Start a new drawing using the **A_LI_MEC** template.
2. Develop and dimension the following drawing as shown.
3. Save the drawing as **C13E6** on your work diskette and then close the drawing.

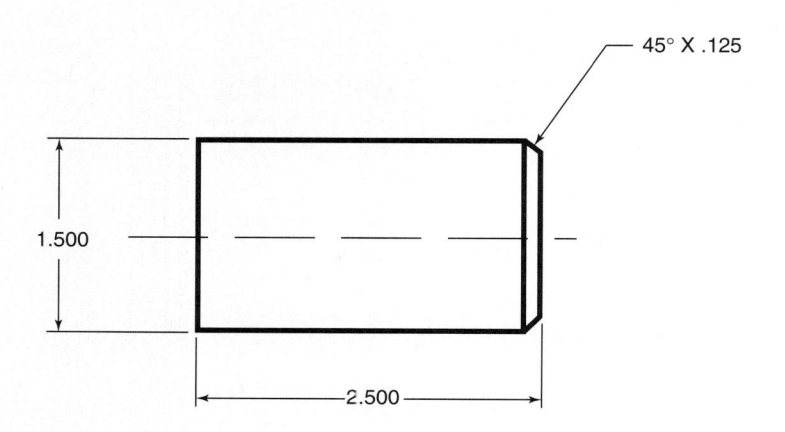

Associativity

AutoSketch dimensions are *associative*. This means that they are attached to the points you pick to place the dimension. When you edit AutoSketch dimensions, such as stretching an object, the dimension and extension lines stretch, and the value of the dimension also changes.

You must include the dimension points in your selection set for associative dimensioning to work properly. *Dimension points* are the points you picked to measure the dimension. The selection set simply means the objects you select to edit. The selection set must include the points you picked for the location of the dimension line, as well as the object you originally dimensioned.

Stretching Dimensions

You can stretch linear and angle dimensions. You must include the dimension point in the crosses/window box. The entire dimension is remeasured and redrawn.

Figure 13-21 shows a triangle dimensioned with several dimensions. The top corner was stretched to a new location. Notice that the affected dimensions changed according to the new location. The vertical dimension still measures vertical distance. The aligned dimension remains aligned. The horizontal dimension was unchanged because no point associated with that dimension was moved.

You can also stretch an angle dimension, Figure 13-22. If you stretch a dimensioned line so that the angular measurement will not fit in the modified space, the dimension is then moved outside.

Figure 13-21.
AutoSketch dimensions are associative. This means that the dimension will stretch with your drawing *if* you include a dimension point in the crosses/window box.

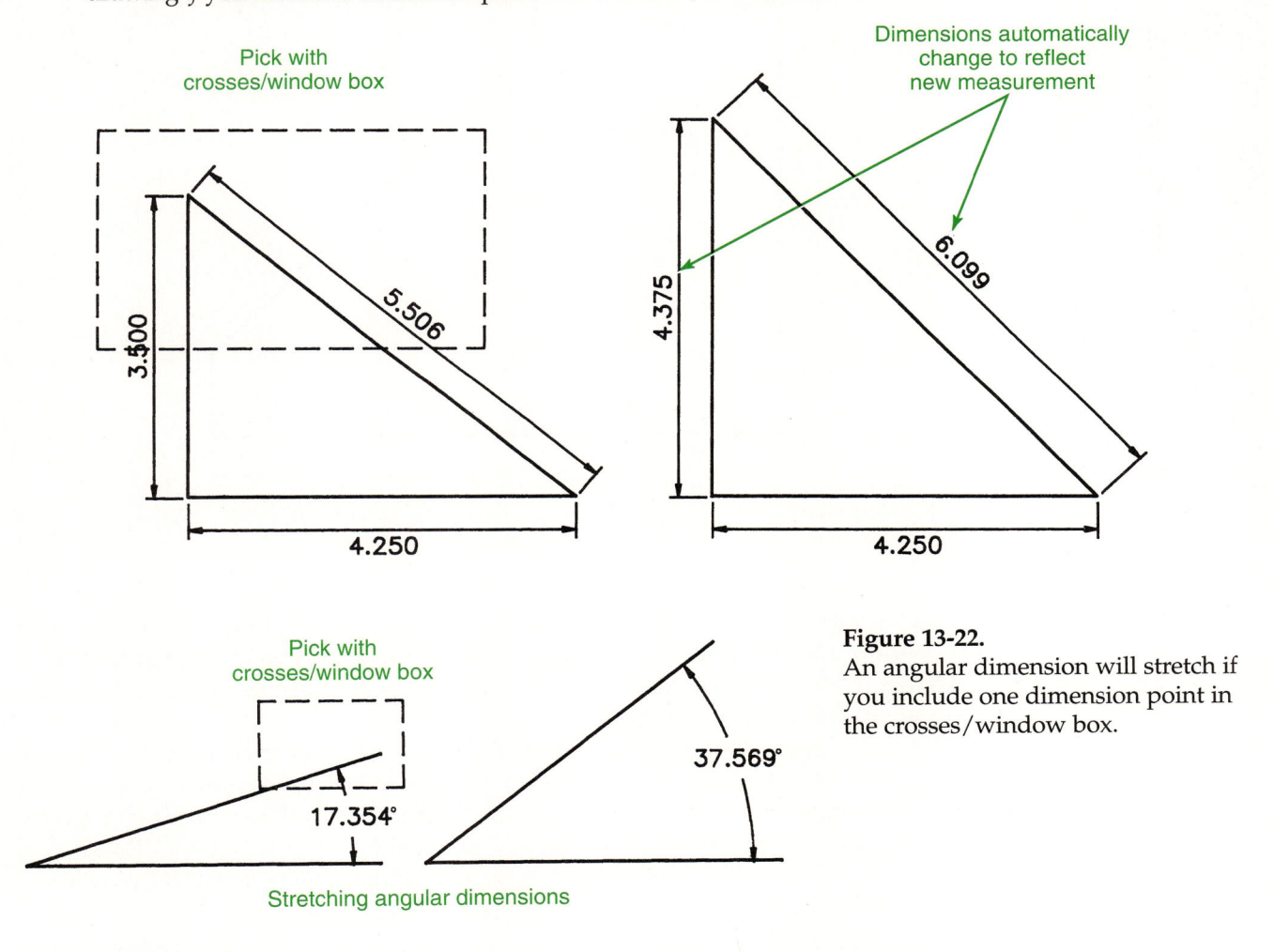

Figure 13-22.
An angular dimension will stretch if you include one dimension point in the crosses/window box.

Stretching angular dimensions

The **Stretch** tool can be used to move dimension lines or leaders. To move a dimension line closer or farther away from the object dimensioned, use a crosses/window selection box to select the dimension line. Do not select the object. This allows you to stretch the extension lines only.

To stretch a leader, use the crosses/window box to select the text and leader tail. Move the cursor to change the length of the leader tail. If you select the leader along with the tail and text, you can change the angle of the leader and the length of the leader. You will notice as you do this that the length of the tail does not change and remains horizontal.

Exercise 13-7

1. Open drawing C13E3 from this chapter.
2. Use the **Stretch** tool to alter the drawing as shown below. Be sure to stretch dimensions when stretching objects.
3. Save the drawing as C13E7 on your work diskette and then close the drawing.

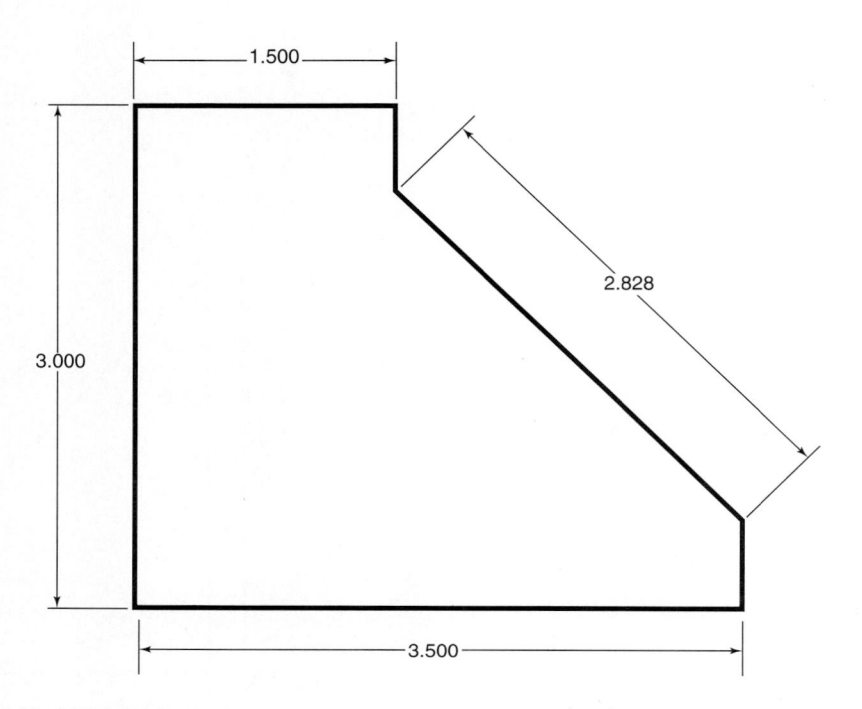

Scaling Dimensions

When scaling dimensioned objects, include the dimensions within the crosses/window box when selecting objects. The dimension will change according to the scale factor. However, the size of the dimension text and arrows are also scaled. Be cautious when using this command.

You can scale just the dimension text. Pick the text only, then a base point and enter a scale factor. This has the same effect as setting the dimension text size in the **Dimension Settings** dialog box and using the **Change Properties** tool to change the properties of the dimension.

Rotating Dimensions

You can rotate dimensions along with objects. Aligned, diameter, or radius dimensions are not affected because they measure an absolute distance or angle.

Rotating objects with horizontal and vertical dimensions has an unusual effect. Refer to the before and after effects of the rectangle dimensioned in Figure 13-23. If you rotate a horizontal dimension 90° or 270°, the dimension text changes to 0 units. This is because what was a horizontal dimension is now a vertical dimension. In the same manner, if you rotate a vertical dimension 90° or 270°, the dimension text changes to 0 because what was vertical is now horizontal. If you plan to rotate objects with linear dimensions, use aligned dimensions instead. In this way, dimension lines and text remain unchanged as you rotate objects, even at 90° and 270°.

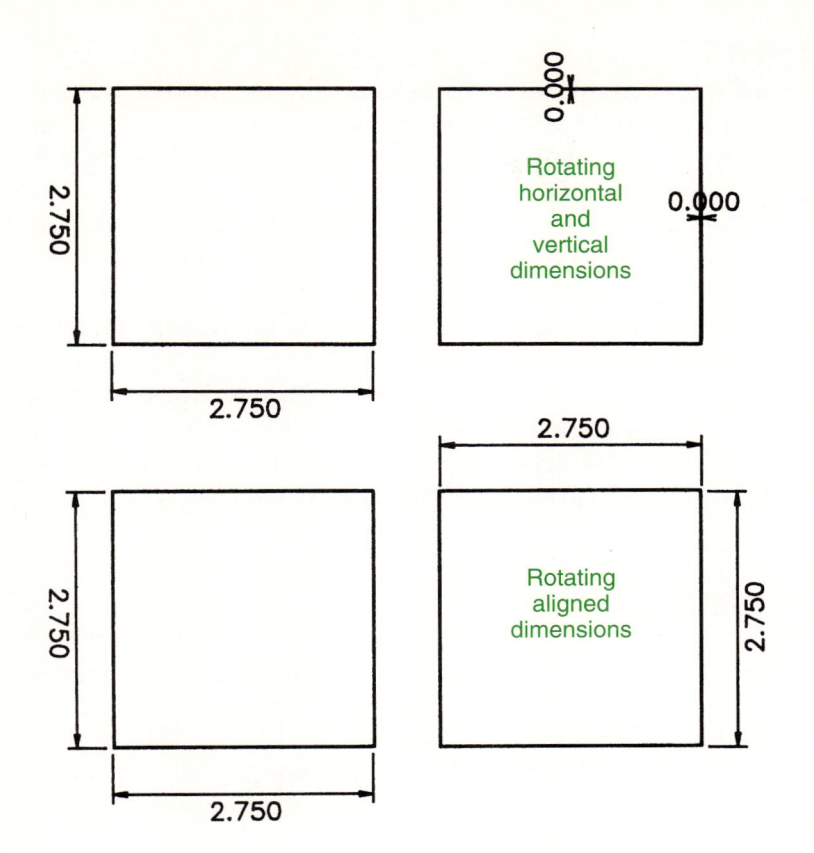

Figure 13-23.
If you plan to rotate dimensioned rectangular objects, use aligned dimensions. Horizontal and vertical dimensions can cause odd results when the object is rotated 90° or 270°.

Angular dimensions work somewhat different. For example, if you rotate one of the lines picked to measure the angle, the angle value will not change. Even if you select one end of the dimension with the line, the dimension will not change. However, if you select both lines and the dimension, the dimension also rotates. The text will remain horizontal, no matter what angle of rotation is used, Figure 13-24.

You can also rotate leaders, or just dimension text. When you rotate a leader, the leader line rotates. However, the leader tail and dimension text remain horizontal. If you select dimension text to rotate, the text will rotate just like any other object.

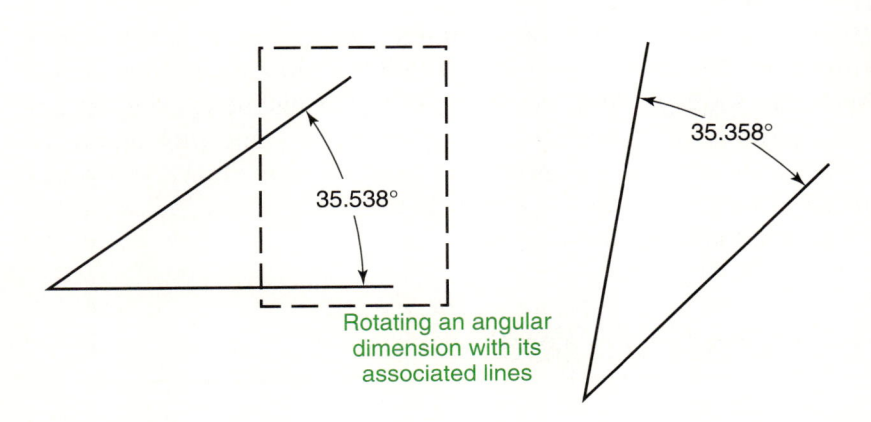

Figure 13-24.
When angular dimensions are rotated, the dimension text remains horizontal.

Restoring Associativity

There is one requirement for a dimension to be associative. You must have the **Measurement** setting turned on in the **Dimension Settings** dialog box when you create the dimension. If you add dimensions without this setting, and later scale or stretch them, the dimension will not change accordingly.

It is also possible to accidentally remove the associativity of a dimension two ways. If you change the properties of the dimension after placing it, and have the **Measurement** setting turned off, the dimension loses associativity. Also, if you use the **Edit Text** tool to delete the %%m character sequence embedded in the dimension text, the dimension will no longer be associative.

To restore associativity to the dimension, you must use the **Change Properties** tool in the **Edit** menu. Double-click on the **Change Properties** tool and select the **Dimension Text** button. This opens the **Dimension Settings** dialog box. Here, turn on **Measurement**. Then, pick **OK** to close the dialog box and pick the dimensions to change.

Editing Dimensions

Dimensions are placed using the current **Units of Measurement** and **Dimension Settings**. At some point, you may want to change these settings. For example, you may want some dimensions measured in millimeters, while others are in inches. Or, you might want some dimensions with solid arrowheads while others end with dots. When you want to place a new dimension with different settings, simply change the **Units of Measurement** and **Dimension Settings.**

Changing Dimension Settings

To change the settings of existing dimensions, first double-click any dimension tool to display the **Dimension Settings** dialog box. Make the needed changes. Then, from the **Edit** menu, double-click on the **Change Properties** tool to display the **Property Settings** dialog box. Check **Dimension Text...** and pick **OK.** Finally, select the dimension to apply the new dimension settings. You can also change the arrow and dimension text placement using this procedure. Just make sure that the setting you want to change is selected in the **Property Settings** dialog box.

Moving Dimension Text

There may be times when you want to move the dimension text to avoid clutter on the drawing. To move just the dimension text, and not the dimension lines, arrow, or extension lines, select the **Modify** menu and **Move** tool. Pick just above the dimension text to move, pick a base point, and then reposition the text and pick again. To move dimension text with a leader, use the **Stretch** tool and select the text and the end of the tail. Make sure the selection box does not include the leader. Then, pick a stretch base point and move the cursor to replace the text.

Changing Dimension Text

You may wish to change dimension text, such as adding a suffix or tolerance to the dimension. Use the **Edit Text** tool in the **Modify** menu. The actual measurement is represented by the %%m characters. Do not delete these or the dimension text loses its associativity. Simply add the text behind the %%m.

General Rules for Dimensioning

There are common rules drafters should follow when placing dimensions. Review the following list and become familiar with these rules.

- Place dimensions on views that show the true shape of the feature being measured.
- Avoid crossing dimension lines. If an extension line must cross a dimension line, most drafters break the extension line around the dimension line. (Note: This is not possible with AutoSketch dimensions.)
- Avoid dimensioning to hidden lines. Dimension features in the view where they appear as object lines.
- The same dimension should not be repeated on different views, unless required to understand the drawing.
- Smaller dimensions are placed nearest the view, while larger, or overall dimensions, are farthest from the view. See the section on stacked dimensions.
- The person reading the drawing should be able to determine all sizes and shapes without using a ruler. (This is especially important when the drawing is not printed or plotted full scale.)
- Locate dimensions together, rather than scattering them around the drawing. This practice makes the dimensions easier to read.
- On manufactured parts, dimensions should be given from finished surfaces and centerlines. On architectural drawings, dimensions should be given from the outside of exterior walls and the centers of interior walls.

Typical Dimensioning Practices

Items to be dimensioned can be broken down into geometric shapes. Here are some typical dimension practices for common geometric shapes. Consult a comprehensive drafting text for additional dimensioning practices.

Rectangular shapes

Dimension both the width and height of rectangular shapes. In a two-view drawing, place the dimension between the two views unless the object is large. Extension lines are extended from the most descriptive view. Square shapes may be dimensioned using the square symbol. See Figure 13-25 for examples.

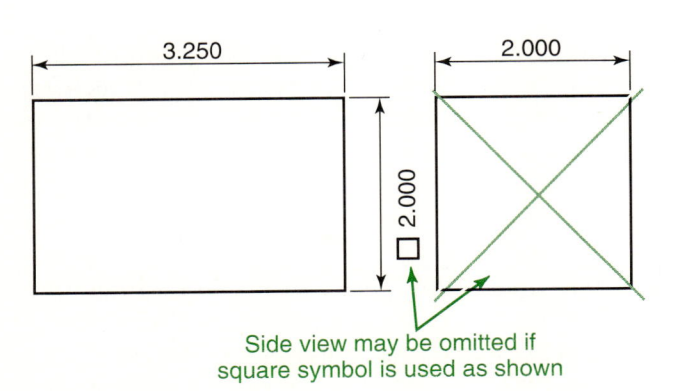

Figure 13-25.
Dimensioning rectangular shapes.

Cylindrical shapes

Dimension cylindrical shapes in the view where the part appears rectangular. For simple cylindrical shapes, the circular view may be omitted. See Figure 13-26.

Figure 13-26. Dimensioning cylindrical shapes.

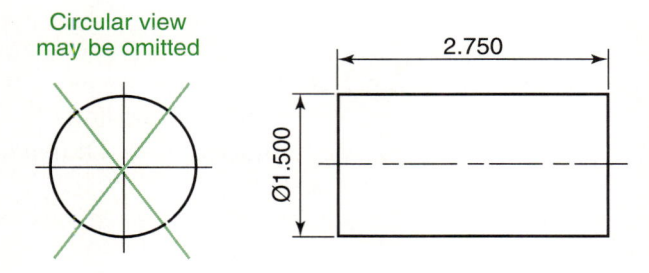

Conical shapes

Conical shapes are dimensioned by the end diameters and length. You can also give the base dimension, taper angle, and length. See Figure 13-27.

Figure 13-27. Dimensioning conical shapes. These shapes can be dimensioned in one of two ways.

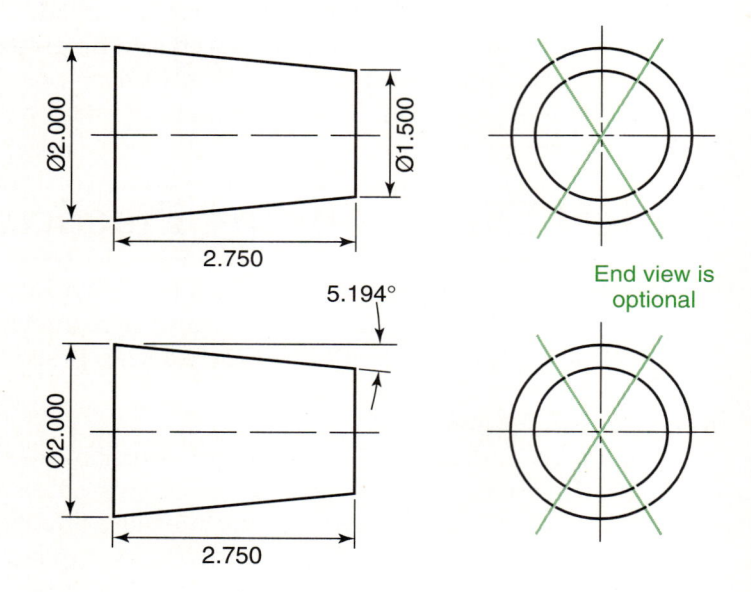

Dimensioning holes

Internal cylinders (or holes) are dimensioned with a diameter dimension in the view where the feature is circular. Holes may be drilled or reamed. They can also be counterbored to a specific depth, countersunk, or spotfaced. This process should be indicated with a local note or symbol. See Figure 13-28. Where it is not clear that a hole goes through the part, the word THRU should follow the dimension.

Slotted holes

Slots provide room for adjustment when fastening parts. They can be milled in solid metal or punched into sheet metal. There are two general methods for dimensioning slotted holes. When punched into sheet metal, the center of the slot is usually located. When milled in thicker stock, the centers of the ending arcs are generally located. See Figure 13-29.

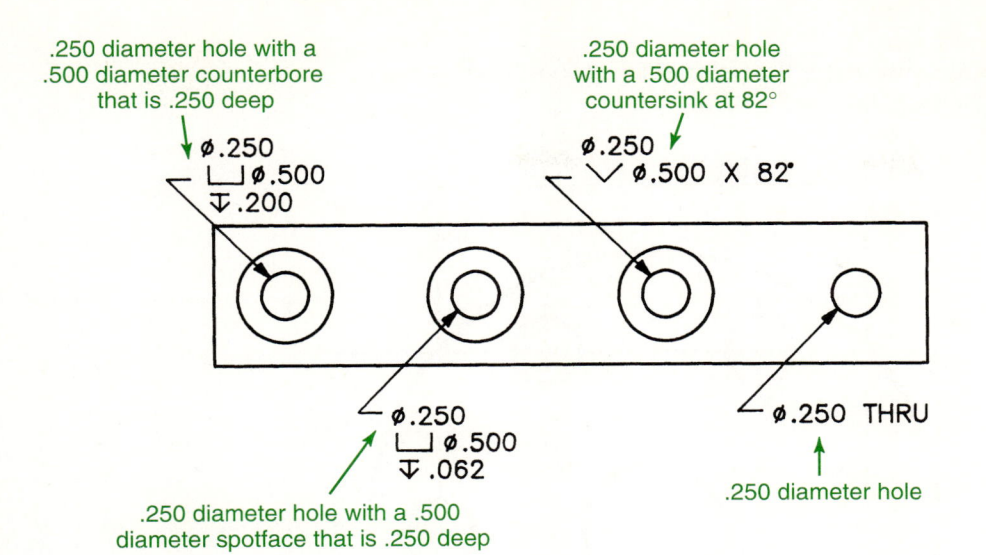

Figure 13-28.
Dimensioning holes for manufactured products. Note how symbols can be used to represent various drilling and machining operations.

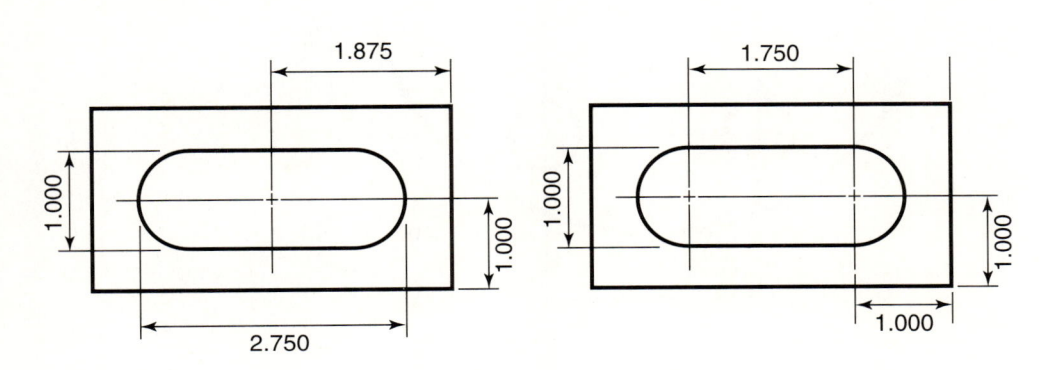

.250 diameter hole with a .500 diameter counterbore that is .250 deep

.250 diameter hole with a .500 diameter countersink at 82°

.250 diameter hole with a .500 diameter spotface that is .250 deep

.250 diameter hole

Figure 13-29.
Dimensioning slotted holes. Dimension the size and slot center for punched slots. For milled slots, dimension the size and both radius centers.

Locating holes and rectangular features

Holes and round parts are always located by their centers, not by the edges. Rectangular features are located by one corner, Figure 13-30. When in a circular pattern, the holes may be dimensioned using rectangular or polar coordinate dimensioning methods, Figure 13-31.

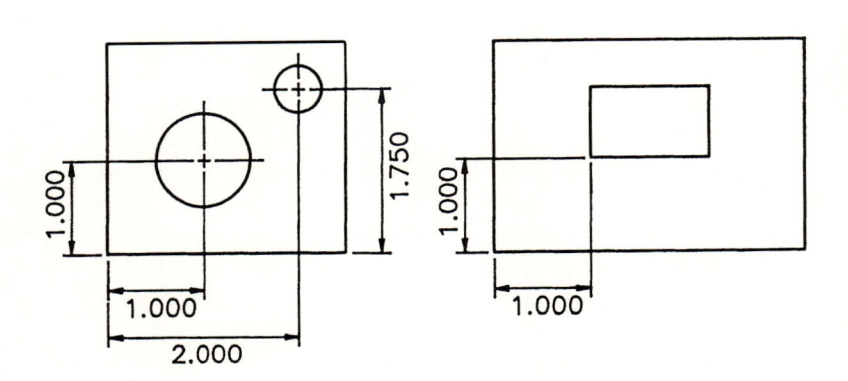

Figure 13-30.
Locate holes by their centers and rectangular features by their corner.

Figure 13-31.
When locating holes in circular pattern, use either the rectangular or coordinate method. Note how the number of holes is indicated with an "X."

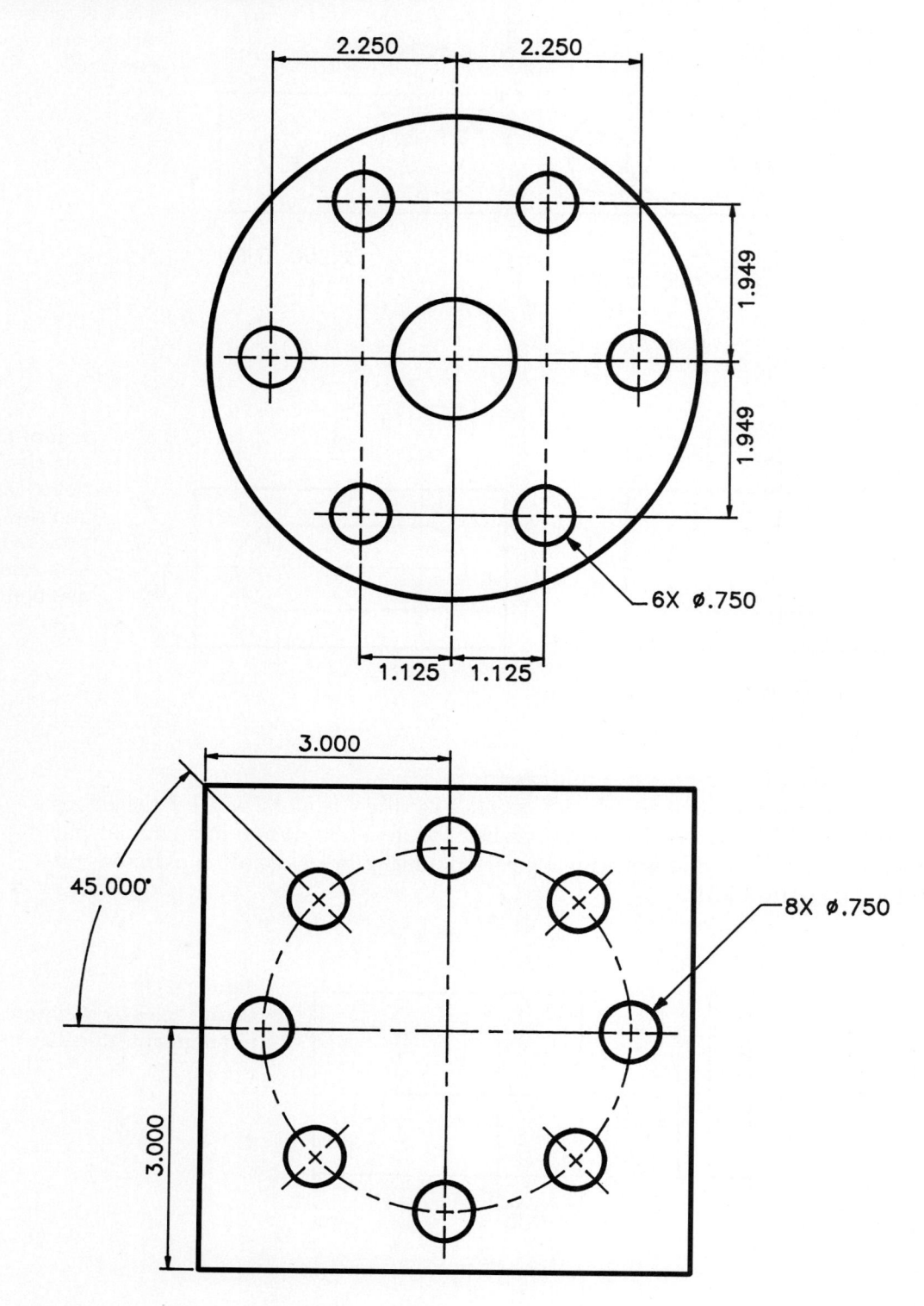

Fillets and rounds

Fillets and rounds are dimensioned by their radius. They can be dimensioned individually, as repetitive features, or with a general note, such as: UNLESS OTHERWISE SPECIFIED, ALL FILLETS AND ROUNDS R.0625.

Angles

Two methods for dimensioning angles are coordinate and angular dimensions. Coordinate dimensioning uses horizontal and vertical linear commands. These dimensions mark the edges of the angle. The angular dimensioning method gives a linear dimension to one corner and the angle in degrees. See Figure 13-32.

Figure 13-32.
Dimensioning angles by linear and angular methods.

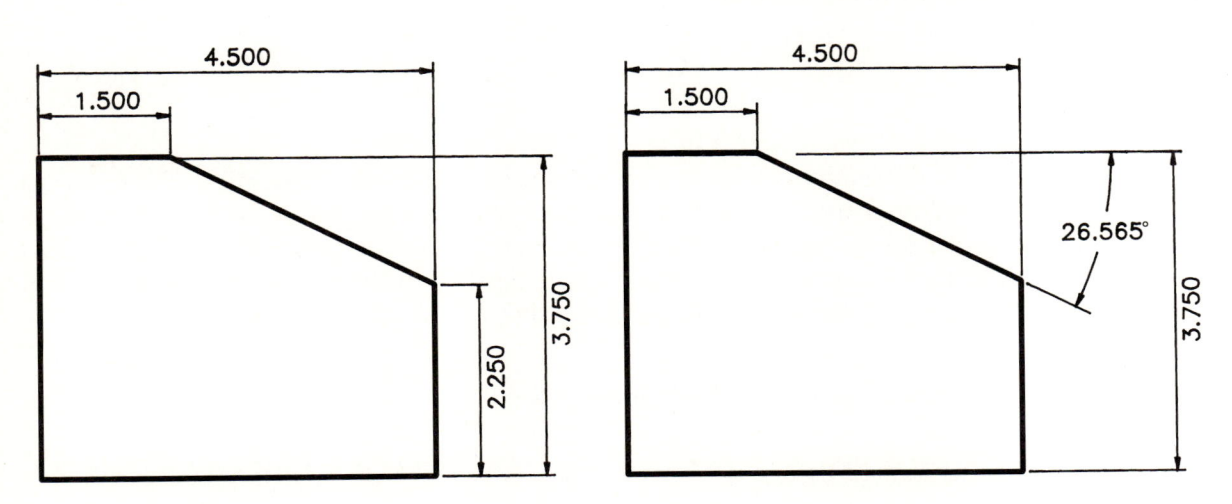

Chamfers

Chamfers can be dimensioned with a leader when they are 45°. Refer to Figure 13-20. Chamfers that are not 45° are dimensioned with two linear measurements, or a linear and angular dimension. (Similar to the angles dimensioned in Figure 13-32.)

Repetitive features

Multiple holes or fillets of the same radius can be dimensioned on one hole or arc. A note giving the number of times the feature is found is added to the dimension. Specify the number of features, then an "X," followed by the size. Refer back to Figure 13-31.

Alternate Dimensioning Practices

A current trend in manufacturing is to use computers to control machining and drilling operations. This has affected some dimensioning practices. Computers move in precise increments measured from a specific place. Therefore, most dimensions are measured from a common point, called a *datum*. Datum dimensioning methods do not necessarily require dimension lines. With *coordinate dimensioning,* dimension text is aligned with the extension lines. The distances of features are given from the datum. Holes are identified with a size symbol, usually a letter. The sizes are usually shown in a table beside or below the drawing. See Figure 13-33.

Figure 13-33.
Coordinate dimensioning system.

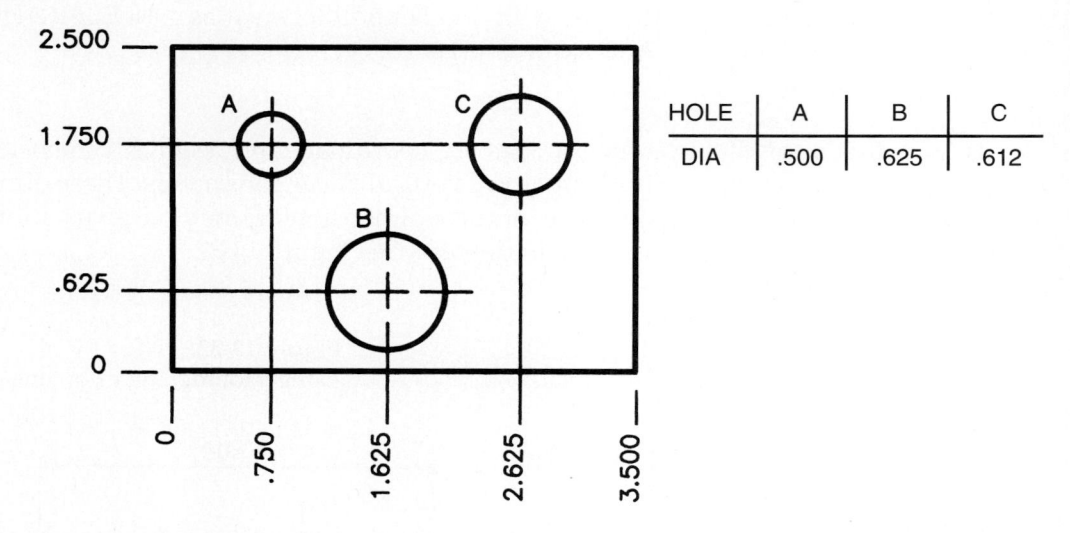

HOLE	A	B	C
DIA	.500	.625	.612

Tabular dimensioning is another technique using the principles of rectangular coordinate dimensioning. Measurements are taken from the X, Y, and Z axes and recorded in a table, Figure 13-34. They are not dimensioned directly on the drawing. Hole diameters or other feature sizes are also given in the table. This method is particularly useful where there are a large number of features to locate.

Figure 13-34. Feature size and location are given in a table with the tabular dimensioning system.

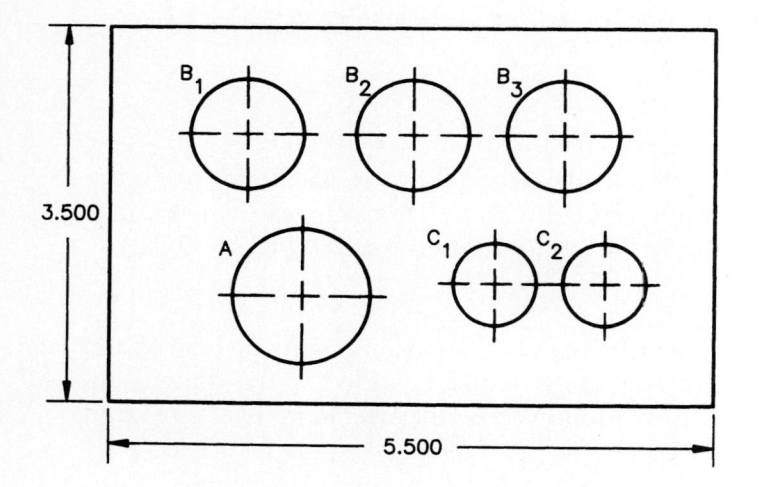

HOLE	SIZE	X	Y	Z
A	⌀.750	2.225	.575	THRU
B_1	⌀.500	1.110	3.010	THRU
B_2	⌀.375	3.235	3.030	THRU
B_3	⌀.425	4.025	2.750	THRU
C_1	⌀.125	3.750	1.000	THRU
C_2	⌀.100	4.700	1.000	THRU

Summary

Dimensions are important aspects of most drawings. Dimensions consist of numbers, lines, symbols, and notes. They provide information about the size and shape of a product. AutoSketch makes the process of applying dimensions to a drawing an easy task.

Dimensions generally describe the size or location of product features. Size dimensions give the width, length, height, diameter, or radius of a feature. Location dimensions give the position of a feature.

Several elements are used when dimensioning a drawing. The most common elements include extension lines, dimension lines, arrows, dimension text, leaders, and notes or symbols. When these elements are combined correctly, they provide exact information about how a product is constructed.

AutoSketch allows you to apply horizontal, vertical, and aligned linear dimensions. Angle dimensions can also be applied to dimension two nonparallel surfaces. If an area that you wish to insert a dimension in is too small, AutoSketch allows you to move the dimension text to a new position.

AutoSketch also allows you to place diameter and radius dimensions either with a dimension line or leader. The unit of measurement can be displayed in many formats, depending on your application. In addition, AutoSketch provides five different types of arrows to meet your needs.

AutoSketch dimensions are associative. This means that the dimensions are "tied" to the points you picked when placing the dimension. When editing drawings with dimensions, you must include the dimensions in your selection set. This allows the dimensions to be changed along with the object you are modifying.

Dimensioning practices vary from industry to industry. However, the basic dimensioning standards are very similar. When dimensioning standards are followed, it allows for consistency and readability from one industry to another.

Important Terms

Aligned dimensions
Angle dimensions
Arrows
Associative
Chained dimensions
Coordinate dimensioning
Datum
Diameter dimensions
Dimension lines
Dimension points
Dimension text
Dimensions
Dual dimensioning
Extension line offset
Extension lines
Feature
General notes
Geometric dimensioning and tolerancing
Horizontal dimensions
Leader
Linear dimensions
Location dimensions
Radius dimensions
Size dimensions
Stacked dimensions
Tabular dimensioning
Unidirectional dimensions
Unit of measurement
Vertical dimensions

New Tools and Commands

Aligned
Aligned Dimension tool
Angle Dimension tool
Arrow Size
Arrow Type
Center Marks
Center Text
Diameter Dimension tool
Dimension Line

Dimension button
Dimension Settings dialog box
Drawing Settings tool
Horizontal Dimension tool
Leader tool
Linear Text Placement
Measurement
Offset with Leader
Radial Placement

Radius Dimension tool
Slide Text
Suffix
Text Size
Units of Measurement dialog box
Units
Vertical Dimension tool
Within Dimension Line

Review Questions

Give the best answer for each of the following questions.

1. Dimensions show _____ and _____.
2. _____ dimensions give the width, length, height, diameter, or radius of a feature.
3. _____ dimensions give the position of a feature or shape.
4. How does AutoSketch make dimensioning an easier process?

5. _____ lines mark the beginning and end of a dimension.
6. List the elements of a typical horizontal dimension.

7. A dimension line should be at least _____ inch(es) away from the surface it measures.
8. Explain the difference between unidirectional and aligned dimension text.

9. Give examples of when leaders are used.

10. Explain the difference between general and specific notes.

11. Where do you set the unit of measurement that determines how dimension text is displayed?
12. True or false. Dimensions with different unit of measurement can be placed on the same drawing.
13. What option(s) in the **Dimension Settings** dialog box affect how the dimension text is placed in a linear dimension?

14. What option(s) in the **Dimension Settings** dialog box affect diameter and radius dimensions? _____

15. What two methods allow you to add additional text behind dimension text?
    A. _____
    B. _____

16. What AutoSketch dimensioning tool(s) allow you to place linear dimensions? _____

17. How do you place a radius or diameter dimension using a leader? _____

18. For most purposes, a leader should be placed at a(n) _____ angle.
19. To stretch a linear dimension, you must include _____ in the selection set.
20. What three ways can a dimension be drawn without, or lose, associativity?
    A. _____
    B. _____
    C. _____
21. True or false. Radius dimensions should be used to dimension holes and cylindrical parts.
22. True or false. The same dimension can be repeated on different views if required to understand the drawing.
23. Dimension cylindrical shapes in the view where the part appears _____ .
24. Two methods to dimension angles are the _____ and _____ methods.
25. With _____ dimensioning, measurements are taken from the X, Y, and Z axes and recorded in a table. Measurements are not shown directly on the drawing.

Activities

Here is a mixture of drawing activities to develop your dimensioning skills. Begin each drawing by selecting one of the templates from Exercise 13-2 and complete the drawing. After saving each activity using the filename given underneath, select the **Close** command from the **File** menu to clear the drawing from the screen.

1 - 5. Create each of the following drawings. Add the proper dimensions following accepted dimensioning practices. (Note: AutoSketch may not allow some of these practices. Follow standard practices when possible.) Place dimensions and object lines on different layers. Also use the colors recommended in this text for the linetypes.

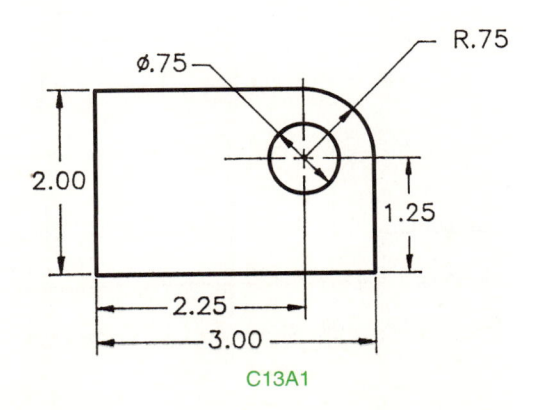

C13A1

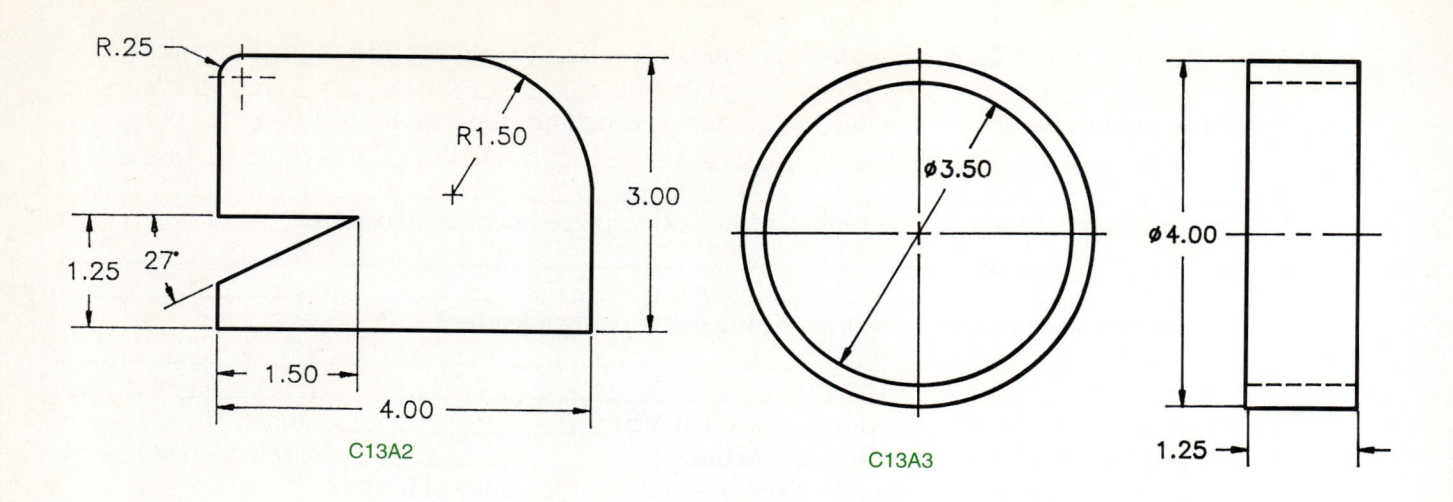

C13A2

C13A3

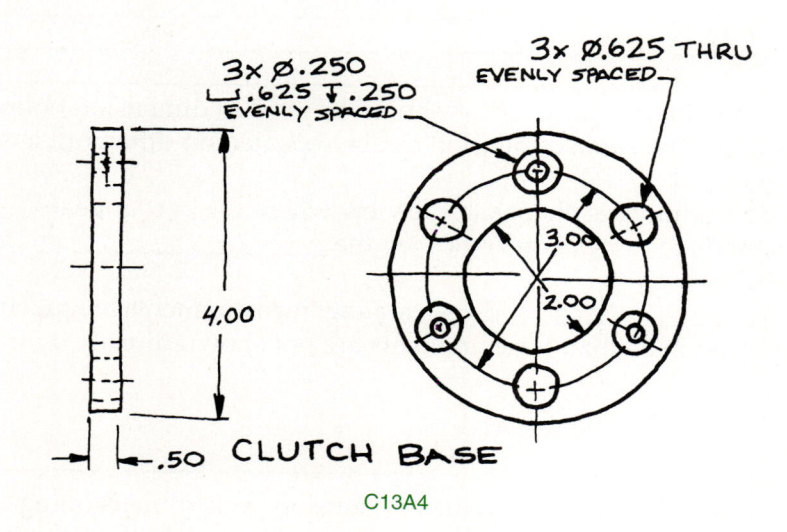

CLUTCH BASE
C13A4

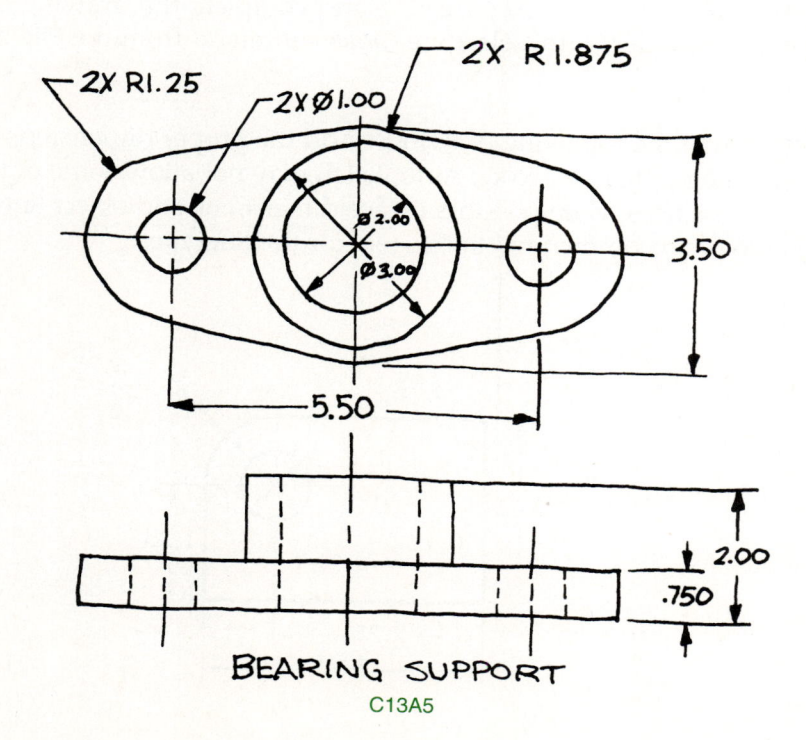

BEARING SUPPORT
C13A5

6 - 10. Create each of the following drawings. Add the proper dimensions following accepted dimensioning practices. (Note: AutoSketch may not allow some of these practices. Follow standard practices when possible.) Place dimensions and object lines on different layers. Also use the colors recommended in this text for the linetypes. Use the .25 grid spacing for size reference.

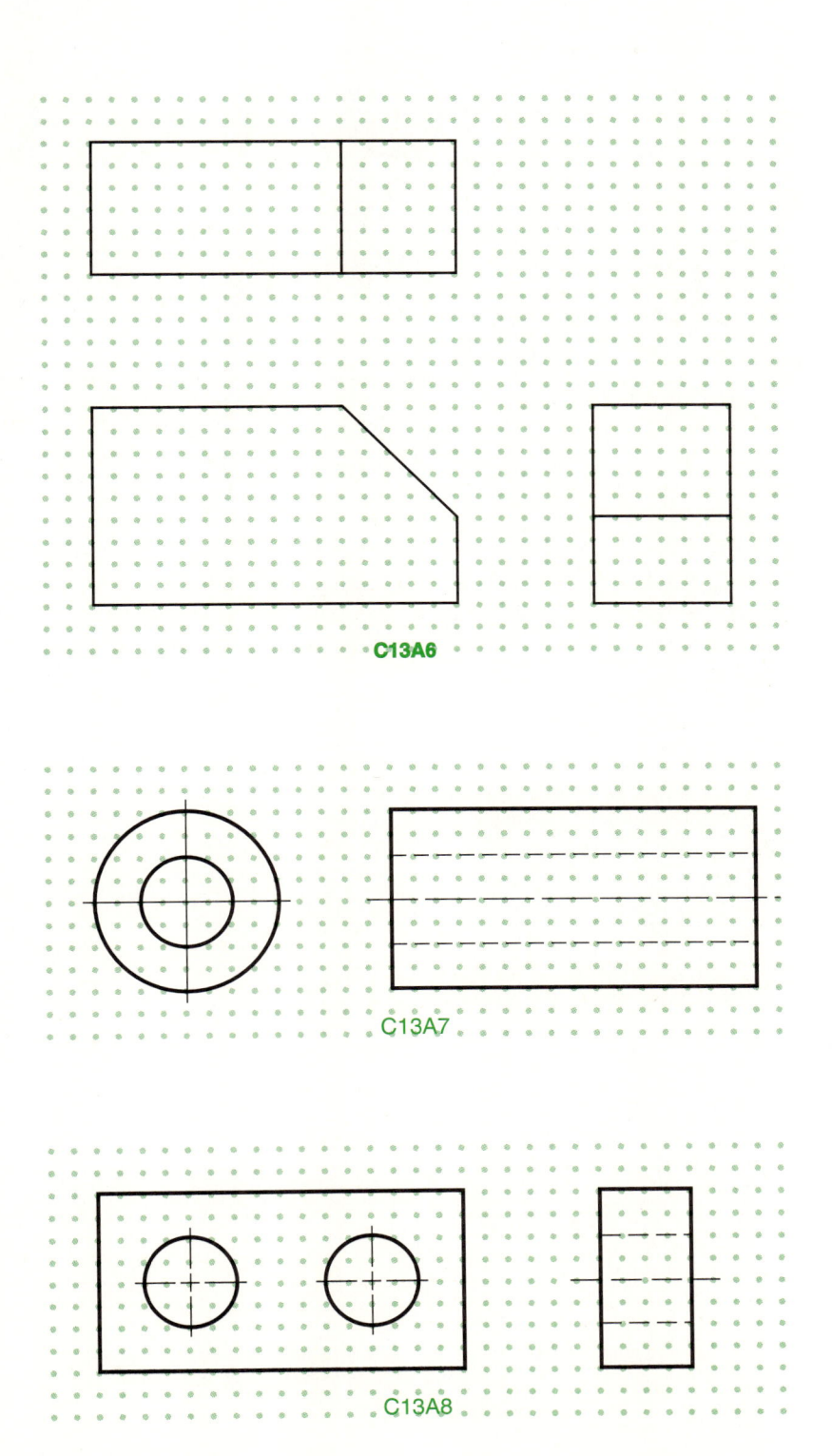

C13A6

C13A7

C13A8

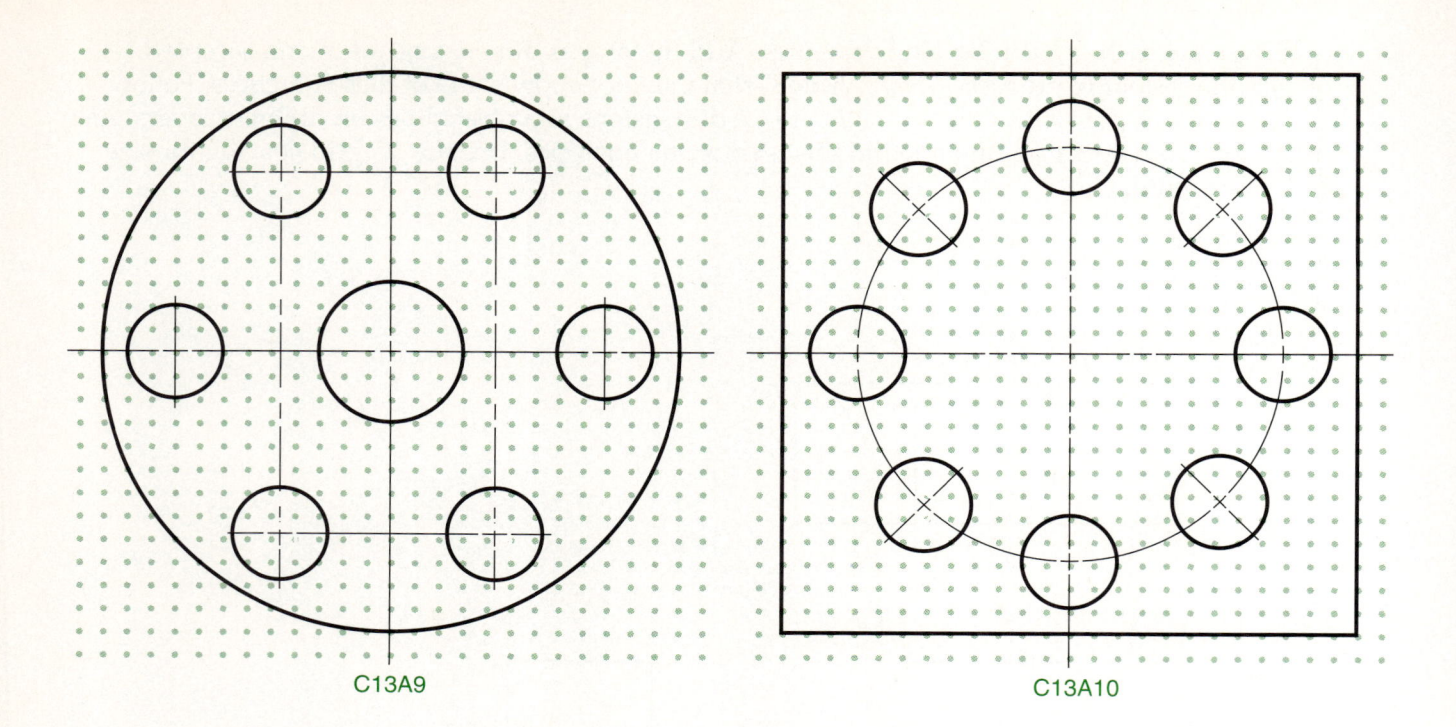

C13A9

C13A10

11 - 20. Select 10 drawings completed in previous chapters. Add the proper dimensions following accepted dimensioning practices. (Note: AutoSketch may not allow some of these practices. Follow standard practices when possible.) Place dimensions and object lines on different layers. Also use the colors recommended in this text for the linetypes. Save the drawings as C13A11 through C13A20.

CHAPTER 14

Creating and Using Symbols

Objectives
After studying this chapter, you will be able to:
- Explain how symbols can save time when creating a drawing.
- Draw and save a symbol using the **Part Clip...** command.
- Insert a symbol into a drawing using the **Part Insert...** command.
- Describe how object properties apply to symbols.
- Organize symbols in a symbols library.

One advantage of a CAD program is that once you draw an item, you should never have to draw it again. Instead, simply copy the item. However, what if you need the same shape in several drawings? Copying objects from one drawing to another is not an easy process. (This will be covered in Chapter 17.) You must have two drawings open. This consumes memory. What do you do? The answer is that AutoSketch allows you to insert an existing drawing into your current drawing. This is done using the **Part Insert...** command found in the **File** menu. The drawing you insert may contain just a few objects, or a complex drawing. Most often you will insert a drawing that contains only a few objects, called a symbol.

Symbols

A *symbol* is a collection of objects that represents a standard component, assembly, or feature. See Figure 14-1. Symbols are used in charts, drawings, diagrams, and schematics. For example, architectural drawings contain symbols that represent doors, windows, appliances, trees, and lights. Electronics drawings contain symbols for resistors, diodes, and transistors. In addition, there are symbols that you use to dimension drawings.

In traditional drafting, symbols are drawn by hand. A plastic template may be used as a guide in some cases. Unfortunately, this method is time-consuming and the quality can vary. Also, there may not be a template made for the unique symbols that you or your company need.

With AutoSketch, you insert symbols using the **Part Insert...** command. AutoSketch uses the term "part" to refer to a symbol. In this text, we will use the term "symbol." Symbols are created in AutoSketch by drawing the symbol and saving it using the **Part Clip...** command. You can create and store an entire library of symbols for future use. You can also use one of hundreds of predrawn symbols that come with AutoSketch. Each one can be inserted as many times as needed in many different drawings.

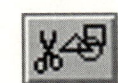

Figure 14-1.
Symbols represent a standard component, assembly, or feature.
(CAD Technology Corp.)

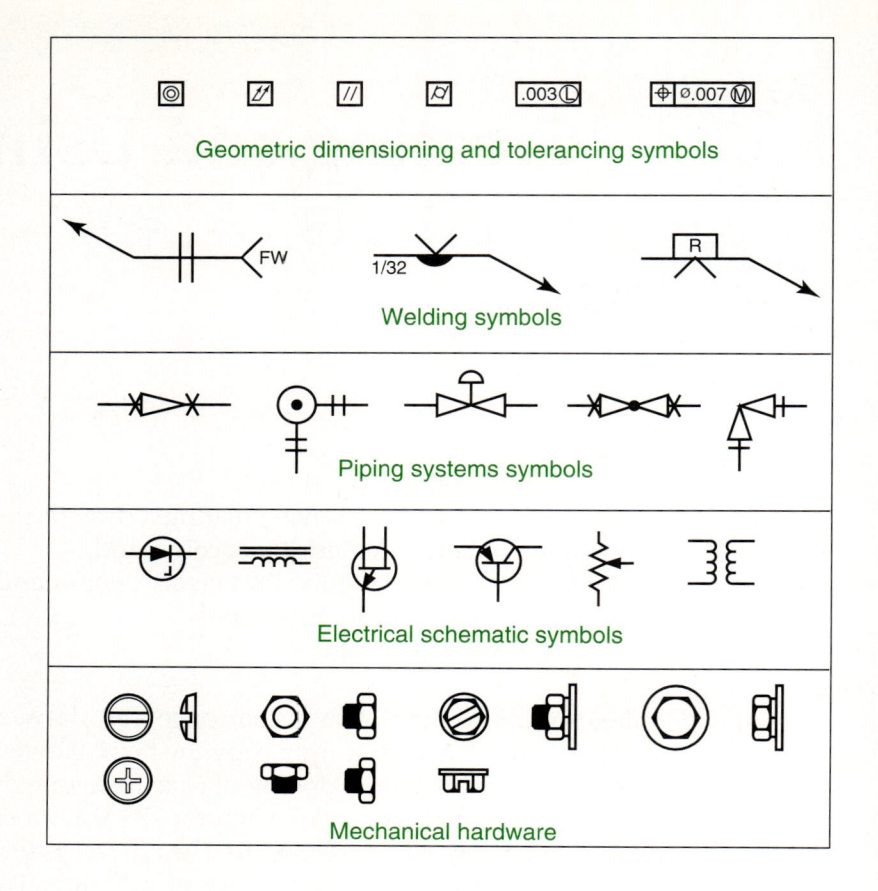

Drawing and Saving a Symbol

Certain steps should be followed when creating symbols. If you work for a company, how you draw the symbol may be a joint decision among product engineers, design engineers, and other drafters. Certain symbol types and sizes that will be used in more than one drawing must be accurate. The steps listed here apply if you are drawing for yourself or drafting for a company.

1. Draw the symbol to scale as a new drawing.
2. Group the objects that make up the symbol.
3. Select the **Part Clip...** command.
4. Select the directory and enter a descriptive symbol filename.
5. Select a part base location for the symbol.
6. Select the objects that make up the symbol.

Drawing a Symbol

Determine the objects that make up the symbol shape before drawing the symbol. There are both standard and custom shapes. Standard symbol shapes are typically set by a *standards organization.* The American Welding Society, for example, governs the size and shape of standard welding symbols, Figure 14-2. Custom symbol shapes are those that are only used by your business. For example, your company may specialize in office layouts. You can make symbols to represent several styles of desks. Keep custom symbols simple when drawing them. Remember, a symbol only represents an item, Figure 14-3. You do not have to fully describe the actual item.

Figure 14-2.
The American Welding Society sets standard welding symbols. These symbols should be used on all welding drawings.

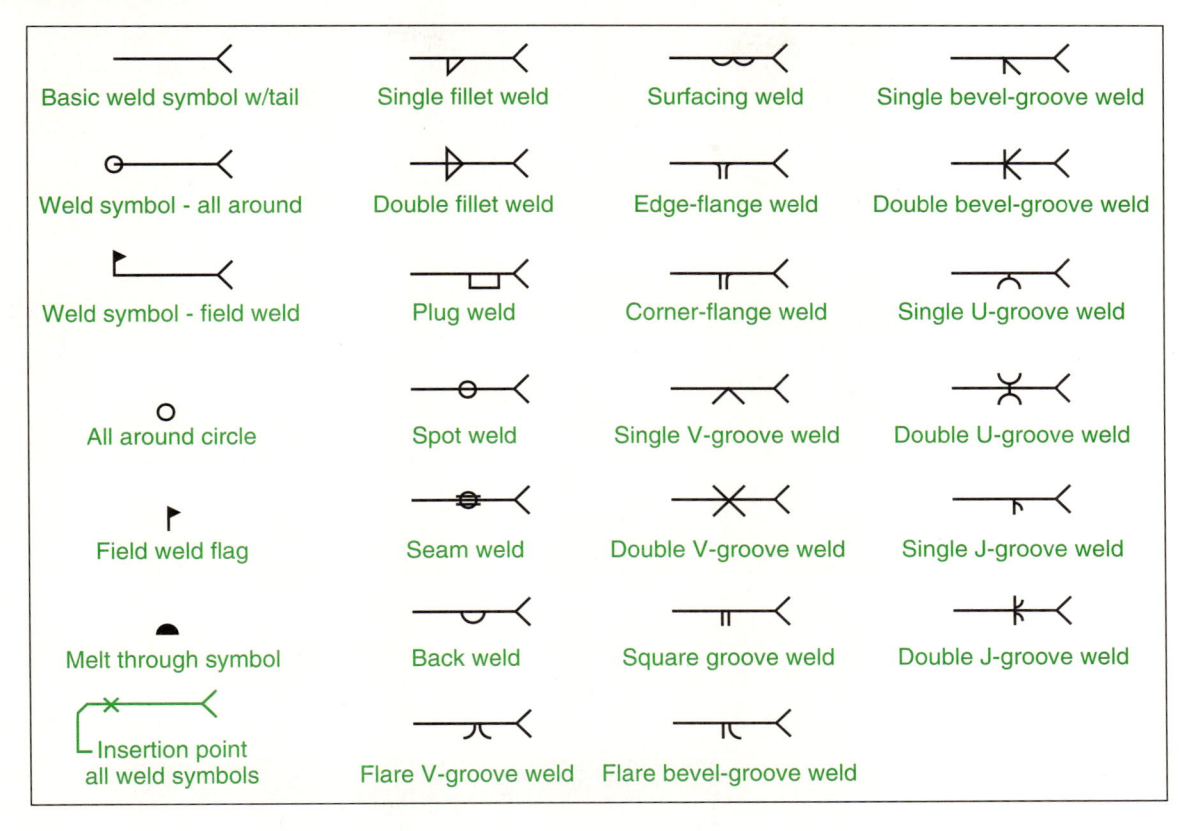

Basic weld symbol w/tail	Single fillet weld	Surfacing weld	Single bevel-groove weld
Weld symbol - all around	Double fillet weld	Edge-flange weld	Double bevel-groove weld
Weld symbol - field weld	Plug weld	Corner-flange weld	Single U-groove weld
All around circle	Spot weld	Single V-groove weld	Double U-groove weld
Field weld flag	Seam weld	Double V-groove weld	Single J-groove weld
Melt through symbol	Back weld	Square groove weld	Double J-groove weld
Insertion point all weld symbols	Flare V-groove weld	Flare bevel-groove weld	

Once you've decided how to draw the symbol, it is best to create it as a new drawing. However, if you've created a symbol in an existing drawing, but later decide it will be used in many drawings, you can simply create a symbol out of those objects. For example, after drawing a house plan you notice that you keep drawing the same window over and over. Group the objects that make up the window and then save it using the **Part Clip...** command. Then, you can insert it as a symbol wherever needed.

Make sure that you accurately draw symbols to scale. Also, match the unit of measurement for drawings that the symbol will be inserted in. For example, suppose you are drawing a window symbol to use in floor plans. During drawing setup, set the scale at 1/4″ = 1′-0″ and select **Feet, inches and fraction** as the units of measurement. However, symbols can be rescaled or rotated as needed after being inserted into another drawing.

Group the Symbol

Group the objects after you draw the symbol. This step is very important. When you insert the symbol into another drawing, the symbol acts like a single object when you move or rotate it. If you need to edit part of the symbol, simply ungroup it. If you are saving objects out of an existing drawing, make sure to group those objects first as well.

Using the Part Clip Command

Once you have drawn and grouped the symbol, select the **Part Clip...** command from the **File** menu. The **Part Clip File** dialog box will appear, Figure 14-4. Here you will need to specify the symbol filename, as well as the drive and directory where the symbol will be stored. The symbol filename should identify the part. For example, the window symbol in Figure 14-3 might be saved as **W-DBL36.** This denotes it as a 36" wide, double-hung window. Naming the symbol WINDOW might not be a good idea since there may be more than one window symbol. Be descriptive.

Figure 14-3.
This symbol represents a window in a floor plan of an architectural drawing. It is simple, but when inserted into a floor plan it will indicate where windows are located.

Figure 14-4.
In the **Part Clip File** dialog box, select the drive and/or directory where you want the part stored. Then enter a descriptive filename.

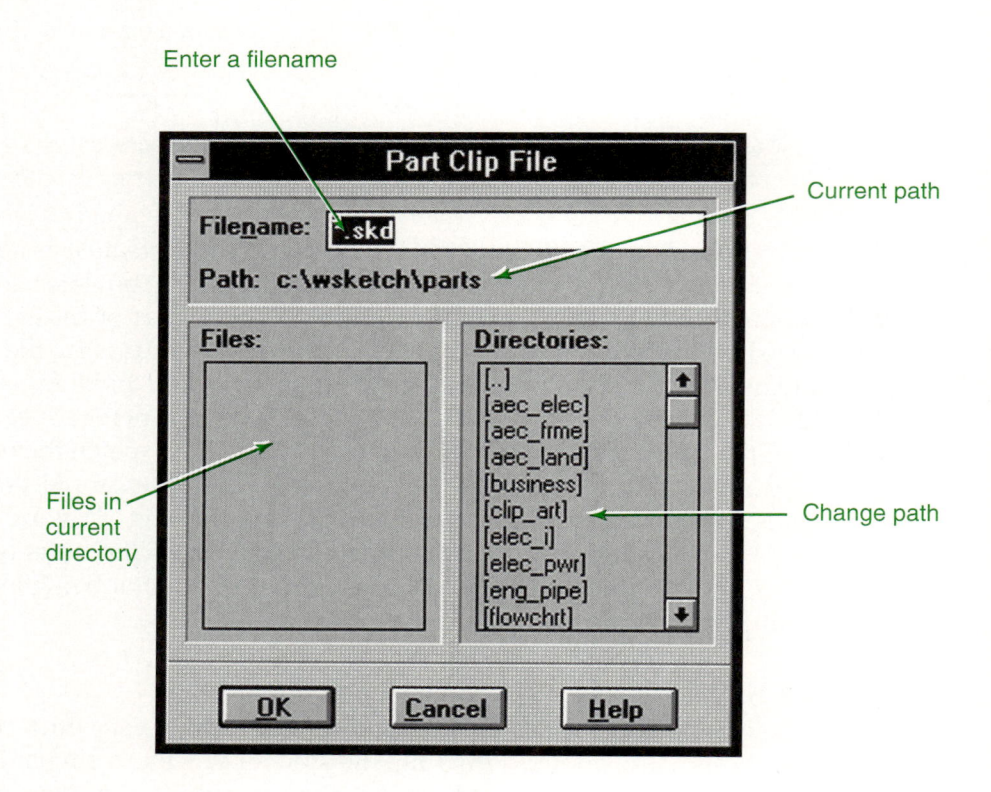

Also select the drive and directory where the symbol should be stored. Symbols are often saved to a special disk, or in a special directory. This is called a *symbol library*. A symbol library is a directory on your disk where similar symbols are stored. For example, all door symbols may be stored in a \DOOR directory. All window symbols may be stored in \WINDOWS. Any directories for your symbols must be created before trying to save to them. Use the Windows File Manager program. These directories keep symbols organized and make finding them easier. To select the directory, simply double-click on the directory. You may need to scroll through the list of available drives and directories. Also, remember that clicking [..] takes you up one level in the directory path. Directories are discussed in Chapter 20.

If you install the symbol libraries that come with AutoSketch they are placed in subdirectories under the \WSKETCH\PARTS directory on your hard disk drive. When you create parts with **Part Clip...** command, AutoSketch stores them in the directory you select. That directory also becomes the new default directory for that drawing.

Choose a Part Base Location

After picking **OK** in the **Part Clip File** dialog box, AutoSketch will prompt you for a part base location.

Part Clip Part base location:

The *part base location* is the place on the symbol where the cursor will "hold" when you insert the symbol. The cursor will drag the symbol around by this location. Refer to Figure 14-5. This point is important because every symbol connects to objects around it in a certain way. A door symbol, for example, connects to a wall by its hinge point. Other symbols might connect by their center or an edge on one object in the symbol. Pick the part base location where the symbol typically connects to features around it. Use **Attach** modes to pick this location at a precise place on an object. You can also enter a coordinate value to select the part base location.

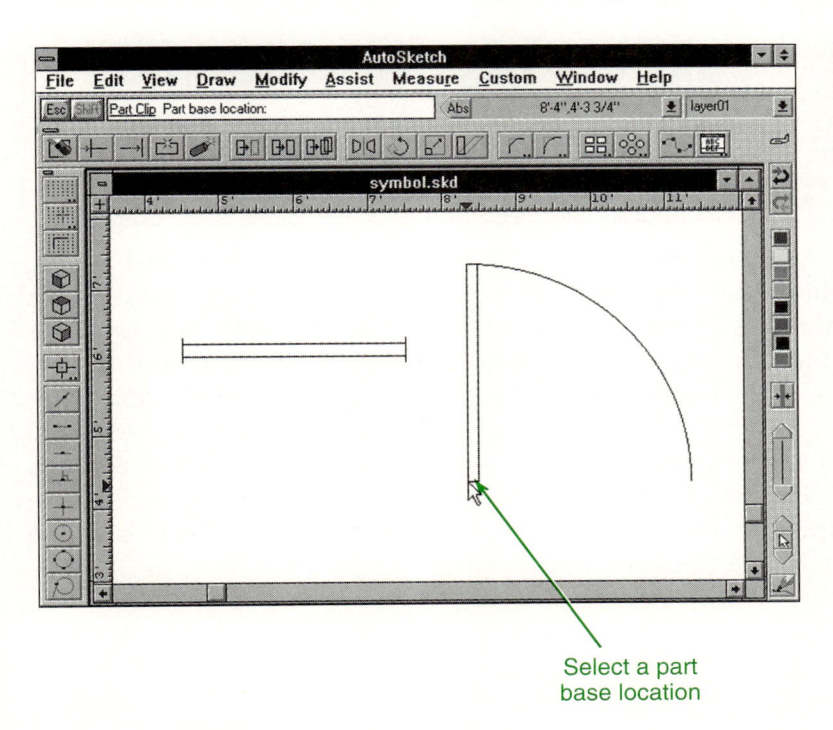

Figure 14-5.
Select a part base location. This is where the cursor will "hold onto" the part when inserting it in a drawing.

Select the Objects

Once you select the part base location, you must tell AutoSketch what will make up the object. The following prompt appears after you select the part base location.

<u>Part Clip</u> Select object:

Since you have already grouped the objects, simply pick one of the objects that makes up the symbol. The entire symbol is selected and saved to disk.

Exercise 14-1

1. Start a new drawing using the **A_LI_ARC** template, but reset the snap for 4". You can also start a new drawing and set up the scale to be 1/4" = 1'-0" with **Feet, inches and fraction** as the units of measurement.
2. Draw the architectural symbols shown below using the 4" grid for reference. Place them on a layer named "floor."
3. Use the **Part Clip...** command to save each symbol to your work diskette using the name given underneath. Consider how the symbol might be inserted when choosing a part base.
4. Save the drawing as C14E1 on your work diskette and then close the drawing.

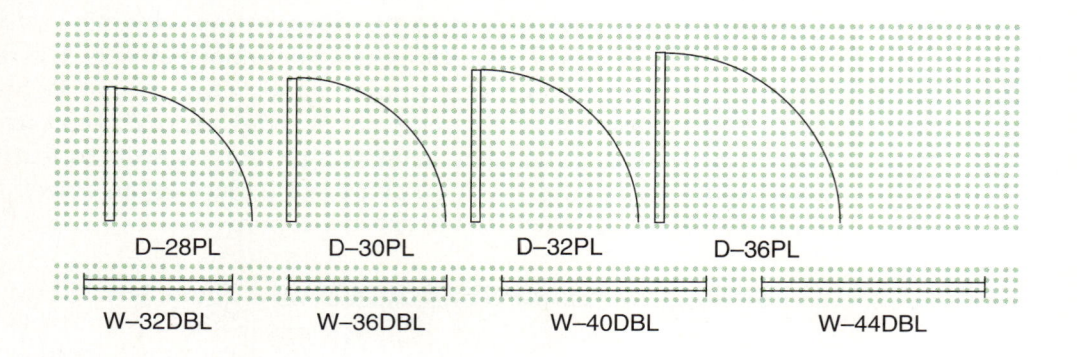

D–28PL   D–30PL   D–32PL   D–36PL

W–32DBL   W–36DBL   W–40DBL   W–44DBL

Saving an Entire Drawing as a Symbol

AutoSketch allows you to insert any existing drawing into your current drawing. Drawings inserted without using the **Part Clip...** command use the origin (0,0) as a part base. The cursor will drag the symbol around by this location. Yet, if the drawing contains small objects far from the origin, you may not see the symbol when inserting it.

You can change the part base for a drawing without using the **Part Clip...** command. First, group the objects in the drawing. Next, select the **Custom** menu and **Drawing Settings** tool. Then, pick the **Part Base...** button to open the **Part Base Settings** dialog box, Figure 14-6. Enter X and Y coordinate location for the part base and choose **OK**. Finally save the drawing. It is best to use the **Save As...** command and save the symbol in the directory where you store your symbols.

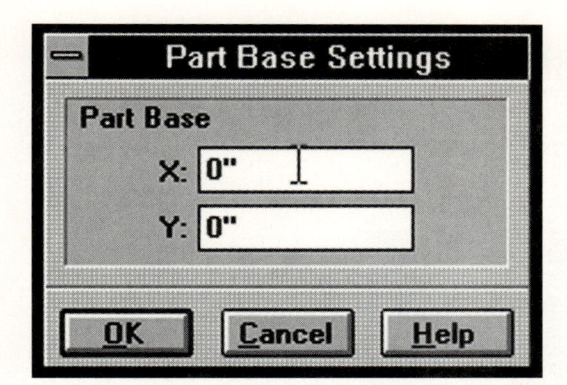

Figure 14-6.
To change the part base of an existing drawing, enter the coordinates in the **Part Base Settings** dialog box.

Exercise 14-2

1. Start a new drawing using the **A_LI_MEC** template.
2. Draw the following basic welding symbol, placing the arrow of the leader at the absolute coordinate 4,4 on the drawing. Do not add the dimensions.
3. Use the **Part Base...** button to set the part base at 4,4.
4. Save the drawing as BASIC on your work diskette and then close the drawing.

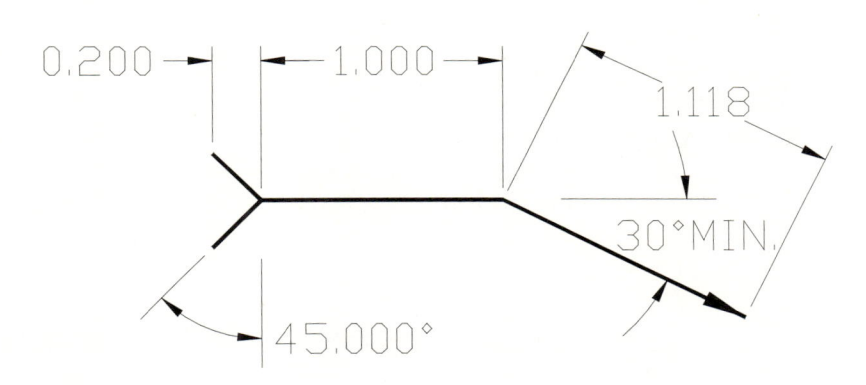

Inserting a Symbol in a Drawing

To insert a symbol in a drawing, select the **Part Insert...** command from the **File** menu. The **Select Part File** dialog box appears, Figure 14-7. Here you chose the symbol you want to insert. Each symbol appears as a small icon. When you click on an icon, the filename of that symbol appears in the **Active Filename** box. If the symbol you need is not listed, select the drive and/or directory where the symbol is stored. To insert the part at its original rotation and size, make sure that **Rotate Part** and **Scale Part** are not selected. Then pick **OK**. The following prompt appears.

Part To point:

Figure 14-7.
You can select the symbol to insert by an icon or filename in the **Select Part File** dialog box. You may need to change the drive and/or directory to find the symbol.

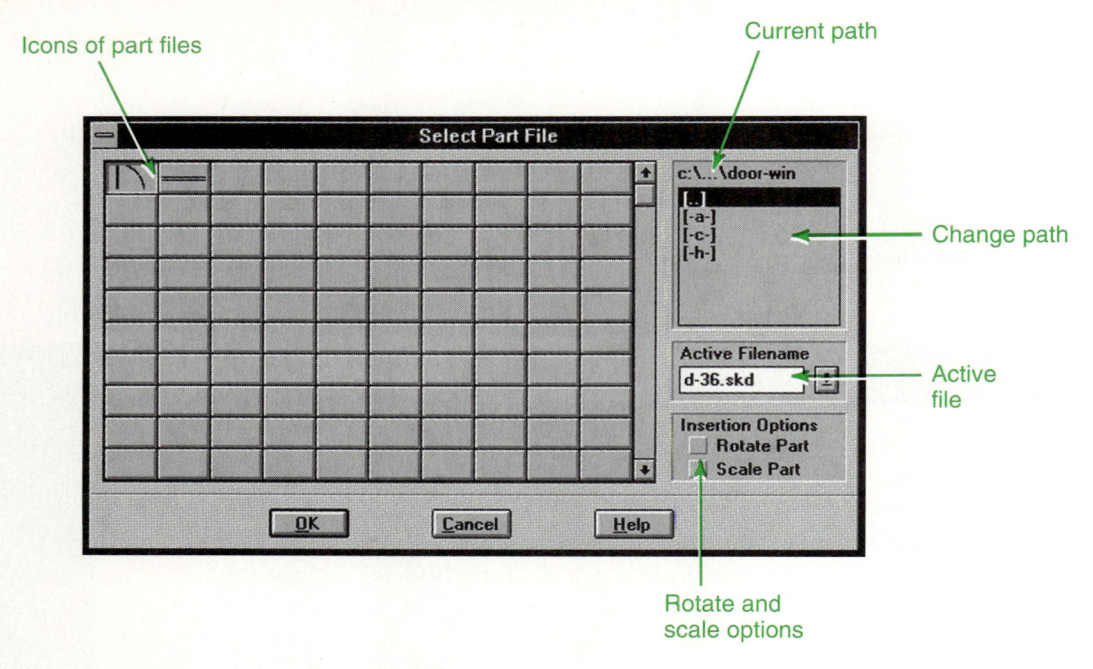

Icons of part files

Current path

Change path

Active file

Rotate and scale options

Pick the location on the drawing where the symbol should be placed. You can visually "drag" the symbol into place, Figure 14-8. The symbol is attached to the cursor by the part base. Use an **Attach** mode if you need to connect the symbol precisely to another object in your drawing. In some cases you might enter coordinate values to locate the part.

Figure 14-8.
Drag the symbol into place. Use an **Attach** mode to connect the symbol precisely to another object. Here, a door symbol is being inserted into a wall.

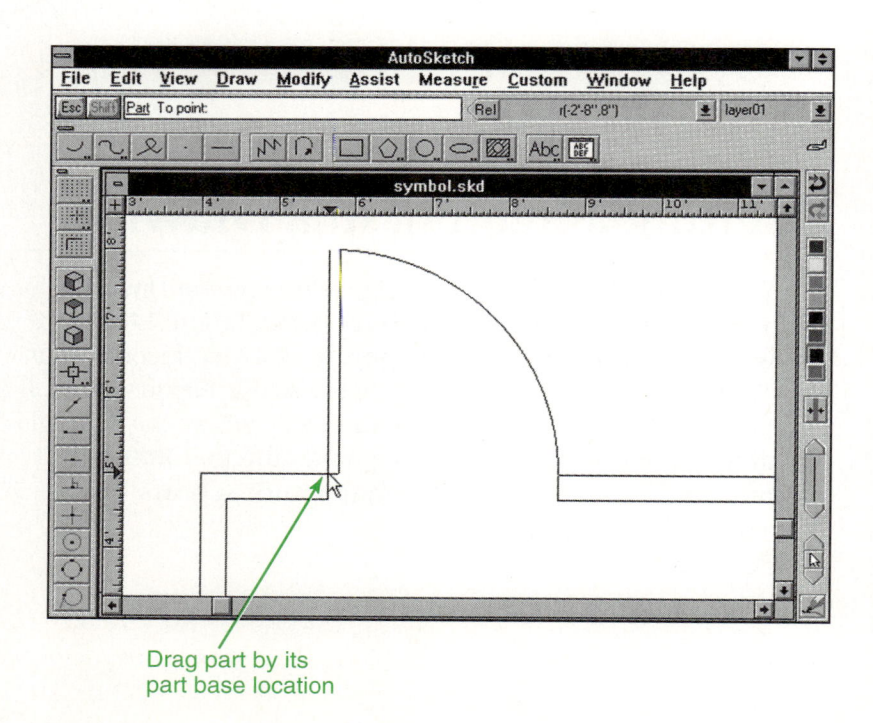

Drag part by its part base location

Exercise 14-3

1. Start a new drawing using the **A_LI_ARC** template.
2. Develop the following drawing. Use the symbols created in Exercise 14-1 to place the door and window.
3. Save the drawing as C14E3 on your work diskette and then close the drawing.

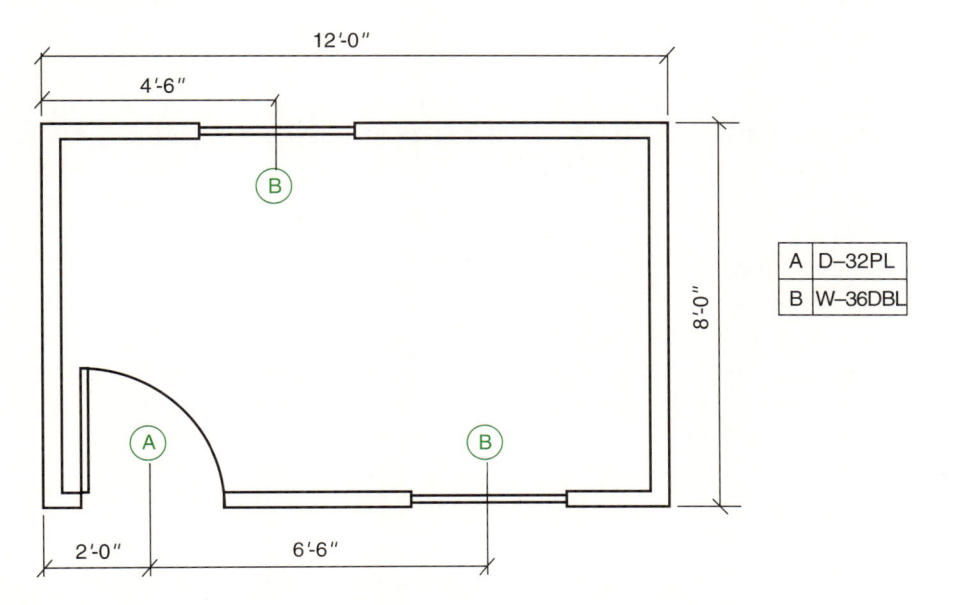

A	D–32PL
B	W–36DBL

Insertion Options

If you need to scale or rotate the part when inserting it, there are two options available. These are **Scale Part** and **Rotate Part.** If either of these options is chosen (a check mark appears), the following is the first prompt you will see.

   Part Reference point:

The reference point serves as a base point for scaled symbols, and a center of rotation for symbols that will be rotated.

Scale

When **Scale Part** is chosen, the reference point serves as a base point. This works exactly like the **Scale** tool. The reference point remains fixed while the symbol is enlarged or reduced around it. Select your base point carefully. Usually, there is one point on the shape that should remain in the same location. The following prompt appears.

   Part Scale factor:

You can either enter an exact scale factor, or use the cursor. To enter an exact scale, type in a value and then press the [Enter] key. A scale factor less than 1 reduces the size of the symbol. A scale factor larger than 1 increases the size of the symbol. Using the cursor, the second point you pick determines the scale factor. Move your cursor in or out to increase or decrease the size in 1/10 increments. A readout in the prompt box shows the scale factor. Click your mouse button when you reach the desired scale factor.

Exercise 14-4

1. Start a new drawing using the **A_LI_MEC** template.
2. Select a symbol from the **\WSKETCH\PARTS\CLIP_ART** directory.
3. Insert the symbol into your drawing two times, but use the **Scale Part** option of the **Part Insert...** command to scale the part both larger and smaller than the original.
4. Save the drawing as C14E4 on your work diskette and then close the drawing.

Rotate

When **Rotate Part** is chosen, the reference point serves as a center of rotation. This works exactly like the **Rotate** tool. Pick the pivot that the symbol should rotate around. The following prompt appears.

Part Rotation angle:

You can either enter the rotation angle, or use the cursor, Figure 14-9. To enter an exact rotation, type in an angle value. When using your cursor, the second point you pick with the cursor determines the angle of rotation. A readout in the prompt box shows the angle of rotation.

Figure 14-9.
When **Rotate** is selected in the **Select Part File** dialog box, you can rotate the part when inserting it. The part will rotate around the part base location.

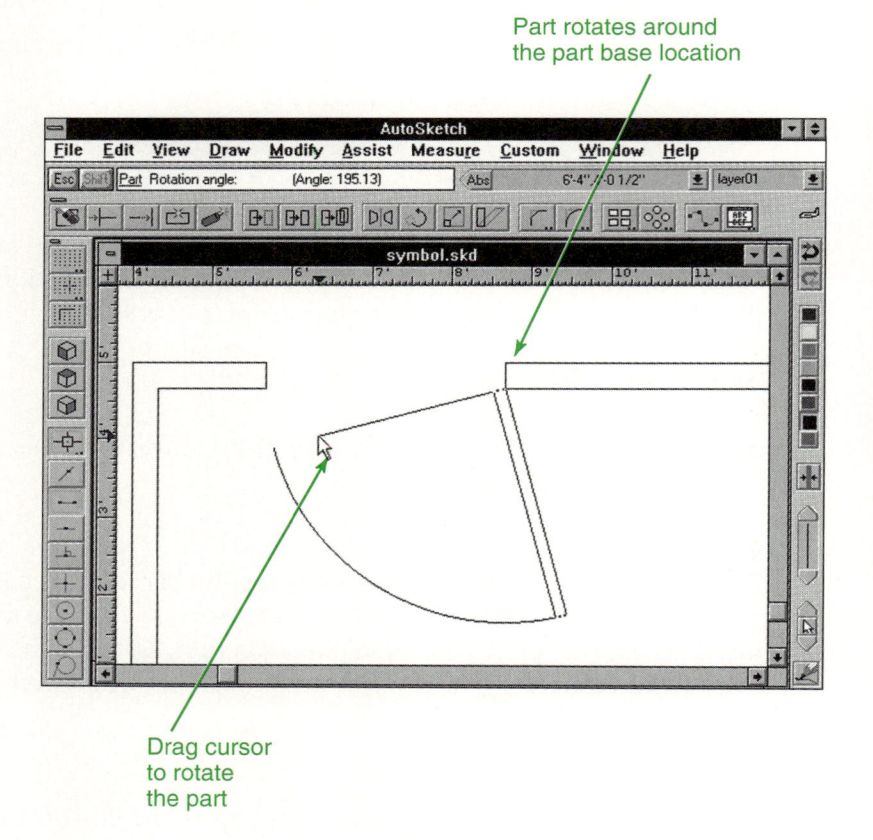

Part rotates around the part base location

Drag cursor to rotate the part

Exercise 14-5

1. Open the drawing C14E3.
2. Add to the drawing as shown below. Use the **Rotate Part** option of the **Part Insert...** command to place the symbols properly. Use the **Break** tool to remove unneeded portions of the walls.
3. Save the drawing as C14E5 on your work diskette and then close the drawing.

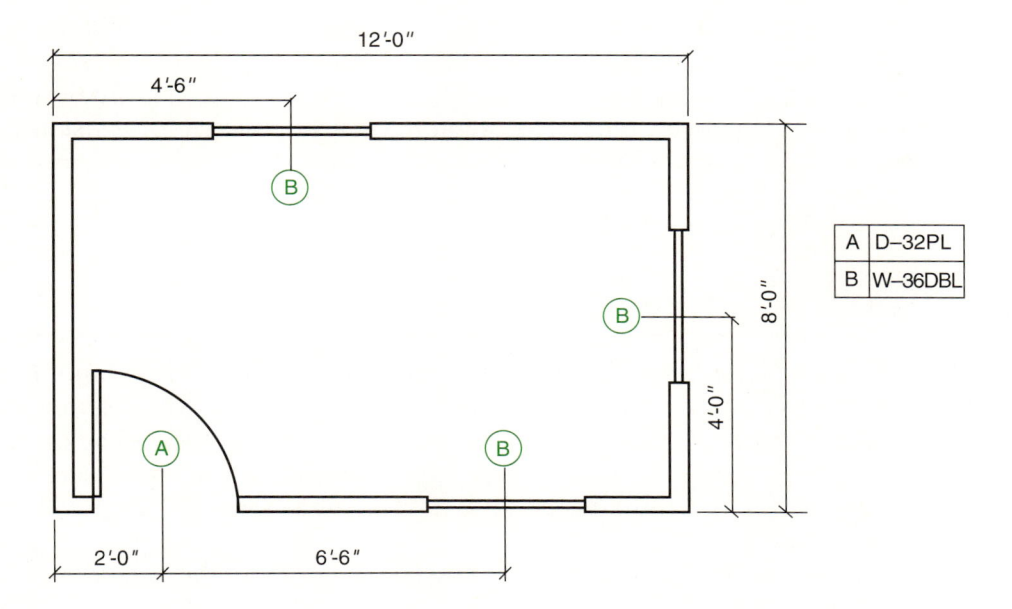

A	D–32PL
B	W–36DBL

Symbol Properties and Layers

When inserted, a symbol retains the object properties that it was created with. It will also retain the layer that it was created on. If you insert the symbol into a drawing that does not have that layer, the layer name will be added. For example, suppose you create a window symbol, and that symbol was drawn on the "floor" layer. Then, you insert the symbol in a different drawing that does not have a "floor" layer. That new drawing will then contain a layer named "floor" after inserting the symbol.

> **SHORTCUT**
> You can access the **Break** tool by pressing the function key [F4].

Editing Symbols

Sometimes, a change may be required after you insert the symbol. You can edit symbols just as you would lines, circles, and arcs. However, a symbol is treated like a single object if you grouped it before saving it. When you pick a grouped symbol to edit, all objects that make up the symbol are affected. Thus, you can move, copy, rotate, or scale the entire symbol. However, to edit individual objects of the symbol, you must first ungroup it. If the changes needed in the symbol are permanent, go back to the original symbol drawing and make the changes.

You can even customize the symbols that come with AutoSketch. Simply open the symbol as a drawing. Make the necessary changes. Then, use the **Save As...** command to save the symbol under a different name. If you save the symbol using the same name, the original symbol provided by AutoSketch is lost.

Symbols Library Management

Keeping track of symbols can be a major task. Symbols may constantly be added if other drafters, engineers, and designers are creating them. A symbol library helps keep symbols organized. In addition, a printed symbol library handbook should be created to keep track of the symbols. This contains all symbols and information relating to each symbol. This book might include a diagram or sketch, dimensions, the symbol name/number, symbol filename, and symbol part base location. Organize the handbook by the products the symbols represent.

Summary

Symbols are individual AutoSketch drawings that are inserted into other drawings. AutoSketch refers to any drawing inserted into another as a "part." Symbols are used to represent a standard component or assembly. A symbol is created by drawing it to scale, grouping it, and selecting the **Part Clip...** command. You then select a part base location and select the group. When saving the symbol, store it in a special drive and/or directory, called a symbol library.

When you insert a symbol using the **Part Insert...** command, the cursor drags the symbol on-screen by the part base location. You can rotate or scale the symbol as it is inserted. You can also edit the inserted symbol just like any group. However, to edit the objects that make up the symbol, you must first ungroup the symbol. Permanent changes should be made to the original symbol file.

Important Terms

Part base location
Standards organization
Symbol
Symbol library

New Tools and Commands

Active Filename box
Part Base... button
Part Base Settings dialog box
Part Clip...
Part Clip File dialog box
Part Insert...
Rotate Part
Scale Part
Select Part File dialog box

Review Questions

Give the best answer for each of the following questions.

1. Explain the difference between copying objects in the same drawing and using symbols.
   _____
   _____
   _____

2. A(n) _____ is a collection of objects that represents a standard component, assembly, or feature.
3. To what scale should symbols be drawn?
   _____
4. Standard symbol shapes are typically set by _____.
5. When drawing a custom symbol, keep it _____ since a symbol only represents the real object.
6. Parts should be drawn to what scale?
   _____
7. After drawing the symbol, you should _____ the objects. How does this affect the symbol when it is inserted?
   _____
8. When would you use the **Part Base...** button rather than the **Part Clip...** command?
   _____
9. When you insert a part, the pointer drags the part on-screen by pulling its _____.
10. What location is best for a part base location point?
    _____
11. Using _____ when placing parts with the pointer makes sure that the part connects precisely to other objects.
12. Where are the symbols that come standard with AutoSketch stored?
    _____
13. When inserting a symbol, what two options allow you to change its appearance?
    A. _____
    B. _____
14. Suppose you create a symbol on the "foundation" layer. You then insert the symbol in a drawing. What layer is the symbol placed on?
    _____
15. How do you edit a symbol that has been inserted into a drawing?
    _____
    _____

Activities

1. - 2. Create the following weld symbols as shown. Use the basic weld symbol called **BASIC** that you created in Exercise 14-2 to start. Save each symbol using the **Part Clip...** command. Then, create each drawing shown below and insert the appropriate symbol.

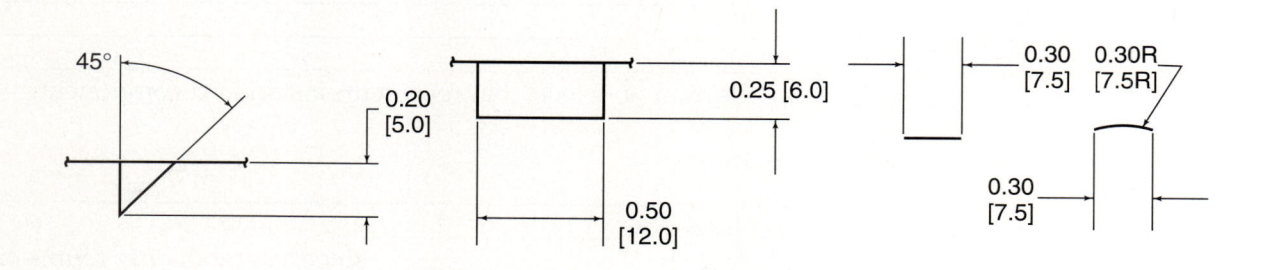

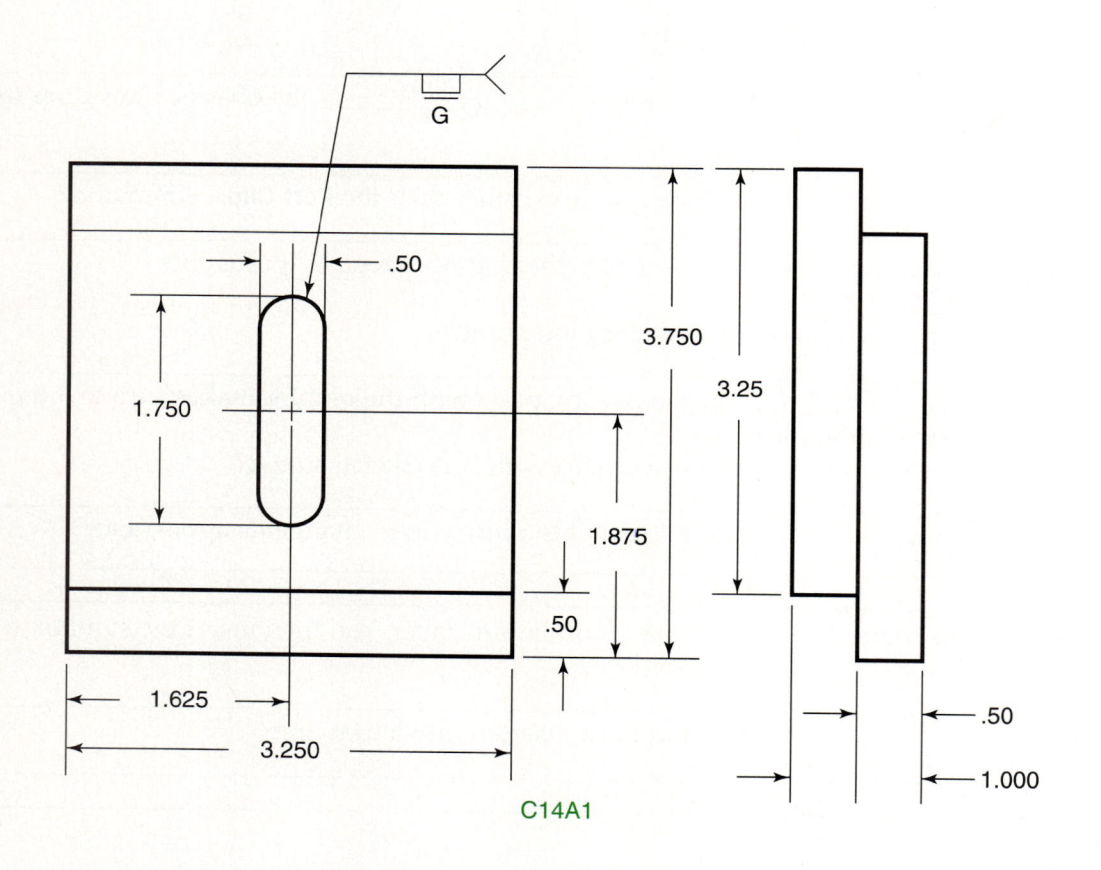

C14A1

C14A2

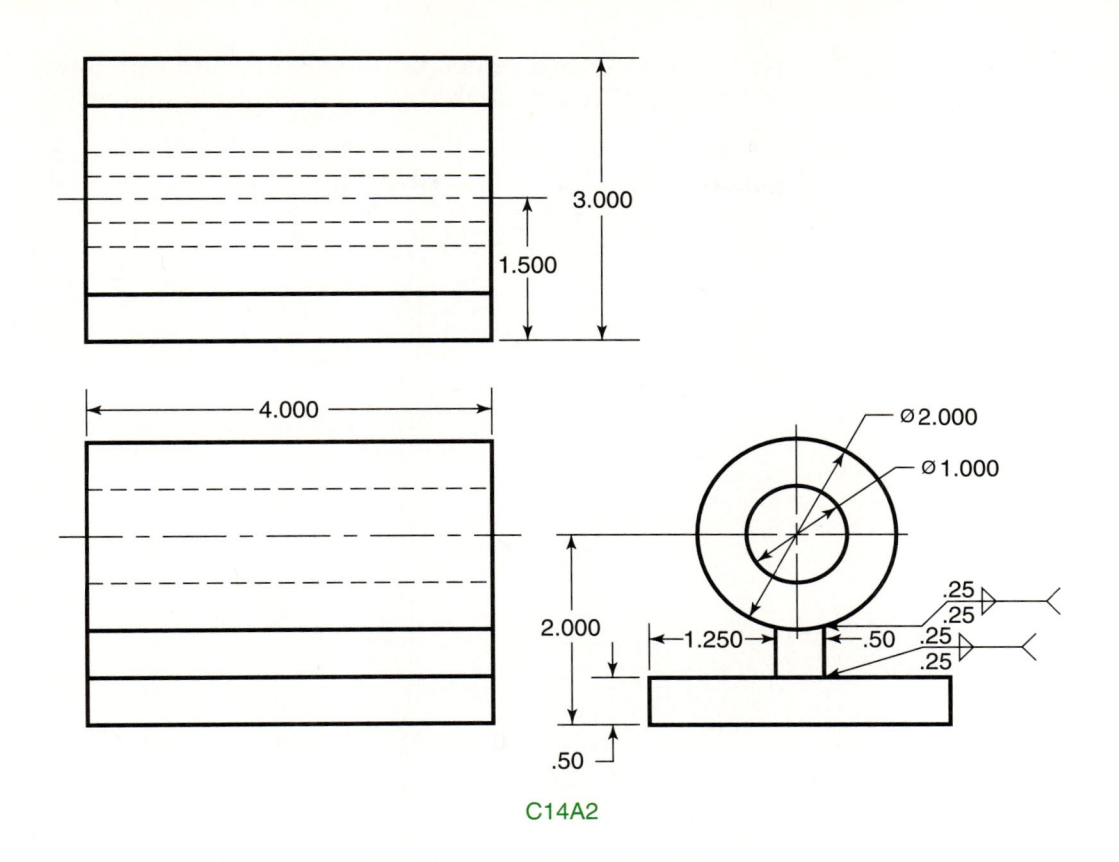

3. Draw parts for the following office furniture: desk, chair, and file cabinet. Then, create several office layouts by inserting the office furniture symbols.

4. Create a landscape plan for a Community Center, similar to that shown here. Develop a part library for the symbols in this drawing. Be sure to include the trees, shrubbery, and cars.

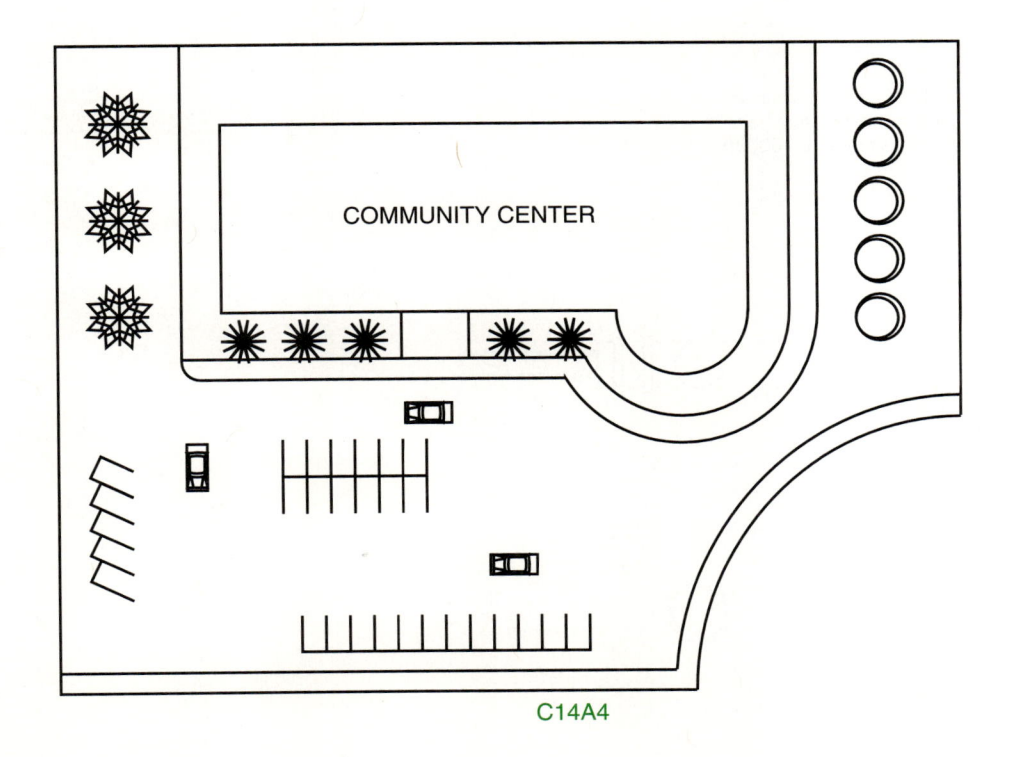

C14A4

5. Create the following floor plan for a recreation center. Develop a furniture part library for the common parts found in this drawing. You can use the trees and shrubs from Activity 4.

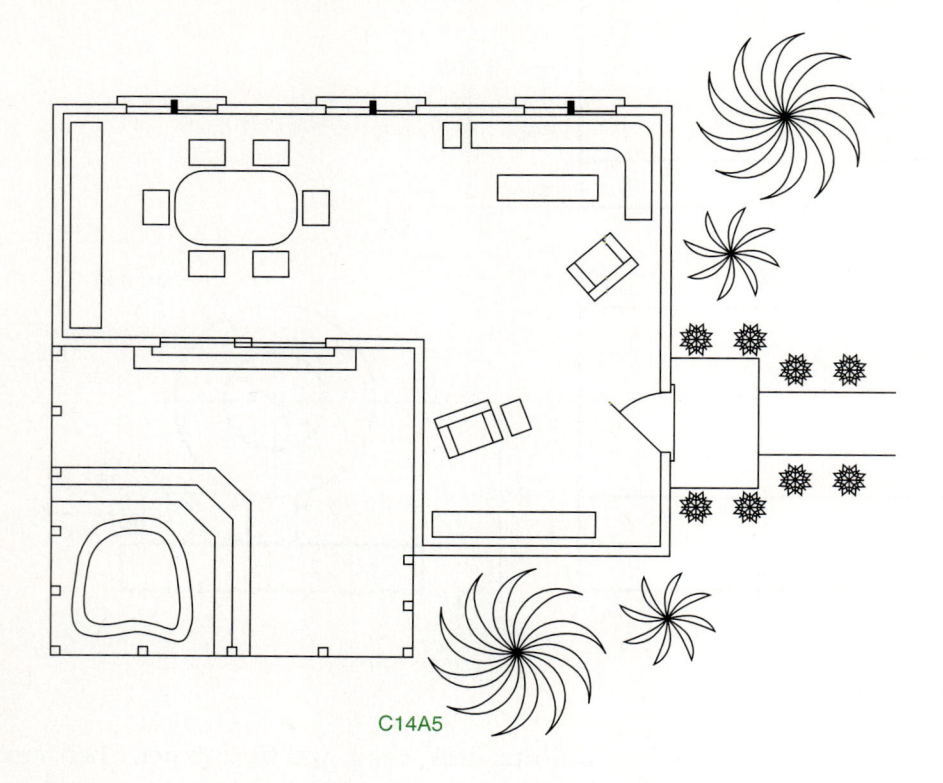

C14A5

6. The object on the left is the symbol for an LED light. Develop the drawing on the right to use in a symbols library handbook.

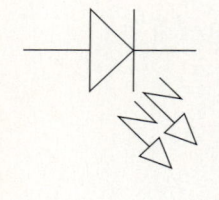

APPROVED SOURCE
OF SUPPLY
VCH INC.
CHEVY CHASE BUS. PARK
1080 Johnson Dr.
Buffalo Grove, IL. 60089
Cage Code:
Part #: CMD5053-200

Only the item described on this drawing when procured from the vendor(s) listed hereon is approved by Barber-Colman Co., for use in the application(s) specified hereon. A substitute item shall not be used without prior testing and approval by Barber-Colman Co., or by the appropriate government activity.

Identification of the approved source(s) of supply hereon is not to be construed as a guarantee of present or continued availability as a source of supply for the item.

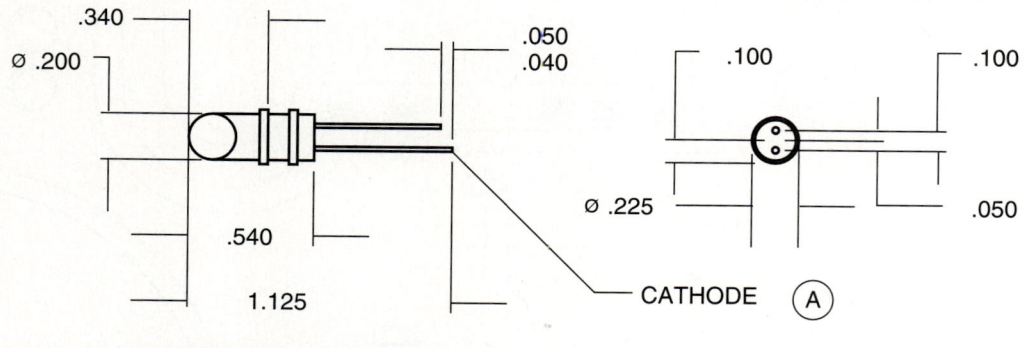

.340
Ø .200
.050
.040
.100
.100
Ø .225
.050
.540
1.125
CATHODE (A)

SPECIFICATIONS
L.E.D. Color: Red
Brightness: 0.5MCD @ 20MA
Power Dissipation: 180MW
Operating Temp.: -55°C to 100°C
Continuous forward current: 100MA
Peak forward current: (IU Sec. Pulse)
(0.3% Duty Cycle) 1.0AMP
Reverse Voltage: 5.0V

CAUTION
Electrostatic sensitive device: Use precautionary procedures when handling and shipping parts.

Dimensions for reference only.

C14A6

(Barber-Colman Co.)

CHAPTER 15

Drawing Pictorial Views

Objectives
After studying this chapter, you will be able to:
- Explain the purpose of oblique, perspective, and isometric views.
- Set left, right, and top isometric drawing planes using AutoSketch.
- Draw an isometric view of a product.
- Create isometric text.
- Demonstrate isometric dimensioning techniques.

Pictorial drawings are three-dimensional views of an object. Pictorial means "like a picture." This refers to many methods of creating a drawing that looks realistic. A pictorial drawing is an excellent way to describe a product's shape or use. The drawing shows the product's height, width, and depth all in one view. Pictorial drawings are often used in technical illustration. These drawings may also show how a product is assembled, Figure 15-1.

There are four popular methods to draw pictorial drawings. These are oblique, isometric, perspective, and three-dimensional (3D) modeling. You can create the first three types of pictorial drawings using AutoSketch. Two-dimensional drawing objects (such as lines and circles) are used to make these drawings *look* three-dimensional. However, AutoSketch does not support 3D modeling. This type of drawing requires special drawing commands that create true three-dimensional objects.

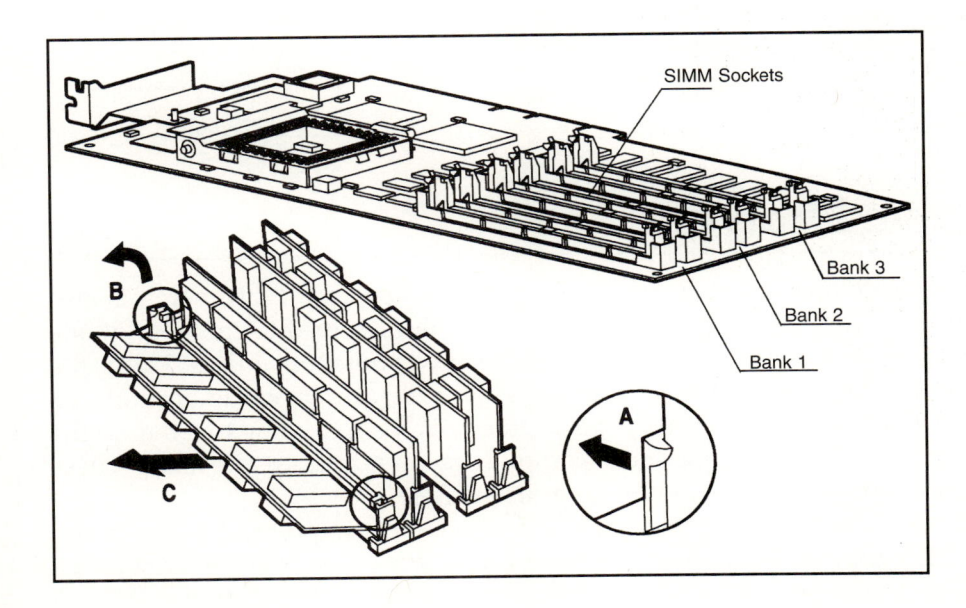

Figure 15-1.
Isometric views are commonly used to show how a product is assembled.

421

Oblique Drawing

An *oblique drawing* shows the front of an object in true size and shape. The depth, however, is represented by lines extending back at an angle between 30° and 60°. See Figure 15-2. When these lines are drawn full length at 45°, the drawing is called a *cavalier oblique drawing*. When these lines are drawn half length at 45°, the drawing is called a *cabinet oblique drawing*. When these lines are drawn three-quarter length at an angle less than 45° (usually 30°), the drawing is called a *general oblique drawing*.

Circular shapes, holes, and curved edges in the front view of an oblique drawing are shown as circles and arcs. However, in the side view they are shown as ellipses, partial ellipses, or curves. To draw these shapes, you must first locate points on the curve and then draw arcs to fit through these points.

When determining the side to be used as the front face of an oblique object, choose the side with the most contour. This will prevent you from having to draw ellipses and curves in the side view. Also, for long objects, place the object so that the longest dimension is horizontal on the drawing and not shown as depth.

Exercise 15-1

1. Start a new drawing using the **B_LI_MEC** template created in Chapter 10.
2. Using the multiview drawing below, create a cavalier oblique, a cabinet oblique, and a general oblique drawing in the same file.
3. Save the drawing as **C15E1** on your work diskette and then close the drawing.

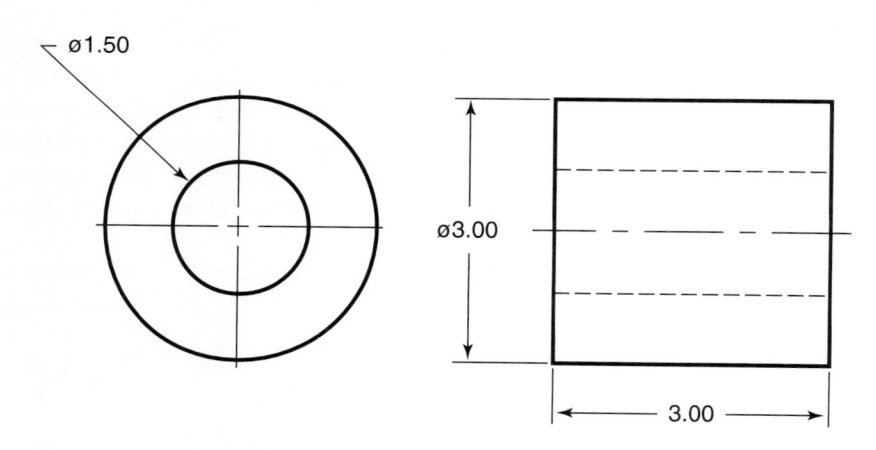

Perspective Drawing

The most realistic form of pictorial drawing is a perspective. However, it is also the most difficult to draw. With a perspective drawing, lines appear to converge at one or more imaginary points in the distance. These are called *vanishing points*. This is the most realistic form of drawing because your eyes naturally see the world around you in perspective. If you look down railroad tracks, the rails seem to come together in the distance. Perspective drawings are most common in architectural drawing. These drawings are also used to some extent in other industries.

The most common types of perspective drawings are one-point and two-point, Figure 15-3. With a *one-point perspective* drawing, one face of the object is drawn true size and shape. The top and side converge at one vanishing point. With a

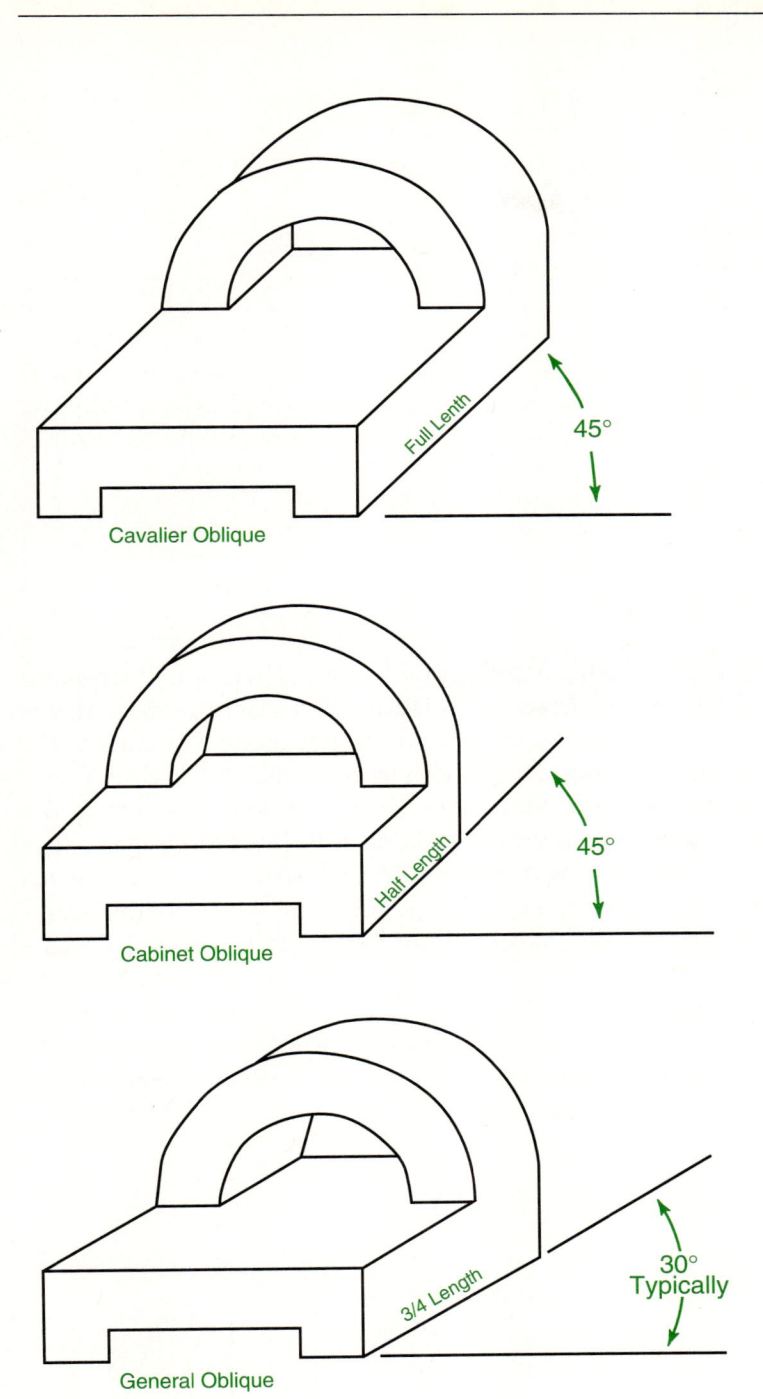

Figure 15-2.
There are three types of oblique drawings.

Cavalier Oblique — Full Length — 45°

Cabinet Oblique — Half Length — 45°

General Oblique — 3/4 Length — 30° Typically

two-point perspective, the object is viewed from an angle. One side of the object converges toward one vanishing point, while the other side converges on a second point.

The steps to draw a perspective drawing are not covered in this book. Perspective drawings are difficult, and require much more information than can be covered in this chapter. In addition, the techniques to draw a perspective drawing using AutoSketch are the same as for manual drafting. It is best that you consult an advanced drafting text if you have a need to construct perspective drawings.

Figure 15-3.
Two most common types of perspective drawings are one-point and two-point.

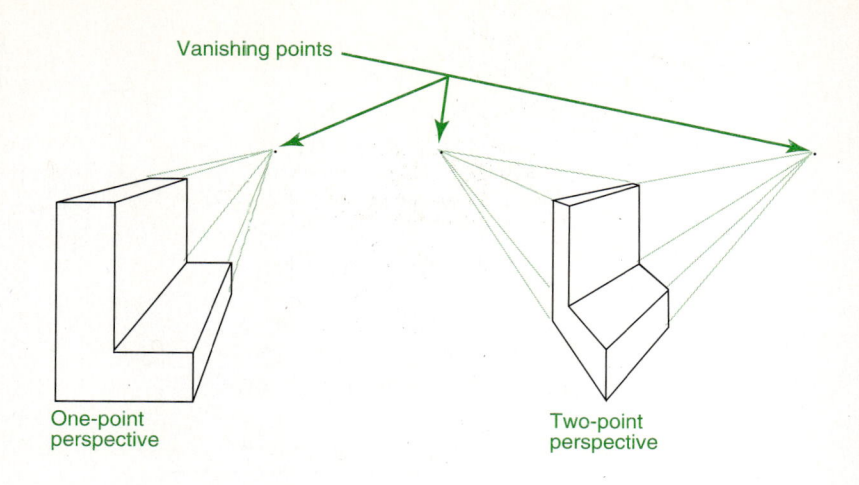

One-point perspective

Two-point perspective

Vanishing points

Isometric Drawing

An *isometric drawing* is a very realistic pictorial drawing. This type of drawing shows the product as if it is tilted toward the viewer, Figure 15-4. An isometric view is not only realistic, but fairly easy to draw. All shapes are drawn using the same scale. Plus, AutoSketch offers a special mode that allows you to create isometric drawings.

Like an oblique drawing, an isometric shows the front, top, and one side of the product. Vertical edges of the object are drawn vertical and full size. Horizontal edges are drawn full size, but at a 30° angle from horizontal in each direction. This creates three axes, Figure 15-5. Lines that are vertical in an orthogonal view (multiview drawing) are placed vertically. Any line parallel to one of the three axes is drawn full scale and can be measured. These are called *isometric lines.* However, any line not parallel to one of the three axes cannot be measured. These lines are called *nonisometric lines.* Refer to Figure 15-5. All holes and other circular shapes are shown as if they are tilted away from the viewer. These shapes are drawn with isometric ellipses. Refer back to Figure 15-4.

Figure 15-4.
An isometric drawing is a realistic pictorial drawing that shows three sides of the object as if the object is tilted toward the viewer.

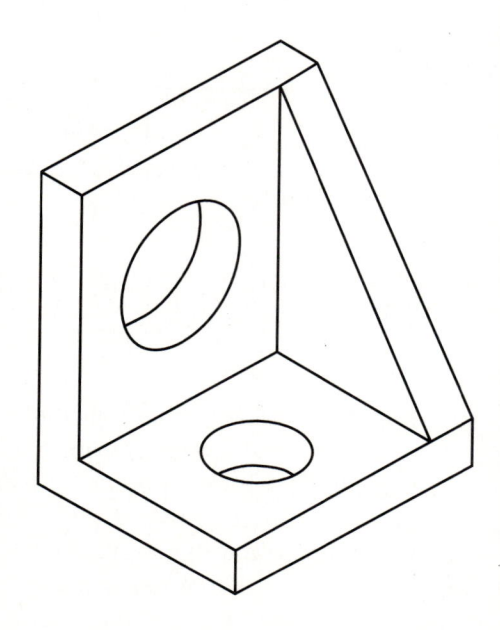

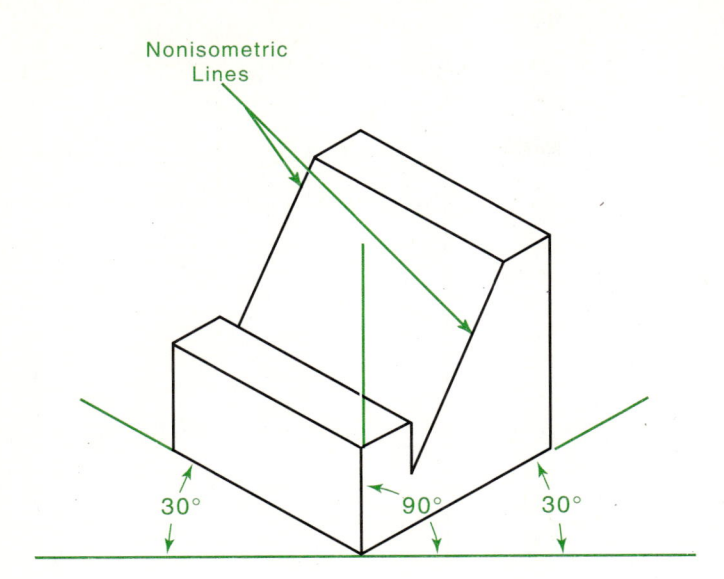

Figure 15-5.
In an isometric drawing, depth and width are shown with lines at 30° from horizontal. Lines that are vertical in an orthographic view are vertical in an isometric view. Any line not parallel to one of the three axes is called a nonisometric line.

Isometric Coordinate System

When drawing in isometric mode, AutoSketch borrows the coordinate system from 3D modeling. It is necessary that you understand this system before attempting to draw an isometric drawing using AutoSketch.

The three dimensional coordinate system begins with a familiar two-dimensional (2D) concept. Remember from Chapter 5 that the Cartesian coordinate system is a method locating objects on a 2D drawing. It consists of two intersecting axes. The horizontal axis is called the X axis. The vertical axes is called the Y axis. Every point, line, or other object can be defined by its position with respect to distance along the X and Y axes.

The three-dimensional coordinate system adds a third axis. This is called the Z axis. See Figure 15-6. The Z axis is perpendicular to the plane defined by the X and Y axes. All three intersect at the origin. Think of it this way. The Z axis now represents height in the isometric drawing. The X axis now represents width or depth. It is placed along the 30° angle of the isometric drawing. The Y axis now represents width or depth. It is placed along the 150° angle of the isometric drawing. All three axes are needed to describe a three-dimensional object.

Having three axes also means that three coordinates are now needed to locate a point. These are the X coordinate, the Y coordinate, and the Z coordinate. Figure 15-7 shows the location of points using the three-dimensional coordinate system. You will find that, like with drawing 2D drawings, you can enter both absolute and relative coordinates.

Exercise 15-2

1. Sketch a three-dimensional coordinate system similar to the one shown in Figure 15-6.
2. Place a dot on the coordinate system to locate the following absolute coordinate points.
   1,1,1
   2,3,1
   4,1,2
   5,5,7

Figure 15-6.
The three-dimensional coordinate system is used when creating an isometric drawing in AutoSketch.

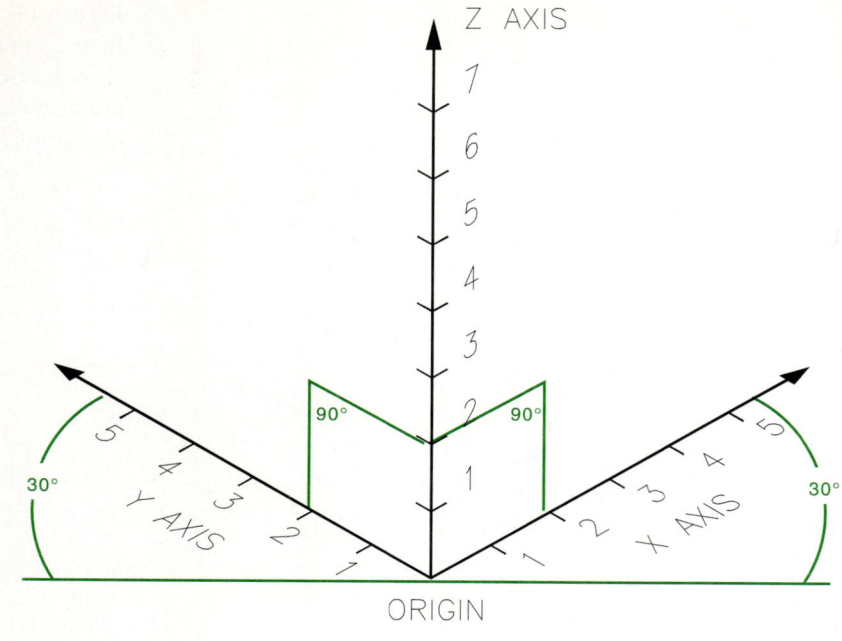

Figure 15-7.
Three coordinates are required to locate a point in the three-dimensional coordinate system.

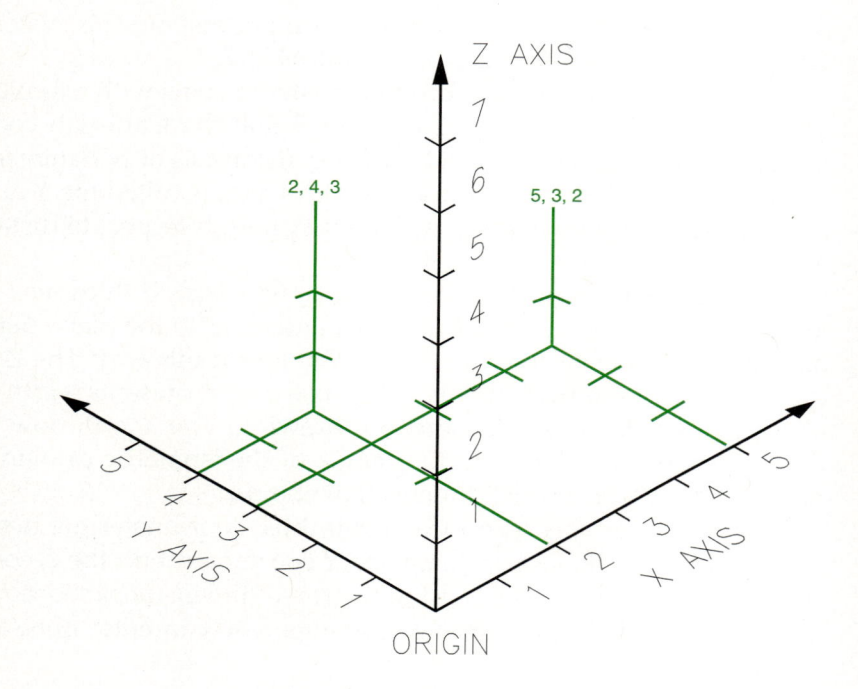

Using the Isometric Mode of AutoSketch

AutoSketch provides an excellent tool for creating isometric drawings. This is the isometric mode. When enabled, it changes the way you draw objects. **Isometric mode** causes AutoSketch to use a three-dimensional coordinate system to represent the three axes of an isometric drawing. The surface formed by any two of these axes is called an **isometric plane.** An object drawn in that plane will appear isometric. This mode is very helpful since you can draw as if you were creating an orthographic (two-dimensional) view. AutoSketch makes all the isometric conversions for you.

Choosing Isometric Planes

AutoSketch allows you to draw in any one surface, or isometric plane, formed by two of the three axes. To select an isometric plane to draw on, choose the **Left Isometric Plane, Top Isometric Plane,** or **Right Isometric Plane** command in the **Assist** menu, Figure 15-8A.

- The left isometric plane is formed by the Y and Z axes. Drawing a box, circle, or arc in this plane makes it appear as if viewed from the left. Refer to Figure 15-8B.

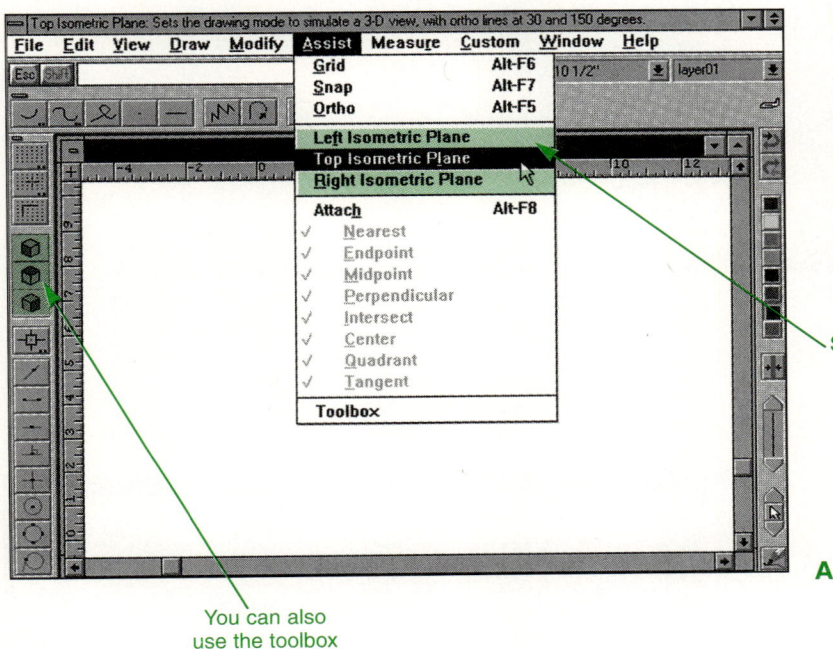

Figure 15-8.
A—Select one of the isometric commands (highlighted here) from the **Assist** menu to draw on that isometric plane. B—A box drawn in an isometric plane appears as a parallelogram. C—A circle drawn in an isometric plane appears as an ellipse. D—An arc drawn in an isometric plane appears as part of an ellipse.

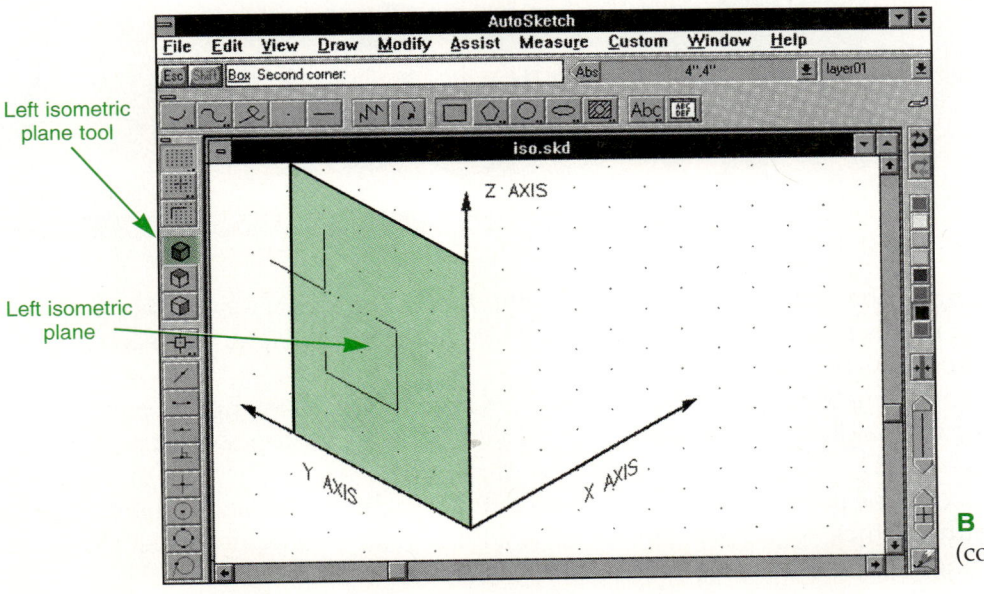

(continued)

Figure 15-8.
(continued)

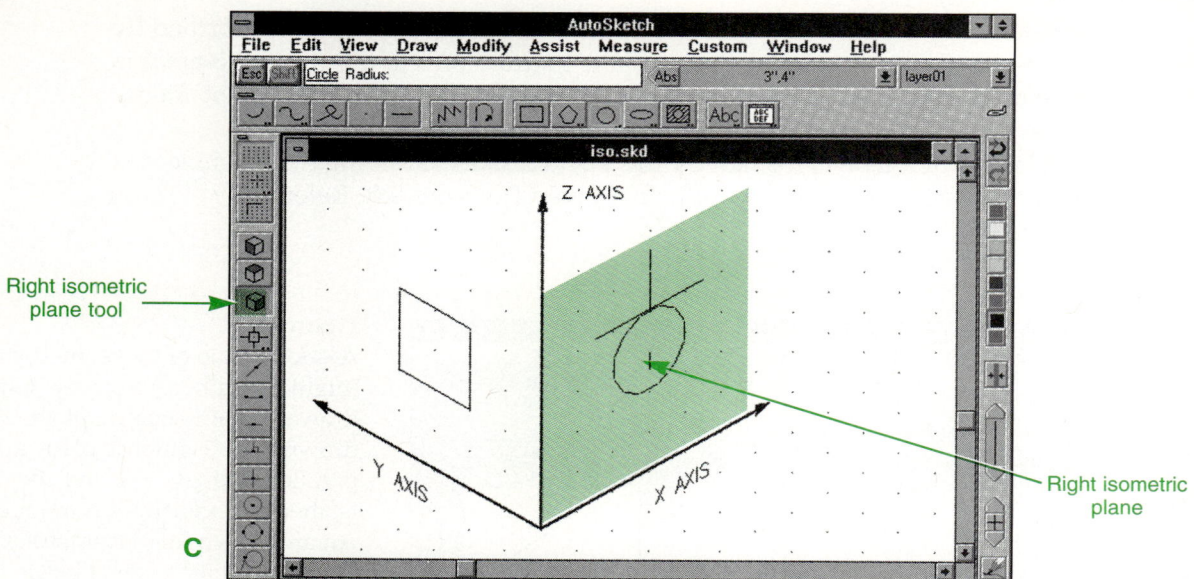

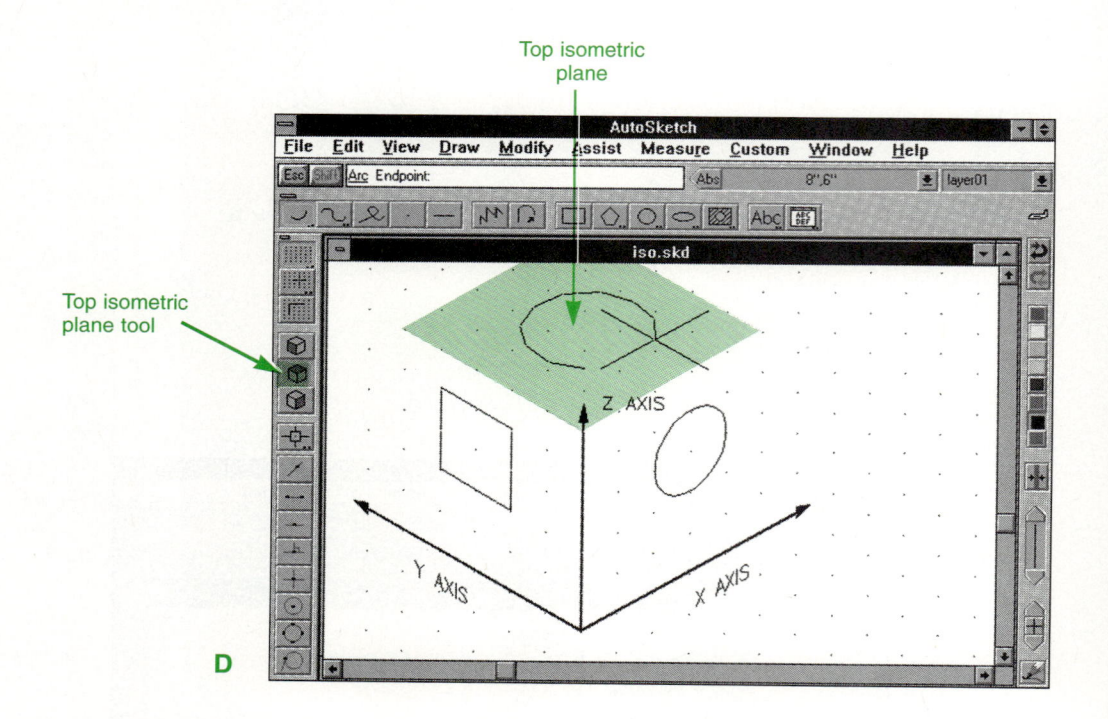

Right isometric plane tool

Right isometric plane

C

Top isometric plane

Top isometric plane tool

D

- The right isometric plane is formed by the X and Z axes. Drawing a box, circle, or arc in this plane makes it appear as if viewed from the right. Refer to Figure 15-8C.
- The top isometric plane is formed by the X and Y axes. Drawing a box, circle, or arc in this plane makes it appear as if viewed from the top. Refer to Figure 15-8D.

The isometric plane affects how AutoSketch draws boxes, circles, and arcs. It does not affect how lines are drawn. You can draw lines in any isometric plane. However, when doing so, it is often helpful to use **Grid, Snap,** or an **Attach** mode.

Isometric Grid, Snap, Ortho, and Cursor

For each isometric plane, AutoSketch changes the appearance of the grid and cursor, and also how **Snap** and **Ortho** work.
- Grid dots will align with the axes of the isometric plane.
- The snap interval will align with the axes of the isometric plane.
- The crosshairs of the cursor will align with the axes of the active isometric plane. (Change the cursor to a crosshairs by picking the **Cursors** tool in the vertical status bar.)
- Ortho restricts certain objects and movements to the axes of the current isometric plane. (When not in isometric mode, the **Ortho** tool restricts lines and movements only to horizontal or vertical directions.)

Constructing an Isometric Drawing

Several of the drawing tools act differently in isometric mode. Boxes are drawn as parallelograms with sides parallel to the axes of the active isometric plane. Refer back to Figure 15-8B. Circles and arcs are drawn as isometric ellipses and partial ellipses. Refer back to Figure 15-8C. Polygons, pattern fills, ellipses, and arc segments of polylines are *not* converted to an isometric perspective when you draw them.

Here are some general steps to take when constructing an isometric drawing of a product, Figure 15-9.
1. Draw three construction lines to represent the front edge of the object. One line is drawn vertical, along the Z axis. The other two are drawn along the X and Y axes. Refer to Figure 15-9A.
2. Draw an isometric box with construction lines to outline the largest dimensions of the product. It is best to use a different color and layer so that you can easily remove these lines later. Refer to Figure 15-9B.
3. Draw the left, right, and top surfaces of the object that lie flat against the surfaces of your isometric box. Refer to Figure 15-9C.
4. Use coordinates or construction lines to locate surfaces that are parallel to, but do not lie flat against, the surfaces of your isometric box. Refer to Figure 15-9D.

SHORTCUT

Remember that drawing aids can be turned on or off (toggled) in the middle of an operation just by picking the tool from the **Assist** toolbox.

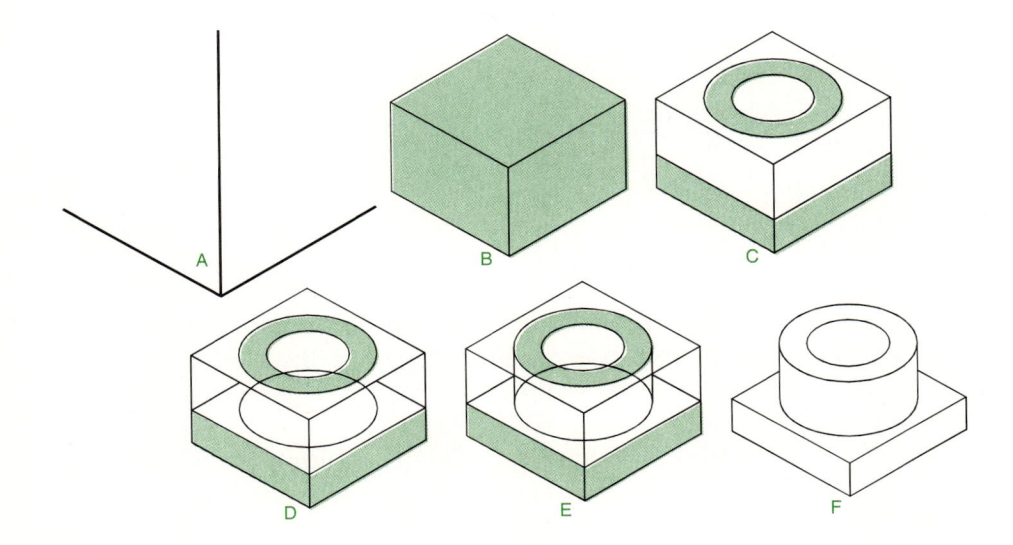

Figure 15-9.
Follow these general steps when constructing an isometric drawing of a product.

5. Construct surfaces that are not parallel to the isometric planes. Locate the edges of these surfaces using surfaces you have already constructed. Refer to Figure 15-9E.
6. Delete all construction lines. Break any parts of objects that are hidden in the isometric view. Refer to Figure 15-9F.

A basic rule when drawing in isometric mode is lines that are parallel in an orthographic view (multiview drawing) must also be parallel in an isometric view.

Exercise 15-3

1. Start a new drawing using the **A_LI_MEC** template created in Chapter 10.
2. Change the cursor type to crosshairs.
3. Turn on the isometric mode by selecting one of the isometric commands.
4. Using the multiview drawing shown below, create an isometric drawing. Remember to change the isometric plane as necessary.
5. Save the drawing as **C15E3** on your work diskette and then close the drawing.

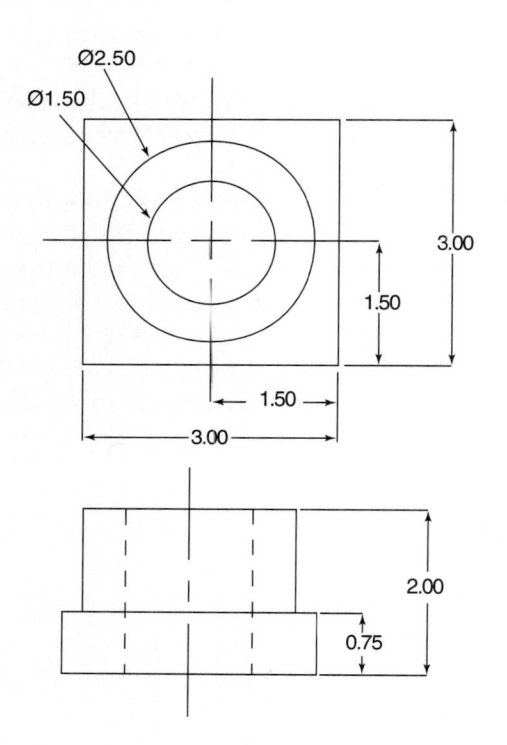

Entering Coordinates in Isometric Mode

AutoSketch allows you to enter three-dimensional coordinates. Both absolute and relative coordinates can be specified. Polar coordinates do not function in isometric mode. The angle is still measured from horizontal, rather than the X or Y isometric axis. Figure 15-10 shows how to enter values to specify absolute or relative coordinates.

If you specify only two coordinate values, AutoSketch assumes that the coordinate not specified is zero (0). In other words, AutoSketch assumes that the object lies *on* the current isometric plane. However, in any isometric plane you can enter all three coordinate values.

Figure 15-10.
When entering three-dimensional coordinates, specify all three values. If the object you are drawing lies on an isometric plane, you can omit the coordinate of that plane (the zero value).

Coordinate Type	Isometric Plane	Isometric Axes of That Plane	Axis Set to 0	Example of Coordinate Entry
Absolute	Any Plane	X,Y,Z		4,3,2
	Top	X,Y	Z	4,3 or 4,3,0
	Right	X, Z	Y	4,2 or 4,0,2
	Left	Y,Z	X	3,2 or 0,3,2
Relative	Any Plane	X,Y,Z		r(4,3,2)
	Top	r(X,Y)	Z	r(4,3) or r(4,3,0)
	Right	r(X,Z)	Y	r(4,2) or r(4,0,2)
	Left	r(Y,Z)	X	r(3,2) or r(0,3,2)

Exercise 15-4

1. Start a new drawing using the **A_LI_MEC** template created in Chapter 10.
2. Change the cursor type to crosshairs.
3. Turn on the isometric mode by selecting one of the isometric commands.
4. Turn off **Grid** and **Snap**.
5. Select the **Polyline** tool and pick the first point of the polyline in the lower center portion of the sheet.
6. Use relative coordinate values to draw the front face of the object shown below using the dimensions given.
7. Use the cursor to draw the remaining sides of the object. Change the isometric plane as necessary.
8. Save the drawing as C15E4 on your work diskette and then close the drawing.

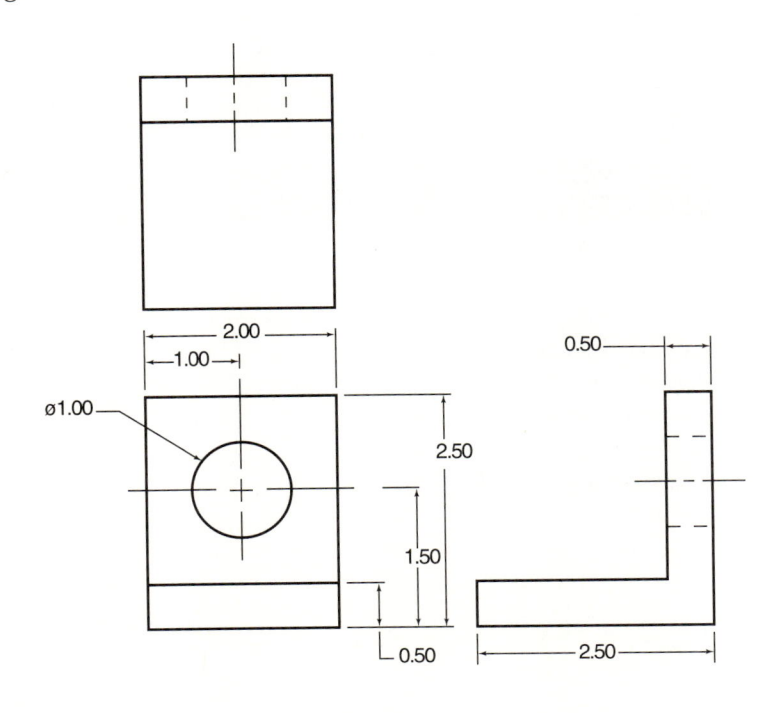

Drawing Isometric Text

Text is not affected by the isometric mode. However, it often looks better if text is placed in the isometric plane of the surface it describes. You can create isometric text by changing the text settings **Oblique Angle** and **Angle** in the **Text Settings** dialog box. You must use an SHX font since the oblique angle (italic) of the text must be at a precise 30° or -30° angle. This is not available with TrueType fonts. The illustration in Figure 15-11 shows the **Oblique Angle** and **Angle** settings for text placed in various positions of all three isometric planes. The SHX font ROMANS was used.

These settings will not work for surfaces that are not parallel with an isometric plane. For example, look at the angled surface in Figure 15-5. It is very difficult to calculate the **Oblique Angle** and **Angle** for these surfaces. It is best to simply try several settings, and "tweak" those settings until the text appears correct. You can also use regular text with a leader pointing to the item or surface the text describes.

Figure 15-11.
By changing the **Oblique Angle** and **Angle** for an SHX font, you can place text on all three isometric planes.

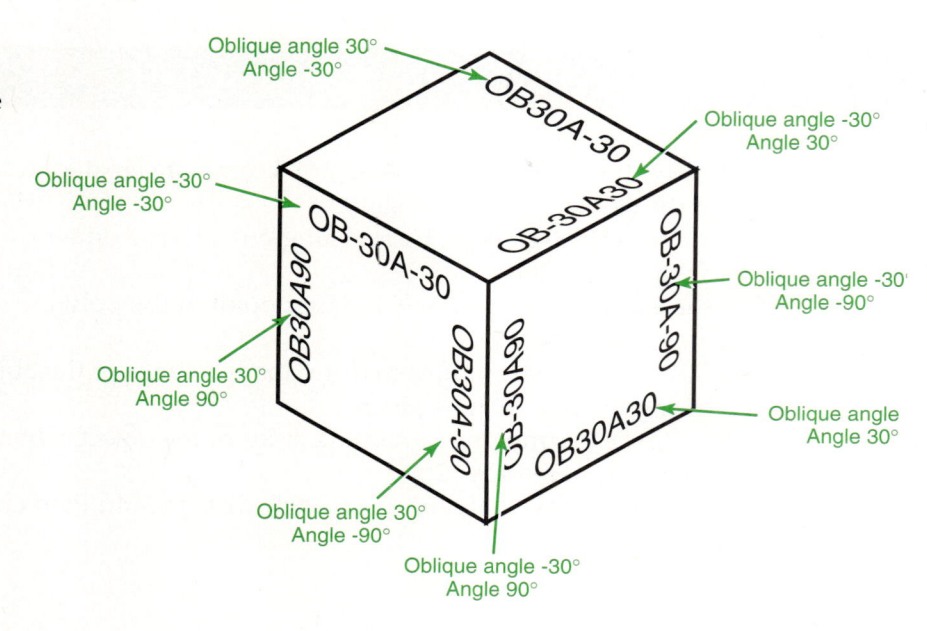

Exercise 15-5

1. Start a new drawing using the **A_LI_MEC** template created in Chapter 10.
2. Change the cursor type to crosshairs.
3. Turn on the isometric mode by selecting one of the isometric commands.
4. Draw the object shown below with the dimensions given. Remember to change the isometric plane as necessary. Use any drawing aids that will help you construct the drawing.
5. Add text to the drawing as shown.
6. Save the drawing as C15E5 on your work diskette and then close the drawing.

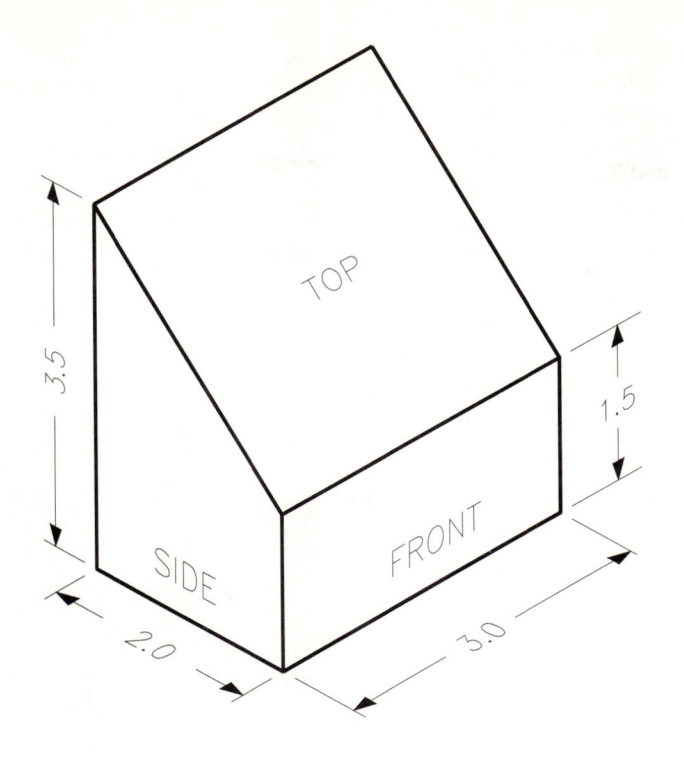

Dimensioning Isometric Views

AutoSketch does not automatically dimension isometric objects. You must manually create the extension lines, dimension lines, arrowheads, and dimension text, Figure 15-12. This is not an easy task compared to dimensioning two-dimensional drawings. Yet, it is easier than manual drafting.

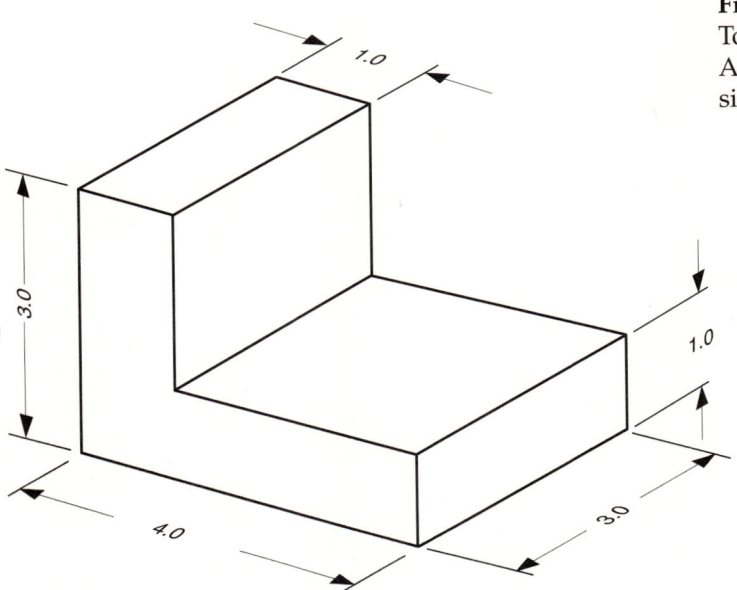

Figure 15-12.
To dimension an isometric drawing in AutoSketch, you must place the dimensions manually.

Creating isometric arrows

You should create a symbol library that contains isometric arrows for the left, right, and top isometric planes. You will need a total of 12 arrows, 4 for each plane, Figure 15-13. The recommended size is .18" long and one-half as wide. If your drawing is larger, remember that you can scale these symbols when inserting them. Use a descriptive name for each arrow symbol. Place the part base location on the tip of the arrow. Some recommended names are shown in Figure 15-14.

Figure 15-13.
You will need to create a symbol library of twelve arrows to dimension isometric drawings in AutoSketch.

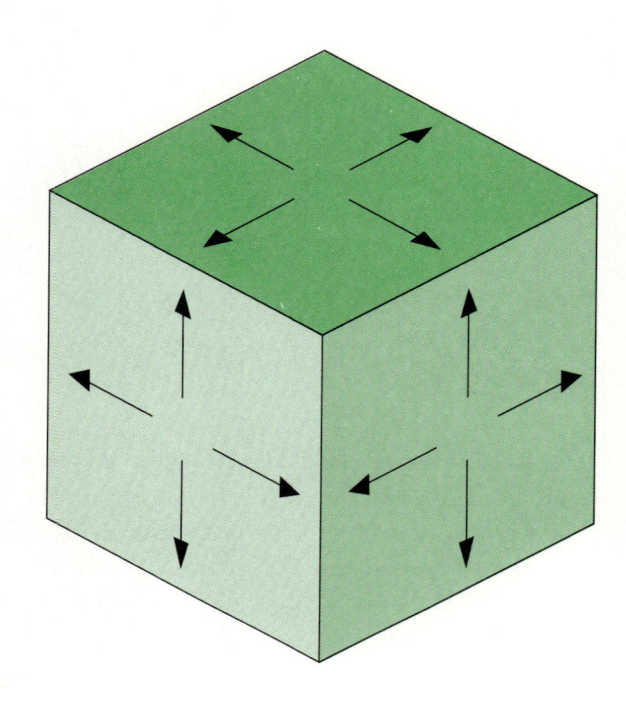

Figure 15-14.
These names are recommended for the arrows you create for isometric dimensioning.

Isometric Plane	Arrow	Symbol Name
Top	Top	ISO-A-TT
	Bottom	ISO-A-TB
	Left	ISO-A-TL
	Right	ISO-A-TR
Left	Top	ISO-A-LT
	Bottom	ISO-A-LB
	Left	ISO-A-LL
	Right	ISO-A-LR
Right	Top	ISO-A-RT
	Bottom	ISO-A-RB
	Left	ISO-A-RL
	Right	ISO-A-RR

Exercise 15-6

1. Start a new drawing using the **Quick Start** template. Set the unit type to **Inches (decimal).**
2. Change the cursor type to crosshairs if it does not already appear as crosshairs.
3. Turn on the isometric mode by selecting one of the isometric commands.
4. Set the snap spacing to .045 and the grid to .09.
5. Draw an arrow using the **Line** tool.
6. Fill the arrow using the **Pattern Fill** tool to create a solid arrowhead.
7. Draw the other arrow heads. (Hint: You can create the other arrows using the **Mirror** and **Rotate** tools.)
8. Use the **Part Clip...** command to save each arrow as a symbol. Remember to place the part base location on the tip of the arrow. Save each arrow symbol to your work diskette using the names given in Figure 15-14.
9. Save the drawing as C15E6 on your work diskette and then close the drawing.

Drawing isometric dimensions

With the arrowheads created in Exercise 15-6, you can now add dimension lines, extension lines, and dimension text to your isometric drawing. Remember to follow proper dimensioning techniques. Refer back to Figure 15-12 for an example of these techniques on a dimensioned isometric part. Remember the following guidelines for placing isometric dimensions.

- Extension lines should extend from the plane being dimensioned.
- Dimension lines should be parallel to the length they measure.
- Dimension arcs should be drawn in the plane they describe.
- Dimension text settings should correspond to the plane being dimensioned.

Exercise 15-7

1. Start a new drawing using the **A_LI_MEC** template created in Chapter 10.
2. Change the cursor type to crosshairs.
3. Turn on the isometric mode by selecting one of the isometric commands.
4. Draw the object shown with the dimensions given. Remember to change the isometric plane as necessary. Use any drawing aids that will help you construct the views.
5. Add dimensions to the drawing. Use the arrow symbols you created in Exercise 15-6.
6. Save the drawing as C15E7 on your work diskette and then close the drawing.

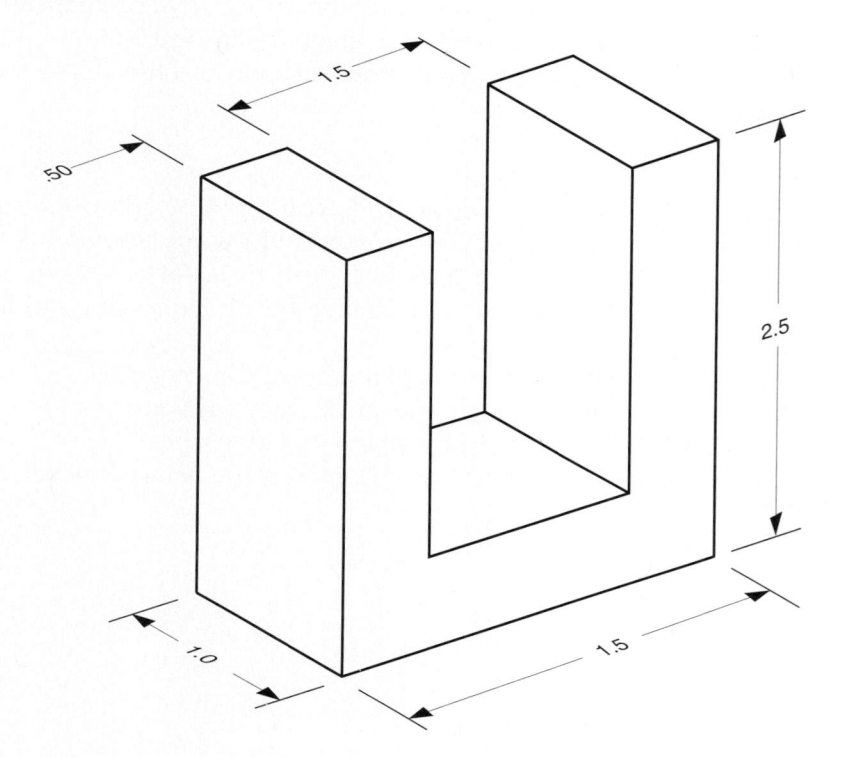

Summary

Pictorial drawings are three-dimensional views of an object using two-dimensional drawing techniques. There are four types of pictorial drawings. These are oblique, perspective, isometric, and 3D modeling. AutoSketch cannot be used to create 3D models.

An oblique drawing shows the front of an object in true size and shape. The depth is shown either full scale or partial scale extending back at a 30° to 60° angle.

An isometric drawing also shows the top, front, and side of an object. Vertical edges of the object are drawn vertical. Horizontal edges are drawn at a 30° angle. AutoSketch provides an isometric mode for creating isometric views. Text can be created for an isometric drawing by modifying the text settings. Dimensions must be created manually. This is more difficult than dimensioning two-dimensional drawings, but easier than manual drafting.

Important Terms

Cabinet oblique drawing
Cavalier oblique drawing
General oblique drawing
Isometric drawing

Isometric lines
Isometric mode
Isometric plane
Nonisometric lines

Oblique drawing
One-point perspective
Two-point perspective
Vanishing points

New Tools and Commands

Left Isometric Plane   **Right Isometric Plane**   **Top Isometric Plane**

Review Questions

Give the best answer for each of the following questions.

1. What is the simplest form of pictorial drawing?

2. When depth lines are drawn half length at a 45° angle, the drawing is called a(n) _____ oblique drawing.

3. In an isometric drawing, circular shapes and holes appear as _____.

4. In an isometric drawing, vertical lines are drawn along the _____ axis.

5. How do you enable isometric mode in AutoSketch?

6. Which isometric plane is formed by the X and Z axes?

7. How does isometric mode affect **Grid** and **Snap**?

8. What type of cursor should you use when drawing an isometric view?

9. For each of the following coordinates, identify which plane the line will be drawn in.

First Point	Second Point	Plane
1,1,1	0,4,5	_____
1,1,1	1,0,7	_____
1,1,1	9,5,0	_____
1,1,1	r(2,0,8)	_____
1,1,1	r(3,1,0)	_____

10. When in isometric mode, which objects does AutoSketch automatically draw in an isometric perspective?

11. Lines that are parallel in an orthographic view must be _____ in an isometric view.

12. What **Oblique Angle** and **Angle** settings are used for text placed horizontally in the right isometric plane?
    A. _____
    B. _____

13. Why should you create symbols for arrows to use when dimensioning an isometric drawing?

14. Which tools in the **Measure** menu will automatically dimension isometric drawings?

15. In an isometric drawing, dimension lines should be _____ to the length they measure.

Activities

Here is a mixture of drawing activities to develop your skills creating pictorial drawings. After saving each activity using the filename given underneath, select the **Close** command from the **File** menu to clear the drawing from the screen.

1. Create two isometric templates for the isometric views you will draw in these activities. Base one template on the **A_LI_MEC** template and the other on the **B_LI_MEC** template created in Chapter 10. Turn on isometric mode and select crosshairs as the cursor. Save the templates as **A_LI_ISO** and **B_LI_ISO**.

2 - 9. Using the isometric templates created in Activity 1, draw the isometric views shown below. The reference grid is .50". Save the drawings on your work diskette using the name given under each view.

10 - 11. Select two isometric views created in Activities 2 - 9. Add dimensions following accepted dimensioning practices. Place dimensions and object lines on different layers. Also, use the colors recommended in this text for the linetypes. Save the drawings on your work diskette using the names C15A10 and C15A11.

12 - 15. Select 4 multiview drawings completed in previous chapters. Draw isometric views and add dimensions following accepted dimensioning practices. Place dimensions and object lines on different layers. Also, use the colors recommended in this text for the linetypes. Save the views on your work diskette using the names C15A12 through C15A15.

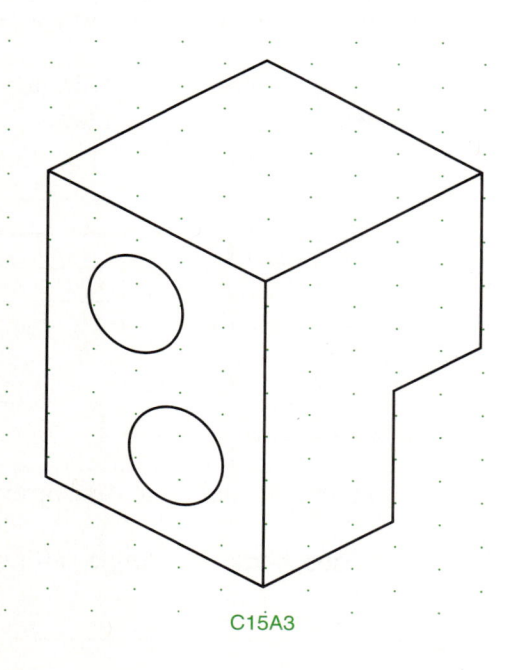

C15A2

C15A3

Chapter 15   Drawing Pictorial Views

C15A4

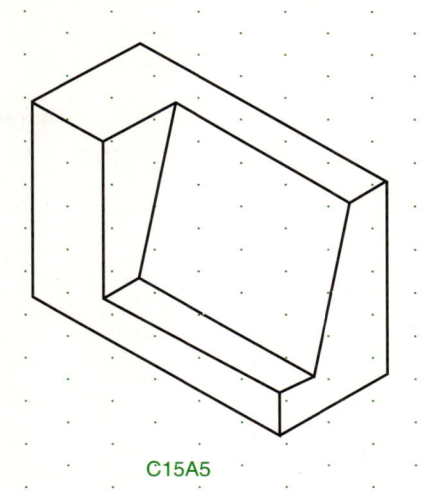

C15A5

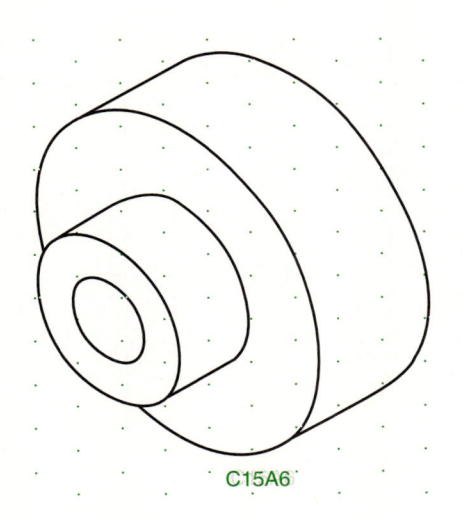

C15A6

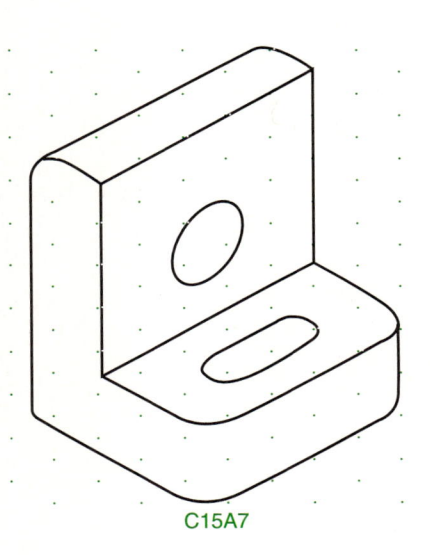

C15A7

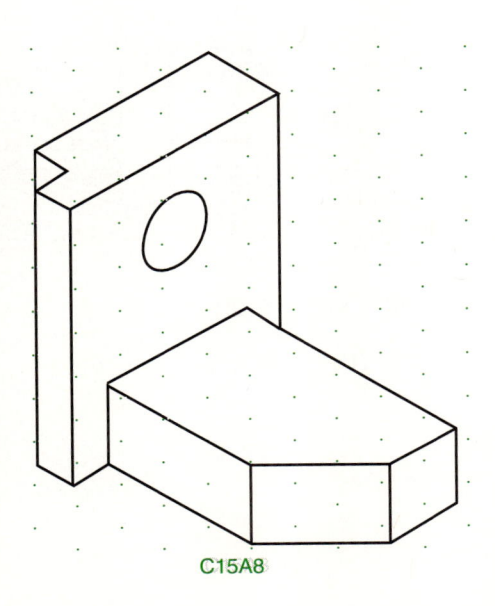

C15A8

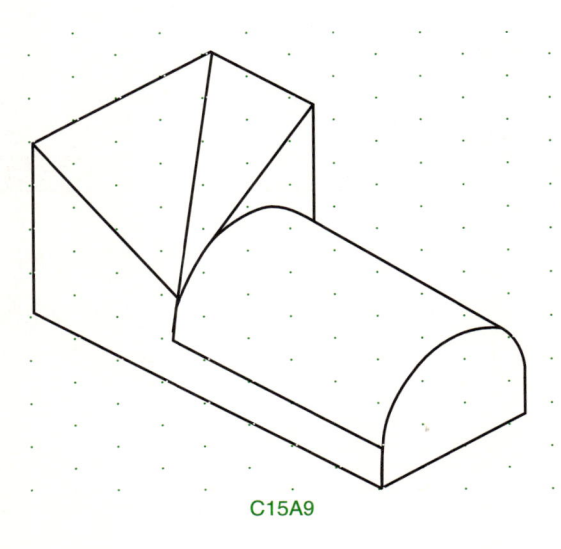

C15A9

This detailed drawing was plotted to be used as part of a written report (Autodesk, Inc.)

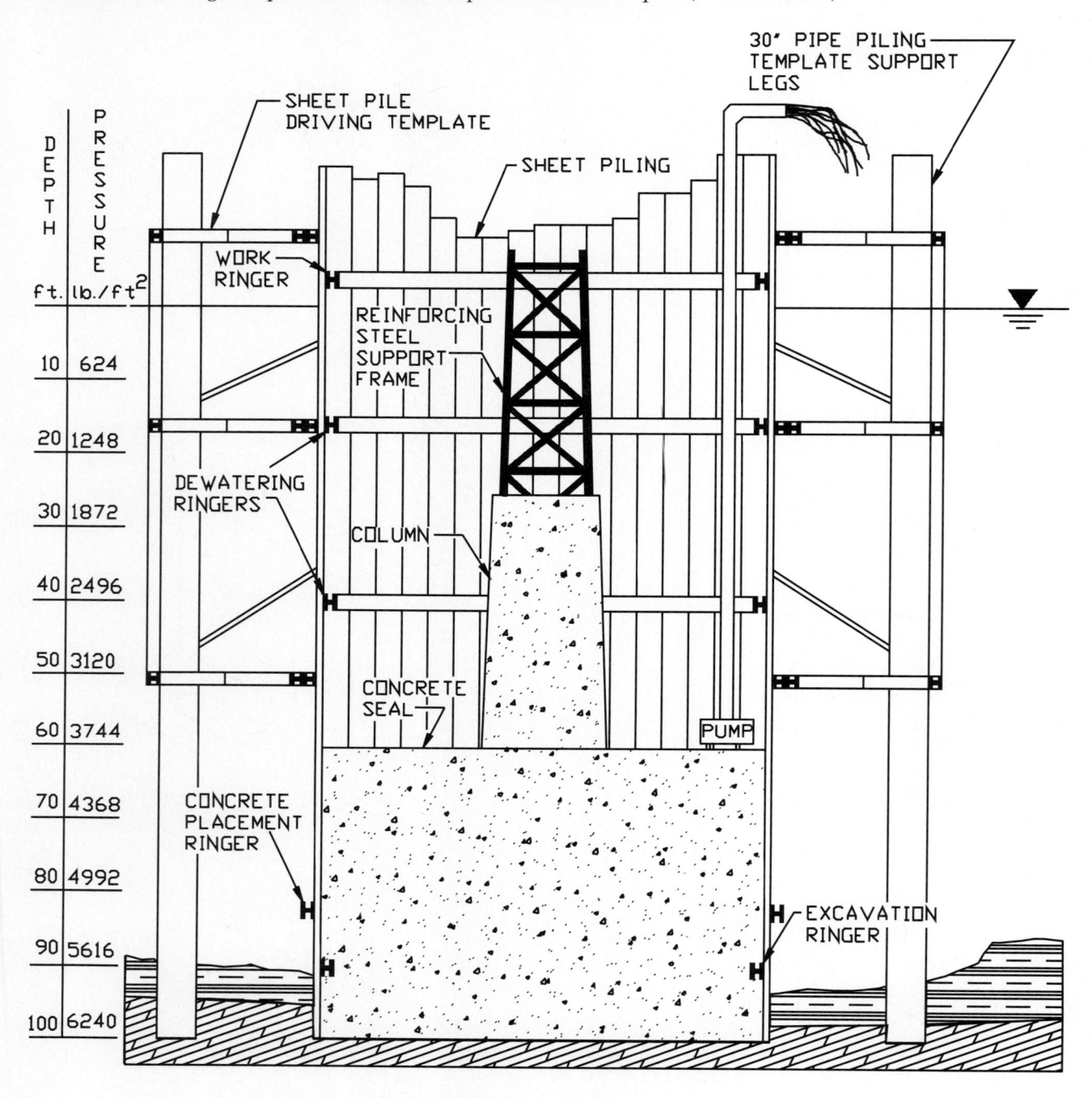

STEPS: CONTINUE CONSTRUCTION IN DRY CONDITION.

NOTE: REINFORCING STEEL NOT SHOWN THIS VIEW.

CHAPTER 16

Printing and Plotting Drawings

Objectives
After studying this chapter, you will be able to:
- Select and configure a printer or plotter.
- Set up sheets to print only a portion of the drawing, or the entire drawing.
- Set view preferences to display sheet text and noncurrent sheets.
- Move and resize sheets to enclose a portion of a drawing to print.
- Preview print settings to determine how the drawing will be printed.
- Send a drawing to a printer or plotter.
- Identify the various types of drawing media and plotter pens.

Printing is the process of sending the drawing on-screen to a printer or plotter to make a paper print of your drawing. During the drawing process, two types of prints are typically made. *Check prints* are made with a dot matrix or laser printer for a quick check on the design progress. *Final plots* are high-quality prints made with a pen plotter or other high-quality output device on larger paper. These are used for copying and distributing the design. AutoSketch also allows you to plot the drawing to a file on disk. The file can later be printed outside of AutoSketch, or imported into a desktop publishing program.

There are four commands in the **File** menu that you will use during the printing process, Figure 16-1. These are listed here in the order you will use them.
- **Print Setup...**—Determines the printer or plotter that you will print to.
- **Sheet Setup...**—Determines what portion of the drawing will be printed.
- **Print Preview...**—Shows how the drawing will look when printed, before actually printing it.
- **Print**—Prints the drawing.

Setting up the printer, sheet, and previewing the print or plot are the most important and time-consuming part of the plotting procedure. The final appearance of the drawing depends on how you set these options. Do not expect to get them correct the first time. You may waste a few sheets of paper until you learn how these settings affect the final output. In addition, before attempting to print any drawing, make sure that the plotter or printer is connected and set up properly. There are special set-up steps for certain types of devices. These procedures are covered thoroughly in the AutoSketch documentation and in your printer or plotter's user manual.

Figure 16-1.
The commands to print drawings are found in the **File** menu or **File** toolbox.

Print Setup

The **Print Setup...** command allows you to pick the printer or plotter to use. You can also change certain settings for the selected device. The most common devices are laser printers and pen plotters. These will be covered in this book.

You should select your printer first. This will later help you with sheet setup. If you always use the same printer or plotter, you will only need to set it up once.

Installing Printer and Plotter Drivers

The correct printer driver for your printer and/or plotter must be installed before using AutoSketch. This is a normal procedure with Windows. If there is a printer or plotter already connected to your system, the driver is most likely already installed.

However, if the correct printer driver is not installed, first select the Control Panel icon from the Main group of the Windows Program Manager. Then, select the Printers icon, Figure 16-2A. Select the Add 》 button. Follow the on-screen prompts. You may need to supply a driver disk. This usually comes with your printer or plotter.

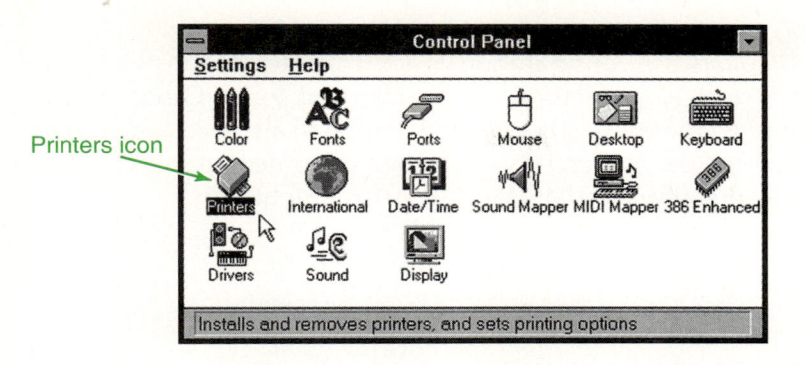

Figure 16-2.
Make sure that the driver for your printer or plotter is installed in Windows. A—To install a driver, first select the Printers icon from Control Panel. B—Then select the Add >> button. Once you add the printer, it appears in the Installed Printers list.

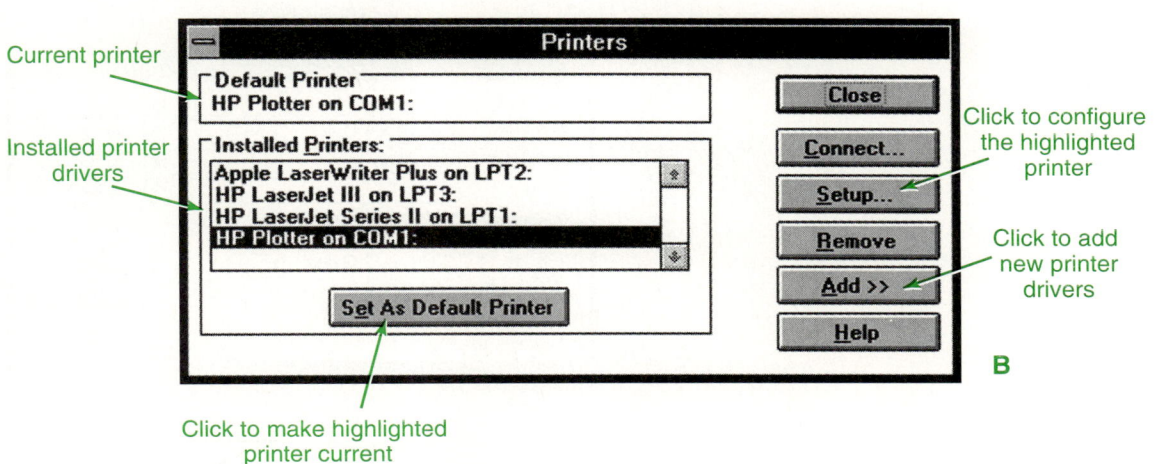

Once added, the printer name will be shown in the Installed Printers: list. See Figure 16-2B. For more detailed information on how to add printers to Windows, see your Microsoft Windows documentation.

Selecting and Configuring a Laser Printer

A laser printer is one of the easiest devices to set up. From the AutoSketch **File** menu, select **Print Setup....** This will display the **Print Setup** dialog box, Figure 16-3. Highlight the laser printer you will use. The **Setup...** and **Margins...** buttons can be used to change the default settings.

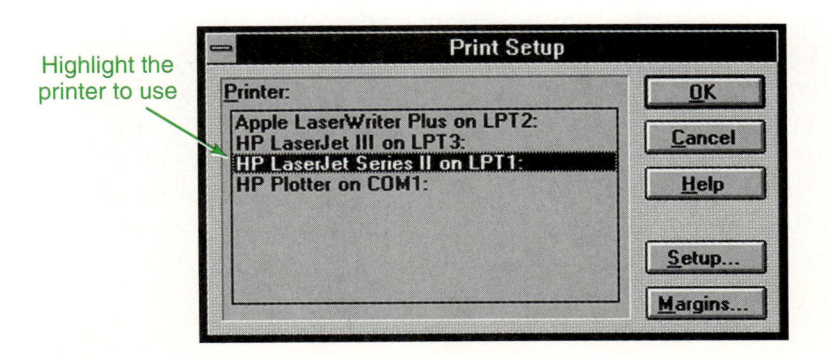

Figure 16-3.
To configure printers while in AutoSketch, pick the **Print Setup...** command from the **File** menu to access the **Print Setup** dialog box.

Setup

The **S̲etup...** button displays the configuration of your printer, Figure 16-4. This includes paper size, paper source, orientation, resolution, and fonts. Other options may be included, depending on the printer. For an explanation of all these values, refer to your Microsoft Windows documentation or your printer documentation. (This configuration dialog box is actually a function of Windows, not AutoSketch.)

Figure 16-4.
This is the setup for a typical laser printer. You usually will only change the paper size and orientation.

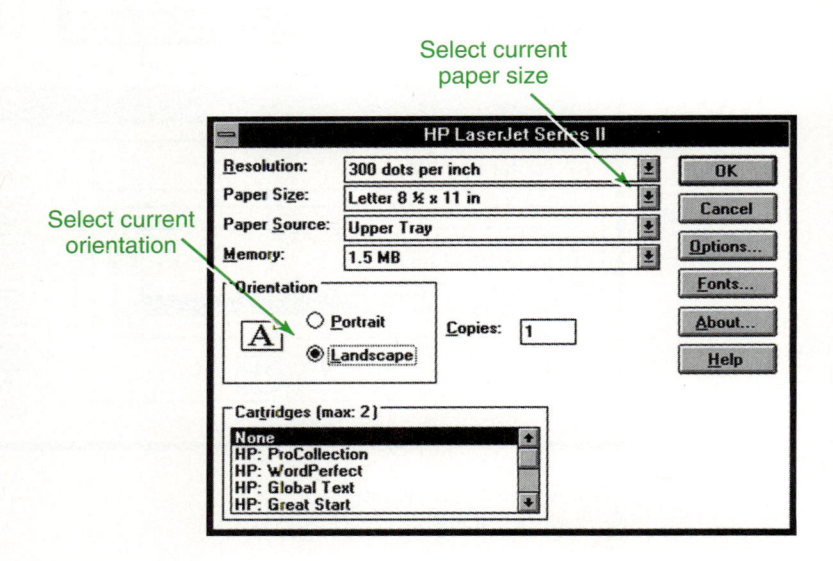

For the activities and exercises in this book, there are two values that you might change. You may want to change the **Paper Si̲ze:** setting. However, not all printers offer more than one paper size setting.

Also, you may need to change the **Orientation** setting. This setting is important when using AutoSketch. Most word processing documents (such as letters and memos) are printed with the **P̲ortrait** setting. This means that when reading the document, it is taller than it is wide. However, most drafting is done with the longest dimension of the paper horizontal. When setting up your drawing, you most likely chose a "wide" sheet. For this, you need to select the **L̲andscape** setting.

Margins

No printer can print all the way to the edge of the paper. Margins identify the outer border on the paper where the printer cannot print. When you select the **Margins...** button, AutoSketch displays the minimum default margins set by the printer driver, Figure 16-5. These are usually correct. The only reason to change the margins is if your printer is cutting off portions of your drawings along the outer edges. This is called *cropping*. In this case, you would increase the margin of the side that is cropped. This tells AutoSketch to print on a smaller area of the paper. Margin values cannot be smaller than the displayed default values. You can switch back to your printer's default margins by choosing the **Default** button.

Figure 16-5.
The default margins for your printer should not result in cropping. However, if your printer is cutting off portions of your drawings, you may need to increase the margins.

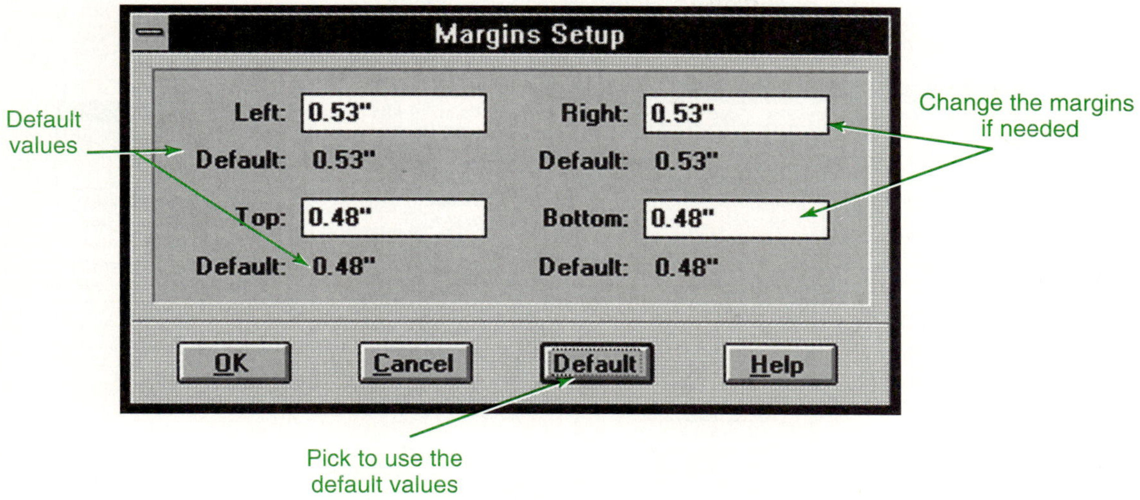
Margins Setup
Left: 0.53"
Default: 0.53"
Top: 0.48"
Default: 0.48"
Right: 0.53"
Default: 0.53"
Bottom: 0.48"
Default: 0.48"
OK
Cancel
Default
Help
Default values
Change the margins if needed
Pick to use the default values

Selecting and Configuring a Plotter

A plotter can be more difficult to set up. Again, from the AutoSketch **File** menu, select **Print Setup...** to display the **Print Setup** dialog box. Highlight the name of your plotter. Like a laser printer, the **Setup...** and **Margins...**buttons can be used to change settings. The **Margins...** button works the same as a laser printer. However, the **Setup...** button has different settings from a laser printer. For further information on plotters, see your Microsoft Windows documentation. Note: Some plotters do not support TrueType fonts.

Figure 16-6 shows the setup dialog box for a Hewlett-Packard series plotter. This is a common type of plotter. Figure 16-6 also shows the various menu options. As with a laser printer, you should select the **Landscape** orientation. The other options are as follows.

- Device—A plotter or printer driver may support more than one model. From the **Device** menu, you can select the specific plotter model you have attached.
- Size—From the **Size** menu, select the size of paper you will plot on. Most plotters use A-size and B-size paper, but some also support up to E-size.
- Paper Feed—Although most plotters require that you load the paper in the plotter manually, some can automatically feed paper from a bin. In the **Paper Feed** section, choose the correct setting for the plotter you are using.

Multipen plotters

There will be additional options if you have a plotter that supports more than one pen. In fact, the advantage of using different colors in your drawing becomes important here.

First, when creating drawings, draw objects using color to represent different line widths. Then, set the pen number at plot time so that the proper width pen (thin or thick) is chosen.

In addition, you may want objects plotted with the same color as they were actually drawn on-screen. Multipen plotters allow you to easily do this.

Figure 16-6.
The setup for a common plotter can be much more involved than a printer. Ensure that you understand the purpose of each option before making any changes. A—The setup dialog box. B—Selecting the device. C—Selecting the paper size. D—Selecting the carousel. E—Selecting the pen color. F—Selecting the pen type. G—Selecting the pen speed.

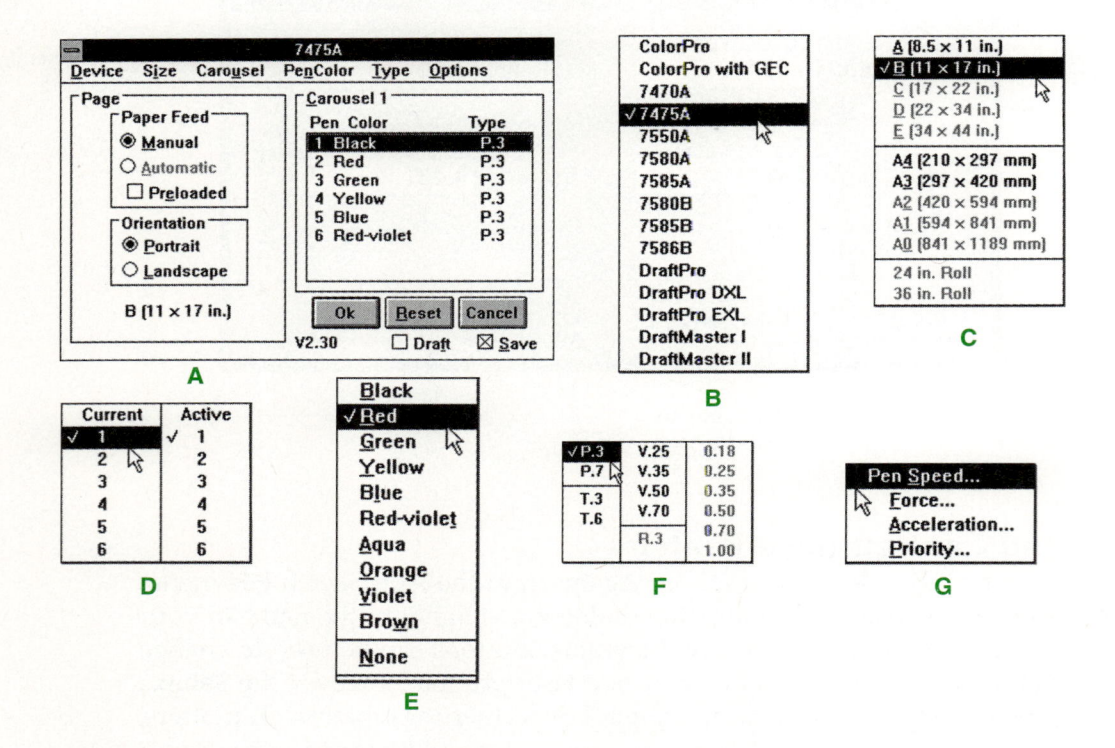

- Carousel—Multipen plotters use a holder called a carousel. From the **Carousel** menu, you can define various carousels that each contain a different mix of pen colors and line widths. The pen number determines which pen the plotter chooses to plot certain color objects.
- Pen Color—When you highlight a pen, the **PenColor** menu option allows you to assign which color on the drawing causes that pen to be selected.
- Type—The **Type** menu allows you to choose the style of pen. There are various types, including specific vendor models of pens, as well as felt-tip, ink, roller ball, and many others.
- Options—The most important command in the **Options** menu is **Pen Speed...** command. This is the speed that the plotter draws with each pen. The speed is often given in centimeters per second. You can set a different speed for each pen. Lines become thin and light at fast speeds. The pen can even skip when plotting too fast. A slower pen speed may prevent some pens from skipping and can darken lines. Generally, set the speed slower for fine-tip pens and faster for broad-tip pens. Setting pen speed is useful only if your plotter supports programmable pen speeds. The **Force...** command sets the pressure of the pen against the paper. This is usually given in grams. The **Acceleration...** command sets how fast the pen reaches the pen speed.

Exercise 16-1

1. Start a new drawing using the **Normal** template.
2. Select the **File** menu and **Print Setup...** command.
3. Highlight your printer in the **Print Setup** dialog box.
4. Select the **Margin...** button and record the default margins for your printer or plotter.
5. Select the **Setup...** button and record the setup options for your printer or plotter.
6 Close the drawing without saving it.

Sheet Setup

Chapter 10 introduced the concept of sheets. A *drawing sheet* is used to tell AutoSketch what part of your drawing to print. After you set up a drawing sheet, solid and dotted boxes appear in the drawing window. The solid line represents the paper size. The dotted line shows the margins. These boxes are not printed, but they are used to determine what portion of the drawing is printed.

At this point, any drawing that you have set up a sheet for in this book can simply be printed using the **Print** command. Chapter 10 recommended that you set up a sheet during drawing setup to define your drawing area. However, there may be times when you want to print only a portion of the drawing.

Defining More than One Sheet

The most common time to define more than one sheet is when you want to print only a portion of a drawing. For example, suppose you have created a mechanical drawing with a multiview and a pictorial view, Figure 16-7. You want to print only the pictorial view. However, your default sheet setup includes the entire drawing. Rather than redefining the default sheet, simply add a second sheet.

To add a second sheet, select the **Sheet Setup...** command from the **File** menu. The **Sheet Setup** dialog box appears, Figure 16-8. Click in the **Sheet Name** box. Type the new name. Then, select **New Sheet** to create a second sheet in addition to the default sheet. However, do not pick **OK** yet. Now that you have a new sheet name, you need to pick the portion of the drawing to print. This can be done one of several ways.

Drag Scale

When you select the **Drag Scale** option and then **OK,** the following prompt appears.

Sheet First corner:

Select the first corner of the area you want to print. AutoSketch then asks for the second corner of the area.

Sheet Second corner:

The sheet that is created will enclose the objects, and the scale will be modified so that the objects fill the sheet size. See Figure 16-9.

Figure 16-7.
This drawing has one sheet set up to print the entire drawing. However, what if you only wanted to print the isometric view?

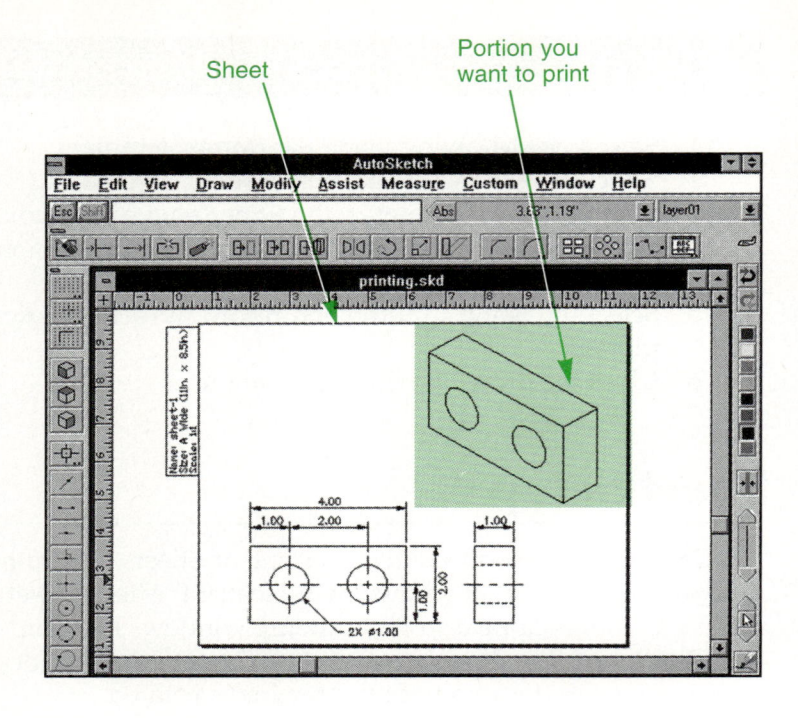

Figure 16-8.
The **Sheet Setup** dialog box is where you can set up additional sheets to print certain portions of your drawing, or the entire drawing with different settings. When your drawing contains different types of views, it is common to print only one of the views.

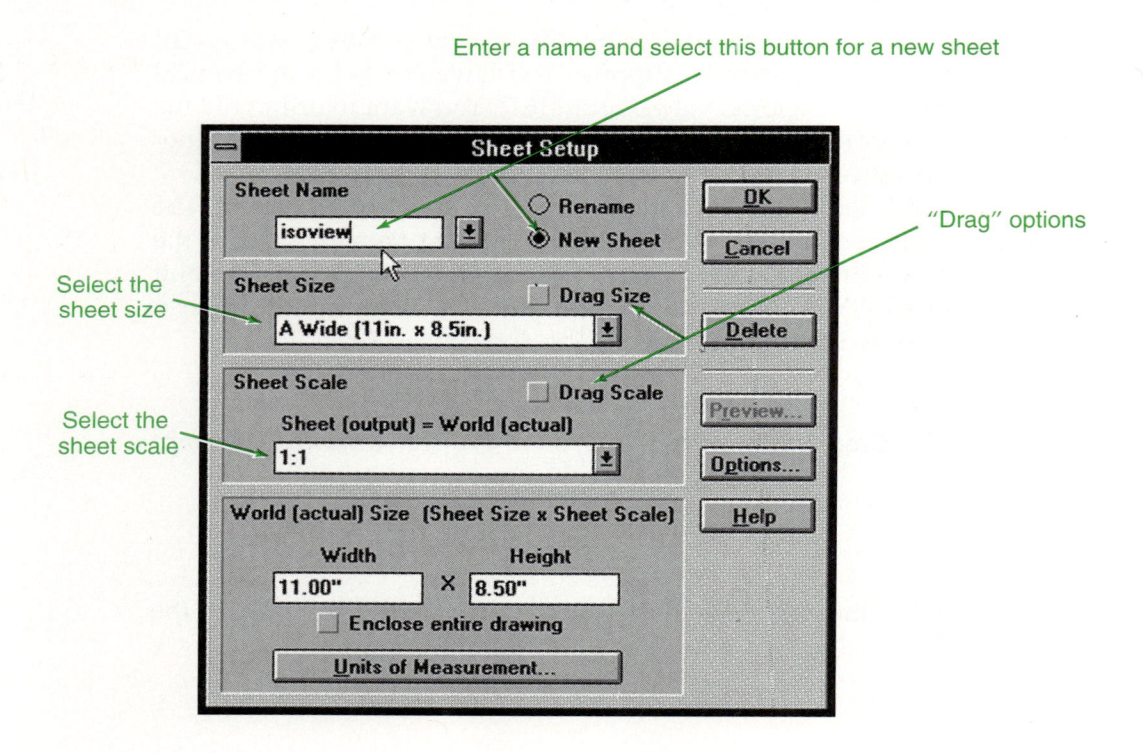

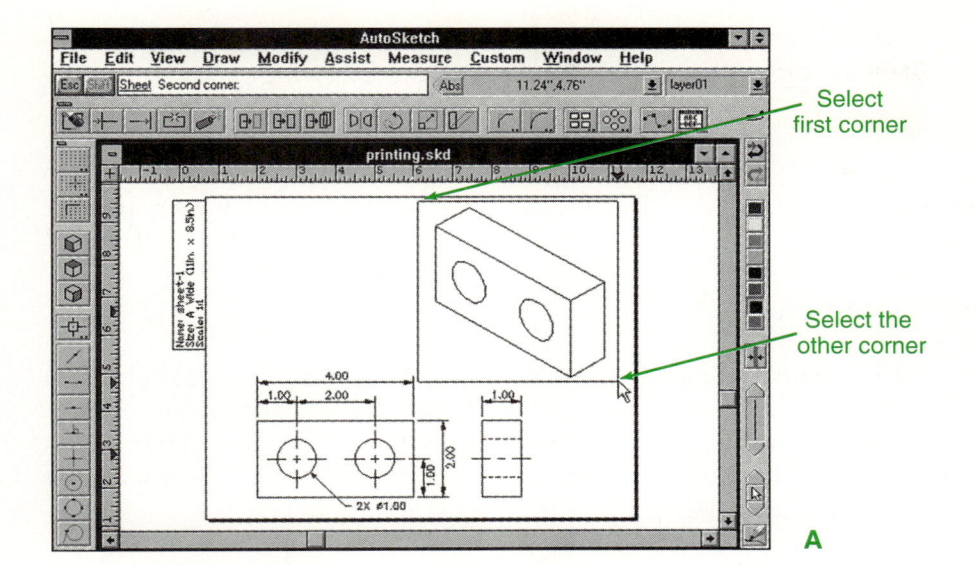

Figure 16-9.
Adding a new sheet with the **Drag Scale** option. A—Pick corners of a sheet around the objects. B—The new sheet appears (highlighted here). The original sheet will remain on-screen if **Hide Current Sheet** is not selected in the **Sheet Options** dialog box. C—If you pick **Sheet Setup** for the new sheet, note that the sheet size remains constant while the drawing scale changes to fit the objects within the sheet.

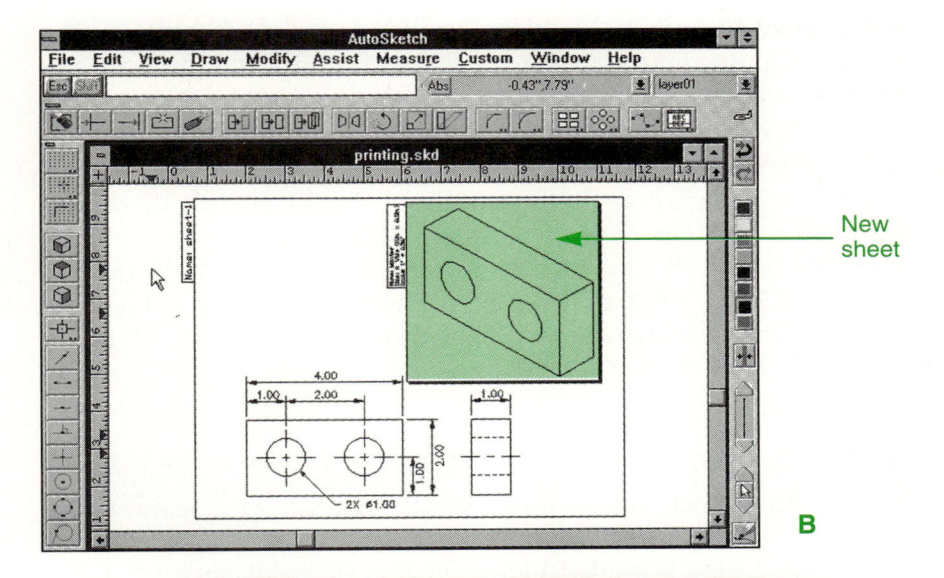

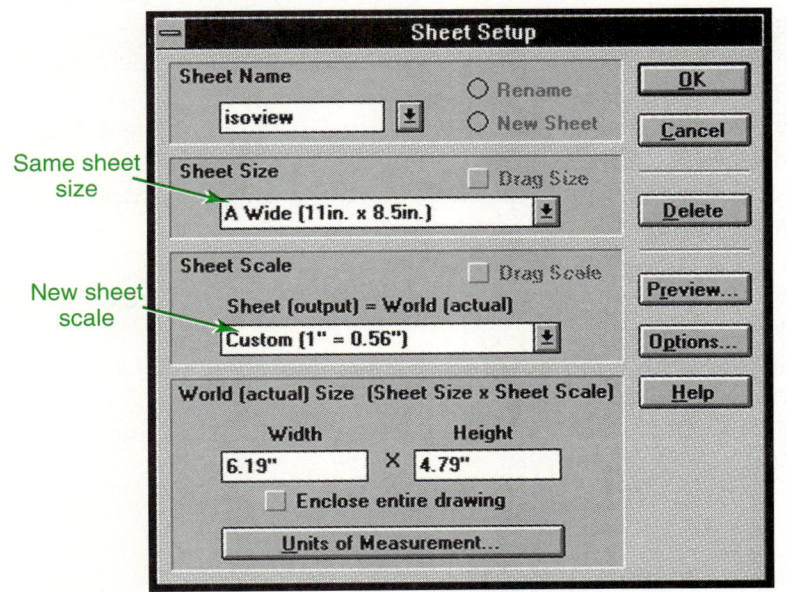

Drag Size

When you select the **Drag Size** option and then **OK,** the following prompt appears.

<u>Sheet</u> First corner:

Select the first corner of the area that you want to print. AutoSketch then asks for the second point.

<u>Sheet</u> Second corner:

The sheet that is created will enclose the objects, and will create a custom sheet size. The size selected should be no larger than the size paper of your printer or plotter. If it is, you may experience tiling when you print. Tiling is discussed later in this chapter. The scale remains the same. See Figure 16-10. If you want a different scale, set the **Sheet Scale** before selecting **OK.**

Enclose entire drawing

The **Enclose entire drawing** option should be used if you want to print all the objects in your drawing to fill the extents of the sheet size. With this option, the **Drag Size** and **Drag Scale** options are not available.

Viewing Sheet Text and Noncurrent Sheets

By default, AutoSketch displays only the current sheet. Also by default, there is no text on-screen to tell you about the current sheet. When working with several sheets, it is best to display all sheets and the text that describes each sheet. To do this, select the **Options...** button in the **Sheet Setup** dialog box. In the **Sheet Options** dialog box, turn off the **Hide Sheet Text** and **Hide Non-current Sheet** options, Figure 16-11. Sheets on an invisible layer are not displayed.

> **SHORTCUT**
> You can also access the **Sheet Options** dialog box by selecting the **View** menu and **View Preferences** tool.

Exercise 16-2

1. Start a new drawing using the **Quick Start** template and select **Inches (decimal)** as the units of measurement.
2. Draw a large square, small circle, small ellipse, and large polygon in each corner of the default sheet.
3. Select the **View Preferences** tool in the **View** menu. Turn off the **Hide Sheet Text** and **Hide Non-current Sheet** options.
4. Create a sheet named "sheet-2" around the circle using the **Drag Scale** option of the **Sheet Setup** dialog box.
5. Use the **Zoom Sheet** tool to make "sheet-1" current.
6. Create a sheet named "sheet-3" around the polygon using the **Drag Size** option of the **Sheet Setup** dialog box.
7. Use the **Print** command to print each of the sheets. Use the **Zoom Sheet** tool to make each sheet current for printing.
8. Save the drawing as C16E2 on your work diskette but leave the drawing on-screen for Exercise 16-3.

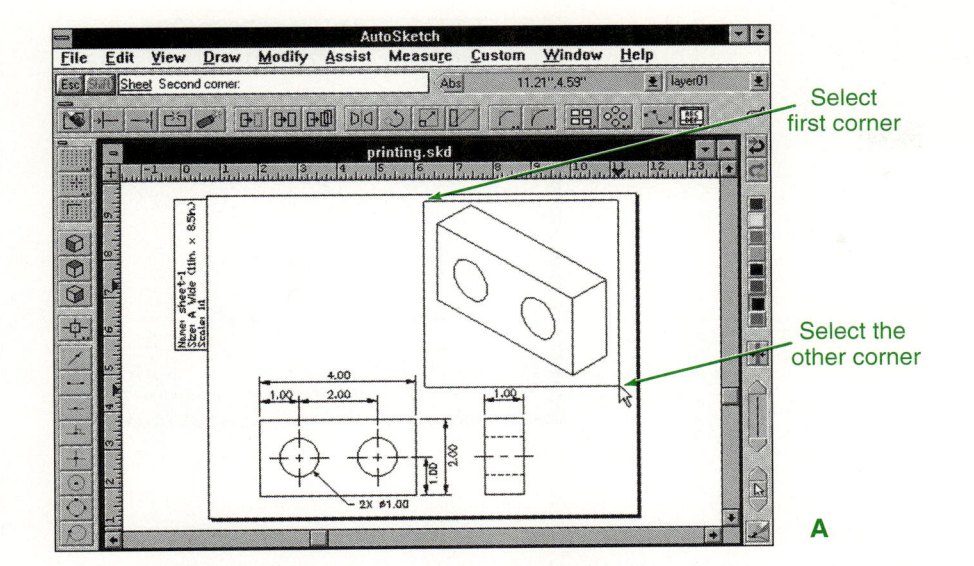

Figure 16-10.
Adding a new sheet with the **Drag Size** option. A—Pick corners of a sheet around the objects. B—The new sheet appears. The original sheet will remain on-screen if **Hide Current Sheet** is not selected in the **Sheet Options** dialog box. C—If you pick **Sheet Setup** for the new sheet, note that the sheet scale remains constant while the sheet size becomes a custom size to match the box you picked.

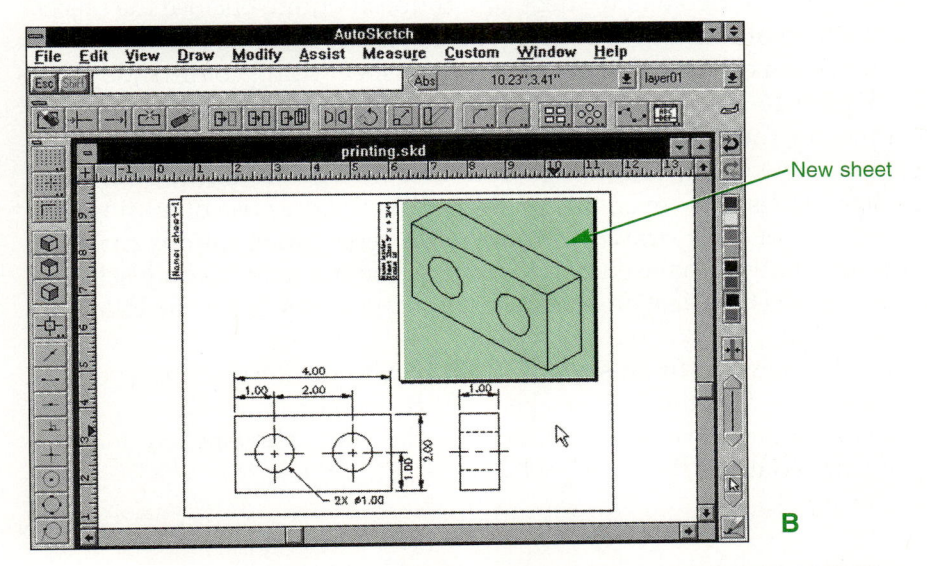

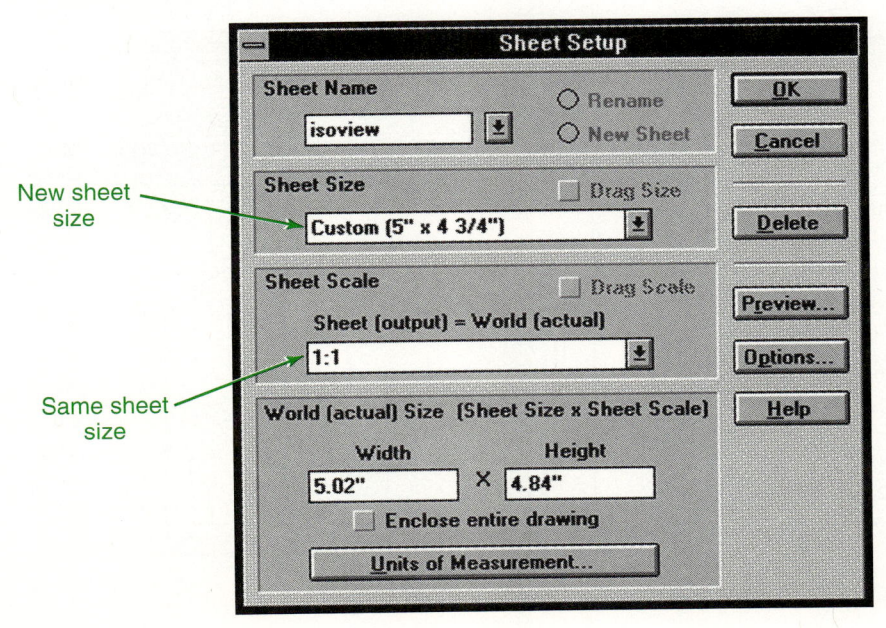

Figure 16-11.
Sheet Option determine which sheets appear in the drawing window, and whether sheet text is displayed to describe each sheet.

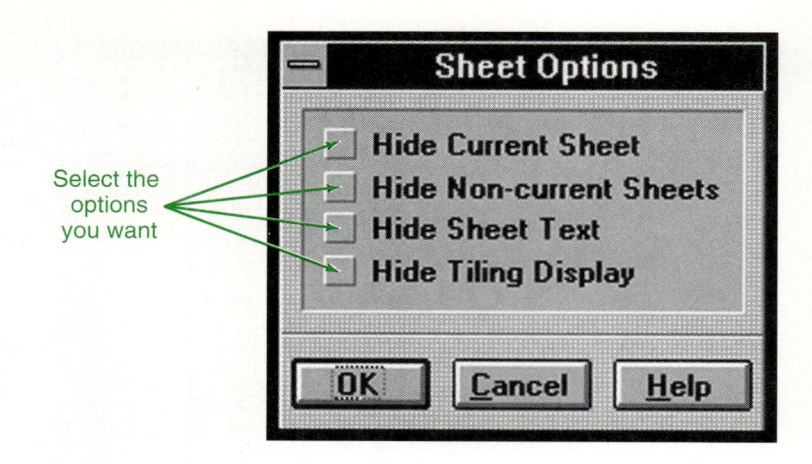

Select the options you want

Working with Existing Sheets

Suppose you have just added a new sheet, and it doesn't quite enclose the objects you want printed. Instead of deleting the sheet, you can move or resize it. A sheet acts much like an object, so you can move, scale, or stretch a sheet using commands in the **Modify** menu. See Figure 16-12. You can even place a sheet on another layer. To select a sheet, you must pick it. You cannot use a crosses/window box to select a sheet.

After you create a new sheet, you can go back and change the **Sheet Size, Sheet Scale,** and **World Size** values. However, this really defeats the purpose of adding new sheets to print a portion of your drawing. You should reset these options carefully, since they will change what you have chosen to print. Sometimes, after you reset these options, you don't even see the sheet on-screen anymore. If so, check for the following common problems.

- The sheet size is too large to fit on-screen. Remedy: Use the **Zoom Sheet** tool from the **View** menu so that the current sheet fills the screen.
- The current layer is not visible. Remedy: Double-click in the **Layers** box and make the layer that contains the sheet visible.

Figure 16-12.
You can scale, stretch, and move sheets. Note here that the smaller sheet from Figure 16-10B (highlighted here) has been moved from the isometric view over the front view.

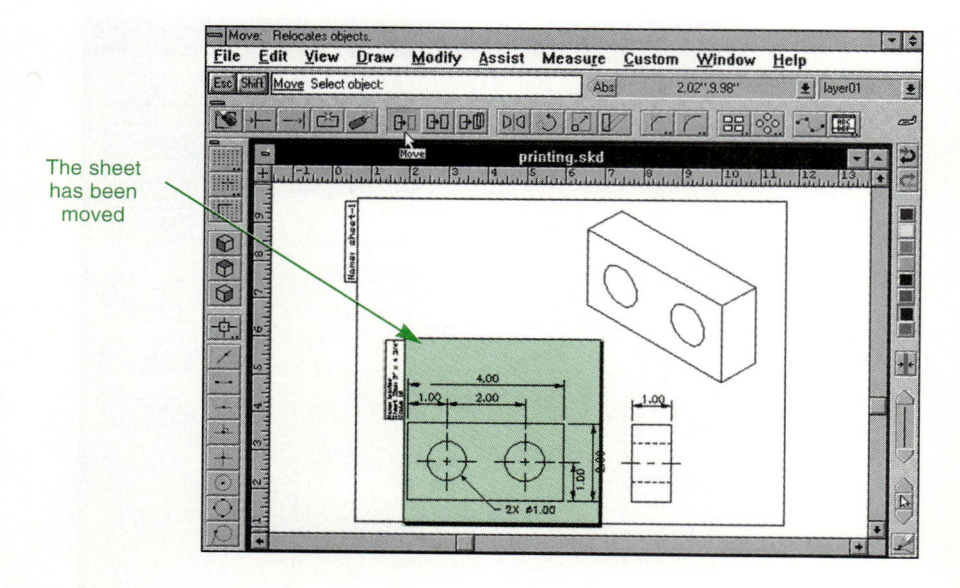

The sheet has been moved

Exercise 16-3

1. Open drawing C16E2 if it is not already on-screen.
2. Move the sheet named "sheet-2" from the circle over the square. Resize the sheet so that it fits around the square.
3. Move the sheet named "sheet-3" from the polygon over the ellipse. Resize the sheet so that it fits around the ellipse.
4. Save the drawing as C16E3 on your work diskette and close the drawing.

Print Preview

Plotter supplies are not cheap. It can also be frustrating to plot a drawing only to find that it is not exactly what you wanted. AutoSketch provides a powerful feature to let you see how the drawing will actually print. This is the **Print Preview...** command. This command is found in the **File** menu.

The **Print Preview** screen displays the drawing area currently selected for printing, Figure 16-13. If what you see is correct, simply pick the **Print** button.

If what you see is not correct, the many options within this screen allow you to change what will be printed. Each time you change an option, the sample display will also change to reflect what your printout will look like. If the sample is not what you expect, change the settings again.

SHORTCUT

You can also choose **Preview...** from the **Sheet Setup** dialog box to access the **Print Preview** screen.

View to Print Options

The **View to Print** options allow you to select which sheet to print. You can also select to print the drawing extents. If you select the **Current Sheet:** option, you can also then select which sheet should be made current, Figure 16-14. If you select the **Drawing Extents** option, AutoSketch allows you to print all the objects in the drawing, regardless of what objects are contained in the current sheet.

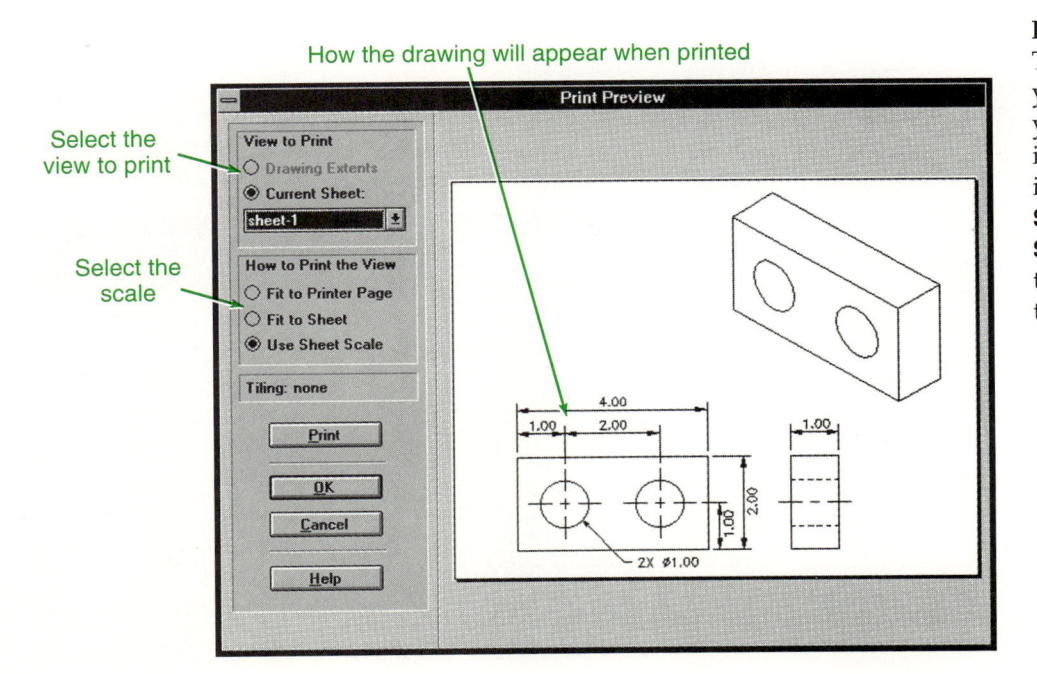

Figure 16-13.
The **Print Preview** screen lets you see the output before you actually print the drawing. Here, the default sheet is chosen as the **Current Sheet** and the **Use Sheet Scale** option is chosen so that printed scale matches the drawing scale.

Figure 16-14.
In this example, the newly created isoview sheet is chosen as the **Current Sheet** and the **Use Sheet Scale** option is chosen so that printed scale matches the drawing scale.

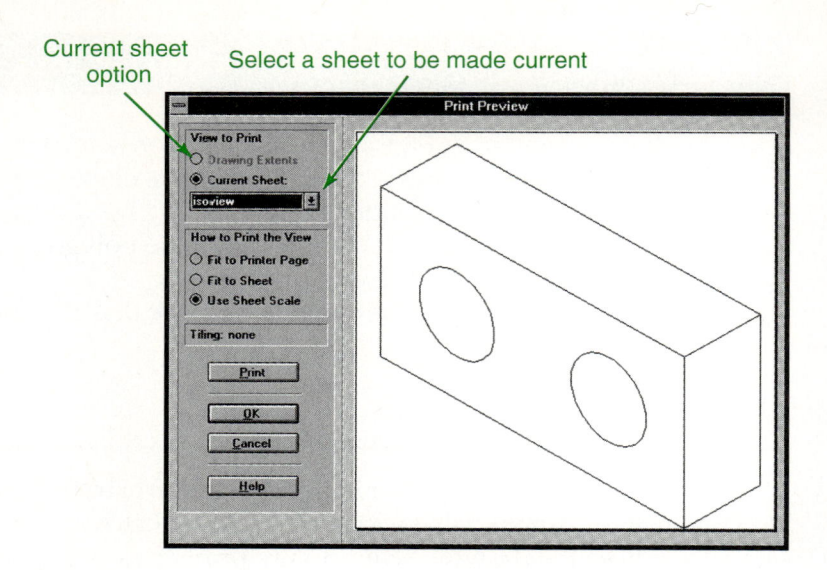

How to Print the View

The **How to Print the View** options allow you to specify how the printed area fits the printer paper. The **Fit to Printer Page** option forces the objects being printed to fill the entire printer page, Figure 16-15. The **Fit to Sheet** option causes the objects being printed to be sized according to the current sheet. The **Use Sheet Scale** makes sure that the objects being printed will not be scaled larger or smaller to fit the paper. They will retain the same scale that they were drawn at. Both the **Fit to Sheet** and **Use Sheet Scale** options can cause the printout to tile.

Tiling

AutoSketch allows you to print a drawing using a scale that makes the printout larger than your paper size. AutoSketch prints the drawing on several sheets of paper. This is called *tiling*. Tiling can be confusing. Remember that a sheet does not have to be the same size as the paper you print on. You can select a B-size sheet when setting up a drawing, yet print on A-size paper. The drawing will be printed on two sheets of paper. The paper can then be taped together.

Figure 16-15.
You can print the entire drawing on the paper size in your printer by selecting **Drawing Extents** and **Fit to Printer Page.**

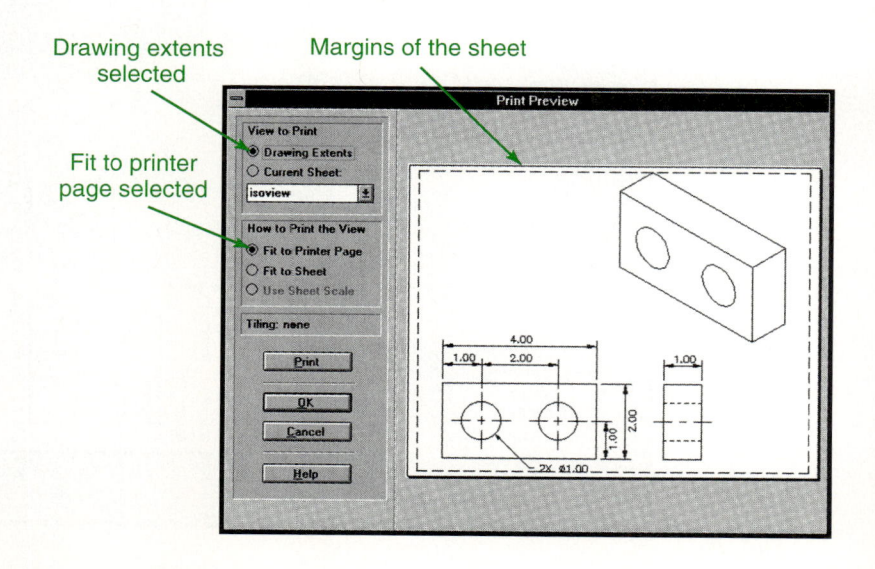

If the sheet is larger than the paper size supported by your printer or plotter, your printer will use several sheets of paper to print the entire drawing. However, if you chose the **Fit to Printer Page** option, AutoSketch will not tile. Instead, the scale is reduced.

When a drawing is tiled when printed, different parts are placed on different pieces of paper. The **Print Preview** screen will show this with lines indicating how the drawing will be split up. In addition, the number of pages used appears in the **Tiling:** box. See Figure 16-16.

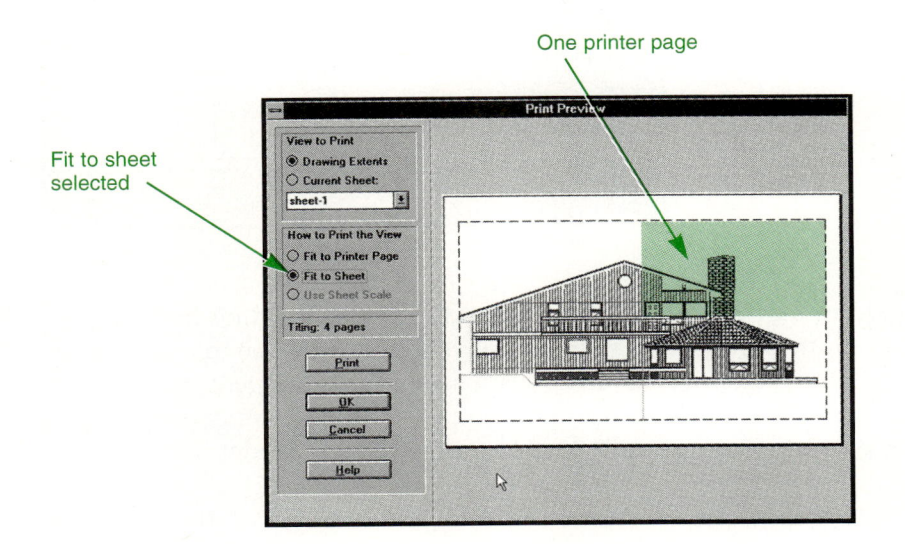

Figure 16-16.
If the sheet does not fit on the paper size supported by your printer, you can select the **Fit to Sheet** option and have AutoSketch print the drawing across several pieces of paper, called tiling. The **Print Preview** screen shows the tiling, and number of pages (one page is highlighted here).

Printing and Plotting Your Drawing

Now that you have completed all the needed setup steps, it's time to actually print or plot your drawing. In most cases, you will want more control of how your drawing is plotted than is provided by the defaults. If you have not gone through the setup procedure discussed in the previous sections, you may need to do so now.

Printing Your Drawing

The setup discussed in the previous sections makes your drawing ready to print to a dot matrix, ink jet, or laser printer. No additional steps are needed, Figure 16-17. Just make sure that paper is loaded in the printer. Then select **Print** from the **File** menu or from the **Print Preview** screen. AutoSketch begins printing your drawing. A message box will also appear that shows the printing status and gives you an option to cancel the job.

Plotting Your Drawing

Plotting your drawing typically requires a few more steps than printing to a printer.

Loading paper

Plotters hold paper differently. Microgrip pen plotters hold the paper with pinch rollers at each end. Paper on a flatbed pen plotter is usually held with tape. Load the paper according to the model you are using.

SHORTCUT
To quickly print while in the drawing editor, simply press the function key [F11].

Figure 16-17.
Having the wrong paper orientation can cause unnecessary tiling. Here, an A-size drawing is tiled because the printer was not configured for Landscape orientation. In this case, select **Cancel** and correct the setup before printing.

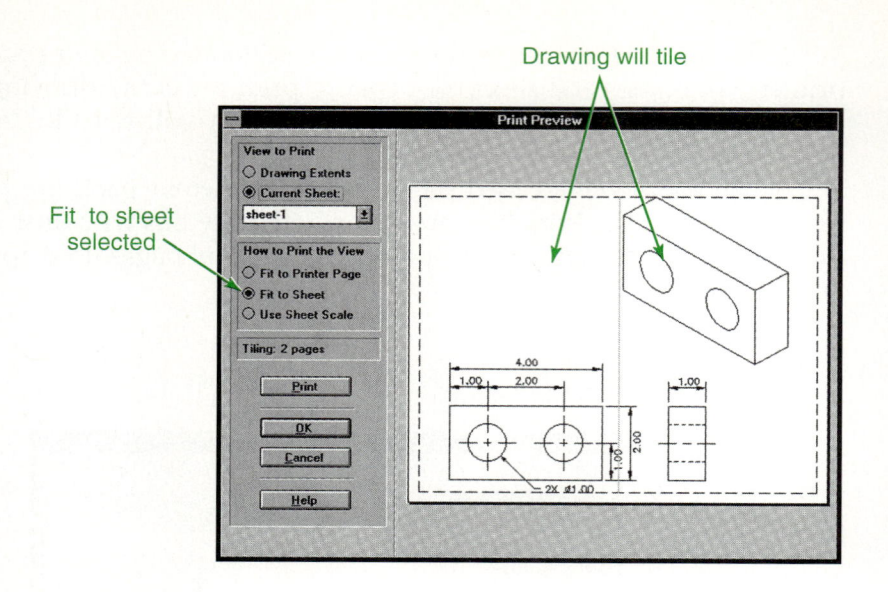

Installing pens

Pen plotters require that you prepare and insert the pens according to the manufacturer's directions. Some plotters also require that you screw the pen into an adapter before inserting it into the plotter. When using disposable liquid ink pens, press the tip into the cartridge. Plastic-tip and fiber-tip pens simply need to be uncapped. Check that you have the proper pens and adapters for your plotter model and drawing media.

Insert the pen or pen/adapter assembly into the plotter. Pens for multipen plotters are inserted into a carousel or pen rack. Make sure the proper pen type is placed into the slot number that corresponds with the value set during setup. When using single-pen plotters, insert pen number one.

Readying the plotter or printer

Plotters and printers have local and remote (on-line) command buttons on the control panel. Make sure the device is on-line. This means it is ready to receive instructions from the computer, not from the control panel. Refer to the owner's manual of your plotter for specific instructions.

Starting the plot

Select **Print** from the **File** menu or from the **Print Preview** screen. The plotting will begin. Single-pen plotters that support multiple pens stop during the procedure for you to insert the next pen. When inserting a new pen, avoid touching the plot with the pen tip. Multipen plotters automatically select the proper pen. When the plot is finished, remove the paper and tightly recap all pens.

Making multipen plots with a single-pen plotter

Some single-pen plotters cannot be used for multipen plots. They will not stop for you to insert a new pen to plot different colors. However, you can still make multipen plots if you follow some basic rules while creating the drawing.

1. Use layers to separate objects. Place all objects that should be plotted with a certain color or pen width on the same layer. For example, place all objects that should be plotted with a wide pen on layer 1. Do this for all colors or pen widths used in the drawing.
2. When you plot, have only one layer visible at a time. Insert the pen width or color associated with objects on that layer and plot the drawing.

> **SHORTCUT**
> To quickly plot a drawing from the drawing editor, simply press the function key [F11] once the plotter is set up.

3. Do not remove the paper when the plot is finished. Instead, press the button that resets the plotter. Then, hide the visible layer and make a different layer visible. Select the pen width or color associated with objects on that layer and plot the drawing again.
4. Repeat Step 3 for each layer. When you are finished, you will have a "multi-pen" plot.

Printing or Plotting a Drawing to File

AutoSketch allows you to print or plot a drawing to a file on disk, rather than to a hardcopy device. You can then do one of two things with the file.

- You can send the file to a hardcopy device outside of AutoSketch (in DOS or Windows, for example). This is helpful if your computer does not have a printer or plotter attached, but you have access to another computer that does.
- You can import the file into a desktop publishing program. (See Chapter 11 and Chapter 17.)

Print/plot to file procedure

A few simple steps must be followed to print or plot your drawing to a file. First, select **Print Setup...** from the AutoSketch **File** menu. Select the **Setup...** button. Then select the **Option...** button.

Printer

If the driver for that printer can print to file, you can select that option. If there is a section to enter a filename, you can do so now. If the driver does not support printing to file, you need to configure the printer in Windows.

Configure in Windows

To configure a printer or plotter in Windows, select the Control Panel icon. This is found in the Main group of the Program Manager. Then, select the Printers icon. Select the printer or plotter you will use and select the **Connect...** button. Scroll down the **Ports:** list and highlight the FILE: option. Finally, click on **OK**. Close the Printers dialog box and the Control Panel application. Now, when you select **Print** from the AutoSketch **File** menu, the **Print To File** dialog box appears. Under **Output File Name:**, enter the path and filename where you want to save the drawing.

> **SHORTCUT**
>
> If there is a filename option in the **Option...** section, it may be easier to leave it blank. AutoSketch then prompts you for a filename when you print.

Outputting a file to a printer/plotter

When sending a file to a printer or plotter, you must send it to the device used to print the file. For example, if you chose a Hewlett-Packard plotter, the drawing file must be copied to a Hewlett-Packard plotter outside of AutoSketch. Plotting to any other device will result in strange images, or nothing at all. A filed drawing is printed or plotted outside of AutoSketch by using the COPY command at the DOS prompt.

C:\> **COPY A:***filename* **LPT***x***:**

This command is used to send a drawing from drive A: to a printer connected to the computer's parallel port. This would typically be an inkjet, laser, or dot matrix printer.

The *filename* refers to the name you gave the file, when you printed it. The *x* in LPT*x* refers to the printer port. This port is usually LPT1:.

Most plotters are connected to a different port, called the serial port. Most computers have at least two serial ports. You must first use the MODE command to set up the communication between computer and serial port to use the serial port. This procedure is covered in your DOS documentation. Then, enter one of the following commands.

COPY A:*filename*.PLT COM1: /B
COPY A:*filename*.PLT COM2: /B

The command that you enter depends on which serial port (COM1: or COM2:) the plotter is connected to. Once you enter the COPY command, the plotter should begin outputting the drawing. If problems arise, consult the AutoSketch documentation.

Plotter Media, Pens, and Tips

There are certain materials and supplies that you will need to plot a drawing. In addition, there are steps you should follow to improve the life and performance of these supplies. These things are covered in the following sections. Also, plotting tips are given to make it easier for you to plot.

Drawing Media

Drawing media includes paper, vellum, and film. Most are cut sheets, except for dot matrix paper. This is often "fan-folded" for tractor-feed printers. Some devices may require special media.

Paper refers to a variety of wood-base and rag-base media. **Bond paper** is a low-cost, opaque, white paper similar to copy paper. **Rag bond** is a cotton-fiber medium that is more durable than bond paper. **Gloss paper** is clay-coated and provides high-contrast plots. It is often used when bright multicolor effects are used for graphics and charts. **Translucent paper** is a semi-transparent medium, similar to tracing paper. It is a low-cost substitute for vellum.

Vellum is a rag-base paper. It is treated with resin to make it transparent, strong, and durable. This media shows ghost marks if you fold or bend it. Most types of pens work on vellum. The paper must be transparent to reproduce prints from it.

Film is a polyester-base transparent media. It is used for high-quality plots. The film may be coated on one or both sides to accept pen ink. A matte finish is a nonglossy coating that accepts liquid ink pens quite well. Clear film is used for overhead transparencies. It has a water-base coating for use with felt- and plastic-tip pens. The combination produces crisp, bright colors.

Plotter Pens

Several types of plotter pens are available. They include liquid ink (refillable and disposable), fiber-tip, plastic-tip, roller ball, and pressurized ballpoint pens.

Liquid ink pens are high-quality pens where a liquid ink flows through a narrow metal tip. They are commonly used for final engineering and architectural plots. Refillable liquid ink pens require careful handling and maintenance. Disposable liquid ink pens do not require as much handling, yet still provide good-quality plots. Stainless steel, jewel, tungsten carbide, or ceramic tips may be used for liquid ink pens. They vary in durability and line quality.

Fiber-tip pens and *plastic-tip pens* have a tip with medium hardness. They are excellent for graphics and charts where an area is to be filled with a solid color. There is a wide selection of bright colors and pen widths. These types of pens are not suitable for final plots because the tip widens with use. Plastic tips wear less, but should still be used with caution.

Fiber-tip and plastic-tip pens are usually available with two types of inks. One is designed for use on paper. The other is designed for use on clear film. If film ink is used on paper, it may "bleed" into the paper.

Roller ball pens and *pressurized ballpoint pens* are used less frequently in industry. Roller ball pens plot well on clear overhead film. They produce darker and more consistent lines than fiber-tip pens. However, they are not available in as many colors as fiber-tip pens. Pressurized ballpoint pens can operate at extremely high speeds and complete many plots during a day. However, the line quality is generally not as good as other pen types.

Maintenance

Proper maintenance of drawing media and pens is important. Store drawing media flat, not vertical or rolled up. It should not be exposed to excessive humidity. Cap the pens after each use. Store them in an airtight plastic container with a damp sponge to keep them in good working condition.

Troubleshooting Plotter Problems

No matter how carefully you select pens for the drawing media, troubles will arise. These may be caused by worn-out pens, unfit plot speeds, improper pen pressure, or environmental problems. The chart in Figure 16-18 provides causes and solutions for common pen plotting problems.

Plotting Tips

Plotting can be time-consuming. Plots may take from a couple minutes for a check plot to over an hour for a final plot of a complex drawing. Remember these helpful hints as you prepare to plot your drawings.
- Use preprinted borders and title blocks when possible. This will save plotting time.
- Make reduced scale plots to avoid plotting drawings on large media. This not only saves time, but also pen ink and paper.
- Photocopy multiple prints. It is not feasible to produce multiple sets of prints with a plotter.
- Plot at the highest speed possible. Use the pen that will plot the fastest, yet still provide the necessary line quality.
- Eliminate unnecessary plots. Never plot a drawing just to see it on paper unless there is a need for a check or approval.
- Use single pen widths and colors when possible. Plotting multiple line widths and colors require that the plotter exchange pens.
- Set up a plotting center, plotting times, or a plotting service in your school or business. A plotter running all the time can disturb drafters and consume valuable computer time.
- Reduce costs by using the least expensive supplies that still produce acceptable results.

Figure 16-18.
Some common plotting problems, their likely cause, and possible solutions. (Koh-I-Noor Rapidograph)

Problem	Cause	Solution
Pen does not write	Pen is clogged because: 1. Ink is dry at tip 2. Pen is out of ink	1. Clean pen, and restart 2. Refill or replace pen
Pen writes by hand, but not on plotter	Pen height set wrong	Adjust pen height
Pen writes on plotter initially, then quits	1. Worn-out pen 2. Pen down-force too high 3. Bad start-up technique 4. Pen/media mismatch 5. Pen ran out of ink 6. Dirty or contaminated surface	1. Replace pen* 2. Reduce pen down-force 3. Change to correct technique 4. Change pen or media 5. Refill or replace pen 6. Clean surface * If this happens often, or too soon: change media
Pen writes on plotter, but lines skip	1. Dirty pen 2. Plotting too fast 3. Ink/media mismatch	1. Clean or replace pen 2. Decrease plotting speed 3. Change ink or media
Line width decreases from start to middle of line	Plotting too fast	Decrease plotting speed
Line width varies	Poor quality media	Change media
Lines feather	1. Ink/media mismatch 2. Plotting too slow	1. Change media or ink 2. Increase plotting speed
Ink beads on surface	Ink/media mismatch	Change media or ink
Lines too wide	Worn out pen	Replace pen
Lines not dark enough	1. Wrong type pen 2. Plotting too fast 3. Wrong type ink 4. Pen down-force too high	1. Use v-groove pen on film 2. Decrease plotting speed 3. Change ink 4. Reduce pen down-force
Line darkness varies	Ink/media mismatch	Change media or ink
Ink smudges on surface	1. Wrong type ink 2. Ink not given enough time to dry 3. Area is too humid	1. Change ink 2. Change program to allow for longer drying time 3. Reduce humidity
Pen leaks	1. Bad start-up technique 2. Dirty pen	1. Change to correct technique 2. Clean or replace pen

Summary

Plotting is the process of making a paper print of a drawing. You can send the drawing data to a printer, plotter, or file. A file can be printed outside of AutoSketch or imported into a desktop publishing program.

The first step in plotting a drawing is selecting your printer. Set up the printer if needed. Then, you may need to set up another sheet if you wish to print only a portion of a drawing. Finally, before printing, you will want to preview the drawing.

Important Terms

Bond paper
Check prints
Cropping
Drawing media
Drawing sheet
Fiber-tip pens

Film
Final plots
Gloss paper
Liquid ink pens
Paper
Plastic-tip pens

Pressurized ball-point pens
Printing
Rag bond
Roller ball pens
Tiling
Translucent paper
Vellum

New Commands

A**cceleration...
Caro**u**sel** menu
Current Sheet:
Default button
Device menu
Drawing Extents
Enclose entire drawing
Fit to Printer Page
Fit to Sheet
Force...
Hide Non-current Sheets

Hide Sheet Text
How to Print the View
Landscape
Margins... button
Options menu
Orientation
Output File Name:
Paper Feed
Paper Siz**e:**
Pen Speed...
PenColor menu

Portrait
Print
Print Prev**iew**
Print Setup...
Print to File dialog box
Setup... button
Sheet Setup...
Size menu
Titling:
Type menu
Use Sheet Scale
View to Print

Review Questions

Give the best answer for each of the following questions.

1. Prints made using a dot matrix, inkjet, or laser printer for review during the design process are called _____.
2. High-quality prints made with a pen plotter for copying and sharing a design are called _____.
3. List the four printing-related commands found in the **File** menu.
   A. _____
   B. _____
   C. _____
   D. _____

4. What steps are needed to change the paper size and orientation? _____
   _____
   _____
   _____

5. What settings should you change if the outer edges of your drawing are cropped on a printout?
   _____

6. How do you change a plotter pen to print a specific color of your drawing?
   _____

7. What three items should you check if not all of the sheets you have set up are displayed?
   A. _____
   B. _____
   C. _____

8. What effect does the **Drag Size** option have on the drawing scale?
   _____

9. When you use the **Scale** or **Stretch** tools to resize an existing sheet, which changes: the sheet size, or the drawing scale? (Hint: You will have to use AutoSketch to find the answer.)

10. Which two options in the **Print Preview** screen allow you to fit your entire drawing on the printer page, regardless of what objects the current sheets contain?
    A. _____
    B. _____

11. _____ will occur if the sheet is larger than the paper size supported by your printer or plotter.

12. How are the steps when printing to a laser printer different than printing to a plotter?
    _____
    _____

13. The _____ value determines which pen AutoSketch selects to plot certain color objects.

14. The _____ value is the speed that the plotter draws with each pen.

15. Name the three general categories of drawing media.
    A. _____
    B. _____
    C. _____

Activities

1. Select a drawing from a previous chapter that will fit full scale on a paper size that your plotter supports. Time how long it takes to plot the drawing.
   Time to plot:_____
2. If you have access to a dot matrix printer and a laser printer or pen plotter, make a check plot and a final plot. Compare the time required to plot each. Also compare the quality.
   Dot matrix—Time to plot:_____ Quality:_____
   Laser/Pen plotter—Time to plot:_____ Quality:_____
3. If you have a multipen plotter, plot a multicolor drawing using a different pen for each color. The drawing should fill the entire paper.
4. If you have a pen plotter and it supports multiple pen speeds, make one plot of a drawing at the slowest pen speed. Make another at the highest pen speed. Time both plots. Compare the times and line quality.
   Slowest pen speed time to plot: _____
   Highest pen speed time to plot: _____
5. Select a B-size or larger drawing from a previous chapter to plot full scale. However, plot on A-size paper. Note how tiling is used.
6. Plot a drawing from a previous chapter using the landscape format. Plot the drawing so that it fills the entire paper. Then, select portrait format. What was the difference in size of the plotted drawing? Why?
   _____
   _____
   _____
7. Plot drawings that you created in previous chapters. Experiment with different settings, as well as various pen types and drawing media. Note which combination of settings creates the best plot. Record these settings below for future use.
   _____
   _____
   _____
   _____

AutoSketch drawings can be shared electronically with other programs, such as Quark XPress.®

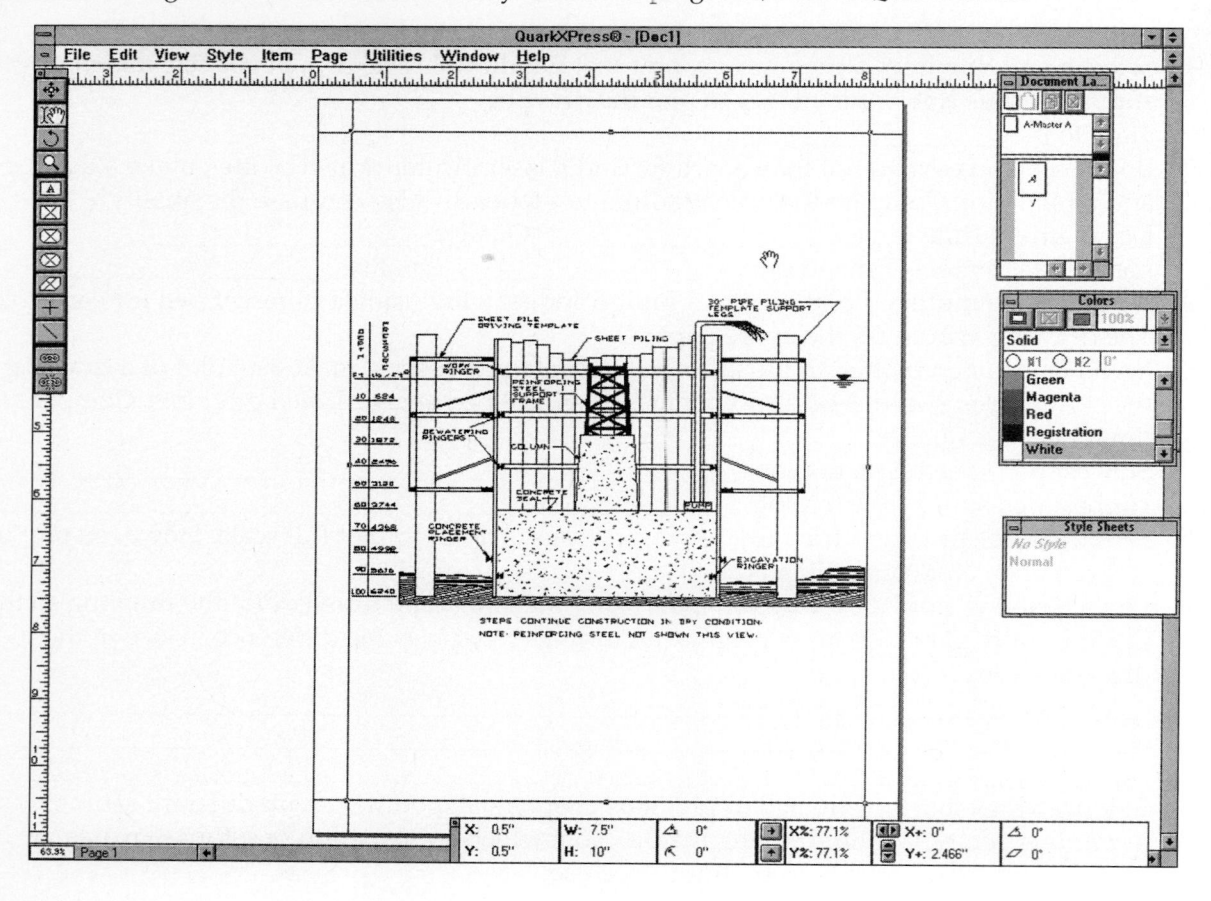

CHAPTER 17

Sharing Drawings with Other Programs

Objectives
After studying this chapter, you will be able to:
- Describe how to share AutoSketch drawings with other applications.
- Export a drawing in AutoCAD DWG format.
- Exporting a drawing in DXF format.
- Import a DXF drawing.
- List restrictions for importing DXF drawings.
- Copy and paste objects using the Windows clipboard.
- Export an Encapsulated PostScript (EPS) file for desktop publishing.
- Export an HPGL file for desktop publishing.
- Make and view slides.

There are a variety of CAD and graphics programs available today. Unfortunately, most CAD programs, graphics software, and desktop publishing programs store data differently. This makes it difficult to share drawings. Several standard file formats have been developed to exchange drawings. Using these formats, you can convert your drawings into a generic format that other programs can understand. Converting drawings into the generic format is called *exporting*. Bringing a file into AutoSketch is called *importing*.

This chapter will discuss four ways to exchange drawings. First, AutoSketch can export drawings in AutoCAD DWG format. Second, AutoSketch supports the Drawing Interchange Format (DXF). This format can be used by many programs. Third, you can use the Windows clipboard to send graphics to other Windows programs. Finally, this chapter will discuss how to modify your printer setup to print files in formats that can be read by other graphics programs.

Exchanging Drawings with AutoCAD

Many engineers and designers use AutoSketch as a preliminary drawing tool. They then send those drawings to the drafting department. However, AutoCAD (also published by Autodesk) is the most widely used production CAD program in the world. Drafting departments across the world use AutoCAD to create final working drawings and three-dimensional models.

In earlier versions of AutoSketch, designers had to export drawings in DXF format. The DXF format was then imported into AutoCAD. However, some details might be lost. With Release 2 of AutoSketch for Windows, you can export an AutoSketch drawing as an AutoCAD DWG file. AutoCAD can read this format more quickly and with greater precision than the DXF format. The DWG files that AutoSketch exports can be used in AutoCAD Release 9 or later.

AutoSketch can only export, not import, DWG files. To import an AutoCAD drawing into AutoSketch, you must use the DXF format. This format is discussed later in this chapter.

Exporting Drawings in DWG Format

To export an AutoSketch drawing to AutoCAD, first open the drawing you want to export. Then, select the **File** menu and the **Export...** command. The **Export** dialog box appears, Figure 17-1. Select the drive and directory where the drawing should be stored. You may need to scroll through the list of available drives and directories. Then, pick the scroll arrow by **List Files of Type:** and select **AutoCAD DWG (*.dwg)**. Last, enter the drawing **Filename** and pick the **Save** button.

Figure 17-1.
The **Export** dialog box. The available options are shown.

SHORTCUT
Select the export file type before you enter a filename. Otherwise, you will need to enter the filename twice.

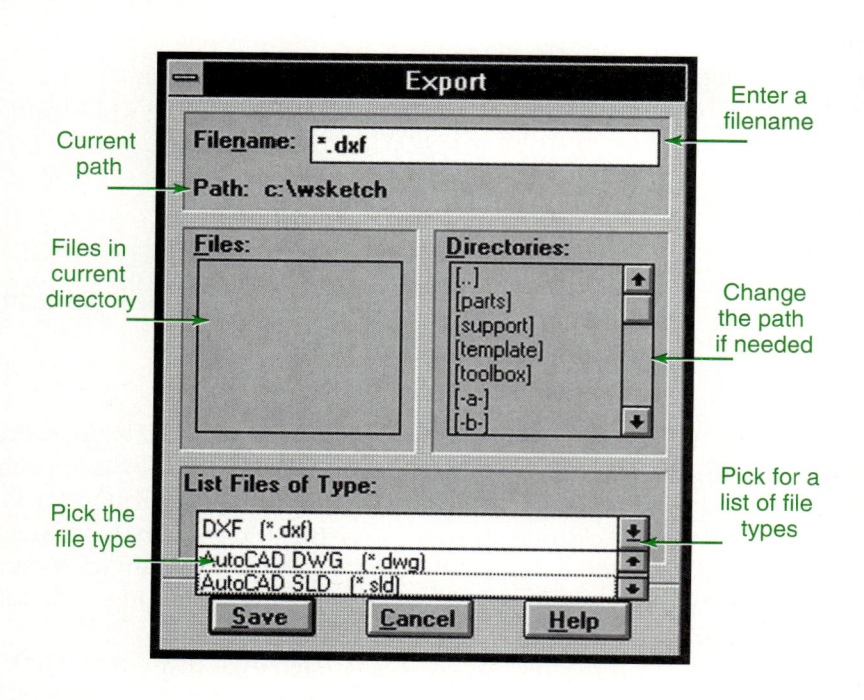

AutoSketch saves the new drawing with a .DWG file extension. You do not have to enter this extension. AutoSketch drawings have the file extension .SKD. Therefore, you can use the same name for the DWG file as your AutoSketch drawing. File extensions are discussed in detail in Chapter 20.

Exchanging Drawings with Other CAD Programs

AutoSketch can export and import drawings using the *Drawing Interchange Format (DXF)* developed by Autodesk, Inc. The DXF format is the most commonly used drawing conversion for personal computer-based CAD programs and computer-aided machining (CAM) programs. In addition, many desktop publishing programs, such as Pagemaker and Quark XPress, can import DXF files as graphics.

Exporting Drawings in DXF Format

To export an AutoSketch drawing in DXF format, first open the drawing you want to export. Select the **File** menu and the **Export...** command. The **Export** dialog box appears. See Figure 17-1. Select the drive and directory where the drawing should be stored. You may need to scroll through the list of available drives and directories. Then, pick the scroll arrow by **List Files of Type:** and select **DXF (*.dxf)**. Finally, enter the drawing **Filename** and pick the **Save** button.

AutoSketch saves the new drawing with a .DXF file extension. You do not have to enter this extension. AutoSketch drawings have the file extension .SKD. Therefore, the DXF file can have the same name as your AutoSketch drawing. File extensions are discussed in detail in Chapter 20.

Importing a DXF Drawing

To import a DXF format drawing, first start a drawing using the **Normal** template. Do not add any objects to your new drawing. If you do not have an empty drawing, the objects imported may overlap existing objects. Next, select the **File** menu and **Import...** command. The **Import** dialog box appears, Figure 17-2. Select the drive and directory where the DXF drawing is stored. You may need to scroll through the list of available drives and directories. Pick the scroll arrow by **List Files of Type** and select **DXF (*.dxf)**. Select the drawing file and pick the **Open** button.

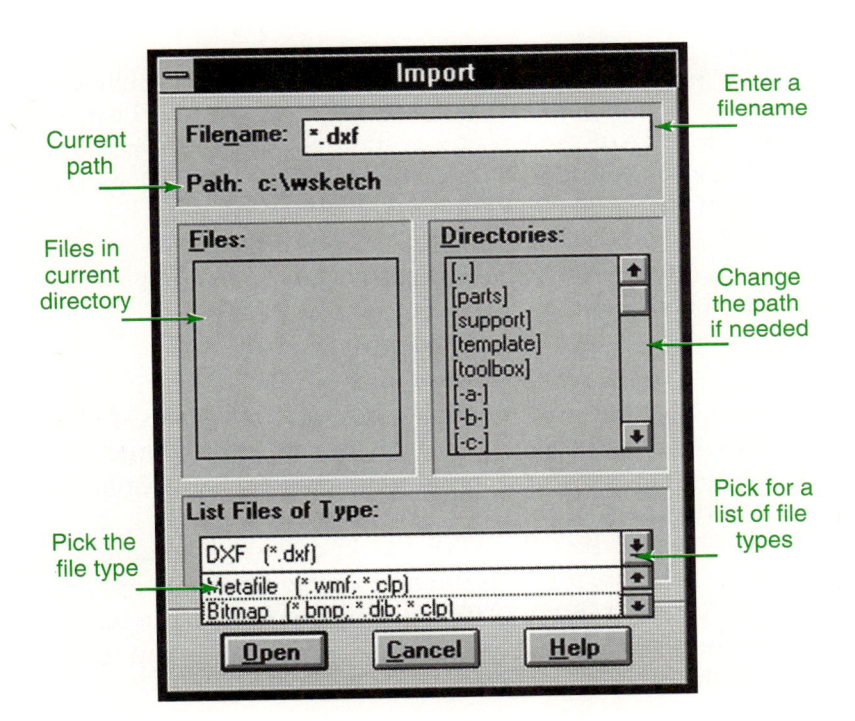

Figure 17-2.
The **Import** dialog box. The available options are shown.

Once you select the **Open** button, the **Import DXF Control Settings** dialog box appears, Figure 17-3. The two options are **Explode Large Blocks** and **Import Visible Attributes.**

Figure 17-3.
The **Import DXF Control Settings** dialog box allows you to include or exclude specific AutoCAD objects.

Select the options you want

Explode Large Blocks

AutoCAD blocks are a cross between an AutoSketch group and a symbol. In AutoCAD, you can save a group of objects by name to use over and over. These are called *blocks* in AutoCAD. Unlike an AutoSketch group, there is no limit to the number of objects in an AutoCAD block. When AutoSketch imports DXF drawings that contains AutoCAD blocks, one of two things can happen.
- AutoSketch imports blocks with 1000 or less objects as a group.
- AutoSketch will not import a block with more than 1000 objects unless the **Explode Large Blocks** option is checked.

Import Visible Attributes

In AutoCAD, you can assign attributes to an AutoCAD block. *Attributes* are text objects that provide information about a specific block. For example, an attribute for a window symbol might be the manufacturer and model number. You can display or hide the attribute text in AutoCAD drawings.

If the attribute text is displayed when the DXF file is created in AutoCAD, AutoSketch can import that text. To do so, select the **Import Visible Attributes** option.

You may not want to import these attributes. Attributes are not really a dimension or note. They are used to generate a bill of materials. If you don't want to import the attributes, deselect the **Import Visible Attributes** option.

Restrictions on Importing DXF Drawings

Remember, AutoCAD can create more complex objects than AutoSketch. Because of this, there are a number of restrictions on what portions of a DXF file AutoSketch will import. The best rule to remember is that AutoSketch ignores DXF objects it cannot recognize.

Refer to the *AutoSketch for Windows User's Guide* for a complete list of restrictions. You will encounter the following most often.

- AutoCAD can draw in true three-dimensional mode. This is not the same as an AutoSketch isometric view. In AutoSketch, there is no Z axis (the third coordinate). Therefore, a straight-line object created with a Z axis value is projected onto the XY plane by stripping the Z coordinate. Circles and arcs are reconstructed on the XY plane using the XY coordinates of the center and radius.
- Objects that have variable widths are reduced to hairline-width polylines. Objects with constant widths are converted to solid pattern fills.
- Quadratic B-splines and fit polylines from AutoCAD are ignored. All other curves are imported.
- Text created in a font that AutoSketch does not have is displayed in the **standard** font.
- Dimensions must be exploded in AutoCAD. AutoSketch will import the components (lines, text). Otherwise, dimensions are discarded.

Copying and Pasting Objects with the Windows Clipboard

Windows provides an easy way to share information between Windows applications. This is the Windows clipboard. The *clipboard* is a temporary holding space. This allows you to switch to another application, and place those objects. This process is commonly called *cutting and pasting* or *copying and pasting*. AutoSketch allows you to copy objects from your drawing to the Windows clipboard.

For example, suppose you draw a map in AutoSketch. Then you want to place the map into a word processing document where you also have written directions. You can copy the objects that make up the map. Then, switch to your Windows-based word processing program. Select the Paste command in that application to paste the objects.

There are five options in the **Edit** menu that allow you to use the Windows clipboard, Figure 17-4. These options are discussed in the next sections.

SHORTCUT
To quickly change between Windows applications, hold down the [Alt] key and press the [Tab] key. Continue pressing the [Tab] key until you reach the program you want. Then release the [Alt] key.

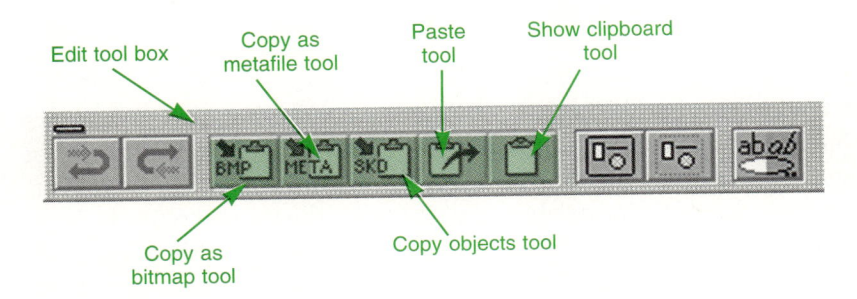

Figure 17-4.
Five tools in the **Edit** menu allow you to copy and paste objects using the Windows clipboard.

Copying Objects as a Bitmap

The **Copy as Bitmap** tool copies an area of the drawing to the clipboard in bitmap format. A *bitmap format* is best compared to a photograph. The image is made up of dots, not geometric objects. In other words, the computer no longer sees a line *as* a line. It sees the line as the *dots* that make up the image.

When a bitmap is pasted, the objects in the image cannot be edited. Also, if you paste the bitmap file back into AutoSketch, it covers any objects under it, Figure 17-5. A bitmap file is treated as a single object. You can move and copy it. However, you cannot explode or edit the pieces that make up the bitmap.

To copy a portion of your drawing as a bitmap, select the **Edit** menu and **Copy as Bitmap** tool. The following prompt appears.

Copy Bitmap First corner:

Select the first corner of a window that will enclose the objects you want to copy. You must use a window, not a crosses box.

Copy Bitmap Second corner:

The bitmap contains the entire area within the box, not just the objects. The bitmap will also include the background color. The image is held on the Windows clipboard until you paste it elsewhere.

Figure 17-5.
Unlike pasting metafiles or AutoSketch objects, pasting a bitmap file (highlighted here) from the clipboard will cover objects underneath it.

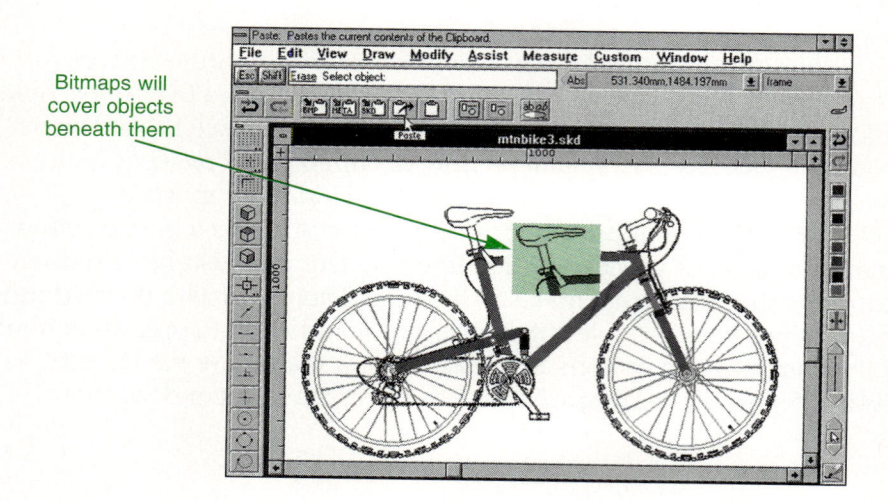

Copying Objects as a Windows Metafile

The **Copy as Metafile** tool copies objects to the Windows clipboard in the *Windows metafile (WMF)* file format. The metafile format retains all the geometric properties of the objects. In other words, the computer still sees a line *as* a line. Therefore, when pasted into another application, the individual objects of the metafile can be edited.

However, do not paste a metafile back into AutoSketch. Although it is inserted as a group of objects, when you ungroup those objects, most do *not* retain their AutoSketch identity. For example, a curve or ellipse is pasted in as individual connected lines. Use the **Copy Objects** tool in the **Edit** menu instead.

To copy a portion of your drawing as a metafile, select the **Edit** menu and **Copy as Metafile** tool. The following prompt appears:

Copy Metafile Select object:

This tool allows you to select objects using all the selection methods for editing AutoSketch objects. Use the **Selection Lock** (**Shift** button) or the crosses/window box to select multiple objects. The metafile contains only the AutoSketch objects you pick. It is held on the Windows clipboard until you paste it elsewhere.

Copying Objects for Another AutoSketch Drawing

The **Copy Objects** tool copies objects to the clipboard in AutoSketch format (SKD). This allows you to paste those objects into another AutoSketch drawing and retain all the object properties. An ellipse is still an ellipse. A curve is still a curve.

Pasting Objects into an AutoSketch Drawing

The **Paste** tool copies the contents of the clipboard into your drawing. Using this tool with the **Copy Objects** tool allows you to copy objects from one drawing into another drawing without using **Part Clip...** command. Pasting objects from other Windows applications is useful for pictures, logos, or designs that you need, but are not easily created with AutoSketch.

Displaying the Clipboard

The **Show Clipboard** tool opens the Windows ClipBook Viewer. This shows what is currently on the clipboard. You can save the contents with the file extension .CLP. The saved information can then be used in another Windows application (or AutoSketch) later.

Importing Metafiles and Bitmaps

You can also import bitmaps and metafiles into AutoSketch. These files can be created in another program, or in a previous AutoSketch drawing session.

To import a bitmap or metafile, select the **File** menu and the **Import...** command. The **Import** dialog box appears. See Figure 17-2. Select the drive and directory where the bitmap or metafile drawing is stored. You may need to scroll through the list of available drives and directories. Pick the scroll arrow by **List Files of Type:** and select either **Metafile (*.wmf;*.clp)** or **Bitmap (*.bmp;*.dib;*.clp)**. The acceptable file extensions are listed after the file type.

Select the file and pick the **Open** button. Like a pasted bitmap, an imported bitmap is treated as a single object. A metafile is imported as grouped objects. To edit the objects in a metafile, you must first ungroup them.

SHORTCUT

A quick way to select multiple objects is by holding the [Shift] and picking the objects. When you release the key, the objects are copied to the clipboard.

SHORTCUT

Remember that clicking [..] takes you up one level in the directory path.

Printing Files to Export to Desktop Publishing

Recall from Chapter 16 that you can print to a file as well as to a printer. There are two types of printer files that can be imported into most desktop publishing programs. These two file types are Hewlett-Packard Graphics Language (HPGL) and Encapsulated PostScript (EPS).

PostScript is a page description language developed by Adobe Systems, Inc. It is widely used to print complex graphics and output from a desktop publishing program. A PostScript file is no more than a text file that "describes" how graphics should be created by a PostScript printer. An *Encapsulated PostScript (EPS)* file is a PostScript file with special header information that allows the file to be imported into a desktop publishing program as a single graphic.

Hewlett-Packard Graphics Language (HPGL) is the language used by Hewlett-Packard plotters. It consists mainly of lines called *vectors*. These vectors are how plotter pens draw. Even text consists of many tiny line strokes. For this reason, HPGL files usually do not reproduce text and complex curves well. HPGL files have the extension .PLT.

You cannot export these types of files from AutoSketch. However, you can easily create HPGL and EPS files with just a few simple configuration steps. These steps are covered in the next sections.

Creating Encapsulated PostScript (EPS) Files

To create an EPS file, you must first install a PostScript printer driver. Then, you must perform some special sheet setup tasks and configure the printer driver to print to an EPS file.

Installing a PostScript driver

The following steps are involved in setting up Windows for a PostScript printer. Refer to Figure 17-6. You may be asked for your Windows installation disks, so have them handy.

Figure 17-6.
Steps involved in adding a PostScript printer driver into Windows. A—Select the Printers icon (highlighted here) from the Control Panel application. B—Highlight the printer and select the Install... button.

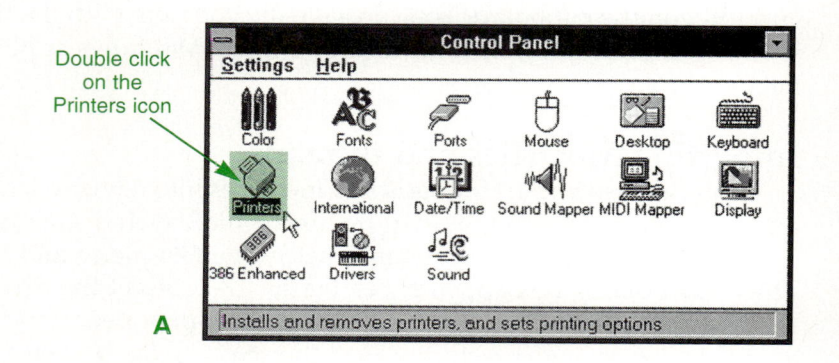

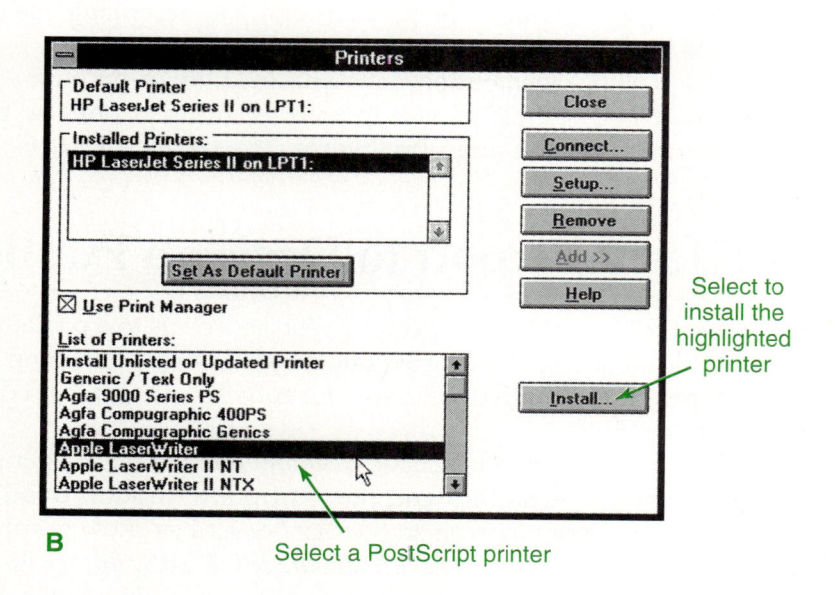

1. Double-click on the Windows Control Panel icon. This is usually found in the Main program group of the Windows Program Manager. This starts the Control Panel application. This application allows you to set up various features of Windows.
2. Double-click on the Printers icon to open the Printers window.
3. Pick the Add>> button.
4. Scroll down the list of Available Printers until you see Apple Laserwriter. This is a common PostScript printer. Highlight it. Note: It doesn't matter if your printer is different, or even if there is no printer connected to your system.
5. Pick the Install... button. Windows may then ask for certain Windows disks. Follow the on-screen prompts. When finished, the printer will appear in the Installed Printers list box.
6. Close the Printers window and the Control Panel application.

Printing to an EPS File from AutoSketch

Once you have installed the printer driver in Windows, printing to file is easy. Use the following steps. Note: You will not need to go through print setup again unless you need to change orientation or use a different printer.

1. Verify that the objects you want to print fit precisely within the current sheet. It is best to select the **Enclose Entire Drawing** option in the **Sheet Setup** dialog box. By doing so, your EPS file will contain exactly the objects in your drawing.
2. Select **Print Setup...** from the **File** menu. The **Print Setup** dialog box will appear.
3. Highlight the PostScript printer in the **Printer:** list box and pick the **Setup...** button.
4. Pick the **Landscape** option if the drawing you are exporting is wider than it is high or **Portrait** if the drawing is higher than it is wide.
5. Pick the **Options...** button to open the **Options** dialog box.
6. Pick the option **Encapsulated PostScript File.** See Figure 17-7. Do not enter a name.

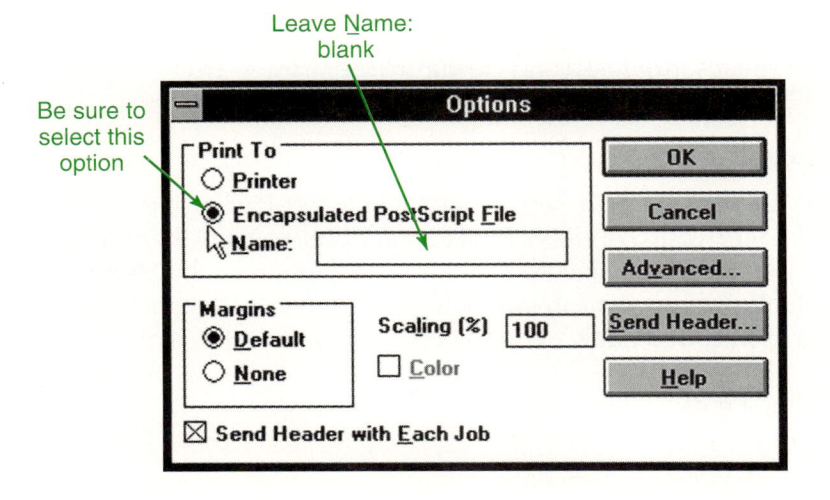

Figure 17-7.
Make sure that the Encapsulated PostScript File option is selected when configuring the printer in the Print Setup dialog box to print an EPS file.

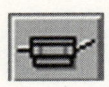

SHORTCUT

You can also print by pressing the function key [F11].

7. Pick the **OK** button to close the **Options** dialog box.
8. Pick the **OK** button again to close the specific printer dialog box.
9. Pick **OK** to close the **Printer Setup** dialog box.
10. Select **Print** from the **File** menu.
11. The **Print to File** dialog box will appear, Figure 17-8. Enter the **Output File Name** and pick **OK**. Use up to 8 characters in the filename. Be sure to add a period and the extension EPS (.EPS) to the name. For example, you might call the file BRACKET.EPS. It is best to also enter a drive and directory path where the file should be stored. Directory paths are discussed in Chapter 20. An example would be C:\GRAPHICS\BRACKET.EPS.
12. Your file is now saved to the drive and directory specified using the filename you entered. The file can be imported into most desktop publishing programs.

Note: If your desktop publishing program supports EPS files, but does not define space for the EPS file on the page, the graphic will print over the entire page even if the image size is smaller. In addition, you cannot print EPS files directly to the printer. To create a printable PostScript file, follow the instructions given in Chapter 16.

Figure 17-8.
Enter the drive, directory, and filename for the EPS file to be printed.

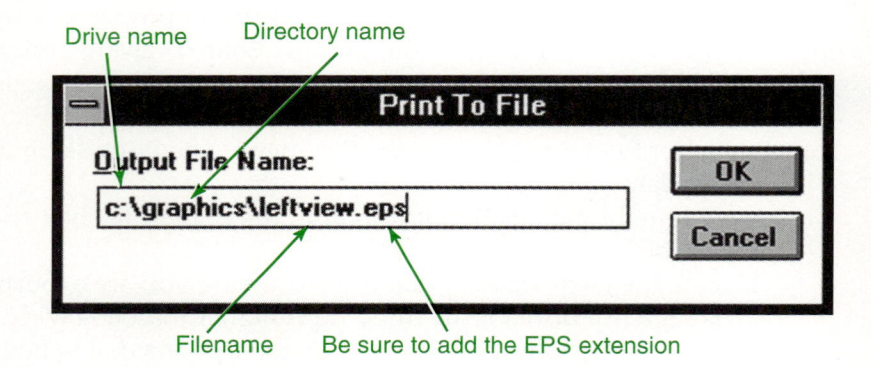

Creating Hewlett-Packard Graphics Language (HPGL) Files

To create an HPGL file, you must install a plotter driver in Windows. You must also perform some special sheet setup tasks and configure the printer driver to plot to a file.

Installing a Plotter Driver

The following steps are involved in setting up a Hewlett-Packard plotter in Windows. You may be asked for your Windows installation disks, so have them handy.

1. Double-click on the Control Panel icon. This is usually found in the Main program group of the Windows Program Manager. This starts the Control Panel application. This application allows you to set up various features of Windows.
2. Double-click on the Printers icon to open the Printers dialog box.
3. Pick the **Add**⟩⟩ button.

4. Scroll down the list of Available Printers until you see HP plotters. There will be a number of models. Highlight HP7475A.
5. Pick the Install... button. Windows may ask certain Windows installation disks. Follow the on-screen prompts. When finished, the plotter will appear in the Installed Printers list box.
6. Highlight the plotter and select the Connect... button.
7. Scroll down the list of Ports: and highlight FILE: in the list of available ports. See Figure 17-9.
8. Click on OK to close the Printers window. Close the Control Panel application.

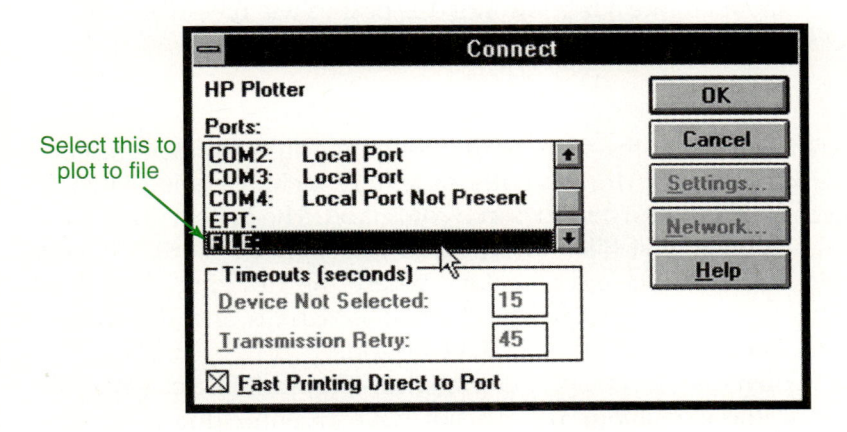

Figure 17-9.
To print an HPGL file, you must configure the printer driver to plot to a file.

Installing a Plotter Driver

Once you have installed and configured the printer driver in Windows, plotting to file is easy. Use the following steps. Note: You will not need to go through print setup again unless you change printers.

1. Verify that the objects you want to print fit precisely within the current sheet. It is best to select the **Enclose Entire Drawing** option found in the **Sheet Setup** dialog box. By doing so, your HPGL file will contain exactly the objects in your drawing.
2. Select the HP plotter in the **Print Setup** dialog box. Click on the **Setup...** button and select the orientation. Close the plotter dialog box and the **Print Setup** dialog box.
3. Select **Print** from the **File** menu.
4. The **Print to File** window will appear. Enter an **Output File Name** and pick **OK.** Make sure to enter an 8 character filename and add the extension .PLT to the name. For example, you might call the file VALVE.PLT. It is best to also enter a drive and directory path where the file should be stored. An example would be C:\GRAPHICS\VALVE.PLT. Directory paths are discussed in Chapter 20.
5. The file is saved to the drive and directory you specified. The file can be imported into graphics or desktop publishing programs.

SHORTCUT
To quickly print, press the function key [F11].

Making and Viewing Slides

An AutoSketch *slide* is a "snap shot" of the current view of your drawing. A slide includes anything on-screen, including sheets and visible layers. The items that are not shown are the grid, objects on invisible layers, toolboxes, and status bar. A slide stores the image of the drawing on-screen, but not the actual drawing data. Thus, slides cannot be edited, printed, zoomed, or otherwise changed in AutoSketch. There are several situations where you might use slides.

- Slides made with AutoSketch can be viewed in many Autodesk programs, including AutoCAD, AutoShade, and Animator Studio. Plus, AutoCAD slides can be viewed in AutoSketch without any conversion process.
- While in AutoSketch, you can view a slide made of another drawing for reference. A slide displays quickly. You can also view a slide while working on another drawing.
- Many desktop publishing and graphics programs can import and print a slide file. However, a slide is much like a bitmap file. It is best to use metafiles, EPS, and HPGL files for desktop publishing.

Making Slides

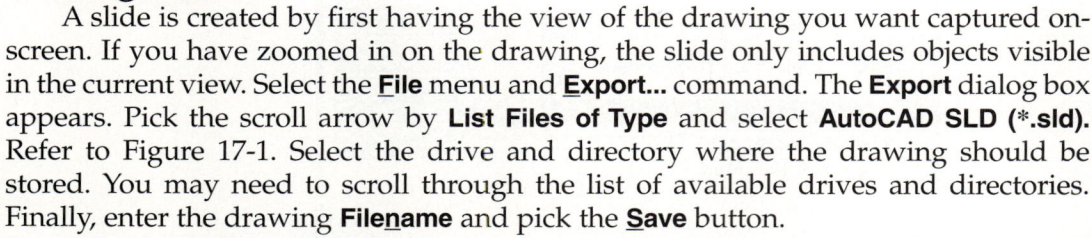

A slide is created by first having the view of the drawing you want captured on-screen. If you have zoomed in on the drawing, the slide only includes objects visible in the current view. Select the **File** menu and **Export...** command. The **Export** dialog box appears. Pick the scroll arrow by **List Files of Type** and select **AutoCAD SLD (*.sld)**. Refer to Figure 17-1. Select the drive and directory where the drawing should be stored. You may need to scroll through the list of available drives and directories. Finally, enter the drawing **Filename** and pick the **Save** button.

You can use the same filename as your AutoSketch drawing. AutoSketch saves the new drawing with a .SLD file extension. You do not have to enter this extension. AutoSketch drawings have the file extension .SKD. File extensions are discussed in detail in Chapter 20.

Viewing Slides

AutoSketch and AutoCAD slides can be viewed in AutoSketch at any time, even while another drawing is on-screen. See Figure 17-10. The slide is shown on top of the current drawing. Select the **View Slide...** command from the **File** menu. The **View Slide File** dialog box appears. Select the drive and directory where the slide is stored. You may need to scroll through the list of available drives and directories. Select the slide file and pick **Open**. The slide file will remain on-screen until you select any tool or option that redraws the display, such as **Redraw** and **Pan**.

You cannot edit slides displayed using the **View Slide...** command. You must edit the original drawing and make another slide to change it. If you save the new slide under a different name, you may want to delete the old slide file to avoid confusion.

Summary

AutoSketch can export and import drawings using a wide variety of formats to communicate with CAD programs and other graphics applications. Using the **Export...** command in the **File** menu you can export AutoCAD format DWG files, DXF format files, and Autodesk SLD slides.

With the clipboard tools in the **Edit** menu, you can copy and paste bitmap files, metafiles, and AutoSketch objects. These tools copy objects to the Windows clipboard. They can then be saved with the .CLP extension or pasted into other applications.

Finally, to send drawings to desktop publishing and non-Windows applications you can print Encapsulated PostScript files or Hewlett-Packard Graphics Language files. To do so, you must configure a printer or plotter in Windows to print to file.

SHORTCUT

Click [..] to go up one level in the directory path.

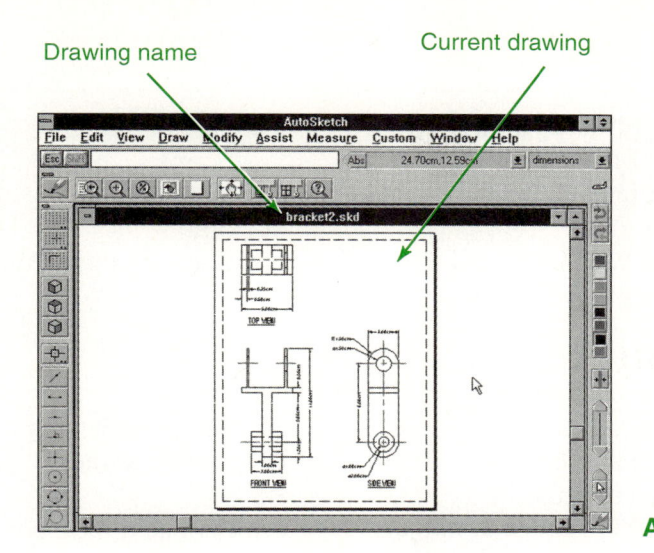

Figure 17-10.
A—You can view a slide with another drawing open. B—Select the slide to view in the **View Slide File** dialog box. C—The slide covers the current drawing until you pick a zoom option, **Redraw**, or **Pan**.

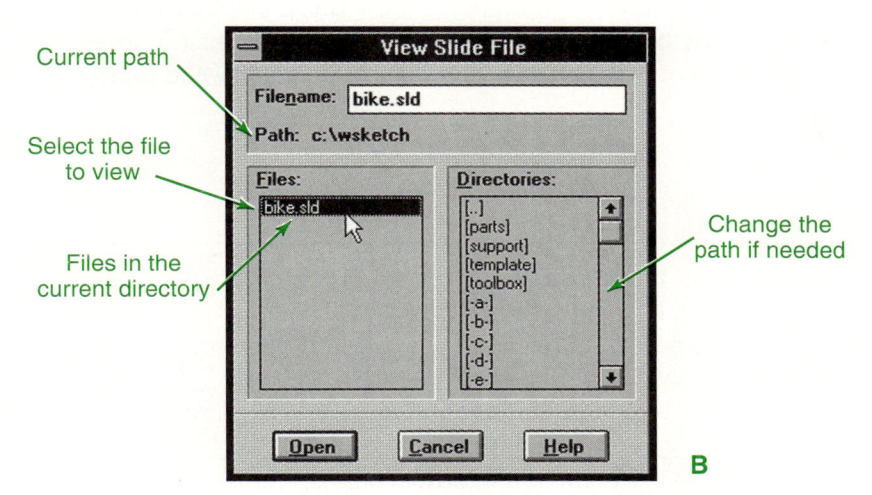

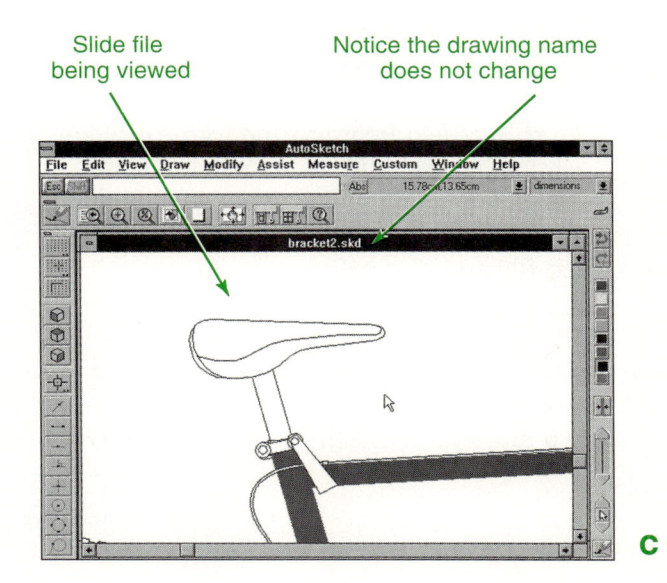

Important Terms

Attributes
Bitmap format
Blocks
Clipboard
Copying and pasting
Cutting and pasting
Drawing Interchange Format (DXF)
Encapsulated PostScript (EPS)
Exporting
Hewlett-Packard Graphics Language (HPGL)
Importing
PostScript
Slide
Vectors
Windows metafile (WMF)

New Tools and Commands

AutoCAD DWG (*.dwg)
AutoCAD SLD (*.sld)
Bitmap (*.bmp;*.dib;*.clp)
Copy as Bitmap tool
Copy as Metafile tool
Copy Objects tool
DXF (*.dxf)
Explode Large Blocks
Export... command
Export dialog box
Import... command
Import DXF Control Settings dialog box
Import Visible Attributes
List Files of Type:
Metafile (*.wmf;*.clp)
Output File Name
Ports:
Print to File dialog box,
Show Clipboard tool
View Slide... command
View Slide File dialog box

Review Questions

Give the best answer for each of the following questions.

1. Since AutoCAD can read both DXF and DWG formats, why should you export drawings from AutoSketch to AutoCAD using the DWG format? _____
2. When exporting drawings and slides, why can you use the same filename without overwriting your AutoSketch drawing file? _____
3. The _____ file format is the most used drawing conversion format for personal computer-based CAD programs.
4. AutoSketch imports AutoCAD blocks with 1000 or less objects as a(n) _____.
5. Because AutoSketch does not support true three-dimensional drawing, imported DXF drawings are stripped of their _____ axis value.
6. Under what conditions can AutoCAD attribute text be imported into AutoSketch using the DXF format?
_____
_____

7. What is the difference between a bitmap file and a metafile?
_____

8. True or false. To import a Windows metafile into AutoSketch, you must use the **Paste** tool.
9. True or false. AutoSketch objects that are copied to the Windows clipboard can be pasted into other Windows-based graphics applications.
10. Explain the difference between pasting a bitmap, a metafile, and objects into AutoSketch from the Windows clipboard.
_____
_____
_____
_____

11. What AutoSketch command is used to create encapsulated PostScript (EPS) and Hewlett-Packard Graphics Language (HPGL) files? _____

12. True or false. You can print an encapsulated PostScript file from a DOS prompt.
13. What is one purpose of making a slide?

14. True or false. A slide can be edited in AutoSketch.
15. After viewing a slide file, how do you remove it from the screen?

Activities

Here is a mixture of activities to develop your skills exporting and importing drawings. Save your drawings on your work diskette using the name given. After saving an activity, select the **Close** command from the **File** menu to clear the drawing from the screen.

1. Export drawing C11A1, C11A2, C11A3, C11A4, or C11A5 from the activities in Chapter 11 in the DWG format. If you have access to an AutoCAD system, bring up this drawing and print it. If not, provide the drawing to a drafter who has an AutoCAD system to print the drawing for you. Record any differences between the exported drawing and the original drawing.

2. Export drawing C11A13, C11A14, or C11A15 from the activities in Chapter 11 to DXF format. Then, import the DXF file back into a new AutoSketch drawing. Record any differences between the exported drawing and the original drawing. Be sure to use the **Show Properties** tool.

3. If you have access to an AutoCAD system, export a drawing in DXF format. If not, ask a drafter who has an AutoCAD system to export an AutoCAD drawing in DXF format. Then, start a new drawing in AutoSketch and import the DXF drawing. Save the drawing as **C17A3**. Record any differences between the imported drawing and the original drawing.

4. Start a new drawing and draw a line, circle, ellipse, and curve. Copy the objects to the clipboard as a bitmap file. Paste the bitmap file back into your AutoSketch drawing. Move the bitmap file over the objects you drew. Explain what happened. Save the drawing as **C17A4**.

5. Start a new drawing and draw a line, circle, ellipse, and curve. Copy the objects to the clipboard as a metafile. Paste the metafile back into your AutoSketch drawing. Use the **Copy** tool to make a copy of the metafile. Move one copy of the metafile over the objects you drew. Explain what happens. Ungroup the other copy of the metafile. Attempt to delete the curve by picking it. Do not use the crosses/window box. Explain what happens. Save the drawing as **C17A5**.

6. Start two new drawings. In one drawing create a line, circle, ellipse, and curve. Use the **Copy Objects** tool to copy the objects to the clipboard. Paste the objects into the other drawing. Erase the curve by picking it. Do not use the crosses/window box. Explain what happens. Do not save either drawing.

_____

_____

7. Select any drawing created in a previous chapter and use the steps necessary to export it as an Encapsulated PostScript file. Save the file as C17A7.EPS. You may need to obtain Windows installation disks if your computer does not have PostScript printer configured for Windows. Import this drawing into any available word processing or desktop publishing application that supports EPS graphics. Examples are Word for Windows, PageMaker, and Quark XPress. Print the page with the EPS file to a PostScript printer using that application. You cannot print the page to a printer that does not support PostScript.

8. Select any drawing created in a previous chapter and use the steps necessary to export it as an HPGL file. Save the file as C17A8.PLT. You may need to obtain Windows installation diskettes if your computer does not have an HP plotter configured for Windows. Import this drawing into any available word processing or desktop publishing application that supports HPGL graphics. Examples are Word for Windows, PageMaker, and Quark XPress. Print the page with the HPGL file to any printer.

9. Select any drawing created in a previous chapter and use the **Zoom Full** tool so that the drawing fills the drawing window. Create a slide named C17A9. Zoom into a small section of your drawing and then view the slide C17A9.SLD. Select **Redraw** to remove the slide from the screen.

CHAPTER 18

Drafting for Computer-aided Machining (CAM)

Objectives

After studying this chapter, you will be able to:
- Define computer-aided machining.
- List machining operations and the tools used for those operations.
- Identify steps of the CAD/CAM/CNC process.
- Explain how CAD is applied in the CAM process.
- Apply guidelines for CAM in creating an AutoSketch drawing.
- Identify key tasks in using CAM software.

What Is CAM?

Computer-aided machining (CAM) is the process of converting data from a computer drawing into code that can be used to control manufacturing machines. CAD software, such as AutoSketch, is used to create the drawing. CAM software is used to convert that drawing into numerical control (NC) code. This code is then loaded into a computer numerical control (CNC) machine. A *computer numerical control (CNC) machine* is able to use NC code to control its motors.

The primary use of CAM is to generate the computer code needed to control milling machines and lathes. However, CAM can also be used for other machining processes, including lasers, sheet metal punches, electrical-discharge machines, cutting torches, and many more. This chapter will discuss CAM as it applies to milling machines and lathes.

Basic Machining Processes

It is important that you understand basic machining processes before drawing parts for CAM and CNC machines. This section gives a brief overview of the two most common CNC machines. These are milling machines and lathes. You should have experience using these machines manually before attempting to machine parts using CAM.

Milling Machines

A *milling machine* uses a revolving cutter to shape the material, Figure 18-1. The material is held in a vise or fixture attached to a table. The material is fed into the cutter by moving the table. Milling machines are widely used for machining flat surfaces, slots, grooves, and keyways. Holes can be drilled by feeding the cutter down into the material. Curved surfaces can also be made. However, this is a complicated process to do by hand. Computer control makes machining curved surfaces on a mill very easy.

Figure 18-1.
In a milling operation, the stock is fixed to a movable table. A rotating cutter mounted in a stationary head shapes the material. (Light Machines Corporation)

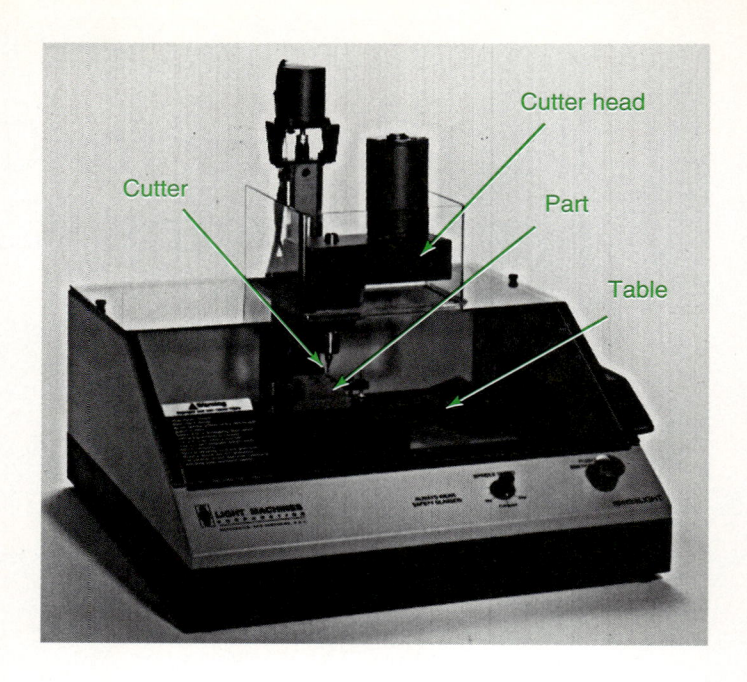

Operations

The following common milling operations are shown in Figure 18-2.
- *Facing* removes material on the top surface of your stock to make it perfectly flat.
- *Contouring* removes material at a given depth along a specified path. This often defines the outline of a part.
- *Drilling* creates holes to a given depth. Drilling may or may not go all the way through the part. The location of a hole is given with a point in your drawing.
- *Pocketing* removes material to a single depth within a continuous boundary. Within a pocket there may be islands where no material is removed.
- A *Surface of Revolution* removes material to create a convex or concave surface defined by revolving a geometry about the X or Y axis. You define a cross section and the computer generates the arcs needed to revolve the cross section.
- *Ruled surfaces* are made by blending two or more contours together to make a surface. The tool moves in straight motions from one contour to the next.
- *Swept surfaces* are made by applying the shape of a cross section contour along another contour.

There are even more sophisticated operations that can be done using CAM software. For example, a swept 3D surface applies a cross section contour over two or more other contours. This is a very complex operation.

Tools

Endmill cutters are the primary tools for milling operations, Figure 18-3. *Flat endmills* can be used for facing, contours, and pockets. A flat endmill is not typically useful for creating curved surfaces. Some flat endmills can only cut on the side of the endmill. However, most endmills can cut on the end of the endmill. This allows them to plunge into material like a drill to start a pocket.

Ball endmills are primarily used for contoured surfaces. They have a round end with a radius the same as the cutter radius.

There are also a number of other cutters that look like miniature saws. These cutters are used for special operations such as T-slots, key seats, face milling, plain milling, side milling, and angle milling.

Figure 18-2.
Common computer-numerical control milling operations. A—Facing. B—Contouring. C—Drilling. D—Pocketing. E—Surface of revolution. F—Ruled surface. G—Swept surface.

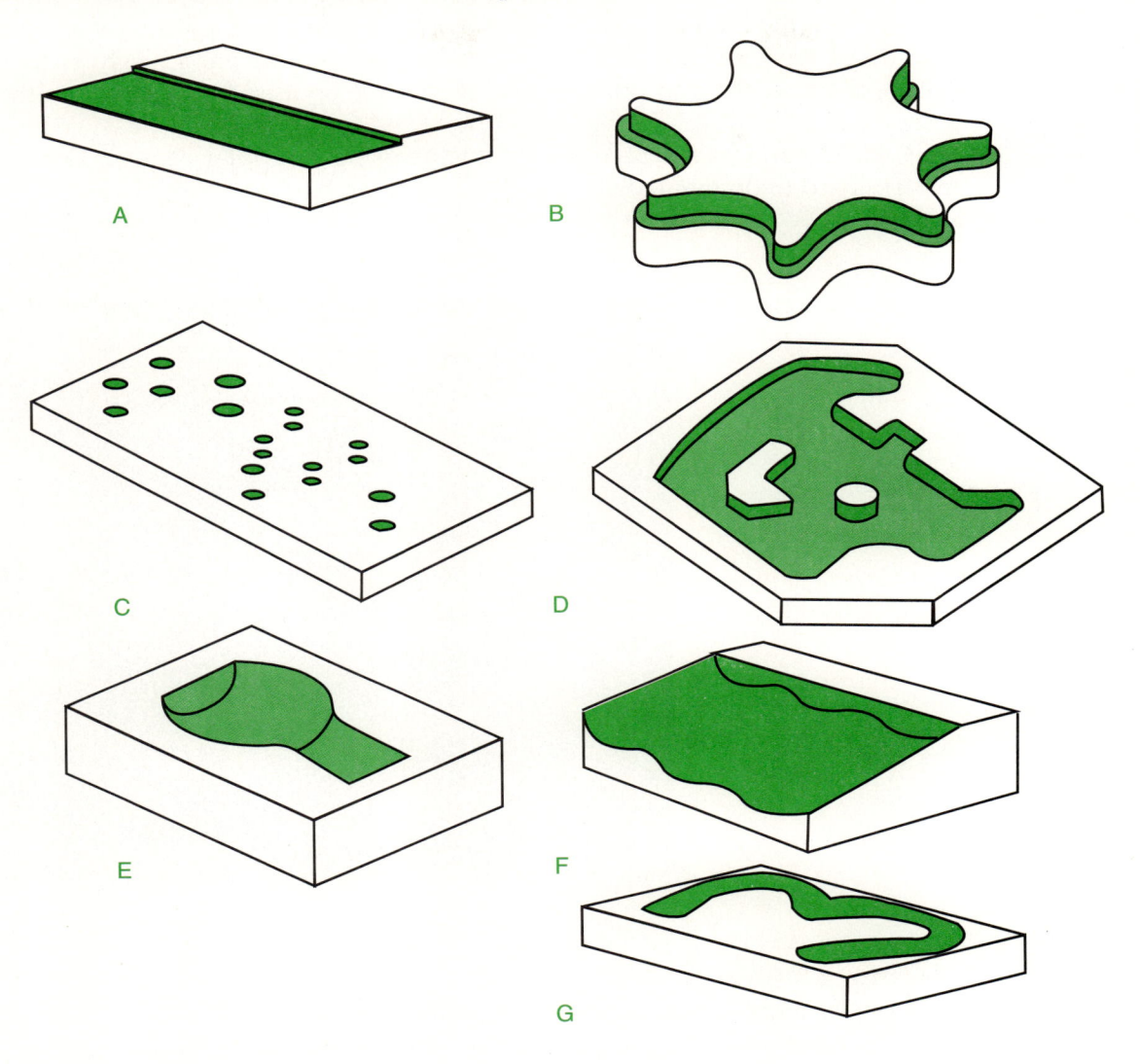

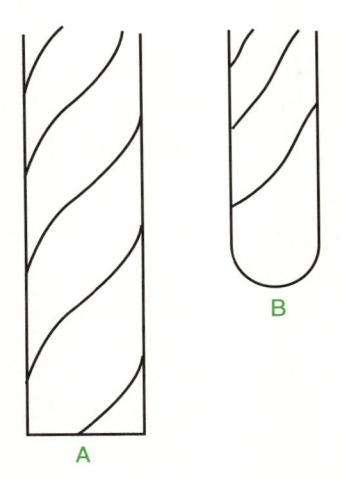

Figure 18-3.
Two common endmill cutters. A—Flat endmill. B—Ball endmill.

Lathe

A *lathe* rotates the material against a stationary cutter to shape the material, Figure 18-4. The material is held in a chuck or between two lathe centers. The stationary cutter is fed into the rotating material by moving the lathe carriage and toolholder.

Lathes are built for making cylindrical and conical parts, such as pins, bolts, screws, shafts, disks, and pulleys. A lathe can also be used for drilling into the end of a part.

Curved surfaces can be made on a lathe using computer control. These surfaces are very hard to do manually.

Figure 18-4.
In a lathe operation, the stock is held in a rotating chuck, or held between centers and rotated. A stationary cutter is fixed in a toolholder. The toolholder can move in and out of the stock, and along its centerline. (Light Machines Corporation)

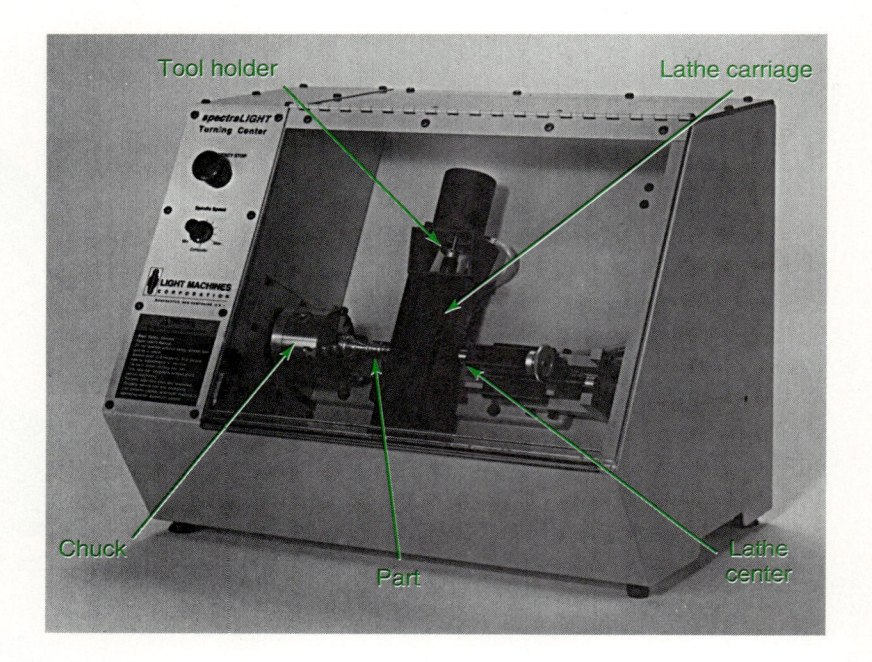

Operations

The following are common lathe operations. See Figure 18-5.
- *Facing* removes material around the circumference of the part to make it perfectly round.
- *Contouring* removes material to one or more depths along the length of the part.
- *Tapering* means to create an angle along the length of the part. The ends of a taper are different diameters.
- *Threading* creates threads, or grooves, so the part can be screwed into (or onto) a receiving threaded part. Threads can be either internal or external.
- *Drilling* and *reaming* create holes in the end of the part held in a chuck.

Tools

The following are tools used for lathe operations.
- *Turning tools* are used for facing and contouring.
- *Radius tools* are used to cut a fillet of a given radius at a shoulder.
- *Cut-off tools*, or *parting tools*, are used for cutting grooves or cutting off stock.
- *Threading tools* create threads for screws and bolts.
- *Knurling tools* create a roughened surface. This type of surface is typically used to help a person grip the part.

Figure 18-5.
Common computer-numerical control lathe operations. A—Facing. B—Contouring. C—Tapering. D—Threading. E—Drilling or reaming.

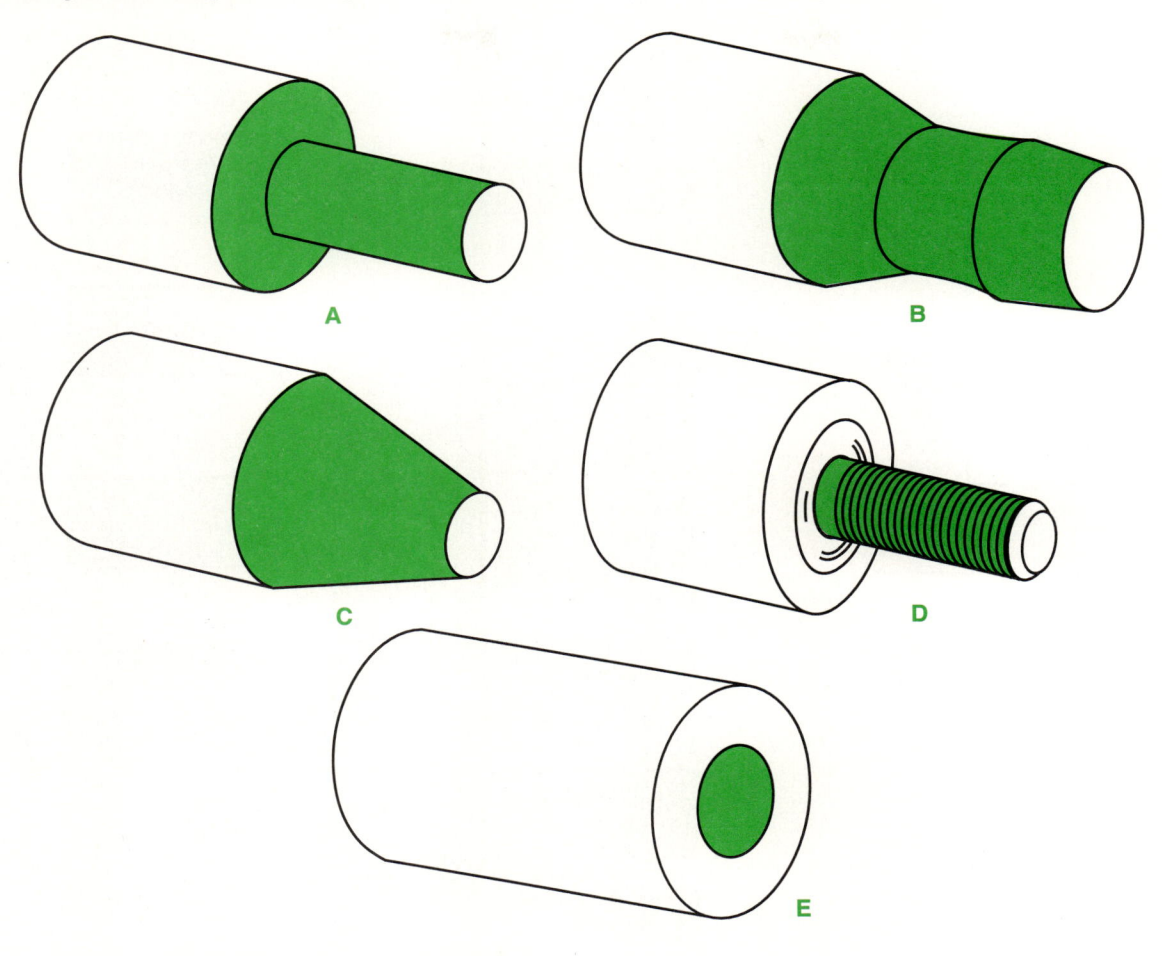

The CAD/CAM/CNC Process

There is a multistep process that links CAD, CAM, and CNC, Figure 18-6. The following steps are involved.

Planning
The first step in the linking process is to plan the part you will machine. It is important to determine up front the material you will use, the general path and sequence that your machine will take to create the part, the type of tool used, and finishing steps. These decisions require that you have a working knowledge of machining techniques.

CAD
The second step in the linking process is to create a drawing of the part using AutoSketch (or other CAD program). The objects you draw to represent the part are commonly called the *geometry.*

Next, save the drawing to a DXF file format. In AutoSketch, you must export the drawing to the DXF format. This common format can be imported by most CAM software. The file contains all of the geometry of the part you wish to machine.

Figure 18-6.
The steps in the CAD/CAM/CNC process.

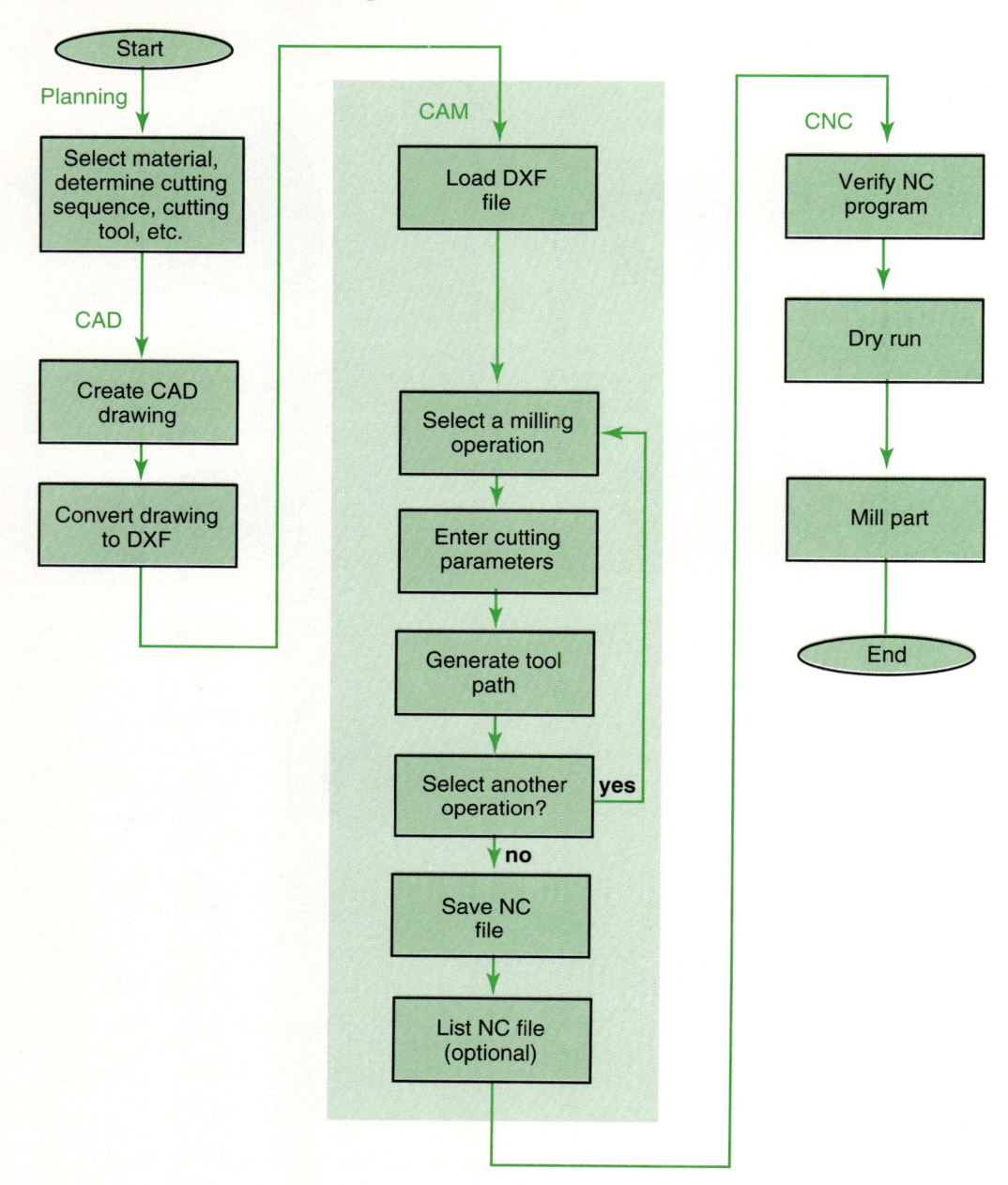

```
Planning: Start → Select material, determine cutting sequence, cutting tool, etc.
CAD: Create CAD drawing → Convert drawing to DXF
CAM: Load DXF file → Select a milling operation → Enter cutting parameters → Generate tool path → Select another operation? (yes → back to Select a milling operation; no → Save NC file → List NC file (optional))
CNC: Verify NC program → Dry run → Mill part → End
```

CAM

The next step in the linking process is to run the CAM program and open the DXF file. The CAM software is used to convert your drawing into a format that can be used by a CNC machine. In addition, some CAM programs allow you to create the part geometry. This eliminates the need for a separate CAD program.

Note: CAM software does not control the machine that will create the part. It only translates the drawing into a different form of information.

Next, select the operation for each geometry of the part, such as facing or pocketing. Most parts cannot be made with one cutting operation, or even one cutting tool.

After that, enter the cutting parameters for each toolpath. For each operation, you must select the tool to use, the speed the tool moves, and the depth of cuts made by the tool. There will likely be different cutting parameters for each toolpath.

Once the parameters are defined, you need to generate the toolpath. A toolpath is created using the combination of cutting parameters, operation, and part geometry. The toolpath represents all the moves the cutting tool will make to create the specific contour.

Save all the toolpaths and cutting parameters as a numerical control (NC) program. The CAM software converts the all toolpaths and cutting parameters to create a single piece of computer code, called a *numerical control (NC) program.* The NC program will be used to control the CNC machine.

CNC

The next step in the linking process is to run the CNC machine control program and open the NC file. The CNC computer, referred to as a *controller,* feeds numerical data to the motors. This computer may load your NC file from another computer or a diskette. Older machines may use punched paper tape.

After that, verify the NC program to check for programming errors. It is important that you verify the code in the NC program before you run it on the machine. This means you must be able to interpret what each line of code will cause the machine to do. An error in the program could cause the tool to crash into your part, or the machine itself. Some CAM programs can accept the NC code and simulate on-screen the entire machining process.

Once you have checked the program for errors, dry-run the NC program to check for an incorrect machine setup. The NC program only tells the tool how to move. It does not load the material in the machine, or align the first tool position. You must do this manually. To check your machine setup, and the NC program itself, you should run the program before you load your material. Or, you might want to machine a sample part using machinable wax.

Finally, run the NC program and machine the part. Once you have created the NC code, checked it for errors, and verified your machine setup, you can machine the parts.

Creating Drawings for CAM

Now that you are familiar with the basic CAM process, and with some of the machines and operations, this section will focus on the drafting process. Computer-aided drafting software, like AutoSketch, can be used to create the complex geometry needed for many machined parts. You can program a numerical-control machine using a keyboard. However, this process is very difficult when compared to having a CAM program convert your drawing into NC code. Plus, the most complex shapes would require hours of manual programming, but only minutes to draw using CAD.

Guidelines for Creating Drawings

Drafting for CAM is quite different than creating working drawings for manual machining or construction. When drafting for CAM, realize that the machine tool will follow every line, arc, and curve you construct. *You* are drawing *the* part. It is not like a house where you create the plan, and someone else builds the house. When drafting for CAM, the data you enter into the computer via AutoSketch is the same data the machine uses to create the part. It follows the path in a sequence. The following are guidelines you must follow when creating your drawing.

Drawing one view

Only one view is used when drawing for computer-aided machining. No other views are required. The single view is imported into the CAM system and depth (the third dimension) is added by assigning the depth of cuts by the milling machine, Figure 18-7. For a turned part (machined on a lathe), only half of the part drawing is needed, Figure 18-8. The material revolves against the cutter to create the entire part.

Figure 18-7. A—One view of a 2D drawing provides the geometry needed to machine a part. B—Once depth is assigned to the toolpath, the part can be machined.

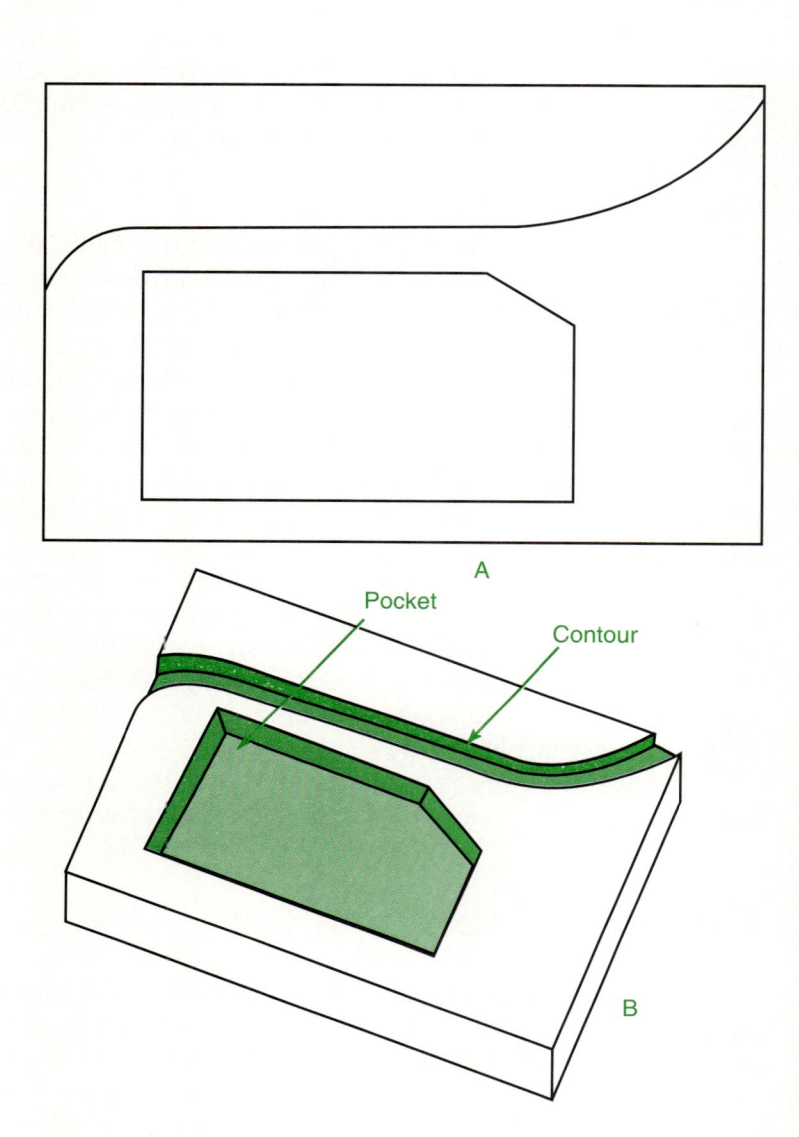

Understand the coordinate system

The machining process begins at a *program reference zero* (**PRZ**) point. This is a starting point where all other movements are measured from. You can also think of this as the origin of your geometry. On a milling machine, the PRZ is typically at the top of the stock at the front, left corner. Movements up, in toward the machine, and to the right are positive values. Movements down, away from the machine, and to the left are negative values. See Figure 18-9. On a lathe, the PRZ is normally set at the front face and centerline of the work piece. Movements into the stock from the face are negative Z values. See Figure 18-10.

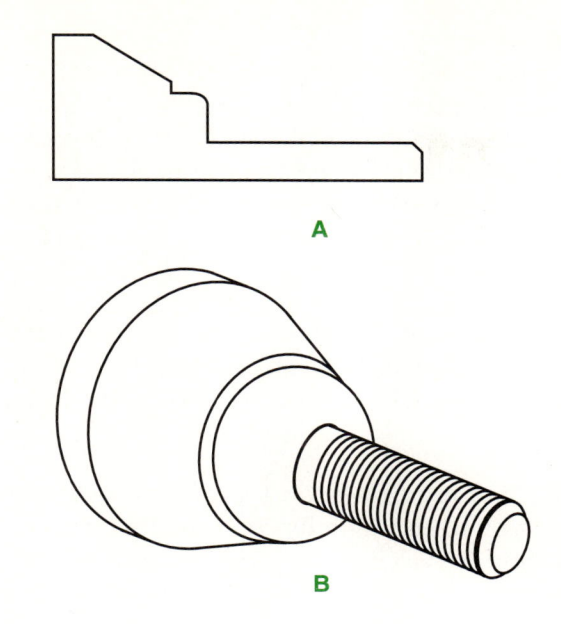

Figure 18-8.
A—For a turned part, only half of the part drawing is needed to cut the part. B—Since the part is rotated on a centerline, the final product is symmetrical.

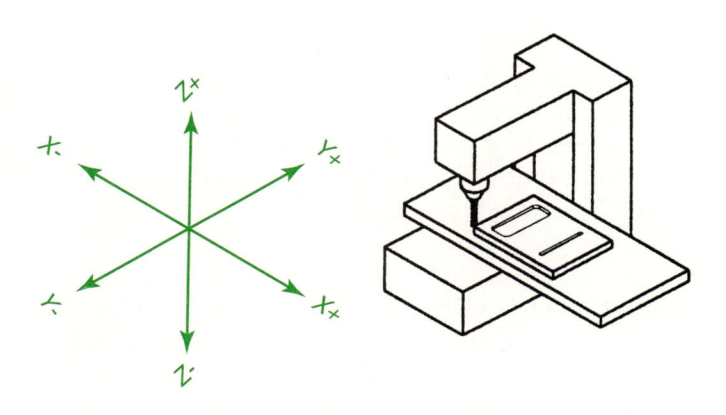

Figure 18-9.
Coordinate axes and values for a milling machine. PRZ for a mill is usually the top of the stock at the front, left corner.

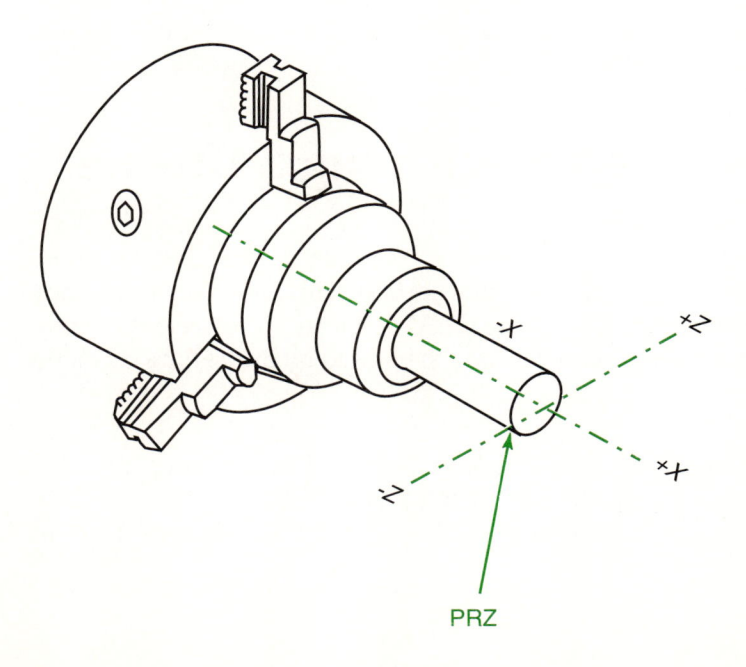

Figure 18-10.
Coordinate axes and values for a lathe. PRZ for a lather is usually the front face of the part centered vertically.

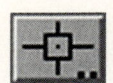

Objects must meet precisely

When objects of a single toolpath meet, they must do so precisely. The objects must touch, but not overlap or cross. Be careful with tangents and endpoints. Make sure that you use **Attach** tools so that the objects must meet precisely. If objects do not touch, the conversion to computer-numerical code will stall. If lines cross, CAM software often has difficulty converting the lines into a toolpath. Typically, the software will go to the end of the first object, then report that no further objects exist.

Do not use polylines

CAD software considers polylines as one continuous line. The software will not recognize the individual segments. Use line and arc objects that have defined endpoints on each segment. The CAM software will be able to interpret the locations.

Delete duplicates and trim overlaps

Check for lines or other objects that are drawn on top of each other and delete them. When copying an object, it is easy to accidentally copy the first object on top of the original. Another hard item to find is two lines in sequence, but where one line crosses the other. One line must be trimmed to meet precisely with the next line at the endpoint. These items will not show up on a printout. You need to search for these types of errors.

Separate toolpaths by layer and color

There are often two or more separate toolpaths, even for a single part. Draw the part geometry for each toolpath on different layers and in different colors. Using different layers allows you to see if you are milling the outer boundary, or possibly an inside pocket. Most CAM software will automatically use the layers to define different toolpaths. The different colors help you see these toolpaths on-screen.

Export only the geometry

When you are ready to export the drawing to DXF format, delete everything that is not part of the geometry of the part to be machined. Geometry refers to the lines that define edges of the part. You should delete dimensions and notes.

You can also select the geometry and copy it to a new AutoSketch drawing. Then, export the geometry from the new drawing. Since there usually are many more dimension and text objects than objects that make up the geometry, this method may be easiest. Remember, you need to export only the part boundaries, nothing else.

Using Computer-aided Machining Software to Generate NC Code

Although every CAM software is different, this section will cover functions common to most programs. Remember, you are using CAM software to convert the drawing geometry made in CAD to numerical control code to be used by the milling machine.

Loading the Geometry

While some CAM programs include a simple drawing program, often you will import geometry from a CAD program. The format you will use most is DXF. However, some CAM software can import AutoCAD (DWG) drawing files. Use the DWG format if possible.

Most CAM software will import layers as separate part geometries. If the objects are drawn on one layer, you will need to separate them manually once imported. There will always be instances where two parts that should touch do not. Or, that the precision of the drawing is not as precise as the CAM software.

Coincidental geometry

Some CAM software has the ability to automatically separate *non-coincidental geometries.* Non-coincidental geometries are geometries that do not touch one another. Some CAM software can generate the different toolpath sequences by interpreting different layers. If endpoints of two geometries are coincident in the drawing, they will not be separated. The CAM program will change the endpoints.

Some CAM programs allow you to specify a tolerance. The CAM software uses this tolerance to chain two lines that are close, but do not exactly touch. Do not rely on such operations. Use **Attach** modes and **Snap** to make sure objects that should touch do so precisely. If objects do not touch, they are non-coincident and considered two separate toolpaths.

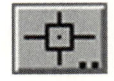

Setup Steps

The following steps are used to set up the process.
- Select the type of equipment that will be used to machine the part. This will determine some of the operations that can be performed. However, most importantly the type of machine determines the type of NC code generated by the CAM software. The part of a CAM program that generates the NC code for a specific machine is called the *post processor.*
- Select the material that you will be machining. The type of material will determine how fast the cutter turns and how much material can be removed with each pass. For example, when machining aluminum, the cutter can turn much faster and remove more material than with steel.
- Select the tool you will use to machine the product. This will also determine how fast the cutter turns and how much material can be removed (how "deep" the cutter can go) with each pass. For example, a 1/4" cutter can go deeper into the material and move much faster than a 1/2" cutter. When selecting a milling cutter, you may also need to specify the diameter, number of teeth, and if it is high speed steel or carbide.

Generating the Toolpath

The *toolpath* represents all the moves the cutting tool will make to create the specific contour. The combination of cutting parameters, operation, and part geometry are used to create a toolpath.

Selecting operations

Most parts cannot be made with one cutting operation, or even one cutting tool. In this step, you select which operation will be performed using which geometry of the drawing. For example, an enclosed polygon might represent a pocket to be milled. A line through the part might represent a slot to be cut.

Entering cutting parameters

Setting the cutting parameters requires a working knowledge of machining practices. *Cutting parameters* include which tool to use, the speed the tool moves, and the depth of cuts made by the tool for each proposed operation. There may be different cutting parameters for each toolpath. You must be able to determine the speed of the tool, feed rate, plunge rate (rate of travel on the Z axis), number of steps to create the depth, and amount of material to remove with each cut. Other parameters determine if coolant should be used while cutting.

Another parameter that must be set is the *cutter offset,* or *cutter compensation.* When the CAM software creates the toolpath, it needs to know where to place the cutter in relation to that path. Keep in mind, if you are using a 1/2" cutter, the tool will cut 1/4" on each side of the path it follows. In other words, you will need to specify if the edge of the cutter is to follow the toolpath, or if the center of the cutter is to follow the toolpath. This is the cutter offset.

Remember, AutoSketch is a two-dimensional drafting program. Yet, machining is a three-dimensional process. This is why you must manually enter any Z-axis, or vertical, movements such as pocketing and drilling. Depending on the CAM program, you may enter a depth, or you may need to enter negative Z-axis values. The negative value is needed for milling, since you are going into the stock from the top.

After you specify these parameters, the CAM program can calculate the three-dimensional toolpath of your geometry. You may even be able to rotate this 3D view to see the part from various angles. On a lathe, X-axis runs vertically, corresponding to diameter of part. Z-axis runs horizontal, along the length of the part. The Y-axis is not used.

Creating the path

Once the operation is chosen and the cutting parameters are entered, the CAM software will typically trace that path on-screen, Figure 18-11. You may need to chain together one or more toolpaths to create the entire part. The cutter must start and have a continuous path, even if this path has the cutter retract from the stock and then plunge back in another location.

Figure 18-11.
The toolpath calculated by the CAM software is shown on-screen.

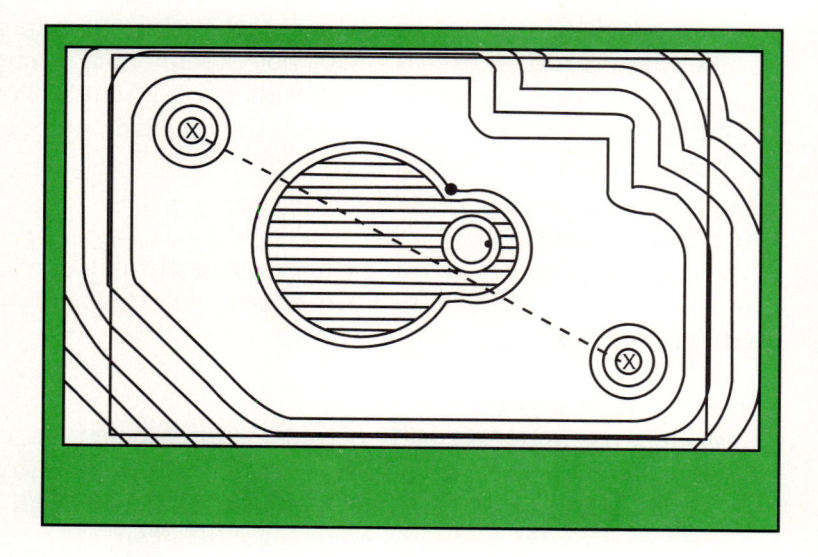

Generating NC Code

Once the toolpath is generated, that information will be converted into the actual NC code. The CAM software combines the toolpaths and cutting parameters to create a single piece of computer code, called a *numerical control (NC) program.* Each sequence of the NC program will control the motors and the tool to cut the material, Figure 18-12.

```
%%
N001G90G80G40G00
N002M06T1H1
N003M03
N004G0X.85Y.375
N005G43H1Z.6
N006G1Z-.1F5.
N007Y2.125F10.
N008X1.475
N009Y.375
N010X.85
N011G0Z.6
N012G28M05
N013M30
%
```

Figure 18-12.
A sample numerical control code program. Each line is numbered. The various letters and number combinations control the cutter and table movements. This program is written in the FANUC language.

Be aware that there are different NC code formats. The most common format is FANUC. Your CAM software can likely export a variety of formats. Choose the NC code format that can be used by the CNC machine controller of the machine you will be using.

It is important that you verify the code in the NC program before you run it on the machine. This means you must be able to interpret what each line of code will cause the machine to do. An error in the program may cause the tool to crash into your part, or the machine itself. There are some CAM programs that allow you to do this before you load the NC program in your CNC machine with a toolpath verification program.

A *toolpath verification program* allows you to verify tool movement. The software is initialized by establishing the stock size and initial tool location. In fact, the toolpath verification can start from any block (or line number) within the NC program. This allows you to verify a specific toolpath within a much larger operation. The display usually shows a graphic of the material and the cuts made by the tool. You can usually tell whether the part will turn out correct from the display.

Machining the Part

Because of the complexity of machining processes, this chapter cannot cover all the processes. However, there are several terms that you should become familiar with before using a computer-numerical control machine.

Program zero reference

The starting position is often called the program reference zero or PRZ. Set the tool manually to locate the PRZ position. The PRZ should be at a point that is easy to locate with all three axes on the milling machine. Typically, this is on the top of the stock and at the front, left corner. All geometry is then created in the proper position. On a lathe, the PRZ is normally set at the front face and centerline of the work piece.

Tool length offset

Before you machine the part, you need to establish the *tool length offsets* between multiple tools. The length of the first tool is used as a reference for other tools. Multiple-tool programs usually require the different tools be mounted in tool holders. This is so that they can be installed in the mill with no variation in length of the tool offsets.

Some mills may have a tool height offset sensor. This is an electronic continuity sensor. The sensor works by indicating electrical continuity from the spindle, through to tool, the sensor, and the table back to the spindle. When the tool touches the material, continuity is made. This eliminates the need for defining the tool length offset. However, the material to be machined must be conductive.

Summary

Computer-aided machining is the process of converting data from a computer drawing into code that can be used to control manufacturing machines. These machines include milling machines and lathes, as well as a number of other manufacturing machines. Many operations could not be done without computers, such as complex curves. A person simply cannot accurately make all the adjustments necessary to create a complex curve.

AutoSketch is an excellent tool for creating the geometry needed for CAM. Drafting for CAM is quite different than creating working drawings for manual machining or construction. Only the lines, arcs, and other objects that define edges of the part are exported to a CAM system. Remember, objects that should touch must do so precisely.

Once the drawing is loaded into a CAM system, there are a number of steps that take place before a part is machined. These include setup steps to select the machine, material, and tool. Also included are steps to define the cutting parameters to use, such as the cutter size, depth of cuts, and speed. Once the geometry is loaded and the cutting parameters are chosen, a toolpath is created. This toolpath is then converted to numerical control code and sent to the machine.

Important Terms

Ball endmills
Computer numerical control (CNC) machine
Computer-aided machining (CAM)
Contouring
Controller
Cut-off tools
Cutter compensation
Cutter offset
Cutting parameters
Drilling
Endmill cutters

Facing
Flat endmills
Geometry
Knurling tools
Lathe
Milling machine
Non-coincidental geometries
Numerical control (NC) program
Parting tools
Pocketing
Post processor
Program reference zero (PRZ)
Radius tools

Reaming
Ruled surfaces
Surface of Revolution
Swept surfaces
Tapering
Threading
Threading tools
Tool length offsets
Toolpath
Toolpath verification program
Turning tools

Review Questions

Give the best answer for each of the following questions.

1. _____ is the process of converting data from a computer drawing into code that can be used to control manufacturing machines.
2. Pocketing is an operation that is done on a(n) _____ machine.
3. A(n) _____ endmill is a cutter used primarily for contoured surfaces.
4. The drawing format used to exchange drawings with AutoSketch and most computer-aided machining programs is _____.
5. What tasks are done with a computer-aided machining program in order to convert a drawing to numerical control code?

6. What two steps should you perform using numerical control code before you actually machine the part?

7. Explain how drafting for computer-aided machining is different than creating a drawing for a part that will be produced manually.

8. Describe why objects that make up the same toolpath must meet precisely in the drawing.

9. Explain why you might separate the geometry of different toolpaths by placing it on different layers.

10. When you export your drawing for CAM, export only the _____, not dimensions or text.
11. The post processor used to generate different types of NC code (such as FANUC) depends on the _____.
12. When setting your CAM software, what three items must you select?

13. The geometry and _____ _____ will determine the toolpath chosen to machine a part using a CNC machine.
14. The point on your stock that you will start the NC code at is called _____.
15. When multiple tools are used to create a single part, they are typically mounted in _____ _____ to be sure of repeatability.

Activities

1. Identify a company in your area that is using computer numerical control machines. Contact the supervisor and visit the company to review the entire process used to take a drawing to a manufactured part. While visiting, determine answers to the following questions.
   A. What is the computer-aided drafting program used by the company?
   B. What is the computer-aided machining program used by the company?
   C. What type of CNC machine is being used and who is the manufacturer?
   D. What type of NC code is used by the machine?
2. Design a tic-tac-toe game board. Use the following steps.
   A. Decide how big you want the game board to be, and the diameter of the holes where the game pieces will sit.
   B. Identify the types of machining operations required to manufacture the product.
   C. Use AutoSketch to create the part geometry.
   D. If you have access to a CAM system, export the part geometry and create the NC code necessary to machine the game board.
   E. Machine the game board out of aluminum or machinable wax.

Note: Consult the directions and safety information for your CAM program and CNC machine before performing any machining operation.

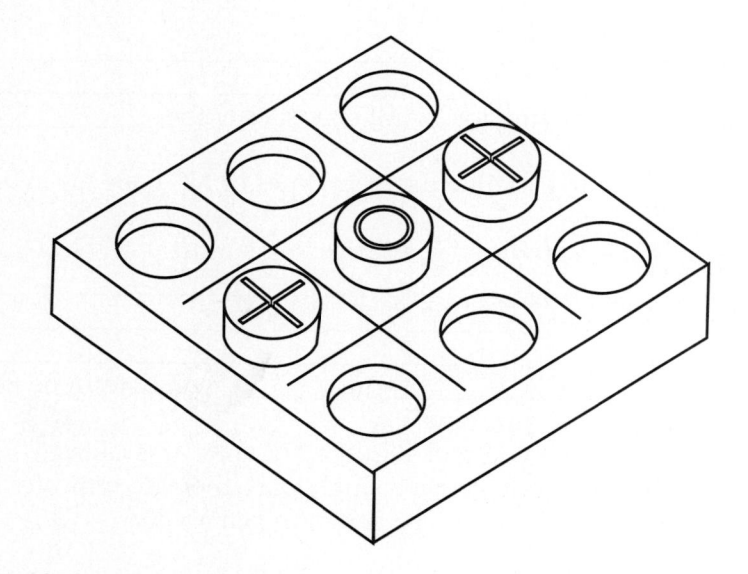

CHAPTER 19

Automating AutoSketch with Macros

Objectives

After studying this chapter, you will be able to:
- Describe the format of macro files.
- Identify the macro tools available to record, play, and add user inputs.
- Perform the steps to record and play, macros.
- List guidelines for recording and playing macros.
- Create and edit custom toolboxes and macro buttons.
- Write and edit macros using the AutoSketch Macro Language.
- Design macros using the proper format for macro language commands.
- Create macros using the proper format for AutoSketch commands.
- Perform math functions within macros.
- Assign and use system variables and user variables in macros.
- Troubleshoot macros.

A *macro* is a stored series of commands. These commands perform specific tasks. While using AutoSketch, you can record a series of actions that can be played back later when needed. You can also write and edit a macro using a text editor. CAD helps automate drafting tasks over manual drafting. Macros further automate drawing and editing tasks of CAD. There are three areas where macros provide help.
- Eliminating repetitive tasks. As you use AutoSketch, you will find that there are tasks you do over and over. For example, suppose you find that you are adding a chart for a bill of materials in each drawing. You might write a macro to draw the chart. Of course, if each chart is the same, use a symbol. However, if the repeated objects have the same general features but are slightly different, a macro may be helpful.
- Conforming to standards. Your drawings should follow certain drafting standards. Macros can help. Remember that when creating a drawing, you should use the hairline line width. However, when finished with the drawing, you must change certain objects to wide line widths to conform to standards. You might write a macro so you can change line widths simply by picking the objects.
- Creating complex objects. Some drafting tasks are very complex, even with AutoSketch drawing tools. A macro might help. For example, suppose you need to draw various size springs. A spring is a difficult drafting task. Instead, you might write a macro that automatically draws various size springs based on values you enter.

Macro Files

The macro itself is simply an ASCII text file that lists AutoSketch commands, dialog boxes, point locations, and objects you pick. An example is shown in Figure 19-1. You can edit the macro with any ASCII text editor or word processor that exports ASCII text. The Windows **Notepad** Program is being used in Figure 19-1.

Figure 19-1.
The example of a macro file can be edited in any ASCII text editor.

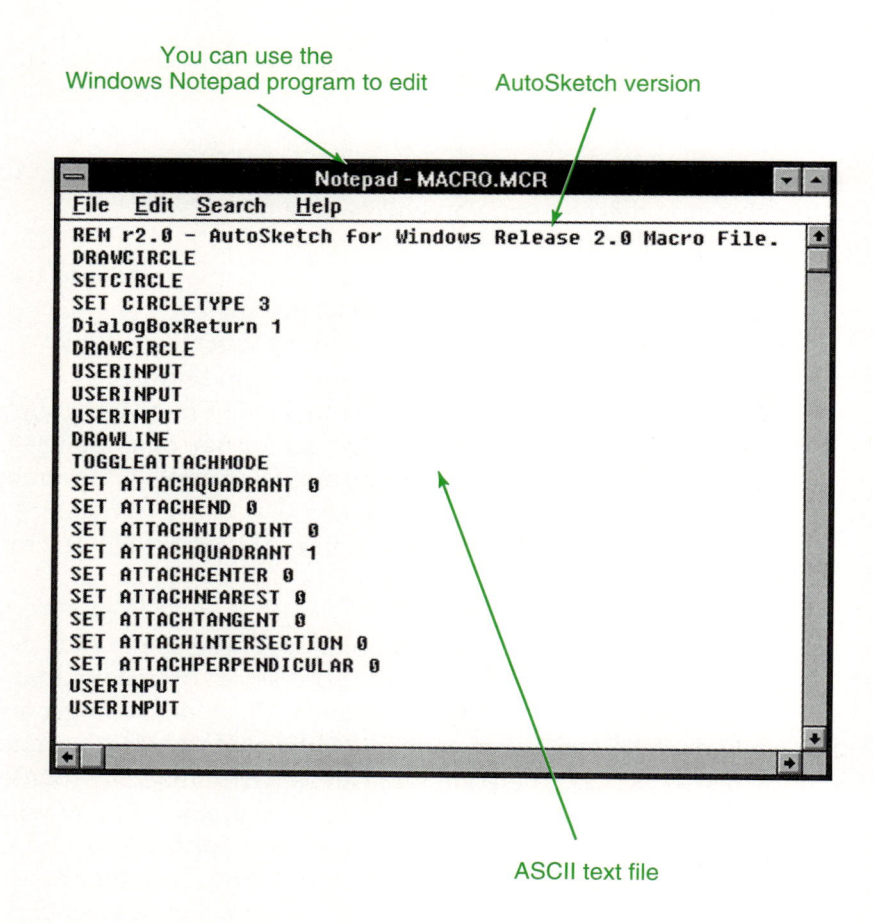

You can use the Windows Notepad program to edit

AutoSketch version

```
Notepad - MACRO.MCR
File  Edit  Search  Help
REM r2.0 - AutoSketch for Windows Release 2.0 Macro File.
DRAWCIRCLE
SETCIRCLE
SET CIRCLETYPE 3
DialogBoxReturn 1
DRAWCIRCLE
USERINPUT
USERINPUT
USERINPUT
DRAWLINE
TOGGLEATTACHMODE
SET ATTACHQUADRANT 0
SET ATTACHEND 0
SET ATTACHMIDPOINT 0
SET ATTACHQUADRANT 1
SET ATTACHCENTER 0
SET ATTACHNEAREST 0
SET ATTACHTANGENT 0
SET ATTACHINTERSECTION 0
SET ATTACHPERPENDICULAR 0
USERINPUT
USERINPUT
```

ASCII text file

However, you may never need to *write* a macro. AutoSketch lets you record your actions as a macro file. The recorded macro is stored to disk with the file extension .MCR. Once recorded, you can use the macro many times in the future for other drawings. For advanced users, the AutoSketch Macro Language and advanced macro writing are covered in detail later in this chapter.

Macro Tools

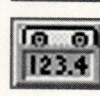

There are three macro tools in the **Custom** toolbox that allow you to record and play macros, Figure 19-2. These tools and their functions are described below.
- The **Record Macro** tool lets you create a macro. Once selected, the **Record Macro** tool changes to the **End Macro** tool.
- The **Play Macro** tool lets you play a recorded macro.
- The **User Input** tool lets you pause the macro to accept user input.

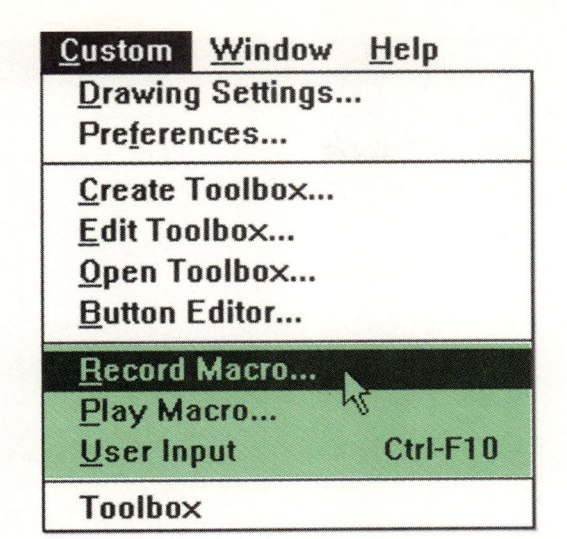

Figure 19-2.
Macro tools (highlighted here) are found in the **Custom** menu or toolbox. The **End Macro** tool appears once the **Record Macro** tool is chosen.

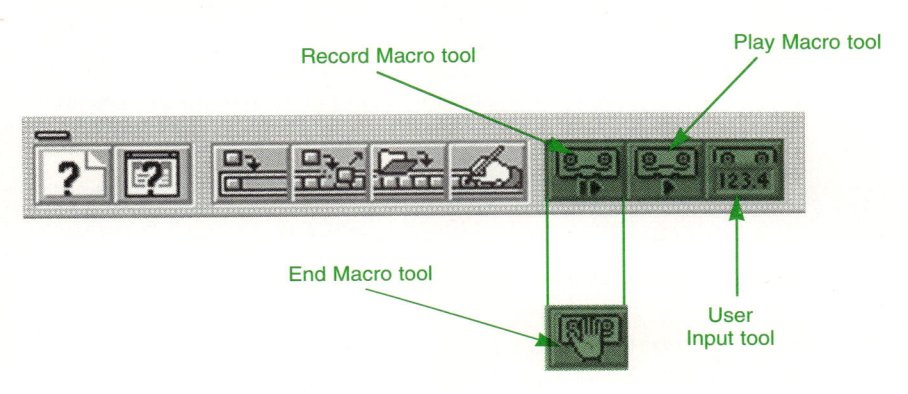

Recording and Playing Macros

In this section, you will create a macro by picking AutoSketch tools, and providing user input.

Recording a Macro

Practice the commands you will use for the macro several times before you actually record it. As you practice your macro, make a note of each time you need to pick an object or location on-screen, or when you need to enter a value in a dialog box. This is called user input and is handled differently.

Select the **Record Macro** tool from the **Custom** menu to begin recording a macro. The **Record Macro File** dialog box appears, Figure 19-3. In the **File name:** box, enter a macro filename of up to eight characters. You do not have to add the .MCR file extension. AutoSketch does that for you. Choose **OK** to close the dialog box. You should save the new macro in the \WSKETCH directory or one of its subdirectories. AutoSketch will look in the \WSKETCH directory first.

Now you are ready to perform the drawing and editing tasks you want to record. Be precise when selecting commands and dialog box options. AutoSketch records your every action, even mistakes.

Figure 19-3.
The **Record Macro File** dialog box. Select the drive and directory, and enter the name of the macro.

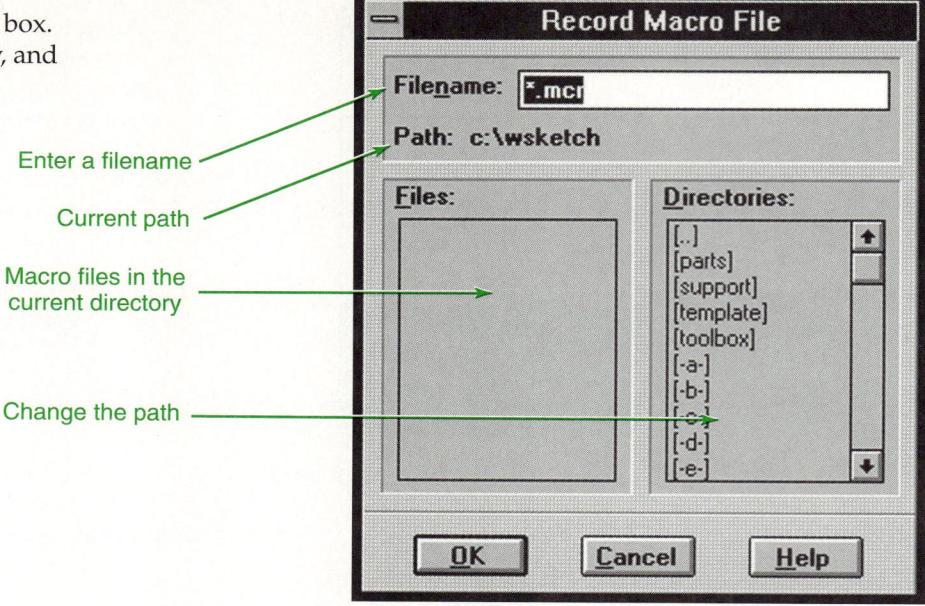

Enter a filename

Current path

Macro files in the current directory

Change the path

Exercise 19-1

1. Start a new drawing using the **A_LI_MEC** template you created in Chapter 10.
2. Use the **Record Macro** tool to create a macro named C19E1. Save your macro on your work diskette. Have the macro do the following.
   A. Draw a border .5" from the edge of the paper.
   B. Place your name in the lower-right hand corner using the **Quick Text** tool.
3. Pick the **End Macro** tool when finished recording.

User input

Most macros will need information from the drafter. This input might be a value entered on the keyboard, or an object or location picked on-screen. For example, a macro that changes line widths requires that you pick the objects to change. To do this, you must instruct the macro to pause to accept *user input.* In this example, the macro must pause for you to pick the object(s) to edit. After you pick the object(s), the macro resumes.

At the point that you need user input, pick the **User Input** tool from the **Custom** menu. You can also press the [Ctrl] key and the function key [F10]. This will cause the macro to pause. Nothing will seem to happen when you do this. However, AutoSketch is waiting for your input. Perform the action that you want the user to perform when running the macro.

The macro will continue to pause until user input ends. User input ends when one of the following four things happens.
- You enter coordinates with the keyboard and press [Enter].
- You pick a point location on-screen.
- You pick an object in the drawing.
- You exit a dialog box using **OK** after entering a value or picking a setting.

The actual input is not part of the macro. For example, if the macro pauses for a point and you enter the coordinate value, AutoSketch doesn't store those values in the

SHORTCUT
You will want to use [Ctrl] and [F10] to insert user input when the dialog box you are in does not allow you to access the **Custom** toolbox.

macro. Instead, it stores a USERINPUT statement. This statement is part of the AutoSketch Macro Language. After you complete the user input, continue recording the macro.

Ending the macro

When finished recording the macro, pick the **End Macro** tool from the **Custom** menu. The **End Macro** tool replaced the **Record Macro** tool after you started recording your macro. AutoSketch saves the macro file under the filename you entered earlier.

Exercise 19-2

1. Start a new drawing using the **A_LI_MEC** template you created in Chapter 10.
2. Use the **Record Macro** tool to create a macro named C19E2. Save your macro on your work diskette. Have the macro do the following.
   A. Set the circle drawing method to **Center, Radius** in the **Circle Settings** dialog box.
   B. Draw a circle that allows for user input to pick each point.
3. Pick the **End Macro** tool when finished recording.

Hints for Recording Macros

The steps to record a macro are very simple. You can almost guarantee that a recorded macro will work as expected in the same drawing during the same drawing session. However, macros are often intended to be used many times, in many other drawings. This can cause some problems.

There are some things to watch for when recording and playing macros. The following sections cover several hints and tips for recording macros.

Paths

During a drawing session, AutoSketch keeps track of all the directories on the disk you are using. This might include directories where files are stored or where symbols are stored. The directory current when recording the macro may not be current when you play the macro. Therefore, you should do one of two things.
- When recording the macro, change to the correct directory. If the correct directory is active, change to another directory and then back to the one you need. This makes sure that your macro will access the correct directory when played later. This method is best.
- Before playing the macro, make sure that the directory paths that should be current *are* current. This is a bit harder to do because you may not remember when directories are used by the macro. Avoid this method if possible.

Save your drawing file

Make sure that the drawing file is saved when you record the macro, and when you play the macro. A macro may not play properly if the drawing file was saved when you recorded the macro, but the file is "untitled" when you play the macro. The same is true if the drawing file was "untitled" when you recorded the macro, but is named when you play the macro. It doesn't matter *what* the filenames are, just that they are both named.

Save before running a macro

There is nothing worse than a macro that "bombs" for some reason. A bad macro may make wrong changes or do wrong operations to your drawing. Always save your drawing before running the macro, just in case you need to recover from a failed macro.

Always enter values

When recording a macro, always enter values in dialog boxes. Never pick **OK** just because the value shown is what you want. Why? The current values of your drawing session take precedence over the values in effect when you recorded the macro.

For example, while recording a macro, you use the **Zoom Percentage** tool. The dialog box appears with the value 75 percent. Since this is what you want, you simply pick **OK**. However, when playing the macro, the zoom percentage part of the macro makes your drawing very tiny. What happened? The answer is that before playing the macro, you probably used the **Zoom Percentage** tool and entered a small value, say 10 percent. This smaller value used earlier in the drawing session, not 75, is used when you play the macro.

To be sure that a specific value is used when you play the macro, change the dialog box settings twice while recording. First, change the setting to any other setting. This sets the value to something different than the current default. Then, enter or select the setting you want to use. When played, the macro will change to the first value, and then to the value you want.

This is only necessary if the setting you wish to use is already displayed. If the default value is *not* what you want, you can simply enter the value you need.

Playing a Macro

To play a recorded macro, pick the **Play Macro** tool from the **Custom** menu. The **Play Macro File** dialog box appears, Figure 19-4. Double-click the macro you want to run. You may need to change to the drive and directory if the macro is stored someplace other than the \WSKETCH directory. Once you double-click on the macro name, AutoSketch performs the recorded tasks.

Figure 19-4.
Play Macro File dialog box. Select the drive and directory, and enter the name of the macro.

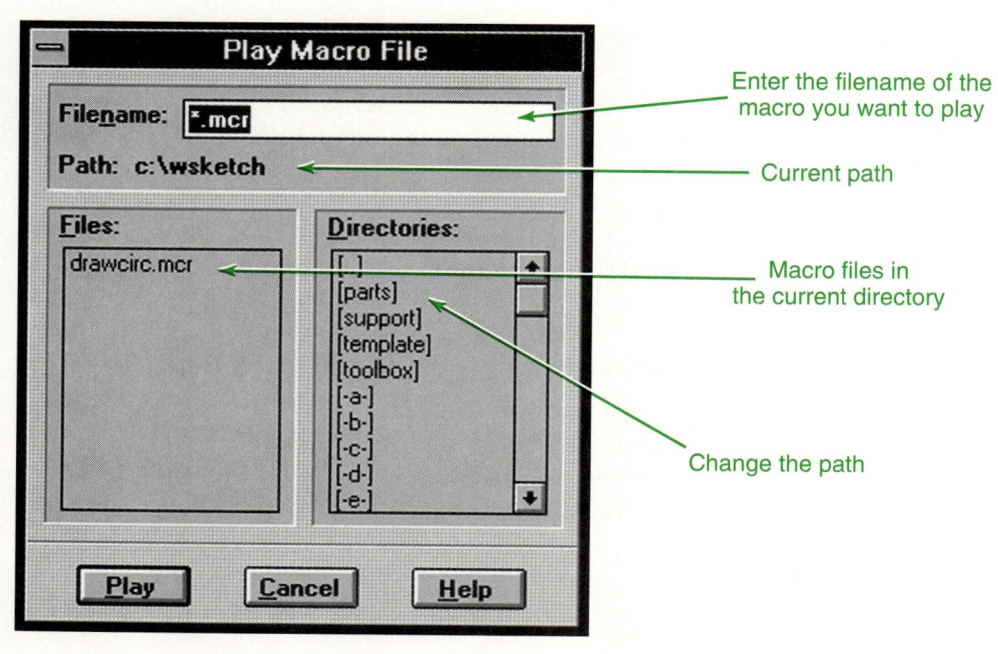

A macro with a pause for user input stops for the user to do one of four things. When you are entering text, pressing [Enter] causes the macro to resume. Picking **OK** when in a dialog box causes the macro to resume. Otherwise, the macro resumes when you pick an object or location with the cursor. Remember to look at the prompt box to determine what information the macro is requesting.

Exercise 19-3

1. Start a new drawing using the **A_LI_MEC** template you created in Chapter 10.
2. Use the **Play Macro** tool to play the macros C19E1 and C19E2.
3. Save the drawing as C19E3 on your work diskette and then close the drawing.

Creating Custom Toolboxes and Macro Buttons

Although a macro automates the drawing process, finding and playing the macro may actually take more time than manually performing the tasks. This is especially true if there is a set of macros that you use quite often.

However, you can create a *custom toolbox* and place your own "tools" in it. These tools consist of custom buttons that activate your macros. This reduces searching for macros to simply clicking on a button. The procedure to create a toolbox is as follows.
1. Create one or more custom buttons.
2. Create a custom toolbox. Assign your macros to your custom buttons and place them into a toolbox.
3. Use the **Edit Toolbox** tool to assign other macros to custom buttons and add them to the new toolbox.

Creating a Custom Button

The first step in creating a custom toolbox is to create *custom buttons.* You can use the sample buttons provided with AutoSketch, Figure 19-5. However, it is better to create your own buttons. A button is simply a picture, and it should represent what the macro does.

To create a button, first pick the **Button Editor** tool from the **Custom** menu. The **Button Editor** dialog box opens, Figure 19-6. You can create a button in three ways.
- Draw the button using the tools in the **Button Editor** dialog box.
- Capture an image from AutoSketch or another application. You can also edit the captured image with the **Button Editor** tool.
- Load a bitmap created with another drawing program. You can also edit the loaded image with the **Button Editor** tool.

The button image is made up of pixels, or small squares. You can edit each pixel to change the color. The total size of the button is 30 pixels wide by 23 pixels tall. This includes the border around the button. The border makes the image appear like a button in the toolbox. The border is a single line of white pixels along the left and top edges, and a double-line of grey pixels along the right and bottom. The border cannot be changed. The area that you can draw in is 27 pixels by 20 pixels.

Figure 19-5.
AutoSketch offers several images for custom buttons. This can save time. However, buttons that you create may be easier to identify.

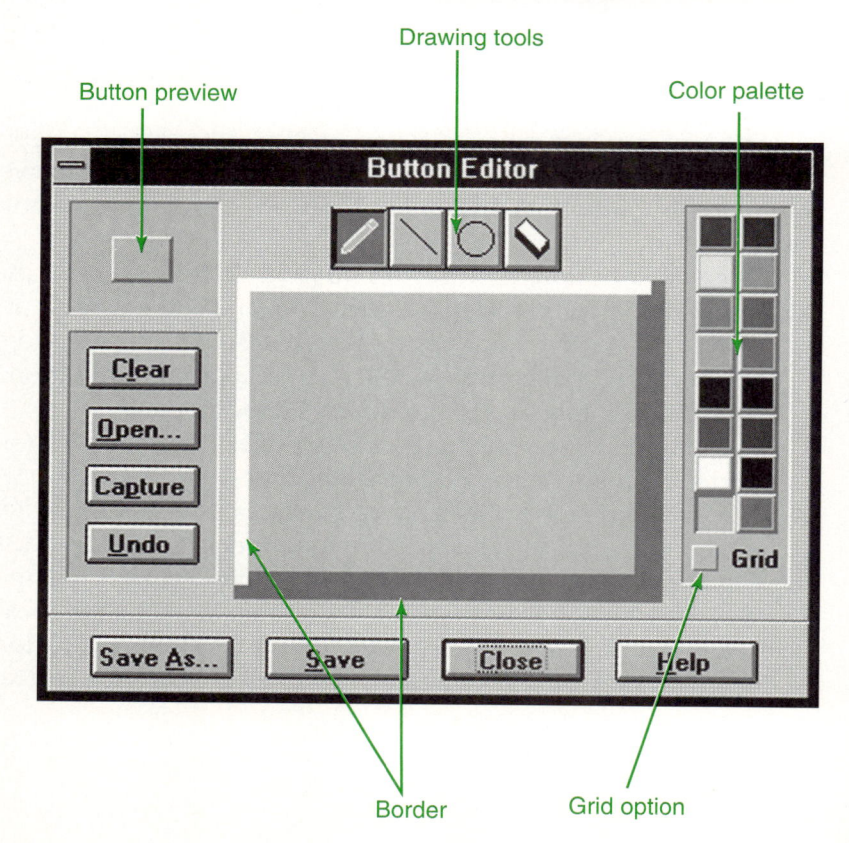

Figure 19-6.
The **Button Editor** dialog box offers you many tools for creating your own buttons.

Using Button Editor tools

The **Button Editor** dialog box has tools for drawing, erasing, changing colors, and displaying a grid. You can also use these tools to edit images from other sources. The following chart explains how to use each of these tools.

Task	Tool	Procedure
Draw Pixels	(pencil icon)	The cursor changes to a pencil. Pick in the drawing area to draw one pixel or hold the mouse pick button and drag the mouse to create a pattern of pixels.
Draw Lines	(line icon)	The cursor changes to a crosshairs cursor. Pick the first endpoint and hold the button. Drag the mouse to draw the line. Release the pick button to finish the line.
Draw Circle/Ellipse	(circle icon)	The cursor changes to a crosshairs cursor. Locate the one side of the circle. Hold the pick button to drag the circle or ellipse to shape. Release the pick button to finish the circle or ellipse.
Erase	(eraser icon)	The cursor changes to an eraser. Pick the pixels you want to erase one at a time or hold down the pick button and drag the eraser over the pixels to erase.
Change Colors	(palette icon)	Pick one of 16 colors in the **Color Palette**. Further pixels or objects are drawn in that color.
Display Grid	Grid	Displays a grid of horizontal and vertical lines in the bitmap drawing area. Each square represents one pixel.

Using Button Editor options

The **Button Editor** has buttons named **C**lear, **O**pen..., **Ca**p**t**ure, **U**ndo, **S**ave **A**s..., **S**ave, **C**lose, and **H**elp. The following chart explains each of these buttons.

Button	Purpose
Clear	Removes all pixels from the drawing area. Select Undo to restore the pixels.
Open...	Load a bitmap (.BMP) file. The file could have been created with the Button Editor, or designed by a program outside AutoSketch. Files larger than 27 by 20 pixels are modified to fit, and the image quality will suffer. The largest bitmap you can load is 300 by 230. To avoid having AutoSketch modify buttons you create outside of AutoSketch, ensure that the buttons are 27 pixels wide by 20 pixels tall. This ensures that they fit exactly within the border.
Capture	Allows you to capture an image from anywhere on screen and inserts it into the Button Editor drawing area. Pick the Capture button and drag the crosshairs to create a box around the area to capture. Release the mouse button to capture the area. If the capture is larger than 27 by 20 pixels, AutoSketch modifies the image to fit.
Undo	Cancels the most recent action.
Save As...	Save your new button to disk or save a modified button under another name and/or directory.
Save	Saves the button to disk using its current filename.
Close	Exits the Button Editor. AutoSketch prompts you to save or discard any changes made to the button.
Help	Displays on-line help.

Exercise 19-4

1. Start a new drawing using the **Normal** template.
2. Use the **Button Editor** tool to create a button named C19E1. The button should represent the macro from Exercise 19-1. Use the drawing tools in the **Button Editor** dialog box to draw the image. Save the button on your work diskette.
3. Use the **Button Editor** tool to create a button named C19E2. The button should represent the macro from Exercise 19-2. Use the capture feature to capture an existing button in the AutoSketch windows. Then, use the drawing tools in the **Button Editor** dialog box to edit the image. Save the button on your work diskette.

Creating a Custom Toolbox

Once you have created some macros and custom buttons, it's time to create a custom toolbox. Once created, you can put your custom "tools" in the toolbox. The process is quite simple. First, select the **Create Toolbox** tool from the **Custom** menu. The **Toolbox Editor** dialog box appears, Figure 19-7.

Pick the **Assign a Macro** button to open the **Assign a Macro** dialog box, Figure 19-8. In the **Macro Path:** box, change the directory if necessary and select a macro. AutoSketch displays the macro name in the **Macro:** display field.

In the **Button Path:** box, select **Files** and the directory path. Then, pick the button you want to use. AutoSketch displays a bitmap image of the button in the **Button:** display field.

Enter the name and a brief description of the new tool in the **Optional Quick Help:** text box. Finally, select the **Add** button. You can do this for as many macros as you wish to add.

Figure 19-7.
The **Toolbox Editor** dialog box allows you to create and change your own toolboxes.

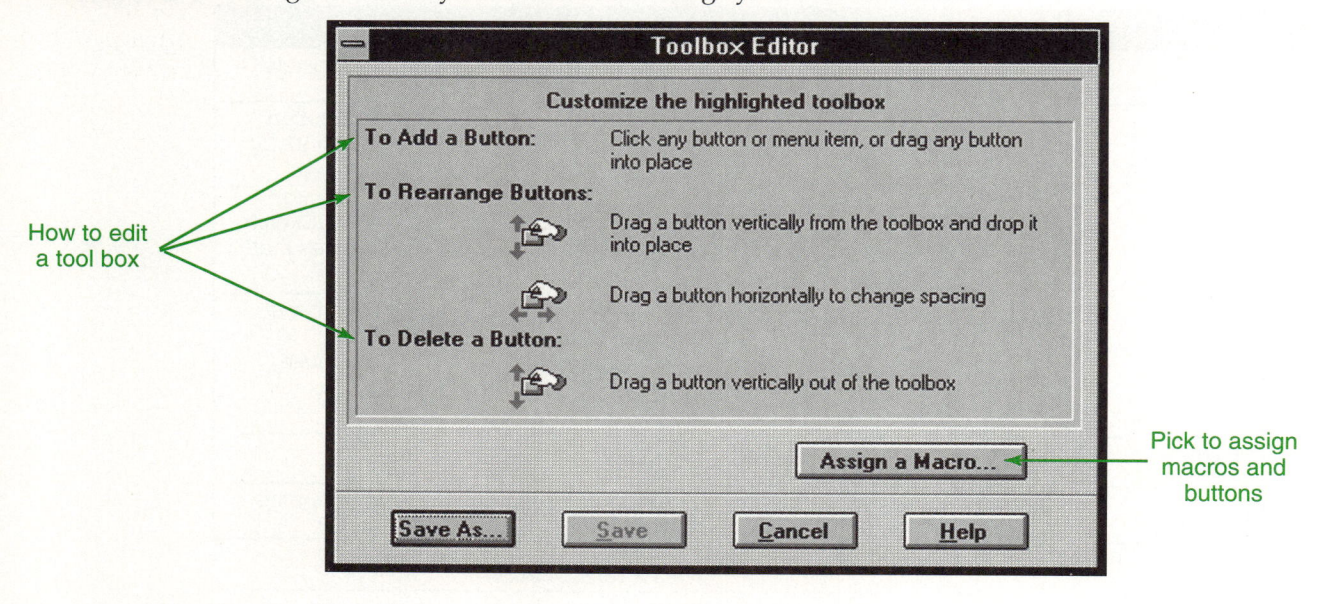

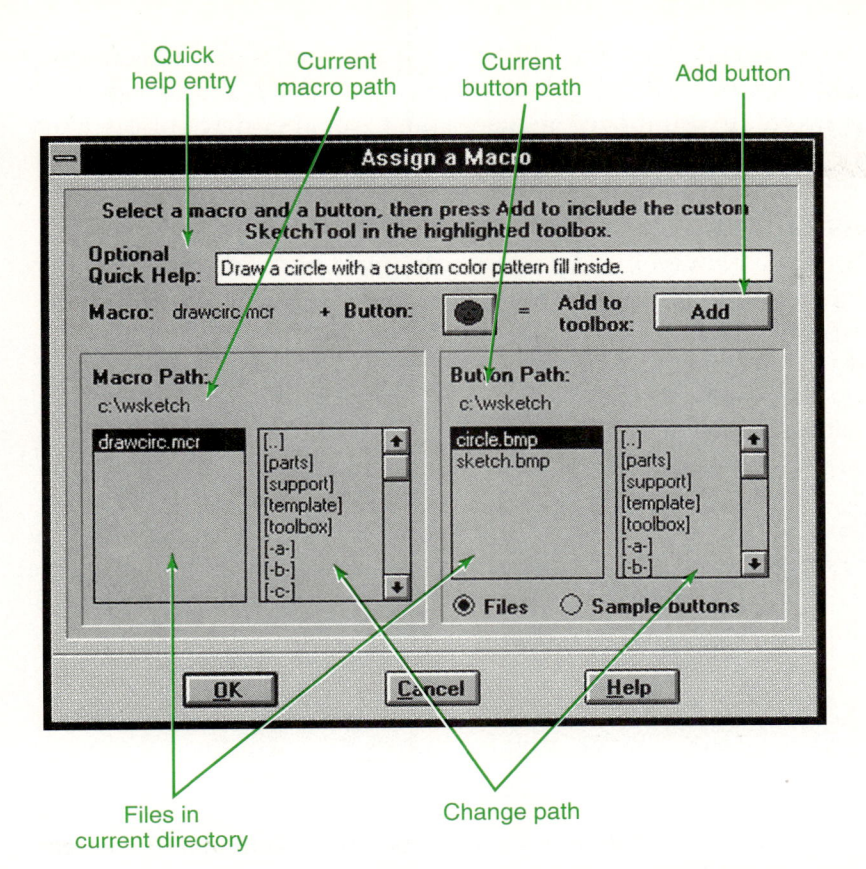

Figure 19-8.
In the **Assign a Macro** dialog box, select a macro file and a button to create a tool in your toolbox.

Callouts: Quick help entry; Current macro path; Current button path; Add button; Files in current directory; Change path

When finished, pick **OK** to close the **Assign a Macro** dialog box. Then, select **Save As...** in the **Toolbox Editor** dialog box and enter a name for your custom toolbox.

Your custom toolbox becomes a part of the AutoSketch program, Figure 19-9. The new toolbox appears in the main toolbox window. The toolbox name also appears in the menu bar.

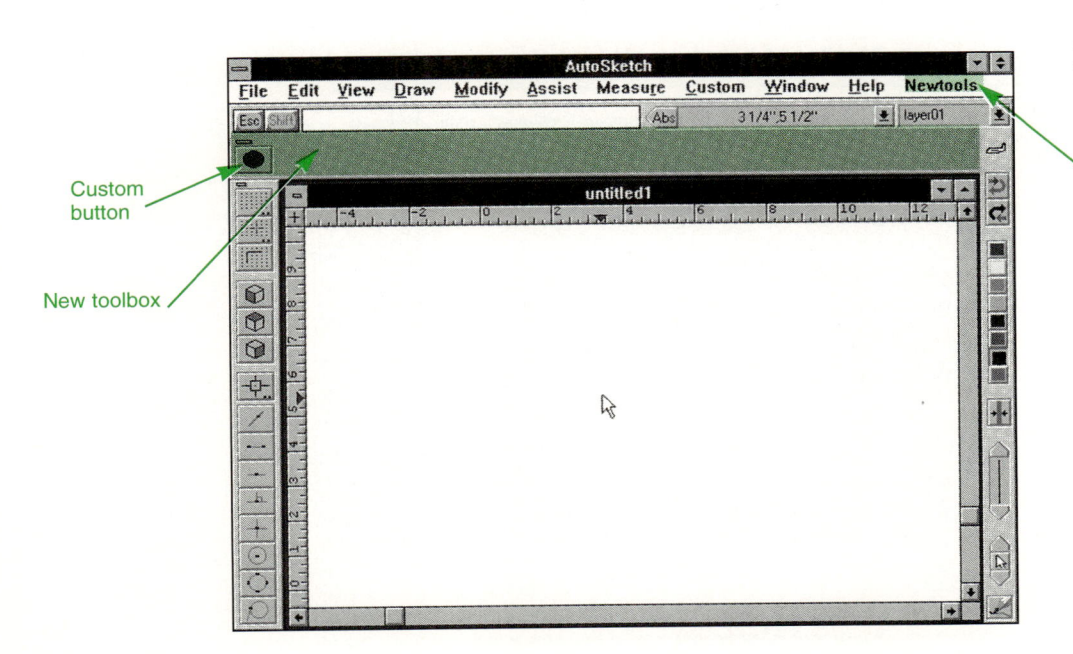

Figure 19-9.
Once you create a toolbox, it becomes part of the AutoSketch menu until you remove it. (Look at the highlighted areas.)

Callouts: Custom button; New toolbox; New menu

Caution! When creating custom tools, AutoSketch stores a copy of the button bitmap and the description, but not the actual macro. The button simply points to where to look for the macro. If you move the macro file to another drive or directory without recreating the tool, your custom tool no longer works.

Editing Your Custom Toolbox

You can modify your custom toolbox by rearranging, adding, or deleting tools. To edit a toolbox, first select it using one of the following methods.
- Double-click any grey area in a non-floating toolbox.
- Open the **Toolbox Control** menu from the toolbox you want to edit, pick **Edit Options**, and then **Edit Toolbox**.
- Pick the **Edit Toolbox** tool from the **Custom** menu. Then, pick the toolbox you want to edit. You may need to pick its name from the menu bar before you can pick the toolbox.

When the **Toolbox Editor** opens, a copy of the toolbox you chose appears below the main toolbox window, Figure 19-10. Perform any editing tasks as shown in the table below. When finished, select **Save** to save the changes. The original toolbox is replaced with the edited toolbox. You can also save it under a new name.

Task	Procedure
Add an existing tool	Drag the tool into the new toolbox.
Delete a tool	Drag the tool anywhere outside the toolbox.
Reposition a tool	Drag the tool to the location you want in the toolbox.
Insert or decrease space between tools	Drag the tool to the left or right.
Create and add more custom tools	Choose the Assign a Macro button and follow the procedure covered earlier to create the first custom tool.

Figure 19-10.
When editing an existing toolbox, a copy of the toolbox appears below the main toolbox window (shown highlighted here).

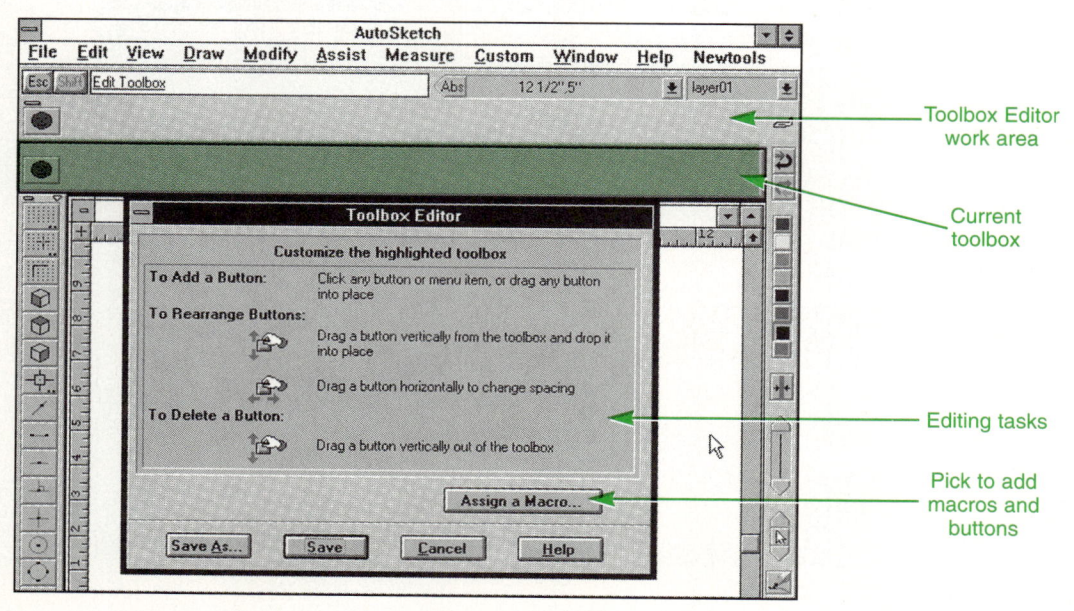

Removing and Restoring Custom Toolboxes

A custom toolbox appears on the menu bar until you remove it. You might chose to remove a custom toolbox if you no longer need the tools it contains. Or, you might want to replace it with a different toolbox. Even though you remove its name from the menu, the toolbox file remains on disk. You can reopen the custom toolbox at any time.

Removing a custom toolbox

To remove a custom toolbox from the menu bar, open the **Toolbox Control** menu. Then, select **Edit Options** and select **Remove Toolbox,** Figure 19-11.

Figure 19-11.
Select **Edit Options** and then **Remove Toolbox** from the control menu to remove a toolbox from the menu bar.

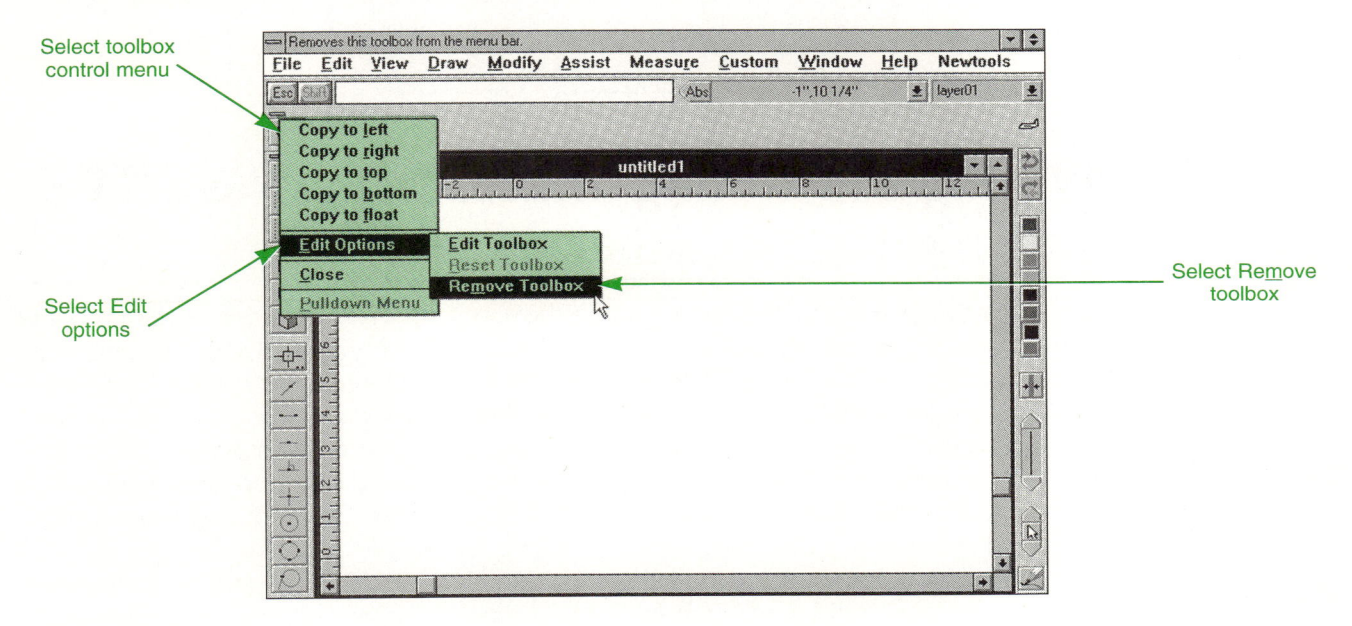

Opening a custom toolbox

To open one of your custom toolboxes, pick the **Open Toolbox** tool from the **Custom** menu. In the **Open Toolbox** dialog box, enter the toolbox filename or select it from the list, Figure 19-12. Select **Open** to open the toolbox. The custom toolbox will appear in the main toolbox window. Its name will also be on the menu bar.

Exercise 19-5

1. Create a custom toolbox for the macros you recorded in Exercises 19-1 and 19-2. Use the buttons you created in Exercise 19-5.
2. Save the custom toolbox on your work diskette as C19E5.

Figure 19-12.
When you pick the **Open Toolbox** tool, the **Open Toolbox** dialog box is displayed. You can then select one of your existing toolboxes to open.

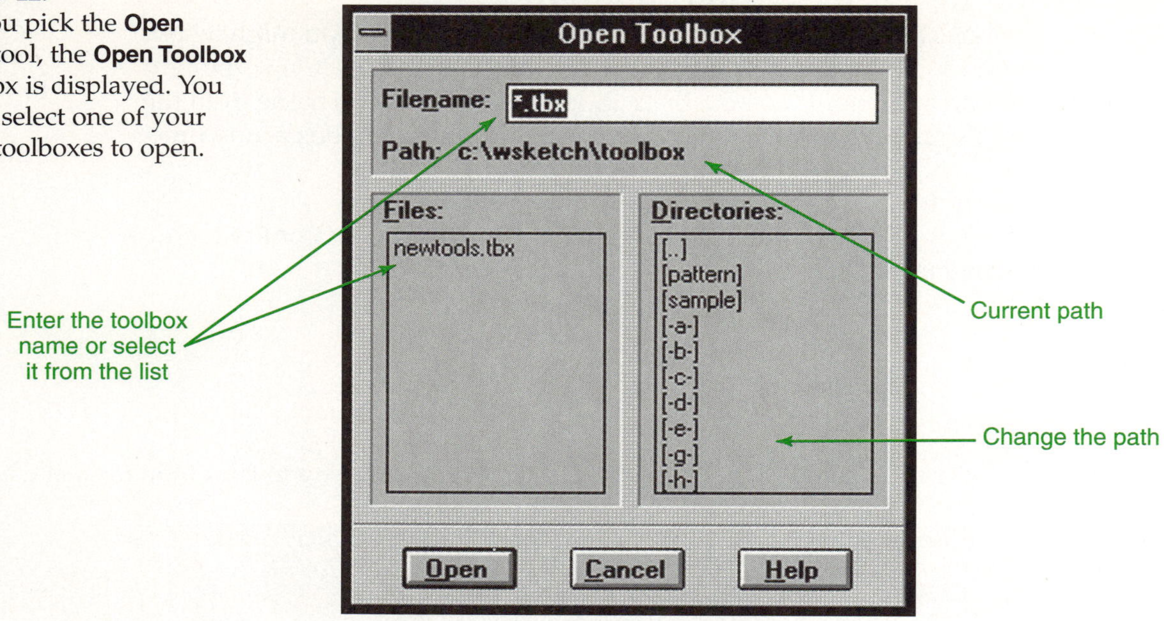

Enter the toolbox name or select it from the list

Current path

Change the path

Writing and Editing Macros Using the AutoSketch Macro Language

Recording and playing macros using the steps just described will satisfy most of your needs. However, there are certain situations where you will want to know the AutoSketch Macro Language. The *AutoSketch Macro Language* is a special programming language used to record and write AutoSketch macros.

For example, suppose you have just created a fairly complex macro that draws 3" circles in blue using a .1 line width. Now, you need a macro that does almost the same thing, but the color should be red. You could rerecord the entire macro from scratch. Or, you can simply edit a copy of the existing macro file to make the change. This requires that you learn more about the AutoSketch Macro Language. The AutoSketch Macro Language will be very familiar if you have done any simple programming in BASIC. If not, don't worry.

Remember that AutoSketch stores macros as text files. The macro file is really just a series of instructions written in ASCII text. You can modify macro files or create new ones using any ASCII text editor.

Figure 19-13 shows a macro file. This macro file is used to draw a 4" line at 30°. It was created using the **Record Macro** tool. The first line of a macro file identifies the AutoSketch version number. Every other line is an instruction or command representing an action. An action may be choosing a tool or entering coordinates. If you look through the statements, they seem very simple. First, the **Line** tool was selected (DRAWLINE). Then, the **User Input** tool was chosen (USERINPUT). This allows the user to pick the first endpoint of the line. Next, the second endpoint of the line was entered with a polar coordinate, not the **User Input** tool. Finally, the **Zoom Full** tool was chosen.

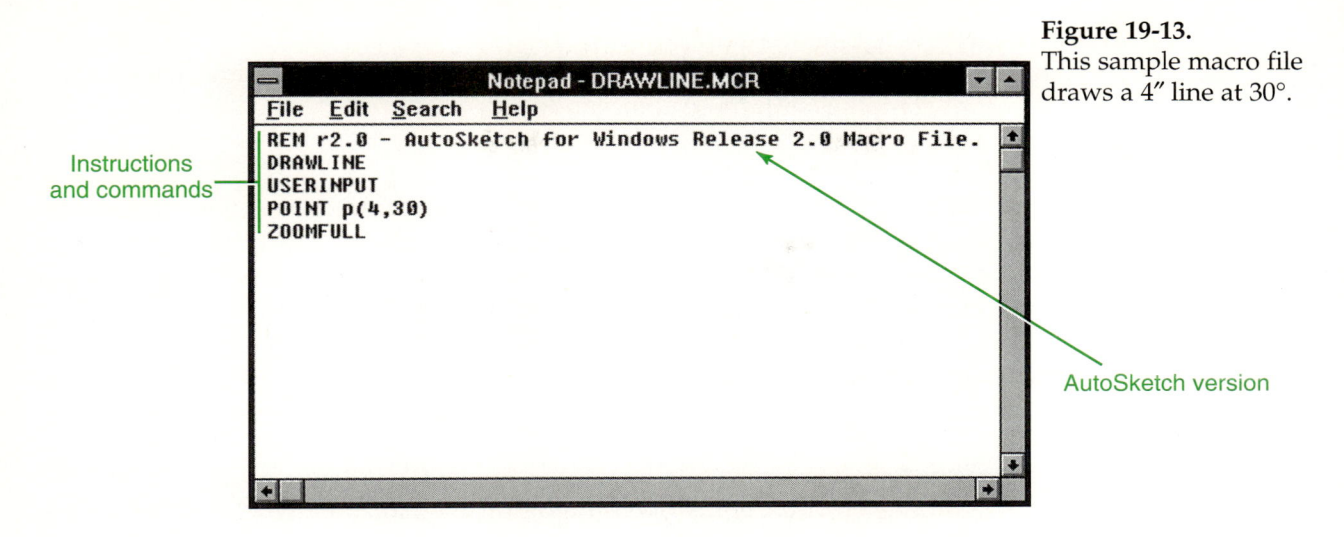

Figure 19-13.
This sample macro file draws a 4″ line at 30°.

```
Notepad - DRAWLINE.MCR
File  Edit  Search  Help
REM r2.0 - AutoSketch for Windows Release 2.0 Macro File.
DRAWLINE
USERINPUT
POINT p(4,30)
ZOOMFULL
```

Instructions and commands — (labels pointing to the lines above)
AutoSketch version — (label pointing to REM line)

Although this is a very simple example, you can also see how easy it would be to edit this macro. For example, you could copy this macro file to a new name. Then, using an ASCII text editor, replace POINT p(4,30) with POINT r(5,6). When played, the new macro will draw a line using a relative coordinate.

Since the AutoSketch Macro Language is a programming language, you can create very complex macros. Some skills are covered here, but you will learn best by recording, writing, and editing your own macros.

Where to Begin?

Instead of writing a macro from scratch, the **Record Macro** tool of AutoSketch gives you a "jump-start." It is very helpful to first record a macro that does something similar to what you want to do. Then, open the macro file in a text editor and modify it to meet your exact needs.

Each macro file should begin with a heading that identifies the version of AutoSketch used to create the macro. It is important to have the version of AutoSketch noted for troubleshooting. Macros may not be completely compatible in different versions of AutoSketch.

The **Record Macro** tool inserts this line automatically. The word **REM** at the beginning of the line tells AutoSketch that the line is a remark. An REM statement is not played. You can use REM statements throughout long macros to identify different sections of the macro.

Understanding Macro Commands

There are two basic types of commands you will see in macros. These are Macro Language commands and AutoSketch commands. *Macro Language commands* are special statements that the macro needs to perform its job and to help you troubleshoot. *AutoSketch commands* in macros are statements that correspond to the AutoSketch tools and commands you pick while drawing.

An example of each of these commands is shown in Appendix E. After reviewing that appendix, you should be able to understand the purpose of a given macro.

Caution! Proper spelling and punctuation are required when writing macros. However, do not rely on a spell checker in your word processor. Many of the macro commands are abbreviations, and also include odd punctuation and spacing. Spell checking your macros can have disastrous results. Pay close attention to the formats given in the following sections.

> **SHORTCUT**
>
> The Windows Notepad program is an ASCII text editor. Once open, you can "[Alt][Tab]" between AutoSketch and Notepad.

Macro Language Commands

The following Macro Language commands are essential to writing macros. You should become familiar with these commands.

REM

REM stands for "remark." This allows you to add comments to your macro. An REM statement is not "played." Instead, REM statements let you break up a macro into sections so that you understand what each part of the macro is doing. Leave a space between **REM** and your comment. The format is as follows.

REM *enter your text here*

DialogBoxReturn

Many AutoSketch commands open dialog boxes that must be closed before the macro can continue. The **DialogBoxReturn** macro command is used to close a dialog box. It is written as follows.

DialogBoxReturn *value*

The *value* refers to the action you would take to accept, decline, or cancel the action of the dialog box. The values are given in the following chart.

DialogBoxReturn 1	Selects Yes or OK to close the dialog box, depending on the dialog box.
DialogBoxReturn 0	Selects Cancel to close the dialog box.
DialogBoxReturn -1	Selects No or Modify, depending on the dialog box.

Note: There is one difference between using dialog boxes in macros and using them on-screen. When you use a Macro Language **SET** command to change a value in a dialog box, the value is changed regardless of what value you provide for **DialogBoxReturn**. **SET** commands are executed immediately and do not require an **OK** or **Yes** response.

Point

The **Point** command enters the coordinate of a location on the drawing. When you draw, you must pick locations or enter coordinates. The **Point** command in the macro does the work for you. The format is entered just as if you type the coordinates in the prompt box.

POINT (*X,Y***)**

The above entry is for an absolute coordinate.

POINT R(*X,Y***)**

The above entry is for a relative coordinate.

POINT P(*D,A***)**

The above entry is for a polar coordinate.

You can also use AutoSketch coordinate or distance system variables. A *system variable* is a word preceded by a front slash (/) that represents a value stored by AutoSketch. These variables can be used instead of a coordinate value whenever you are entering a point location. Refer to the following chart.

/langle	The last angle measured using the Angle or Bearing tool in the Measure toolbox.
/larea	The last area measured using the Area tool in the Measure toolbox.
/ldist	The last distance measured using the Distance tool or one of the dimension tools in the Measure toolbox. This variable also stores the perimeter value derived from using the Area tool in the Measure toolbox.
/lpoint	The last coordinate of the last point entered.
/lx	The X coordinate of the last point entered.
/ly	The Y coordinate of the last point entered.
/pi	The value of pi.

The following are two examples using these variables.

DRAWLINE
POINT (/LX,/LY)
POINT (/LX,3)

These three statements will draw a line that begins from the previous point you picked, and with a second endpoint that has the same X coordinate but a Y coordinate of 3.

DRAWLINE
USERINPUT
POINT P(/LDIST/LANGLE)

These three statements allow you to draw a line where you pick the first endpoint. The second endpoint is placed using the distance and angle last measured with the **Distance** and **Angle** (or **Bearing**) tools.

Note: AutoSketch records only the system variable name, not a specific value. When you run the macro, the values used are taken from the last values before you ran the macro, or ones entered during macro operation. When using system variables, it is better to let the macro itself set these values rather than rely on values set during your drawing session.

SET

The **SET** command changes an AutoSketch system variable value. System variables include the following.
- Values set in AutoSketch dialog boxes, such as the method to draw a circle.
- Toggle functions, such as grid, snap, and attach, that are turned on and off.

The format is as follows.

SET *variable value*

The *variable* is the name of the AutoSketch system variable to be changed. The *value* is the new value for the variable. For example, the following macro command turns on snap.

SET SNAP 1

There are three types of values you will use. These are integer, real, and Boolean.

An *integer value* is a whole number, and usually refers to a setting dialog box with multiple options. Refer to the **ARCTYPE** variable in Appendix E.

A *real value* is any decimal number, and usually refers to a distance. For example, look at the variables **GRIXY** and **GRIDX** in Appendix E. These are used to set the grid spacing.

Finally, a *Boolean value* is either true or false. The number 1 is used for true, and 0 is used for false. Boolean values are used for variables that are either on or off. Refer to the **ATTACH** variables in Appendix E.

Stop

Placing the word **Stop** as a line in your macro immediately stops macro execution. This statement is very useful for troubleshooting long macros, or only using the first portion of a long macro.

String

The **String** command enters text as if it is typed with the keyboard. This command is used in response to prompts for text, such as with the **Quick Text** tool. The format is as follows.

STRING *your text*

> **SHORTCUT**
> One ASCII value you will use often is \013. This is the code for the [Enter] key.

The *your text* is the text you want to enter. You can use any printable ASCII characters found on the keyboard. You can also encode other keys and characters in the string using a backslash (\) followed by the three-digit number. The number represents the ASCII value of the key. Consult your DOS manuals for a complete list of ASCII codes.

In the following line, the macro places the quick text "General Tolerances" and ends the line (with the code for the [Enter] key).

STRING General Tolerances\013

UserInput

The **UserInput** command pauses the macro to allow the user to supply input. This is the entry that AutoSketch places in the file when you pick the **User Input** tool while recording a macro.

Restart

The **Restart** command tells the macro to repeat until the user picks another tool. This command is very useful because it makes your custom buttons function like AutoSketch drawing tools that repeat.

AutoSketch Commands in Macros

In addition to Macro Language commands, macro files will also include AutoSketch commands. See Appendix E. Notice that all the commands are one word (no spaces).

Exercise 19-6

1. Start recording a macro named C19E6 using the **Record Macro** tool. Store the macro on your work diskette. The macro should do the following.
   A. Set a .02 custom line width.
   B. Change the color to blue.
   C. Draw a box that allows user input to select the corners.
   D. Finishes with the **Zoom Full** command. Remember to select **End Macro** when finished.
2. Test the macro to make sure that it works properly.
3. Use the Notepad program in Windows to copy the macro C19E6 to C19E6Band edit the copy so that it performs the following functions.
   A. Sets a custom line width of .03.
   B. Changes the color to yellow.
   C. Draws a circle.
   D. Uses the **Radius, Diameter** option to draw a circle and allows user input to select the center and point on the circle. Remember to save the edited macro as C19E6B on your work diskette.
4. Test the macro again to make sure that it works properly.

Advanced Macro Language Topics

The following sections offer additional information on functions supported in the AutoSketch Macro Language. Review all of these sections before attempting to use any of the advanced functions.

Math Functions

AutoSketch contains a built-in calculator that you can use from the prompt box or any real number input field in a dialog box. The calculator supports math operations as well as variables and AutoSketch system variables. Refer to the following chart.

+	Add
-	Subtract
*	Multiply
/	Divide
ABS( )	Absolute value
ATAN( )	Arc Tangent (returns degrees)
CEIL( )	Ceiling (round a real number up to the next integer)
COS( )	Cosine (argument in degrees)
COSH( )	Hyperbolic Cosine (argument in hyperbolic radians)
EXP( )	Scientific notation e^x
FLOOR( )	Floor (round a real number down to the next integer)
LOG( )	Natural logarithm
LOG10( )	Base 10 logarithm
SIN( )	Sine (argument in degrees)
SINH( )	Hyperbolic Sine (argument in hyperbolic radians)
SQRT( )	Square Root
TAN( )	Tangent (argument in degrees)
TANH( )	Hyperbolic Tangent (argument in hyperbolic radians)

When using math functions, use parentheses to group operations. Multiplication and division take precedence over addition and subtraction. Evaluation is from left to right. The following example will return the value 3 for grid spacing.

SetGrid
This opens the **Grid/Snap Settings** dialog box.

SET GRIDX (4+2).5
This sets the grid spacing to 3.

SET GRID 1
This turns on the grid.

DialogBoxReturn 1
This selects **OK** to close the dialog box.

Note: When using architectural units, the subtraction function (minus sign) must be handled differently. For example, AutoSketch reads 2'-6" as 30" (2 feet, 6 inches), not 18" (2 feet minus 6 inches). This is because the "-" is assumed to be a dash, not a minus sign. To avoid confusion, insert spaces around this math function. For the previous example, 2' - 6" is read as 18".

User Variables

A powerful feature of the AutoSketch Macro Language is user variables. Remember that system variables of AutoSketch can hold the value of a coordinate, or the spacing of the grid. However, you can also define variables to hold your own values. To set a variable, you can use one of the following two formats.

LET *myvalue=something*
myvalue=something

The only difference in these two formats is that statements beginning with **LET** do not appear in the prompt box when the macro is run. *Myvalue* is the name you want to give your variable. Also, *something* can be just about any numerical value, equation, or system variable. For example, you could assign a variable to a mathematical equation or to a distance.

LET MYVALUE=/DIST+6

At this point, **MYVALUE** is a number equaling the most recently measured distance plus six. You can use this user variable in any equation as a real number. For example, you can enter a point location using a macro with the following expression.

POINT MYVALUE,sqrt(*myvalue***)**

An example

The best way to illustrate is with an example. The following macro allows you to set the snap spacing by picking two point locations. Four user variables store the X and Y coordinate values of the two points picked. The coordinate values are pulled from AutoSketch system variables (**/LX** and **/LY**) discussed earlier in the chapter. The macro also uses a math equation to set the grid to twice the snap spacing. The information in italic type is *not* in the macro file, but provided here for explanation.

REM r2.0 - AutoSketch for Windows Release 2.0 Macro File.	
REM *** Part 1. Turn off the grid or snap in case it is one. ***	
REM *** This will make it easier to pick the distance. ***	
SETSNAP	*Opens the* **Grid/Snap Settings** *dialog box*
SET SNAP 0	*Turns off snap*
SET GRID 0	*Turns off grid*
DialogBoxReturn 1	*Closes the* **Grid/Snap Settings** *dialog box*
REM *** Part 2. Pick two points and set user variables to those X and Y coordinates. ***	
REM *** The Measure Distance tool is used since you can pick two points without drawing. ***	
MEASUREDISTANCE	*Activates the* **Measure Distance** *tool*
USERINPUT	*Pick the first point to measure*
LET MYX1=/LX	*Sets a user variable equal to the X coordinate value of the first point*
LET MYY1=/LY	*Sets a user variable equal to the Y coordinate value of the first point*
USERINPUT	*Pick the second point to measure*
LET MYX2=/LX	*Sets a user variable equal to the X coordinate value of the second point*
LET MYY2=/LY	*Sets a user variable equal to the Y coordinate value of the second point*
DialogBoxReturn 0	*Closes the distance information box*
REM *** Part 3. Set the grid and snap spacing using a math equation and the user variables. ***	
SETSNAP	*Opens the* **Grid/Snap Settings** *dialog box*
SET SNAPX MYX2-MYX1	*Sets the snap spacing equal to the distance between the first and second X coordinates measured*
SET SNAPY MYY2-MYY1	*Sets the snap spacing equal to the distance between first and second Y coordinates measured*
SET GRIDX (MYX2-MYX1)*2	*Sets the grid spacing equal to twice the distance between X coordinates measured*
SET GRIDY (MYY2-MYY1)*2	*Sets the grid spacing equal to twice the distance between Y coordinates measured*
SET SNAP 1	*Turns on snap*
SET GRID 1	*Turns on the grid*
DialogBoxReturn 1	*Closes the dialog box by picking* **OK**

User variable guidelines

When you create variables for macros, follow the guidelines below. Most of these aspects are included in the above example.

- Variable names must consist of letters (A - Z) and numbers (0 - 9). All other characters are not allowed.
- Variable names must begin with a letter.
- Variable names cannot exceed 16 characters.
- Uppercase and lowercase letters are considered the same.
- You cannot give your variables the same name as AutoSketch system variable names (such as **/LX**, **GRIDX**, and **CURVETYPE**).
- You can use variables and equations in **SET** statements.
- You can assign your user variables the value of a system variable. However, remember that AutoSketch records only the system variable name, not a specific value. When you run the macro to assign your user variable, the value used is

value. When you run the macro to assign your user variable, the value used is taken from the last value assigned the system variable before you ran the macro, or values entered during macro operation.
- You can enter user variables in dialog text boxes that require a position, or in the prompt box when AutoSketch prompts you for a point.

Exercise 19-7

1. Use the Notepad program in Windows to enter the example macro from this section. Name the macro C19E7 and save it on your work diskette.
2. Test macro C19E7 to make sure that it works properly.
3. Edit macro C19E7 and add the commands necessary to draw a box whose width is the X spacing and height is the Y spacing.
4. Test the macro again to make sure that it works properly.

Troubleshooting Your Macros

Your macro will seldom work on the first try. That's why you should save your drawing before running a new macro. You may need to discard the changes and reopen your drawing to test the macro again. Below are some tips and some ways AutoSketch helps you with troubleshooting.
- When a macro you obtain from someone else doesn't run right, first check the version. This should be the first line in the macro file. Different versions of AutoSketch have different features. You may need to check the macro to make sure it conforms to the macro language in Release 2 for Windows.
- If there is a syntax error in your macro file, AutoSketch will display a message, and the line number of the error, Figure 19-14. This is a very helpful feature. A *syntax error* might be a misspelled command or incorrect spaces and punctuation. In fact, sometimes you simply forget to put a **REM** statement in front of your notes in the macro.
- Use the **STOP** command in very long macros to check your work. For example, if the first section of your macro is supposed to set the color, and it doesn't, place a **STOP** command after that section. Next, work on that section of the macro until it changes the color correctly. Then, remove the **STOP** command and work on the next section of your macro.
- If all else fails, go back to basics. Perform the macro steps without recording a macro. Make a note of every tool you pick, text and coordinates you enter, or location you pick. Compare this against a printout of your macro file.

Figure 19-14.
AutoSketch displays an error message if you have a misspelling or other syntax problem in your macro. The message identifies the path and the filename, and the line where the error is. Note: When this message is displayed, it does not mean that the line given is the *only* line with an error. This is just the first line that AutoSketch encountered with an error.

Summary

This chapter covered a broad range of topics. Some of these topics are very technical. The following is a recap of hints that will help you record and write macros, and avoid troubleshooting.

- Keep track of drawing and editing tasks that you do often. These may be easily automated with a macro.
- Use a familiar word processor or text editor to write and edit macros. It should have the features search, replace, cut, and paste. It must also be able to save files in ASCII text format.
- Assign names to your macros that reflect what the macro does.
- Never assume that a particular setting, property, or directory needed for the macro is current at the time you play the macro. Always set them within your macro.
- When possible, first record a macro that does something similar to what you want your written macro to do. Then, open the macro file in a text editor and modify it to meet your needs.
- Test your macros before adding them to a custom toolbox. Building your custom toolboxes over time is much easier than creating an entire untested toolbox.
- Create multiple custom toolboxes that group together similar macros. You can open and float multiple toolboxes on-screen. Putting macro buttons that are not similar into a single toolbox may be confusing.
- Macros and toolboxes work best when placed in the default directories displayed by AutoSketch. However, in a school setting, you may need to place them on a separate diskette.
- Practice, practice, practice. The exercises and activities in this chapter will give you a chance to write several macros. Your competence and creativity in writing macros will increase with each one you develop.

Important Terms

AutoSketch commands	Example	REM
AutoSketch Macro Language	Integer value	Syntax error
Boolean value	Macro	System variable
Custom buttons	Macro Language commands	User input
Custom toolbox	Real value	

New Tools and Commands

Add button
Assign a Macro button
Assign a Macro dialog box
Button:
Button Editor dialog box
Button Editor tool
Button Path:
Capture button
Clear button
Create Toolbox tool
DialogBoxReturn
Edit Options

Edit Toolbox tool
End Macro tool
Files
LET
Macro:
Macro Path:
Open Toolbox dialog box
Open Toolbox tool
Optional Quick Help:
Play Macro File dialog box
Play Macro tool

Point
Record Macro File dialog box
Record Macro tool
REM Remove Toolbox
Restart
SET
Stop
String
Toolbox Editor dialog box
User Input tool
UserInput

Review Questions

Give the best answer for each of the following questions.

1. What are the three most common uses for macros?

2. True or false. A macro file can be viewed and edited with a ASCII text editor.
3. What tool do you use to begin recording a macro?

4. List four times when you will want to use the **User Input** tool.

5. How do you find the **End Macro** tool?

6. Always _____ your drawing before running a macro.
7. Explain how a custom toolbox that contains your macros can enhance your productivity.

8. List three ways to create a custom button in the **Button Editor** dialog box.

9. Identify the maximum button size (in pixels) and what happens when you open a bitmap file that is larger than this size.

10. Describe how the grid used in the **Button Editor** dialog box is different than the grid used for drawing objects.

11. Explain why you would need to recreate a custom tool if you move the macro file to another drive or directory? _____

12. True or false. Once you create a custom toolbox, you cannot remove it from the menu bar.

13. What AutoSketch Macro Language command allows you insert comments in your macro file? _____

14. What AutoSketch Macro Language command picks location on-screen using coordinate values? _____

15. What is the syntax for setting an AutoSketch variable? _____

16. What AutoSketch Macro Language command pauses the macro to accept user input? _____

17. What is the macro syntax of the AutoSketch command to draw a polyline? _____

18. When inserting math functions into a macro statement, what is the difference between 1'-4" and 1' - 4"? _____

19. What AutoSketch Macro Language command is used to help troubleshoot very long macros? _____

20. What type of filename is recommended when saving macro files? _____

Activities

Here is a mixture of activities to develop your skills recording and writing macros. Save your macros, buttons, toolboxes, and AutoSketch drawings on your work diskette using the name given. When instructed, select the **Close** command from the **File** menu to clear the drawing from the screen.

1. Write a macro that places NOTES: at a location you pick on-screen. The text should be the ROMANS font at a height of .25". Save the macro on your work diskette as C19A1.
2. Write a macro that changes the current layer to layer01. Save the macro on your work diskette as C19A2.
3. Write a macro that changes any object you pick to the current layer. Save the macro on your work diskette as C19A3.
4. Write a macro that changes all text on-screen to the current text features. Save the macro on your work diskette as C19A4.
5. Develop a macro that deletes all objects on a layer named CONSTRUCT, and then deletes that layer. This is a good macro to remove construction lines used while constructing a drawing. Save the macro on your work diskette as C19A1.
6. Write a macro that trims the ends of two lines to meet precisely. Use a 0 radius fillet to do the job. The macro should set the fillet radius back to its current value when done.
7. Develop a macro that places two parallel lines on opposite sides of an imaginary line you pick with two points. The macro should also allow the user to set the distance between the parallel lines. This macro can be used to draw walls in an architectural floor plan. Save the macro on your work diskette as C19A2.
8. Write a macro that draws a right triangle with two sides of equal length. Create it so that you only have to pick the endpoints of the first line. The macro should draw the second line and a third line that connects the endpoints.

9. Create a macro that places an "X" between the opposite corners of a box you pick on-screen. You will need to assign user variables to the X and Y coordinates of both points. Then, use the **DRAWLINE** command to draw two lines between the points.
10. Write a macro that inserts centerlines on circles that are not dimensioned. The interior intersecting lines should be .25" long. The lines that cross the circle should begin .125" away from the interior lines, and extend .25" from the circle circumference.

CHAPTER 20

Managing Files in Windows and DOS

Objectives

After studying this chapter, you will be able to:
* Manage files using both Windows **File Manager** and the DOS prompt.
* Explain directories, filenames, and file extensions.
* Create, delete, move, rename, and copy directories.
* List and search for files.
* Format and copy diskettes.
* Move, copy, rename, and delete files.

The Disk Operating System (DOS) prompt can be intimidating to organize your drawing files, disk drives, and other information. Windows 3.1 and Windows for Workgroups 3.11 provide a graphical program called **File Manager** to do the same task. With it you can organize files and directories, move and copy files, format diskettes, and start applications much easier than with the DOS prompt. This chapter will show you how to use **File Manager.** Then, you will also learn how to do many of the same tasks from the DOS prompt. But first, you need to learn how the computer structures files and directories on your hard drive and diskettes.

Note: For specific information on managing files in Windows 95, see Chapter 21 *AutoSketch and Windows 95.*

Understanding Directories, Files, and Extensions

One function of the operating system allows you to organize information on your hard drive and diskettes. Think of your hard drive or diskette as a filing cabinet, Figure 20-1. The following sections look at how the hard drive is organized.

Directories

Consider the drawers of your file cabinet to be directories. *Directories* are used to organize files on the drive into smaller, more manageable sections. You can place all your papers (files) in a drawer (directory). Or, you can insert folders (subdirectories) in the drawer to better manage your information. A *subdirectory* further divides the structure of your hard drive. Now, open a folder and pull out a piece of paper. That is a file. A *file* is a group of data saved under a single name. The file might be a word processing document or an AutoSketch drawing file.

Figure 20-1.
Think of the hard disk as a filing cabinet with drawers (directories), folders (subdirectories), and papers (files).

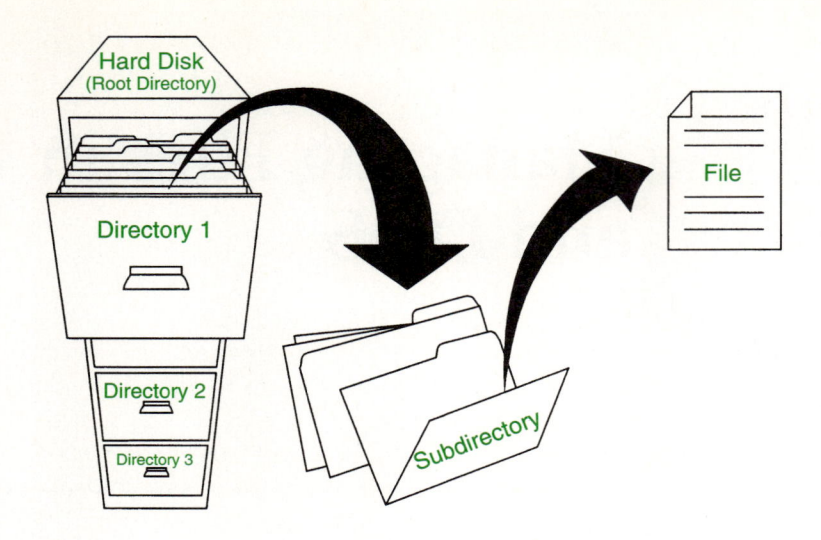

Filenames and Extensions

You have been working with AutoSketch drawing files throughout this text. When saving a new drawing, you entered a filename of up to eight characters. When opening a drawing, you picked the drawing from a list of files stored on disk. However, this was done entirely within AutoSketch. AutoSketch shows you only part of the filename as it is stored on disk. A computer thinks of filenames a bit differently.

In DOS, a *filename* includes up to eight characters and up to a three-character optional extension separated by a period. For example, FILENAME.EXT is a valid DOS filename. The name must start with either a letter or number. It can contain any uppercase or lowercase characters. However, Windows and DOS do *not* recognize the difference between upper and lower case letters. The filename also *cannot* contain any of the following characters.

period	.
quotation mark	"
slash	/
backslash	\
brackets	[ ]
colon	:
semicolon	;
vertical bar	|
equal sign	=
comma	,
space	

In addition, the following names are reserved and cannot be used for files or directories: CON, AUX, COM1, COM2, COM3, COM4, LPT1, LPT2, LPT3, PRN, and NUL.

Note: The rules governing filenames have changed in Windows 95. For specific information on filenames in Windows 95, see Chapter 21 *AutoSketch and Windows 95*.

When saving a drawing, AutoSketch automatically adds the extension .SKD to the filename. AutoSketch also adds .DXF and .SLD file extensions when you export drawings and work with slides. However, when you copy, delete, or rename files *outside* of AutoSketch, you must enter the filename plus the period and extension.

When copying or renaming drawing files, be sure to include the extension. If you do not include the extension, programs like AutoSketch and AutoCAD will not recognize their files. Also, do not erase or otherwise alter files with extensions such as .COM, .EXE, .SYS, .BAT, or .OVL. These are program files. They are usually important to the operation of the computer. Deleting these files might cause serious problems. In fact, it is best that you not delete a file unless you are absolutely sure what that file contains, and that it is not needed.

Using Windows File Manager

File Manager is an application that comes with Windows 3.1 and Windows for Workgroups 3.11. It allows you to manage the files on your disks. You can search for files, create directories, copy and move files, and many other tasks.

Note: There are three different versions of Windows commonly used today. These versions are Windows 3.1, Windows for Workgroups 3.11, and Windows 95. One of the primary differences between Windows 3.1 and Windows for Workgroups 3.11 (WFW) is additional options in the WFW **File Manager**. One of the options is a customizable toolbar. Throughout the rest of this chapter, the toolbar icons for a command are located in the margin. These icons are printed in green and have the abbreviation WFW below them. Keep in mind, these buttons are *only* found in **File Manager** with Windows for Workgroups. Also, in Windows 95, **File Manager** is replaced by **Windows Explorer**. This chapter will only look at **File Manager**. For specific information on **Windows Explorer**, see Chapter 21 *AutoSketch and Windows 95*.

The **File Manager** icon is by default in the **Main** program group. The features of the **File Manager** program are shown in Figure 20-2 and explained below.

Figure 20-2.
The features of **File Manager**.

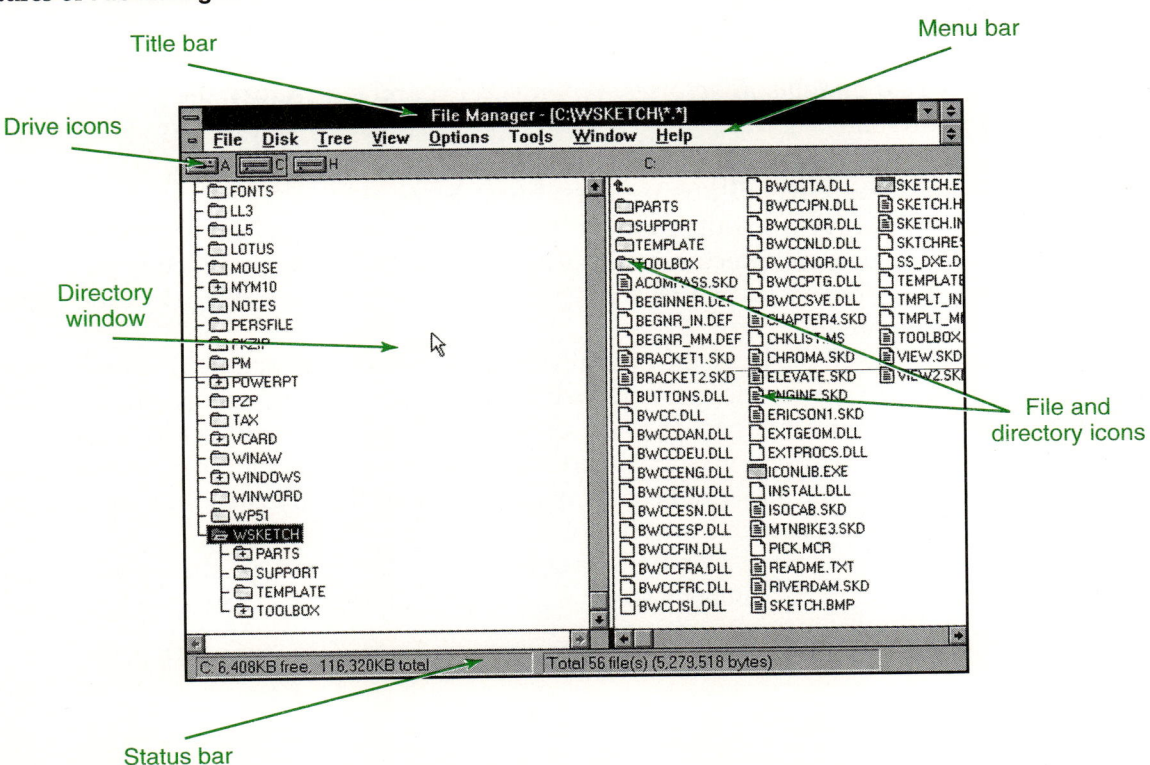

- Menu bar. The menu bar provides pull-down menus listing all commands and options. See Figure 20-3 for a display of all pull-down menus.
- Directory window. In each directory window, the directory tree is displayed on the left, and the contents of the current directory is displayed on the right. The current directory is highlighted, and appears as an open folder. The *directory tree* simply means that the drive and directories are shown as connected branches of a tree.
- File and directory icons. Each filename and directory has an icon next to it, indicating what it is. Figure 20-4 shows the icons and what they mean.
- Drive icons. In the top left corner of each directory window are icons for each drive you currently have access to. These icons are called drive icons. Different icons represent the different types of drives on your computer: hard disk drives, floppy disk drives, network drives, RAM drives, and CD-ROM drives. You can change to a different drive by picking its drive icon.
- Status bar. At the bottom of **File Manager** is a status bar that shows the free storage available on the current drive, total storage, and the number of files in the current directory. When files are selected, the status bar shows how many files are selected and their total size.
- Title bar. The title bar shows the path of the current selected directory. The path simply means a text string that gives the drive letter, a colon, and separates the directories by backslashes. Paths are discussed further in the section on managing files and directories from the DOS prompt.

Displaying the Directory Tree

When **File Manager** is first started, usually only one directory window appears. You can open more than one directory window. This is discussed later.

Remember that the drive, directories, and subdirectories are called a tree. The **Tree** menu provides several options to determine how many "branches" **File Manager** shows. However, the first option to enable in the **Tree** menu is **Indicate Expandable Branches.** This option may make **File Manager** run a bit slower, but it is very helpful. This option marks directories that have subdirectories with a plus sign (+). These are called *expandable directories.* When you expand a directory, the plus sign changes to a minus sign (–).

SHORTCUT
To quickly expand one level, double-click on the directory icon. To collapse an expanded level, double-click on the directory icon.

Expanding and collapsing directory views

To display the contents of a directory, select it in the directory tree. If the directory you need is a subdirectory and does not appear, expand the directory. To see one more levels of directories under an expandable directory (+ sign), select the **Expand One Level** command, Figure 20-5.

To see all the levels of a certain directory, select the **Expand Branch** command. You will see all subdirectories, and their subdirectories, of the current directory. This command is a quick way to open up an entire "branch" of the directory tree.

To see all directories and subdirectories on the entire hard drive, select **Expand All.** This command may not be the best to use since the directory tree can become quite long and hard to move around in.

Finally, to close a branch, double-click on the directory shown with a minus (–) sign, or select the **Collapse Branch** command. The subdirectories no longer appear.

Figure 20-3.
The pull-down menus and their commands in **File Manager**. (Note that the **Window** and **Options** pull-down menus are from Windows 3.11 Windows for Workgroups.)

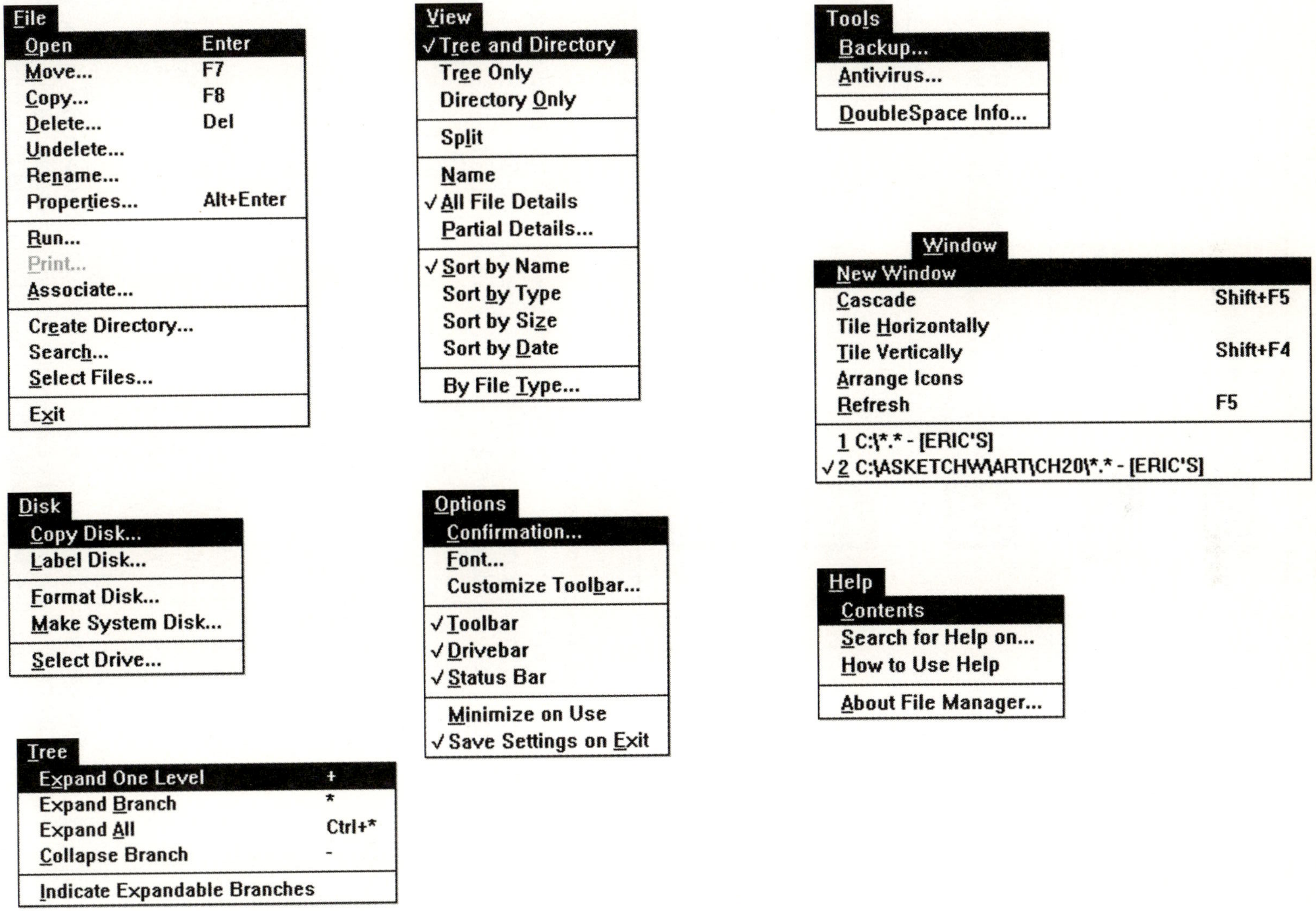

File	
Open	Enter
Move...	F7
Copy...	F8
Delete...	Del
Undelete...	
Rename...	
Properties...	Alt+Enter
Run...	
Print...	
Associate...	
Create Directory...	
Search...	
Select Files...	
Exit	

View
√ Tree and Directory
Tree Only
Directory Only
Split
Name
√ All File Details
Partial Details...
√ Sort by Name
Sort by Type
Sort by Size
Sort by Date
By File Type...

Tools
Backup...
Antivirus...
DoubleSpace Info...

Window	
New Window	
Cascade	Shift+F5
Tile Horizontally	
Tile Vertically	Shift+F4
Arrange Icons	
Refresh	F5
1 C:*.* - [ERIC'S]	
√ 2 C:\ASKETCH\WART\CH20*.* - [ERIC'S]	

Options
Confirmation...
Font...
Customize Toolbar...
√ Toolbar
√ Drivebar
√ Status Bar
Minimize on Use
√ Save Settings on Exit

Disk
Copy Disk...
Label Disk...
Format Disk...
Make System Disk...
Select Drive...

Tree	
Expand One Level	+
Expand Branch	*
Expand All	Ctrl+*
Collapse Branch	−
Indicate Expandable Branches	

Help
Contents
Search for Help on...
How to Use Help
About File Manager...

Figure 20-4.
The icons in **File Manager** refer to various types of files.

Icon	Meaning
📁	Directories
📁	Program files, PIFs, and batch files. These files start applications.
📄	Document files. These files are associated with an application.
📄	System or hidden files.
📄	All other files.

Hiding the directory tree

If you do not want to see the directory tree at all, select **Directory Only** from the **View** menu. Only the current directory will appear, Figure 20-6. To move around in the directory tree, double-click on ⬆... at the top to move up a level, or double-click on a directory icon to move down a level.

Changing the width of the directory tree

If your directory tree is especially deep, all the subdirectories may not fit on-screen. Use the **Split** command in the **View** menu. After you select this command, the cursor drags a vertical line across screen for you to select the dividing line, Figure 20-7.

Displaying Directory Contents

Once you select the directory, **File Manager** is very flexible in how it displays the contents. There are several options.

Displaying file details

File Manager will show as much, or as little, detail as you like. In the **View** menu, select **Name** to display only the names of files and directories. Select **All File Details** to show the name, size, last modification date and time, and file attributes for each file. Or, select **Partial Details...** and pick the items you want to see, Figure 20-8.

> **SHORTCUT**
> You can adjust the split of the directory tree simply by moving your cursor over the dividing line. The cursor will change to the vertical line just as with the **Split** command.

Figure 20-5.
Double-click on an expandable directory icon to see its subdirectories (shown in B highlighted).

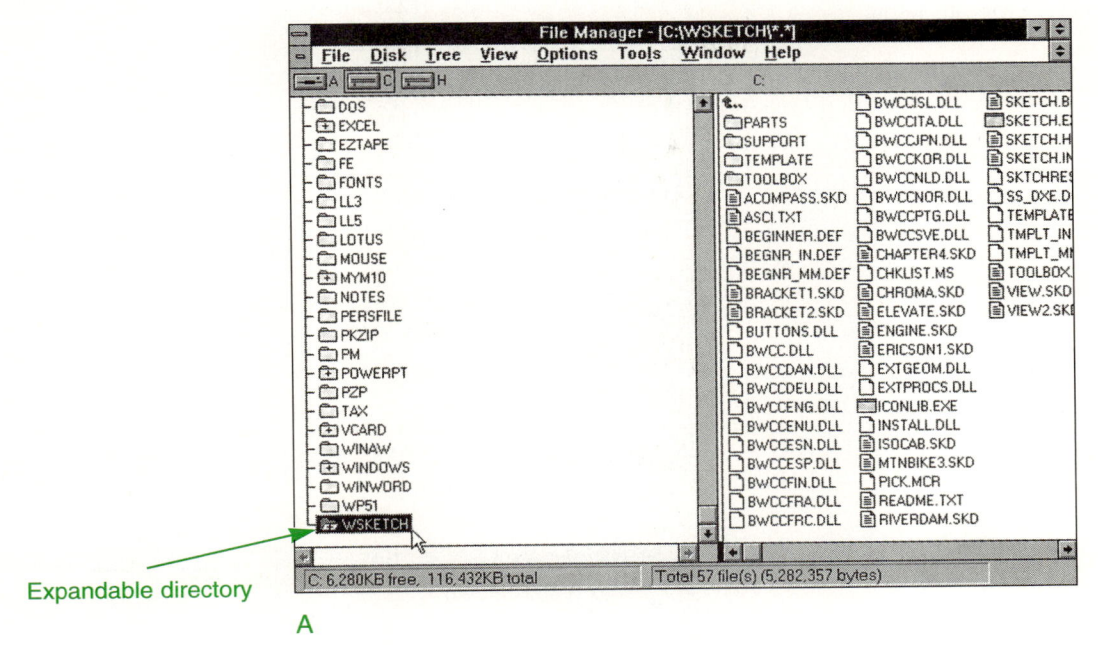

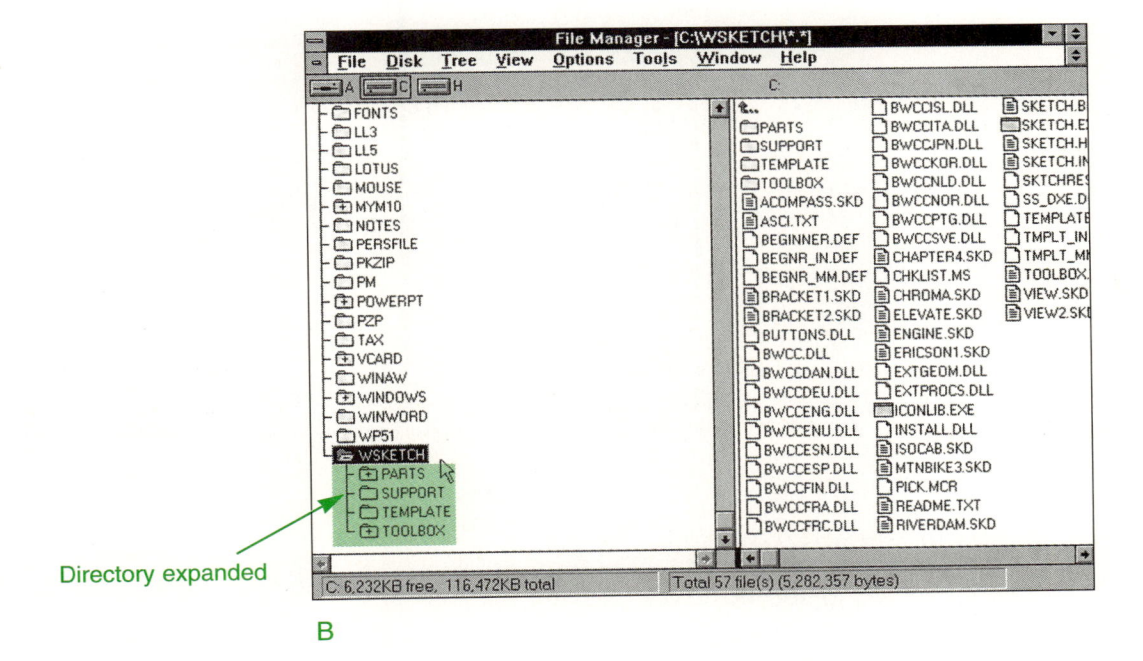

Expandable directory

A

Directory expanded

B

Sorting files

In some situations, you may want to sort files by something other than the name. For example, you may want to sort files by size to move large files to another directory. Or, you might sort files by date to easily see old drawing files. Select one of the available sort options in the **View** menu.

Figure 20-6.
You can display only the directory contents, or only the tree, or both. Here, only the directory contents are shown.

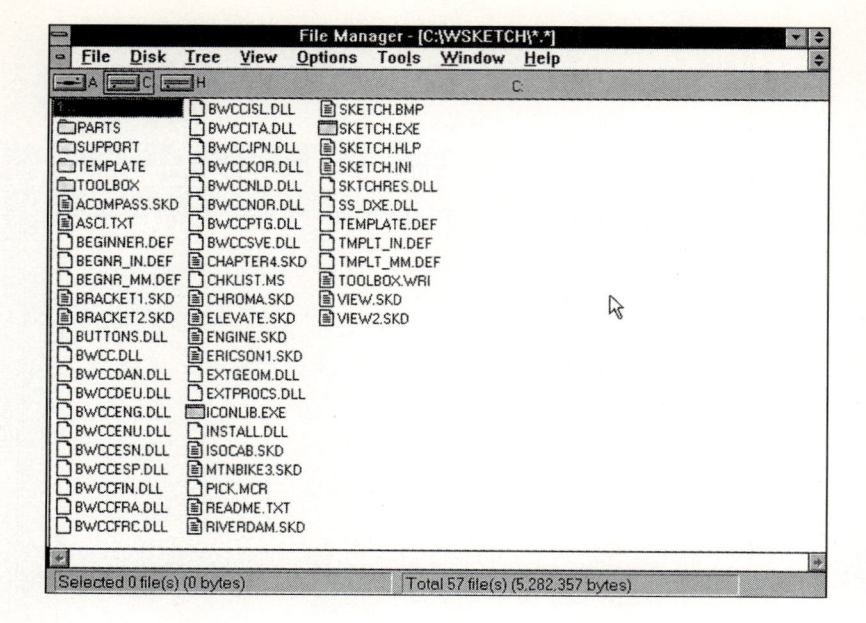

Figure 20-7.
You can change the width of the directory tree and contents. When doing so, the cursor changes to vertical lines with two arrows.

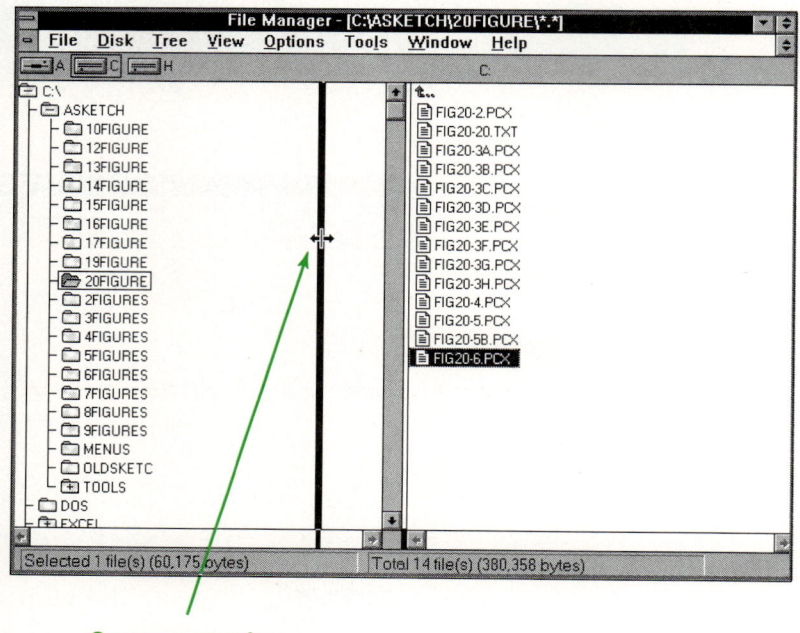

Screen cursor for adjusting the screen split

WFW

Displaying groups of files

In some situations, you may want to display only certain files by name, extension, or those associated with applications. Select the **By File Type...** command in the **View** menu to open the **By File Type** dialog box, Figure 20-9. You can display only directories or programs by checking the box next to these options. Select documents to display those files which are associated with Windows applications. The other category is to display all files that do not fit into one of the above categories.

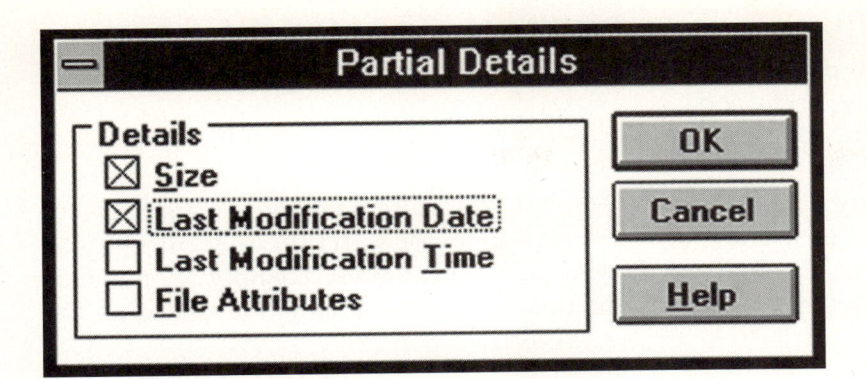

Figure 20-8.
To display only certain file information, use the **Partial Details...** command. Then, select the information you want displayed.

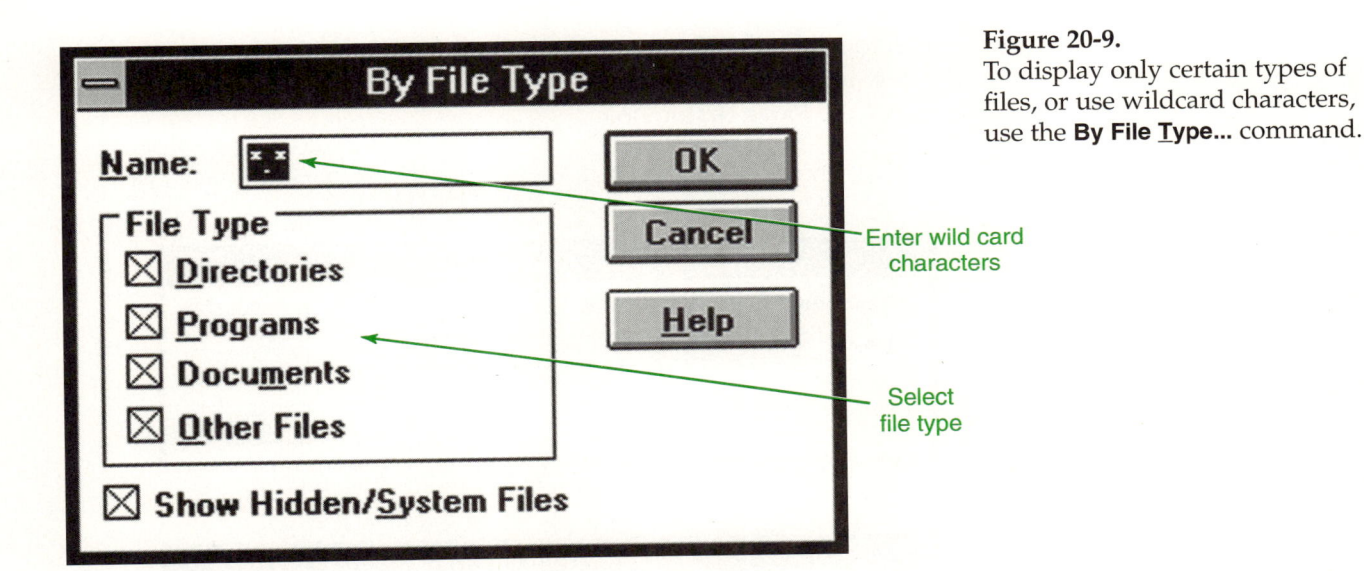

Figure 20-9.
To display only certain types of files, or use wildcard characters, use the **By File Type...** command.

Enter wild card characters

Select file type

You can use this command another way, too. Leave all the options checked. But, in the **Name:** box, use wildcard characters. *Wildcard characters* are characters that can represent one or more letters. An asterisk (*) means any group of letters, while a question mark (?) identifies a single letter. To display all files with similar names or extensions, use wildcards in the **Name:** box.

For example, to display only files that have the extension .TXT, type *.TXT. To display all drawing files that start with DRAW, enter DRAW*.SKD. To display the files that begin with HOUSE, end with PL, and have an .SKD extension, enter HOUSE?PL.SKD. Since the question mark represents only a single character, drawings named HOUSE1PL.SKD and HOUSE2PL.SKD would be shown, but one named HOUSEPLA.SKD would not be. When the default (*.*) is specified, all files are displayed.

Managing Drives and Directories

Managing drives and directories involves selecting drives and creating, deleting, moving, and renaming directories.

SHORTCUT
To change drives, you can also press the [Ctrl] key and the letter of the drive.

Changing drives

You may want to display the directory tree and contents of a different drive. To change drives, pick the drive icon.

Creating a directory

To create a directory, first select the drive and directory where you want to create the new directory. If you want it to be a main directory, select the root directory at the very top of the tree. Then, select the **File** menu and **Create Directory...** command. Enter the name of the new directory in the **Create Directory** dialog box and select the **OK** button.

Moving a directory

You can move a directory to another drive, or make it a subdirectory of another directory on the current drive. All subdirectories of the moved directory become part of the tree.

To move a directory, use the **Move...** command in the **File** menu. In the **Move** dialog box, enter the new location. You can also highlight the directory and press the function key [F7] to access the **Move** dialog box.

SHORTCUT
To quickly move a directory on the same drive, simply select it in the directory tree and drag to the new location. If you are moving the directory to another drive, press the [Shift] key while dragging it.

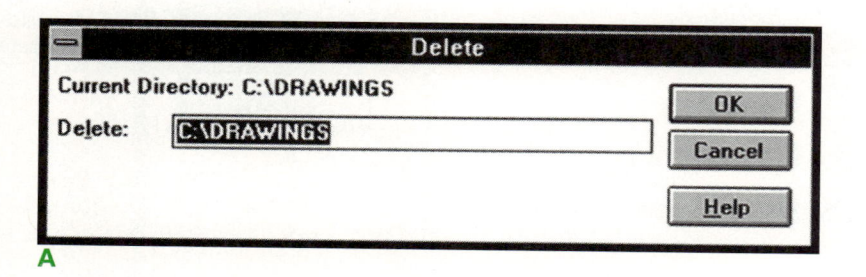

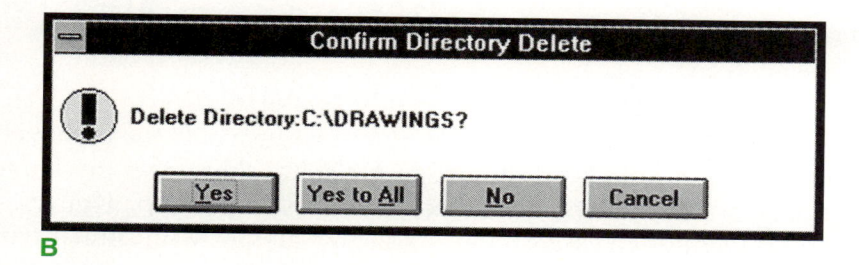

Figure 20-10.
A—The **Delete** dialog box allows you to enter what files and directories to delete. What appears in the boxes will be related to the currently selected item. B—If a directory is being deleted, this message appears. C—If a file is being deleted, this message appears.

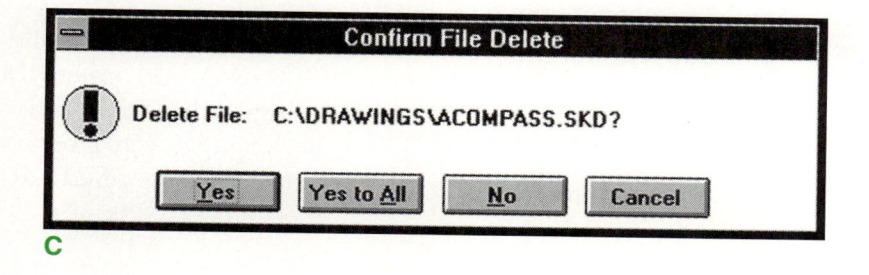

Deleting a directory

If you have a directory that is no longer needed, delete it. This frees space on your hard drive for other files. When you delete a directory, all of its subdirectories and files are deleted. To delete a directory, select the directory and then choose **Delete...** from the **File** menu. In the **Delete** dialog box, the selected directory is listed, Figure 20-10A. Finally, pick **OK**. Before you delete a directory, a message appears asking whether you want to delete it, Figure 20-10B. A similar message appears when deleting a file, Figure 20-10C.

You can turn off confirmation messages using the **Confirmation...** command in the **Options** menu, Figure 20-11. However, it is not recommended to do so.

Copying a directory

You can copy a directory to another drive, or to a subdirectory of another directory on the current drive. All of the copied directory's subdirectories become part of the tree. To copy a directory, select the **Copy...** command in the **File** menu. You can also highlight the directory and press the function key [F8]. The **Copy** dialog box will then appear. Enter the location.

Renaming a directory

You may wish to change the name of a directory. Select the directory and then the **Rename...** command from the **File** menu. The **Rename** dialog box appears, with a **From:** box that shows the currently selected directory, Figure 20-12. Enter the new name in the **To:** box.

Copying disks

The **Copy Disk...** command in the **Disk** menu allows you to copy the contents of one diskette to another diskette. Use this command to make a backup of a disk. If you have more than one "floppy" disk drive, this command will ask for the source and destination drives.

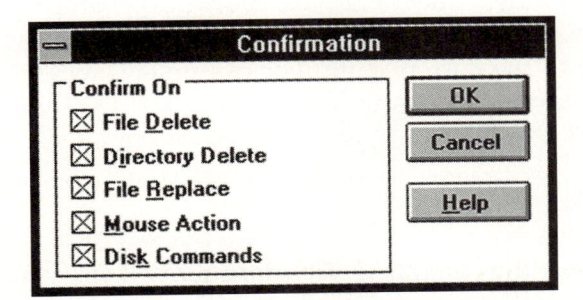

SHORTCUT
To copy a directory on the same drive, hold the [Ctrl] key and drag the directory to the location. If you are copying the directory to another drive, you do not need to use the [Ctrl] key while dragging it.

WFW

WFW

Figure 20-11.
You can turn off confirmation messages using the **Confirmation ...** command.

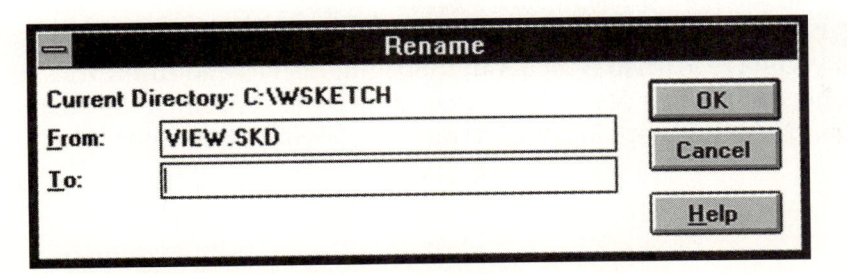

Figure 20-12.
The **Rename** dialog box allows you to change the name of a file or directory.

Formatting a floppy disk

Before you can use a disk to store files, you must format it. Formatting prepares the disk for use with your computer. It also deletes any information stored on the diskette. To format a diskette, insert a diskette in a drive. Select **Format Disk...** from the **Disk** menu. The **Format Disk** dialog box appears, Figure 20-13. Specify the drive letter, disk capacity, volume label, and whether or not to make it bootable (**Make System Disk...**).

Figure 20-13.
The **Format Disk** dialog box allows you to select the drive and density. You can also label the disk, or make it a system disk.

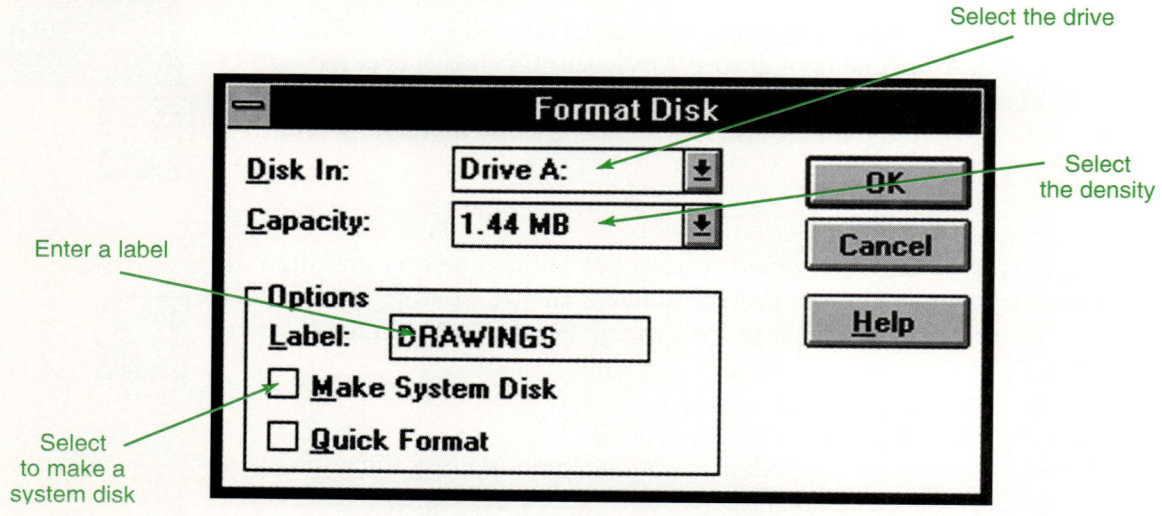

- Select the drive
- Select the density
- Enter a label
- Select to make a system disk

Labeling a disk

You can relabel a disk using the **Label Disk...** command in the **Disk** menu. The disk label is simply a name that is electronically imprinted on the disk. When that disk is the current disk in **File Manager**, the label will appear at the top of the directory tree next to the current path. For example, C:\WSKETCH*.* HARD-DRIVE may appear if you have labeled your hard drive as HARDDRIVE and the current directory is the \WSKETCH directory.

Making a bootable disk

You can also make a disk bootable using the **Make System Disk...** command in the **Disk** menu. When you select the command, **File Manager** will ask you if you want to copy the system files to the drive. Pick the **Yes** button.

Managing Files

Now that you have learned a bit about managing drives and directories, this section discusses how to manage files within those directories. This includes several ways to select files, and moving, copying, renaming, and deleting those files.

Selecting files

There are several ways to select files. If the files are shown in the directory window, simply select them with the mouse. To select two or more files in a row, select the first file. Then, hold the [Shift] key and select the last file in the group you want. To select or deselect individual files, hold down the [Ctrl] key while you pick each file. See Figure 20-14.

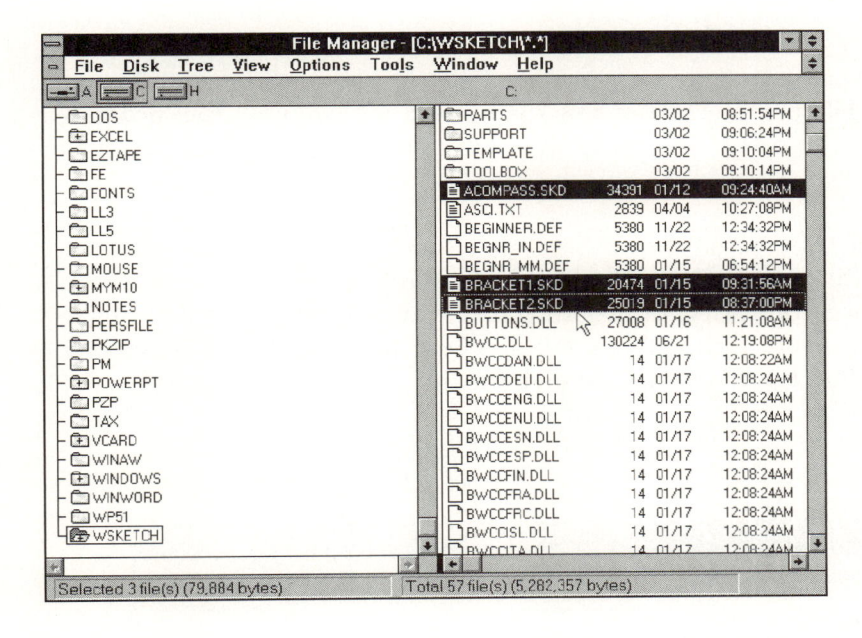

Figure 20-14.
Select sequential files by holding the [Shift] key when picking files, and select or deselect individual files by holding the [Ctrl] key.

Another method to select files is with the **Select Files...** command in the **File** menu. You can use this to select similar files in a directory, or use wildcard characters.

In the **Select Files** dialog box, type the name of the file you want to select. Then pick the **Select** button. You can also use wildcards to select a group of files. For example, to select all AutoSketch drawings in the current directory, enter *.SKD. The **Deselect** button deselects the files specified.

Moving a file

You may need to move a file from one directory or drive to another. To move a file to a directory on the same drive, simply select the file and drag it to the new location. If you are moving the file to another drive, press the [Shift] key while dragging it. Either way, you will notice that the "file" icon in the current directory disappears. Also, there will be no plus symbol on the "file" icon that you are dragging. You can also use the **Move...** command in the **File** menu to do the same thing.

Deleting a file

If you have a file that is no longer needed, delete it. This frees space on your hard drive for other files. To delete a file, select the file and then chose **Delete...** from the **File** menu.

Before you delete a file, a message appears asking if you want to delete it. You can turn off confirmation messages using the **Confirmation...** command in the **Options** menu.

SHORTCUT
You can quickly delete a selected file (or group of files) by pressing the [Delete] key and then [Enter]

Copying a file

You may need to copy a file to another drive or directory. To copy a file to another directory on the same drive, hold the [Ctrl] key and drag the file to the location. If you are copying the file to another drive, you do not need to use the [Ctrl] key while dragging it. You will notice that the "file" icon will remain in the current directory. Also, a plus symbol will appear on the icon that you are dragging. You can also use the **Copy...** command in the **File** menu to do the same thing.

Renaming a file

You may wish to change the name of a file. Select the file and then the **Rename...** command from the **File** menu. The **Rename** dialog box appears with two boxes. The **From:** box shows the selected file. Enter the new name and any extension in the **To:** box.

Searching for files

There are often times when you need to find a file, but don't know where it is. Searching through each directory and subdirectory can be time-consuming. Instead, use the **Search...** command in the **File** menu. The **Search Dialog** box appears, Figure 20-15. In the **Search For:** box, enter the name of the file.

Figure 20-15.
The **Search** dialog box allows you to define what to search for.

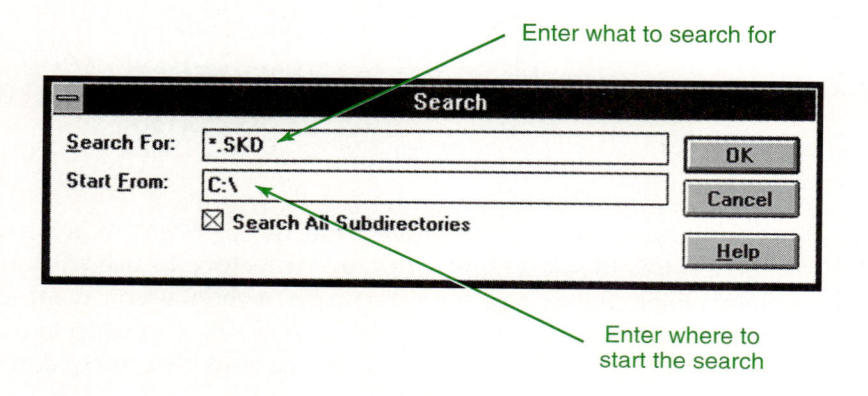

If you only know part of the name, use wildcard characters. For example, to search for an AutoSketch drawing file that starts with the letter D, enter D*.SKD.

The **Start From:** box determines where to start searching. The default entry will be the current selected directory in the directory tree. The **Search All Subdirectories** option determines whether to search only the **Start From:** directory, or to include its subdirectories. For example, to search the entire hard drive, enter C:\ for **Start From:** and select **Search All Subdirectories.**

After you pick the **OK** button, **File Manager** will search for all instances of the search criteria and display that information in a search window, Figure 20-16. To show the search window and the directory window(s) side by side, select **Tile** from the **Window** menu, Figure 20-17. In Windows for Workgroups (Windows 3.11), the tile commands are **Tile Horizontally** and **Tile Vertically.** Now, you can manage the files just as you would files within a directory window.

Figure 20-16.
The **Search Results:** window shows the files and their paths of anything matching the search criteria. The window can be tiled, cascaded, or maximized (as shown here).

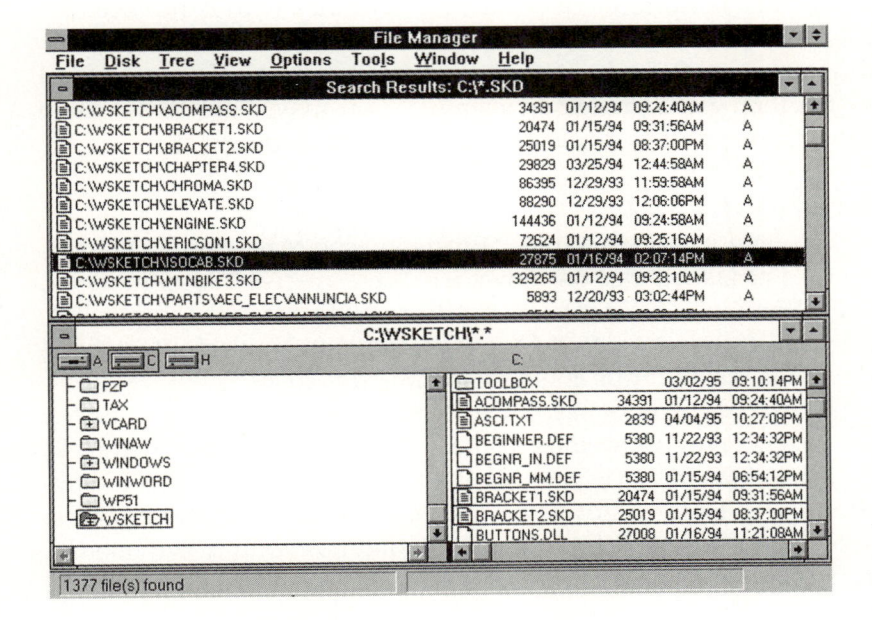

Figure 20-17.
You can organize the search window and other directory windows using the **Tile** and **Cascade** commands in the **Window** menu. The two windows shown here are tiled.

Opening Drawings from File Manager

File Manager has a unique feature called *association*. This allows you to automatically open a file from **File Manager**. When you double-click on a file, **File Manager** looks at the extension and determines whether there is a program to run that file.

For example, during AutoSketch installation, the extension .SKD is associated with AutoSketch. When you double-click on a file with the .SKD extension, **File Manager** starts AutoSketch and opens the drawing.

Some programs will not allow you to start more than one copy of the program. AutoSketch is one of these programs. If AutoSketch is already running when you double-click on an .SKD file, a warning appears, Figure 20-18. In this case, you must switch to AutoSketch and open the drawing.

SHORTCUT
To quickly switch between open applications, press and hold the [ALT] key, and press the [Tab] key until the name of the program you want is displayed. Then, release the [ALT] key and that program is made active.

Figure 20-18.
Some programs do not allow you to start more than one copy of the program. If you double-click on an associated file and that program is already open, this message may appear. If so, you must switch to the program and open the file there.

Associating file extensions

Not all programs automatically associate their file extension during installation. You may also want to associate an extension with a different program. You can manually associate file extensions in **File Manager**.

To associate the .SKD extension with AutoSketch, select the **File** menu and the **Associate...** command. In the **Associate** dialog box, enter SKD for **Files with Extension:** (notice that you do not have to include the period). Then, pick the **Browse...** button and select SKETCH.EXE from the \WSKETCH directory (or the directory where you installed AutoSketch). Finally, pick the **OK** button to associate all .SKD files with AutoSketch.

This same procedure can be used to associate any file extension with any program. For example, the extension .TXT is by default associated with Window's **Notepad** program. If you have another word processor installed, you may want to associate .TXT files with it. To do so, enter TXT and select the program filename.

SHORTCUT

When you associate file extensions with a program, be sure that the program you pick can open that type of file.

Windows for Workgroups 3.11

File Manager in Windows for Workgroups 3.11 (often called Windows 3.11) has a few extra features over the **File Manager** in Windows 3.1. There are two **Tile** commands, as opposed to the one in Windows 3.1. There is also a drop-down list in the upper, left-hand corner that allows you to select a drive. However, the most powerful feature of **File Manager** in Windows for Workgroups is the toolbar, Figure 20-19.

The toolbar is similar to the toolbars used in AutoSketch. You can customize the toolbar to your needs by choosing **Customize Tool&ar...** from the **Options** menu. In addition, many third-party "add-on" programs for Windows make use of this feature by providing tool buttons. This feature allows you to place the tools that you use most often together for easy access, even if they are from different pull-down menus. Therefore, the toolbar makes **File Manager** much easier to use.

Using DOS Commands

While **File Manager** is an excellent way to manage your hard drive, there are times when you will also do these tasks from the DOS prompt. When you boot the computer, a *system prompt* appears when the operating system has loaded. Many computers boot directly into Windows. In that case, you will need to exit Windows to see the DOS prompt. To restart Windows from a DOS prompt, simply type WIN.

The DOS prompt will look like C:\>. The system (DOS) prompt lets you know that the computer is ready for your next command. The letter (C:) shows the disk drive the computer looks to when you enter a command.

Figure 20-19.
The toolbar in Windows for Workgroups **File Manager** is a powerful and useful feature. The toolbar is customizable, so you can change it to meet your own needs.

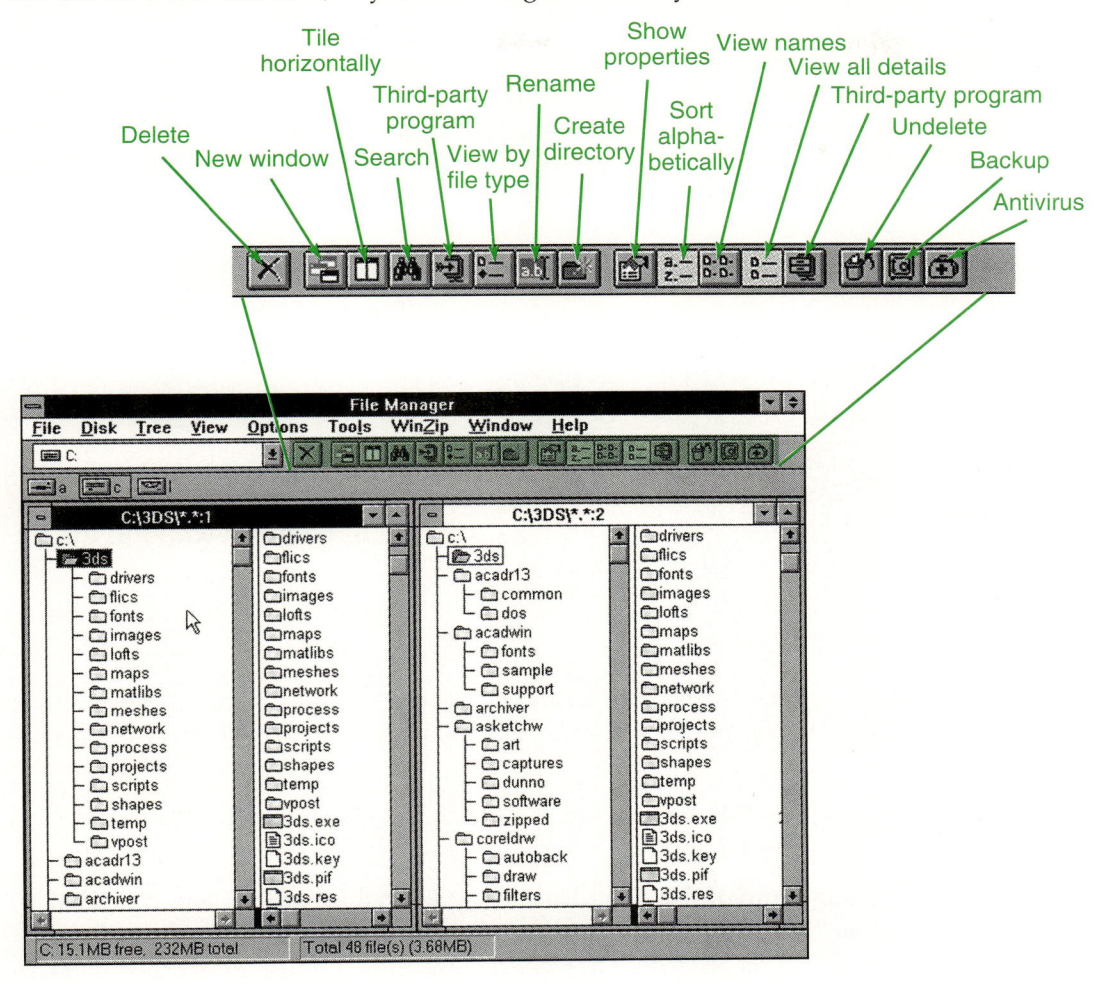

Toolbar labels: Delete, New window, Tile horizontally, Search, Third-party program, View by file type, Rename, Create directory, Show properties, Sort alphabetically, View names, View all details, Third-party program, Undelete, Backup, Antivirus

Moving around the Hard Drive

When using **File Manager**, you enter directories simply by picking them. It is different when working from the DOS prompt.

Changing the active drive

The *active drive* is where the computer looks for files when you enter a DOS command. You always know the active drive by the letter that appears in the system prompt. The first and second floppy disk drives are generally called the A: and B: drives. The hard disk is usually called the C: drive. A CD-ROM drive is usually called the D: drive. (These are the same names used in the Windows **File Manager**.)

Change the active drive of the computer by typing its name (the letter of that drive followed by a colon). The prompt will change to indicate the active drive.

C:\>**A:**↵
A:\>**C:**↵
C:\>

> **SHORTCUT**
>
> You can access the DOS prompt through Windows by using the **MS-DOS Prompt** icon. This is usually found in the **Main** program group. To return to Windows, type **EXIT**. Note: There may be certain commands that will not work using the **MS-DOS Prompt** icon (such as **DEFRAG**). For these commands, you will need to exit Windows.

Notice how the prompt changes to reflect the active drive. The first entry changed the active drive to the A: drive. The second entry changed it back to the C: drive.

Changing directories

You've already learned about the disk drive structure, directories, and files. This same knowledge is needed at the DOS prompt. When you first start up the computer, or exit Windows, you are in the root directory. Compare this to standing in front of a file cabinet with all of the drawers closed. Entering a directory is like opening a drawer. To change to the DOS directory, type the following command.

C:\>**CD\DOS**↵

CD means "change directory." The backslash (\) tells the computer that the tree starts at the root directory.

Making directories

DOS allows you to make new directories, or "drawers," where you can store files. For example, you may want to make a new directory \DRAWINGS to store your AutoSketch drawings. First, exit any directory you might be in by entering the following.

C:\>**CD**↵

The backslash without a directory name means to exit any directory (close all file drawers). After the **CD** command, you will be in the root directory. Then, enter the following command.

C:\>**MD DRAWINGS**↵

DRAWINGS is the name of the new directory you wish to create. You can think of the **MD** command as "Make Directory."

The directory name can be from 1 to 8 characters. You can use letters, numbers, or symbols. You can also add a period and a three character extension. However, the naming guidelines given earlier in the chapter for files also applies to directories. Remember that you can move between directories with the **CD** command. For example, to move between directories on the hard disk, enter the following.

C:\>**CD\WSKETCH**↵
C:\WSKETCH>**CD\DRAWINGS**↵
C:\DRAWINGS>**CD**↵

The first entry changed the current directory to the \WSKETCH directory. The second entry changed the current directory to the \DRAWINGS directory. The last entry removed you from all directories and placed you at the root directory. Notice that a backslash precedes a directory name in all cases. Also notice that the system prompt changed to reflect the current directory. This is done with the **PROMPT PG** command in your computer's startup file (AUTOEXEC.BAT). Most computers come standard with this configuration. However, you can also enter this command at the DOS prompt.

SHORTCUT

You can go down a level into a subdirectory by typing CD, a space, and the subdirectory name. For example, if you are in the \WINDOWS directory and want to go to the \WINDOWS\SYSTEM subdirectory, type CD SYSTEM.

Removing directories

Remove a directory by first deleting all files in that directory. You can delete the files individually, or delete them all at once using a wildcard character. Then, exit that directory using the **CD** command. Finally, enter the **RD** command and the directory name. You can think of **RD** as "Remove Directory."

Suppose you have copied all of your AutoSketch drawings from the \DRAWINGS directory to a floppy disk, and have erased them from the hard disk. The \DRAWINGS directory can then be removed by entering the following command.

C:\>**RD DRAWINGS.**↵

Notice that you do not need to precede the directory name with a backslash when removing the directory, but you do need to include a space after **RD.**

Understanding Paths

A *path* represents the location of a file within a directory and disk drive. You may need to enter paths to tell AutoSketch where to store or look for files. The path is the combination of drive, directory, and a filename. Look at the DOS directory listing in Figure 20-20. The location of the LEADGLAS.SKD file is C:\WSKETCH\PARTS\HOME\DOOR_WIN. You can see a graphic representation of this location by reviewing the **File Manager** directory window.

Suppose you wanted to delete the AutoSketch drawing HOUSE.SKD stored in the \DRAWINGS directory of the hard disk drive. From the root directory, you must enter the following.

C:\>**DELC:\DRAWINGS\HOUSE.SKD.**↵

You could also change to the \DRAWINGS directory first. Then, entering the path is not required.

C:\>**CD\DRAWINGS.**↵
C:\DRAWINGS>**DEL HOUSE.SKD.**↵

```
Volume in drive C
Volume Serial Number is 250F-16E6
Directory of  C:\WSKETCH\PARTS\HOME\DOOR_WIN

.               <DIR>           11-17      2:30p
..              <DIR>           11-17      2:30p
LEADGLAS SKD            10,199  12-20      3:16p
        3 file(s)       10,199 bytes
                    30,892,032 bytes free
```

Figure 20-20.
Note the path showing where the file is located.

Managing Files from the DOS Prompt

DOS includes commands that allow you to copy disks, list files stored on a disk, and delete files, in addition to managing files in your computer. You should know some basic DOS commands to make your work with the computer more enjoyable and productive.

Listing files

When you wish to copy or rename files, you should first display a list of files stored on the drive. The **DIR** (DIRectory) command lists the files stored in the current disk drive and directory. The **DIR** command also shows size of the file (in bytes, or characters) and the date and time the files were created or last edited. The **DIR** command also displays the "total bytes free" at the bottom of the list.

Type DIR at the system prompt and press [Enter] to list the files. A list similar to the one shown in Figure 20-21 appears.

C:\>**DIR**↵

Figure 20-21.
A display of files using the DOS **DIR** command.

```
Volume in drive C
Directory of C:\WSKETCH2\PARTS

.            <DIR>          06-06      3:01p
..           <DIR>          06-06      3:01p
DOOR30_P SKD            5,473 12-19    12:43a
REFRI36P SKD            6,100 12-19    12:26a
ROVEN30P SKD            7,309 12-19    12:26a
SINKD36P SKD            7,521 12-19    12:28a
TBLC36_P SKD            7,573 12-19    12:29a
TREE     SKD           14,604 12-19    12:30a
WIN36_P  SKD            5,898 12-19    12:31a
WIN60_P  SKD            5,898 12-19    12:32a
CLIP_ART     <DIR>          06-06      3:01p
       11 file(s)      60,376 bytes
                   15,872,000 bytes free
```

There are three other variations of the **DIR** command. The command **DIR /P** tells the computer to pause when a full screen of files is displayed. Press any key to see the next full screen of files. The **DIR /W** command shows a wide, or windowed, version of the file list. This command omits the date and size of files so that more files can fit on-screen. The **DIR /S** command displays the current directory and all subdirectories. This is especially useful when looking for a specific file, but you do not know which directory it is in.

You can list files stored on another disk drive or directory without first making it current by simply entering the path name. Suppose you are working in the A: drive, but want to list all drawing files stored on the hard disk in the \WSKETCH\PARTS subdirectory on the hard drive. Enter the following (if your hard drive is the C: drive).

A:\>**DIR C:\WSKETCH\PARTS*.SKD**↵

Formatting new disks

As you learned earlier, diskettes must be formatted before you can store files on them. Formatting prepares the surface of new disks into a format readable by the computer. To do so, enter the format command and diskette drive letter.

C:\>**FORMAT A:**↵

Remember that formatting is designed to prepare a new disk. Formatting a disk erases any data stored on the disk. Make sure you perform a **DIR** command on the disk before formatting it to make sure it does not contain needed files.

Copying files

The **COPY** command allows files to be copied from the hard drive to a diskette or from a diskette to the hard drive. It also allows you to change the filename as you copy it. When copying files to a diskette, the diskette must already be formatted.

The simplest **COPY** command might be entered to copy an AutoSketch drawing from the hard drive to a diskette drive.

    C:\DRAWINGS>**COPY NETWORK.SKD A:**↵

This command copies the AutoSketch drawing NETWORK.SKD from the \DRAWINGS directory to the A: diskette drive. (You must be in the \DRAWINGS directory.)

In addition, a filename can be changed when copying a file from one drive to another. If you want the drawing placed under a different name on the B: drive, enter the following.

    C:\DRAWINGS>**COPY NETWORK.SKD B:\OLDNET.SKD**↵

In this example, the NETWORK drawing on the hard drive is copied to the B: drive and given the name OLDNET.

Another command might copy a file from the A: drive into a directory on the hard disk. Here is one example.

    C:\>**COPY A:MYDRAW.SKD C:\WSKETCH\ASSIGN1.SKD**↵

This command copies the AutoSketch drawing MYDRAW from the A: disk drive to the C: hard drive, and places the drawing in the \WSKETCH directory under the name ASSIGN1. This is probably the most complex copy command you would ever enter.

Another command might copy a drawing between hard disk directories.

    C:\>**COPY C:\WSKETCH\HOUSE.SKD C:\DRAWINGS**↵

This command copies the AutoSketch drawing HOUSE from the \WSKETCH directory of the C: drive into the \DRAWINGS directory of the C: drive.

Renaming files

Renaming files may be desired to change the name of a drawing without creating a copy. When necessary, you can rename a file by using the **REN** (REName) command.

    C:\>**REN A:OLDNAME.SKD NEWNAME.SKD**↵

Do not include the drive name and directory for the new name. DOS automatically places it in the same drive and directory. Now look at this more complex renaming task.

    C:\>**REN C:\DRAWINGS\HOUSE.SKD NEWHOUSE.SKD**↵

In this case, the AutoSketch drawing HOUSE.SKD in the \DRAWINGS directory was renamed NEWHOUSE.SKD and placed in the same directory. Notice that you do not place the C:\DRAWINGS\ path preceding the new name. DOS does not need it.

Deleting files

The **DEL** (DELete) command is used to remove files from disk. Use this command with care. It is difficult, and often impossible, to restore an erased file. If you accidentally erase a file, stop and get help. There are utility programs, possibly an undelete program on your computer, that allow you to retrieve deleted files. However, you must not save anything else on the disk before using the utility program. Use the command as follows.

A:\>**DEL MYDRAW.SKD**↵

This command removes the drawing MYDRAW from the A: drive. Remember that you must enter the drive name (and directory, if any) to delete files stored somewhere other than the current drive and directory. If you want to delete the file ENGINE.SKD from the \WSKETCH directory of the C: drive when you are currently in the A: drive, enter the following.

A:\>**DEL C:\WSKETCH\ENGINE.SKD**↵

Using wildcard characters

Just as in Windows **File Manager**, DOS allows you to use wildcard characters. Wildcard characters represent a single character, or any number of characters. They are best used when copying and deleting files. The asterisk wildcard represents any number of characters. Look at the following example.

C:\>**COPY A:*.SKD C:**↵

This command copies all files on the A: drive with an .SKD extension to the C: drive into the current directory.

C:\>**COPY A:DR*.SKD C:**↵

This command copies all files on the A: drive that begin with DR and have an .SKD extension from the A: drive to the C: drive. You can also use the **DEL** command to delete, rather than copy, the drawings.

Another wildcard character is the question mark. It represents a single character. Look at the following example.

C:\>**DEL A:HOUSE?PL.SKD**↵

This command deletes only those drawings on the A: drive that begin with HOUSE, end with PL, and have an .SKD extension. Since the question mark represents only a single character, drawings named HOUSE1PL.SKD and HOUSE2PL.SKD would be deleted, but one named HOUSEPLA.SKD would not be deleted.

Copying entire disks

DOS allows you to copy an entire diskette at one time with the **DISKCOPY** command. This command is useful to make a back-up copy of the disk where your

AutoSketch drawings are stored. If you have a hard disk and two diskette drives, enter DISKCOPY A: B: at the C:\> prompt. If you have only one diskette drive, use the following command.

C:\>**DISKCOPY A: A:**↵

The computer first copies the contents of one disk into memory. You then insert the target disk and the data is copied from memory to the backup disk. The target disk does not need to be formatted. This command automatically formats the target disk.

Note! You cannot use the **DISKCOPY** command if the A: and B: drives are different sizes or densities. Also, the **DISKCOPY** command will copy over everything stored on your target disk. Make sure the target, or backup, disk is new or does not contain needed files.

Summary

Managing files is an important part of computer-aided drafting. You must organize your files in a way that makes it easy to locate them. To do this you must understand how to divide the hard disk space into directories and subdirectories. The **File Manager** program that comes with Windows is an excellent tool for this. With it, you can graphically display the contents of your drive. Using the mouse, you can drag and drop files and directories to move or copy them. You can also use **File Manager** to delete and rename files and directories, and format diskettes.

You can do the same tasks of **File Manager** from the DOS prompt. However, this requires that you fully understand DOS paths, filenames and extensions. The filename is up to eight characters followed by a period and up to three characters. To help work with filenames, you can use wildcard characters to display or manipulate groups of files. DOS commands are somewhat cryptic, so you will need to remember the command names and syntax.

Important Terms

Active drive
Association
Directories
Directory tree

Expandable directories
File
Filename
Path

Subdirectory
System prompt
Wildcard characters

New Tools and Commands

All File Details
Associate...
Associate dialog box
Browse... button
By File Type...
By File Type dialog box
CD\
Collapse Branch
Confirmation...
Copy...
Copy Disk...
COPY
Copy dialog box
Create Directory...
Create Directory dialog box
Customize Toolbar...
Delete...
DEL
Deselect button
DIR
DIR /P
DIR /S
DIR /W
Directory Only
DISKCOPY
Disk menu
Expand All
Expand Branch
Expand One Level
File menu
File Manager
Files with Extension:
Format Disk...
Format Disk dialog box
From: box
Indicate Expandable Branches
Label Disk...
Make System Disk...
MD
Move...
Name
Name: box
Options menu
Partial Details...
PROMPT PG
RD
Rename...
Rename dialog box
REN
Search...
Search All Subdirectories
Search Dialog box
Search For: box
Select Files...
Select Files dialog box
Select button
Split
Start From: box
Tile
Tile Horizontally
Tile Vertically
To: box
Tree menu
View menu
Window menu

Review Questions

Give the best answer for each of the following questions.

1. What is the purpose of a directory?
   _____

2. Which of the following filenames are valid DOS filenames?
   A. BACK/23.DOC
   B. FLANG.TXT
   C. SLIDE
   D. VALVE0001.SKD

3. AutoSketch adds the extension _____ to your drawing names.

4. The directory structure shown in **File Manager** is called the directory _____.

5. What is an expandable branch? How is it shown in the directory window?
   _____
   _____

6. To see one more level of directories under an expandable directory in **File Manager**, what must you do? _____
   _____

7. What command can you select in **File Manager** to display only the name and size of files?
   _____

8. What command can you select in **File Manager** to display only AutoSketch drawing files (*.SKD)?
   _____

9. How do you use the cursor to move a directory to another location on the same drive?
   _____

10. How do you use the cursor to copy a directory to another location on the same drive?

11. How do you use the cursor to move a directory to another location on a different drive?

12. **File Manager** will warn you before you copy or delete files unless you use the _____ command found in the _____ pull-down menu.

13. To select multiple files in **File Manager** that are not in order, hold down the _____ key while you pick each file.

14. What would you enter in the **Search** dialog box of **File Manager** to search for all AutoSketch drawing files on the hard drive (C:)?
    S̲earch For: _____
    Start F̲rom: _____
    S̲earch All Directories _____

15. True or false. You can launch AutoSketch by double-clicking on a drawing file within **File Manager**.

16. The _____ of DOS lets you know that the computer is ready for your next command.

17. What is the command at a DOS prompt to change the active drive from C: to A:?
    C:\> _____

18. What is the command at a DOS prompt to change the current directory from the root to the directory \DRAWINGS?
    C:\> _____

19. What is the command at a DOS prompt to create a directory called \MECH under the directory \DRAWINGS?
    C:\> _____
    C:\> _____

20. Given the path C:\DRAWINGS\HOUSE.SKD, on what drive and in what directory is the AutoSketch drawing HOUSE located? _____

21. Suppose you are the root directory C:\. From this location, what single command could you enter to delete the drawing FLOOR1.SKD from the subdirectory \DRAWINGS\ARCH?
    C:\> _____

22. Suppose you are in the root directory C:\. What command would you enter to search for a drawing called FOUND.SKD anywhere on the hard drive?
    C:\> _____

23. What does the following DOS command do?
    COPY A:HOUSE.SKD C:\DRAWINGS

24. What does the following DOS command do?
    COPY C:\WSKETCH*.SKD A:*.OLD

25. What does the following DOS command do?
    DEL C:\WSKETCH\PROJECT?.SKD

Activities

Here is a mixture of activities to develop your skills with managing files. Do not save files on disk drives where you are not authorized to do so.

1. Use either **File Manager** or a DOS command to make a backup copy of your work diskette.
2. Using **File Manager**, create directories on your work diskette for each chapter in this book. Move your drawings for each chapter into the appropriate directory.
3. Using **File Manager**, create a directory on the hard drive using your name. Copy the entire directory tree from your work diskette into that directory.
4. Delete the directory tree from the hard drive that you created in Activity 3.
5. Using **File Manager**, search for AutoSketch backup files (*.BAK) and delete them.

CHAPTER 21

AutoSketch and Windows 95

Objectives
After completing this chapter, you will be able to:
- Identify the differences between Windows 95 and Windows 3.1.
- Install AutoSketch for Windows in the Windows 95 environment.
- Use AutoSketch for Windows in Windows 95.
- Explain how to manage files in using Windows 95.

Windows 95 is an operating system designed to take advantage of the 32-bit architecture of 386, 486, and Pentium (Intel or compatible) computers. Windows 95 is very different from Windows 3.1 and Windows for Workgroups 3.11. This chapter will look at three aspects that will directly affect you when installing and using AutoSketch in the Windows 95 environment.

First, this chapter looks at how Windows 95 is different. Secondly, how to install AutoSketch and how it functions in the Windows 95 operating system are examined. Finally, managing files with Windows 95 is explained.

Note: Windows 3.1 and Windows for Workgroups 3.11 operate very similar to each other. Therefore, this chapter will explain how Windows 95 and Windows 3.1 relate to each other. However, in certain instances where Windows 3.1 and Windows for Workgroups 3.11 have different features or operate differently, how Windows 95 relates to Windows for Workgroups 3.11 will be explained as well.

How Is Windows 95 Different?

There are many features in Windows 95 that are new or operate differently than in Windows 3.1. While this text does not look at every new feature, the next sections give you an overview of the important changes you need to know to install and use AutoSketch, and to manage files.

Operating System
One of the biggest changes in Windows 95 is that it is an operating system, like DOS. What this means is that Windows 95 provides the basic instructions to the computer. With a DOS system, even though Windows 3.1 may start automatically when you turn on your computer, all of the basic instructions to the computer are provided by DOS. Windows 3.1 is simply a graphic interface running on top of DOS.

Windows 95 has a DOS compatibility box that allows you to run DOS programs. In fact, most DOS programs run faster in Windows 95. You also have the option of shutting down your computer and restarting in DOS mode.

A Different Look

The first change that you may actually *notice* is that Windows 95 *looks* different, Figure 21-1. There is no longer the **Program Manager** found in Windows 3.1. Instead, there is the **Task Bar** located along the bottom edge of the screen. Also, located in the top-left corner of the screen are two icons. These are called **My Computer** and **Recycle Bin**. **My Computer** has taken over some of the functions of **Program Manager** found in Windows 3.1. **Recycle Bin** is the undelete function of Windows 95. The **Task Bar**, **My Computer**, and **Recycle Bin** are covered later in this chapter.

Note: The names of the icons given in this chapter are the default names after installing Windows 95. If you have changed the names on your system, the icons will have the new names.

Figure 21-1.
Windows 95 looks and operates much different from Windows 3.1. Located along the bottom of the screen is the **Task Bar**. This is used to start or switch between programs. The **My Computer** icon performs other functions of **Program Manager** in Windows 3.1.

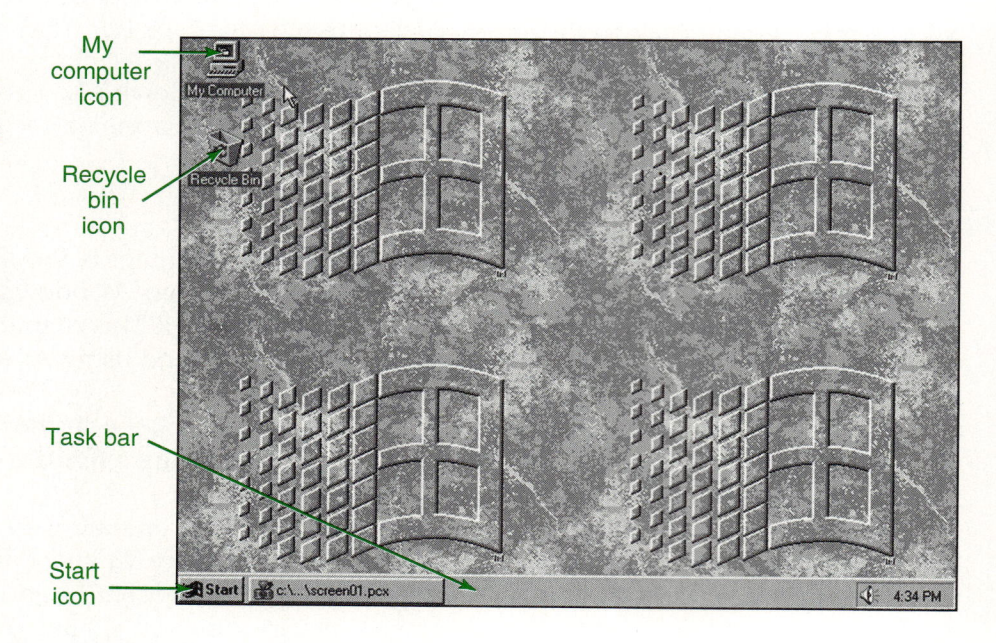

Managing Files and Installing Programs

Another change is that **File Manager** from Windows 3.1 has been replaced by **Windows Explorer**. This is explained in more detail later in the section called *Managing Files Using Windows 95*. Another change, as mentioned earlier, is how programs are installed in Windows 95. This is looked at in more detail in the section called *Installing AutoSketch in Windows 95*.

Starting Programs

A very big change in Windows 95 is how programs are launched (started). In Windows 3.1, you open the program group that contains the program icon and double-click on the icon. However, in Windows 95 you will notice that there are no program groups. To start a program, click on **Start** located on the **Task Bar** and select the appropriate item. This feature is looked at in more detail in the section called *Using AutoSketch in Windows 95*.

Multitasking

Another important change is in how you perform multitasking. *Multitasking* means running two or more programs at the same time. For example, you have AutoSketch and a desktop publishing program running. You create a simple illustration in AutoSketch and want to place it in a desktop layout. You "cut" the object from

AutoSketch, then switch to the desktop publishing program and "paste" the object. This is an example of multitasking.

In Windows 3.1, to switch between programs, hold down the [Alt] key and press the [Tab] key until the name of the program you want is displayed. Then, release the [Alt] key and that program is shown in the foreground.

However, switching between programs in Windows 95 is different. As mentioned earlier, there is the **Task Bar** located along the bottom edge of the screen. The **Task Bar** is always shown no matter what program is active, Figure 21-2. When a program is started, a button appears on the **Task Bar**. The button for the currently active program appears "pushed in," Figure 21-3. Other programs that are running also have a button on the **Task Bar**. These programs are referred to as *minimized*. This means that they are running, but are not the currently active program. To switch between programs in Windows 95, simply click on the appropriate button on the **Task Bar.** That program is made the currently active program.

The "[Alt][Tab]" function still exists in Windows 95. However, it operates differently from Windows 3.1. When you press and hold the [Alt] key, a box appears in the middle of the screen with an icon for each open program. The icon for the current program has a small box, or frame, around it. As you press the [Tab] key, the small box moves between the icons. When the box is around the icon for the program that you want, release the [Alt] key.

Figure 21-2.
The **Task Bar** located along the bottom of the screen (shown here highlighted) is always shown.

Figure 21-3.
The button on the **Task Bar** for the currently active program is "pushed in." To switch to another program that is minimized, simply click on its button. The **Start** icon is used to start other programs.

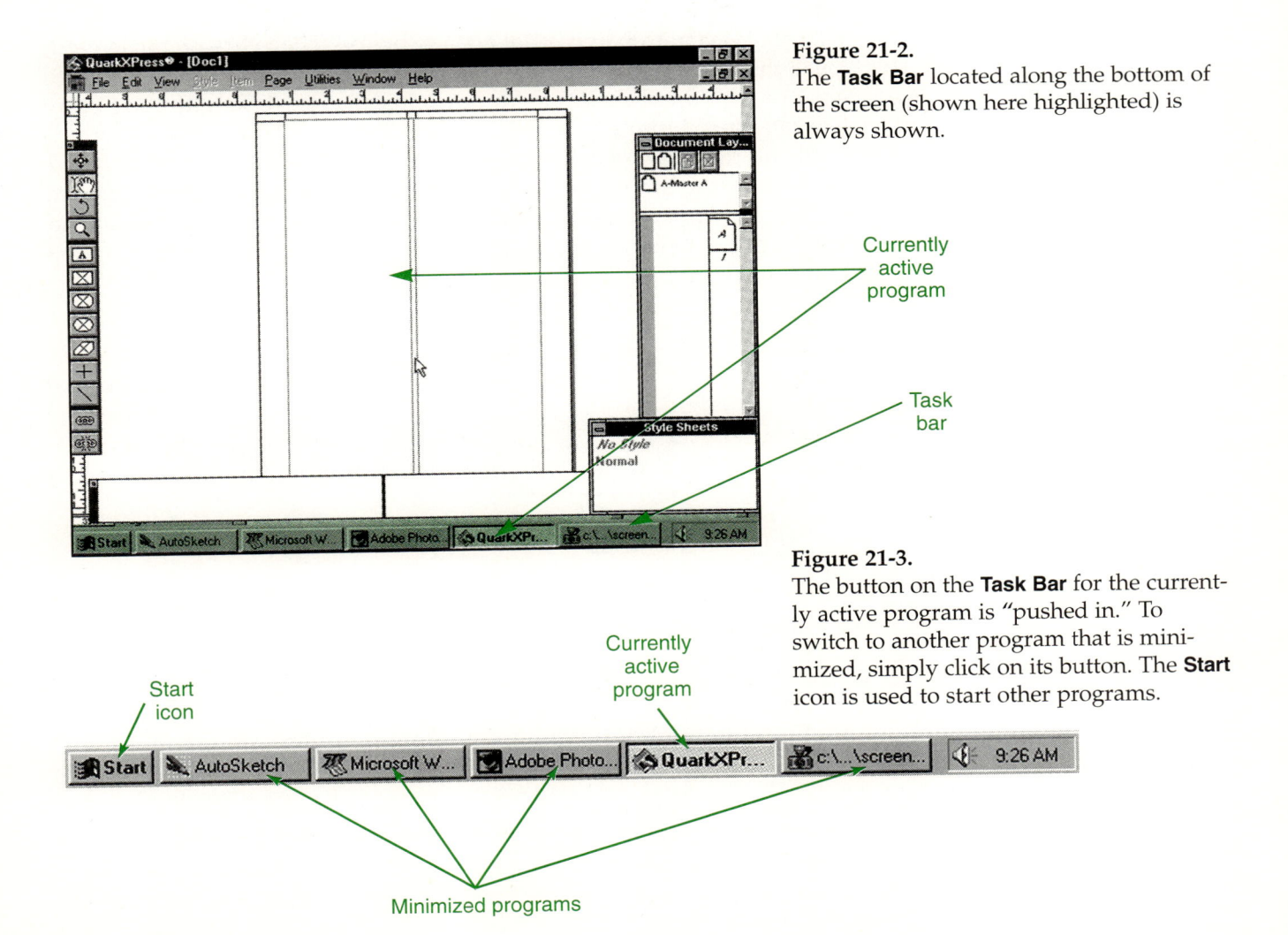

Installing AutoSketch in Windows 95

Installing new programs in Windows 95 is very different from Windows 3.1. Though the basics of "telling" the computer to install a new program from a floppy disk (or CD-ROM) remains, *how* you tell the computer to do this has changed. The following installation procedure is used for any program, including AutoSketch and the Symbol Library.

Installing a Program

To install a new program, first double-click on the **My Computer** icon found in the upper-left corner of the screen. This opens the **My Computer** window, Figure 21-4. In this window, you will notice several icons. (Icons are sometimes called *objects* in Windows 95.) There is **Control Panel, Printers, 3 1/2 Floppy [A:]**, **Hard drive [C:]**, and **[D:]**. Note: Your system may have different icons, depending on the setup.

Next, double-click on the **Control Panel**. This opens the **Control Panel** window, Figure 21-5. This window is similar to the **Control Panel** in Windows 3.1. The **Control Panel** is where you can change various system settings, such as the display, system sounds, and adding/removing printers.

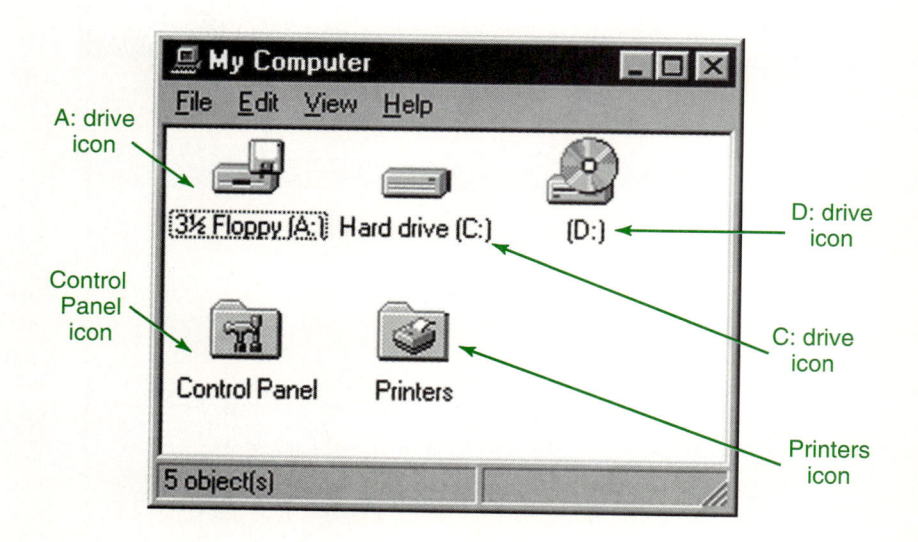

Double-click on the **Add/Remove Programs** icon. This opens the **Add/Remove Programs Properties** window, Figure 21-6. In this window, you will see three tabs along the top labeled **Install/Uninstall**, **Windows Setup**, and **Startup Disk**.

Click on the **Install/Unistall** tab, if it is not already on top. Then click on the **Install...** button. This opens the **Install Program From Floppy Disk or CD-ROM** window, Figure 21-7. Insert Disk 1 of the AutoSketch for Windows installation disks into the floppy drive and click on the **Next)** button.

Windows will automatically search for the installation program. When it finds the program, the path and program name are displayed in the **Run Installation Program** window, Figure 21-8. If this information is incorrect, click on the **Browse...** button and locate the correct program. Once the information is correct, click on the **Finish** button and follow the on-screen prompts.

Figure 21-4.
The **My Computer** window.

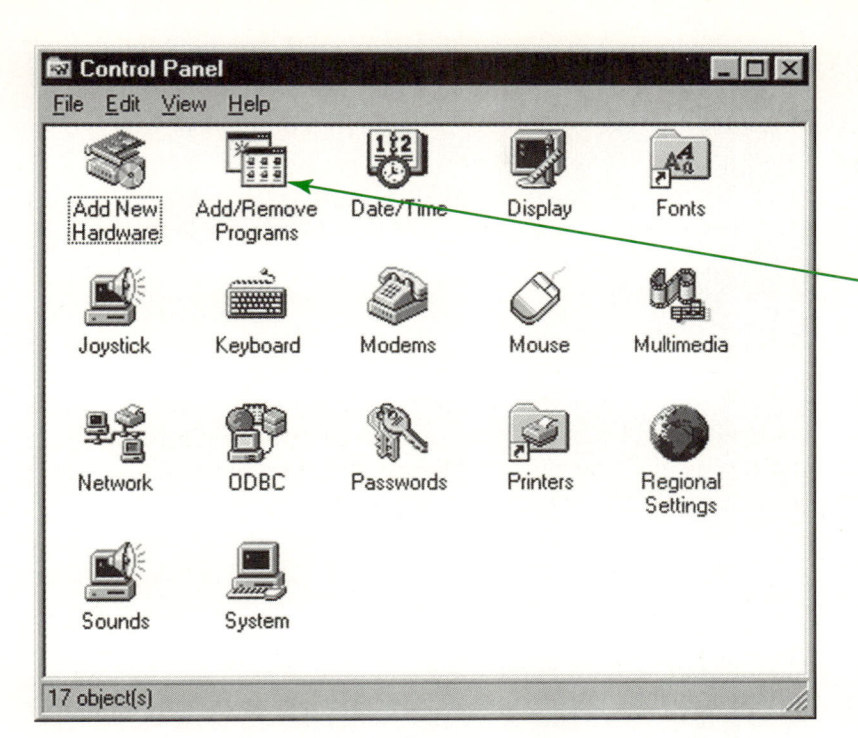

Figure 21-5.
Double-click on the **Add/Remove Programs** icon found in **Control Panel** to install a new program.

Add/remove programs icon

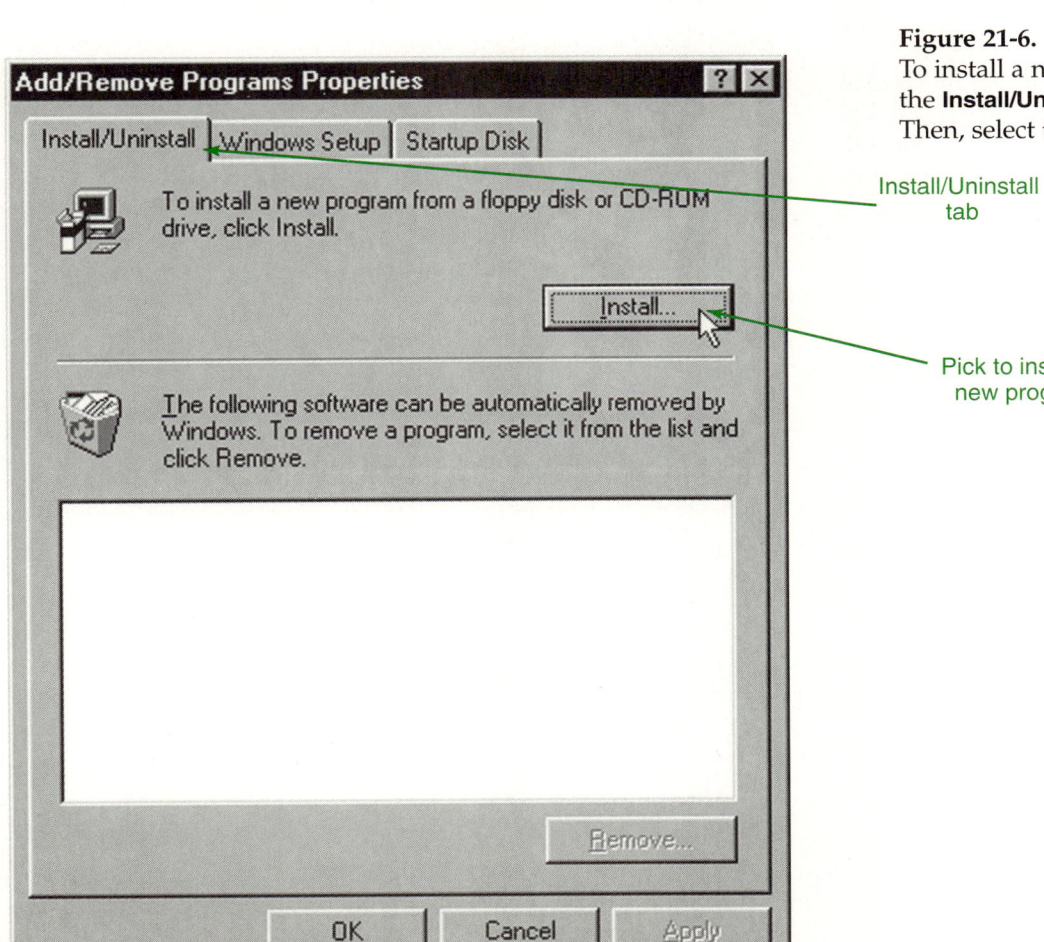

Figure 21-6.
To install a new program, make sure the **Install/Uninstall** tab is on top. Then, select the **Install...** button.

Install/Uninstall tab

Pick to install a new program

Figure 21-7.
The **Install Program From Floppy Disk or CD-ROM** window prompts you to insert the first installation disk. Then, pick the **Next...** button.

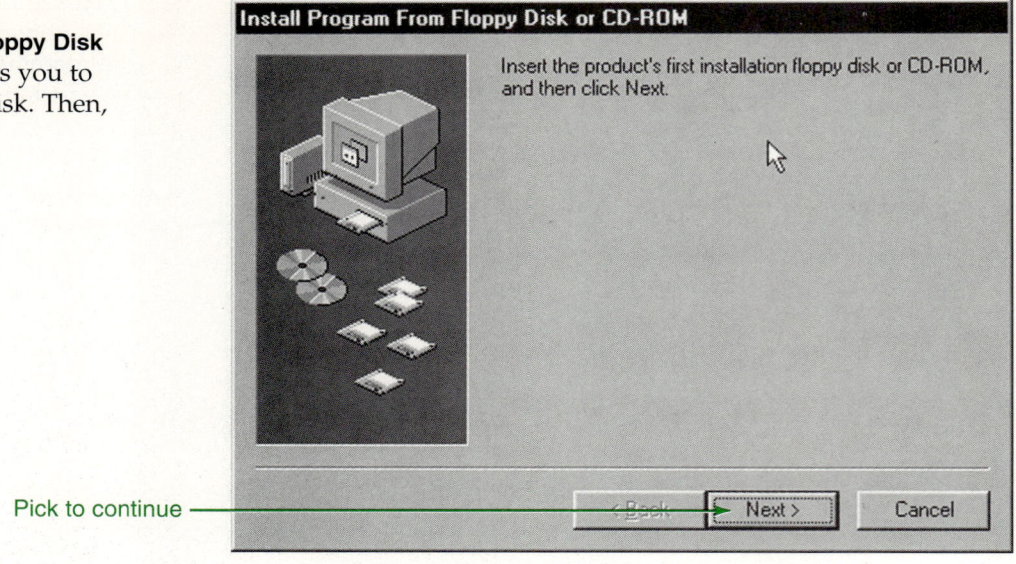

Pick to continue

Figure 21-8.
Windows 95 automatically searches the floppy and CD-ROM drive for an installation program. If the program it finds, shown in the **Command line for installation program:** box, is incorrect, pick the **Browse...** button and locate the correct program. Otherwise, pick the **Finish** button.

Installation program name and path

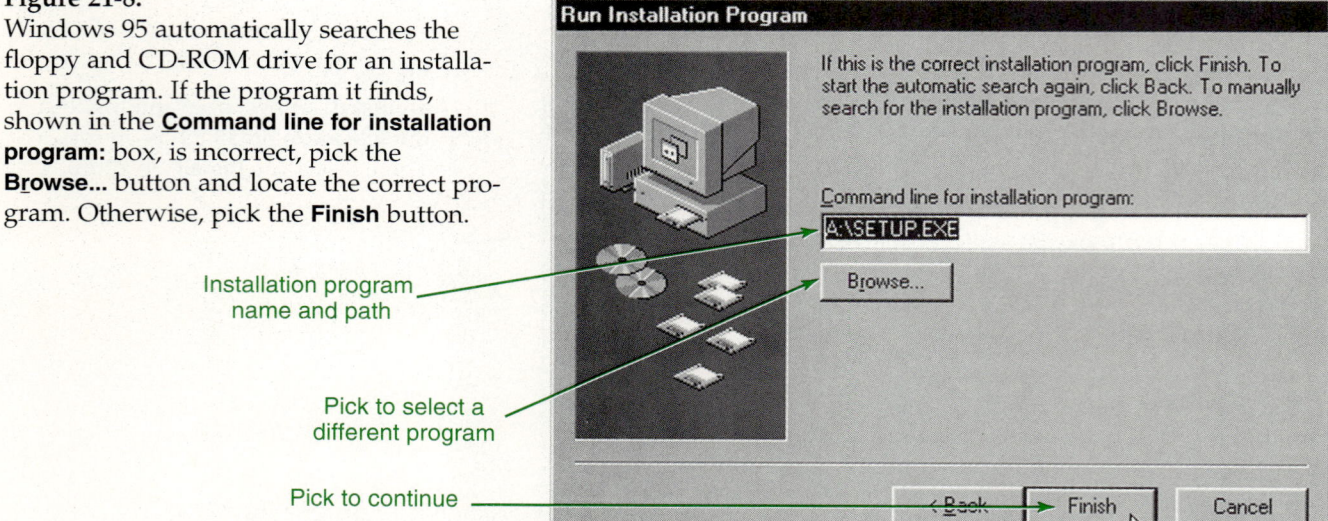

Pick to select a different program

Pick to continue

Differences

To install a program in Windows 3.1, you first have to select the **File** menu, then the **Run...** command. Then in the **Run** dialog box, you must manually enter the correct path and filename, or pick the **Browse...** button and manually search for the program.

Windows 95 automatically searches for the installation program. If Windows 95 has located the wrong program, you must search for the program manually. However, the "browsing" feature has changed for Windows 95, Figure 21-9. Select the proper drive and file type. Once the file is located, highlight it and pick the **Open** button.

Figure 21-9.
This window appears when you pick the **Browse...** button in the **Run Installation Program** window. Use the **Browse** window to locate the correct installation program. When you have located the correct program, highlight it and pick the **Open** button.

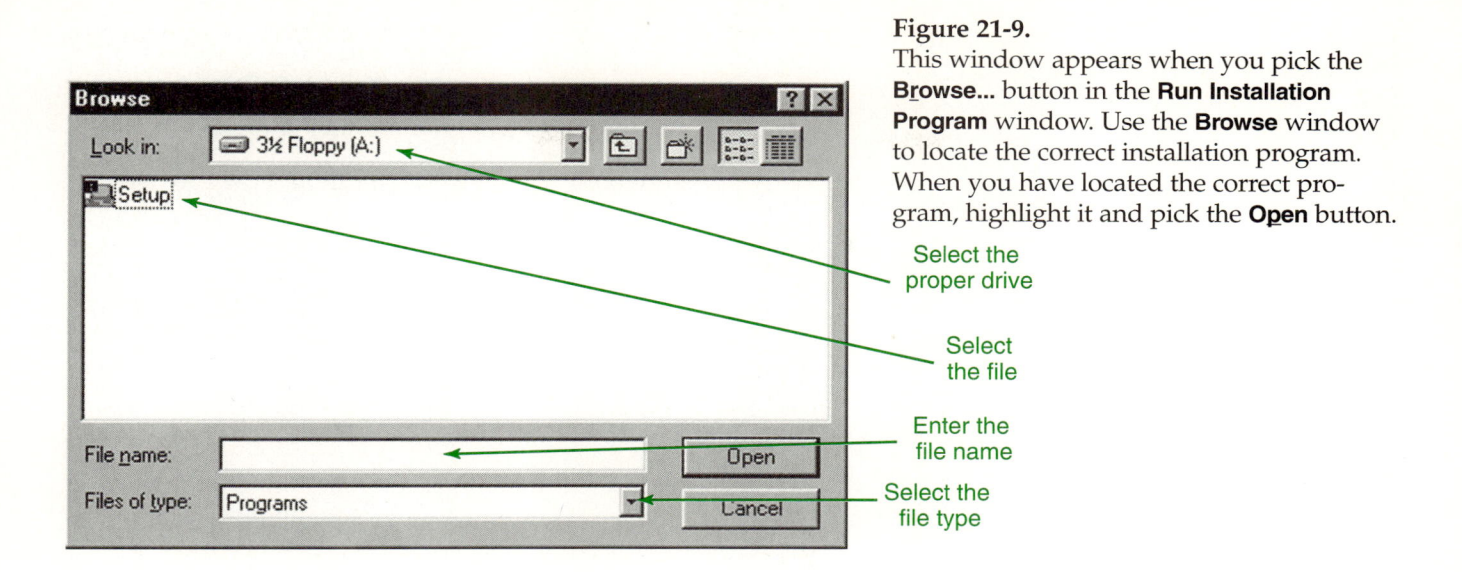

Using AutoSketch in Windows 95

The way that programs are *launched* (started) is different from Windows 3.1 Though the following section tells you how to start AutoSketch in Windows 95, the same procedure is used to start any program. In addition, Windows 95 slightly changes how AutoSketch works. This is covered in the section *How AutoSketch Operates in Windows 95*.

Starting AutoSketch in Windows 95

To start AutoSketch once it is installed, click on **Start** located on the **Task Bar**, Figure 21-10. This opens a pop-up menu with several selections, Figure 21-11. Some of the menu items include **Programs**, **Shut Down...**, and **Help**.

Figure 21-10.
To start a program, click on the **Start** icon located at the left of the **Task Bar.**

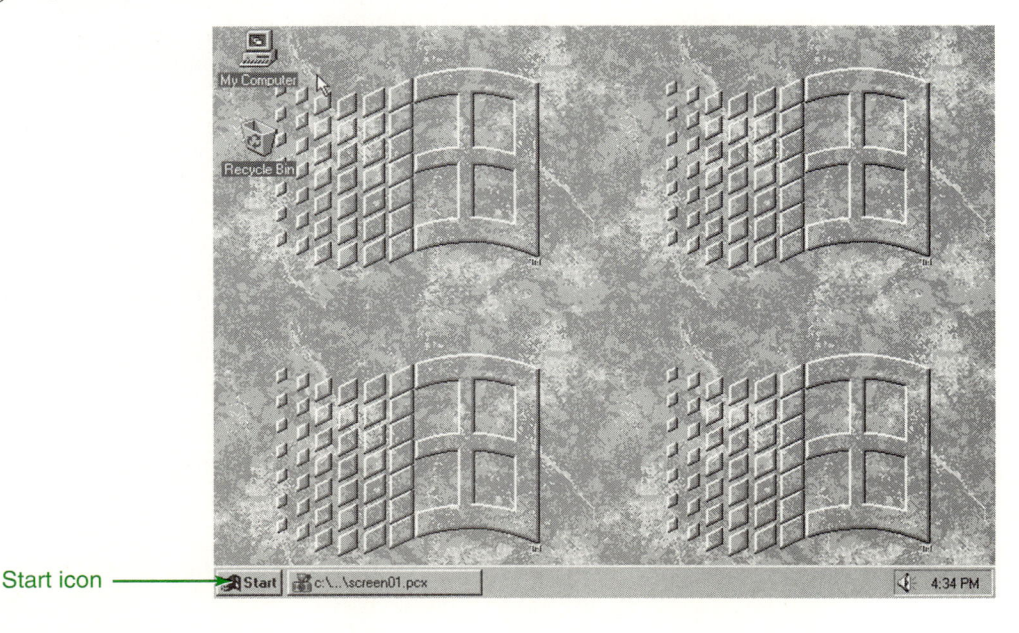

Figure 21-11.
After you click on the **Start** icon, this pop-up menu appears. To start a program, choose the **Programs** selection.

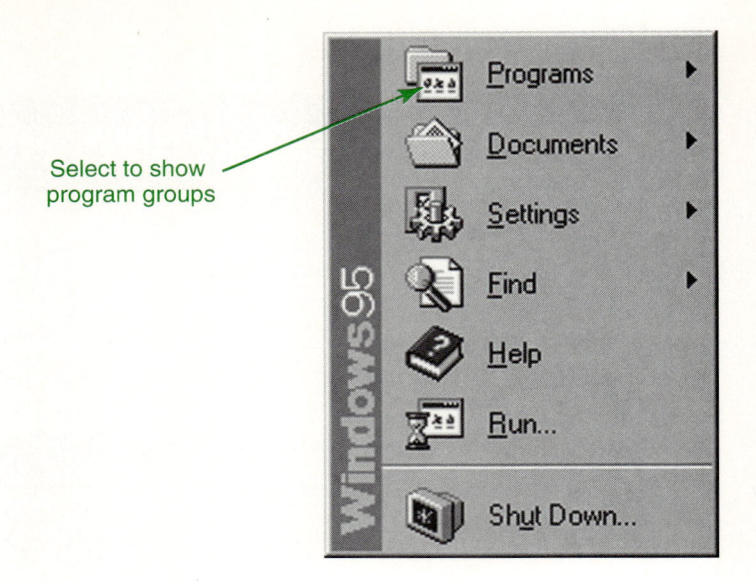

Select to show program groups

Figure 21-12.
This pop-up menu appears when you select **Programs** from the pop-up menu shown in Figure 21-11. The selections in this pop-up menu are similar to the program groups in Windows 3.1.

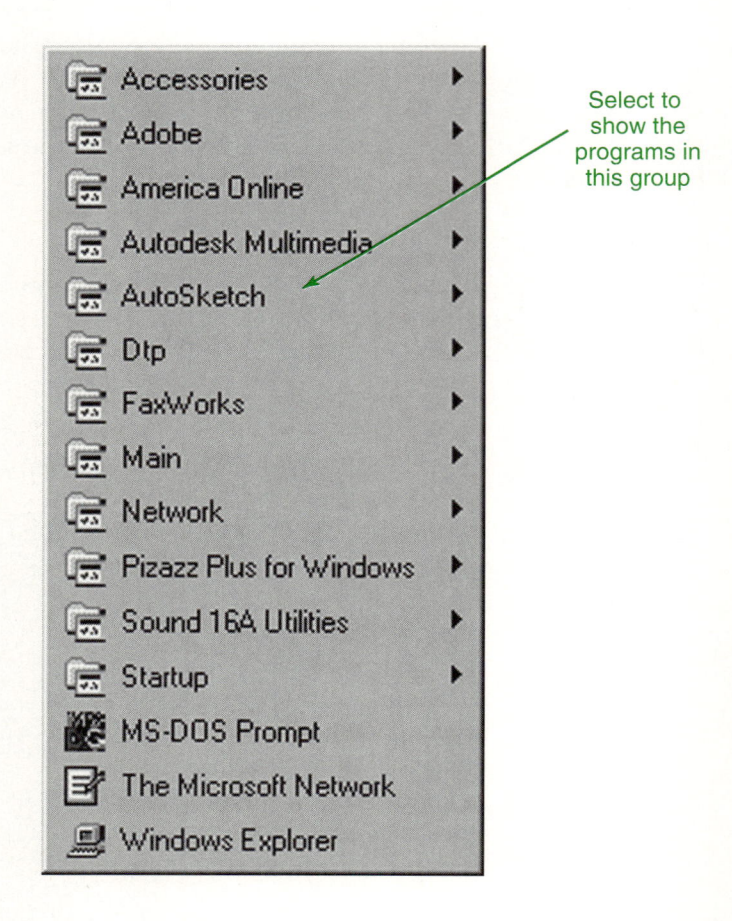

Select to show the programs in this group

Move your cursor to the **Programs** selection. You can click on the selection, or wait a second for Windows to automatically highlight it. Another pop-up menu appears, Figure 21-12. The items in this list are similar to the program groups in Windows 3.1. In fact, if you installed Windows 95 over an existing copy of Windows 3.1, the names of the program groups are carried over to Windows 95 and appear in this list.

Move your cursor to the **AutoSketch** selection, and another list appears, Figure 21-13. This list contains the names of the programs "under" the selection highlighted in the previous menu. In the case of the **AutoSketch** selection, there is only one program, so only one item is listed.

To start AutoSketch, move your cursor to the AutoSketch icon (selection) and click. You only have to click once. In Windows 3.1, you have to double-click on an icon to start the program.

Differences

To start a program in Windows 3.1, you have to first open the program group that contains the program's icon. Then, you must double-click on the program's icon to launch it.

To start a program in Windows 95, click on **Start** located on the **Task Bar** and select **Programs**. Then select the appropriate menu item and single-click on the appropriate program selection.

Figure 21-13.
When you choose a "group" in the pop-up menu shown in Figure 21-12, the programs in that group are shown. Single-click on the program that you want to start.

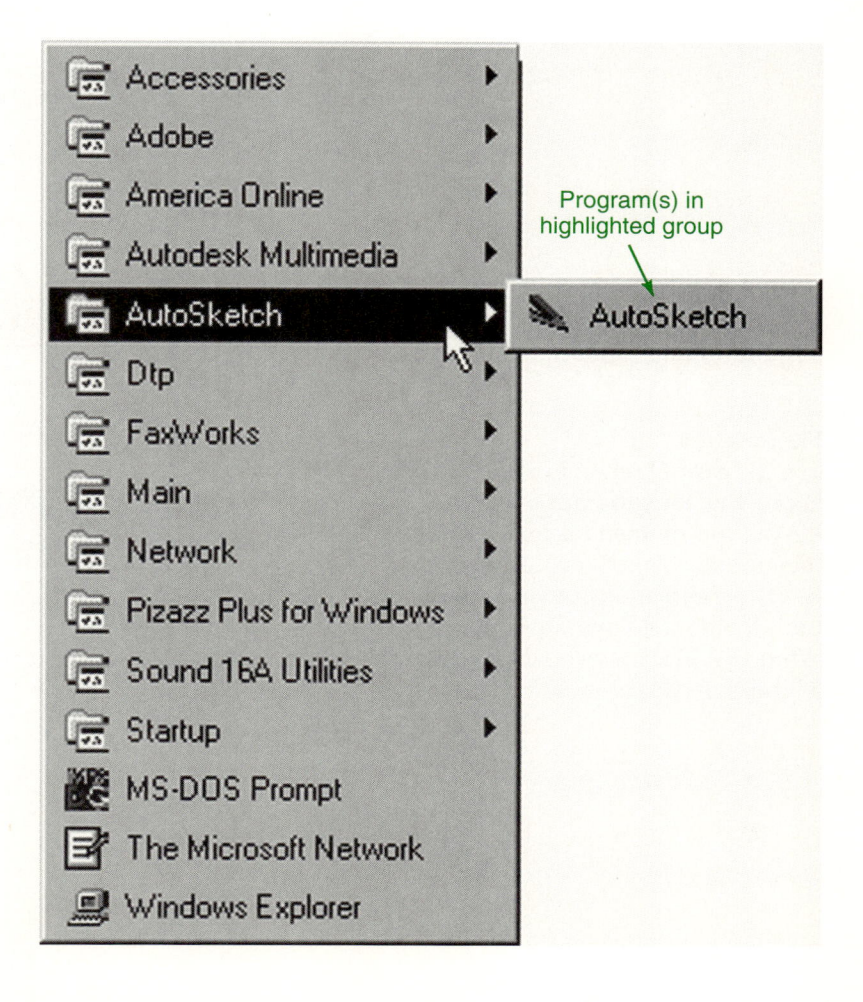

How AutoSketch Operates in Windows 95

AutoSketch operates nearly the same in Windows 95 as it does in Windows 3.1. In fact, the only real differences are in the visual appearance and a few extra tools that Windows 95 provides.

One of the first things you will notice when AutoSketch is started in Windows 95 is that the titles are aligned to the left, not centered, Figure 21-14. The next thing you will likely notice right away is the **Task Bar** at the bottom of the screen.

You will also notice that the file control menu appears as the AutoSketch icon, Figure 21-15. When you activate the menu, notice that the options are the same as those found in Windows 3.1, except that **S**w**itch To...** command has been removed. This is due to the addition of the **Task Bar** in Windows 95. The commands in the file control menu all operate the same. However, the menu looks slightly different. The file control menu for the drawing window also appears as the AutoSketch icon, Figure 21-16.

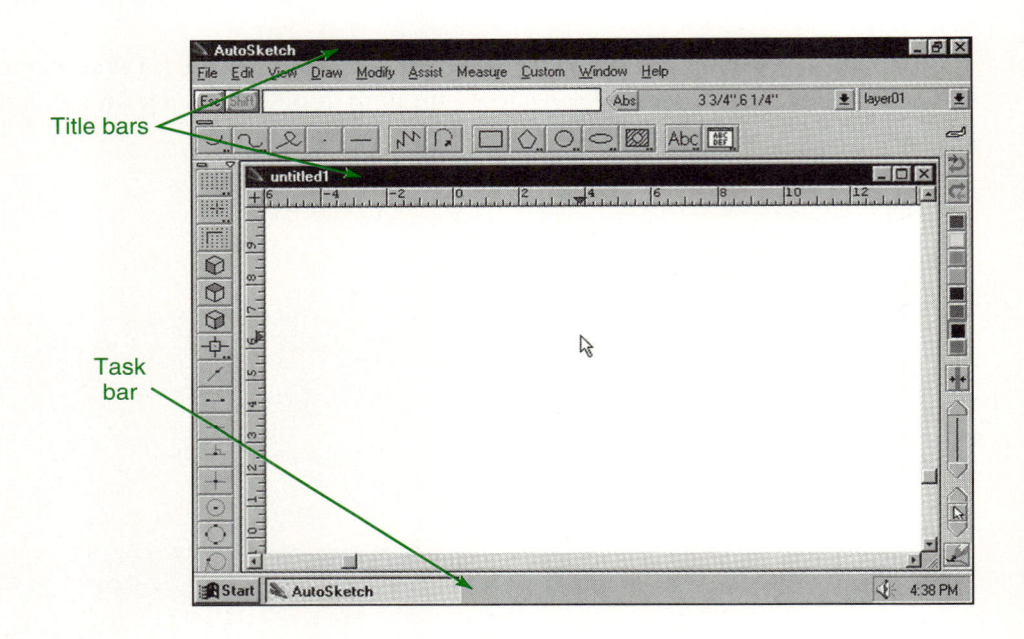

Figure 21-14.
AutoSketch performs the same in Windows 95 and Windows 3.1. However, there are additional tools provided by Windows 95, and the appearance is slightly different.

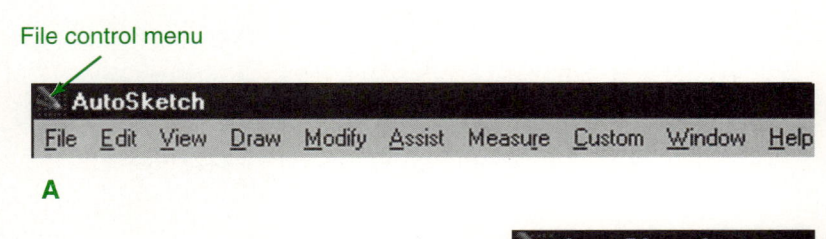

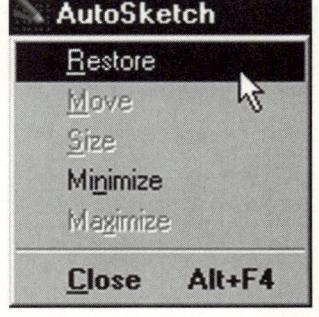

Figure 21-15.
The file control menu in Windows 95 appears as the program's icon. A—Here, the file control menu is for AutoSketch, therefore the AutoSketch icon is shown. B—The commands in the file control pull-down menu operate the same as in Windows 3.1. However, the menu looks a bit different.

The final thing to point out is the three buttons on the right side of the AutoSketch title bar. These are the minimize, maximize/window, and close application buttons, Figure 21-17. The minimize button has a short horizontal line on it. The maximize/window button has two overlapping windows on it. The close application button has an X on it.

By picking the maximize/window button, you can switch between a full screen and a window of AutoSketch. When AutoSketch is minimized, the AutoSketch button on the **Task Bar** no longer appears "pushed in." By single-clicking on the close application button, the program is closed.

Another aspect of Windows 95 is the ability to use long filenames. (Long filenames are discussed later in this chapter.) However, AutoSketch does not allow you to enter long filenames. When naming your drawing, you are limited to eight characters and the .SKD extension, just as in Windows 3.1.

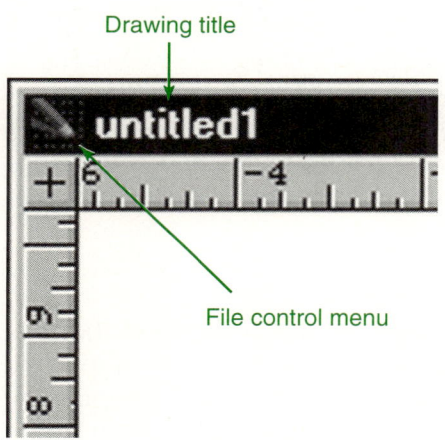

Figure 21-16.
The file control menu for the AutoSketch drawing windows also appears as the AutoSketch icon.

Figure 21-17.
The three buttons on the right side of the title bar are used to minimize, maximize/window, or close the application.

Differences

In Windows 3.1, the file control menu appears as a square with a horizontal line in it. On the right side of the AutoSketch title bar, there are two buttons. One is to minimize AutoSketch and the other is to maximize/window AutoSketch. In addition, the titles for Windows 3.1 are centered.

In Windows 95, the file control menus appear as the AutoSketch icon. Also, there are three buttons on the right side of the AutoSketch title bar for minimizing/maximizing/closing AutoSketch. Finally, titles are aligned to the left of the title bar in Windows 95.

Placing AutoSketch on Your Desktop

The *desktop* is what appears when Windows 95 starts. The **Task Bar**, **My Computer** icon, and **Recycle Bin** icon are all on the desktop. You can also add shortcuts to your desktop. A *shortcut* simply tells the computer to open a file. A shortcut can appear in **Windows Explorer** as a file, or it can be placed on the desktop as an icon. The following procedure can be used to create a shortcut for any program. Placing a shortcut on your desktop is also explained.

Creating a shortcut

First, start **Windows Explorer** using the **Start** button located on the **Task Bar.** To create a shortcut for AutoSketch, highlight SKETCH.EXE from the \WSKETCH directory. Then, select **Create Shortcut** from the **File** menu. A file appears in the \WSKETCH directory with the name Shortcut to WSKETCH.EXE. If you double-click on this file, AutoSketch is launched.

Placing a shortcut on the Desktop

To place a shortcut on your desktop, first highlight the file in **Windows Explorer**. Then, drag the shortcut to the Desktop in **Windows Explorer** (Desktop is the top of the tree in **Windows Explorer**). When you close **Windows Explorer**, an icon with the name Shortcut to WSKETCH.EXE is on the desktop. When you double-click on this icon, AutoSketch is launched.

Renaming a shortcut

If you wish to rename the shortcut, you can select it in **Windows Explorer** and rename it using the **Rename** command in the **File** menu. This can be done either before or after you have "dragged" the shortcut to the Desktop.

Managing Files Using Windows 95

The **File Manager** in Windows 3.1 has been replaced in Windows 95 with **Windows Explorer**. **Windows Explorer** operates much differently than **File Manager**. The next sections cover how to manage your files using **Windows Explorer**.

Overview of Windows Explorer

To start **Windows Explorer**, click on **Start** on the **Task Bar**. Then select **Programs** and **Windows Explorer** (located at the bottom of the list), Figure 21-18. The **Windows Explorer** window opens with the title **Exploring** and the name of the current drive or directory, Figure 21-19. Also, an **Exploring** button appears on the **Task Bar**. Note: Directories are often called *folders* in Windows 95.

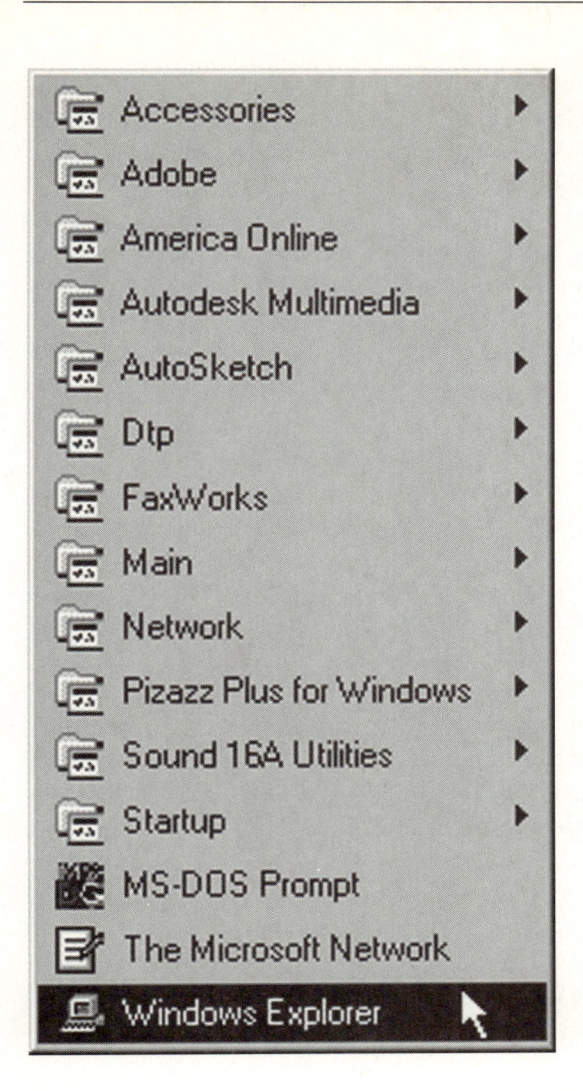

Figure 21-18.
To launch **Windows Explorer**, select **Start**, **Programs**, then **Windows Explorer**.

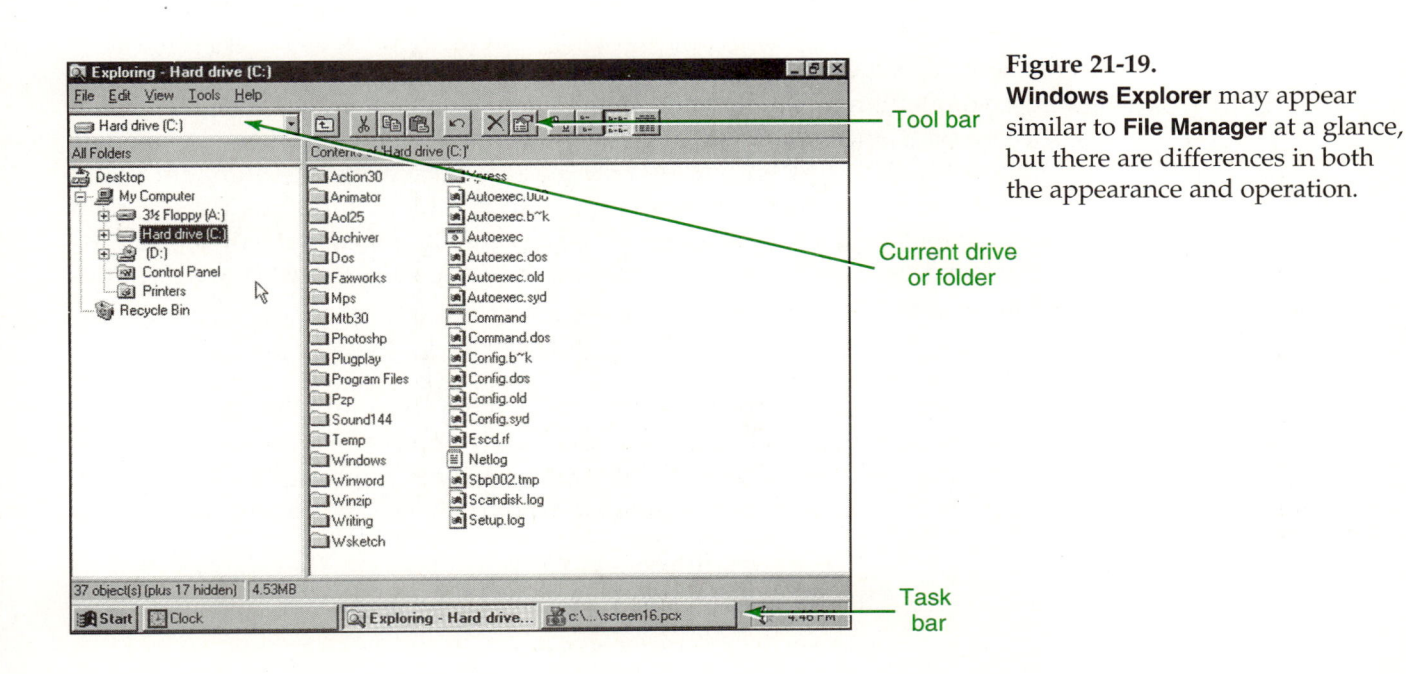

Figure 21-19.
Windows Explorer may appear similar to **File Manager** at a glance, but there are differences in both the appearance and operation.

You will notice that there is a tree similar to the one found in **File Manager**. However, where the top of the tree in **File Manager** is the root directory of a drive (such as the hard drive), in **Windows Explorer** the top of the directory is called **Desktop**. Also, the first "branches" are **My Computer** and **Recycle Bin**.

Under **My Computer**, the next branches are the disk drives installed on your computer. The floppy drive, hard drive, and CD-ROM drive (if installed) are listed. You can double-click on the drive you want to view to expand the directories on that drive.

You will also notice a toolbar along the top of the window, much like the one found in Windows for Workgroups 3.11. On the toolbar, you will find a button with a file folder having an up arrow on it. This can be used to move up one level in the tree.

Differences

In Windows 3.1, **File Manager** shows a directory tree for the current drive. At the top of the tree is the root directory. Different drives can be viewed at the same time by opening multiple windows.

In Windows 95, **Windows Explorer** also shows a tree, but the top of the tree is **Desktop**. The tree contains all of the disk drives installed on your computer. You can expand the tree on a drive by double-clicking on its icon.

Selecting Files

To select a single file, simply highlight that file with your cursor. To select several files in a series, highlight the first file, hold down the [Shift] key, and then select the last file in the series. The entire series is highlighted. To select several individual files, hold down the [Ctrl] as you select the files.

Differences

Files are selected in the same way in both Windows 3.1 and Windows 95. There have been no changes to this function. The [Ctrl] and [Shift] keys are still used to select groups of files.

Copying and Moving Files

To copy a file to a different location on the same drive, highlight the file and hold down the [Shift] key as you drag it to the location where you want the copy. To copy a file to a different drive, simply highlight the file and drag it to the location on the other drive where you want it. In either case, you will notice a small square with a plus in it next to the cursor as you drag the file.

To move a file to a different location on the same drive, simply highlight it and drag it with the cursor to the new location. To move a file to a different drive, hold down the [Shift] key as you drag it. In either case, you will notice that as you drag the file, only the filename and icon are dragged with the cursor. There is no box with a plus in it that follows the cursor.

Differences

The major differences between copying/moving files in **File Manager** and **Windows Explorer** is in the visual appearance and menu options. In addition, formatting a disk is not done in **Windows Explorer**, but in **Control Panel.** For complete information on the menu options, see the documentation that comes with Windows 95. Formatting a disk is covered later in the section called *Formatting a Disk.*

Deleting and Undeleting Files

To delete a file, highlight the file and click on the **Delete** button on the toolbar. You are then asked Are you sure you want to send *filename* to the Recycle Bin?. Pick the **Yes** button. You can also select a group of files and delete them all at the same time.

To undelete a file, minimize **Windows Explorer** by selecting the button in the upper-right corner with a horizontal bar on it. Then, double-click on the **Recycle Bin** icon on the main Windows 95 screen. This opens the **Recycle Bin** window, Figure 21-20. Highlight the file or files you want to undelete and select **Restore** from the **File** menu. The file is restored to its original location and removed from the **Recycle Bin.**

Deleted files are not kept in the **Recycle Bin** forever. If you "empty" the **Recycle Bin**, the files are permanently erased. Also, if your hard drive has most of the space used, new files that are saved may overwrite files in the **Recycle Bin**. In addition, if you defragment your hard drive, the files in the **Recycle Bin** will be lost.

Differences

The undelete function in Windows 3.1 is located in the **File** menu of **File Manager**. In Windows for Workgroups 3.11, the undelete function can also be placed as a tool on the toolbar. To use this command, simply select the file(s) and select the command. You will then be asked to provide the first letter of the filename, as DOS substitutes a ? for the first letter. In addition, you can only delete files that were in the current directory.

The undelete function in Windows 95 is separate from **Windows Explorer** and called **Recycle Bin**. To undelete a file, open **Recycle Bin.** Then select the file(s) and pick **Restore** from the **File** menu. You do not need to enter the first letter of the filename, as Windows 95 retains the entire filename and location.

SHORTCUT
To quickly delete a file or group of files, simply highlight them and press the [Delete] key.

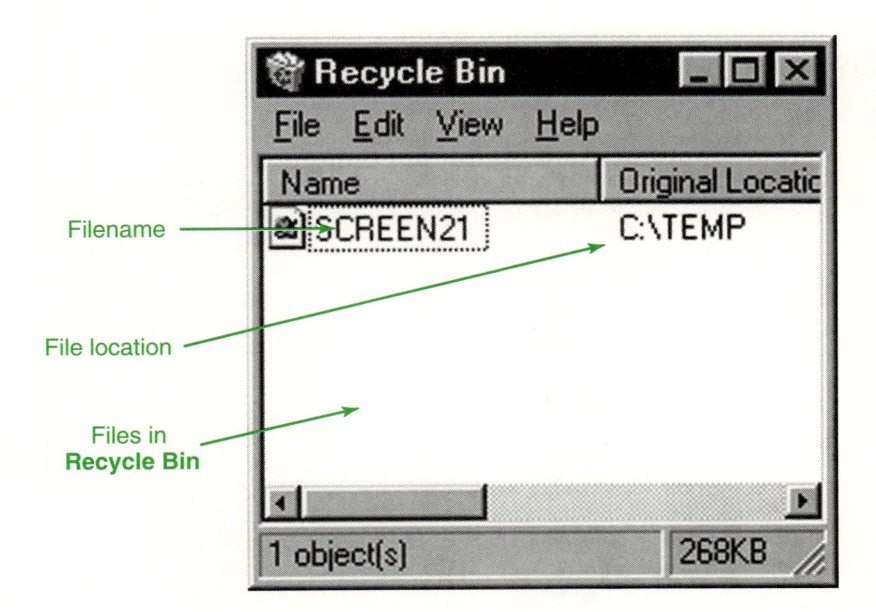

Figure 21-20.
The undelete function in Windows 95 is **Recycle Bin**. All deleted files are moved here, no matter the original location. **Recycle Bin** keeps the entire filename and location.

Formatting a Disk

To format a disk in Windows 95, single-click on **3 1/2 Floppy Drive [A:]** found in My Computer (if the disk is in the A: drive). The icon is highlighted. Then, select **Format...** from the **File** menu. This opens the **Format - 3 1/2 Floppy [A:]** window (if you are using the A: drive), Figure 21-21.

Select the density (**Capcity:**) and format type. You can also add a label to the disk or make it a system disk (**Copy system files**). When you have made the appropriate selections, click on the **Start** button. A meter appears at the bottom of the window showing the progress of the format.

> **SHORTCUT**
> If you don't know what is on the disk, double-click on the drive icon. This will show any files that may be on the disk. If there are no files on the disk, this window will be blank. Then, close the window and proceed with the format.

Differences

To format a disk in Windows 3.1, select **Format Disk...** from the **Disk** menu in **File Manager.** Then, specify the drive and density of the disk. You also have the option of labeling the disk or making it a system disk.

To format a disk in Windows 95, first open **Control Panel**. Then single-click on the drive icon where your disk is. Select **Format...** from the **File** menu. Finally, select the appropriate options and start the formatting.

Filenames

In Windows 95, you can use long filenames. *Long filenames* can contain up to 255 characters, including spaces. In addition, the file extension can also be 255 characters long. However, a filename cannot contain \ ? : " < > | as these are reserved characters.

Note: AutoSketch does *not* make use of long filenames in Windows 95. When naming a drawing, you still are limited to eight characters and the .SKD extension.

Differences

Since Windows 3.1 uses DOS as the operating system, the rules that apply to DOS filenames also apply to Windows 3.1. In DOS, a filename can only contain eight characters, a period, and a three character extension. In addition, a filename cannot contain \ ? : " < > | as these are reserved characters.

Figure 21-21.
When formatting a disk, select appropriate options and click on the **Start** button.

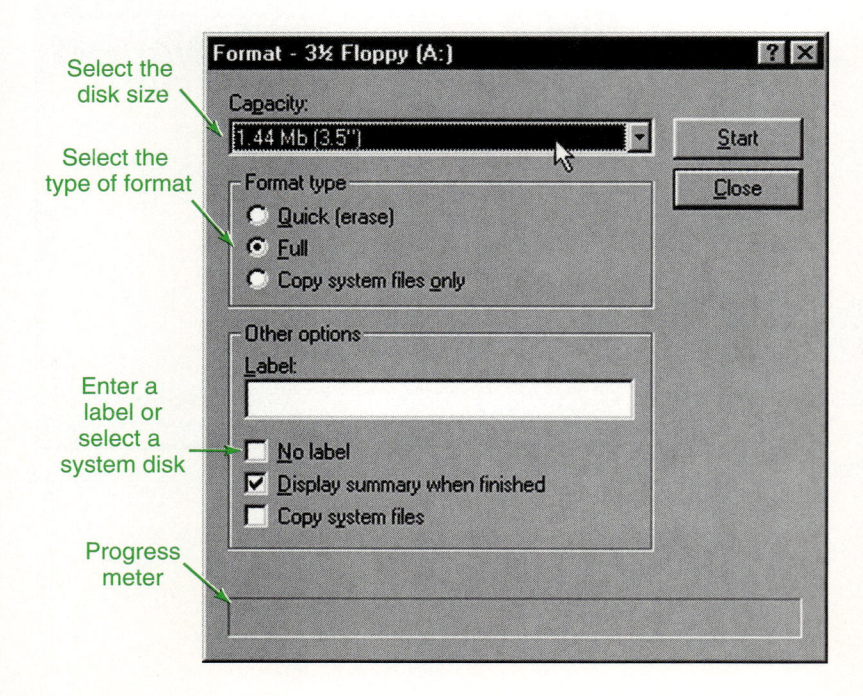

Summary

Windows 95 is an operating system designed to take advantage of the 32-bit architecture of most computers. Since Windows 95 is an operating system, it provides the basic commands to the computer just as DOS does.

There are many features that are new or operate differently in Windows 95. The **Task Bar** and **Start** button have replaced **Program Manager**. The **File Manager** in Windows 3.1 is replaced in Windows 95 with **Windows Explorer.** Window control buttons have a different appearance. Also, to switch tasks you only need to click on the application button on the **Task Bar**. The procedure for installing new programs in Windows 95 has changed. Finally, a big boost for file management in Windows 95 is the ability to use long filenames.

Important Terms

Desktop
Folders
Launched

Long filenames
Minimized
Multitasking

Objects
Shortcut

New Tools and Commands

3 1/2 Floppy Drive [A:]
Add/Remove Programs
  Properties window
Add/Remove Programs
Browse... button
Cap*c*ity:
Control Panel
Copy system files
Delete button
Desktop

Finish button
Form*a*t...
Format - 3 1/2 Floppy [A:]
  window
Install... button
Install Program From Floppy
  Disk or CD-ROM window
Install/Uninstall tab
My Computer
Open button

Programs
Recycle Bin
Restore
Run Installation Program
  window
Start
Startup Disk tab
Task Bar
Windows Explorer
Windows Setup tab

Review Questions

Give the best answer for each of the following questions.

1. What is multitasking? _____

2. True or false. When using Windows 95, DOS provides the basic commands to the computer.
3. Where is the **Install/Remove Programs** icon located? _____

4. True or false. When installing a program, Windows 95 will automatically look for an installation program on your floppy or CD-ROM drive.

5. How do you launch a program in Windows 95? _____
6. What program is used in Windows 95 to manage files? _____
7. Explain the two ways that can be used to multitask in the Windows 95 environment.
   A. _____

   B. _____

8. How do you format a disk in Windows 95? _____

9. How many characters can a filename have in Windows 95? _____
10. How do you undelete a file in Windows 95? _____

Activities

Here is a mixture of activities to develop your skills using Windows 95. Do not save files on disk drives where you are not authorized to do so. This section assumes that you have Windows 95 installed on your computer.

1. Find a program that you can install on your computer (such as the AutoSketch Symbol Library). Install that program in Windows 95. Note how the Windows 95 installation process is different from Windows 3.1.
2. Using Windows 95, format an unformatted disk, or a disk with files you no longer need.
3. Using **Windows Explorer**, create a directory on the hard drive using your name. Copy the entire directory tree from your work diskette into that directory.
4. Using **Windows Explorer**, create directories on the diskette from Activity 2 for each chapter in this book. Move your drawings for each chapter into the appropriate directory.
5. Delete the directory tree from the hard drive that you created in Activity 3.

Appendix A

Function Keys and Key Combinations

This chart shows different function keys and key combinations and the resulting action. These keys and combinations are also identified throughout the text in the "Shortcuts."

Key Combination	Function	Key Combination	Function	Key Combination	Function
F1	Help	Alt + F1	Line	Ctrl + F1	Arc Mode toggle
F2	Undo	Alt + F2	Redo	Ctrl + F2	Box Array
F3	Erase	Alt + F3	Arc	Ctrl + F3	Mirror
F4	Break	Alt + F4	Exit	Ctrl + F4	Close window
F5	Move	Alt + F5	Ortho toggle	Ctrl + F5	Rotate
F6	Copy	Alt + F6	Grid toggle	Ctrl + F6	Switch windows
F7	Stretch	Alt + F7	Snap toggle	Ctrl + F7	Box
F8	Pan	Alt + F8	Attach toggle	Ctrl + F8	Ellipse
F9	Zoom Last	Alt + F9	Group	Ctrl + F9	Pattern Fill
F10	Zoom Box	Alt + F10	Ungroup	Ctrl + F10	Macro-User Input
F11	Print				
F12	Save				

Appendix B

AutoSketch Fonts

This chart shows the .SKD fonts supplied with AutoSketch. TrueType fonts can also be used with AutoSketch. The TrueType fonts available will vary from system to system. These fonts are supplied by Windows and other applications, and not AutoSketch. Double-click on the Fonts icon located in the Windows **Control Panel** to see what TrueType fonts are installed on your system.

COMPLEX				CYRILLIC				CYRILTLC				GOTHICE				GOTHICG				GOTHICI			
A	a	1	[	А	а	1	Ъ	А	а	1	Ъ	A	a	1	[	A	a	1	[	A	a	1	[
B	b	2	]	Б	б	2	ь	Б	б	2	ь	B	b	2	]	B	b	2	]	B	b	2	]
C	c	3	}	В	в	3	Ч	В	ч	3	ь	C	c	3	}	C	c	3	}	C	c	3	}
D	d	4	{	Г	г	4	ь	Д	д	4	Ъ	D	d	4	{	D	d	4	{	D	d	4	{
E	e	5	\	Д	д	5	Ы	Е	е	5	Ы	E	e	5	\	E	e	5	\	E	e	5	\
F	f	6	|	Е	е	6	ы	Ф	ф	6	ы	F	f	6	|	F	f	6	|	F	f	6	|
G	g	7	;	Ж	ж	7	;	Г	г	7	;	G	g	7	;	G	g	7	;	G	g	7	;
H	h	8	:	З	з	8	:	Х	х	8	:	H	h	8	:	H	h	8	:	H	h	8	:
I	i	9	'	И	и	9	'	И	и	9	'	I	i	9	'	I	i	9	'	I	i	9	'
J	j	0	"	Й	й	0	"	Щ	щ	0	"	J	j	0	"	J	j	0	"	J	j	0	"
K	k	·	,	К	к	Я	,	К	к	Я	,	K	k	·	,	K	k	·	,	K	k	·	,
L	l	~	.	Л	л	э	.	Л	л	э	.	L	l	~	.	L	l	~	.	L	l	~	.
M	m	!	<	М	м	!	ю	М	м	!	ю	M	m	!	<	M	m	!	<	M	m	!	<
N	n	@	>	Н	н	@	я	Н	н	@	я	N	n	@	>	N	n	@	>	N	n	@	>
O	o	#	/	О	о	#	/	О	о	#	/	O	o	#	/	O	o	#	/	O	o	#	/
P	p	$	?	П	п	$	?	П	п	$	?	P	p	$	?	P	p	$	?	P	p	$	?
Q	q	%		Р	р	%		Ц	ц	%		Q	q	%		Q	q	%		Q	q	%	
R	r	^		С	с	Э		Р	р	Ю		R	r	^		R	r	^		R	r	^	
S	s	&		Т	т	&		С	с	&		S	s	&		S	s	&		S	s	&	
T	t	*		У	у	*		Т	т	*		T	t	*		T	t	*		T	t	*	
U	u	(		Ф	ф	(		У	у	(		U	u	(		U	u	(		U	u	(	
V	v	)		Х	х	)		В	в	)		V	v	)		V	v	)		V	v	)	
W	w	–		Ц	ц	–		Ш	ш	–		W	w	–		W	w	–		W	w	–	
X	x	_		Ч	ч	Ю		Ж	ж	э		X	x	_		X	x	_		X	x	_	
Y	y	=		Ш	ш	=		Й	й	=		Y	y	=		Y	y	=		Y	y	=	
Z	z	+		Щ	щ	+		З	з	+		Z	z	+		Z	z	+		Z	z	+	

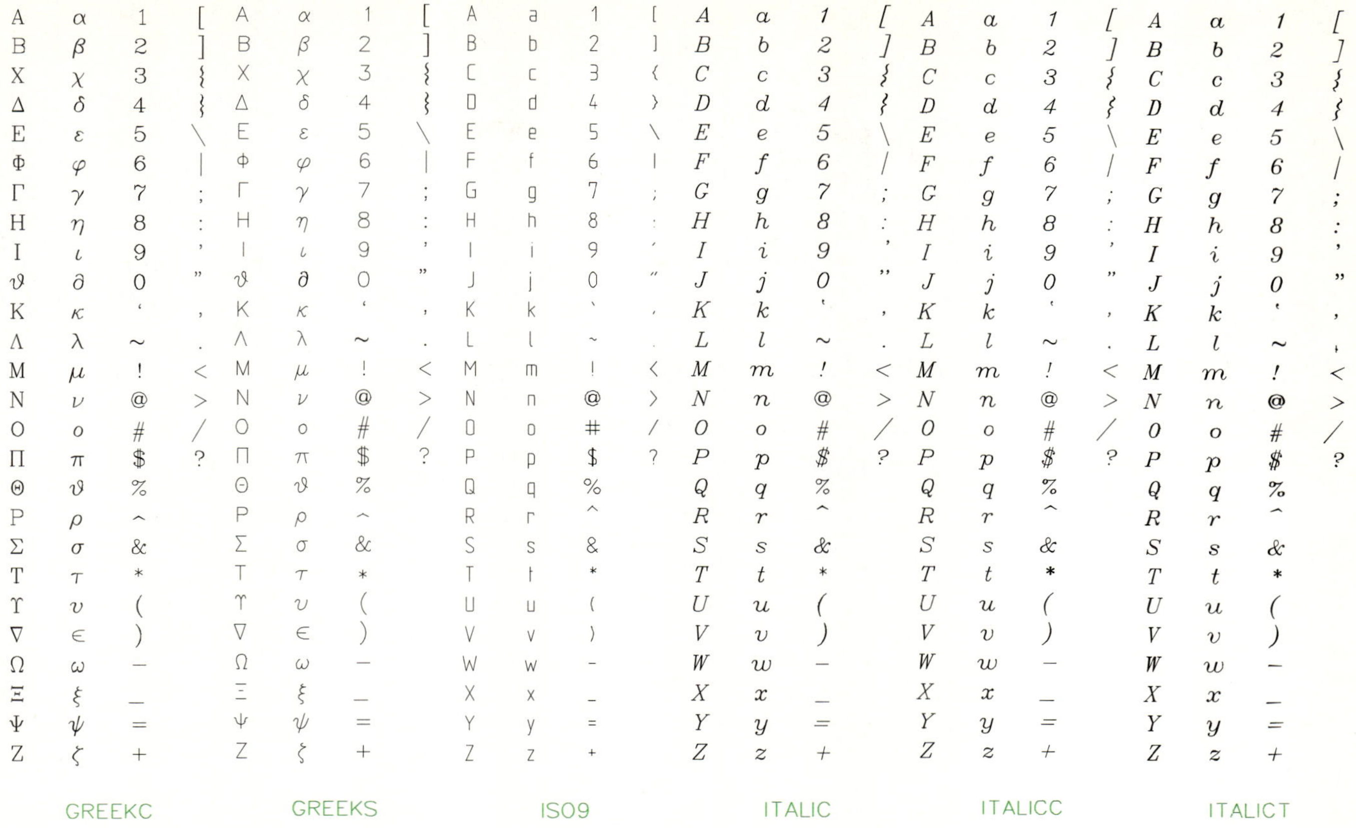
GREEKC
GREEKS
ISO9
ITALIC
ITALICC
ITALICT

MONOTEXT
ROMANC
ROMAND
ROMANS
ROMANT
SCRIPTC

SCRIPTS SIMPLEX STANDARD SYASTRO SYMAP

SYMATH SYMETEO SYMUSIC TXT

Appendix C

AutoSketch Pattern Fills

These are the pattern fills supplied with AutoSketch.

five	flemish	flemishx	flex	geol1	geol2	geol3	geol4
flgstone	flortile	four	garden	geol5	geol6	geol7	geol8
glasbloc	glass	glass2	grass	griddot	grtstone	hex	honey
grass2	grate	gravel	grecodek	hound	insul	lath	lattice1
lattice2	lava	line	log-log	mosaic	mudst	mudstone	net
log	logend	melbury	moncello	net3	nine	octagons	octasqr
one	panel	parquet	parquet1	parquet6	parquet7	paving2	pavingbr
parquet2	parquet3	parquet4	parquet5	phantsqr	pipe	plast	plasti
plywood	randtile	rattan	rigidins	scallop1	scallop2	scrnbloc	secgrill
rock	rubble	sacncr	sand	semi-log	seven	shadeblk	shingles
siding	six	skirting	soil	square	stagger	stars	steel
solid	spanish	spantile	sqshngle	stones	swamp	texture	textured
three	tile1	trans	treadplt	weave	woodface	zero	zigzag
triang	two	vbricks	water				

Appendix D

AutoSketch Pull-down Menus

The arrangement of the pull-down menus for AutoSketch, and the commands found in each menu.

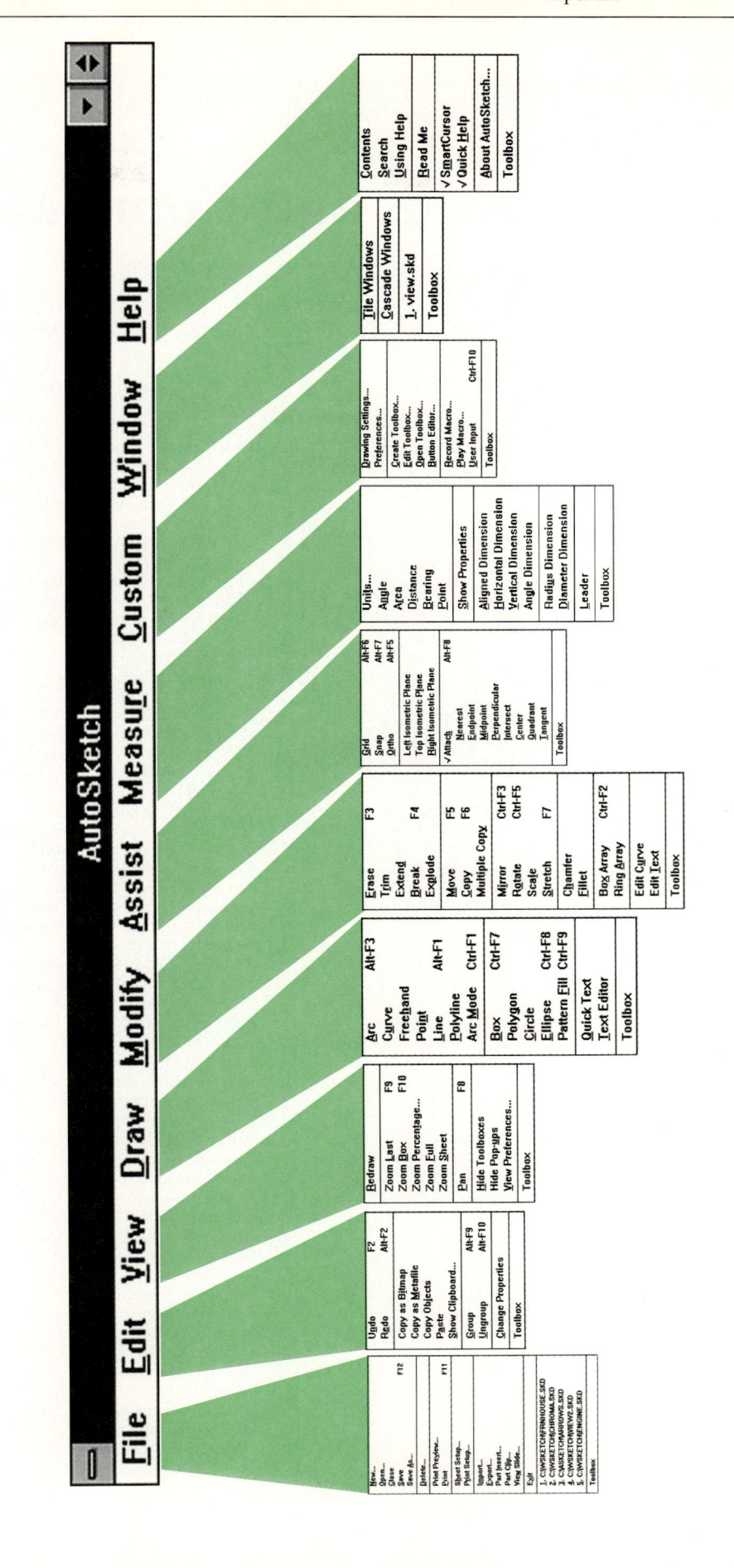

AutoSketch

File	Edit	View	Draw	Modify	Assist	Measure	Custom	Window	Help
New...	Undo F2	Redraw	Arc Alt-F3	Erase F3	Grid Alt-F6	Units...	Drawing Settings...	Tile Windows	Contents
Open...	Redo Alt-F2	Zoom Last F9	Curve	Trim	Snap Alt-F7	Angle	Preferences...	Cascade Windows	Search
Close		Zoom Box	Freehand	Extend	Ortho Alt-F5	Area	Create Toolbox...	1. view.skd	Using Help
Save	Copy as Bitmap	Zoom Percentage...	Point	Break F4		Distance	Edit Toolbox...	Toolbox	Read Me
Save As... F12	Copy as Metafile	Zoom Full F10	Line Alt-F1	Explode	√ Attach Alt-F8	Bearing	Open Toolbox...		√ SmartCursor
Delete...	Copy Objects	Zoom Sheet	Polyline		Nearest	Point	Button Editor...		√ Quick Help
Print F11	Paste		Arc Mode Ctrl-F1	Move F5	Endpoint		Record Macro...		About AutoSketch...
Print Preview...	Show Clipboard...	Pen F8		Copy F6	Midpoint	Show Properties	Play Macro...		Toolbox
Sheet Setup...		Hide Toolboxes	Box Ctrl-F7	Multiple Copy	Perpendicular	Aligned Dimension	User Input Ctrl-F10		
Print Setup...	Group Alt-F9	Hide Pop-ups	Polygon		Intersect	Horizontal Dimension	Toolbox		
	Ungroup Alt-F10	View Preferences...	Circle	Mirror Ctrl-F3	Center	Vertical Dimension			
Import...	Change Properties	Toolbox	Ellipse Ctrl-F8	Rotate Ctrl-F5	Quadrant	Angle Dimension			
Export...	Toolbox		Pattern Fill Ctrl-F9	Scale	Tangent				
Part Insert...			Fillet	Stretch F7	Toolbox	Radius Dimension			
Part Clip...				Chamfer		Diameter Dimension			
View Slide...			Quick Text						
Exit			Text Editor	Box Array Ctrl-F2		Leader			
				Ring Array		Toolbox			
1. C:\WSKETCH\FRMHOUSE.SKD			Toolbox	Edit Curve					
2. C:\WSKETCH\DIPOMA5.SKD				Edit Text					
3. C:\WSKETCH\ARROWS.SKD									
4. C:\WSKETCH\WREWZ.SKD				Toolbox					
5. C:\WSKETCH\ENGINE.SKD									
Toolbox									

Appendix E

Macro Variables and Commands

The first chart gives set variables for AutoSketch macros. The second chart gives AutoSketch command names and what their functions are.

Set Variables

Variable	Type	Description
ANSIDXF	Boolean	Read and write ANSI DXF and DWG files
ANSIMCR	Boolean	Read and write ANSI macro files
ANSITXT	Boolean	Read and write ANSI text files
ARCHPREC	Integer	Architectural precision: 0 = 1", 1 = 1/2", 2 =1/4", 3 = 1/8", 4 = 1/16", 5 = 1/32", 6 = 1/64"
ARCMODE	Boolean	Arc Mode on/off
ARCTYPE	Integer	Type of arc to be drawn: 0 = Start point, Point, Endpoint, 1 = Start point, Endpoint, Point 2 = Start point, Center, Endpoint, 3 = Center, Start point, Endpoint
ARRAYAROUND	Real	Ring Array angle to fill
ARRAYCDIST	Real	Box Array column distance
ARRAYCP	Boolean	Set Ring Array center point by pointing on/off
ARRAYCW	Boolean	Ring array clockwise on/off
ARRAYCX	Real	Ring array center X coordinate
ARRAYCY	Real	Ring array center Y coordinate
ARRAYFILLC	Boolean	Divide column distance by number of columns on/off
ARRAYFILLR	Boolean	Divide row distance by number of rowson/off
ARRAYICP	Boolean	Set Ring Array pivot point by pointing on/off
ARRAYNOCOLUMNS	Integer	Box array, number of columns
ARRAYNOITEM	Integer	Ring array, number of items
ARRAYNOROWS	Integer	Box array, number of rows
ARRAYPOINTC	Boolean	Prompt for array column distance on/off
ARRAYPOINTR	Boolean	Prompt for array row distance on/off
ARRAYRDIST	Real	Box array row distance
ARRAYROT	Boolean	Rotate objects during ring array on/off
ARRAYROTANG	Real	Ring Array rotation angle
ARROWSIZE	Real	Dimension arrow size
ATTACH	Boolean	Attach mode on/off
ATTACHCENTER	Boolean	Attach to Center on/off
ATTACHEND	Boolean	Attach to Endpoints on/off
ATTACHINTERSECTION	Boolean	Attach to Intersection on/off
ATTACHMIDPOINT	Boolean	Attach to Midpoint on/off
ATTACHNEAREST	Boolean	Attach nearest on/off
ATTACHNODE	Boolean	Attach to Node point on/off
ATTACHPERPEDICULAR	Boolean	Attach to Perpendicular on/off
ATTACHQUADRANT	Boolean	Attach to Quadrant points on/off
ATTACHTANGENT	Boolean	Attach to Tangent on/off
AVIEWCENTERX	Real	Center of Aerial View window, X coordinate
AVIEWCENTERY	Real	Center of Aerial View window, Y coordinate
AVIEWMODE	Integer	Aerial View mode number 0 = View Extents, 1 =View Limits, 2 = View Box
AVIEWSIZE	Real	Aerial View window size
BACKGROUNDCOLORMODE	Integer	Background color type: 0 = White background, 1 = Black background, 2 = Light gray back ground, 3 = Blue background, 4 = Windows system background
BITMAPPATFILENAME	String	Bitmap fill pattern filename
BUTTONDIR	String	Location of button bitmap files
BUTTONFILENAME	String	Button filename
BUTTONHIGHLIGHTCOLOR	Integer	Button highlighting color
BUTTONINVERT	Boolean	Invert button color on/off

Apendix 575

Set Variables

Variable	Type	Description
CAPTIONHELP	Boolean	Caption Help on/off
CHAMFERA	Real	Chamfer distance A
CHAMFERB	Real	Chamfer distance B
CHPCOLOR	Boolean	Change properties: change color
CHPDIMARROW	Boolean	Change properties: change dimension arrow
CHPDIMTXTPT	Boolean	Change properties: change dimension text placement
CHPDIMUNITS	Boolean	Change properties: change dimension units
CHPFONT	Boolean	Change properties: change text font
CHPLAYER	Boolean	Change properties: change layer
CHPLINETYPE	Boolean	Change properties: change linetype
CHPPATFILL	Boolean	Change properties: change fill pattern
CHPPOLYWIDTH	Boolean	Change properties: change polyline width
CHPTEXT	Boolean	Change properties: change text size, alignment, etc
CIRCLETYPE	Integer	How circle is to be drawn: 0 = Center, Radius, 1 = Center, Diameter, 2 = Two-point, 3 =Three-point
COLOR	Integer	Color number
COORDTYPE	Integer	Coordinate display type: 0 = Absolute coordinates, 1 = Polar coordinates, 2 = Relative coordinates, 3 = Last Point coordinates, 4 = No coordinates
CROSSHAIRS	Boolean	Crosshairs on/off
CROSSHAIRSIZE	Integer	Crosshair size index
CURVECLOSED	Boolean	Closed curve on/off
CURVESTYLE	Integer	Style of curve to be drawn: 1 = B-spline, 2 =Bezier
CURVETYPE	Integer	How curve is to be drawn: 1 = Polyline frame, 2 = Fitted to points
CUSTOMUNITFACTOR	Real	Scale of custom units to meters
CUSTOMUNITTYPE	Integer	Unit type of custom unit factor
DECIPREC	Integer	Decimal precision
DEFAULTTEMPLATEFILENAME	String	Default template filename
DIMARROW	Integer	Dimension arrow style: 0 = Standard, 1 = Solid, 2 = Tick, 3 = Dot, 4 = None
DIMARROWUNITTYPE	Integer	Unit type for dimension arrow size
DIMNUMBER	Boolean	Numerical Dimension on/off
DIMRADIALCENTERMARK	Boolean	Center mark for radial dimension on/off
DIMRADIALINTERIOR	Boolean	Draw radial dimension line as well as leader
DIMRADIALNOCENTERMARK	Boolean	Suppress center mark for radial dimension
DIMRADIALNOTED	Boolean	Suppress radial dimension line
DIMTEXT	String	Dimension text
DIMTXTABOVE	Boolean	Align dimension text above the dimension line on/off
DIMTXTALIGN	Boolean	Align dimension text with dimension line on/off
DIMTXTHEIGHT	Real	Dimension text height
DIMTXTHEIGHTUNITTYPE	Integer	Unit type for dimension text height
DIMTXTHORIZ	Boolean	Align dimension text horizontal on/off
DIMTXTLEADER	Boolean	Place dimension text on a leader line on/off
DIMTXTWITHIN	Boolean	Align dimension text within the dimension line on/off
DIMUNIT	Boolean	Display units in dimensions on/off
DISPLAYUNIT	Boolean	Display units on/off
DRAWINGDESCRIPTION	String	Drawing description
DRAWINGDIR	String	Drawing directory name
DRAWINGFILENAME	String	Drawing filename
DRAWNFROM	Integer	Which side a wide entity is to be drawn from: -1 = left side, 0 = center, +1 = right side
DXFATTRIMPORT	Boolean	Import visible attributes from DXF file on/off
DXFDIR	String	DXF directory name
DXFEXPLODE	Boolean	Explode large blocks during DXF reading on/off
DXFFILENAME	String	DXF filename
DXFPRECISION	Integer	Precision for next ASCII DXF file exported
ELLIPSETYPE	Integer	How ellipse is to be drawn: 0 = Center and Both Axes, 1 = Axis and Planar Rotation, 2 = Two Foci and Point
ENTHILIGHTON	Boolean	Highlight selected entities on/off
FILLETRAD	Real	Fillet radius
FILLMODE	Boolean	Fill/Outline pattern fill entities on/off
GRID	Boolean	Grid on/off
GRIDX	Real	Grid X spacing
GRIDY	Real	Grid Y spacing

Set Variables

Variable	Type	Description
INSBASEX	Real	Insertion base point X coordinate
INSBASEY	Real	Insertion base point Y coordinate
ISOMODE	Boolean	Isometric mode on/off
ISOPLANE	Integer	Data type is Short Isometric plane: 0 = Left, 1 = Top, 2 = Right
LASTOLDFONTFILENAME	String	Filename for the last font of the noncurrent font type
LEADERARROW	Integer	Leader arrow style: 0 = Standard, 1 = Solid, 2 = Tick, 3 = Dot, 4 = None
LIMITXMAX	Real	Maximum X drawing limit value
LIMITXMIN	Real	Minimum X drawing limit value
LIMITYMAX	Real	Maximum Y drawing limit value
LIMITYMIN	Real	Minimum Y drawing limit value
LINESCALE	Real	Scale of all linetypes
LINETYPE	Integer	Line style: 0 = Solid, 1 = Dashed, 2 = Hidden, 3 = Center, 4 = Phantom, 5 = Dot, 6 = Dashdot, 7 = Border, 8 = Divide, 9 = Dots
LOCKGRIDTOSNAP	Boolean	Lock grid size to snap size
MACRODIR	String	Macro directory name
MACROFILENAME	String	Macro filename
MEASUREMESSAGES	Boolean	Display measured values in alerts on/off
OLDDWGUNITTYPE	Integer	Last drawing unit type
OLDFONTDIR	String	Last font directory name
ORTHO	Boolean	**Ortho** tool on/off
PALETTECOLOR1	Integer	Top color palette box color number
PALETTECOLOR2	Integer	Second color palette box color number
PALETTECOLOR3	Integer	Third color palette box color number
PALETTECOLOR4	Integer	Fourth color palette box color number
PALETTECOLOR5	Integer	Fifth color palette box color number
PALETTECOLOR6	Integer	Sixth color palette box color number
PALETTECOLOR7	Integer	Seventh color palette box color number
PALETTECOLOR8	Integer	Eighth color palette box color number
PARTDIR	String	Part directory
PARTFILENAME	String	Part filename
PATALIGN	Boolean	Set fill pattern alignment point by pointing on/off
PATALIGNPTX	Real	Fill pattern alignment point X coordinate
PATALIGNPTY	Real	Fill pattern alignment point Y coordinate
PATANGLE	Real	Fill pattern angle
PATDIR	String	Fill pattern directory
PATFILENAME	String	Fill pattern filename
PATSCALE	Real	Fill pattern scale
PFBLANK	Boolean	Blank fill polylines on/off
PFDBLHATCH	Boolean	Double-hatch pattern fill
PFPATFILL	Boolean	Pattern fill polylines on/off
PFSOLID	Boolean	Solid fill polylines on/off
PICKBOX	Boolean	Show Selection Area Box on/off
PICKBOXSIZE	Integer	Pickbox size index
PICKDELTA	Integer	Size of Selection Area Box as a percentage of screen height
PLINEWIDTH	Real	Polyline width
POLYGONSIDES	Integer	Number of sides in a polygon
POLYGONTYPE	Integer	How polygon is to be drawn: 1 = Inscribed, 2 = Circumscribed, 3 = Edge
POPUPS	Boolean	Show/Hide pop-up toolboxes on/off
PRINTDIR	String	Print file directory
PRINTROTATED	Boolean	Rotate printout 90 degrees on/off
PRINTSIZE	Integer	How to print the view: 0 = Fit to sheet, 1 = Fit to printer page, 2 = Use sheet scale
PRINTVIEWTYPE	Integer	View to print: 0 = Drawing Extents, 1 = Current Sheet
QUICKSELECTLOCK	Boolean	Shift-style selection editing on/off
ROTATEPART	Boolean	Rotate part during insertion on/off
RULERS	Boolean	Display Rulers on/off
SAVETOTEMPLATE	Boolean	Save drawing as a template on/off
SCALEPART	Boolean	Scale Part during insertion on/off
SELECTLOCK	Boolean	Selection set editing on/off
SHEETCUSTOMSIZE	Real	Custom sheet width in meters
SHEETCUSTOMXUNITTYPE	Integer	Custom sheet width unit type

Apendix 577

Set Variables

Variable	Type	Description
SHEETCUSTOMYSIZE	Real	Custom sheet height in meters
SHEETCUSTOMYUNITTYPE	Integer	Custom sheet height unit type
SHEETNAME	String	Sheet name
SHEETPAGETILINGX	Integer	Number of page tiles in width required by current sheet
SHEETPAGETILINGY	Integer	Number of page tiles in height required by current sheet
SHEETPAPERTYPE	Integer	Sheet size table index
SHEETSCALE	Real	Sheet scale
SHEETSCALEINDEX	Integer	Sheet scale table index
SHEETSCALESHEETUNITS	Real	Custom sheet units in meters
SHEETSCALESHEETUNITTYPE	Integer	Custom sheet unit type
SHEETSCALETYPE	Integer	Standard or custom sheet scale: 1 = Standard scale, 2 = Custom scale, 3 = Drawing Extents scale
SHEETSCALEWORLDUNITS	Real	Custom world units in meters
SHEETSCALEWORLDUNITTYPE	Integer	Custom world unit type
SHEETSIZETYPE	Integer	Standard or custom sheet size: 1 = standard size, 2 = custom size
SHEETSIZEX	Real	Sheet width
SHEETSIZEY	Real	Sheet height
SHOWALLSHEETS	Boolean	Display all sheets on/off
SHOWPAGETILING	Boolean	Display page tiling marks on/off
SHOWSHEETTEXT	Boolean	Display sheet information text on/off
SHOWTEMPLATEDIALOG	Boolean	Display Select Template dialog box after New on/off
SHXFONTDIR	String	Text font directory
SHXFONTFILENAME	String	Font filename
SLIDEDIR	String	Slide directory
SLIDEFILENAME	String	Slide filename
SMARTCURSOR	Boolean	**SmartCursor** on/off
SNAP	Boolean	Snap on/off
SNAPX	Real	Snap X spacing
SNAPY	Real	Snap Y spacing
SPLINEFRAME	Boolean	Show Curve frame on/off
STANDARDUNIT	Integer	Last noncustom sheet unit type
STARTUPSCREEN	Boolean	Display startup screen on/off
TEMPLATEDIR	String	Template directory name
TEXTANGLE	Real	Angle to draw line of text
TEXTEDITDIR	String	**Text Editor** import/export directory
TEXTFONTTYPE	Integer	Current font type: 1 = SHX font, 2 = TrueType font
TEXTHEIGHT	Real	Text character height
TEXTHEIGHTUNITTYPE	Integer	Unit type for text height
TEXTINPOINTS	Boolean	Text size is specified in points
TEXTJUSTCENTER	Boolean	Center text on/off
TEXTJUSTLEFT	Boolean	Left justify text on/off
TEXTJUSTMIDDLE	Boolean	Middle justify text on/off
TEXTJUSTRIGHT	Boolean	Right justify text on/off
TEXTMODE	Boolean	Draw/outline text entities on/off
TEXTOBLIQUEANG	Real	Angle to slant SHX text characters
TEXTTTBOLD	Boolean	Boldface TrueType text on/off
TEXTTTITALIC	Boolean	Italicize TrueType text on/off
TEXTTTSTRIKEOUT	Boolean	Strike out TrueType text on/off
TEXTTTUNDERLINE	Boolean	Underline TrueType text on/off
TEXTWIDTH	Real	Text character width
TOOLBOXDIR	String	Toolbox directory
TOOLBOXES	Boolean	Show/hide all toolboxes on/off
TOOLBOXFILENAME	String	Toolbox filename
TTFONTFILENAME	String	TrueType font filename
UNDODIR	String	Undo file directory
UNDOON	Boolean	Undo on/off
UNITS	Integer	Units style: 0 = Decimal units, 1 = Fractional units
WIDTH1UNITTYPE	Integer	Unit type of width palette entry #1
WIDTH2UNITTYPE	Integer	Unit type of width palette entry #2
WIDTH3UNITTYPE	Integer	Unit type of width palette entry #3
WIDTH4UNITTYPE	Integer	Unit type of width palette entry #4

Set Variables

Variable	Type	Description
WIDTH5UNITTYPE	Integer	Unit type of width palette entry #5
WIDTH6UNITTYPE	Integer	Unit type of width palette entry #6
WIDTHSIZE1	Real	Size of width palette entry #1
WIDTHSIZE2	Real	Size of width palette entry #2
WIDTHSIZE3	Real	Size of width palette entry #3
WIDTHSIZE4	Real	Size of width palette entry #4
WIDTHSIZE5	Real	Size of width palette entry #5
WIDTHSIZE6	Real	Size of width palette entry #6
WIDTHTYPE	Integer	Entity width index
WINDOWCENTERX	Real	Drawing X coordinate for center of current view
WINDOWCENTERY	Real	Drawing Y coordinate for center of current view
WINDOWSIZE	Real	Height of current view in drawing units
XANGLE	Real	Crosshatching angle
XSPACE	Real	Crosshatching spacing
ZOOMX	Real	Zoom factor

AutoSketch Commands

Command Name	Function
BoxArray	Activates the **BoxArray** tool
Break	Activates the **Break** tool
ButtonEditor	Displays a **Button Editor** dialog box
CascadeWindows	Selects cascading window mode
Chamfer	Activates the **Chamfer** tool
ChangeProperty	Activates the **Change Properties** tool
CloseFile	Close the current window
Copy	Activates the **Copy** tool
CopyBitmap	Activates the **Copy as Bitmap** tool
CopyMetafile	Activates the **Copy as Metafile** tool
CopyObject	Activates the **Copy Object** tool
CreateToolbox	Displays the **Toolbox Editor** dialog box
DeleteFile	Opens the **Delete File** dialog box
DrawAlignedDimension	Activates the **Aligned Dimension** tool
DrawAngularDimension	Activates the **Angle Dimension** tool
DrawArc	Activates the **Arc** tool
DrawBox	Activates the **Box** tool
DrawCircle	Activates the **Circle** tool
DrawCurve	Activates the **Curve** tool
DrawDiameterDimension	Activates the **Diameter Dimension** tool
DrawEllipse	Activates the **Ellipse** tool
DrawFreehand	Activates the **Freehand** tool
DrawHorizontalDimension	Activates the **Horizontal Dimension** tool
DrawLeader	Activates the **Leader** tool
DrawLine	Activates the **Line** tool
DrawPart	Activates the **Part** tool
DrawFillArea	Activates the **Pattern Fill** tool
DrawPoint	Activates the **Point** tool
DrawPolygon	Activates the **Polygon** tool
DrawPolyline	Activates the **Polyline** tool
DrawQuickText	Activates the **Quick Text** tool
DrawRadiusDimension	Activates the **Radius Dimension** tool
DrawTextEditor	Activates the **Text Editor** tool
DrawVerticalDimension	Activates the **Vertical Dimension** tool
EditCurve	Activates the **Curve Edit** tool
EditText	Activates the **Edit Text** tool
Edittoolbox	Activates the **Toolbox Editor** tool
Erase	Activates the **Erase** tool
Explode	Activates the **Explode** tool
Export	Creates a DXF, DWG, or SLD file of the current drawing
ExportSlide	Creates a Slide file of the current view
Extend	Activates the **Extend** tool
Fillet	Activates the **Fillet** tool
Group	Activates the **Group** tool
Import	Opens the **Import** dialog box
Information	Displays the **About AutoSketch** information
MeasureAngle	Activates the **Angle** tool
MeasureArea	Activates the **Area** tool
MeasureBearing	Activates the **Bearing** tool
MeasureDistance	Activates the **Distance** tool
MeasurePoint	Activates the **Point** tool
Mirror	Activates the **Mirror** tool
Move	Activates the **Move** tool
MultipleCopy	Activates the **Multiple Copy** tool
NewFile	Activates the **New** command
OpenFile	Activates the **Open** command
OpenToolBox	Displays the **Open Toolbox** dialog box, but can't record actions taken in the dialog box
PartClip	Activates the **Part Clip** tool
PasteClipboard	Activates the **Paste** tool
Preferences	Opens the AutoSketch **Preferences** dialog box
Print	Prints the drawing
PrintPreview	Opens the **Print Preview** dialog box
PrintSetup	Opens the **Print Setup** dialog box
Quit	Exits AutoSketch
Redo	Reverses the effect of the last **Undo** command
Redraw	Redraws the screen
RingArray	Activates the **Ring Array** tool
Rotate	Activates the **Rotate** tool
SaveFile	Saves the current drawing
SaveFileAs	Saves the current drawing to a new filename
Scale	Activates the **Scale** tool
ScrollDown	Activates scroll bar Down
ScrollLeft	Activates scroll bar Left
ScrollPgDown	Activates scroll bar Page Down
ScrollPgLeft	Activates scroll bar Page Left
ScrollPgRight	Activates scroll bar Page Right
ScrollPgUp	Activates scroll bar Page Up
ScrollRight	Activates scroll bar Right
ScrollUp	Activates scroll bar Up
SetArc	Opens the **Arc Settings** dialog box
SetAttach	Opens the **Attachment Settings** dialog box

AutoSketch Commands

Command Name	Function
SetBoxArray	Opens the **Box Array Settings** dialog box
SetChamfer	Opens the **Chamfer Settings** dialog box
SetCircle	Opens the **Circle Settings** dialog box
SetColor	Opens the **Color Settings** dialog box
SetCurve	Opens the **Curve Settings** dialog box
SetDimension	Opens the **Dimension Settings** dialog box
SetDwgSettings	Opens the **Drawing Settings** dialog box
SetEllipse	Opens the **Ellipse Settings** dialog box
SetFillet	Opens the **Fillet Setting** dialog box
SetGrid	Opens the **Grid/Snap Settings** dialo box
SetLayer	Opens the **Layer Settings** dialog box
SetLineType	Opens the **Line Type Settings** dialog box
SetOrigin	Activates the **Change Origin** tool
SetPartBase	Opens the **Part Base Settings** dialog box
SetPattern	Opens the **Pattern Settings** dialog box
SetPolygon	Opens the **Polygon Settings** dialog box
SetPolyline	Opens the **Polyline Settings** dialog box
SetProperty	Opens the **Property Settings** dialog box
SetRingArray	Opens the **Ring Array Settings** dialog box
SetSelectionArea	Opens the **Selection Area Settings** dialog box
SetSheet	Opens the **Sheet Settings** dialog box
SetSnap	Opens the **Grid/Snap Settings** dialog box
SetText	Opens the **Text Settings** dialog box
SetUnits	Opens the **Units of Measurement** dialog box
SetViewPreferences	Opens the **View Preferences** dialog box
SetWidth	Opens the **Width Settings** dialog box
ShowProperties	Opens the **Show Properties** dialog box
Stretch	Activates the **Stretch** tool
TileWindows	Selects tiling window mode
ToggleArcMode	Toggles **Arc** mode
ToggleAttachMode	Toggles **Attach** mode
ToggleCaptionHelp	Toggles **Caption Help** on/off
ToggleCrosshair	Toggles the Crosshairs cursor
ToggleFillMode	Toggles **Fill** mode
ToggleFrame	Toggles the **Curve Frame** display
ToggleGrid	Toggles the **Grid** display
ToggleIsoLeft	Switch to **Left** isometric mode or back to orthogonal drawing
ToggleIsoRight	Switch to **Right** isometric mode or back to orthogonal drawing
ToggleIsoTop	Switch to **Top** isometric mode or back to orthogonal drawing
ToggleOrthoMode	Toggles **Ortho** mode
TogglePopups	Shows or hides pop-up toolboxes
ToggleRuler	Toggles the **Rulers**
ToggleSelectLock	Toggles the **Selection Lock** tool
ToggleSmartCursor	Toggles the **SmartCursor** on/off
ToggleSnap	Toggles **Snap** mode
ToggleTextMode	Toggles between displaying text entities and their outlines
ToggleToolboxes	Shows or hides all toolboxes
Trim	Activates the **Trim** tool
Undo	Undoes the last operation
Ungroup	Activates the **Ungroup** tool
UpdateLink	Update an OLE link
ViewLastPrintBox	Activates the **Last Print Box** tool
ViewLastView	Activates the **Last View** tool
ViewPan	Activates the **Pan** tool
ViewSlide	Displays a slide file
Window1	Switch to the first window
Window2	Switch to the second
Window3	Switch to the third window
Window4	Switch to the fourth window
Window5	Switch to the fifth window
ZoomBox	Activates the **Zoom Box** tool
ZoomFull	Activates the **Zoom Full** tool
ZoomPercent	Activates the **Zoom Percent** tool
ZoomSheet	Activates the **Zoom Sheet** tool

Glossary

A

Absolute Coordinates: X and Y coordinates that mark an exact position measured from the origin of the Cartesian coordinate system.

Aligned Dimensions: Dimensions that measure the true length of a surface drawn at an angle. The dimension line is parallel to the angled surface.

Alphabet of Lines: Guidelines set by the American National Standards Institute (ANSI) that assign meanings to various line patterns and widths.

Alphanumeric Keys: Keyboard keys that represent letters of the alphabet, numbers, and symbols such as $, %, &. These keys allow the user to add text to a drawing.

Angle: Figure formed by two lines extending from the same point.

Angle Dimensions: Dimensions that measure the angle formed by nonparallel surfaces.

Apprentice Drafter: Redraws, revises, and repairs existing drawings. They may also develop simple drawings under the close supervision of a drafter.

Architectural Drafting: Involves preparing drawings that describe the structure and materials for residential or commercial buildings.

Arcs: Partial circles containing a center point, radius, start point, endpoint, and an included angle.

Area: Surface enclosed by a shape and expressed in square units.

Arrows: Marks placed at the ends of dimension lines to show the extent of the dimension. In architectural drafting, dimension lines may end with arrows, tick marks, or dots. AutoSketch for Windows has five types of arrows: the standard arrow, solid arrow, tick, dot, and no arrow.

Associative: Dimensions are attached to the points picked to place the dimension. When edited, such as stretching an object, the dimension and extension lines stretch and the value of the dimension also changes.

Attributes: AutoCAD text objects that provide information about a specific block.

AutoSketch Macro Language: Programming language used to record and write AutoSketch macros.

B

Backup File: A reserve file containing the previously saved version of work, making it possible for the user to retrieve some or all of accidentally deleted material.

Bezier Curve: A curve that bends more freely around the control points than other types of curves (refers to mathematical equation that determines the curve's shape).

Bitmap Format: An image made up of dots (like a photograph), not geometric objects. The computer no longer sees a line *as* a line. It sees the line as the *dots* that make up the image.

Blocks: An AutoCAD application where the user can save a group of an unlimited number of objects by name for use over and over.

Bond Paper: Low-cost, opaque, white printing medium.

Boolean Value: Either true or false. The number one is used for true, and zero is used for false. Boolean values are used for variables that are either on or off. Refer to the **Attach** variables in Appendix E.

Booting: Process of turning on the computer and loading the operating system.

Border: Solid, wide line that surrounds the drawing sheet, providing a margin between the edge of the paper and the drawing.

Boundary: Object that defines where to cut an object. Also, the object that trimmed or extended objects meet.

Brainstorming: Group problem-solving technique that involves the spontaneous contribution of ideas from all members of the group.

Break Points: Reference positions where a portion of an object will be removed.

B-spline Curve: A curve that is very tightly controlled by the control points that define the curve. (Refers to the mathematical equation that determines the curve's shape).

Byte: Memory needed to store one character. Memory is usually added in increments of one million bytes. One million bytes is one megabyte (MB).

C

Cabinet Oblique Drawing: A pictorial drawing where one surface is drawn true size and shape, and depth lines are drawn half length at 45°

CAD Program: A drawing program with special functions needed for drafting, including the capacity to add dimensions and draw lines, circles, boxes, ellipses, and complex curves.

Card: Circuit board added to the inside of the computer to expand its functions.

Cartesian Coordinate System: Horizontal (X) and vertical (Y) axes that intersect at 90° at the origin. Coordinates locate points at a distance from the origin along the X and Y axes.

Cavalier Oblique Drawing: A pictorial drawing where one surface is drawn true size and shape, and depth lines are drawn full length at 45°.

Center of Rotation: Point that objects will revolve around when rotated or arrayed.

Centerlines: Thin lines made up of alternating long and short dashes showing the centers of holes, cylindrical objects, and symmetrical objects. Along an object's length, it notes the axis extending lengthwise through the object's center. In an end view, they cross to show the center.

Chained Dimensions: Also called point-to-point dimensions, they continue a linear dimension from the second extension line of the previous dimension. Chained dimensions are less accurate since each dimension depends on others in the chain.

Chamfering: Beveling done on outside corners to relieve the stress of machining.

Check Prints: Paper copies of a drawing made with a dot matrix or laser printer for a quick check on the design progress.

Checker: Experienced drafter who checks prepared drawings and looks for errors that were not caught by the original drafter.

Chief Drafter: Supervises the work of the drafting department. This person, who usually has production experience as well as drafting skills, may also develop complex working drawings.

Circle: A closed curve with every point equidistant from a fixed center point within the curve. Often used in a drawing to show holes and round objects.

Circumference: Perimeter of a circle.

Circumscribed: An object drawn on the outside of another object.

Civil Drafting: The process of making drawings that describe land terrain, road systems, and utility systems. Surveys of property are typical civil drawings.

Click: To press the button on the input device and choose a tool, command, menu selection, object, or location on a drawing.

Clipboard: A temporary holding space in Windows that allows the user to "cut" or "copy" objects and paste them later.

Computer: An electronic device that stores, retrieves, and processes data.

Computer-aided Drafting (CAD): The process of creating a drawing using a computer and drafting software.

Computer-aided Machining: The process of converting data from a computer drawing into code that can be used to control manufacturing machines.

Computer-integrated Manufacturing (CIM): A system that uses computers to control machines and the production process.

Control Points: Connections that define an object's size and shape. For example, the four corners of a box are control points.

Coordinate Dimensioning: Technique where dimension are given from a datum.

Copying and Pasting: Process of copying objects from a drawing to the Windows clipboard and placing them in the same or different Windows application.

Cropping: Changing the margins on the page set-up. This is only necessary when the printer is cutting off portions of the drawing.

Current Layer: The layer that objects will be added to.

Cursor: Indicates on-screen position and allows the user to choose menus, tools, to draw, and to select items to edit. A cursor may appear as an arrow, crosshairs, dot, I, or underscore.

Cursor Keys: Keyboard keys labeled with arrows and the words [Home], [End], [PgUp], and [PgDn]; they can be used to move the cursor around the screen.

Curve: A smooth-bending line.

Custom Buttons: When creating a macro, the user may design or choose icons to represent the macro's function.

Custom Toolbox: When creating a macro, you may also create a toolbox with "tools" that activate macros.

Cutting and Pasting: Process of taking objects from an application, placing them on the clipboard, and inserting them later.

Cutting-plane Line: Designates an imaginary cut used to identify the portion of an object "cut away" to create a section view.

D

Data Storage Devices: Devices such as hard drives, disk drives, and tape drives where electronic information is stored.

Datum: A common point or surface that dimensions are measured from.

Default: The selection automatically used by a computer program unless you enter or select another value.

Design Criteria: Varying guidelines that provide focus and set limits for problem solving.

Design Process: Six steps that lead to solving a design problem. 1. Identify the problem or need. 2. Set guidelines. 3. Gather information. 4. Develop possible design solution. 5. Select the final design solution. 6. Evaluate the solution.

Designer: A person who prepares sketches and writes specifications to develop new products.

Desktop: The initial appearance of Windows 95. The desktop includes the **My Computer** and **Recycle Bin** icons, and the **Task Bar**. Shortcuts can be added to the desktop.

Detailer: A person who develops working drawings from specifications received from sketches, from notes made by an engineer or designer, and/or from verbal instructions.

Dialog Boxes: Windows that appear on-screen to request or provide information.

Diameter: The distance as measured through the center of circles, cylinders, holes, and other circular shapes.

Digitizer: Input device composed of a rectangular plastic tablet and a movable stylus or puck. Unlike a mouse, a digitizer allows the tracing of an existing drawing.

Dimension Lines: Lines that show the direction and extent of a dimension. The line is straight for a linear dimension and curved for an angular dimension.

Dimension Points: Points picked on an object to create a dimension.

Dimension Text: Numeric value of the measurement plus any needed symbols or notes.

Dimensions: Show the size and location of shapes, and may consist of numbers, lines, symbols, and notes.

Disk Drive: A piece of computer hardware that has a read/write head that saves, or "writes," data by charging magnetic particles on the disk's surface. This device loads data into computer memory by "reading" the charges of the particles. Can refer to either a floppy disk drive or a hard disk drive.

Diskettes: Round flat plates, enclosed in a vinyl jacket or plastic case, coated with magnetic particles, that data is stored on. Also referred to as a "disk" or a "floppy."

Dot Matrix Printer: A computer hardware device where images are formed by 9 or 24 steel wires mounted in a print head impacting an ink ribbon onto paper. The print head moves from side-to-side as the paper moves up through the printer one line at a time.

Double-click: To quickly press the pick button on the mouse twice.

Drafter: A person who develops working drawings from specifications received from sketches, from notes made by an engineer or designer, and/or from verbal instructions.

Drafting: A language used to communicate ideas with drawings.

Draw Programs: Object-oriented computer programs that store each object mathematically as a single item as opposed to a dot pattern.

Drawing Aids: Tools and functions that help locate a position on a drawing or existing object more accurately.

Drawing Files: Computer files for storing work created using AutoSketch.

Drawing Exchange Format (DXF): Developed by Autodesk, Inc. The DXF format is commonly used to exchange drawings between CAD programs and computer-aided machining (CAM) programs. In addition, many desktop publishing programs, such as Pagemaker and Quark XPress, can import DXF files as graphics.

Drawing Media: Printing medium, such as paper, vellum, and film.

Drawing Plan: A standard form for laying out a design and keeping track of its progress.

Drawing Setup: Actions taken to prepare the drawing file before adding objects. This may include entering values for the unit of measurement, sheet size, drawing aids, linetype, width, and color.

Drawing Sheet: A defined area that can be used to help create a drawing or determine what portion of the drawing is printed. After setting up a drawing sheet, solid-line and dotted-line boxes appear in the drawing window. The solid-line box represents the paper size. The dotted-line box shows the margins. These boxes are not actually part of your drawing. Drawing sheets are usually set to standard paper sizes.

Drawing Window: The working space on the computer monitor.

Dual Dimensioning: A dimensioning practice that uses both customary (inch/feet) and metric measurements.

E

Editing: Process of changing existing objects.

Electrical Plans: Architectural drawings that show the location of electrical fixtures and switches. Also explains types of fixtures, switches, wiring, and other items necessary to distribute electricity.

Electronics Drafting: Involves the design and layout of circuits, parts, and wiring for electrical and electronic products.

Electrostatic Plotter: A hardcopy device that forms images with a pattern of small dots. A writing head moves across the paper to apply tiny electrostatic charges and places the charges on paper in the dot pattern. The paper then passes through toner that sticks to the paper where the charges were placed.

Elevation Plans: Architectural drawings that show how the exterior of a structure looks.

Ellipse: A closed curve generated by a point moving so that the sum of the distance from two fixed points is a constant. When a circle is viewed at an angle, it will appear as an ellipse.

Encapsulated PostScript (EPS): A PostScript file with special header information that allows the file to be imported into a desktop publishing program as a single graphic.

Engineer: A person who develops ideas into practical designs, including the processes used to manufacture a product.

Entities: Items added to a drawing. Also called objects.

Expansion Slots: Vacant slots inside the computer where circuit boards may be inserted, allowing the addition of more devices to the computer.

Exporting: Converting drawings into a format different from the normal type of file used by a program. For example, AutoSketch saves files as SKD files. However, it can also export DXF, DWG, and SLD file types.

Extension Line Offset: The distance between the object or feature and the beginning of the extension line.

Extension Lines: Also known as "witness lines," they mark the beginning and end of a dimension. These are not used when the dimension is located within a shape, such as a circle.

F

Feature: Refers to any distinct part of an object. This may include a surface, edge, hole, slot, wall, window, door, appliance, or cabinet.

Feedback: Evaluation of information gathered from customers and from brainstorming sessions.

Fiber-tip Pens: A pen where liquid ink flows through a medium-hard tip. These pens are used in pen plotters.

Fillet: Rounding of an inside corner to relieve the stress of machining.

Film: A polyester-base transparent printing media used for high-quality plots.

Final Plots: High-quality prints made with a pen plotter or other high-quality output device.

First-angle Projection: The method used in Europe to show the relationship of views in a multiview drawing. The front view is the central view. The projection technique for the remaining views creates "shadow" images. The left side view is on the right. The right side view is on the left. In a similar manner, the top view is below and the bottom view is above the front. The rear view is shown beside the right side view.

Fixed Disk Drives: Also known as "hard drives," these disk drives are fixed inside the computer and can store many times the amount of data that a floppy disk can store.

Flatbed Pen Plotter: A hardcopy output device where pens move across paper that is taped to a flat surface.

Floor Plans: Architectural drawings showing the division of space for a building. Included are the locations of walls, doors, and windows.

Folders: The term used in Windows '95 to describe directories.

Font: The style of typeface (lettering).

Formatting: An operation that prepares the surface of new disks so that it can be read by the computer.

Foundation Plans: Drawings describing the concrete, block, beams, and other materials needed to support the structure.

Frame: The imaginary lines that connect control points and are used to form a curve.

Function Keys: Keyboard keys that provide instant access to important commands. These keys are labeled F1 to F12 and are located along the top of the keyboard. Keys located either on the right or left side of the keyboard labeled **Alt** (alternate), **Ctrl** (control), and **Shift** change the meaning of character keys and function keys when held.

G

General Notes: Text notes that apply to the entire drawing and appear in the lower-left or upper-left corner of the drawing area, or next to the title block.

General Oblique Drawing: A pictorial drawing where one surface is drawn true size and shape, and where depth lines drawn three-quarter length at an angle less than 45°.

Geometric Dimensioning and Tolerancing: An advanced dimensioning technique that uses symbols to describe the form, position, texture, or tolerance of features on a product.

Glass Box: A technique used by drafters to develop multiview drawings.

Gloss Paper: A clay-coated printing medium that provides high-contrast plots. It is often used when bright multicolor effects are used for graphics and charts.

Graphics Controller: A circuit board or chip integrated into the computer that controls the monitor. This device determines the resolution of the monitor and the number of colors supported.

Grid: A visual reference composed of a pattern of dots (resembling graph paper). The grid is often set up using units that match coordinates in the Cartesian coordinate system.

Group: A collection of objects treated as one item.

H

Hairline: The thinnest line that can be drawn in AutoSketch.

Hard Disk Drives: Standard mass storage devices that store more information (from 20 to over 4000 megabytes) and retrieve it faster than floppy disks. It is sealed in an airtight container. A small light on the front of the computer indicates when it is storing or searching for information. Often referred to as the hard drive. A hard drive is usually labeled the C drive.

Hard Drives: The common term used to describe a hard disk drive.

Hardcopy Device: A computer hardware devices that takes a drawing from computer memory and copies it onto paper (or other printing medium). Popular hardcopy devices include laser printers, pen plotters, electrostatic plotters, inkjet printers, thermal printers, and dot matrix printers.

Hardware: The physical parts of a computer system.

Heating, Ventilation, and Air Conditioning (HVAC) Plans: Architectural drawings that show how the building environment (temperature, humidity, and air flow) will be controlled.

Hewlett-Packard Graphics Language (HPGL): A language used by Hewlett-Packard plotters to print graphics. This language uses vectors, or lines, to reproduce images on paper.

Hidden Lines: Lines that show edges of an object that are not visible in a view. Hidden lines are drawn with a dashed, thin linetype.

Horizontal Dimensions: Dimensions that measure a horizontal linear distance between two points.

Horizontal Status Bar: Part of the AutoSketch screen that provides a prompt box, the location of your cursor on the drawing, and the current layer.

Horizontal Toolbox: Toolbox menu that contains the tools in the selected menu. These tools can be used during a drawing session.

Hot Keys: Underlined characters in pull-down menu names and command names. To access a menu, press the [Alt] key and type the underlined letter. To execute a command, press the underlined character of the command once the menu is open.

I

Icon: A graphic symbol that represents an available function. Selecting an icon with the mouse activates that function.

Importing: Bringing a file into a program that is a different format from the program's normal format. For example, AutoSketch saves files with an SKD extension. Files with DXF, DWG, or SLD extensions must be imported.

Included Angle: The angle between two lines.

Informational Dialog Box: A pop-up window that provides information or informs the user of an error.

Ink Jet Printer: An output device that forms a high-quality images by shooting ink droplets onto paper using a dot pattern.

Input Devices: Parts of the computer that allow you to enter commands, text, or draw on-screen.

Inscribed: An object that is drawn on the inside of another object.

Insertion Point: Location that marks the point on the drawing where text or other object (such as a symbol/part) will be inserted.

Integer Value: A whole number usually referring to a setting dialog box with multiple options. Refer to the **ARCTYPE** variable in Appendix E.

Isometric Drawing: A realistic, pictorial drawing showing the product as if it is tilted toward the viewer, displaying the front, top, and one side.

Isometric Lines: Lines parallel to one of the three axes in an isometric drawing. These lines are drawn full scale and can be measured.

Isometric Mode: An AutoSketch tool for creating isometric drawings. This tool causes AutoSketch to use a three-dimensional coordinate system to represent the three axes of an isometric drawing. The surface formed by any two of these axes is called an isometric plane.

Isometric Plane: The surface formed by two of the three axes in an isometric drawing.

J

Joystick: A hand-held cursor control device with a pick button and movable shaft. Pressing the pick button selects commands and points.

K

Keyboard: A series of keys, resembling a typewriter, used to operated a computer. A typical personal computer keyboard has alphanumeric keys, function keys, and cursor keys.

L

Landscape Plans: Architectural drawings that show the layout of trees, shrubs, and other ground cover surrounding a building.

Laser Printer: A high-resolution printer for computer output that prints an image formed by a laser. These printers work much like a photocopy machine.

Launch: To start a program. This term is usually used with Windows '95.

Layer: Similar to a clear sheet of paper that you can add objects to. However, unlike in manual drafting, these "sheets" can be shown together in any combination, so objects on all layers can be combined.

Layering Scheme: A plan that determines what details are to be placed on what layer. In a mechanical drawing, dimensions, object lines, title block and border, hidden lines and centerlines, and specifications typically all have their own layer. Architectural drawings have the different plans on different layers.

Leader: A thin line leading from a note or dimension to where it applies. Consists of an optional arrow, a leader line, a tail, and a note.

Linear Dimensions: Dimensions that measure straight distances. The dimension can be placed vertically, horizontally, or aligned (parallel) to an object at an angle.

Line: Straight object of set width and pattern, drawn by picking two endpoints.

Linetype: Line pattern including solid, dashed, or a combination of solid and dashed segments.

Liquid Ink Pens: High-quality pens used in pen plotters. Liquid ink flows through a narrow metal tip and onto the paper.

Location Dimensions: Dimensions that give the position of a feature or shape that is measured from an edge, center, surface, or another feature.

Long Break Lines: Lines used to shorten the view of a long part. The section of the part not shown must be the same as the section shown.

Long Filenames: Can contain up to 255 characters, including spaces. In addition, the file extension can also be 255 characters long. However, a filename cannot contain \ ? : " < > | as these are reserved characters.

M

Macro: Stored series of commands that perform specific tasks and may be created or edited by the user for personal use.

Macro Language Commands: Statements that a macro needs to perform its job and to help you troubleshoot.

Magnification Value: The factor used with the **Zoom Percent** tool to determine if the view will be enlarged or reduced, and by how much.

Manual Drafting: Process of creating a drawing on paper using manual drafting instruments.

Math Coprocessor: A computer chip that speeds up the computer by performing math functions normally performed by the central processing unit.

Maximizing: Enlarging the drawing window so that it takes up the entire workspace.

Mechanical Drafting: Involves making drawings that describe the shape and size of manufactured products.

Memory: The place where computer information (your drawing) is stored while you work. Memory is often called Random Access Memory (RAM). RAM is a collection of computer chips that is able to hold information *while the computer is on.* Work must be saved to either a floppy disk or to the hard drive, or it will be lost when the computer is turned off. It is important to remember that *memory* is not the same as *disk space.*

Menu: A list of text or icon commands where you can select an operation for the computer program to perform.

Menu Bar: A line located at the top of the screen containing a list of menus. When you choose a menu name, it displays either a pull-down menu or toolbox. Each pull-down menu or toolbox offers a list of commands.

Microgrip Pen Plotter: An output device where the pens and paper move to draw the image. The pens move in one direction along a rail and the pinch rollers, which hold the paper in place, move the paper to produce the other direction.

Microprocessor: A single integrated circuit that is the "brain" of the computer.

Minimizing: Reducing the current window, file, or application to an icon at the bottom of the workspace.

Mirror Line: An invisible line used with the **Mirror Tool** that objects are reflected about and copied.

Monitor: A display screen where you view the drawing, menus, and commands.

Mouse: A hand-held device that is rolled around a flat surface to move the cursor on-screen. A mouse shows movement, not absolute position.

Multitasking: Running more than one program at the same time.

Multiview Drawing: A drawing giving true shape and size by showing an object as viewed straight-on from different positions (views).

N

Nested Group: A group that is part of another group.

Network: Individual computers connected to share disk storage or peripherals (such as a printer).

Nonisometic Lines: Lines not parallel to one of the three axes in an isometric drawing. These lines are not drawn full scale and cannot be measured.

O

Object Lines: Solid, thick lines drawn to show edges and contours of the object visible in a view. They should clearly contrast with other lines on the drawing. All object lines should be the same color.

Object Properties: The settings for color, linetype, and width of an object created in AutoSketch.

Objects: Lines, boxes, polygons, circles, arcs, and other elements that AutoSketch can create and add to a drawing. Also refers to icons in the Windows '95 environment.

Oblique Drawing: Shows the front of an object in true size and shape, and the depth at an angle.

One-point Perspective Drawing: A drawing where one face of the object is drawn true size and shape. The top and sides converge at one vanishing point.

On-line Help: A command or program that provides information and directions on tools, menus, and features.

Operating System: A program that instructs the computer how to interact with hardware and software.

Origin: The intersection of the horizontal (X) and vertical (Y) axes in the Cartesian coordinate system.

Orthogonal: Perpendicular, or intersecting at right angles.

Orthographic: Refers to right (90°) angles.

Orthographic Projection: Process where a multiview drawing is created. Each face of the object is projected onto an imaginary glass surface and appears true size and shape.

P

Projection: The process of reproducing an object onto a plane.

Paint Programs: A computer program that creates and stores images as dot patterns.

Pan Destination: A point that marks the distance and direction you want to move your view of the drawing.

Pan Reference: A point that marks a "handle" you hold to "pull" the drawing into, or across, your viewing window.

Panning: Allows you to move around the magnified view of a drawing.

Paper: Variety of wood-base and rag-base printing media.

Part Base Location: The place on a symbol where the cursor will "hold" to move the symbol around when inserted into a drawing.

Pen Plotter: An output device that moves a pen across a drafting medium such as paper, vellum, or film to create a hard copy of your drawing

Percentage Meter: Shows how much of the 2048 character limit of the **Text Edit Box** has been used.

Perimeter: The distance around a shape.

Peripheral Device: Hardware used with the computer including printers, modems, and scanners.

Personal Computer: A computer that uses a single computer chip to process data.

Perspective Drawing: The most realistic and most difficult form of a pictorial drawing to create. Lines appear to converge at one or more imaginary points in the distance (vanishing points).

Phantom Lines: Lines that show the alternate positions of moving parts.

Pick: To select something on-screen, such as a tool or command, by moving the cursor into position and pressing a button on the input device.

Pick Button: Usually the left-hand button on the mouse. Used to select on-screen items.

Pictorial Drawing: A three-dimensional drawing that looks like a picture, but does not show the true size, shape, and relationship of features.

Pixels: Abbreviation for picture elements. These are small dots or rectangles used to create an image on a monitor. Also, the unit of measure used to determine the resolution of a monitor.

Planar Rotation: The angle that a circle is viewed at, making the circle appear as an ellipse.

Plans: Architectural drawings that show various aspects of a building. Several plans make up a set of working drawings.

Plastic-tip Pens: A pen used in pen plotters where a liquid ink flows through a medium hard tip.

Platters: Magnetic coated, aluminum disks that are stacked to form a hard drive. Several read/write heads store and load information between RAM and the platters.

Plot: Paper copy of a drawing created by hardcopy devices such as pen plotters.

Plot Plans: Architectural drawings describing the parcel (piece) of land, its perimeter, elevations, and the building location.

Plotting: The process of taking a drawing from computer memory and outputting it on paper or into a file.

Plumbing Plans: Architectural drawings showing location of plumbing fixtures and tracing the flow of water and waste for a building.

Point: Marks an exact position on the drawing. It has no size, only location. Points are helpful as a reference for placing other objects. Also, a macro command that enters the coordinate of a location on the drawing just as if you typed the coordinates in the **Prompt box**.

Pointing Devices: An input device, such as a mouse or digitizer, used to move the cursor and pick commands, objects, or locations.

Polar Coordinates: Coordinates that measure position from the previous point by distance and the angle from the X axis.

Polygon: A closed shape made with three or more connected straight lines.

Polyline: A line that continues from the last point when drawn until the command is canceled.

Pop-up Windows: Small windows that remain on-screen to help in your drawing session.

Ports: Connections on the back of the computer that can be used to connect peripheral devices.

PostScript: A page description language developed by Adobe Systems, Inc. It is widely used to print complex graphics and output from a desktop publishing program. A PostScript file is no more than a text file that "describes" how graphics should be created by a PostScript printer.

Precedence of Lines: The rule determining whether an object line, hidden line, or centerline should be drawn when two or more linetypes coincide on a drawing.

Pressurized Ball-point Pens: A pen used in pen plotters that can operate at extremely high speeds, plot well on clear overhead film, and produce darker and more consistent lines than fiber-tip pens. The line quality is generally not as good as other pen types.

Print: Paper copy created by such hardcopy devices as laser or ink jet printers.

Printing: Taking drawings from computer memory and putting them on paper.

Projection Plane: One of the "sides" of the glass box used in orthographic projection. The view of the object is projected at right angles onto the plane.

Prompt: A message or question requesting information from the user that AutoSketch displays in the prompt box.

Prompt Box: Window displayed on-screen containing a message, request for information, or providing information to the user.

Puck: An input device that looks like a small box with a plastic extension. Inside the plastic extension are metal crosshairs that sense position on a digitizer. One button on the puck is the pick button. The other buttons may perform other commands.

R

Radius: Distance from the center to a point on an arc or circle.

Radius Dimensions: Dimensions that specify the size of arcs and circles from the center to a point on the arc or circle.

Rag Bond: A cotton-fiber printing medium that is more durable than bond paper.

Real Value: Any decimal number usually referring to a distance. For example, look at the variables **GRIDX** and **GRIXY** in Appendix E. These are used to set the grid spacing.

Regular Polygons: Closed shapes with equal sides and equal internal angles, such as squares, hexagons, and octagons.

Relative Coordinates: Coordinate points that are measured position from the previous point.

REM: Stands for "remark." Allows addition of comments to a macro. An REM statement is not "played." Instead, REM statements let you break up a macro into sections so that you understand what each part of the macro is doing.

Resolution: Refers to the sharpness of an image measured by the number of pixels horizontally and vertically on the monitor. Resolution of a printer is the number of dots placed per inch. The more dots per inch, the sharper the image will be.

Roller Ball Pens: Pens used in pen plotters that can operate at extremely high speeds, plot well on clear overhead film, and produce darker and more consistent lines than fiber-tip pens. Line quality is generally not as good as other pen types.

Rounding: Done on outside corners of a part to relieve the stress of machining.

Rulers: Strips along top and left of the drawing area showing measurements and the current position of the cursor on the drawing.

S

Scroll Bars: Strips along the bottom and right edges of the drawing window that let you move horizontally and vertically across your drawing. Each scroll bar has a scroll arrow at each end and a scroll box.

Scrolling: Moving around the magnified view of a drawing. The same as panning, but using the scroll bars.

Section Lines: Indicate surfaces in a section view where material was "cut away" to show interior detail.

Selection Set: Temporary group of objects that can be modified or edited at the same time. Unlike using the crosses/window box, this set may be built by picking objects from different areas and does not have to contain all objects in a given area.

Senior Drafter: A person who supervises the work of the drafting department. This person, who usually has production experience as well as drafting skills, may also develop complex working drawings.

Server: The main computer of a computer network.

SET: A macro command used to change an AutoSketch system variable value.

Short Break Lines: Lines used in a single section view to show where material was "cut away."

Shortcuts: A file or icon that tells Windows '95 to open a particular file.

SHX Fonts: Fonts that come with AutoSketch and AutoCAD.

Single Inline Memory Modules (SIMMs): Small memory boards that fit into special sockets in the computer. SIMMs are added to increase the RAM of the computer.

Size Dimensions: Dimensions that give the width, length, height, diameter, or radius of a feature.

Slide: A "snap shot" of the current view of your drawing. This will include anything on-screen, including sheets and visible layers. Items not shown are the grid, objects on invisible layers, toolboxes, and status bar. A slide stores the image of the drawing on-screen, but not the actual drawing data. Thus, slides cannot be edited, printed, zoomed, or otherwise changed in AutoSketch.

Software: Electronic instructions (program) for a computer.

Specific Notes: Refer to particular features on a drawing and are attached to the feature by a leader line. Notes are always placed horizontally, even if dimension text follows the aligned system.

Stacked Dimensions: Also called datum dimensions, they continue linear dimensions from a common edge or surface, called a datum.

Standards Organization: A group of individuals who are experts in a particular field. They create a set of "rules" or guidelines to follow for that field. These rules are called a standard.

Storage Device: Computer hardware that stores and retrieves programs and files.

Stylus: A pen-like device used with a digitizing tablet to move the cursor and select commands and point locations.

Symbol Library: A special directory or disk where symbols are stored.

Symbols: Objects used in charts, drawings, diagrams, and schematics that represent a standard component, assembly, or feature. Standard shapes are typically set by a standards organization.

Symmetrical: One-half of an object is a reflection of the other half.

Syntax Error: An error in a macro file. This error indicates that one (or more) line in the macro is incorrectly stated. This error might be a misspelled command, incorrect spaces, or punctuation in a macro file.

System Variable: A word preceded by a front slash (/) that represents a value stored by AutoSketch. These variables can be used in macros instead of a coordinate value whenever you are entering a point location.

T

Tabular Dimensioning: A dimensioning technique using the principles of the coordinate system. Measurements are taken from the X, Y, and Z axes and recorded in a table placed on the drawing. Dimensions are not placed directly on the object.

Tangent: An object touching another object at only one point, but not intersecting the other object.

Technical Drafting: The process of preparing working drawings for products that are manufactured or constructed.

Technical Illustration: The process of making two-dimensional or three-dimensional (pictorial) drawings for assembly or presentation.

Technical Illustrator: A person who prepares technical illustrations. This may involve adding colors to drawings to make them look more realistic. A technical illustrator has both technical and artistic skills.

Template: A model drawing with a border and a title block used to begin new drawings. It is set up with the proper sheet size, text features, linetypes, and standard symbols.

Text: Letters, numbers, words, and sentences added to a drawing to explain features that cannot be described with graphics.

Text Features: The appearance of text, including font, height, width, oblique angle, and angle.

Text Justification: The alignment (left, right or center) of text in relation to the point you pick to place the text.

Thermal Printers: A printer that works by melting ink from a ribbon or transfer sheet and applying it to paper. It is possible to have ribbons or sheets that carry several colors. The colored ink is deposited in several layers to produce a multicolor image. Thermal printers use a dot pattern to form images.

Third-angle Projection: A method used in the United States to show the relationship of views in a multiview drawing. The front view is the central view. The left side view is on the left of the front view, and the right side view is on the right. The top view is above and the bottom view is below the front view. The rear view, if needed, is shown beside the left side view.

Tiling: Printing drawings that are larger than the printer paper using multiple sheets of paper. The printed sheets can then be assembled to form the entire drawing.

Title Block: An area on the drawing that gives important information about the product and progress of the project, as well as providing space for the designer's name and the drawing name.

Toggle Modes: Functions that are either on or off while using other functions.

Tool: Graphic symbol, or icon, of a command.

Toolbox: A menu of icons that represent commands.

Trackball: A stationary input device that can be described as an "upside-down mouse." You move the ball with your thumb and the buttons are located on the top or side of the base.

Translucent Paper: A semi-transparent printing medium that is similar to tracing paper and a low-cost substitute for vellum.

TrueType Fonts: Fonts that are used by most Windows applications. There are many of these fonts available. The point size ("height") of these fonts can be changed, however, the angle cannot.

Two-point Perspective Drawing: A drawing where the object is viewed from an angle. One side of the object converges toward one vanishing point, while the other side converges on a second point.

U

Unidirectional Dimensions: A dimensioning system where dimension text is placed horizontally. This allows all dimensions to be read from the bottom of the drawing.

Unit of Measurement: Determines the real-world distance between each unit of the Cartesian coordinate system. AutoSketch measures size, distance, and location of objects you place on the drawing. You can select a standard unit (such as inches, feet, and millimeters) or specify a custom unit of measurement.

User Input: Information that you need to enter during a command.

V

Vanishing Points: Imaginary points in the distance where lines appear to converge.

Vectors: Straight lines used by a plotter to create objects.

Vellum: A rag-base printing medium that is treated with resin to make it transparent, strong, and durable. This media shows ghost marks if you fold or bend it. Most types of pens work on vellum.

Vertex: Point on an object where lines or sides meet.

Vertical Dimensions: Dimensions that measure the vertical linear distance between two points.

W

Width: The thickness of lines.

Windows Metafile File Format (WMF): A file format that retains all the geometric properties of objects cut and pasted between Windows applications.

Working Drawings: Plans showing the information needed to build, machine, assemble, install, or service a product.

X

X Axis: The horizontal axis in the Cartesian coordinate system.

Y

Y Axis: The vertical axis in the Cartesian coordinate system.

Index

A

Absolute coordinates, 134, 135
    drawing with, 134, 135
    using unit notation, 135
Absolute, relative, and polar coordinates, combining, 140, 141
Active drive, 539
Active Pattern box, 345
Adding and editing text, 255-274
    benefits of CAD text, 255, 256
    editing text, 267-269
    placing text on drawing, 262-267
    selecting text features, 256-262
Advanced editing and measuring tools, 209-255
    breaking objects, 209-211
    drawing a box array, 225-232
    drawing a ring array, 232-238
    drawing fillets and chamfers, 217-222
    editing curves, 238-242
    exploding objects, 211, 212
    extending objects, 215-217
    measuring distance, angle, and area, 243-249
    stretching objects, 222-225
    trimming objects, 213-215
Aerial View button, 55
Aerial View pop-up window, 197
Aerial view, 197-200
    moving the **Aerial View** window, 198
    using the **Aerial View** as a **View** tool, 198, 199
    using the **Aerial View** control menu, 199, 200
Aligned dimensions, 368, 376
Aligned linear dimensions, placing, 376, 377
Aligned section, 358
Alphabet of lines, 315
Alphanumeric keys, 35
Angle dimensions, 379
Applying AutoSketch, 307-342
    architectural drafting, 320-334
    desktop publishing, 334-338
    mechanical drafting, 307-320
Apprentice drafter, 39
Arc Mode tool, 123
Arc Settings, 106, 107
    Start, Center, End, 107
    Start, End, Point, 107
    Start, Point, End, 106, 107
Architects, 40
Architectural drafting, 21, 22, 307, 320-334
    space planning, 321-324
Arcs, 106
Area, 244
Assembly view, 358
Assign a Macro dialog box, 507
Assist toolbox, 150
Association, 537
Associative, 388
Associativity, 388-392
    restoring, 392
    rotating dimensions, 390, 391
    scaling dimensions, 390
    stretching dimensions, 388-390
Attach options, 150, 151
    Center, 150
    Endpoint, 150
    Intersect, 151
    Midpoint, 151
    Nearest point, 151
    Perpendicular, 151
    Quadrant, 151
    Tangent, 151
Attach tools, 178
    turning on and off, 151, 152
Attach, 149
Attaching to objects, 149-152
    drawing precisely with coordinates and drawing aids, 149-152
    turning **Attach** tools on and off, 151, 152
Attachment Settings dialog box, 151
Automating AutoSketch with macros, 497-522
    advanced Macro Language topics, 515-518
    creating custom toolboxes and macro buttons, 503-510
    macro files, 498
    macro tools, 498, 499
    recording and playing macros, 499-503
    writing and editing macros using AutoSketch Macro Language, 510-515
AutoSketch, basic drawing with, 62-66
    starting, 49, 50
    why use?, 47-49
AutoSketch and Windows 95, 549-566
    how is Windows 95 different?, 549-551
    installing AutoSketch in Windows 95, 552-555
    managing files using Windows 95, 560-564
    using AutoSketch in Windows 95, 555-560
AutoSketch commands, 511
    in macros, 514
AutoSketch control menus, 53
AutoSketch fonts, 568-570
AutoSketch Macro Language, 510-515
    macro language commands, 512-514
    understanding macro commands, 511
    where to begin?, 511
AutoSketch pattern fills, 571, 572
AutoSketch pull-down menus, 573
AutoSketch system, 49
AutoSketch and Windows 95, 549-566
AutoSketch vs. other CAD programs, 47, 48
AutoSketch vs. paint and draw software, 48, 49
AutoSketch window layout, 50-55
    horizontal status bar, 51
    main window, 51
    menu bar, 51
    quick help, 51
Axis and planar rotation, 113

B

B-Spline curve, 118
Background drawings, 331, 332
Backup file, 80
Ball endmills, 482
Baseline Angle, 226
Basic machining processes, milling machines, 481-483
Bezier curve, 118
Bitmap format, 469
Bitmap pattern, 346
Bond paper, 458
Boolean value, 514
Booting, 49
Border, 294
Boundary, 213
Box array, drawing, 225-232
Box Array Settings, 225
Box Array tool, 168, 225
Brainstorming, 277
Break points, 209
Break tool, 168, 209
Breaking objects, 209-211
Broken-out section, 356
Business graphics, 24
Button Editor dialog box, 503
Button Editor options, using, 505, 506
Button Editor tools, 503
    using, 505
Byte, 27

C

Cabinet oblique drawing, 422
CAD, 485
CAD drafters, 40
CAD operators, 40
CAD programmer, 40
CAD programs, 49
CAD text, benefits, 18, 255, 256
CAD, tools of, 24-39
CAD/CAM/CNC process, 485-487
CAM software to generate NC code, 490-494
    generating, 493
    generating the toolpath, 491, 492
    loading the geometry, 490, 491
    machining the part, 493, 494
    setup steps, 491
CAM, 486
Cartesian coordinate system, 133, 134
    locating points, 133, 134
Cascade drawings, 80
Cavalier oblique drawing, 422
Center and both axes, 113
Center of rotation, 175
Center point and diameter, 96
Center point and radius, 96
Centerlines, 316
Chained and stacked linear dimensions, 377
Chained dimensions, 377
Chamfer on objects, effect, 220-222
Chamfer tool, 219
Chamfering, 217

Change Origin, 142
Change Properties tool, 180, 182, 299
Changing views of your drawing, 189-208
    Aerial View, 197-200
    changing what is seen with **View Preferences,** 201-205
    panning and scrolling, 195-197
    redrawing the view, 200, 201
    using **View** tools with other commands, 201
    Zoom tools, 190-195
Check prints, 441
Checker, 39
Chief drafter, 39
Circle Setting dialog box, 96
Circles, 95
Circumference, 95
Circumscribed, 102
Civil drafting, 22, 23
Clicking, 35
Clipboard, 469
    displaying, 471
Closed Curve option, 119
CNC, 487
Coincidental geometry, 491
Color Palette, 55
Color Settings dialog box, 89
Commands, repeating, 60
Computer-aided drafting, introduction, 15-46
Computer-integrated manufacturing (CIM), 19
Computers, 26-28
    memory, 27, 28
    math coprocessor, 27
    microprocessor, 26
    ports, 28
Contouring, 482, 484
Control menu, 199
Control points, 117, 222
Controller, 487
Conventional break, 357
Coordinate dimensioning, 397
Coordinate Display box, 54
Coordinate Display Pop-up Window, shrinking and closing, 144, 145
Coordinate Display, 143-145
    changing the **Coordinate Type,** 143, 144
    displaying all **Coordinate Types,** 144
    displaying certain **Coordinate Types,** 144, 145
Coordinate locations, typing in, 134 -141
Coordinate Type, changing , 143, 144
Coordinate Types, displaying, 144, 145
Coordinate values, format for entering, 139
Coordinate Window button, 144
Coordinates and drawing aids, drawing precisely, 133-162
Coordinates in isometric mode, entering, 430, 431
Copy as Bitmap tool, 469
Copy as Metafile tool, 470
Copy Objects tool, 471
Copy tool, 162-175
Copying and pasting objects with Windows clipboard, 469-471
Copying and pasting, 469
Copying objects as a bitmap, 469, 470
Copying objects as a Windows metafile, 470, 471
Copying objects for another AutoSketch drawing, 471
Copying objects, 172-175
    making multiple copies, 174, 175
Creating and using symbols, 405-420
    drawing and saving a symbol, 406-410
    editing symbols, 415, 416
    inserting a symbol in drawing, 411-415
    saving an entire drawing as a symbol, 410, 411
    symbols library management, 416
Creating custom toolboxes and macro buttons, 503-510
Creating drawings for CAM, 487-490
    guidelines, 487-490
Cropping, 444
Crosses box, 127
Crosses/window box, 126
Crosses/window, 126
Current Layer box, 54, 293
Current layer, 291
Current LineType button, 55
Cursor keys, 35
Cursor, 52
    using, 55, 56
Cursors button, 55
Curve, 117
Curve Settings dialog box, 120
Curve Settings, 117
Curve type, 118
Custom button, creating, 503-508
Custom buttons, 503
Custom toolbox, 503, 506-510
    creating, 506-508
    editing, 508
    opening, 509
    removing and restoring, 509, 510
Customizing widths, setting object color, width, and linetype, 90, 91
Cut-off tools, 484
Cutter compensation, 492
Cutter offset, 492
Cutting and pasting, 469
Cutting parameters, 491
Cutting plane, 343
Cutting-plane line, 318, 343

D

Data storage devices, 28-32
    care of diskettes, 31, 32
    diskettes, 29-31
    formatting diskettes, 31
    hard disk drives, 28, 29
    tape drives, 32
Datum, 397
Default, 51
Delete dialog box, 532
Deleting a drawing, 79, 80
    deleting a backup drawing file, 79, 80
    deleting a drawing file, 79
    retrieving a backup drawing file, 80
Design criteria, 276
Design process (chart), 278
Design process, 276-279
    develop possible solutions, 277
    evaluate solution, 278
    gather information, 276, 277
    identify problem or need, 276
    select final solution, 277, 278
    set guidelines, 276
Designer, 40
Desktop publishing (DTP), 334
Desktop publishing, 334-338
    other ways of using CAD with DTP, 337, 338
    overview of DTP programs, 334, 335
    working with Drawing Exchange Format (DXF) files, 335, 336
    working with plot files, 336, 337
    working with PostScript files, 336
Desktop, 560
Detailer, 39
Developing new drawings and templates, 275-306
    defining unit of measurement, design process, 276-279
    setting object properties, drawing aids, and text, 294, 295
    setting up drawing sheet, 284-290
    setting up new drawing, 282
    sketching designs, 279-281
    sorting information with layers, 290-294
    tips for developing drawings, 298, 299
    using templates, 296-298
Dialog boxes, 60
Diameter dimensions, 383
Diameter, 95
Digitizer, 36
Dimension lines, 367
Dimension points, 388
Dimension Settings dialog box, 368, 372
Dimension text, 368, 393
    changing, 393
Dimensioning, 330
Dimensioning circles and arcs, 382-386
Dimensioning a diameter with a leader, 384
Dimensioning an arc with a leader, 385, 386
Dimensioning angles, 379-382
Dimensioning circles and arcs, placing diameter dimensions, 383, 384
Dimensioning drawings, 365-404
    associativity, 388-392
    changing dimension text, 393
    dimensioning angles, 379-382
    dimensioning circles and arcs, 382-386
    dimensioning linear distance, 374-377
    dimensioning with AutoSketch, 371-374
    editing dimensions, 392
    elements in dimensioning, 367-371
    general rules for dimensioning, 393
    leaders, 387, 388
    linear dimensions with leaders, 378, 379
    size and location dimensions, 365, 366
    typical dimensioning practices, 393-398
Dimensioning isometric views, creating isometric arrows, 434
    drawing isometric dimensions, 435
Dimensioning linear distance, 374-377
    chained and stacked linear dimensions, 377
    placing aligned linear dimensions 376, 377
    placing horizontal linear dimensions, 374, 375
    placing vertical linear dimensions, 375, 376
Dimensioning lines, 318
Dimensioning practices,
    alternate dimensioning practices, 397, 398
    angles, 397
    chamfers, 397
    cylindrical shapes, 394
    fillets and rounds, 397
    holes, 394
    locating holes and rectangular features, 395, 396
    rectangular shapes, 393
    repetitive features, 397
    slotted holes, 394
Dimensioning with AutoSketch, 371-374
    setting **Dimension Settings,** 372-374
    setting the unit of measurement, 372
Dimensioning with circles, 383
Dimensions, 365
Directories, 523
Directory contents,
    displaying, 528-531
    displaying file details, 528, 529
    displaying groups of files, 530, 531
    sorting files, 529
Directory tree,
    changing the width, 528
    displaying, 526-528
    hiding, 528
Disk drive, 29
Disk space, 27
Diskettes, 29-32
    care, 31, 32
Display **Unit Type,** 284

Displaying the frame, 239-242
    deleting a control point, 241
    inserting a control point, 241, 242
    moving a control point, 240
Document, 335
DOS commands, using, 538-545
Dot matrix printer, 38, 39
Double-click, 50, 55
Drafter, 39
Drafting, 15
Drafting environment, 41
Drafting for computer-aided machining (CAM), 481-496
    basic machining processes, 481-485
    CAD/CAM/CNC process, 485-487
    creating drawings for CAM, 487-490
    using computer-aided machines software to generate NC code, 490-494
    what is CAM?, 481
Drafting positions and qualifications, 39-42
    drafting environment, 41
    job outlook, 41, 42
    positions related to CAD, 40, 41
    traditional drafting occupations, 39, 40
    training, 41
Drafting, design, and application of computers, 18-24
    business graphics, 24
    graphic design, 23, 24
    technical drafting, 19-23
    technical illustration, 23
Drag Scale option, 447, 449
Drag Scale, 450
Draw programs, 48
Drawing a basic section view with AutoSketch, 345-350
Drawing precisely with coordinates and drawing aids, attaching to objects, 149-152
Drawing a simple ring array, 234-238
Drawing a box array, 225-232
    picking distances, 227-229
    simple box array, 226, 227
    setting **Box Array Settings**, 225, 226
    using **Fit and Baseline** angle, 229-232
Drawing a circle using **Center, Diameter**, 97, 98
Drawing a circle using **Center, Radius**, 96, 97
Drawing a curve, 120
Drawing a polyline with line and arc segments, 123-125
    starting with a line segment, 123, 124
    starting with arc segment, 124, 125
Drawing a polyline with only line segments, 122, 123
Drawing a ring array, 232-238
    drawing a simple ring array, 234-238
    setting **Ring Array Settings**, 232, 233
Drawing a **Three-point Circle**, 100, 101
Drawing a **Two-point Circle**, 99, 100
Drawing aids, 143
Drawing, opening a stored, 76-78
Drawing and erasing objects, 87-132
    drawing ellipses, 112-117
    drawing arcs, 106-112
    drawing boxes, 94, 95
    drawing circles, 95-101
    drawing curves, 117-121
    drawing lines, 93, 94
    drawing points, 92, 93
    drawing polygons, 101-106
    drawing polylines, 122-125
    erasing objects, 126-128
    setting object color, width, and linetype, 88-92
    undoing and redoing your commands, 129
Drawing and saving a symbol, 406-410
    choose a part base location, 409
    drawing, 406, 407
    group the symbol, 407
    select the objects, 410
    using the **Part Clip...** command, 408, 409
Drawing arcs, 106-112
    arc settings, 106
    drawing an arc using **Center, Start, End**, 111, 112
    drawing an arc using **Start, Center, End**, 109, 110
    drawing an arc using **Start, End, Point**, 107, 109
    drawing an arc using **Start, Point, End**, 107, 108
Drawing boxes, 94, 95
Drawing circles, 95-101
    drawing a circle using **Center, Diameter**, 97, 98
    drawing a circle using **Center, Radius**, 96, 97
    drawing a **Three-point Circle**, 100, 101
    drawing a **Two-point Circle**, 99, 100
    setting **Circle Settings**, 95, 96
Drawing curves, 117-121
    drawing a curve, 120
    drawing **Freehand** curves, 120, 121
    setting **Curve Settings**, 117
Drawing ellipses, 112-117
    drawing an ellipse using **Axis and Planar Rotation**, 114, 115
    drawing an ellipse using **Center and Both Axes**, 113, 114
    drawing an ellipse using **Two Foci and Point**, 116
    setting **Ellipse Settings**, 113, 114
Drawing files, 71
Drawing fillets and chamfers, 217-222
    drawing chamfers, 219, 220
    drawing fillets, 217, 218
    effect of chamfer on objects, 220-222
    effect of, 218, 219
Drawing fillets, 217, 218
Drawing Interchange Format (DXF), 335, 466
Drawing lines, 93, 94
Drawing media, 458
Drawing method, 118, 119
Drawing pattern fills and section views, 343-364
    drawing a basic section view with AutoSketch, 345-350
    editing pattern fills, 351-354
    elements of a section view, 343-345
    other uses for pattern fills, 359-361
    types of sections, 354-359
Drawing pictorial views, 421-440
    isometric drawing, 424-426
    oblique drawing, 422
    perspective drawing, 422-424
    using isometric mode of AutoSketch, 426-436
Drawing plan, 281
Drawing points, 92, 93
Drawing polygons, 101-106
    drawing a **Circumscribed Polygon**, 103, 104
    drawing a polygon using **Endpoints**, 104-106
    drawing an **Inscribed Polygon**, 102, 103
    setting **Polygon Settings**, 101, 102
Drawing polylines, drawing a polyline with line and arc segments, 123-125
    drawing a polyline with only line segments, 122, 123
Drawing precisely with coordinates and drawing aids, 133-162
    Cartesian coordinate system, 133, 134
    changing origin of drawing, 142
    Coordinate Display, 143-145
    Grid and **Snap**, 145-148
    Ortho tool, 149
    rulers, 152, 153
    typing in coordinate locations, 134-141
    using **/lpoint** to enter location, 141
    with cursor, 143
Drawing session, ending, 83
Drawing setup, 282
Drawing sheet, 284, 447
Drawing window, 51
Drawings, recent, 77
    from different versions of AutoSketch, 77, 78
Drilling, 482, 484
Drives and directories,
    changing drives, 532
    copying a directory, 533
    copying disks, 533
    creating a directory, 532
    deleting a directory, 533
    formatting a floppy, 534
    labeling a disk, 534
    making a bootable disk, 534
    managing, 531-534
    moving a directory, 532
    renaming, 533
Dual dimensioning, 369

E

Edit and **Modify** tools, 351-353
Edit B-spline, 239
Edit Bezier dialog box, 239
Edit Curve tool, 238
Edit Toolbox tool, 503
Editing curves, 238-242
    displaying the frame, 239-242
Editing dimensions, 392
    changing dimension settings, 392
    moving dimension text, 392
Editing pattern fills, 351-354
    changing scale and angle of patterns, 353, 354
    creating custom patterns, 354
    View tools, 354
Editing symbols, 415, 416
Editing text, 267-269
    changing text features, 267
    changing text with other editing commands, 268, 269
    changing words and letters, 267
Editing, 163
Electrical plans, 21, 331
Electronics drafting, 22
Electrostatic plotter, 37, 38
Elements in dimensioning, 367-371
    arrows, 368
    dimension lines, 367
    dimension text, 368
    dual dimensioning, 369
    extension lines, 367
    leader, 369
    notes and symbols, 369-371
    position of dimension text, 368, 369
    unidirectional and aligned dimension text, 368
    unit of measurement, 368
Elevation plans, 22, 332
Ellipse, 112
Encapsulated PostScript (EPS), 471
Enclose entire drawing option, 450
Endmill cutters, 482
Endpoints, 102
Engineer, 40
Entities, 49
EPS file from AutoSketch, printing, 473, 474
EPS files, creating, 472-474
Erase tool, 126, 351
Erasing multiple objects with **Crosses/Window** box, 126
    erasing objects within and crossing the box, 126, 127

Erasing objects, 126-128
    erasing multiple objects with
        Crosses/Window box, 126, 127
    erasing single objects, 126
    selecting objects using a selection set,
        127, 128
Erasing single objects, 126
Esc button, 53
Exchanging drawings with other CAD
    programs,
    exporting drawings in DXF format, 467
    importing DXF drawing, 467, 468
    restrictions on importing DXF
        drawings, 468, 469
Expandable directories, 526
Expansion slots, 28
Explode tool, 211
Exploding objects, 211, 212
Export dialog box, 466
Exporting, 465
Exporting drawings in DWG format, 466
Exporting drawings in DXF format, 467
Extend tool, 215
Extending objects, 215-217
Extension line offset, 367
Extension lines, 367

F

Facing, 482, 486
Feedback, 278
Feature, 365
Fiber-tip pens, 459
File extensions, associating, 538
File Manager,
    features, 525
    opening drawings from 537, 538
File, 523
Filename, 524
Filenames and extensions, 524, 525
Files,
    copying, 536
    deleting, 535
    managing, 534-537
    moving, 535
    renaming, 536
    searching for, 536
    selecting, 535
Fillet, 217
Fillet command, 219
Fillet on objects, effect of, 218, 219
Fillet Setting dialog box, 218
Fillet tool, 217, 219
Film, 458
Final plots, 441
First-angle projection, 311
Fitted option, 118
Fixed disk drives, 28
Flat endmills, 482
Flatbed pen plotter, 37
Floor plans, 21, 325
Floppy drive, 29
Folders, 560
Font, 256-258
    SHX fonts, 257
    TrueType fonts, 258
    selecting, 258
Formatting, 31
Formatting diskettes, 31
Foundation plans, 21
Frame, 117, 335
Framed option, 118, 119
Freehand curves, drawing, 120, 121
Full section, 355
Function keys and key combinations, 567
Function keys, 35, 58

G

General notes, 369
General oblique drawing, 422

Generating the toolpath, 491, 492
    creating the path, 492
    entering cutting parameters, 491, 492
    selecting operations, 491
Geometric dimensioning and tolerancing, 371
Geometry, 485
Getting help, 60-62
    information about a tool, 61
    moving around in help, 61
    search for information, 61
Glass box, 309
Gloss paper, 458
Graphic design, 23, 24
Graphics controller, 34
Grid, 145
Grid = Snap option, 147
Grid and **Snap**, 145-148
    guidelines for using **Grid**, 147, 148
    guidelines for using **Snap**, 148
    spacing setting, 145-147
    turning **Grid** and **Snap** on and off, 147
Grid and **Snap** boxes, 147
Grid/Snap Settings dialog box, 145
Group tool, 166
Group, 165
Guidelines for creating drawings, 487-490
    delete duplicates and trim overlaps, 490
    do not use polylines, 490
    drawing one view, 488
    export only the geometry, 490
    objects must meet properly, 490
    separate toolpaths by layer and color, 490
    understand the coordinate system,
        488, 489

H

Hairline width, 89
Half section, 355
Hard disk drives, 28, 29
Hard drive, moving around, 539-541
Hardcopy device, 26
Hardcopy devices, 36-39
    dot matrix printer, 38, 39
    electrostatic plotter, 37, 38
    ink jet printers, 38
    laser printers, 37
    pen plotters, 37
    thermal printers, 38
Hardware, 26
Hatch pattern, 344, 346
Heating, ventilation, and air conditioning
    (HVAC), 21
Hewlett-Packard Graphics Language
    (HPGL), 336, 472
Hidden lines, 316
Hide Fill, 201
Hide Pop-ups, 204
Hide Text option, 202
Hide Toolboxes, 203
Hide/Show Toolboxes tool, 204
Hiding features that slow performance, 201, 202
Hiding other features, 203-205
    pop-up windows, 204
    rulers, 205
    toolboxes, 203
Hiding Sheet Options, 202, 203
    current sheet, 203
    non-current sheets, 203
    sheet text, 203
    tiling display, 203
Horizontal dimensions, 374
Horizontal linear dimensions, placing, 374, 375
Horizontal status bar, 51, 53
Horizontal toolbox, 52
Hot keys, 59
How to Print the View options, 454
HPGL files,
    advantages, 337

creating, 475
disadvantages, 337
importing, 337
installing a plotter driver, 475

I

Icons, 77
Import dialog box, 467
Imported, 335
Importing DXF drawing, 467, 468
    explode large blocks, 468
    import visible attributes, 468
Importing metafiles and bitmaps, 471
Importing, 465
Included angle, 110
Informational dialog boxes, 60
Ink jet printers, 38
Input devices, 26, 34-36
    alphanumeric keys, 35
    Ctrl, Alt, Shift, 35
    cursor keys, 35
    digitizer, 36
    function keys, 35
    joystick, 36
    keyboard, 34
    mouse, 35
    trackball, 35
Inscribed, 102
Insertion options, 413, 414
Installing AutoSketch in Windows 95, 552-555
Integer value, 514
Introduction to AutoSketch, 47-70
    basic drawing with AutoSketch, 62-66
    getting help, 60-62
    selecting menus, tools, and pull-down
        menus, 56-60
    starting AutoSketch, 49, 50
    using the cursor, 55, 56
    why use AutoSketch?, 47-49
    window layout, 50-55
    your AutoSketch system, 49
Introduction to computer-aided drafting, 15-46
    drafting positions and qualifications,
        39-42
    drafting, design, and application of
        computers, 18-24
    tools of CAD, 24-39
    what is computer-aided drafting?, 16-18
Islands, 349
Isometric drawing, 424-426
    isometric coordinate system, 425
Isometric grid, **Snap, Ortho,** and cursor, 429
Isometric lines, 424
Isometric mode, 426
Isometric mode of AutoSketch, using, 426-436
Isometric plane, 426-428
    choosing, 427, 428
Isometric text, using, 432
Isometric view, dimensioning, 433-435

J

Job outlook, 41, 42
Joystick, 36

K

Keyboard, 34
Knurling tools, 484

L

Landscape plans, 22
Laser printer, configuring, 443-445
Laser printers, 37
Lathe, 484, 485
Launched, 555
Layer Settings dialog box, 291
Layering schemes, 293, 294
Layers,
    adding, renaming, and deleting named,
        291, 292
    sorting information, 290-294

Leader, 369
Leaders, 387, 388
Line, 93
Line Settings dialog box, 91
Line tool, 134
Linear dimensions, 374
Linear dimensions with leaders, 378, 379
Linetype, 91
Liquid ink, 458
Loading the geometry, CAM software to generate NC code, 490, 491
Location dimensions, 366
Long filenames, 564
Long-break lines, 318

M

Machining the part,
    program zero reference, 493, 494
    tool length offset, 494
Macro, 497
Macro files, 498
Macro Language commands, 511-514
    DialogBoxReturn, 512
    Point, 512
    REM, 512
    Restart, 514
    SET, 513
    Stop, 514
    String, 514
    UserInput, 514
Macro Language topics, 515-518
    math functions, 515, 516
    user variable guidelines, 517, 518
    user variables, 516
Macro tools, 498, 499
Macro variables and commands, 574-579
Macros,
    always enter values, 502
    ending the macro, 501
    hints for recording, 501, 502
    playing, 502, 503
    recording and playing, 499-503
    recording, 499-502
    save before running, 502
    troubleshooting, 518
    user input, 500, 501
Magnification value, 190
Managing files from DOS prompt, 541-545
    copying entire disks, 544, 545
    copying files, 543
    deleting, 544
    formatting new disks, 542, 543
    listing files, 542
    renaming files, 543, 544
    using wildcard characters, 544
Managing files in Windows and DOS, 523-548
    understanding directories, files, and extensions, 523-525
    using DOS commands, 538-545
    using Windows File Manager, 525-538
Managing files using Windows 95, 560-564
    copying and moving, 562
    deleting and undeleting, 563
    filenames, 564
    formatting, 564
    Windows Explorer, 560-562
Manual drafting, 16
Math coprocessor, 27
Math functions, Macro Language topics, 515, 516
Maximize a drawing, 82
Maximizing, 82
Measure Angle tool, 244
Measure Area tool, 244
Measure Bearing tool, 246
Measure Distance tool, 243
Measure Point tool, 245
Measuring bearing, 246-249
Measuring distance, angle, and area, 243-249
    displaying point coordinates, 245, 246
    measuring an angle, 244
    measuring bearing, 246-249
    measuring distance, 243
    measuring perimeter and area, 244, 245
Mechanical drafting, 19, 20, 307-320
    developing a multiview mechanical drawing, 318-320
    glass box, 309
    multiview drawings, 308
    relationship of views, 311, 312
    selecting views to draw, 312-314
    showing features in mechanical drawings, 315-318
    sketching your ideas, 309, 310
Memory, 27, 28
Menu, 56
Menu bar, 51
Menus, tools, and pull-down menus, selecting, 56-60
Metafiles and bitmaps, importing, 471
Microgrip pen plotter, 37
Microprocessor, 26
Milling machines, 481-483
Minimize a drawing, 82
Minimize, 82
Minimized, 551
Mirror line, 177
Mirror tool, 177, 352
Mirroring objects, 177, 178
Modifying and editing objects, 163-188
    choosing objects to edit, 163-169
    copying objects, 172-175
    mirroring objects, 177, 178
    modifying object properties, 181-184
    moving objects, 170-172
    rotating objects, 175-177
    scaling objects, 179, 180
Modifying object properties,
    changing object properties, 182, 183
    showing object properties, 181, 182
Monitor, 26, 32-34
    graphics controller, 34
    resolution, 34
    types, 33
Mouse, 35
Move tool, 170
Moving around the hard drive, changing directories, 540
    changing the active drive, 539, 540
    making directories, 540
    removing directories, 541
Multipen plotters, 445, 446
Multiple Copy tool, 175, 225
Multitasking, 49, 550
Multiview drawing, 307
Multiview mechanical drawing, 318-320

N

Nested group, 166-168
Network, 18
Non-coincidental geometries, 491
Nonisometric lines, 424
Notes and symbols, 369-371
Number of sides, 102
Numerical control (NC) program, 487

O

Object lines, 316
Object properties, 88
Objects, 49, 87
Objects to edit, choosing, 163-169
    selecting objects with groups, 165-168
    selecting objects with selection sets, 163-165
    why use selections sets and groups?, 169
Objects with groups, selecting, 165-168
Objects with selection sets, selecting, 163-165
Oblique Angle, 260
Oblique drawing, 422
Offset section, 356
Offset with Leader option, 378
On-line help, 61
One-point perspective, 422
Open Drawing File dialog box, 77
Open Toolbox dialog box, 509
Operating system, 49
Origin of drawing, changing, 142
Origin, 133
Ortho tool, 149, 175
Orthogonal, 149
Orthographic, 308
Orthographic projection, 308
Outputting a file to printer/plotter, 457, 458
Overlay drafting, 332

P

Paint programs, 48
Panning, 195, 198
Panning and scrolling, 195-197
    using scroll bars to pan , 195-197
    using the **Pan** tool, 195
Paper, 458
Part base location, 409
Part Base Settings dialog box, 411
Part Clip File dialog box, 79, 408
Part Clip... command, 405, 408, 409
Part Insert... command, 405
Parting tools, 484
Paste tool, 471
Pasting objects into AutoSketch drawing, 471
Path, 541
Pattern Alignment Point, 348
Pattern Fill command, 348
Pattern Fill tool, 361
Pattern Options, 347
Pattern Settings dialog box, 345
Pattern Settings, choosing, 345-348
Pattern, placing the, 348
Pen plotters, 37
Perimeter and area, measuring, 244, 245
Perimeter, 244
Peripheral device, 26
Personal computer, 26
Perspective drawing, 422-424
Phantom lines, 318
Pick and click, 55
Pick button, 34, 35
Pictorial drawing, 308
Pixels, 34
Placing text on drawing,
    drawing special characters, 266, 267
    using **Quick Text,** 262, 263
    using the **Text Editor,** 263, 264
Planar rotation, 115
Plans, 21
Plastic-tip pens, 459
Platters, 29
Play Macro File dialog box, 502
Play Macro tool, 498, 502
Plot, 26
Plot files, working with, 336, 337
Plot plan, 21, 324
Plotter media, pens, and tips, 458-460
    drawing media, 458
    maintenance, 459
    plotter pens, 458, 459
    plotting tips, 460, 461
    troubleshooting plotter problems, 459
Plotter pens, 458, 459
Plotter, configuring, 445-447
Plotting, 18
Plotting problem, cause, solution, 461
Plotting your drawing, 455-457
    installing pens, 456
    loading paper, 455, 456
    making multipen plots with single-pen plotter, 456, 457

readying the plotter or printer, 456
  starting the plot, 456
Plumbing plans, 22
Pocketing, 482
Point, 92
Point coordinates, displaying, 245, 246
Polar coordinates, 137
  drawing with, 138
Polygon, 101
Polyline, 122
Polyline tool, 140
Pop-up windows, 204
Ports, 28
Post processor, 491
PostScript, 336, 471
PostScript drive, installing, 472, 473
PostScript files, working with, 336
Precedence of lines, 317
Preferences tool, 152
Pressurized ballpoint pens, 459
Print, 26
Print preview, 453-455
  How to Print the View options, 454
  View to Print options, 453, 454
Print setup, 442-447
  installing printer and plotter drivers, 442, 443
  selecting and configuring a laser printer, 443-445
  selecting and configuring a plotter, 445-447
Printer and plotter drivers, installing, 442, 443
Printing, 18, 442
Printing and plotting drawings, 441-464
  plotter media, pens, and tips, 458-460
  print preview, 453-455
  print setup, 442-447
  sheet setup, 447-453
  your drawing, 455-458
Printing and plotting your drawing, 455-458
  plotting, 455-457
  printing, 455
  to file, 457, 458
Printing files to export to desktop publishing, 471-475
Printing or plotting a drawing to file, 457, 458
  configure in Windows, 457
  print/plot to file procedure, 457
  printer, 457
Program reference zero (PRZ), 488
Projection, 308
Projection plane, 309
Prompt box, 53
Prompt, 50
Prompts, 60
Property line, 324
Property Settings dialog box, 182
Puck, 36
Pull-down menus, 58

Q

Quick Text tool, 262, 263
  more lines, 262

R

Radius, 95
Radius dimensions, 384-386
Radius tools, 484
Rag bond, 458
Real value, 514
Reaming, 484
Record Macro File dialog box, 499
Record Macro tool, 498
Redo tool, 129
Redraw button, 55
Redraw tool, 200
Regular polygons, 101
Relative coordinates, 136, 137, 140, 141

drawing, 136, 137
  when to use, 140, 141
REM, 512
Removed section, 357
Rename dialog box, 533
Resolution, 34, 37
Revolved section, 356
Ring Array Settings, setting, 232, 233
Ring Array tool, 168, 232, 352
Ring Array, drawing, 232-238
Riser, 328
Roller ball pens, 459
Rotate tool, 175
Rotate, 352
Rotating dimensions, 390, 391
Rotating objects, 175-177
Rounding, 217
Ruled surfaces, 482
Rulers, 52, 152, 153

S

Save As... command, 416
Saving a drawing under new name, 78
Saving a drawing, 78, 79
Saving part of drawing, 78, 79
Scale and angle of patterns, changing, 353, 354
Scale tool, 179, 351
Scaling dimensions, 390
Scaling objects, 179, 180
Scroll bars, 53, 195
Scrolling, 195
Search dialog box, 536
Section lines, 318
Section view, 343
  elements, 343-345
Sectioning assemblies, 358, 359
Sections, types, 354-359
Select Part File dialog box, 414
Select Template dialog box, 297
Selecting commands from pull-down menus, 58
Selecting commands using function keys, 58, 59
Selecting commands using toolboxes and tools, 57, 58
Selecting commands with hot keys, 59
Selecting menus, tools, and pull-down menus, dialog boxes,57-60
  prompts, 60
  repeating commands, 60
  selecting commands from pull-down menus, 58
  selecting commands using function keys, 58, 59
  selecting commands using toolboxes and tools, 57, 58
  selecting commands with hot keys, 59
Selecting objects using a selection set, 127, 128
Selecting objects with groups,
  building a group, 166
  nesting groups, 166-168
  ungrouping objects, 168
  using editing commands on groups, 168
Selecting text features, 256-262
  angle, 259, 260
  color, 261, 262
  font, 256-258
  height, 258, 259
  oblique angle, 260, 261
  text justification, 261
  TrueType options, 261
  width factor, 259
Selection set, 163
Selection sets and groups, why use?, 169
Senior drafter, 39
Server, 18
Setting **Box Array Settings**,
  Number of items, 226
  Baseline Angle, 226
  Fit, 226
  Point, 225

Setting **Circle Settings**, **Center point and diameter**, 96
  Center point and radius, 96
  Three points, 96
  Two points, 96
Setting **Curve Settings**, curve type, 118
  drawing method, 118, 119
Setting object color, width, and linetype,
  changing the color palette, 89
  customizing widths, 90, 91
  setting object color, 89
  setting object linetype, 91, 92
  setting object width, 89-91
Setting object linetype, 91, 92
Setting object properties, drawing aids, and text, 294, 295
Setting **Ring Array Settings**, **Center Point of Array**, 233
  Degrees between Items, 233
  Draw Clockwise, 233
  Included Angle, 233
  Number of Items, 232, 233
  Pivot Point, 233
Setting that affect objects (chart), 88
Setting up drawing sheet, **Sheet Scale** and **World Size**, 286, 287
  sheet name, 285
  sheet size, 285, 286
Settings dialog box, 60
Sharing drawings with other programs, 465-480
  copying and pasting objects with Windows clipboard, 469-471
  exchanging drawings with AutoCAD, 465, 466
  exchanging drawings with other CAD programs, 466-469
  making and viewing slides, 475, 476
  printing files to export to desktop publishing, 471-475
Sheet name, 285
Sheet Options dialog box, 450
Sheet Options, 202
  hiding, 202, 203
Sheet setup, 447-453
  defining more than one sheet, 447-450
  viewing sheet text and noncurrent sheets, 450
  working with existing sheets, 452
Sheet Setup dialog box, 447, 448
Sheet size, 285, 286
Sheet Size, Drawing Scale, and **World Size**, combinations, 287-289
Shift button, 53
Short-break lines, 318
Show Clipboard tool, 471
Show Properties tool, 181
Show Rulers box, 153
Show Rulers option, 205
SHX fonts, 257
Simple box array, drawing, 226, 227
Single Inline Memory Modules (SIMMs), 28
Size and location dimensions, 365, 366
Size dimensions, 365
Sketching designs, 279-281
  drawing plan, 281
Slide, 475
Slides, making and viewing, 475, 476
  making, 476
  viewing, 476
SmartCursor, 52
Snap, 145
  grid, 145
Software, 26
Sorting information with layers, 290-294
  adding, renaming, and deleting named layers, 291, 292
  layering schemes, 293, 294
  making a layer current, 293

making layers visible and invisible, 293
moving objects to new layer, 293
Space diagram, 322
Space planning, 321-324
Specifications, 334
Stacked dimensions, 377
Standard welding symbols, 407
Standards organizations, 15, 406
Start, Center, End, 107
Start, End, Point, 107
Start, Point, End, 106, 107
Starting and saving drawings, 71-86
    closing a saved drawing, 76
    deleting a drawing, 79, 80
    ending your drawing session, 83
    opening a stored drawing, 76-78
    saving a drawing, 78, 79
    saving a new drawing, 74-76
    starting a new drawing, 73, 74
    working with multiple files on-screen, 80-82
Storage device, 26
Stretch, how objects are affected by, 224, 225
Stretch base, 223
Stretch tool, 168, 179, 222, 351
Stretching dimensions, 388-390
Stretching objects, 222-225
    how objects are affected by **Stretch,** 224, 225
Stringer, 328
Subdirectory, 523
Surface of Revolution, 482
Swept surfaces, 482
Symbol, 405
Symbol in drawing, inserting, 411-415
Symbol library, 409
Symbol properties and layers, 415
Symbols library management, 416
Symmetrical, 177
Syntax error, 518
System prompt, 538
System variable, 513
Systems manager, 40

T

Tabular dimensioning, 398
Tape drives, 32
Tapering, 484
Technical drafting, 19-23
    architectural drafting, 21, 22
    civil drafting, 22, 23
    electronics drafting, 22
    mechanical drafting, 19, 20
Technical illustration, 23
Technical illustrator, 40
Template Description dialog box, 297
Template, 18, 50, 73, 296
Templates, 282
    using, 296-298
Text, 255
Text edit box, 264
Text Editor dialog box, 264, 265, 387
Text Editor tool, 256
Text features, selecting, 256-262
Text justification, 261, 264
Text on drawing, placing, 262-267
Text Setting dialog box, 256
Thermal printers, 38
Thin section, 357
Third-angle projection, 311
Threading, 484
Threading tools, 484
Tile drawings, 80
    tiling, 454, 455
Tips for developing drawings, avoid switching properties often, 299
    begin at reference points, 298
    construct minimum number of objects, 298, 299
    plan your commands, 299
    use hairline width, 299
Title block, 294
Toggle modes, 143
Tool, 57
Toolbox, 52
Tool length offsets, 494
Toolpath, 491
Toolpath verification program, 493
Tools of CAD, 24-39
    computers, 26-28
    data storage devices, 28-32
    hardcopy devices, 36-39
    input devices, 34-36
    monitors, 32-34
Trackball, 35
Traditional drafting occupations, 39, 40
Training, 41
    drafting skills, 41
    hardware skills, 41
    software skills, 41
Translucent paper, 458
Tread, 328
Trim tool, 213
Trimming objects, 213-215
TrueType fonts, 258
TrueType options, 261
Turning tools, 484
Two Foci and Point option, 113, 116
Typing in coordinate locations, absolute coordinates, 134, 135
    combining absolute, relative, and polar coordinates, 140, 141
    drawing with polar coordinates, 138
    format for entering coordinate values, 139
    polar coordinates, 137
    relative coordinates, 136, 137

U

Undo and **Redo** buttons, 55
Undo command, 129
Undoing and redoing your commands, 129
Ungroup tool, 168
Ungrouping objects, 168
Unidirectional dimensions, 368
Unit notation, using, 135
Unit of measurement, 282-284, 368
Unit Type, 283
Units Display, 284
Units of Measurement dialog box, 282, 368
Units of Measurement settings, 283, 284
User Input tool, 498
Using **/lpoint** to enter location, 141
Using AutoSketch in Windows 95, how AutoSketch operates, 558-560
    placing AutoSketch on your desktop, 560
    starting, 555-557
Using DOS commands, 538-545
    managing files from DOS prompt, 541-545
    moving around the hard drive, 539-541
    understanding paths, 541
Using isometric mode of AutoSketch, 426-436
    choosing isometric planes, 427, 428
    constructing isometric drawing, 429, 430
    dimensioning isometric views, 433-435
    drawing isometric text, 432
    entering coordinates in isometric mode, 430, 431
    isometric grid, **Snap, Ortho,** and cursor, 429
Using templates, saving, 297
    starting drawing with template, 297, 298
Using Windows File Manager, displaying directory contents, 528-531
    displaying the directory tree, 526-528
    managing drives and directories, 531-534
    managing files, 534-537
    opening drawings from File Manager, 537, 538

V

Vanishing points, 422
Vellum, 458
Vertical dimensions, 375
Vertical Status Bar, 53, 54
Vertical toolbox, 52
View box, 198
    creating new, 198, 199
View Preferences, changing what is seen with, 201-205
View to Print options, 453, 454
View tools, 201, 354
View tools with other commands, using, 201
Views to draw, selecting, 312-314

W

Wall diagram, 323
What is computer-aided drafting?, 16-18
    benefits of CAD, 18
Why use AutoSketch?, 47-49
    AutoSketch vs. other CAD programs, 47, 48
    AutoSketch vs. paint and draw software, 48, 49
Width button, 55
Width Settings dialog box, 90
Width, 89
Wildcard characters, 531
Window box, 126
Windows 95, 549-551
    different look, 550
    managing files and installing programs, 550
    multitasking, 550, 551
    operating system, 549
    starting programs, 550
Windows environment, 49
Windows File Manager, using, 525-538
Windows for Workgroups 3.11, 538
Windows metafile, 470
Word wrap, 264
Work triangle, 321
Working drawings, 15
Working with multiple files on-screen, 80-82
    arranging drawing windows, 80, 81
    working with duplicate windows, 82
World size, (chart), 288
World Size, more, 287

X

X axis, 133
X: value, 146

Y

Y axis, 133
Y: value, 146

Z

Zoom Box tool, 190
Zoom Full tool, 192
Zoom in, 198
Zoom Last tool, 194
Zoom out, 198
Zoom Percent, 190-192
Zoom Percentage dialog box, 192
Zoom tools, 190-195
    Zoom Full, 192, 193
    Zoom Last, 194, 195
    Zoom Percent, 190-192
    Zoom Sheet, 193, 194